Mineral Resource Reviews

Mineral Resource Reviews is the official book series of the Society for Geology Applied to Mineral Deposits (SGA). It publishes selected conference and symposium proceedings, monographs, Short Course Volumes, In-depth Reports and Field Guides by researchers and professionals in the fields of mineral resources; their study, development and environmental impacts. The series will focus on providing in-depth descriptions, analyses and reviews of deposit types, metallogenic regions, ore forming processes as well as the analytical/investigative techniques needed to study them. High quality scientific contributions from experts in mineral deposit geology are invited for publication in this series.

David Huston • Jens Gutzmer
Editors

Isotopes in Economic Geology, Metallogenesis and Exploration

Editors
David Huston
Geoscience Australia
Canberra, ACT, Australia

Jens Gutzmer
Helmholtz Institute Freiberg
for Resource Technology
Freiberg, Saxony, Germany

ISSN 2365-0559 ISSN 2365-0567 (electronic)
Mineral Resource Reviews
ISBN 978-3-031-27899-0 ISBN 978-3-031-27897-6 (eBook)
https://doi.org/10.1007/978-3-031-27897-6

This Springer imprint is published by the registered company Springer Nature Switzerland AG
The registered company address is: Gewerbestrasse 11, 6330 Cham, Switzerland

Contents

Isotopes in Economic Geology, Metallogeny and Exploration—An Introduction

David L. Huston, Ian Lambert, and Jens Gutzmer

Abstract

Although (Soddy, Nature 92:399–400, 1913) inferred the existence of isotopes early last century, it was not until the discovery of the neutron by (Chadwick, Nature 129:312, 1932) that isotopes were understood to result from differing numbers of neutrons in atomic nuclei. (Urey, J Chem Soc 1947:562–581, 1947) predicted that different isotopes would behave slightly differently in chemical (and physical) reactions due to mass differences, leading to the concept of isotopic fractionation. The discovery that some elements transformed into other elements by radioactive decay happened even before the recognition of isotopes (Rutherford and Soddy, Lond Edinb Dublin Philos Mag 4:370–396, 1902), although the role that different isotopes played in this process was discovered later. The twin, and related, concepts of isotopes and radioactive decay have been used by geoscience and other scientific disciplines as tools to understand geochemical processes such as mineralization, and also the age and duration of these processes. This book is a review of how isotope geoscience has developed to better understand the processes of ore formation and metallogenesis, and thereby improve mineral system models used in exploration.

D. L. Huston (✉) · I. Lambert
Geocience Australia, GPO Box 378, Canberra ACT 2601, Australia
e-mail: David.Huston@ga.gov.au

J. Gutzmer
Helmholtz-Institute Freiberg for Resource Technology, Helmholtz-Zentrum Dresden-Rossendorf, Chemnitzer Str. 40, 09599 Freiberg, Germany

1 Isotopic Research in the Geosciences

After the discovery of radiogenic decay and recognition of isotopes, largely by physicists and chemists, other disciplines—such as geology and biology—took up process-oriented research and began to apply isotopic studies to natural systems led largely by universities and government research organizations.

1.1 Radioactivity and Geochronology—Determining the Timing and Duration of Mineralization

Within a decade of the discovery of radiation by Henri Becquerel in 1896, Strutt (1905) first used radioactive decay (in this case the production of He (alpha particles) by Th decay) to estimate the age of a thorianite sample from Ceylon (present day Sri Lanka) at ca 2000 Ma. The dating of ore minerals followed soon after: Boltwood et al.

D. Huston and J. Gutzmer (eds.), *Isotopes in Economic Geology, Metallogenesis and Exploration*, Mineral Resource Reviews, https://doi.org/10.1007/978-3-031-27897-6_1

(1907) used the U–Pb system to date uraninite samples, yielding uraninite ages of between 410 and 2200 Ma for different localities in Norway and North America. Holmes (1946) and Houtermans (1946) independently developed a method (in effect two-point isochrons) to estimate the age of lead-rich minerals. The old ages indicated by these studies were a key to understanding Earth's evolution, indicating for the first time the extreme antiquity of Earth.

Although some of these early geochronology studies have been discredited for various reasons, they demonstrated the potential of radiogenic isotope systems to provide rigorous absolute ages for geological processes such as mineralization. Advances in analytical techniques and understanding of radiogenic isotope systems have continually improved so that for many mineral systems, absolute ages can be readily determined, and durations of mineralizing events can, in many cases, be robustly estimated.

1.2 Stable Isotopes—Tracers for Mineralizing Processes

Within two decades of the recognition of isotopes and within a decade of the theoretical prediction of isotopic fractionation, Epstein et al. (1953) calibrated a geothermometer based on measured carbon isotope fractionations from molluscs grown at different temperatures. Engel et al. (1958) first used stable isotopes in economic geology when they documented the effects of hydrothermal alteration on carbon and oxygen isotope characteristics of limestone in the Leadville district, Colorado, USA.

By the mid-1960s light stable isotopic studies had become a mainstream tool for understanding processes and sources of components in ore systems, and the variety and precision of analytical techniques and isotopic systems has since only increased. Isotopic studies were critical for the development of many modern ore genesis models, for example, porphyry (Sheppard et al. 1969, 1971; Field and Gustafson 1976; Wilson et al. 2007) and volcanic-hosted massive sulfide (VHMS) (Sangster 1968; Beaty and Taylor 1982; Ohmoto et al. 1983; Cathles 1993) systems. Light stable isotopes provide important constraints on the origin of, and interaction, between ore and ambient fluids, sulfur sources and chemical processes such as mixing, boiling, disproportionation and redox reactions.

As an example of the application of isotope research to mineral and other geological systems, Box 1 presents a summary of research undertaken at the Baas Becking Geobotanical Laboratory (BBL), a research laboratory supported by the Bureau of Mineral Resources (BMR, now Geoscience Australia), Commonwealth Scientific and Industrial Research Organisation (CSIRO) and industry. Although this laboratory was involved in many aspects of isotopic research, it is particularly recognized for having produced ground-breaking research on sulfur isotope variability in sedimentary rocks and mineral deposits in the Archean to Proterozoic. Although trends originally noted by the BBL have held up to subsequent research, the interpretation of some of these trends has changed because of continued acquisition of similar data and the availability of new types of data.

The development of inductively coupled plasma-mass spectrometry in the late 1970s to early 1980s (Houk et al. 1980) allowed analysis of new isotopic systems in the 1990s, including metallic, stable isotopes (Halliday et al. 1995; Zhu et al. 2002; Albarède 2004). Initially this work concentrated on major metals of economic interest—iron, copper and zinc (Zhu et al. 2000; Albarède, 2004; Mason et al. 2005; Johnson and Beard 2006)—but other work has studied the isotopic behaviour of minor metals such as antimony, molybdenum and silver (Rouxel et al. 2002; Arnold et al. 2004; Mathur et al. 2018), among others. The advantage of metallic stable isotopes is that they can track sources of and processes that affect ore metals themselves and do not require the frequent implicit assumption that there is a genetic relationship metals and the isotopic system (e.g., O, H, C or S) being analyzed. A challenge of metallic stable isotopes is that the degree of fractionation is much smaller than that of light stable isotopes, making the range in isotopic values much less and, therefore, more challenging to resolve analytically.

Box 1 Baas Becking Laboratory

The Baas Becking Geobiological Laboratory (BBL), which brought together geologists, geochemists, chemists, microbiologists and biochemists from the Bureau of Mineral Resources (now Geoscience Australia) and the Commonwealth Scientific and Industrial Research Organisation (CSIRO), operated from 1963 to 1987. It was established, with mineral industry backing, to investigate the role of microorganisms in geological processes, particularly the formation of metal sulfide deposits such as sediment-hosted Zn–Pb–Ag deposits like Mount Isa, McArthur River and Broken Hill. Research at the BBL included studies that contributed to (1) determining sulfur isotope fractionation associated with bacterial sulfide reduction (BSR) in modern marine sediments, (2) defining and interpreting sulfur isotope trends in the Archean and Proterozoic, and (3) understanding changes in the carbon cycle leading up to the Cambrian explosion of life.

Sulfur isotope characteristics of modern marine sediments

Studies of sulfur isotope fractionation in modern marine sediments provided insights into the signatures of biological activity in metal sulfide ores in sedimentary strata. In modern sedimentary environments where sulfate availability is not constrained, biogenic pyrite exhibits a wide range of negative $\delta^{34}S$, up to 60‰ less than $\delta^{34}S$ of source sulfate. In systems, where sulfate is not/slowly replenished, $\delta^{34}S$ values of biogenic sulfides trend to a range of positive values as sulfate is progressively depleted (Trudinger et al. 1972; Chambers and Trudinger 1979). The Permian Kupferschiefer base metal deposit in black shales in Germany, is a classic example of mineralisation exhibiting a typical biogenic S isotope signature.

Sulfur isotope trends in the Archean and Proterozoic

Documenting sedimentary sulfur isotopic trends through the Precambrian helped to define some major trends in biogeochemical evolution as well as providing insight into potential sulfur sources in mineral systems. The vast majority of iron sulfide minerals in Archaean metasedimentary rocks, collected near nickel, gold and iron ore deposits, have $\delta^{34}S$ in a narrow range of − 4 to 4‰, similar to the nearby mineral deposits (Donnelly et al. 1978; Lambert and Donnelly 1991). The $\delta^{34}S$ values are consistent with derivation directly or indirectly from magmatic sulfur, consistent with high levels of igneous activity. There were a few sites sampled, however, with more variability. Although these results had been interpreted previously (Goodwin et al. 1976; Ohmoto and Felder 1987) to indicate BSR, analyzed minerals were from veins and massive lenses and are closely associated with volcanics, leading to the conclusion that the anomalous $\delta^{34}S$ values were more likely generated by hydrothermal activity, and that there was no clear evidence of BSR, nor significant levels of sulfate, in the Archean hydrosphere. The low concentrations of sulfate in Archean seawater have largely been substantiated by later studies, although with some complexities (see below, Huston et al. 2023a, b and references therein).

This research also had implications for genesis of and, potentially, exploration for mineral deposits. In contrast with most Archean orogenic gold deposits, which have $\delta^{34}S_{sulfide}$ mostly in the range 1–4‰, Kalgoorlie samples have values of − 7 to − 1‰, interpreted as resulting from gold transport as reduced sulfur complexes and its precipitation as a result of sulfidation of magnetite-bearing host rocks, which resulted in partial oxidation of the ore fluids (Lambert et al. 1983). Based on this example, it was suggested that Archean epigenetic gold deposits of unknown size are likely to be very large if they have extensive alteration zones containing abundant pyrite with negative $\delta^{34}S$—a conclusion supported by further studies (Phillips et al. 1986).

Although Archean oceans are generally thought to be sulfate-poor (Lambert and Donnelly 1991), there are localised occurrences of sulfate minerals in Paleoarchaean provinces, particularly the Pilbara (Western Australia) and Kaapvaal (South Africa) cratons. In the Pilbara, BBL studied bedded and discordant barite from the North Pole (Dresser) deposit, and barite from the volcanic-hosted massive sulfide (VHMS) Big Stubby deposit (Lambert et al. 1978). Barite from North Pole had $\delta^{34}S$ mostly of 3–4‰, tailing to weakly negative values; associated sulfide minerals had similar values. The barite was interpreted to have formed from Ba-bearing hydrothermal fluids introduced into a restricted (evaporative?) basin in which microbiological activities oxidised reduced sulfur, leading to barite deposition. In contrast, Big Stubby barite was found to have $\delta^{34}S$ values of 11–13‰, while associated sulfides had values in the range − 1 to − 5‰. This was interpreted in terms of the incorporation of sulfate-bearing basin waters into barium and base metal bearing hydrothermal exhalations related to felsic volcanism.

Research by the BBL contributed to the recognition of a sustained change in the sulfur isotope record from the early Paleoproterozoic, which was related to the Great Oxidation Event (GOE). Overall, Proterozoic sulfides in (meta)sedimentary rocks have wide ranges of negative to positive $\delta^{34}S$ values, with mean values well above both those of magmatic sulfur and modern marine sulfides (Hayes et al. 1992). Uncommon sulfate deposits have $\delta^{34}S$ values of 10–20‰. The $\delta^{34}S$ variations can be accounted for by sulfate becoming a stable component in low concentrations in the hydrosphere around the Archean-Proterozoic boundary, and its enrichment in ^{34}S as a result of bacterial and hydrothermal reduction processes (Hayes et al. 1992). The rising sulfate levels likely paralleled the stabilisation of oxygen in the atmosphere during the GOE, and proliferation of sulfate-reducing bacteria (Lambert and Groves 1981; Lambert and Donnelly 1991; Hayes et al. 1992). The predominance of positive $\delta^{34}S$ values in Mid to Late Proterozoic sulfides in stratiform base metal deposits and unmineralized strata were interpreted to indicate extensive reduction of sulfate-limited systems—restricted intracratonic marine or non-marine basins, possibly on supercontinents, where ^{34}S enriched sulfides formed as biological sulfate reduction proceeded.

Changes in the carbon cycle leading up to the Cambrian explosion of life

Carbon isotopic signatures of carbonate and organic carbon in strata straddling the Neoproterozoic-Cambrian boundary in China and Svalbard were studied in collaboration with the Proterozoic Paleobiology Research Group (Knoll et al. 1986; Lambert et al. 1987; see figure). These data and data from other sites around the world defined a trend of carbonate $\delta^{13}C$ from the latest Proterozoic that were more positive than younger marine carbonates. Taken together with organic carbon $\delta^{13}C$ data, this was interpreted to indicate major burial of organic carbon through this period, in some cases in closed environments and in others in open marine systems (Knoll et al. 1986; Lambert et al. 1987). This would likely have been accompanied by a major

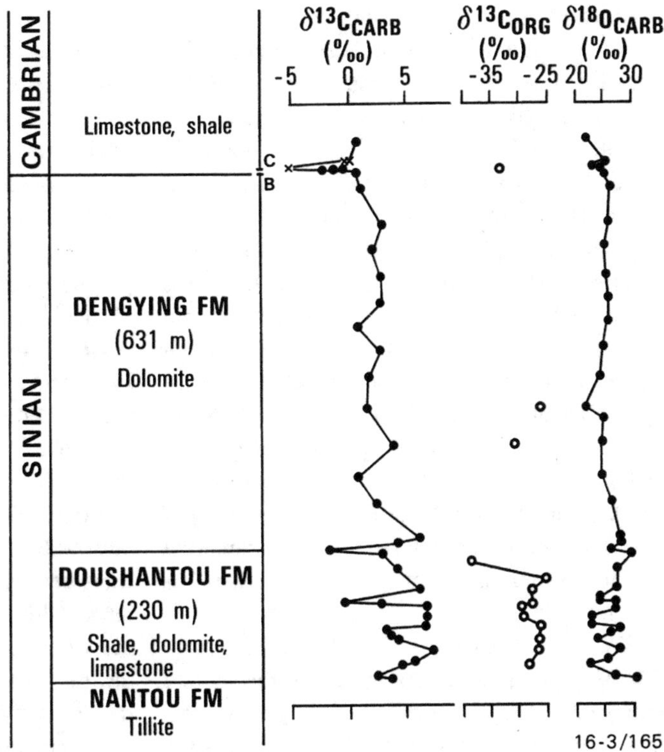

Fig. 1 Isotopic data for the carbonate-rich Upper Proterozoic to Early Cambrian succession from the Yangtze Gorges near Ichang, China. Reproduced with permission from Lambert et al. (1985); Crown copyright 1985 Geoscience Australia. The original figure used data summarized by Hsu et al. (1985)

increase in atmospheric O_2, the Neoproterozoic Oxidation Event (Canfield 2005). The transition to the Cambrian was accompanied by a trend toward negative carbonate $\delta^{13}C$, indicating decreases in rates of accumulation and/or increased oxidation of organic matter—the latter would be expected as a result of the evolution of animals that effectively bioturbated the sediments, allowing ingress of oxygenated waters (Fig. 1).

2 Isotopes in Economic Geology, Metallogeny and Exploration

This book provides a review of the use of isotopes to understand mineralizing processes at scales ranging from microscopic to continental, including the timing of duration of these processes, regional controls on the localization of ore formation, sources of ore-forming components, and chemical and physical processes of ore deposition. The book is set out in four parts. Part I describes the use of radiogenic isotopes to determine the absolute timing and duration of mineralizing processes. Part II documents the use of radiogenic isotopes to determine metal sources, fingerprint deposit types and map tectonic and metallogenic provinces. Part III describes how light stable isotopes have been used to determine fluid and sulfur sources and to establish ore forming reactions and processes. Part IV examines the utility of stable metallic isotopes for the robust determination of metal sources and for understanding geochemical processes of mineralization like redox reactions.

2.1 Part I—Radiogenic Isotopes and the Age and Duration of Mineralization

Ever since the first attempt to date ore minerals by Boltwood (1907), the main use of radiogenic isotopes in economic geology, and in the geosciences in general, has been determination of the age of geological events. Largely because of the lack of analytical methods and uncertainties in the rates of radioactive decay, geochronology did not develop as a discipline until the latter half of the twentieth century, with the development and widespread adoption of thermal ionization mass spectrometry (TIMS) and secondary ion mass spectrometry (SIMS). The development and application of ICP-MS, particularly in conjunction with in situ sampling of individual minerals by laser ablation (LA-ICP-MS) from the 1990s to now, has revolutionized geochronology by enabling access to inexpensive and rapid analysis. This part of the book presents an overview of the use of radiogenic isotopes and geochronology to understand the timing and duration of mineralization.

The first chapter in Part I, by Chiaradia (2023), presents an overview of the theory and radiogenic isotope systems used in geochronology, including U–Th–Pb, Re–Os, K–Ar (Ar–Ar), Rb–Sr and Sm–Nd. An important aspect of geochronology here is the concept of closure, when a mineral ceases to undergo isotopic exchange with its surroundings. Potential problems related to the lack of closure were recognized from the start: Strutt (1905) observed that the amount of helium would give a minimum age of a Th-bearing mineral because "the helium may not have been all retained". Closure depends upon many factors, including the mineral, the isotopic system, temperature and strain. Of these, the most important is temperature. Closure temperature is the temperature below which the mineral locks in isotopic composition. This varies not only with the mineral, but also the isotopic system (mainly limited by diffusion) and physical variables such as cooling rate and crystal size. Chiaradia (2023) presents closure temperatures for a range of minerals/isotope systems for which ages are commonly determined. Consideration of closure temperatures allows an interpretation of the significance of different ages (ages determined from mineral systems do not necessarily reflect primary ages of mineralization if the mineral remains open for an extended period of time or re-opens later due to processes such as heating or deformation), but also allows an understanding of the rates of cooling of mineral systems and possible constraints on subsequent events. Another important aspect of geochronology addressed by Chiaradia (2023) is the duration of mineralizing processes: individual mineralizing pulses associated with porphyry copper deposits last at most a few tens of thousands of years. Individual mineralizing pulses, however, can overprint each other, incrementally building up to total endowment to form world-class deposits.

Chelle-Michou and Schaltegger (2023) provide a description of the U–Pb isotopic system and how this system has been the main, and best understood, isotopic system to determine ages in most geological systems, including mineral systems. This overview provides information on theory, data presentation, processes that can disturb the system and complicate data interpretation, analytical methods, guidelines for data interpretation, minerals that can be dated as well as case studies. In addition, they give a brief summary of the Th–Pb system. Even though the U–Th–Pb isotopic system is well understood, there are many impediments to obtaining robust ages, and, because ore minerals are rarely dated directly, geological relationships between the dated mineral and the ore assemblage must be considered in data interpretation. Despite this, the U–Th–Pb system has been the workhorse for mineral system geochronology, and the high precision obtainable from this system has enabled robust studies on the duration of mineral systems.

The third chapter (Norman 2023) describes the theory and application of the Re–Os and Pt–Os systems to dating mineral systems. The Re–Os system is probably second in importance for mineral deposit geochronology after the U–Th–Pb system. In contrast to the U–Th–Pb system, many minerals dated by the Re–Os systems are ore minerals (e.g. molybdenite) or closely related to ore minerals (e.g. arsenopyrite and pyrite). The Pt–Os system, which can also be used to date ore minerals, is much less used owing the lower abundance of suitable minerals and analytical challenges. Molybdenite Re–Os dating has been particularly useful in age dating of mineral systems as, in addition to being an ore mineral in many systems, it is resistant to resetting. Despite these advantages, several processes can complicate data interpretation. Norman (2023) also describes how the Re–Os system can be used in understanding processes in orthomagmatic systems such as metal sources and partitioning of platinum group elements between sulfide and silicate melts during magma evolution.

Isotopic systems not described in detail in this book include K–Ar (Ar–Ar), Rb–Sr and Sm–Nd. Although the use of the latter two systems is limited in mineral deposit studies, the K–Ar (Ar–Ar) system is commonly used, particularly in dating K-bearing alteration minerals. Owing to complexities in data interpretation, low closure temperatures for many datable minerals (e.g. Chiaradia 2023) and uncertainties in relating alteration minerals to mineralizing events, care must be taken in interpreting the significance of K–Ar and Ar–Ar ages in mineral systems. More detailed descriptions and applications of the K–Ar (Ar–Ar) system can be found in Richards and Noble (1998), Vasconcelos (1999) and Kelley (2002).

2.2 Part II—Radiogenic Isotopes: Metal Source, Deposit Fingerprinting and Tectonic and Metallogenic Mapping

The second part of this book presents uses of radiogenic isotopic systems—Sm–Nd, Pb–Pb and Lu–Hf—to determine metals sources and to map tectonic and metallogenic provinces using variations in isotopic ratios. Although demonstrated as a method last century (Zartman 1974; Bennett and DePaolo 1987), recent advances in geographic information systems (GIS) and increases in the amount of data available have enabled isotopic mapping at the province to continental scales.

Champion and Huston (2023) and Huston and Champion (2023) discuss isotopic mapping using the Sm–Nd and Pb–Pb isotopic systems, respectively. Both papers present the theory behind the isotopic systems as well as analytical methods prior to describing applications of the data to metallogenic studies. The availability of large isotopic datasets combined with GIS visualization and contouring tools has enabled mapping of variations in isotopic parameters at the province to continental scales. Champion and Huston (2023) illustrate that variations in parameters derived from Sm–Nd isotopic data commonly coincide with boundaries between tectonic and metallogenic provinces. Moreover, they show that some types of mineral systems prefer different types of crust (i.e. isotopically juvenile versus evolved). Huston and Champion (2023) show that maps based on parameters derived from lead isotope analyses of ores (e.g., $\mu = {}^{238}U/{}^{204}Pb$) also define tectonic and metallogenic provinces that can be used to predict the metallogenic potential of mineral provinces. Moreover, they review previous literature that shows that lead isotope data can be used to determine the lead sources and to fingerprint deposit types in regions of complex metallogeny.

Waltenberg (2023) discusses the theory and analytical methods of the Lu–Hf isotopic system prior to considering its utility in metallogenic studies, particularly isotopic mapping. Although in many ways similar to the Sm–Nd system, Lu–Hf isotopic data, which are mostly derived from zircon analyses, also provide information on changes in isotopic patterns through time, a characteristic not available with the Sm–Nd or Pb–Pb systems. Like the Sm–Nd system, parameters derived from Lu–Hf data define tectonic and metallogenic provinces or boundaries, in this case at specific times. Understanding of the Lu–Hf system is evolving rapidly, hence the utility of the Lu–Hf system to metallogenic studies will increase in the future.

2.3 Part III—Light Stable Isotopes: Fluid and Sulfur Sources and Mineralizing Processes

Since their early application to metallogenic studies (Engel et al. 1958), light stable isotopes (initially H, C, O and S, but expanding to include B, Mg, Si, etc.) have played an important role in developing models for ore deposit and then mineral systems. Part III of this book presents the theory, analytical methods and processes that fractionate the most used light stable isotope systems (H, C, O, S and, in some cases, B) and then illustrates how these data have been used to develop mineral system models for VHMS, orogenic gold, shale-hosted Zn–Pb and iron ore deposits.

Huston et al. (2023a) present an overview of the most used stable isotopes in mineral systems studies. This chapter describes the general theory and conventions of these data, documents analytical techniques, including recent developments in microanalytical capabilities and interpretation of multiple sulfur isotope ratios, and discusses how light stable isotope data can be used to constrain possible fluid and sulfur sources, how isotopic patterns can be used to infer geochemical and geophysical processes, and, potentially, exploration. Huston et al. (2023a) stress that the interpretation of isotopic data is commonly ambiguous in that different sources and different processes can produce similar isotopic characteristics.

Volcanic-hosted massive sulfide deposits, the ancient analogues of modern black smoker deposits, form at or close to the seafloor in submarine volcanic successions. Because of this environment, VHMS mineral systems involve a complex interplay between component sources and processes. Huston et al. (2023b) review how stable isotope data, particularly oxygen, hydrogen and sulfur data, have been used to determine (or at least to delimit) fluid and sulfur sources, to infer processes, and to demonstrate that sources and processes have evolved with geological time. Oxygen-hydrogen data indicate that the main fluid was (evolved) seawater although magmatic-hydrothermal fluids are thought to have been present in some systems. Moreover, these data, particularly $\delta^{18}O$, define reasonably consistent patterns that have been used in exploration and discovery. Sulfur isotope data (including $\delta^{34}S$, $\delta^{33}S$ and $\delta^{36}S$) indicate that sulfur sources are a mixture between seawater sulfur (reduced sulfate) and igneous (leached or magmatic-hydrothermal) sulfur, with the proportion of seawater-derived sulfur increasing with geological time. Detailed deposit-scale sulfur isotope studies using in situ microanalysis demonstrate a complex mixture of igneous sulfur and that produced by thermochemical and biogenic sulfate reduction (TSR and BSR).

Orogenic gold deposits form in orogenic belts commonly intruded by igneous rocks, are mostly in the form of veins or stockworks. The association of these deposits with both orogenic belts and granitic rocks has led to controversies over the origin of ore fluids and sulfur. Quesnel et al. (2023) address these controversies using a database of over 8000 oxygen, hydrogen, carbon, sulfur, nitrogen, boron and silicon isotope analyses collected from deposits of all ages around the world. Based upon this dataset, Quesnel et al. (2023) conclude that the isotopic data for these deposits are most consistent with a dominant metamorphic fluid having a temperature of 360 ± 76 °C (1σ) and that this broad fluid type did not change significantly over geological time. Moreover, isotopic arrays can be interpreted to indicate mixing between this deep metamorphic fluid and upper crustal fluids. Secular variations were noted for nitrogen and sulfur isotopes. Quesnel et al. (2023) interpret the changes in nitrogen to reflect secular variations in $\delta^{15}N$ values, whereas the changes in $\delta^{34}S$ ultimately reflect secular changes in $\delta^{34}S$ of seawater.

Williams (2023) also used an isotopic dataset compiled from clastic-dominated (shale-hosted or SEDEX) Zn–Pb deposits in the Paleo-Mesoproterozoic North Australian Zinc Belt and in the Paleozoic Northern Cordillera of Canada and the United States to determine sulfur and carbon sources and assess competing syngenetic and diagenetic/epigenetic timings of mineralization. An important conclusion of the study is that earlier bulk sulfur isotope studies using conventional analytical techniques blurred the isotopic pattern that became sharper with the advent of microanalytical methods beginning in the late 1980s. Carbon–oxygen data from carbonates suggest that the ore fluids were warm (> 150 °C) basinal brines. Microanalytical sulfur isotope data in combination with observations of mineral textures suggest that the earliest, barren pyrite incorporated sulfide produced by BSR, whereas later-formed sphalerite and galena incorporated sulfide derived from TSR. The sulfur isotope and paragenetic data are most consistent with a diagenetic/epigenetic timing for mineralization, although in some deposits (e.g., Red Dog, Alaska) the data are indicative of protracted mineralization, including early syngenetic sulfide deposition.

Hagemann et al. (2023) document isotopic changes in the complex iron ore mineral system in which iron formation protore was progressively enriched via hypogene and supergene processes to form iron ore. Data from iron ore deposits in Australia, South Africa and Brazil, showed a consistent decrease in $\delta^{18}O$ from values of 4–9‰ in iron formation protore to values as low as − 10‰ in high-grade iron ore. Hagemann et al. (2023) interpret this $\delta^{18}O$ shift to indicate the incursion of ancient meteoric fluids along fault and fracture zones. They also interpret that upgrading involved magmatic fluids in the Carajás (Brazil), basinal brines in the Hamersley (Australia) and deep crustal (metamorphic or magmatic) fluids in the Quadrilátero Ferrífero (Brazil) provinces. The systematic decrease of $\delta^{18}O$ values in iron oxides from the early to late paragenetic stages and from the distal to proximal alteration zones, including the ore zone, may be used as a geochemical vector. In this case, oxygen isotope analyses on iron oxides would provide a potential exploration tool (Hagemann et al. 2023).

2.4 Part IV—Metallic Stable Isotopes: Metal Sources and Mineralizing Processes

The development of ICP-MS has enabled stable isotope studies to be extended from the traditional light stable isotopes into transition metals such as copper, iron and zinc and other ore and related elements such as antimony, molybdenum and silver. Of these, studies of iron, copper and zinc in this part are reviewed in chapters by Lobato et al. (2023), Mathur and Zhao (2023) and Wilkinson (2023).

Lobato et al. (2023) compile existing data and present new data on the variability of iron isotopes in banded iron formation-hosted iron ore deposits from Australia and Brazil. This analysis found that although some characteristics of iron isotopes varied by metallogenic province, others, such as the association of hypogene ores with lower $\delta^{56}Fe$ and supergene ores with higher $\delta^{56}Fe$, persist in all metallogenic provinces. In the Quadrilátero Ferrífero district (Brazil), iron isotopic characteristics in hematite ore differ between deformational domains: deposits in domains with low deformation have lower $\delta^{56}Fe$ than those in high-strain domains, possibly reflecting different fluid characteristics, such as temperature. In the Corumbá region (Brazil), $\delta^{56}Fe$ characteristics reflect the interplay of primary seawater, microbial activity and supergene alteration. In the Carajás district (Brazil), hypogene magnetite and hematite iron ores have lower $\delta^{56}Fe$ than their iron formation protores. In the Hamersley Province (Australia), $\delta^{56}Fe$ and $\delta^{18}O$ values appear correlated during greenschist metamorphism and hypogene upgrading, but negatively correlated during subsequent supergene upgrading. Lobato et al. (2023) highlight that hypogene iron ores tend to have lower $\delta^{56}Fe$ values, whereas supergene overprints result in higher $\delta^{56}Fe$ values, although they note that more data are required to confirm these observations.

Mathur and Zhao (2023) review the development of copper isotopes over the last two decades in ore genesis, environmental monitoring and exploration. Compared to other metallic stable isotopes, $\delta^{65}Cu$ has a large range of 10‰, possibly due to copper being involved in redox reactions, during both hypogene and supergene mineralizing processes. After documenting analytic methods and reporting conventions, Mathur and Zhao (2023) provide an overview of the natural variability of copper isotopes followed by a discussion of how copper isotopes fractionate and evolve in mineral systems including magmatic Ni–Cu, porphyry copper (and related skarns), VHMS, sediment-hosted copper and supergene-enriched systems. They document that the greatest fractionations occur during low temperature redox reactions, such as supergene enrichment, and how this information can be used during exploration of, for instance, leached caps that form over copper-rich deposits at depth or in groundwaters.

For zinc isotopes, Wilkinson (2023) reviews analytical methods and reporting conventions, and then describes natural variability in rocks and mineral deposits. Compared to $\delta^{56}Fe$ and $\delta^{65}Cu$, the variability observed in $\delta^{66}Zn$ (< 2‰) is small, probably because in zinc on Earth has only one valence state and, hence is not involved in redox reactions. Because of the newness of analytical methods, zinc isotope data from mineral deposits are uncommon. One exception is sediment-hosted zinc deposits, including both carbonate-hosted and shale-hosted deposits. Wilkinson (2023) demonstrates a reasonably consistent enrichment of $\delta^{66}Zn$ from the core to the peripheries of these deposits. For other deposit types for which data are available (VHMS, veins and porphyry and related deposits) currently available data are simply insufficient to establish consistent patterns.

Although not discussed in this book, significant natural variations have been observed in other metallic and semi-metallic isotopes, including antimony, molybdenum and silver (Rouxel et al. 2002; Arnold et al. 2004; Mathur et al. 2018). Because these elements are extracted from some ores or are closely associated with ore metals, they have potential to further understand ore genesis.

3 Conclusion

Isotope geochemistry has played an essential role in the development of mineral deposit and metallogenic models at all scales, providing constraints on sources (fluid, sulfur and metal), processes and ages. In particular, the data have provided critical information for mineral systems models for porphyry copper/epithermal, VHMS, orogenic gold, iron ore, sediment-hosted Zn–Pb and Cu, and orthomagmatic Ni–Cu deposits, among others. The development of rapid, inexpensive and robust geochronological methods for dating ore and ore-related minerals has provided the fourth dimension to economic geology and metallogenic research. Radiogenic isotopes also provide information on the sources of ore metals, including lead and rare earth elements; in addition, mapping of isotopic ratios and derived parameters provides new insights into tectonic and metallogenic provinces, including fertility and metal endowment. Light stable isotopes have been a workhorse in ore genesis studies since the mid-1960s, providing important constraints on fluid and sulfur sources as well as ore forming processes. Metallic stable isotopes provide information on metal sources and mineralizing processes, particularly redox processes. This isotopic research, like isotope mapping, is in the early stage of evolution and, when combined with other major development in isotopic research, such as the analysis of multiple isotope ratios in the oxygen ($\delta^{18}O$ and $\delta^{17}O$) and sulfur ($\delta^{34}S$, $\delta^{33}S$ and $\delta^{36}S$) systems, will undoubtedly provide further invaluable insights into ore genesis.

Acknowledgements The editors (DLH and JG) thank John Slack for proposing this book as part of the Society for Geology Applied to Mineral Deposits (SGA) Mineral Resource Reviews book series. We also thank the authors for their contributions and, particularly, their patience in seeing this book through to publication. DLH notes the support of his employer, Geoscience Australia, who allowed GA authors to contribute and provided financial support for this book to be openly accessible. The authors thank Andrew Cross for review, and the contribution is published with permission of the Executive Director, Geoscience Australia.

References

Albarède F (2004) The stable isotope geochemistry of copper and zinc. Rev Mineral Geochem 55:409–427

Arnold GL, Anbar AD, Barling J, Lyons TW (2004) Molybdenum isotope evidence for widespread anoxia in Mid-Proterozoic oceans. Science 304:87–90

Beaty DW, Taylor HP Jr (1982) Some petrologic and oxygen isotope relationships in the Amulet mine, Noranda, Quebec, and their bearing on the origin of Archean massive sulfide deposits. Econ Geol 77:95–108

Bennett VC, DePaolo DJ (1987) Proterozoic crustal history of the western United States as determined by neodymium isotopic mapping. Geol Soc Am Bull 99:674–685

Boltwood BB (1907) On the ultimate disintegration products of the radio-active elements. Part II. The disintegration products of uranium. Am J Sci 23:77–88

Canfield DE (2005) The early history of atmospheric oxygen: homage to Robert M. Garrels. Ann Rev Earth Planet Sci 33:1–36

Cathles LM (1993) Oxygen isotope alteration in the Noranda mining district, Abitibi greenstone belt, Quebec. Econ Geol 88:1483–1511

Chadwick J (1932) Possible existence of a neutron. Nature 129:312

Chambers LA, Trudinger P (1979) Microbiological fractionation of stable sulfur isotopes: a review and critique. Geomicrobiology 1:249–293

Champion DC, Huston DL (2023) Applications of Nd isotopes to ore deposits and metallogenic terranes; using regional isotopic maps and the mineral systems concept. In: Huston DL, Gutzmer J (eds), Isotopes in economic geology, metallogensis and exploration. Springer, Berlin. This volume

Chelle-Michou C, Schaltegger U (2023) U–Pb dating of mineral deposits: from age constrains to ore-forming processes. In: Huston DL, Gutzmer J (eds) Isotopes in economic geology, metallogensis and exploration. Springer, Berlin. This volume

Chiaradia M (2023) Radiometric dating applied to ore deposits: theory and methods. In: Huston DL, Gutzmer J (eds) Isotopes in economic geology, metallogensis and exploration. Springer, Berlin. This volume

Donnelly TH, Lambert IB, Oehler DZ, Hallbert J, Hudson DR, Smith JW, Bavinton OA, Golding LY (1978) A reconnaissance study of stable isotope ratios in Archaean rocks from the Yilgarn Block, Western Australia. J Geol Soc Austr 24:409–420

Engel AEJ, Clayton RN, Epstein S (1958) Variations in isotope composition of oxygen and carbon in Leadville Limestone (Mississippian, Colorado) and in its hydrothermal and metamorphic phases. J Geol 66:374–393

Epstein S, Buchsbaum HA, Lowenstam H, Urey HC (1953) Revised carbonate-water isotope temperature scale. Geol Soc Am Bull 64:1315–1326

Field CW, Gustafson LB (1976) Sulfur isotopes in the porphyry copper deposit at El Salvador, Chile. Econ Geol 71:1533–1548

Goodwin AM, Monster J, Thode HG (1976) Carbon and sulfur isotope abundances in Archean iron formations and early life. Econ Geol 71:870–891

Hagemann S, Hensler A-S, Figueiredo e Silva RC, Tsikos H (2023) Light stable isotope (O, H, C) signatures of BIF-hosted iron ore systems: implications for genetic models and exploration targeting. In: Huston DL, Gutzmer J (eds) Isotopes in economic geology, metallogensis and exploration. Springer, Berlin. This volume

Halliday AN, Lee D-C, Christensen JN, Walder AJ, Freedman PA, Jones CE, Hall CM, Yi W, Teagle D (1995) Recent developments in inductively coupled plasma magnetic sector multiple collector mass spectrometry. Int J Mass Spectrom Ion Proc 146:21–33

Hayes JM, Lambert IB, Strauss H (1992) The sulfur-isotope record. In: Schopf JW, Klein C (eds) The Proterozoic biosphere: a multidisciplinary study. Cambridge University Press, Cambridge, pp 129–132

Holmes A (1946) An estimate of the age of the earth. Nature 157:680–684

Houk RS, Fassel VA, Flesch GD, Svec HJ, Gray AL, Taylor CE (1980) Inductively coupled argon plasma as an ion source for mass spectrometric determination of trace elements. Anal Chem 52:2283–2289

Houtermans FG (1946) Die Isotopen-Haufigkeiten im naturlichen Blei und dasAalter des Urans. Naturwissenschaften 33:185–187

Hsu KJ, Oberhänsli H, Gao JY, Sun S, Chen H, Krahenbuhl U (1985) 'Strangelove Ocean' before the Cambrian explosion. Nature 268:209–213

Huston DL, Champion DC (2023) Applications of lead isotopes to ore geology, metallogenesis and exploration. In: Huston DL, Gutzmer J (eds) Isotopes in economic geology, metallogensis and exploration. Springer, Berlin. This volume

Huston DL, Trumbull RB, Beaudoin G, Ireland T (2023a) Light stable isotopes (H, B, C, O and S) in ore studies —methods, theory, applications and uncertainties. In: Huston DL, Gutzmer J (eds) Isotopes in economic geology, metallogensis and exploration, Springer, Berlin. This volume

Huston DL, LaFlamme C, Beaudoin G, Piercey S (2023b) Light stable isotopes in volcanic-hosted massive sulfide ore systems. In: Huston DL, Gutzmer J (eds) Isotopes in economic geology, metallogensis and exploration. Springer, Berlin. This volume

Johnson CM, Beard B (2006) Fe isotopes: an emerging technique in understanding modern and ancient biogeochemical cycles. GSA Today 16:4–10

Kelley S (2002) K–Ar and Ar–Ar dating. Rev Mineral Geochem 47:785–818

Knoll AH, Hayes JM, Kaufman AJ, Swett K, Lambert IB (1986) Secular variation in carbon isotope ratios from Upper Proterozoic successions of Svalbard and East Greenland. Nature 321:832–838

Lambert IB, Donnelly TH (1991) Atmospheric oxygen levels in the Precambrian: a review of isotopic and geological evidence. Palaeogeogr Palaeoclimatol Palaeoecol (global and Planetary Change Section) 97:83–89

Lambert IB, Groves DI (1981) Early earth evolution and metallogeny. In: Handbook of stratabound and stratiform ore deposits 8, pp 339–447

Lambert IB, Donnelly TH, Dunlop JSR, Groves DI (1978) Stable isotope studies of early Archaean sulphate deposits of probable evaporitic and volcanogenic origins. Nature 276:808–811

Lambert IB, Phillips GN, Groves DI (1983) Stable isotope compositions and genesis of Archaean gold mineralization. In: Foster RP (ed) Gold 82. Balkema, pp 373–387

Lambert IB, Donnelly TH, Walter MR, Southgate PN, Olley J, Zeilinger I, Zang W, Lu S, Ma G, Perkins DJ, Shergold JH, Jackson MJ (1985) Stable isotope trends and their palaeoenvironmental significance. Baas Becking Geobiol Lab Ann Rep 1985:21–31

Lambert IB, Walter MR, Zang W, Lu S, Ma G (1987) Palaeo-environment and carbon isotope stratigraphy of Upper Proterozoic carbonates of the Yangtze Platform. Nature 235:140–142

Lobato LM, Figueiredo e Silva RC, Angerer T, Mendes M, Hagemann S (2023) Fe isotopes applied to BIF-hosted iron ore deposits. In: Huston DL, Gutzmer J (eds) Isotopes in economic geology, metallogensis and exploration. Springer, Berlin. This volume

Mason TFD, Weiss DJ, Chapman JB, Wilkinson JJ, Tessalina SG, Spiro B, Horstwood MSA, Spratt J, Coles BJ (2005) Zn and Cu isotopic variability in the Alexandrinka volcanic-hosted massive sulfide (VHMS) ore deposit, Urals, Russia. Chem Geol 221:170–187

Mathur R, Zhao Y (2023) Copper isotopes used in mineral exploration. In: Huston DL, Gutzmer J (eds) Isotopes in economic geology, metallogensis and exploration. Springer, Berlin. This volume

Mathur R, Arribas A, Megaw P, Wilson M, Stroup S, Meyer-Arrivillaga D, Arriba I (2018) Fractionation of silver isotopes in native silver explained by redox reactions. Geochim Cosmochim Acta 224:313–326

Norman MD (2023) The ^{187}Re-^{187}Os and ^{190}Pt-^{186}Os radiogenic isotope systems: techniques and applications to metallogenic systems. In: Huston DL, Gutzmer J (eds) Isotopes in economic geology, metallogensis and exploration. Springer, Berlin. This volume

Ohmoto H, Felder RP (1987) Bacterial activity in the warmer, sulfate-bearing Archean oceans. Nature 328:244–246

Ohmoto H, Mizukami M, Drummond SE, Eldridge CS, Pisutha-Arnond V, Lenagh TC (1983) Chemical processes in Kuroko formation. Econ Geol Mon 5:570–604

Phillips GN, Groves DI, Neall FG, Donnelly TH, Lambert IB (1986) Anomalous sulfur isotope compositions in the Golden Mile, Kalgoorlie. Econ Geol 81:2008–2016

Quesnel B, Scheffer C, Beaudoin G (2023) The light stable isotope (H, B, C, N, O, Si, S) composition of orogenic gold deposits. In: Huston DL, Gutzmer J (eds) Isotopes in economic geology, metallogensis and exploration. Springer, Berlin. This volume

Richards JP, Noble SR (1998) Applications of radiogenic isotope systems to the timing and origin of hydrothermal processes. Rev Econ Geol 10:195–233

Rouxel O, Luden J, Fouquet Y (2002) Antimony isotope variations in natural systems and implications for the use as geochemical tracers. Chem Geol 200:25–40

Rutherford E, Soddy F (1902) The cause and nature of radioactivity—Part I. Lond Edinb Dublin Philos Mag 4:370–396

Sangster DF (1968) Relative sulphur isotope abundances of ancient seas and strata-bound sulphide deposits. Proc Geol Assoc Can 19:79–91

Sheppard SMF, Nielsen RL, Taylor HP Jr (1969) Oxygen and hydrogen isotope ratios of clay minerals from porphyry copper deposits. Econ Geol 64:755–777

Sheppard SMF, Nielsen RL, Taylor HP Jr (1971) Hydrogen and oxygen isotope ratios in minerals from porphyry copper deposits. Econ Geol 66:515–542

Soddy F (1913) Intra-atomic charge. Nature 92:399–400

Strutt RJ (1905) On the radio-active minerals. Proc R Soc Lond A 76:88–101

Trudinger PA, Lambert IB, Skyring GW (1972) Biogenic sulphide ores: a feasibility study. Econ Geol 67:1114–1127

Urey HC (1947) The thermodynamics of isotope substances. J Chem Soc 1947:562–581

Vasconcelos PM (1999) K–Ar and ^{40}Ar/^{39}Ar geochronology of weathering processes. Ann Rev Earth Planet Sci 27:183–229

Waltenberg K (2023) Application of Lu-Hf isotopes to ore geology, metallogenesis and exploration. In: Huston DL, Gutzmer J (eds) Isotopes in economic geology, metallogensis and exploration. Springer, Berlin. This volume

Williams N (2023) Light-element stable isotope studies of the clastic-dominated lead-zinc mineral systems of northern Australia and the North American Cordillera: implications for ore genesis and exploration. In: Huston DL, Gutzmer J (eds) Isotopes in economic geology, metallogensis and exploration. Springer, Berlin. This volume

Wilkinson JJ (2023) The potential of Zn isotopes in the science and exploration of ore deposits. In: Huston DL, Gutzmer J (eds) Isotopes in economic geology, metallogensis and exploration. Springer, Berlin. This volume

Wilson AJ, Cooke DR, Harper BJ, Deyell CL (2007) Sulfur isotopoic zonation in the Cadia district, southeastern Australia: exploration significance and implicates for the genesis of alkalic porphyry gold-copper deposits. Mineral Deposita 42:465–487

Zartman RE (1974) Lead isotope provinces in the Cordilleran of the western United States and their geological significance. Econ Geol 69:792–805

Zhu XK, O'Nions RK, Guo Y, Belshaw NS, Rickard D (2000) Determination of natural Cu-isotope variation by plasma-source mass spectrometry; implications for use as geochemical tracers. Chem Geol 163:139–149

Zhu XK, Guo Y, Williams RJP, O'Nions RK, Matthews A, Belshaw NS, Canters GW, de Waal EC, Weser U, Burgess BK, Salvato B (2002) Mass fractionation processes of transition metal isotopes. Earth Planet Sci Lett 200:47–62

Radiometric Dating Applied to Ore Deposits: Theory and Methods

Massimo Chiaradia

Abstract

Metallic ore deposits have contributed to the development of the human society since pre-historic times and nowadays are one of the pillars of unprecedented technological developments. In order to understand how metallic ore deposits form and thus construct genetic models that may serve as exploration guides, determining the age of an ore deposit is one of the most important pieces of information needed. More recently it has also become evident that determining the temporal duration of mineralizing events can offer valuable information on how certain deposits form and thus improve genetic models. Radiometric dating of ore minerals or of other minerals that are demonstrably associated in space and time with mineralization is the most accurate and precise tool to date an ore deposit. This Introductory Chapter summarizes basic concepts on why ore deposit dating is important and how this can be achieved through different methods. It illustrates basic differences among different methods and serves as an introduction to the more detailed descriptions of specific dating methods presented in the following Chapters.

M. Chiaradia (✉)
Department of Earth Sciences, University of Geneva, Rue des Maraîchers 13, 1205 Geneva, Switzerland
e-mail: Massimo.Chiaradia@unige.ch

1 Introduction

Metallic ore deposits are anomalous accumulations of metal commodities within a few kilometres of the Earth's surface, which have been concentrated significantly above their average crustal contents and in mineralogical forms that make them economically and metallurgically exploitable (e.g., Skinner 1997). Although various geological processes are implicated in concentrating metals in the Earth's crust, metals in most ore deposits, except some types, like orthomagmatic deposits, are precipitated by aqueous solutions displaying a range of temperature and pH conditions (e.g., Seward and Barnes 1997). These aqueous solutions may derive directly from cooling and/or rising magmas or may be modified seawater, meteoric or connate water, or a mixture of all the above and other fluids, depending on the typology of the ore deposit formed (e.g.,Giggenbach 1997; Scott 1997; Mckibben and Hardie 1997). Precipitation of metals from these aqueous solutions is usually accompanied by an alteration process of the rocks adjacent to the fluid path (e.g., Reed 1997). This is due to chemical disequilibrium between the fluid and the rocks it interacts with. Therefore, a mineralization process results ultimately in the precipitation of metal-bearing ore minerals

D. Huston and J. Gutzmer (eds.), *Isotopes in Economic Geology, Metallogenesis and Exploration*, Mineral Resource Reviews, https://doi.org/10.1007/978-3-031-27897-6_2

(usually sulfides and/or oxides) and metal-barren gangue minerals (e.g., carbonates, sulfates, silicates) directly from hydrothermal fluids or as replacement of pre-existing mineral phases by fluid-rock reactions.

Dating the time in the past when metals in the Earth's crust have been concentrated is instrumental for a complete understanding of the geological processes that form mineral deposits. Determining the timing of mineralization allows us to associate it with magmatic, metamorphic, climatic, biologic, tectonic, and/or geodynamic events that are the drivers of the mineralization processes. Many researchers have studied the distribution of mineral deposit types through time highlighting their link with major events of the Earth's history (e.g., supercontinent assembly, oxidation events, climatic changes) and their temporal recurrence or uniqueness (Watson 1978; Veizer et al. 1989; Barley and Groves 1992; Goldfarb et al. 2001; Frimmel 2005; Heinrich 2015). Geochronology of ore deposits is fundamental to pinpoint these relationships, to evaluate their repeatability through time, and eventually to offer a "predictive" tool for exploration of different mineral deposits within specific geochronological windows.

In addition, establishing the timescales of mineralizing events improves our understanding of the physico-chemical processes that lead to metal precipitation, especially the rate at which metal precipitation occurs, which is a function of the energy of the system and how this is distributed through time (Weis et al. 2012; Chiaradia et al. 2013, 2014; Chelle-Michou et al. 2017; Chiaradia and Caricchi 2017; Chiaradia 2020).

Dating of mineral deposits can be accomplished through "indirect" stratigraphic methods (e.g., paleomagnetism, fossil assemblages, isotope chemostratigraphy using C and Sr isotopes) or "direct" absolute radiometric dating methods (e.g., U–Pb, Re–Os, $^{40}Ar/^{39}Ar$, K/Ar, Rb–Sr). It should be emphasized that ages obtained through "indirect" stratigraphic methods are ultimately determined by absolute radiometric dating that establish the geological time scale of the sedimentary sequences deposited on the surface of our planet (Gradstein et al. 2020). Such "indirect" stratigraphic methods can only be applied to deposits that are interpreted or inferred to be syn-diagenetic or syn-depositional, e.g., VHMS deposits, potentially, but not necessarily, sediment-hosted base metal deposits and carbonate-hosted Pb–Zn deposits (Symons and Sangster 1994; Leach et al. 2001). In the case of deposits that are not syn-sedimentary, these stratigraphic methods only allow us to establish upper and/or lower limits to the age of mineralization. The recognition of the syn-diagenetic or syn-depositional nature of a mineralization is a difficult task especially where deposits of older ages are overprinted by subsequent geological events that may mask primary textural relationships (e.g., see controversy on the age of the Zambia copper belt: Sillitoe et al. 2017a, b; Hitzman and Broughton 2017; Muchez et al. 2017).

The most accurate and precise way of determining the age of an ore deposit is radiometric dating, while recognizing that careful textural constraints on the minerals used for dating are essential in order to date ore or alteration minerals that are genetically associated with the mineralization (e.g., Chiaradia et al. 2013). Accuracy refers to how close a measurement of an age is to the "real" value, which is not trivial to assess because a priori we do not know what the "real" age is. Estimation of the accuracy can be done through measurement of standards with known, certified ages and evaluating how far our measure is from the certified value. However, such measurements do not guarantee that sample ages are accurate, if, for instance, unknown samples are affected by open system behavior. Several parameters may affect the accuracy of an age determination as it will be discussed below. Precision indicates the uncertainty that we can attach to an age that we have measured and usually depends on limitations of the analytical tools used.

The goal of this chapter is to provide basic information on minerals, techniques, accuracy and precision of radiometric methods used for dating metallic mineral deposits.

2 Principles of Radiometric Dating

2.1 Radioactive Decay

Isotopes are nuclides of the same element and as such are characterized by the same number of protons but different numbers of neutrons. Radioactivity is the process through which naturally unstable isotopes of elements emit sub-atomic particles and energy resulting in changes of the number of protons and neutrons which lead to the transformation of the initial isotope into an isotope of another element (Fig. 1) (Rutherford and Soddy 1902a, b). When a radioactive isotope is incorporated into a mineral at the time of its formation it becomes a natural clock that may allow the determination of the time in the past when such incorporation, or the mineral formation, occurred. Strutt (1905) was the first to use radioactivity as a geochronological tool. Although the formation of a mineral may not be instantaneous at the scale of human life it is so at the scale of most geological events. The way the radioactive isotope clock works is that in several specific minerals radioactive isotopes of some elements can be incorporated at levels sufficient to be measured with available analytical techniques and instruments (mass spectrometers). Through time these radioactive isotopes (called parents) decay, emitting various forms of energy and sub-atomic particles and transforming into isotopes of other elements (called radiogenic daughters) which may also be measurable if enough time has occurred since the formation of the mineral to allow sufficient decay of the radioactive isotope (Fig. 1). Several types of radioactive decay exist (e.g., α and β decay: Fig. 1) and apply to different parent radioactive isotopes (see specific literature for more details: e.g., Faure and Mensing 2005; Dickin 2005; Table 1). The products of radioactive decay may be stable (i.e., non-radioactive) daughter isotopes (a typical example is the decay of ^{187}Re to stable ^{187}Os by β^- decay or the decay of ^{147}Sm to ^{143}Nd by α decay) or daughter isotopes that are themselves radioactive parents and decay to another daughter isotope a typical example of the latter is the decay chain of ^{238}U to ^{206}Pb, which occurs through a series of intermediate radioactive daughter/parent isotopes formed by α and β decay processes, e.g.,

$$^{238}U \rightarrow^{\alpha} {}^{234}Th \rightarrow^{\beta} {}^{234}Pa \rightarrow^{\beta} {}^{234}U \rightarrow^{\alpha} {}^{230}Th \quad (1)$$

and so on until stable ^{206}Pb.

The use of radioactivity as a geological clock is possible because the decay rate of a large population of a particular radioactive isotope is statistically constant (and for this reason it is called the decay constant, identified by the Greek letter λ), can be measured and quantified, and does not change with changes of physico-chemical parameters like temperature, pressure and composition of the phase hosting the radioactive isotopes (e.g., Faure and Mensing 2005). Different radioactive parent isotopes have widely differing rates of decay encompassing > 10 orders of magnitude (Table 1), which allows determination of events that range in age from billions of years ago to decades ago depending on the chosen isotopic decay system. Thus, by measuring the amounts of radioactive parent isotopes and daughter products present today in a mineral, and knowing the constant rate at which the parents have decayed into the daughters, we can determine the time elapsed over which the mineral has been accumulating the daughter isotope since its formation. Time, rate of disintegration and atomic abundances of parents and daughters are linked together through a simple exponential equation

$$D^* = N\left(e^{\lambda t} - 1\right) \quad (2)$$

where D* and N are the numbers of radiogenic daughter and radioactive parent isotopes, respectively, measured in the system today, λ is the decay rate (known as decay constant) and t is the time since the system (mineral in geological applications) has formed incorporating the radioactive parent isotope. This equation can be solved in order to obtain the time

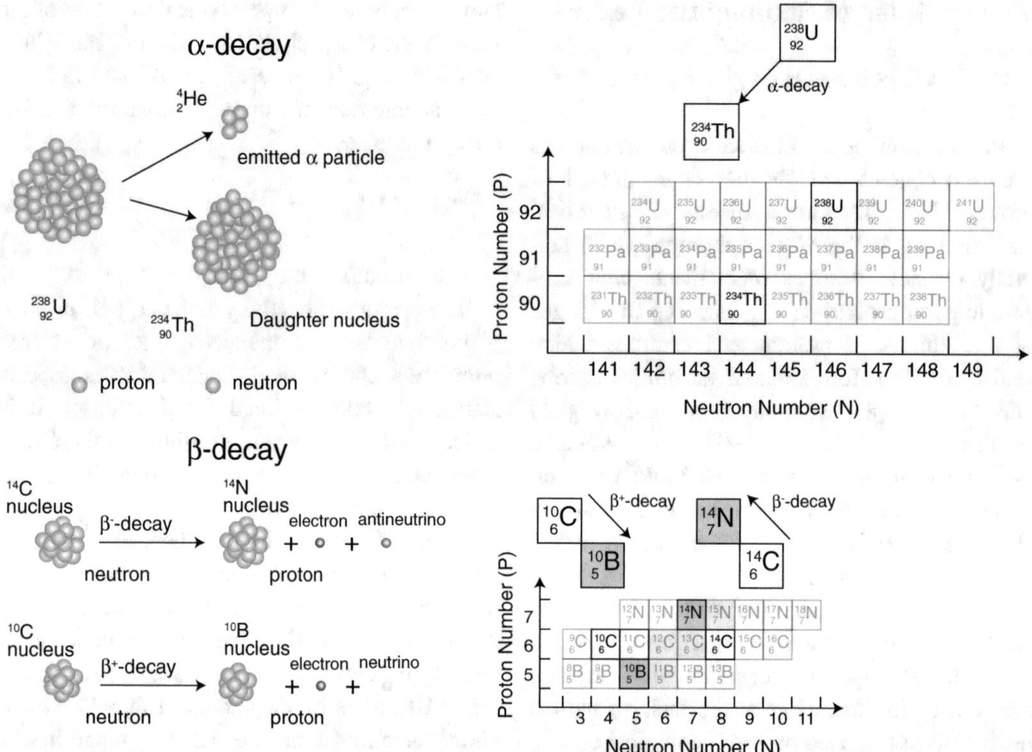

Fig. 1 Graphic examples of α and β decay processes and their expression in the pertinent portions of the nuclide table

$$t = \frac{1}{\lambda}\ln\left(\frac{D^*}{N} + 1\right). \tag{3}$$

2.2 Conditions

There are several conditions that must be met in order to use Eq. (3) to obtain a meaningful age. First of all, the mineral must behave as a closed system since its formation, not allowing any escape or any ingression of either parent or daughter isotope, which would modify the D^*/N ratio produced by the radioactive decay.

This is a condition that is difficult to be realized because: (i) all elements occurring in minerals, including the radioactive isotopes, tend to diffuse in order to re-equilibrate with the surrounding environment under the drive of chemical and thermal gradients; (ii) radioactive decay creates damage to the crystalline lattice of minerals, which favors the escape or ingression of elements; (iii) the minerals hosting the radioactive parents may be subject to interactions with fluids during their geological life which will cause a release or an ingression of radiogenic and parent isotopes (the reader may refer to Mattinson 2005, to see how the problem of crystal lattice damage of zircons can be tackled in U–Pb geochronology). Among these three processes, thermally activated diffusion is a ubiquitous process, by which elements diffuse within crystals as a function of atomic size, crystal structure and size, and temperature. For a particular element in a particular crystal structure diffusion becomes increasingly fast with increasing temperature, through an exponential function known as the Arrhenius law.

All the points above bear on the fundamental question of what exactly are we dating when we apply a radiometric clock to a mineral? In ideal cases, where the mineral has remained a closed system since its formation, the answer is that we

Table 1 Some radioactive system decay schemes and their applications to dating mineral deposits

Decay scheme	Decay constant	Half-life	Types of datable minerals	Datable minerals
$^{238}U \rightarrow {}^{206}Pb$ $^{235}U \rightarrow {}^{207}Pb$	1.55125×10^{-10} 9.8485×10^{-10}	4.468×10^{9} 0.704×10^{9}	Magmatic and hydrothermal ore-associated minerals ± ore minerals (e.g., cassiterite)	zr, tit, rt, bd, xen, mon, all, cs, gar
$^{232}Th \rightarrow {}^{208}Pb$	4.9475×10^{-11}	14.010×10^{9}	Magmatic and hydrothermal ore-associated minerals ± ore minerals	all, mon
$^{187}Re \rightarrow {}^{187}Os$	1.66×10^{-11}	4.17×10^{10}	Ore minerals (sulfides)	mol, various sulfides
$^{40}K \rightarrow {}^{40}Ar$	$\sim 5.81 \times 10^{-11}$	1.193×10^{10}	Gangue hydrothermal minerals ± fluids contained in ore and gangue minerals	K-bearing minerals (micas, amph, Kf, al, ….) and fluids
$^{87}Rb \rightarrow {}^{87}Sr$	1.42×10^{-11}	4.88×10^{10}	Gangue hydrothermal minerals ± fluids contained in ore and gangue minerals	micas, Kf, sl, and Rb-bearing fluids (as inclusions in, e.g., quartz)
$^{147}Sm \rightarrow {}^{143}Nd$	6.54×10^{-12}	1.06×10^{11}	Certain gangue and ore minerals (e.g., scheelite)	cpx, gar, sch
$^{176}Lu \rightarrow {}^{176}Hf$	1.94×10^{-11}	3.57×10^{10}	Certain gangue minerals	gar
$^{234}U \rightarrow {}^{230Th}$	2.829×10^{-6}	2.45×10^{5}	Young zircons and carbonates	zr, carbonates
$^{230}Th \rightarrow {}^{226}Ra$	9.1929×10^{-6}	7.54×10^{4}	Young zircons and carbonates	zr, carbonates
$^{226}Ra \rightarrow {}^{222}Rn$	4.33×10^{-4}	1.6×10^{3}	Young carbonates (e.g., travertines)	carbonates
$^{210}Pb \rightarrow {}^{210}Bi$	3.06×10^{-2}	2.26×10^{1}	Lake sediments, snow/ice	bulk sediments, snow/ice

Decay constant and half-life values are from Faure and Mensing (2005) and are purely indicative. Continuous refining of these values is proposed in the literature and the reader should refer to the latest proposed values for geochronological applications

Abbreviations: zr = zircon, tit = titanite; rt = rutile; bd = baddeleyte; xen = xenotime; mon = monazite; all = allanite; cs = cassiterite; mol = molybdenite; amph = amphibole; Kf = K-feldspar; al = alunite; cpx = clinopyroxene; gar = garnet; sch = scheelite; sl = sphalerite

are dating the time elapsed since the mineral crystallised. In other cases, the radiometric age may reflect the time since the mineral cooled to a temperature below which diffusion of the parent and daughter isotopes becomes so slow that the mineral acts as a closed system. This temperature is known as closure temperature (Dodson 1973). Because the closure temperature depends on crystal structure, grain size of the crystal, and atomic size of the diffusing isotope, there is a wide range of closure temperatures (from > 900 to < 100 °C) for the different radiometric dating systems applied to different minerals (Fig. 2).

In cases where a mineral has grown at temperatures below its closure temperature for the isotopic system of interest, an isotopic age from this system is likely to reflect the time of formation of that mineral. In contrast, where a mineral grows at temperatures above its closure temperature, the isotopic age will be younger than the timing of mineral formation. The time at which a mineral passes below its closure temperature can be variably younger than the time of its formation and reflects the thermal history of the rock that contains the mineral.

These considerations highlight the need to carefully assess the geological meaning of a radiometric age on a case by case basis, requiring as much knowledge as possible of the geological context of the sample.

2.3 Systems not Incorporating Initial Daughter or with Known Initial Daughter

A rock or mineral can be dated using Eq. (3) only if it does not incorporate any significant amount of daughter isotope from the surrounding environment during its formation or if the amount and isotopic composition of the incorporated daughter isotope are known. In such cases either all the daughter isotopes measured in the mineral can be attributed to in-situ radioactive decay or they can be corrected for the incorporation of the initial daughter isotopes. Daughter isotopes of any parent-daughter system are continuously produced everywhere on Earth and therefore the possibility to use Eq. (3) depends on the ability of a mineral to selectively incorporate a radioactive parent and exclude the radiogenic daughter at the time of mineral formation. Several commonly-dated mineral and isotopic pairs satisfy this requirement, for example, the U–Pb system in zircon, the Re–Os system in molybdenite, and, to some extent, the K–Ar system in micas and feldspars. In the latter case it is assumed that initial argon present in a mineral has the isotopic composition of atmospheric argon, which is known and is assumed to have not changed through time (see below): this allows the subtraction of the initial atmospheric Ar.

The K–Ar method of dating of K-bearing minerals is more complex than U–Pb dating of zircon and Re–Os dating of molybdenite for at least two additional physical and methodological issues. The first issue is that the relatively low closure temperatures of Ar diffusion in commonly dated K-bearing minerals (Fig. 2), due to the high diffusivity of Ar, implies that K–Ar dates record the time at which the minerals pass below their Ar closure temperatures. The latter, for most K-bearing minerals, fall in the lower range of hydrothermal temperatures (Fig. 2) and therefore may be lower than the dated mineral crystallization temperatures. In such a case, the K–Ar age recorded by the mineral will be younger than the mineral crystallization age. In other cases, the crystallization of the K-bearing minerals may occur at temperatures that are similar or lower than their closure temperature. In such a case the $^{40}Ar/^{39}Ar$ date will yield the crystallization age of the mineral. Careful petrographic studies should be carried out to evaluate the meaning of K–Ar ages of K-bearing minerals associated with ore deposits. This is especially true in multi-pulsed systems like porphyry deposits that are characterized by long-lived thermal anomalies above the closure temperature of commonly dated K-bearing minerals (e.g., Chiaradia et al. 2014). In contrast, the high closure temperatures of zircon (Cherniak and

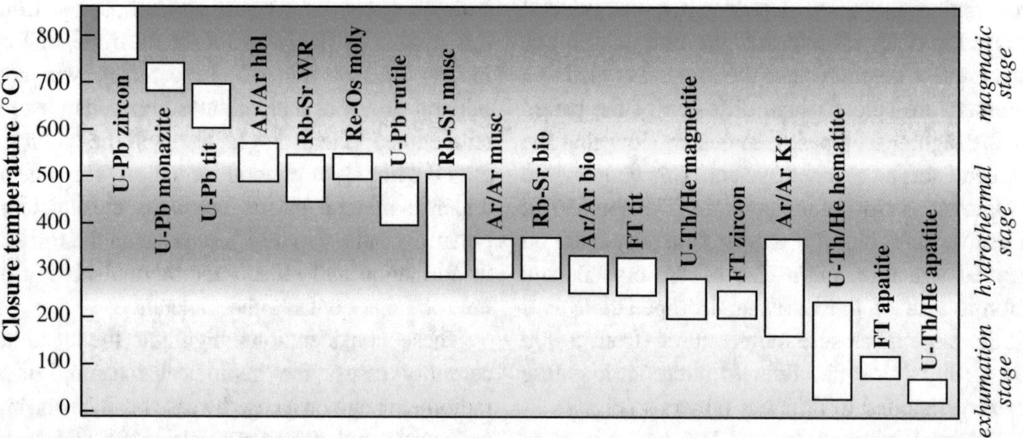

Fig. 2 Range of closure temperatures for various geochronometers. Reproduced with permission from Chiaradia et al. (2013); Copyright 2013 Society of Economic Geologists

Watson 2001) and molybdenite (Stein et al. 2001) for the corresponding U–Pb and Re–Os systems, most of the time higher than crystallization temperatures of these minerals, imply that dates obtained on these minerals are crystallization ages. The second issue is that the K–Ar method requires independent measurements of K and Ar by distinct analytical methods on different sample aliquots, which poses a problem if the sample aliquots are not homogeneous.

The $^{40}Ar/^{39}Ar$ method of dating is a derivation of the K–Ar method, which has the advantage that a date can be obtained through a single measurement of the same sample aliquot on a noble gas mass spectrometer, thanks to the conversion of ^{39}K of the sample to ^{39}Ar by neutron irradiation in a nuclear reactor. Since ^{39}K and ^{40}K occur in a fixed ratio, ^{39}Ar becomes a proxy of ^{40}K measurement. This method requires the use of monitors with a known age for the quantification of the conversion of ^{39}K to ^{39}Ar in the reactor, and is therefore an indirect dating method relying on the accuracy of the monitors' ages.

Another advantage of the $^{40}Ar/^{39}Ar$ method is that it allows the acquisition of a series of dates on the same sample by incrementally heating it, which causes the release of Ar aliquots at each different heating step. The dates so obtained correspond to the release of Ar from increasingly retentive (i.e., higher closure temperatures) parts or crystallographic domains of the mineral. This method allows us to draw an age spectrum diagram using the dates obtained for each step of argon released during the mineral heating process. In ideal cases, several steps (usually the higher temperature ones) provide statistically identical ages forming a so-called plateau age. Younger ages, usually pertaining to the lower temperature steps, correspond to parts and/or crystallographic domains of the mineral that have undergone variable Ar loss. Sometimes, steps with older ages than a plateau may also occur and can be associated with other problems of the $^{40}Ar/^{39}Ar$ method like Ar excess and Ar recoil. The reader is invited to consult the specific literature (e.g., Faure and Mensing 2005; Dickin 2005) for detailed information on the K–Ar and $^{40}Ar/^{39}Ar$ dating methods.

2.4 Systems with Initial Daughter—The Isochron Method

For other cases, where a significant and not constrainable amount of daughter isotope may be incorporated in the mineral at the time of its formation

$$D = D_i + D^* \tag{4}$$

where D is the total amount of daughter isotope resulting from the sum of the daughter isotope incorporated from the surrounding environment at the time of formation of the mineral (D_i) plus the daughter isotope produced within the mineral after its formation by decay of the incorporated radioactive parent (D^*).

Thus, Eq. (4) above becomes, by substituting Eq. (2) into it,

$$D = D_i + N\left(e^{\lambda t} - 1\right) \tag{5}$$

and Eq. (5) can be solved for t

$$t = \frac{1}{\lambda}\ln\left(\frac{D - D_i}{N} + 1\right) \tag{6}$$

This equation, to be solved for t, requires knowledge of D_i, which cannot be known if t is not known, except in some cases like discussed above for the K–Ar system, in which all initial Ar (D_i) is attributed to atmospheric Ar. However, a solution to Eq. (6) is given by the observation that

$$\frac{D - D_i}{N} \tag{7}$$

is the slope of a straight line in the space [$D–D_i$] versus N, which corresponds to an isochron. All the terms of Eq. (7) are usually normalized to a common reference isotope, like a stable non-radiogenic isotope of the daughter element.

This solution allows us to solve the age Eq. (6) by determining the slope of a linear regression between daughter radiogenic and parent radioactive isotopes, which form as a consequence of the radioactive decay.

In order to obtain a linear regression at least two points should be obtained, but, in reality, two and three point regressions have very limited statistical significance (e.g., Ludwig 2001). Therefore, the reliability and precision of linear regressions lies in the number of points by which they are formed and in the statistical dispersion around the regression (Fig. 3).

Robustness and precision of the regression depend on: (i) the spread of the points along the isochron (the wider the spread the lower the uncertainty on the isochron age), (ii) the analytical uncertainty of the isotope ratios of each analysis (the lower this uncertainty the lower the uncertainty on the age), and (iii) the number of points defining the regression (Fig. 3). In many applications of isochron methods in geochronology, the different points of a linear regression are represented by different but cogenetic minerals. The use of different minerals ensures that they have a broad range of parent/daughter ratios which is a prerequisite to obtain a spread of points along the regression. In some cases (e.g., Pb–Pb isochrons) a single mineral isochron can be obtained by sequentially leaching the mineral and extracting different proportions of radiogenic lead from it (e.g., Frei and Kamber 1995).

The use of Eq. (6) requires that the minerals (or mineral leachate fractions) used for the regression have remained closed systems since their formation and that they have formed coevally incorporating the radiogenic isotope from the formation environment (fluid) in the same ratio to the non-radiogenic isotope, i.e., they must be formed by an isotopically homogeneous hydrothermal fluid, in the case of hydrothermal ore deposits. It should be noted that, for the reasons discussed above and summarized in Fig. 3, the precision of the isochron dating method is usually poorer than dating of minerals which incorporate very little initial daughter isotope, e.g., U–Pb dating of zircon, Re–Os dating of molybdenite, $^{40}Ar/^{39}Ar$ (K–Ar) dating of K-bearing minerals (under the reasonable assumption that all initial Ar is atmospheric).

3 Dating Methods

Table 1 reports the most used radiometric dating systems in ore geology. The usefulness of these methods depends on the occurrence of datable minerals in any particular ore system.

Each method has different ranges of applicability, but a first-level evaluation of the radiometric dating methods concerns what type of ore or ore-related minerals they can date. From this perspective we can distinguish three main classes:

1. Methods that may directly date ore minerals (e.g., sulphides, oxides);
2. Methods that may date hydrothermal minerals that are paragenetically associated with ore minerals as a result of fluid-rock interaction or co-precipitation from the ore fluid;
3. Methods that allow upper and lower age bracketing of the ore event.

3.1 Methods Directly Dating Ore Minerals

Re–Os is the most precise and accurate method for directly dating ore minerals (Stein et al. 2001; Norman 2023). Rhenium is a siderophile to chalcophile element, which is incorporated to variable extents into sulfide minerals. This method allows high precision dating of molybdenite (MoS_2) through Eq. (3) and a variety of sulfides (pyrite, chalcopyrite, etc.; Morelli et al. 2007; Saintilan et al. 2018) through the isochron method (Eq. (6)). Because of the widespread occurrence of molybdenite and other sulfides datable by the Re–Os method in a great variety of geological environments this method is very versatile and allows dating of a broad range of ore deposits (e.g., porphyry systems, stratiform copper deposits, iron oxide copper–gold (IOCG), volcanic-hosted massive sulfide (VHMS), Mississippi Valley-type (MVT) deposits: Li et al. 2017; Saintilan et al. 2018; Liu et al. 2015; Requia et al. 2003; Nozaki et al. 2014).

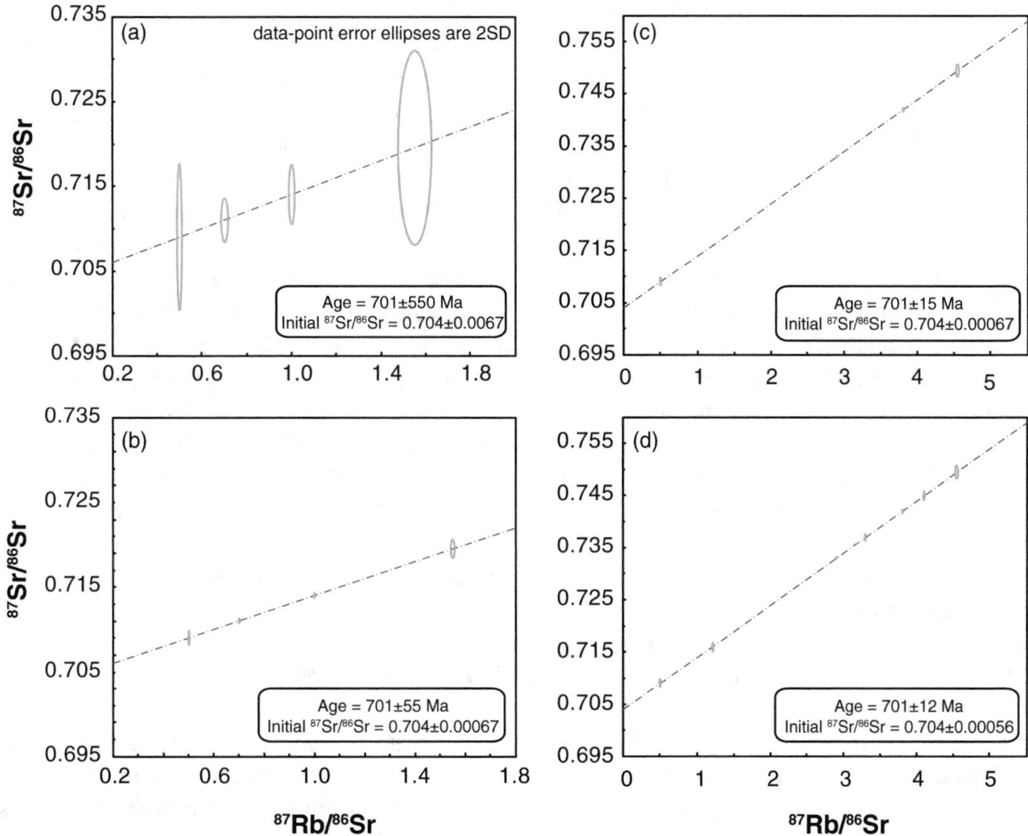

Fig. 3 Examples of different synthetic Rb–Sr isochrons with common natural ranges of $^{87}Sr/^{86}Sr$ and $^{87}Rb/^{86}Sr$ values and with a slope corresponding to an age of 701 Ma: **a** 4-point isochron with large internal uncertainties of single point analyses (up to 1.3% for $^{87}Sr/^{86}Sr$) yielding a large uncertainty on the age (~80%); **b** same 4-point isochron with 10 times smaller internal uncertainties of single point analyses yielding a 10 time better precision on the age ~8%); **c** 4-point isochron with larger spread of $^{87}Rb/^{86}Sr$ values yielding a much smaller uncertainty on the age (~2%); **d** 7-point isochron with same spread of $^{87}Rb/^{86}Sr$ values yielding an even smaller uncertainty on the age (< 2%)

Another method that has been used for dating directly an ore mineral (sphalerite) is the Rb–Sr isochron method. This method allows direct dating of sphalerite from MVT deposits (Nakai et al. 1993; Christensen et al. 1997). U–Pb dating of cassiterite is another method that allows direct dating of ores containing this mineral (e.g., Yuan et al. 2011).

Other methods that have been used in particular cases for directly dating ore minerals are Pb-Pb isochrons on sulfides and oxides (Frei and Kamber 1995; Requia et al. 2003), $^{40}Ar/^{39}Ar$ dating of pyrite (Smith et al. 2001), sphalerite (Qiu and Jiang 2007), and U-Th/He dating of Fe-oxides (Wernicke and Lippolt 1994). $^{40}Ar/^{39}Ar$ dating of hydrothermal quartz fluid inclusions (Kendrick et al. 2001) and Rb–Sr isochron dating of quartz-hosted fluid inclusions (Li et al. 2008) are methods that have been applied to date ore-related fluids. In addition, Sm–Nd isochron dating has been applied to date scheelite ($CaWO_4$) as the main ore mineral (Eichhorn et al. 1997) or associated with gold mineralization of variable age and type (e.g., Anglin et al. 1996; Zhang et al. 2019). These methods may represent the only possibility to date the mineralization in some types of deposits and may return successful results.

3.2 Methods Dating Hydrothermal Alteration Associated with Mineralization

$^{40}Ar/^{39}Ar$ dating of K-bearing phases is the most applied method for dating hydrothermal alteration associated with ore processes. This is due to the fact that $^{40}Ar/^{39}Ar$ dating yields high precision dates and age spectra that indicate the thermal history of the dated mineral (see above). K-bearing alteration minerals (ranging from amphibole to biotite, adularia, muscovite, illite, alunite and various clay minerals) are almost ubiquitously associated with porphyry system mineralization and with low sulfidation epithermal deposits, but also with other types of mineralization (orogenic gold, Carlin-type, VHMS) and therefore they have been widely used for dating hydrothermal activity associated with various types of mineralization (e.g., Henry et al. 1997; Marsh et al. 1997; Arehart et al. 2003). As discussed above, the $^{40}Ar/^{39}Ar$ system, due to its low closure temperature in the greatest majority of datable minerals (Fig. 2), records the time at which a certain mineral has passed below its closure temperature. Closure temperatures for the most commonly dated K-bearing minerals mentioned above range between 550° and 200 °C (Fig. 2), which is a range of temperatures for a wide variety of geological processes. Therefore, this method is sensitive to later geologic events during which temperatures may rise above the closure temperatures of these minerals, but even to sustained temperature in the lifetime of the same mineralization event which may be characterized by a protracted thermal anomaly induced by multiple subsequent magma intrusions (Chiaradia et al. 2013). In such a case the time recorded by the minerals will be the time at which the temperature drops below the closure temperatures of each mineral which may be variably later than its time of formation. In other cases, in contrast, the temperature of formation of the K-bearing mineral can be similar to its closure temperature and/or the cooling rate of the mineral can be sufficiently rapid that the cooling age virtually coincides with the crystallization age of the mineral.

Rb–Sr dating of hydrothermal minerals (biotite, muscovite) has the same problems of closure temperatures as $^{40}Ar/^{39}Ar$ (Fig. 2) and additionally it is less precise due to the fact that a date in such a case is obtained through the isochron method (see above).

U–Pb dating of hydrothermal titanite, zircon, rutile, monazite, xenotime, allanite, and U–Pb isochron dating of garnet have been applied in several studies and may allow dating of different mineralization types like skarns, VHMS, and orogenic gold-type deposits among others (e.g., Vielreicher et al. 2003; Schaltegger 2007; Chiaradia et al. 2009; Fielding et al. 2017; Wafforn et al. 2018; Schirra and Laurent 2021). U–Pb isochrons have been applied to date hydrothermal carbonates associated with MVT mineralization (Grandia et al. 2000).

3.3 Bracketing

An alternative approach to direct dating of either ore or hydrothermal minerals associated with the mineralization is that of dating geological events bracketing the mineralization. The most common case is dating magmatic events that, through stratigraphic or cross-cutting relationships, are demonstrably pre- and post-mineralization.

For instance, U–Pb dating of zircon has been widely used to bracket age and duration of magmatic hydrothermal activity associated with porphyry-type mineralization by dating pre- to syn- and post-ore porphyry intrusions (e.g., von Quadt et al. 2011). Figure 4 shows an idealized example of such a situation in which the age of mineralized veins can be bracketed by U–Pb zircon ages of subsequently emplaced porphyritic intrusions. The same method, applied to dating volcanic rocks occurring in the stratigraphic footwall and hanging-wall of strata-bound VHMS mineralization, has also been used to bracket the age of VHMS mineralization (e.g., Barrie et al. 2002). Of course the closer in age the bracketing events the more useful the resulting age for constraining the timing of mineralization. Unfortunately, this cannot be known a priori, although both in the porphyry and in the VHMS environment genetic models suggest that magmatism used for bracketing is closely spaced in time (to a maximum of few 100 s of thousands of years in most cases). In the absence of alternative methods this remains a powerful method, principally when it is based on the presently most accurate and precise dating method, which is U–Pb dating of zircon.

3.4 Other Methods Applicable to Dating of Ore Deposits and Associated Processes

Low-temperature thermochronological methods like the U-Th/He and fission track dating methods have been used in combination with geochronological methods (e.g., U–Pb zircon dating) to determine timing and duration of porphyry mineralization processes, rate of exhumation and erosion of intrusive-related ore deposits and comparative preservation potential (e.g., McInnes 2005).

4 Successful Dating of Ore Deposits

The possibility to date reliably and precisely ore-forming events depends on the occurrence of suitable minerals in the mineralization. Ore deposits intimately associated with intermediate-felsic magmatism (e.g., porphyry deposits, high-sulfidation epithermal deposits, skarn deposits) are probably the most reliably and precisely datable. This is so because these deposits can be dated by the three most accurate and precise dating methods available. U–Pb zircon ages provide a close-to-mineralization upper temporal limit or, in the case of multiple intrusive events showing cross-cutting relationships, also a tight temporal bracketing (Fig. 4). In addition, such deposits very often contain molybdenite, which allows direct Re–Os dating of the mineralization and its comparison with zircon U–Pb ages of the magmatic pulses associated with it (Fig. 4). Further constraints on these types of deposits come from the possibility of dating K-bearing alteration minerals by the $^{40}Ar/^{39}Ar$ method. There is a large amount of literature showing the successful combined use of U–Pb dating (zircon, titanite), Re–Os (molybdenite), and $^{40}Ar/^{39}Ar$ (various K-bearing alteration phases) in porphyry Cu-Au, high-sulfidation epithermal, and skarn deposits (e.g., Maksaev et al. 2004; Barra et al. 2013; Deckart et al. 2013; Burrows et al. 2020). Figure 5 shows an example of the gold skarn of Nambija (Ecuador), which has been dated using U–Pb in zircon of the causative porphyry intrusions, U–Pb dating of hydrothermal titanite from the prograde skarn, and Re–Os dating of molybdenite from veins of the retrograde skarn stage, cross-cutting the gold mineralization (Chiaradia et al. 2009). All three minerals returned undistinguishable ages, around 145 Ma, within uncertainty indicating that magma intrusion, skarn formation and gold deposition occurred in a short time that is not resolvable with the available precision of the most precise radiometric techniques. The occurrence of sericite in the same paragenetic association (Fig. 5) could also be used for $^{40}Ar/^{39}Ar$ dating of this mineral.

Ore deposits for which a genetic and spatial association with magmatic activity is more elusive rely on the occurrence of datable ore and/or gangue minerals. Examples of these deposits are low sulfidation epithermal deposits that have been dated by $^{40}Ar/^{39}Ar$ on K-bearing minerals (Love et al. 1998; Hames et al. 2009), IOCG deposits that have been dated using Re–Os ages on molybdenite (Requia et al. 2003), U–Pb ages on hematite and U–Th–Pb dating of monazite (Zhou et al. 2017; Courtney-Davies et al. 2019, 2020), and orogenic gold deposits that have been dated using a variety of methods (Re–Os isochrons on sulfides, U–Pb on hydrothermal monazite and xenotime, $^{40}Ar/^{39}Ar$ on white mica).

Low-temperature hydrothermal deposits hosted by sedimentary sequences, e.g., Carlin-type deposits, MVT deposits, sediment-hosted massive sulfides, stratiform copper deposits, are

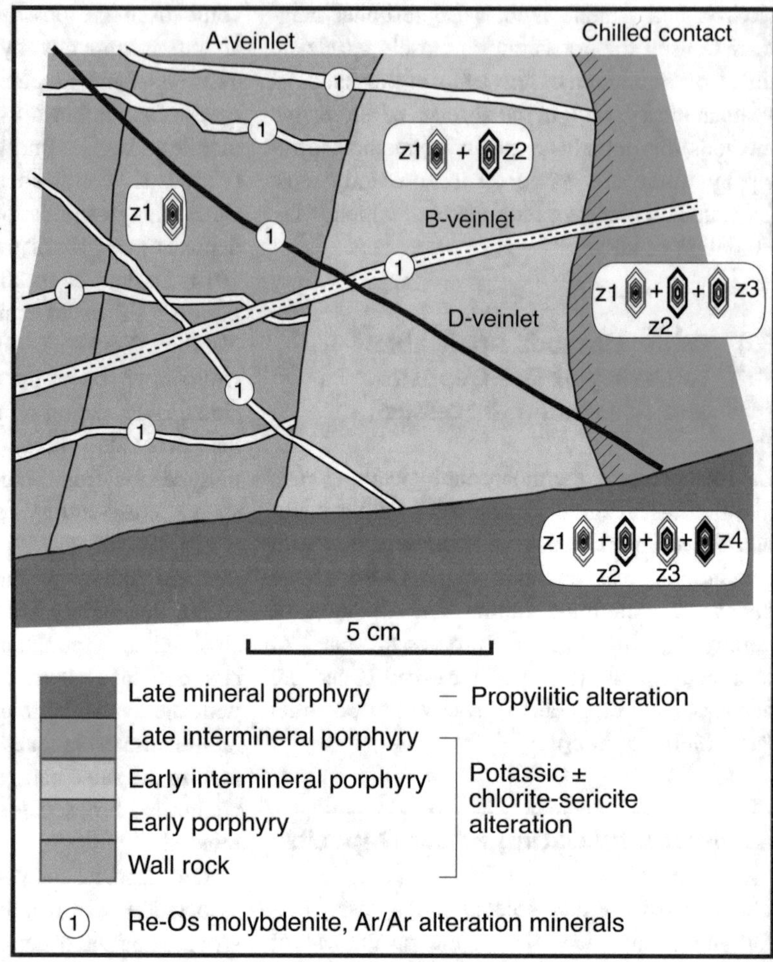

Fig. 4 Schematic crosscutting relationships between early porphyry (immediately premineral), intermineral, and late mineral porphyry phases in porphyry Cu stocks and wall rocks (modified from Sillitoe 2010 and Chiaradia et al. 2014). U–Pb zircon, Re–Os molybdenite, and $^{40}Ar/^{39}Ar$ dating can be applied to magmatic phases and hydrothermal veins. With younging ages, each magmatic unit may present an increasing range of zircon types inherited from protracted crystallization in the deeper parental magma chamber

probably the most difficult to date due to the paucity of datable minerals with reliable and precise methods and the small size of the minerals which make them more prone to open-system behavior (e.g., Rb–Sr on galkhaite, Sm–Nd on fluorite and calcite, Re–Os on Cu–Co sulphides: Tretbar et al. 2000; Saintilan et al. 2017, 2018; Tan et al. 2019).

In summary, whenever possible, a multi-method approach carried out on minerals encompassing different stages of the mineralization process (magmatism, ore deposition, alteration) and coupling in situ dating of petrographically constrained grains with higher precision bulk grain or sub-grain methods on the same minerals should be viewed as the ideal approach for a successful dating of ore deposits.

5 Timing and Duration of Mineralization Processes

It has been highlighted in the Introduction that radiometric dating may provide two types of information: (i) the time in the past when the mineralization occurred and possibly (ii) how long the mineralizing process lasted.

Mineral deposit formation is invariably linked to specific triggering processes, which in turn depend on large-scale geodynamic, climatic, and biologic events that may have recurrent or cyclic occurrence (e.g.,Bekker et al. 2010; Richards 2013; Wilkinson 2013; Heinrich 2015). Determining the time in the past when a mineralization occurred allows us to place the mineralizing

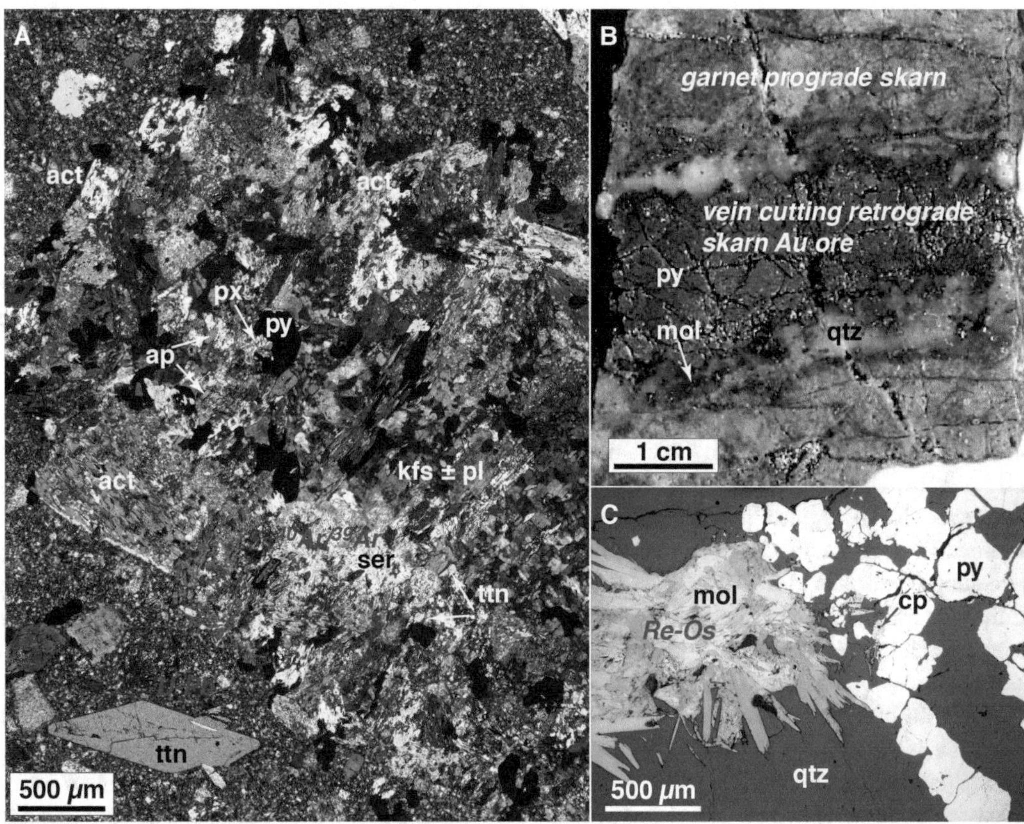

Fig. 5 Skarn and associated ore and alteration mineral assemblages from the Nambija gold skarn (Ecuador) showing various minerals that have been dated with different techniques (modified from Chiaradia et al. 2009). **a** Photomicrograph of the endoskarn assemblage with hydrothermal titanite (ttn) that has been dated by the U–Pb method (transmitted light, crossed nicols). **b** Molybdenite (mol) dated by Re–Os method in a sulfide-rich vein cutting through garnet prograde skarn and post-dating retrograde skarn gold mineralization. **c** Photomicrograph of molybdenite in the sulfide-rich vein dated by Re–Os (reflected light). In the skarn assemblage there are also K-bearing minerals (e.g., K-feldspar, sericite) that could be eventually dated by the $^{40}Ar/^{39}Ar$ method. U–Pb titantite and Re–Os molybdenite dates have returned undistinguishable ages ($\sim$ 145 Ma) also with respect to U–Pb zircon dates of the causative porphyritic intrusions (see Chiaradia et al. 2009 for details). Other abbreviations: act = actinolite, ap = apatite, cp = chalcopyrite, kfs = K-feldspar, pl = plagioclase, px = pyroxene, py = pyrite, qtz = quartz, ser = sericite

event in relationship with geological processes that have triggered and caused the mineralization. This is the case for instance of dating the metallogenic belts of the Andes, which has revealed an initial trenchward shift of the metallogenic porphyry belts and a later shift away from the trench towards the back-arc through time (Sillitoe 1988). Also dating of orogenic gold deposits has highlighted their links with major geodynamic changes through Earth's history (Goldfarb et al. 2001). Such information may help us to understand the possibilities to find similar deposits in specific time intervals as a result of geodynamic events and/or of preservation and is therefore essential for mineral exploration strategies.

In recent years, ore geologists have also started to appreciate the importance of determining the duration of a mineralizing process. This can in fact provide valuable information on the rate at which metals are deposited and, therefore, a better understanding of the processes that govern metal deposition and the formation of a mineral deposit (e.g.,Chelle-Michou et al. 2017;

Chiaradia and Caricchi 2017, 2022). Porphyry deposits have been the most investigated from this point of view because the most accurate and precise dating methods (U–Pb, Re–Os, $^{40}Ar/^{39}Ar$) can all be applied to this type of mineralization. We have now acquired, also in combination with thermodynamic (Cathles et al. 1997) and numerical modelling (Weis et al. 2012), and element diffusion studies (Mercer et al. 2015), a quite good understanding of how long a single mineralizing pulse and the overall mineralizing event can last to form a porphyry deposit. The high precision dating obtained in these deposits (see summary in Chiaradia et al. 2013, 2014; Chiaradia and Caricchi 2017; Chiaradia 2020) agrees with the durations obtained through thermodynamic and numerical modelling (Cathles et al. 1997; Weis et al. 2012) or through mass balance calculations carried out using metal concentrations and fluid fluxes of active geothermal fields (e.g., Simmons and Brown 2006). According to these data, single magmatic pulses and associated hydrothermal-ore activity last for a maximum of a few 10 s of thousands of years. However, in order to form the largest deposits these single pulse events must be repeated several times through a process that leads to a progressive step-wise increment of the metal endowment of the deposit over time intervals that can span up to more than one million years (Ballard et al. 2001; Deckart et al. 2013; Chiaradia and Caricchi 2017; Chiaradia 2020).

6 Instrumentation

High-precision radiometric dating is carried out by mass spectrometry (the reader may consult Faure and Mensing 2005 and Dickin 2005 for more detailed introductions to mass spectrometry). Currently the following types of mass spectrometry techniques are the most used for radiometric dating of ore deposits:

1. Isotope Dilution Thermal Ionization Mass Spectrometry (ID-TIMS) (Chelle-Michou and Schaltegger 2023): This technique is routinely used for U–Pb, Re–Os, Rb–Sr and Sm–Nd dating of various minerals. Single or multiple grains, but even portions of single grains (Fig. 6), are dissolved in acids and the elements of interest are separated from each other through column chemistry, and then measured on a mass spectrometer.
2. Noble gas mass spectrometry: this is the technique used for $^{40}Ar/^{39}Ar$ dating. Previous irradiation of the sample to be analysed to convert ^{39}K to ^{39}Ar, the sample is outgassed under vacuum using a furnace or a laser, to liberate stepwise the Ar gas contained in it (see above). At each step Ar isotope ratios are obtained that can be converted into dates yielding an age spectrum (see above).
3. LA-(MC)-ICPMS (Chelle-Michou and Schaltegger 2023): Laser ablation coupled to a multi- or single collector mass spectrometer is used for dating zircons (mostly) but also other minerals (e.g., titanite, cassiterite, monazite, hematite, calcite). The material analysed is ablated by a laser inside an ablation cell and transferred to the mass spectrometer where it is ionized and isotope ratios are measured. The ablated material corresponds usually to circular pits of 30–60 μm in diameter and a few μm in depth for a total ablated volume in the order of 10^3–10^4 μm^3 compared to $\sim 10^5$ μm^3 for TIMS analyses (Sylvester 2008) (Fig. 6).
4. SIMS (Chelle-Michou and Schaltegger 2023): Secondary Ion Mass Spectrometry is a dating technique largely used for U–Pb dating of minerals that is conceptually similar to LA-ICPMS but differs in the way by which the analysed elements are extracted from the mineral. In SIMS an ion beam (usually of oxygen ions for U–Pb dating) hits the mineral surface and sputters it producing secondary ions that are accelerated and focused into the mass spectrometer. The beam size can be as small as 15 μm thus providing the highest spatial resolution of all dating techniques, but also resulting in smaller volumes of analysed material ($\sim 10^2$ μm^3; Sylvester 2008) (Fig. 6).

The major differences among these techniques in terms of precision and accuracy of the obtained ages are two: techniques 1 and 2 above, by virtue of the larger amounts of material analysed, yield higher-precision whereas in situ techniques (3 and 4 above) yield a spatial resolution that is not obtainable with bulk mineral techniques (Fig. 6). Also, in situ techniques have a much higher throughput and overall lower operational costs.

It should be highlighted that the "bulk" minerals (1 and 2) and in situ (3 and 4) methods should be viewed as complementary and that the choice of one or the other depends on the actual goals of the dating. High precision dating by ID-TIMS is the only technique to precisely and accurately constrain the timescales of hydrothermal events where minerals datable by high-precision U–Pb and Re–Os techniques are available, whereas in-situ techniques are the choice for regional studies and for dating minerals with complex textures indicating several growth episodes. High precision CA-ID-TIMS dating of zircon may also benefit from previous dating of the same zircons by in-situ methods in order to allow a selection of the most appropriate zircon grains to be subsequently dated by the more labor intensive CA-ID-TIMS technique (Chelle-Michou et al. 2014).

A limitation in all radiometric dating is the physical limitation of signal detectors in mass spectrometers to precisely and accurately measure very low signals. This places limits on precision and accuracy of ages in very young geological systems in which not enough time has elapsed to allow the formation of measurable signals of radiogenic daughter isotopes (Chiaradia et al., 2013). These limits are being progressively improved via technological advances in ion detection. Accuracy and precision of all dating methods are also dependent on minimization of 'blanks' and background signals within the measurement system, that is, the presence of interfering isotopic signals related to sample preparation and/or instrument cleanliness.

7 Intra-method Age Reproducibility, Inter-laboratory Comparison, Inter-method Comparisons and Other Caveats

From the above discussion, it is clear that radiometric dating of an ore deposit can be conducted through different methods. Various analytical techniques are available at different laboratories throughout the world. The quality of the data output of a laboratory is monitored by continuous analysis of standards (natural or synthetic) with known (ideally certified) isotopic compositions, intercalated with the measurement of the unknown samples. There are two main parameters associated with the age obtained at a laboratory that can be used to evaluate the quality of the measurement: (i) the uncertainty of the measured value (precision) and (ii) the accuracy of the obtained value. Uncertainty depends mostly on repeatability of procedural methods (including blanks), counting statistics of the measurements and on mass instrumental bias among the different isotopes. Counting statistics depend on the size of the sample analyzed (Chiaradia et al. 2013), which translates into higher or lower signals (counts) of the analyzed isotopes in the mass spectrometer. Mass instrumental bias may represent the largest amount of the uncertainty associated with an age (e.g., U–Pb: Schmitz and Schoene 2007) and can be overcome by using appropriate doping solutions (spikes) to the unknown samples. These solutions consist of mixtures of isotopes in known ratios which allow the determination of an instrumental fractionation factor associated with the measurement that can be used to correct isotope ratio measurements in the mass spectrometer of the radiometric systems (U–Pb, Re–Os: Markey et al., 2003; Schmitz and Schoene, 2007). Additional sources of uncertainties derive from uncertainties in the calibration of spikes, uncertainties of the decay constants, and also uncertainties in the ages of secondary standards in the $^{40}Ar/^{39}Ar$ method (e.g., Min et al. 2000; Kuiper et al. 2008; Renne et al. 2010).

a

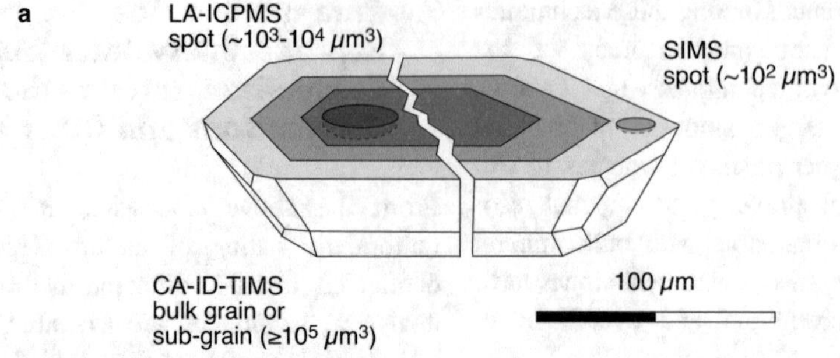

b

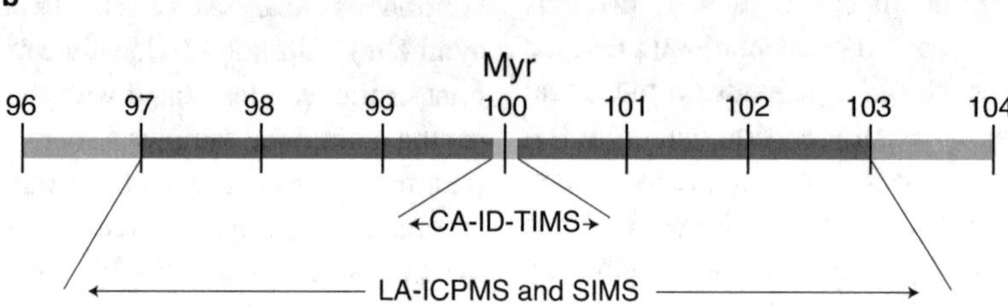

Fig. 6 Analytical techniques of high-precision U–Pb geochronology. **a** CA-ID-TIMS: Chemical-Abrasion, Isotope-Dilution Thermal Ionization Mass Spectrometry can be applied to bulk single grains or portions of them. LA-ICP-MS (Laser Ablation Inductively Coupled Plasma Mass Spectrometry) and SIMS (Secondary-Ion Mass Spectrometry) are in situ techniques allowing a much higher spatial resolution but a smaller volume of analyzed material. **b** Different precisions of the CA-ID-TIMS versus the in situ LA-ICPMS and SIMS methods resulting from the different volumes of zircon analysed by these methods. The 2SD error bars refer to single spot or single/fraction grain a zircon 100 Myr old assuming typical precisions associated with the 2 methods (i.e., 0.1% for CA-ID-TIMS and 3% for LA-ICPMS and SIMS). See text for further discussion

The accuracy of a measurement is how close the value of such a measurement is to the "real" value. A way to assess the accuracy of measurements in a laboratory is by measuring certified standards in order to see if the measurements return the expected values of the certified standard. Accuracy depends on systematic uncertainties of various origins (Chiaradia et al. 2013; Schaltegger et al. 2015).

Age data obtained within the same laboratory using the same radiometric method can be compared confidently with each other using the "internal" reproducibility that is associated with the measurements at that laboratory. In such a case, uncertainties associated with spike calibration and decay constants can be disregarded because they apply in the same way to all sample analyses (since the same lab will use the same spike solution and for a given radiometric system the decay constant uncertainties are fixed). These "internal" uncertainties are very low (< 0.1–0.2%, 95% confidence level) for the main dating systems (i.e., ID-TIMS U/Pb dating of zircons, ID-NTIMS Re–Os dating of molybdenite, $^{40}Ar/^{39}Ar$ dating of K-rich minerals; Chiaradia et al. 2013). This means that the uncertainty on a mineral that is 10 Ma old is only 10,000 years! However, if one wants to compare ages obtained using the same radiometric method at two different laboratories, then "external" uncertainties in the spike calibration must be taken into account, if the two laboratories do not use the same calibrated spike solution. Uncertainty on the calibration of the spikes, which is an essential part of the high precision obtainable by ID-(N) TIMS dating techniques (Chiaradia et al. 2013), can be considerable (Chiaradia et al. 2013). For this reason, there is an increased awareness in the geochronological community to use common calibrated spike solutions which allow the achievement of reproducible ages of standard minerals from different laboratories at the same level of uncertainty obtained by dating within the same laboratory (e.g., Condon et al. 2015).

When we want to compare ages from different radiometric systems (e.g., U–Pb versus Re–Os and $^{40}Ar/^{39}Ar$ for instance), uncertainties in the decay constants must also be considered. Currently, the decay constants of ^{238}U and ^{235}U are known to a higher precision level than those of ^{40}K and ^{187}Re (see discussion in Chiaradia et al. 2013) and this is one of the main reasons why the U–Pb system is used as a reference system. Therefore, there is a progressive increase of the uncertainties of an age measurement when one wants to compare such a measurement with another measurement using the same radiometric system in the same laboratory ("internal" uncertainty associated mostly with counting statistics and instrumental bias), or with another measurement obtained with the same radiometric system but in another laboratory ("internal" uncertainty plus "external" uncertainty associated, with uncertainty in the spike calibration, if not using the same calibrated solution), or with a measurement carried out using another radiometric system ("internal" plus "external" uncertainties, plus uncertainties in decay constants).

Accuracy of the ages of standards in secondary dating methods (like $^{40}Ar/^{39}Ar$) must be also considered and may result in significant discrepancies of the dates of the same geological event using different geochronometers (see discussions in Chiaradia et al. 2013, 2014).

7.1 Geological Factors

An important geological consideration about radiometric dating is that the older the mineral dated the more likely is the possibility that it has been affected by post-formation geological processes which might have variably disturbed its closed-system behavior, a fundamental requirement to be a reliable mineral for dating. Superimposed geological processes may also cause remobilization and formation of new ore minerals at a different time from that of the initial mineralization, complicating the interpretation of geochronologic data.

8 Conclusion

Radiometric systems provide the only quantitative tool for dating ore deposits. Dating ore deposits is an essential step in the formulation of metallogenetic models because it allows us to relate the formation of an ore deposit to a specific geological cause or trigger. In certain cases, like porphyry systems, it also allows us to determine the duration of the mineralizing event which provides essential information on the amount of energy, fluid and magma needed to form such deposits, and which can be used to model the magmatic system features needed to comply with such timescales.

Dating ore deposits using radiometric methods is nonetheless not an easy task and require specialized laboratories and trained personnel. One should keep in mind that dating minerals from an ore deposit should be the final step of a detailed geological and petrographic work that already provides us with an idea of the relative timing of the events we intend to date with absolute ages. Discordancy between a clear-cut geological and petrographic sequence of events and absolute radiometric dating should warn us that some issues may have affected radiometric dating.

Successful radiometric dating of ore deposits is best accomplished using a multi-method approach, through which minerals belonging to different stages of the ore deposit (e.g., causative magmatic intrusion, ore deposition, alteration) are dated with different methods. This, on the other hand, requires careful consideration and tackling of the issues that comparing dates obtained with different radiometric clocks implies, as discussed above. Also coupling in situ dating techniques with "bulk" (single grain or sub-grain) dating is a complementary approach that should be considered for an optimal use of the different advantages that these different methods offer.

Analytical and technological developments over the last 10–15 years and continuously ongoing have resulted in increased accuracy and precision of radiometric dating and this is opening new avenues for the application of geochronology to the understanding of how mineral deposits form.

Acknowledgements I would like to thank Geoff Fraser for his review that contributed to improve this work and David Huston for inviting me to write this overview chapter and for his comments on a previous version of the manuscript.

References

Anglin CD, Jonasson IR, Franklin JM (1996) Sm–Nd dating of scheelite and tourmaline: implications for the genesis of Archean gold deposits, Val d'Or, Canada. Econ Geol 91:1372–1382

Arehart GB, Chakurian AM, Tretbar DR, Christensen JR, McInnes BA, Donelick RA (2003) Evaluation of radioisotope dating of Carlin-type deposits in the Great Basin, western North America, and implications for deposit genesis. Econ Geol 98:235–248

Ballard JR, Palin JM, Williams IS, Campbell IH (2001) Two ages of porphyry intrusion resolved for the super-giant Chuquicamata copper deposit of northern Chile by ELA-ICP-MS and SHRIMP. Geology 29:383–386

Barley ME, Groves DI (1992) Supercontinent cycles and the distribution of metals through time. Geology 20:291–294

Barra F, Alcota H, Rivera S, Valencia V, Munizaga F, Maksaev V (2013) Timing and formation of porphyry Cu–Mo mineralization in the Chuquicamata district, northern Chile: new constraints from the Toki cluster. Miner Depos 48:629–651

Barrie TC, Amelin Y, Pascual E (2002) U–Pb Geochronology of VMS mineralization in the Iberian Pyrite Belt. Miner Depos 37:684–703

Bekker A, Slack JF, Planavsky N, Krapez B, Hofmann A, Konhauser KO, Rouxel OJ (2010) Iron formation: the sedimentary product of a complex interplay among mantle, tectonic, oceanic, and biospheric processes. Econ Geol 105:467–508

Burrows DR, Rennison M, Burt D, Davies R (2020) The Onto Cu–Au discovery, eastern Sumbawa, Indonesia: a large, Middle Pleistocene lithocap-hosted high-sulfidation covellite-pyrite porphyry deposit. Econ Geol 115:1385–1412

Cathles LM, Erendi AHJ, Barrie T (1997) How long can a hydrothermal system be sustained by a single intrusive event? Econ Geol 92:766–771

Chelle-Michou C, Schaltegger U (2023) U–Pb dating of mineral deposits: from age constrains to ore-forming processes. This volume

Chelle-Michou C, Chiaradia M, Ovtcharova M, Ulianov A, Wotzlaw JF (2014) Zircon petrochronology reveals the temporal link between porphyry systems and the magmatic evolution of their hidden plutonic roots (the Eocene Coroccohuayco deposit, Peru). Lithos 198:129–140

Chelle-Michou C, Rottier B, Caricchi L, Simpson G (2017) Tempo of magma degassing and the genesis of porphyry copper deposits. Sci Rep 7:40566

Cherniak DJ, Watson EB (2001) Pb diffusion in zircon. Chem Geol 172:5–24

Chiaradia M (2020) Gold endowments of porphyry deposits controlled by precipitation efficiency. Nat Commun 11:248. https://doi.org/10.1038/s41467-019-14113-1

Chiaradia M, Caricchi L (2017) Stochastic modelling of deep magmatic controls on porphyry copper deposit endowment. Sci Rep 7:44523

Chiaradia M, Caricchi L (2022) Supergiant porphyry copper deposits are failed large eruptions. Res Sq, https://doi.org/10.21203/rs.3.rs-802055/v1

Chiaradia M, Vallance J, Fontboté L, Stein H, Schaltegger U, Coder J, Richards J, Villeneuve M, Gendall I (2009) U–Pb, Re–Os, and $^{40}Ar/^{39}Ar$ geochronology of the Nambija Au skarn and Pangui porphyry-Cu deposits, Ecuador: implications for the Jurassic metallogenic belt of the Northern Andes. Miner Depos 44:371−387

Chiaradia M, Schaltegger U, Spikings R, Wotzlaw JF, Ovtcharova M (2013) How accurately can we date the duration of magmatic-hydrothermal events in porphyry systems? An invited paper. Econ Geol 108:565–584

Chiaradia M, Schaltegger U, Spikings R (2014) Time scales of mineral systems—advances in understanding over the past decade. SEG Spec Publ 18:37–58

Christensen JN, Halliday AN, Leigh KE, Randell RN, Kesler SE (1997) Direct dating of sulfides by Rb–Sr: a critical test using the Polaris Mississippi Valley-type Zn–Pb deposit. Geochim Cosmochim Acta 59:5191–5197

Condon DJ, Schoene B, McLean NM, Bowring SA, Parrish RR (2015) Metrology and traceability of U–Pb isotope dilution geochronology (EARTHTIME Tracer Calibration Part I). Geochim Cosmochim Acta 164:464–480

Courtney-Davies L, Ciobanu CL, Verdugo-Ihl MR, Dmitrijeva M, Cook NJ, Ehrig K, Wade BP (2019) Hematite geochemistry and geochronology resolve genetic and temporal links among iron-oxide copper gold systems, Olympic Dam district, South Australia. Precamb Res 335:105480

Courtney-Davies L, Ciobanu CL, Tapster SR, Cook NJ, Ehrig K, Crowley JL, Verdugo-Ihl MR, Wade BP,

Condon DJ (2020) Opening the magmatic-hydrothermal window: high-precision U–Pb geochronology of the Mesoproterozoic Olympic Dam Cu–U–Au–Ag deposit, South Australia. Econ Geol 115:1855–1870

Deckart K, Clark AH, Cuadra P, Fanning M (2013) Refinement of the time-space evolution of the giant Mio-Pliocene Río Blanco-Los Bronces porphyry Cu–Mo cluster, Central Chile: New U–Pb (SHRIMP II) and Re–Os geochronology and $^{40}Ar/^{39}Ar$ thermochronology data. Miner Depos 48:57–79

Dickin AP (2005) Radiogenic isotope geology, 2nd edn. Cambridge University Press, Cambridge

Dodson MH (1973) Closure temperature in cooling geochronological and petrological systems. Contrib Miner Petrol 40:259–274

Eichhorn R, Höll R, Jagout E, Schärer U (1997) Dating scheelite stages: a strontium, neodymium, lead approach from the Felhertal tungsten deposit, Central Alps, Austria. Geochim Cosmichim Acta 61:5005–5022

Faure G, Mensing TM (2005) Isotopes: principles and applications. Wiley, New York

Fielding IOH, Johnson SP, Zi J-W, Rasmussen B, Muhling JR, Dunkley DJ, Sheppard S, Wingate MTD, Rogers JR (2017) Using in Situ SHRIMP U–Pb monazite and xenotime geochronology to determine the age of orogenic gold mineralization: an example from the Paulsens Mine, Southern Pilbara Craton. Econ Geol 112:1205–1230

Frei R, Kamber BS (1995) Single mineral Pb–Pb dating. Earth Planet Sci Lett 129:261–268

Frimmel HE (2005) Archaean atmospheric evolution: evidence from the Witwatersrand gold fields, South Africa. Earth-Sci Rev 70:1–46

Giggenbach WF (1997) The origin and evolution of fluids in magmatic-hydrothermal systems. In: Barnes HL (ed) Geochemistry of hydrothermal ore deposits, 3rd edn. Wiley, New York, pp 737–795

Goldfarb RJ, Groves DI, Gardoll S (2001) Orogenic gold and geologic time: a global synthesis. Ore Geol Rev 18:1–75

Gradstein F, Ogg JG, Schmitz M, Ogg G (2020) The geologic time scale 2020, 2nd edn. Elsevier, Amsterdam

Grandia F, Asmerom Y, Getty S, Cardellach E, Canals À (2000) Pb dating of MVT ore-stage calcite: implications for fluid flow in a Mesozoic extensional basin from Iberian Peninsula. J Geocheml Explor 69:377–380

Hames W, Unger D, Saunders J, Kamenov G (2009) Early Yellowstone hotspot magmatism and gold metallogeny. J Volcanol Geotherm Res 188:214–224

Heinrich CA (2015) Witwatersrand gold deposits formed by volcanic rain, anoxic rivers and Archaean life. Nat Geosci 8:206–209

Henry CD, Elson HB, McIntosh WC, Heizler MT, Castor SB (1997) Brief duration of hydrothermal activity at Round Mountain, Nevada, determined from 40Ar/39Ar geochronology. Econ Geol 92:807–826

Hitzman MW, Broughton DW (2017) Discussion: "Age of the Zambian copperbelt" by Sillitoe et al (2017). Miner Depos 52:1273–1275

Kendrick MA, Burgess R, Pattrick RAD, Turner G (2001) Halogen and Ar–Ar age determinations of inclusions within quartz veins from porphyry copper deposits using complementary noble gas extraction techniques. Chem Geol 177:351–370

Kuiper KF, Deino A, Hilgen FJ, Krijgsman W, Renne PR, Wijbrans JR (2008) Synchronizing rock clocks of Earth history. Science 320:500–504

Leach DL, Bradley D, Lewchuk MT, Symons DT, Marsily G, Branon J (2001) Mississippi Valley-type lead–zinc deposits through geological time: implications from recent age-dating research. Miner Depos 36:711–740

Li Q-L, Chen F, Yang J-H, Fan H-R (2008) Single grain pyrite Rb–Sr dating of the Linglong gold deposit, eastern China. Ore Geol Rev 34:263–270

Li Y, Selby D, Feely M, Costanzo A, Li X-H (2017) Fluid inclusion characteristics and molybdenite Re–Os geochronology of the Qulong porphyry copper-molybdenum deposit. Tibet. Miner Depos 52:137–158

Liu Y, Qi L, Gao J, Ye L, Huang Z, Zhou J (2015) Re–Os dating of galena and sphalerite from lead-zinc sulfide deposits in Yunnan Province, SW China. J Earth Sci 26:343–351

Love DA, Clark AH, Hodgson CJ, Mortensen JK, Archibald DA, Farrar E (1998) The timing of adularia-sericite-type mineralization and alunite-kaolinite-type alteration, Mount Skukum epithermal gold deposit, Yukon Territory, Canada: $^{40}Ar–^{39}Ar$ and U–Pb geochronology. Econ Geol 93:437–462

Ludwig KR (2001) Isoplot v. 2.2—a geochronological toolkit for Microsoft Excel. Berkeley Geochron Cent Spec Publ 1a

Maksaev V, Munizaga F, McWilliams M, Fanning M, Mathur R, Ruiz J, Zentilli M (2004) New chronology for El Teniente, Chilean Andes, from U–Pb, $^{40}Ar/^{39}Ar$ Re–Os, and fission-track dating: implications for the evolution of a supergiant porphyry Cu–Mo deposit. SEG Spec Publ 11:15–54

Markey R, Hannah JL, Morgan JW, Stein HJ (2003) A double spike for osmium analysis of highly radiogenic samples. Chem Geol 200:395–406

Marsh TM, Einaudi MT, McWilliams M (1997) $^{40}Ar/^{39}Ar$ geochronology of Cu–Au and Au–Ag mineralization in the Potrerillos district, Chile. Econ Geol 92:784–806

Mattinson JM (2005) Zircon U–Pb chemical abrasion ("CA-TIMS") method: combined annealing and multi-step partial dissolution analysis for improved precision and accuracy of zircon ages. Chem Geol 220:47–66

McInnes BIA (2005) Application of thermochronology to hydrothermal ore deposits. Rev Mineral Geochem 58:467–498

McKibben MA, Hardie LA (1997) Ore-forming brines in active continental rifts. In: Barnes HL (ed) Geochemistry of hydrothermal ore deposits, 3rd edn. Wiley, New York, pp 877–935

Mercer CN, Reed MH, Mercer CM (2015) Time scales of porphyry Cu deposit formation: insights from titanium diffusion in quartz. Econ Geol 110:587–602

Min K, Mundil R, Renne PR, Ludwig KR (2000) A test for systematic errors in $^{40}Ar/^{39}Ar$ geochronology through comparison with U–Pb analysis of a 1.1-Ga rhyolite. Geochim Cosmochim Acta 64:73–98

Morelli RM, Creaser RA, Seltmann R, Stuart FM, Selby D, Graupner T (2007) Age and source constraints for the giant Muruntau gold deposit, Uzbekistan, from coupled Re–Os–He isotopes in arsenopyrite. Geology 35:795–798

Muchez P, André-Mayer A, Dewaele S, Large R (2017) Discussion: age of the Zambian copperbelt. Miner Depos 52:1269–1271

Nakai S, Halliday AN, Kesler SE, Jones HD, Kyle JR, Lane TE (1993) Rb-Sr dating of sphalerites from Mississippi valley-type (MVT) ore deposits. Geochim Cosmochim Acta 57:417–427

Norman M (2023) The $^{187}Re–^{187}Os$ and $^{190}Pt–^{186}Os$ radiogenic isotope systems: techniques and applications to metallogenic systems. This volume

Nozaki T, Kato Y, Suzuki K (2014) Re–Os geochronology of the Hitachi volcanogenic massive sulfide deposit: the oldest ore deposit in Japan. Econ Geol 109:2023–2034

Qiu H-N, Jiang Y-D (2007) Sphalerite $^{40}Ar/^{39}Ar$ progressive crushing and stepwise heating techniques. Earth Planet Sci Lett 256:224–232

Reed MH (1997) Hydrothermal alteration and its relationship to ore fluid composition. In: Barnes HL (ed) Geochemistry of hydrothermal ore deposits, 3rd edn. Wiley, New York, pp 303–365

Renne PR, Mundil R, Balco G, Min K, Ludwig KR (2010) Joint determination of ^{40}K decay constants and $^{40}Ar^*/^{40}K$ for the Fish Canyon sanidine standard, and improved accuracy for $^{40}Ar/^{39}Ar$ geochronology. Geochim Cosmochim Acta 74:5349–5367

Requia K, Stein H, Fonboté L, Chiaradia M (2003) Re–Os and Pb–Pb geochronology in the Archean Salobo iron oxide copper-gold deposit, Carajás mineral province, northern Brazil. Miner Depos 38:727–738

Richards JP (2013) Giant ore deposits formed by optimal alignments and combinations of geological processes. Nat Geosci 6:911–916

Rutherford E, Soddy F (1902a) The cause and nature of radioactivity. Pt I Philos Mag 4 (Ser 6):370−396

Rutherford E, Soddy F (1902b) The cause and nature of radioactivity. Pt II Philos Mag 4 (Ser 6):569−585

Saintilan NJ, Creaser RA, Bookstrom AA (2017) Re–Os systematics and geochemistry of cobaltite (CoAsS) in the Idaho cobalt belt, Belt-Purcell Basin, USA: evidence for middle Mesoproterozoic sediment-hosted Co–Cu sulfide mineralization with Grenvillian and Cretaceous remobilization. Ore Geol Rev 86:509–525

Saintilan NJ, Selby D, Creaser RA, Dewaele S (2018) Sulphide Re–Os geochronology links orogenesis, salt and Cu–Co ores in the Central African Copperbelt. Sci Rep 8:14946

Schaltegger U (2007) Hydrothermal zircon. Elements 3 (1):51–79

Schaltegger U, Schmitt AK, Horstwood MSA (2015) U-Th–Pb zircon geochronology by ID-TIMS, SIMS, and laser ablation ICP-MS: recipes, interpretations, and opportunities. Chem Geol 402:89–110

Schirra M, Laurent O (2021) Petrochronology of hydrothermal rutile in mineralized porphyry Cu systems. Chem Geol 581:120407

Schmitz MD, Schoene B (2007) Derivation of isotope ratios, errors, and error correlations for U–Pb geochronology using $^{205}Pb–^{235}U–(^{233}U)$-spiked isotope dilution thermal ionization mass spectrometric data. Geochem Geophys Geosys 8:Q08006. https://doi.org/10.1029/2006GC001492

Scott SD (1997) Submarine hydrothermal systems and deposits. In: Barnes HL (ed) Geochemistry of hydrothermal ore deposits, 3rd edn. Wiley, New York, pp 797–875

Seward TM, Barnes HL (1997) Metal transport by hydrothermal ore fluids. In: Barnes HL (ed) Geochemistry of hydrothermal ore deposits, 3rd edn. Wiley, New York, pp 435–485

Sillitoe RH (1988) Epochs of intrusion-related copper mineralization in the Andes. J South Am Earth Sci 1:89–108

Sillitoe RH (2010) Porphyry copper systems. Econ Geol 105:3–41

Sillitoe RH, Perelló J, Creaser RA, Wilton J, Wilson AJ, Dawborn T (2017a) Age of the Zambian copperbelt. Miner Depos 52:1245–1268

Sillitoe RH, Perelló J, Creaser RA, Wilton J, Wilson AJ, Dawborn T (2017b) Reply to discussions of "Age of the Zambian Copperbelt" by Hitzman and Broughton and Muchez et al. Miner Depos 52:1277–1281

Simmons SF, Brown KL (2006) Gold in magmatic hydrothermal solutions and the rapid formation of a giant ore deposit. Science 314:288–291

Skinner BJ (1997) Hydrothermal mineral deposits: what we do and don't know. In: Barnes HL (ed) Geochemistry of hydrothermal ore deposits, 3rd edn. Wiley, New York, pp 1–29

Smith PE, Evensen NM, York D, Szatmari P, de Oliveira DC (2001) Single-crystal $^{40}Ar–^{39}Ar$ dating of pyrite: no fool's clock. Geology 29:403–406

Stein HJ, Markey RJ, Morgan JW, Hannah JL, Scherstén A (2001) The remarkable Re–Os chronometer in molybdenite: How and why it works. Terra Nova 13:479–486

Strutt RJ (1905) On the radioactive minerals. Proc R Soc London Ser A 76:88–101

Sylvester PJ (2008) LA-(MC)-ICP-MS trends in 2006 and 2007 with particular emphasis on measurement uncertainties. Geostand Geoanal Res 32:469–488

Symons DTA, Sangster DF (1994) Palaeomagnetic methods for dating the genesis of mississippi valley-type Lead-Zinc deposits. In: Fontboté L, Boni M (eds) Sediment-hosted Zn–Pb Ores. Spec Publ Soc Geol Appl Miner Depos 10:42−58

Tan Q, Xia Y, Xie Z, Wang Z, Wei D, Zhao Y, Yan J, Li S (2019) Two hydrothermal events at the Shuiyindong Carlin-type gold deposit in southwestern China: insight from Sm–Nd dating of fluorite and calcite. Minerals 9:230

Tretbar DR, Arehart GB, Christensen JN (2000) Dating gold deposition in a Carlin-type gold deposit using Rb/Sr methods on the mineral galkhaite. Geology 28:947–950

Veizer J, Laznicka P, Jansen SL (1989) Mineralization through geologic time: recycling perspective. Am J Sci 289:484–524

Vielreicher NM, Groves DI, Fletcher IR, McNaughton NJ, Rasmussen B (2003) Hydrothermal monazite and xenotime geochronology: a new direction for precise dating of orogenic gold mineralization. SEG Newsl 53(1):10–16

Von Quadt A, Erni M, Martinek K, Moll M, Peytcheva I, Heinrich CA (2011) Zircon crystallization and the lifetimes of ore-forming magmatic-hydrothermal systems. Geology 39:731–734

Wafforn S, Seman S, Kyle RJ, Stockli D, Leys C, Sonbait D, Cloos M (2018) Andradite garnet U–Pb geochronology of the big Gossan skarn, Ertsberg-Grasberg mining district, Indonesia. Econ Geol 113:769–778

Watson JV (1978) Ore deposits through geological time. Proc R Soc London 362:305–328

Weis P, Driesner T, Heinrich CA (2012) Porphyry-copper ore shells form at stable pressure-temperature fronts within dynamic fluid plumes. Science 338:1613–1616

Wernicke RS, Lippolt HJ (1994) Dating of vein specularite using internal (U+Th)/4He isochrons. Geophys Res Lett 21:345–347

Wilkinson JJ (2013) Triggers for the formation of porphyry ore deposits in magmatic arcs. Nat Geosci 6:917–925

Yuan S, Peng J, Hao S, Li H, Geng J, Zhang D (2011) In-situ LA-MC-ICP-MS and ID-TIMS U–Pb geochronology of cassiterite in the giant Furong tin deposit, Hunan Province, South China: new constraints on the timing of tin-polymetallic mineralization. Ore Geol Rev 43:235–242

Zhang Z, Xie G, Mao J, Liu W, Olin P, Li W (2019) Sm-Nd dating and in-situ LA-ICP-MS trace element analyses of scheelite from the Longshan Sb–Au deposit, Xiangzhong metallogenic province, south China. Minerals 9:87

Zhou H, Sun X, Wu Z, Liao J, Fu Y, Li D, Hollings P, Liu Y, Lin H, Lin Z (2017) Hematite U–Pb geochronometer: insights from monazite and hematite integrated chronology of the Yaoan gold deposit, Southwest China. Econ Geol 112:2023–2039

U–Pb Dating of Mineral Deposits: From Age Constraints to Ore-Forming Processes

Cyril Chelle-Michou and Urs Schaltegger

Abstract

The timing and duration of ore-forming processes are amongst the key parameters required in the study of mineral systems. After more than a century of technical developments, innovations and investigation, the U–Pb system arguably is the most mature radioisotopic system in our possession to conduct absolute dating of a wide range of minerals across geological environments and metallogenic processes. Here, we review the basics of U–Pb geochronology, the key historic developments of the method, and the most commonly used analytical techniques (including data reduction, Pb-correction, uncertainty propagation and data presentation) and minerals while pointing out their respective advantages, weaknesses and potential pitfalls. We also highlight critical aspects that need to be considered when interpreting a date into the age of a geological process (including field and petrographic constraints, open-system behavior, handling and interpretation of uncertainties). While U–Pb geochronology is strongly biased toward zircon dating, we strive to highlight the great diversity of minerals amenable to U–Pb dating (more than 16 mineral species) in the context of mineral systems, and the variety of geological events they can potentially date (magmatism, hydrothermal activity, ore-formation, cooling, etc.). Finally, through two case studies we show (1) how multi-mineral geochronological studies have been used to bracket and decipher the age of multiple geological events associated with the world-class Witwatersrand gold province, and (2) how rather than the absolute age, the duration and rate of the mineralizing event at porphyry copper deposits opens new avenues to understand ore-forming processes and the main controls on the size of such deposits. The improving precision, accuracy and spatial resolution of analyses in tandem with high-quality field and petrographic observations, numerical modelling and geochemical data, will continue to challenge paradigms of ore-forming processes and contribute significant breakthroughs in ore deposit research and potentially to the development of new exploration tools.

1 Introduction

The knowledge of the timing and duration of ore-forming processes are perhaps one of the most desirable pieces of information that geologists

C. Chelle-Michou (✉)
Department of Earth Sciences, ETH Zurich, Clausiusstrasse 25, 8092 Zurich, Switzerland
e-mail: cyril.chelle-michou@erdw.ethz.ch

U. Schaltegger
Department of Earth Sciences, University of Geneva, Rue des Maraîchers 13, 1205 Geneva, Switzerland

D. Huston and J. Gutzmer (eds.), *Isotopes in Economic Geology, Metallogenesis and Exploration*, Mineral Resource Reviews, https://doi.org/10.1007/978-3-031-27897-6_3

require to draw a complete picture of the deposit and to put its genesis into a coherent regional or even global geological framework. In many cases, it represents an essential parameter for establishing detailed genetic models, and can critically impact on exploration strategies. This necessarily requires a reliable, precise and accurate geochronometer.

In the past two decades, U–Pb dating has seen a remarkable success across the Earth Sciences to become the most commonly used absolute isotopic geochronometer. This great success results from considerable improvements in the analytical techniques and in advances of our understanding of the U–Pb system in the geological environment. The paramount advantage of U–Pb dating relies on the coexistence of two chemically identical but isotopically distinct radioisotopes of U (^{238}U and ^{235}U), both of which have their very own decay chain and decay rates. Furthermore, their half-lives are particularly suitable for geologically relevant ages. This allows the determination of two independent dates of which equivalence (concordance) can usually be taken as a sign of the meaningfulness of the date, while discordant dates can be either geologically irrelevant or may be extrapolated to a meaningful date if the cause(s) of this discordance can be identified.

The recent success of U–Pb geochronology is the result of numerous stepwise improvements over the last decades (see detailed history in Davis et al. 2003; Corfu 2013; Mattinson 2013), but has experienced a boost due to coordinated community efforts (EARTHTIME for isotope dilution analysis: http://www.earth-time.org; PLASMAGE for laser ablation analysis: http://www.plasmage.org).

Geochronology was born out of the U–Pb system. Radioactivity was discovered at the dawn of the nineteenth century by H Becquerel, M and P Curie in their work with various uranium compounds (U-salts, U-metal, pitchblende) (Becquerel 1896a, b; Curie et al. 1898; Curie and Skolodowska Curie 1898; Skolodowska Curie 1898). Soon after, E Rutherford first suggested that the Pb/U ratio of geological materials could be used to date them (Rutherford 1906). The next year, B Boltwood applied this method to 43 uranium ore samples and obtained the first absolute total-U and total-Pb ages ranging from 410 to 2200 Ma (Boltwood 1907). This revolution conclusively supported the suggestion made by Charles Darwin half a century prior, that the earth was several hundred million years old, and was about to provide absolute age calibrations for the geological timescale of A Holmes (1911, 1913). However, it was not until the turn of 1930 that the existence of two radioactive U isotopes and their respective Pb daughter isotopes was recognized in U ores (Rutherford 1929; Aston 1929; von Grosse 1932), paving the way for modern U–Pb geochronology. Ever since, improvements in mass spectrometry, laboratory procedures and advances in nuclear physics have permitted the analysis of increasingly smaller quantities of U and Pb with improved precision and accuracy. This in turn, enabled a switch from the analysis of U ore minerals, to low-U bearing minerals such as zircon, titanite and apatite in the second half of the last century (Larsen et al. 1952; Tilton et al. 1955, 1957; Webber et al. 1956). However, dating still involved multigrain mineral fractions which typically show discordance between ^{206}Pb/^{238}U and ^{207}Pb/^{235}U dates, and render their interpretation subjected to debate, assumption and uncertainty. The 1970s to 1980s period arguably marks the turning point of U–Pb geochronology. At that time, the development of low blank single grain zircon dating (Mattinson 1972; Krogh 1973; Krogh and Davis 1975; Lancelot et al. 1976; Michard-Vitrac et al. 1977; Parrish 1987), air-abrasion techniques (Krogh 1982) and in-situ ion probe dating (Hinthorne et al. 1979; Hinton and Long 1979; Froude et al. 1983) concurred to routinely produce concordant U–Pb ages and triggered an expansion in the range of application of U–Pb dating across various minerals, geological terrains and planetary materials. The 1990s saw the advent of the chemical abrasion technique (Mattinson 1994) and of laser-ablation inductively coupled plasma mass spectrometry (Fryer et al. 1993; Horn et al. 2000) that are now common practices in many laboratories around the world. This is the time when U–Pb dating was embraced by the Earth Sciences community, and became an essential tool of geological mapping and mineral exploration. Perhaps as a sign of a mature discipline, the last decade has seen U–Pb practitioners around the world collaborating in a community driven effort to push precision, accuracy and inter-laboratory reproducibility of dates toward unprecedented limits, the EARTHTIME initiative (http://www.earthtime.org).

This century of development of U–Pb dating has left us with a powerful tool for ore deposit studies. While zircon is arguably the most commonly used and understood mineral due to its robustness and minimal amount of Pb it can incorporate in its lattice during crystallization (so-called "common" Pb), a number of other U-bearing minerals are amenable to U–Pb dating (e.g., titanite, apatite, monazite, xenotime, rutile, baddeleyite, perovskite, columbo-tantalite, cassiterite, allanite, calcite, etc.). While most minerals can date their crystallization, a handful of them (e.g., apatite, rutile, titanite) actually date their arrival below their respective closure temperature for the U–Pb system. This diversity of minerals allows a variety of ore deposit types and related geological processes (magmatic, hydrothermal, metamorphic, sedimentary and supergene) to be dated. As we write, U–Pb dates have been published on almost the full spectrum of deposit types and an increasing number of minerals are being tested and improved for U–Pb geochronology. However, the systematics of the U–Pb system are only really well-known in zircon and possibly monazite, followed by titanite, apatite, rutile, baddeleyite, and xenotime.

Geochronology can illuminate the apparent geological chaos at some deposits or districts, as well as support, refute or generate hypotheses for ore-forming processes. Nevertheless, only in rare cases does the dated mineral directly date the ore itself (e.g., columbo-tantalite, cassiterite, uraninite). As examples, zircon from a porphyry stock dates magma intrusion and not the cross-cutting copper mineralization, titanite in a skarn dates the high temperature metasomatism and not the deposition of the polymetallic ore at lower temperature. Some minerals may date magmatic crystallization (e.g., zircon, baddeleyite), or metamorphic reactions (e.g., monazite, titanite) and some may date their precipitation from hydrothermal fluids (e.g., monazite, xenotime, calcite, uraninite). In fact, the meaning of any date remains deeply anchored into proper field observations and sample characterization. Some minerals and dating methods (e.g., fission tracks in apatite and zircon, $^{40}Ar/^{39}Ar$ in micas and K-feldspar, etc.) can also record low-temperature events that that post-date ore formation, allowing a fuller understanding of the coupled temperature–time evolution of mineral systems.

While U–Pb geochronology has been extensively used to determine the age of geological events, it remains to current and future generations of scientists to give increasingly more added value to increasingly more precise and accurate dates, feeding quantitative and numerical models or ore-forming processes. For example, when combined with numerical models, the duration of magmatic-hydrothermal events or the probability density distribution of a population of dates may be interpreted in terms magmatic-hydrothermal flux and volume (Caricchi et al. 2014; Chelle-Michou et al. 2017). This will be a critical step if we want to uncover the processes at play during ore formation, and provide mineral exploration professionals with innovative and efficient tools that may help locating a distant or deeply buried deposit, or that could provide early information on the potential size of the explored deposit (e.g., Chelle-Michou et al. 2017).

This chapter reviews the basics of the U–Pb geochronology and the most commonly used dating techniques and minerals while pointing out their respective advantages, weaknesses and potential pitfalls. Through a series of case studies, we illustrate the various usages of U–Pb dating for the study of mineral deposits. Admittedly, U–Pb geochronology is a field that is strongly biased toward the use of zircon and this chapter is not an exception. Nevertheless, we will also shed light on U–Pb dating applied to less commonly encountered and dated minerals.

2 Basics of U–Pb Geochronology

2.1 The U–Pb System

On first approximation, both naturally occurring long-lived parent uranium isotopes (^{238}U and ^{235}U) decay to stable lead isotopes (^{206}Pb and ^{207}Pb, respectively) at distinct rates, and thus have different half-lives and decay constants (λ_{238} and λ_{235}). Details of the U decay to Pb are actually more complex and involve a long chain of alpha or beta decays with the production of a number of intermediate daughter isotopes (Fig. 1a). This allows the formulation of two generalized age equations:

$$\left(\frac{^{206}\mathrm{Pb}}{^{204}\mathrm{Pb}}\right) = \left(\frac{^{206}\mathrm{Pb}}{^{204}\mathrm{Pb}}\right)_0 + \left(\frac{^{238}\mathrm{U}}{^{204}\mathrm{Pb}}\right)\left(e^{\lambda_{238}t} - 1\right), \quad (1)$$

$$\left(\frac{^{207}\mathrm{Pb}}{^{204}\mathrm{Pb}}\right) = \left(\frac{^{207}\mathrm{Pb}}{^{204}\mathrm{Pb}}\right)_0 + \left(\frac{^{235}\mathrm{U}}{^{204}\mathrm{Pb}}\right)\left(e^{\lambda_{235}t} - 1\right), \quad (2)$$

where ^{204}Pb is the only non-radiogenic isotope of Pb and the subscript 0 indicate the initial isotopic composition of lead at the time (t) when the system closed. In cases where the proportion of initial to radiogenic Pb is negligible, which is common for zircon, monazite, and xenotime, Eqs. (1) and (2) can be simplified:

$$\left(\frac{^{206}\mathrm{Pb}^*}{^{238}\mathrm{U}}\right) = e^{\lambda_{238}t} - 1, \quad (3)$$

$$\left(\frac{^{207}\mathrm{Pb}^*}{^{235}\mathrm{U}}\right) = e^{\lambda_{235}t} - 1, \quad (4)$$

where the superscript * indicate the amount of radiogenic Pb that has formed since the system closed. If the system has remained closed since the mineral crystallized, the ^{206}Pb/^{238}U and ^{207}Pb/^{235}U dates should be identical. Dividing Eqs. (1) and (2) yield a third age equation:

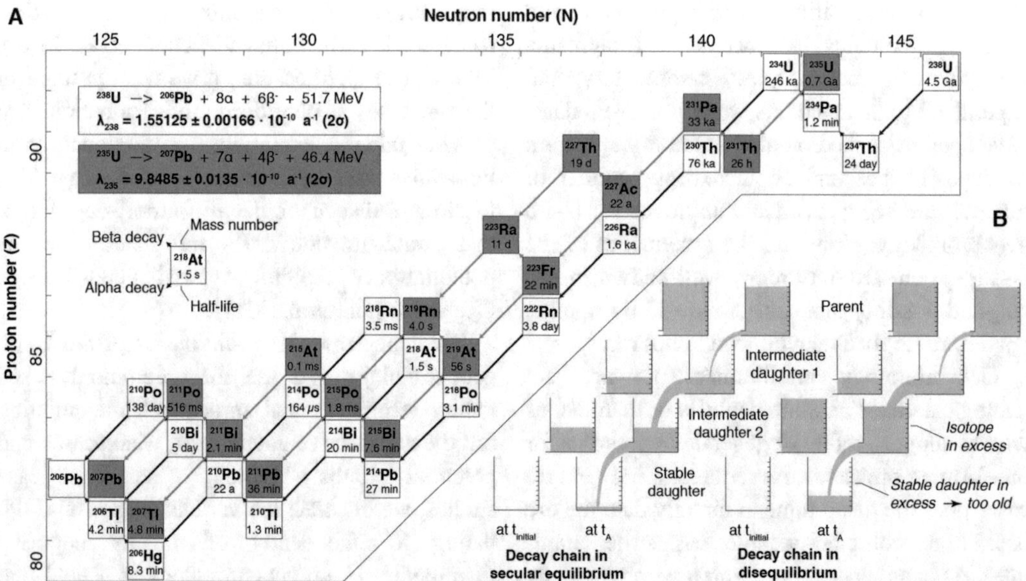

Fig. 1 **a** Decay chains of ^{238}U and ^{235}U with the approximate half-live indicated for each radionuclide. **b** Cartoon illustrating the difference between a decay chain in secular equilibrium and one in disequilibrium. $t_{initial}$ and t_A refer to the time immediately after and some time after mineral crystallization, respectively

$$\frac{\left(\frac{^{207}Pb}{^{204}Pb}\right) - \left(\frac{^{207}Pb}{^{204}Pb}\right)_0}{\left(\frac{^{206}Pb}{^{204}Pb}\right) - \left(\frac{^{206}Pb}{^{204}Pb}\right)_0} = \left(\frac{^{207}Pb}{^{206}Pb}\right)^* = \left(\frac{^{235}U}{^{238}U}\right)\frac{\left(e^{\lambda_{235}t} - 1\right)}{\left(e^{\lambda_{238}t} - 1\right)}. \quad (5)$$

This equation has the advantage that the determination of the age does not require measurement of the U isotopes because the present-day $^{238}U/^{235}U$ ratio is mostly constant in U-bearing accessory minerals and equal to 137.818 ± 0.045 (2σ; Hiess et al. 2012). However, in practice, $^{207}Pb/^{206}Pb$ dates are relevant only for ages older than ca. 1 Ga (see below). The constancy of this ratio and the low abundance of ^{235}U further allow the measurement of the ^{235}U to be neglected, which is common practice in many laboratories.

Decay constants for ^{238}U and ^{235}U are by far the most precisely determined ones among those used in geochronology. Recommended values are those determined by Jaffey et al. (1971) and are $\lambda_{238} = 1.55125 \pm 0.00166 \cdot 10^{-10}$ a^{-1} and $\lambda_{235} = 9.8485 \pm 0.0135 \cdot 10^{-10}$ a^{-1} (2σ) (Schoene 2014). However, these constants have been suggested to be slightly inaccurate (Schoene et al. 2006; Hiess et al. 2012), but always within their reported 2σ uncertainties. More accurate values may be available in the future providing further counting experiments are done.

2.2 Data Presentation

The trinity of age equations presented above (Eqs. 3–5) has promoted the emergence of U–Pb specific plots, the concordia diagrams, that provide a convenient and elegant representation of the data. By far, the most common visual representations of U–Pb data use either the Wetherill concordia plot (Fig. 2a; Wetherill 1956) or the Tera-Wasserburg concordia plot (Fig. 2b; Tera and Wasserburg 1972a, b). These concordia diagrams are bivariate plots where each axis corresponds to one of the three isotopic ratios used in eqs. 3–5 or their inverse (i.e., $^{206}Pb/^{238}U$, $^{238}U/^{206}Pb$, $^{207}Pb/^{235}U$ and $^{207}Pb/^{206}Pb$). On each diagram, the curve represents the line where both isotopic ratios (in abscissa and ordinate) correspond to the same dates, it is the so-called concordia curve. The curvature of the concordia simply reflects the contrasted decay rates of ^{238}U and ^{235}U. If the U–Pb system has remained closed since the crystallization of the mineral and no common Pb is present, the three dates will be the same and plot on the Concordia line, meaning they are be concordant.

For both diagrams (Fig. 2a, b), each analysis is represented by an ellipse where the center is the measured isotopic ratios and the size of the ellipse depicts the analytical uncertainties at a given level of confidence (usually 2σ). Additionally, uncertainties of isotopic ratios plotted on both axis of the concordia diagram are not fully independent from each other and often correlated (e.g., York 1968; Ludwig 1980). This is either due to the use of the ^{206}Pb measurement on both ratios of the Tera-Wasserburg plot or to the use of ^{238}U to calculate ^{235}U for the Wetherill diagram. Thus, the orientation of the uncertainty ellipse reflects the correlation (or covariance) of the errors.

For data that are concordant, it is also convenient to use only the most precise of the three isotopic dates (usually the $^{206}Pb/^{238}U$ or $^{207}Pb/^{206}Pb$ date) and plot them as ranked bars of which the center represents the date and the length reflect the associated uncertainty (Fig. 2c). For a population of dates, the same information can also be presented as a probability density function (Fig. 2c) or a kernel density estimate. The latter is particularly suitable for detrital studies (e.g., Vermeesch 2012).

Because the production of these specific diagrams can be quite labor intensive and calculations in geochronology involve advanced statistical methods, it is recommended to use available software packages dedicated to isotopic geochronology. The most popular and versatile package is the Isoplot Microsoft Excel VBA add-in of K Ludwig (Ludwig 2012) that has served isotope geochronologists for nearly two decades. However, Isoplot is no longer being updated for later versions of Microsoft Excel (last versions

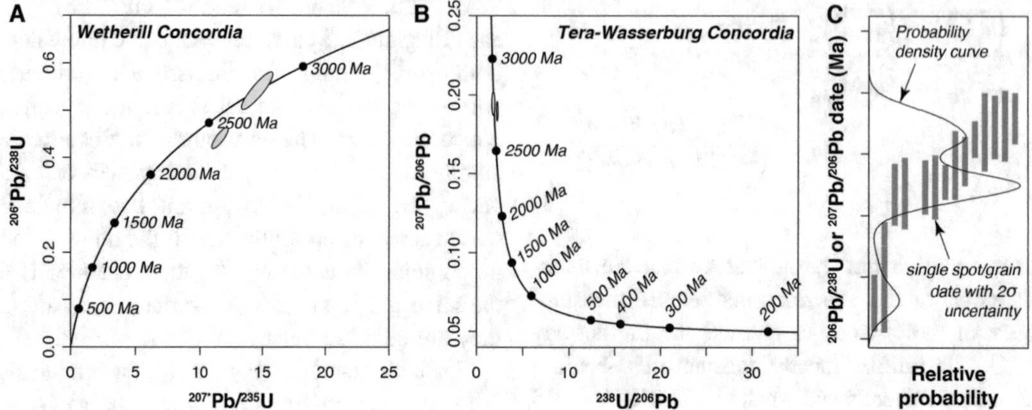

Fig. 2 Classical plots used to present U–Pb geochronological data. **a** Wetherill concordia plot with one concordant and one discordant analysis shown as example, **b** Tera-Wasserburg concordia plot with the same analyses, **c** ranked isotopic date plot for synthetic concordant data together with the corresponding probability density curve. Note that the while the y-axis is valid for both the data bars and the density curve, the x-axis labelled "relative probability" is only relevant for the probability density curve. Single spot/grain dates are ranked only to facilitate the reading of the figure

working on Excel 2010 on PC and Excel 2004 on Mac). This was the incentive for the development of the multiplatform replacement geochronological application IsoplotR. IsoplotR is a package developed for the R statistical computing and graphics software environment by P. Vermeesch (University College London, UK) and can be used through the command line in R or as an online RStudio Shiny applet at http://isoplotr.london-geochron.com (Vermeesch 2018).

2.3 Causes of Discordance

Since the beginning of isotopic dating, discordance has been the main concern of U–Pb geochronologists. Ultimately, understanding the causes of discordance and trying to eliminate it has been the most powerful driving force to advance U–Pb dating during the second half of the twentieth century (Corfu 2013). It is now established that discordance can have a number of origins including: mixing of various age domains, Pb-loss during physical and chemical changes in the crystal lattice (partially opened system), initial intermediate daughter isotopic disequilibrium, incorrect or no correction for non-radiogenic Pb, or a combination of these (Fig. 3). Nevertheless, one should keep in mind that the recognition of some dates as being discordant is intimately tied to the uncertainty of the data. Indeed, low-precision data might appear perfectly concordant, while high-precision ones would actually reveal otherwise (e.g., Moser et al. 2009). This means that any method is blind to discordance at a degree that is inferior to the best age resolution of that method. Below we present the classical causes of discordance and the most appropriate ways to avoid, mitigate or value them.

2.3.1 Mixing Multiple Age Domains

A number of minerals (e.g., zircon, monazite, xenotime) often record multiple growth events. The recognition of different growth zones is crucial for the analysis and interpretation of any dating result. Imagery using transmitted and reflected light together with cathodoluminescence (CL) and back-scattered electron (BSE) microscopy greatly aids in this process but is not always definitive. These images can reveal that a mineral grain can be made up of a sequence of growth zones starting in the center, and mantled by sequential zones towards the rim, all of which can have distinct U–Pb ages. Bulk (whole grain) dating of such multi-domain

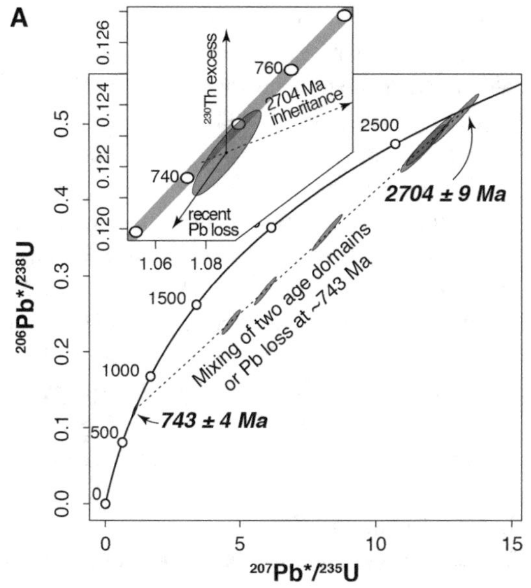

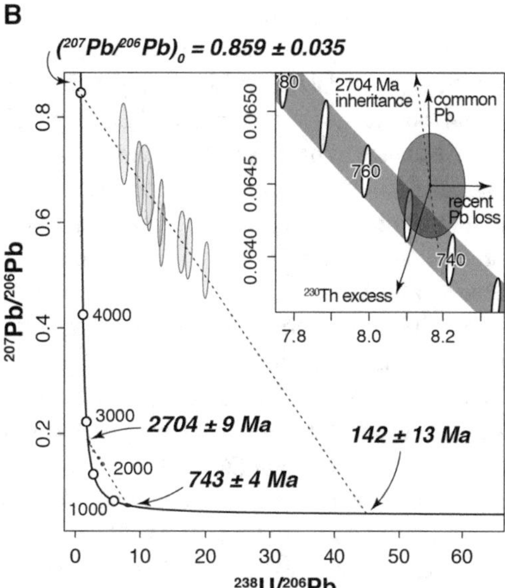

Fig. 3 Main causes of discordance plotted on **a** Wetherill concordia diagram and **b** on a Tera-Wasserburg concordia diagram. Discordance of the red ellipses group is caused by either mixing of two age domains (one at 2704 ± 9 Ma and one at 743 ± 4 Ma) or by Pb-loss of 2704 ± 9 Ma minerals at 743 ± 4 Ma. Discordance of the yellow ellipses group is caused by the presence of common lead in minerals crystallized at 142 ± 13 Ma (Pb_c uncorrected data). Insets shows the possible vectors of discordance

mineral grains could result in discordant dates, if the age differences are sufficiently large. A similar effect can arise from dating multigrain mineral fractions if they include grains with different isotopic ages. In the case of a simple two component mixture of two different age domains, several analyses could plot along a linear array (a so-called discordia line) in concordia diagrams, of which the lower and upper intercept dates would correspond to the respective ages of the two components (red ellipses on Fig. 3). However, multicomponent mixtures may show more scattered distribution or even plot along artificial, and often poorly correlated discordia arrays of which the upper and lower intercept dates have no geological significance, therefore inhibiting meaningful interpretation of the data.

In order to avoid problems arising from mixing several age domains, imagery of the minerals has become a necessary prerequisite to any dating (either in-situ or whole grain) in order to accurately place the spot of the analysis (for in-situ dating) or to select only those grains (or grain fragment) that have one age domain (for whole grain dating). However, small cores or domains with distinct ages can still go unrecognized if they are present below the imaged surface or have a similar chemistry to the surrounding zones. This effect may be monitored on the time-resolved signal for in-situ measurements (changing isotopic ratio) but would hinder the interpretation of whole grain dates.

2.3.2 Open System Behavior

It has long been recognized that the crystallographic lattice of minerals can, under certain conditions, behave as an open system with respect to the U–Pb system (e.g., Holmes 1954; Tilton 1960) through the partial or complete loss of radiogenic Pb. Radiogenic intermediate daughter products of the U decay chains experience a recoil during ejection of the highly energetic alpha particle. The final radiogenic Pb^{2+} is thus situated in a decay-damaged area with enhanced fast pathway diffusion characteristics and could tend to leave this site when appropriate conditions are met. Mechanisms of Pb-loss have been studied extensively, but no simple process can be universally put forward to explain it. Leaching of metamict (radiation-damaged) crystal domains, metamorphic recrystallization, crystal plastic deformation and thermally activated volume diffusion are the most commonly advocated causes of Pb-loss, in decreasing order of importance (see Corfu 2013; Schoene 2014 and references therein). At the sample scale, all these processes will result in discordance of the $^{206}Pb/^{238}U$ and $^{207}Pb/^{235}U$ dates if the age difference is large enough. By calculating by a linear regression through a series of discordant analyses, upper and lower intercepts ages can be reconstructed, corresponding to the age of crystallization of the mineral and to the age of the Pb-loss event, respectively (Fig. 3). Multiple Pb-loss events are notoriously difficult to unravel and may present as excess data scatter or even spurious discordia lines. Furthermore, highly metamict crystal domains may also experience U loss or U gain that would result in inversely (i.e., above the Wetherill concordia) or normally discordant data, respectively. In such cases, no age interpretation can be made. Complete recrystallization of a grain may lead to complete loss of all accumulated radiogenic Pb and reset the age to zero. The extremely low diffusion constants for Pb and U in zircon (Cherniak et al. 1997; Cherniak and Watson 2001, 2003) means that volume diffusion is a very inefficient process to remove radiogenic Pb from an undisturbed zircon lattice. It is for this reason that cases of U–Pb system survival have been reported in granulite facies rocks (e.g., Möller et al. 2003; Kelly and Harley 2005; Brandt et al. 2011; Kröner et al. 2015).

Open-system-related discordance is caused by several distinct processes that cause fast diffusion pathways in the zircon lattice, and such discordant data may be difficult to interpret. Features like multiple growth zones, overgrowth rims, dissolution-reprecipitation textures, or metamorphic recrystallization can be recognized in BSE or CL images (Geisler et al. 2007). Furthermore, recrystallized domains have distinct trace element compositions that can be identified by in-situ chemical analysis (Geisler et al. 2007). Pb-loss through fluid leaching of metamict domains can result in the deposition of minute amounts of 'exotic' elements that normally would not be able to enter the mineral structure (e.g., Fe or Al in zircon; Geisler et al. 2007). Additionally, the degree of metamictization, crystal ordering and ductile crystal reorientation can be evaluated with Raman spectroscopy, electron backscatter diffraction (EBSD), and transmission electron microscopy (TEM), respectively. Finally, for the specific case of zircon, the chemical abrasion technique (Mattinson 2005) has proven to be a powerful method for removing zircon domains that have suffered Pb-loss due to fission tracks, metamictization or other fast diffusion pathways.

2.3.3 Common Pb

Common Pb is a generic name for the fraction of Pb that is not radiogenic in origin and results from a mixture of initial Pb (i.e., Pb incorporated during mineral crystallization) and/or Pb contamination (both in nature and in the lab). The measurement of ^{204}Pb (the only non-radiogenic Pb isotope) undoubtedly pinpoints the presence of common Pb. However, ^{204}Pb measurement can be very challenging for low concentrations of common Pb, or may be prone to isobaric interference with ^{204}Hg, inherent to the LA-ICPMS technique (see analytical methods). On a Tera-Wasserburg plot, analyses containing common Pb typically display a linear array of discordant ellipses defining an upper intercept date older than 4.5 Ma which points to the $^{207}Pb/^{206}Pb$ common Pb composition on the ordinate axis, and a lower intercept providing the age of the mineral (2D isochron; Fig. 3b). If $^{204}Pb/^{206}Pb$ can be measured, it can be plotted on a third axis and the data regressed to estimate the common Pb composition, the age of the mineral and to evaluate the relative contributions of common Pb and Pb-loss on the cause of discordance (3D isochron; Wendt 1984; Ludwig 1998). This approach has been shown to provide better precision for the common Pb composition than the 2D isochron method (Amelin and Zaitsev 2002; Schoene and Bowring 2006). Another Pb-correction practice in LA-ICPMS and SIMS analysis consists of deducing the common Pb correction from measurement of ^{208}Pb (stable decay product of ^{232}Th) and by assuming concordance of the U and Th systems. However, these correction methods may result in overcorrection of some data that are discordant for reasons other than common Pb only. When possible, it is therefore ideal to apply a more robust correction based on the direct measurement of the sample ^{204}Pb. The Pb isotopic composition from laboratory contamination ("blank") is also an important consideration in high-precision U–Pb geochronology using isotope-dilution TIMS, and is obtained through repeated measurement of blank aliquots.

The isotopic composition of initial Pb incorporated during the crystallization of a mineral is best obtained from measurements of cogenetic low-U minerals such as feldspars, galena or magnetite. Alternatively, initial Pb compositions for a known age may be estimated from bulk Earth evolution models (Stacey and Kramers 1975). However, this last approach is less reliable compared to the measurement of a cogenetic low-U mineral (Schmitz and Bowring 2001; Schoene and Bowring 2006). Finally, for the specific case of zircon where the presence of common Pb is essentially limited to inclusions, fractures and metamict domains (see Sect. 6.1), the chemical abrasion technique (Mattinson 2005) has proven to be a powerful method for removing initial Pb from the crystal, leaving only the need for a laboratory blank correction.

2.3.4 Intermediate Daughter Disequilibrium (^{230}Th and ^{231}Pa)

The age equations presented above (Eqs. 1–5) are valid under the assumption that the decay chains are in secular equilibrium, that is, one atom of Pb is created for every decay of one atom of U (Fig. 1b). However, elemental fractionation during mineral crystallization or partial melting would likely disrupt a previously established secular equilibrium (Fig. 1b). This effect should ideally be accounted for in geochronology. Nevertheless, most intermediate decay products of the U series have half-lives of many orders of magnitude smaller (microseconds to years) than the half-lives of U (Ga; Fig. 1a) and potential disequilibrium would have negligible effect on the U–Pb dates even at the best of current analytical capabilities (i.e., 0.5‰ uncertainty on the date). However, intermediate daughters ^{230}Th (^{238}U decay chain) and ^{231}Pa (^{235}U decay chain) have half-lives that are long enough (75.6 ka and 32.8 ka, respectively; Fig. 1a; Robert et al. 1969; Schärer 1984; Parrish 1990; Cheng et al. 2013) to critically impact on the accuracy of the calculated date if disequilibrium is not accounted for (Schärer 1984; Parrish 1990; Anczkiewicz et al. 2001; Amelin and Zaitsev 2002; Schmitt 2007). For example, during monazite crystallization, Th (of which ^{230}Th) is preferentially incorporated into the crystal lattice compared to U, thus resulting in excess ^{206}Pb (e.g., Fig. 1b) and in erroneously old ^{206}Pb/^{238}U dates if the excess ^{230}Th is not accounted for (Figs. 3, 4a). In turn, the Th-uncorrected ^{207}Pb/^{206}Pb date for the same crystal would be too young (Fig. 4b). Conversely, zircon preferentially incorporates U over Th, rendering ^{230}Th-uncorrected ^{206}Pb/^{238}U dates typically too young (Fig. 4a). Similarly, the ^{207}Pb/^{235}U isotopic system is potentially affected by ^{231}Pa excess as has been reported for zircon (e.g., Anczkiewicz et al. 2001).

The magnitude of the correction that needs to be applied to correct the isotopic dates for initial ^{230}Th and ^{231}Pa disequilibrium depends on the distribution coefficient of Th/U and Pa/U between the dated mineral and the liquid from which it crystallized (a melt or an aqueous fluid), respectively (Schärer 1984). For the ^{207}Pb/^{206}Pb date, it also depends on the age of the mineral (Parrish 1990). Figure 4 shows the effect of initial ^{230}Th and ^{231}Pa disequilibrium has on the ^{206}Pb/^{238}U, ^{207}Pb/^{206}Pb and ^{207}Pb/^{235}U dates. It shows that for low mineral/liquid distribution coefficients ($D_{Th}/D_U < 1$) date offsets converge to a minimum of − 109 ka and − 47 ka for the ^{206}Pb/^{238}U and ^{207}Pb/^{235}U dates, respectively. However, if the distribution coefficients are high (> 1), excess ^{206}Pb/^{238}U and ^{207}Pb/^{235}U dates up to few Ma can be expected. Conversely, Th/U distribution coefficient < 1 causes excess ^{207}Pb/^{206}Pb dates of few ka to ca. 0.5 Ma (depending on the age of the mineral), and distribution coefficient > 1 causes a deficit

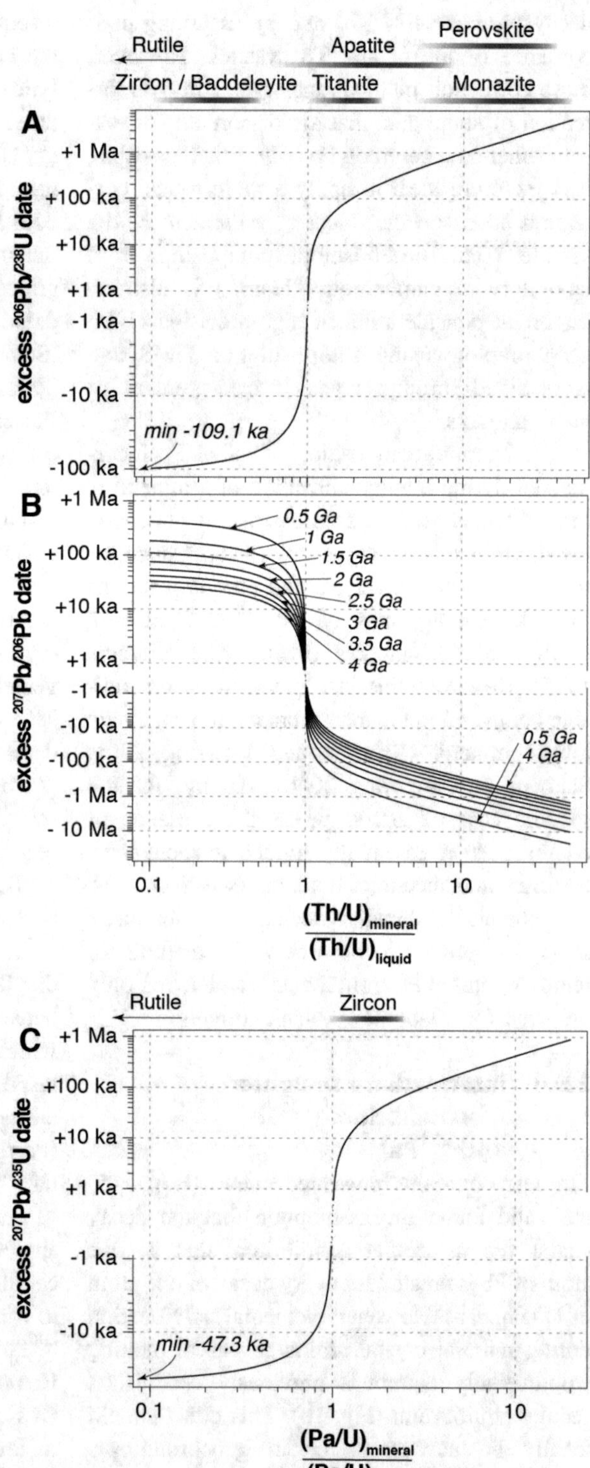

Fig. 4 Excess in **a** $^{206}Pb/^{238}U$ and **b** $^{207}Pb/^{206}Pb$ dates due to initial ^{230}Th disequilibrium, and **c** excess in $^{207}Pb/^{235}U$ date due to initial ^{231}Pa disequilibrium as a function of Th/U and Pa/U mineral/liquid distribution coefficients, respectively (modified after Schärer 1984; Parrish 1990). Typical ranges of mineral/melt distribution coefficients for commonly dated minerals are shown for reference

in $^{207}Pb/^{206}Pb$ dates up to few Ma for Precambrian samples (Fig. 4b).

In practice, the Th/U ratio of the mineral is measured as $^{232}Th/^{238}U$ or estimated from the measured amount of its stable daughter isotope ^{208}Pb by assuming concordance of the U–Pb and Th–Pb dates. For minerals crystallized from a melt, available Th/U mineral-melt distribution coefficients (Fig. 4a) can then be used to reconstruct the Th/U of the melt needed for the Th-disequilibrium correction (e.g., adopting the values from Tiepolo et al. 2002; Klemme and Meyer 2003; Prowatke and Klemme 2005, 2006; Klemme et al. 2005; Rubatto and Hermann 2007; Stepanov et al. 2012; Beyer et al. 2013; Chakhmouradian et al. 2013; Stelten et al. 2015). Alternatively, direct measurement of melt inclusions hosted in the dated mineral, of glass or of whole rock Th/U ratio are also commonly used. Choosing the most appropriate estimate of the melt Th/U ratio at the time of mineral crystallization (using partition coefficient or direct measurement on whole rock or melt inclusions) should be done at the light of all possible information concerning the crystallization conditions of the dated mineral (e.g., temperature, crystallinity, co-crystallizing Th-bearing mineral phases, etc.; see examples in Wotzlaw et al. 2014, 2015).

In essence, ^{230}Th- and ^{231}Pa-corrections are based on the assumption that the dated mineral crystallized from a liquid in secular equilibrium with respect to the U-series. While this might be an acceptable assumption for some magmatic systems (at least for ^{238}U and ^{230}Th) (Condomines et al. 2003), it should not be regarded as a rule, especially for hydrothermal systems in which Th and U have distinct solubilities (Porcelli and Swarzenski 2003; Drake et al. 2009; Ludwig et al. 2011). Indeed, the contrasted partitioning behavior of U and Th into a hydrothermal fluid causes isotopic disequilibrium in the fluid (^{230}Th excess or deficit). In cases where the existence of this fluid is very short (e.g., for magmatic-hydrothermal systems) no time is given for radiogenic ingrowth in the fluid which would remain out of secular equilibrium. Finally, the fractionation of U and Th promoted by the crystallization of U- and Th-bearing hydrothermal minerals may further enhance isotopic disequilibrium. In such cases, the Th-correction (or Pa) should aim at determining the Th/U ratio of the last medium where the decay chain was in secular equilibrium before the crystallization of the mineral. This equates to determining the bulk source (in secular equilibrium) to sink (dated mineral) distribution coefficient of Th/U, regardless of the intermediate process(es), assuming short transport timescales and a unique source of U and Th. For example, Chelle-Michou et al. (2015) used the Th/U ratio of the porphyries (same as for magmatic zircons; Chelle-Michou et al. 2014) to correct the dates obtained on hydrothermal titanite from the Coroccohuayco skarn deposit. In this case, the U-series elements (mainly U and Th) were likely sourced from the magma which was assumed to be in secular equilibrium and transported to the site of deposition by a magmatic fluid in a short period of time.

2.4 A Note on Th–Pb Geochronology

Although less commonly used than U–Pb geochronology, Th–Pb dating may, in some cases, be advantageous and complementary to U–Pb dating. Due to comparable ionic radii of U and Th and similar valence (tetravalent except for oxidized systems where U in mostly hexavalent), most minerals hosting U into their structure will also incorporate Th (if it is available in the system), and vice versa. The single long-lived isotope of Th, ^{232}Th, decays to ^{208}Pb through a chain of alpha and beta decays. The Th–Pb decay offers the possibility of a third independent geochronometer embedded within the mineral allowing for a further assessment of the robustness and meaningfulness of the obtained date. In addition, the nearby masses of ^{235}U, ^{235}U and ^{232}Th on one side, and of ^{204}Pb, ^{206}Pb, ^{207}Pb, and ^{208}Pb on the other side, allows for simultaneous measurement of U–Th–Pb isotopes from the same volume of analyte (ablated volume or dissolved grain). The generalized age equation writes as follow:

$$\left(\frac{^{208}\mathrm{Pb}}{^{204}\mathrm{Pb}}\right) = \left(\frac{^{208}\mathrm{Pb}}{^{204}\mathrm{Pb}}\right)_0 + \left(\frac{^{232}\mathrm{Th}}{^{204}\mathrm{Pb}}\right)\left(e^{\lambda_{232}t} - 1\right), \quad (6)$$

where λ_{232} is the ^{232}Th decay constant. If common Pb is negligible Eq. (6) can be simplified to:

$$\left(\frac{^{208}\mathrm{Pb}^*}{^{232}\mathrm{Th}}\right) = e^{\lambda_{232}t} - 1. \quad (7)$$

The ^{232}Th decay constant is much smaller to that of ^{235}U (half-life of 14 Ga) and is commonly considered to be $4.947 \pm 0.042 \cdot 10^{-11}$ a^{-1} (2σ; Holden 1990). Despite a good accuracy of the ^{232}Th decay constant as suggested by the common concordance of Th–Pb and U–Pb dates (e.g., Paquette and Tiepolo 2007; Li et al. 2010; Huston et al. 2016), its precision is an order of magnitude lower than those of ^{238}U and ^{235}U. This can represent the main source of systematic uncertainty on Th–Pb dates and the main limitation of this system when working below the percent precision level. However, unlike uranium, intermediate daughter isotopes of the ^{232}Th decay chain have short half-lives such that any isotopic disequilibrium formed during mineral crystallization will fade within few decades only. Therefore, the ^{232}Th decay chain can be considered to have remained in secular equilibrium on geological timescale. It results that on cases where U–Pb dates require a large initial ^{230}Th-disequilibrium correction and parameters required for this correction are difficult to estimate (e.g., hydrothermal minerals), Th–Pb dates may be much more accurate than U–Pb ones (but often of lower precision).

Due to the very long half-live of ^{232}Th, the optimal use of Th–Pb geochronology (highest analytical precision) is achieved for old sample and/or minerals with high Th concentrations so that large amount of ^{208}Pb have been accumulated. In the case of Th-rich minerals (e.g., monazite and perovskite, and, to a lesser extent, xenotime, apatite, titanite and allanite), thorogenic ^{208}Pb (i.e., ^{208}Pb*) would typically be so abundant than common Pb correction may not introduce significant uncertainties into the computed ^{208}Pb*/^{232}Th ratio or may even be neglected.

^{208}Pb/^{232}Th dates are most commonly presented in rank-order plots such as Fig. 2c, the center of each bar representing the date and the length reflecting the associated uncertainty. To evaluate the concordance of the Th–Pb and U–Pb systems, concordia diagrams (^{208}Pb*/^{232}Th vs. ^{206}Pb*/^{238}U or ^{207}Pb*/^{235}U) offer a convenient graphical representation of the data.

3 Analytical Methods (Including Data Reduction, Pb-Correction, Uncertainty Propagation and Data Presentation)

Currently, three methods are commonly used to measure isotopic ratios necessary for U–Pb geochronology: (1) laser ablation-inductively coupled plasma mass spectrometry (LA-ICPMS); (2) secondary ion mass spectrometry (SIMS); and (3) isotope dilution-thermal ionization mass spectrometry (ID-TIMS). Each of these methods have particular strengths and weaknesses (see summary in Table 1). In most cases, U–Pb geochronology involves the separation of the mineral of interest through gravimetric and magnetic techniques (e.g., heavy liquids, Wilfley shaking table, Frantz magnetic separator) and the selection of individual grains (picking) under binocular microscope. However, in-situ dating with LA-ICPMS and SIMS can also be done directly on polished thin section, thus preserving the petrographic context of the dated mineral, which may be key for the interpretation of the data in some cases.

The main difference between these three techniques resides in the way the dated material is prepared, ionized and introduced into the mass spectrometer. Below, we present an overview of the main aspects of the state-of-the-art procedures for these methods, while highlighting their respective advantages and disadvantages and the handling of uncertainties. For more details on the technical aspects of mass spectrometry, the interested reader is referred to a number of good textbooks and papers (e.g., Ireland and Williams 2003; Parrish and Noble 2003; Gehrels et al.

Table 1 Comparison of the three analytical techniques used for U–Pb dating

	LA-ICPMS	SIMS	CA-ID-TIMS
Spatial resolution	Spot diameter typically of 10–50 μm, depth of 15–40 μm	Spot diameter typically of 10–15 μm, depth of 1–2 μm	Whole mineral grain or grain fragment. Mixing of age domains is hard to avoid
Standardization	External with a known reference material and accuracy controlled with a secondary standard	External with a known reference material and accuracy controlled with a secondary standard	Internal with tracer solution (preferably double Pb—double U isotope tracer)
Sample preparation	Mineral separate mount or thin section, Imagery (CL, BSE, …)	Mineral separate mount or thin section, Imagery (CL, BSE, …)	Mineral separation, imagery, chemical abrasion (for zircon only) and washing, digestion, column chemistry
Time required for sample preparation	Few days for mineral separation, sample mount preparation and imagery	Few days for mineral separation, sample mount preparation and imagery	Few days for mineral separation and imagery; 1 day for chemical abrasion of zircon; $\geq$ 3 days for acid digestion; 1 day for chemical separation of Pb and U
Time required for one analysis (sample or standard)	2–3 min	15–30 min	3–4 h
Analytical precision (reference for typical zircon: see Fig. 5)	2–5% on single spot date and ~0.2–2% on weighted mean date	1–5% on single spot date and ~0.1–1% on weighted mean date	0.1–0.05% on single grain $^{206}Pb/^{238}U$ date and ~0.02% on weighted mean date
Accuracy	~1–5%	~1–5%	0.03–0.3%; fully traceable to SI units
Preferred geologic application	Large scale survey, detrital geochronology, in-situ dating, minerals with inherited cores	In-situ dating, complexely zoned minerals	Used when highest temporal resolution or highest accuracy are necessary
Limitations	Imprecise common Pb correction, matrix matched standard material	Matrix matched standard material required for $^{206}Pb/^{238}U$ and $^{207}Pb/^{235}U$ dates, but not required for $^{207}Pb/^{206}Pb$ dates	Only very limited spatial resolution (microsampling)

2008; Arevalo et al. 2010; Arevalo 2014; Carlson 2014; Ireland 2014; Schoene 2014; Schaltegger et al. 2015).

3.1 Laser Ablation-Inductively Coupled Plasma Mass Spectrometry (LA-ICPMS)

LA-ICPMS is an efficient U–Pb dating technique that allows high spatial resolution and high sample throughput. Analysis is done directly from a thin section or from polished grains mounted in epoxy resin that have been imaged by transmitted and reflected light, CL and/or BSE techniques prior to analysis. Typical analytical uncertainties for zircon dates are on the order of 3–5% for single spot and of 0.2–2% for the weighted mean dates (Fig. 5). However, accuracy may not be better than 3% (Klötzli et al. 2009; Košler et al. 2013), which should be considered when comparing LA-ICPMS U–Pb dates

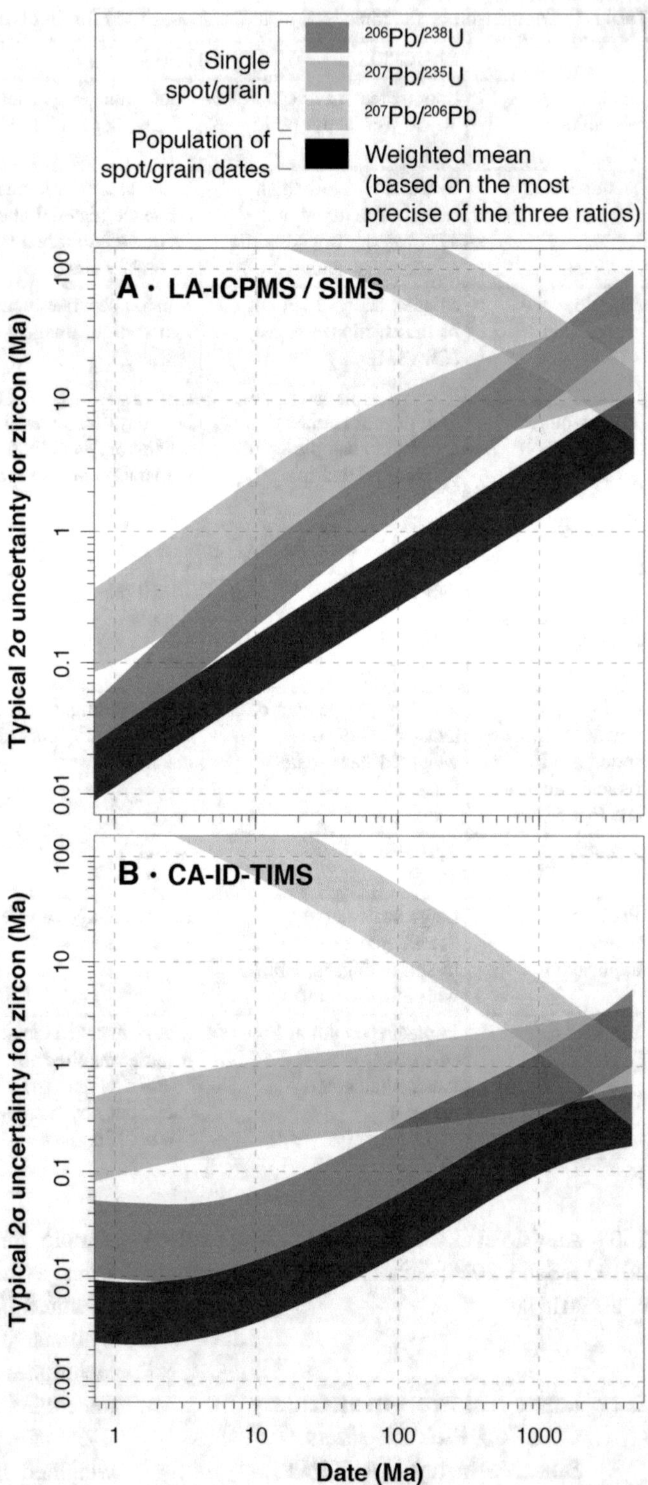

Fig. 5 Typical analytical uncertainties for zircon ^{206}Pb/^{238}U, ^{207}Pb/^{235}U, ^{207}Pb/^{206}Pb single spot/grain dates for modern **a** LA-ICPMS, SIMS and, **b** CA-ID-TIMS dating techniques. Weighted mean dates refers to the weighted mean of a set of statistically equivalent single spot/grain dates based the most precise isotopic ratio (typically ^{206}Pb/^{238}U for dates younger than ca. 1 Ga and ^{207}Pb/^{206}Pb for dates older than 1 Ga)

from different studies or with dates from other isotopic systems.

The LA-ICPMS setup consists of a laser of short wavelength in the UV range (typically 193 nm), an ablation cell and an ICPMS instrument. The sample is placed into the ablation cell along with several standards. During ablation, repeated laser pulses are focused on the surface of the dated mineral. The resulting ablated aerosol is subsequently transported by a carrier gas (usually He $\pm$ Ar $\pm$ N_2) toward the Ar-sourced plasma torch at the entry of the mass spectrometer where it is ionized and transferred into the ion optics of the mass spectrometer. LA-ICPMS U–Pb dating is mostly carried out on single-collector sector-field ICP-MS instruments that offer sequential measurement of individual Pb and U isotopes in a mixed ion-counting – Faraday cup mode.

The spot size used for LA-ICPMS geochronology mainly depends on target size and the U concentration of the dated mineral. As a reference, 25–35 μm spots are commonly used for zircon and can be as low as 5 μm for monazite (Paquette and Tiepolo 2007). Crater depth for a 30–60 s analysis is on the order of 15–40 μm depending on the fluence of the laser and on the ablated material. However, laser-induced U–Pb fractionation increases with crater depth during ablation, which negatively impacts on the analytical uncertainty of the measured Pb/U ratio. Ultimately, this is an important limiting factor for precision and accuracy in LA-ICPMS geochronology (Košler et al. 2005; Allen and Campbell 2012). The technique requires a laser setup that yields reproducible ablation with small particles (subsequently more efficiently ionized in the plasma torch) and that limits crater depth to no more than the spot diameter by minimizing the laser fluence (e.g., Günther et al. 1997; Horn et al. 2000; Guillong et al. 2003).

Another important limitation of LA-ICPMS U–Pb dating is the imprecise common Pb correction due to the difficulty of precisely measuring common ^{204}Pb caused by an isobaric interference with ^{204}Hg (traces of Hg are contained in the Ar gas). Common Pb correction protocols using ^{208}Pb may be employed and are preferred over simple rejection of discordant analyses. It results that age interpretation of minerals with elevated common Pb contents (e.g., titanite, rutile) may be hampered by large age uncertainties due, in part, to the large uncertainties associated with the common Pb-correction.

LA-ICPMS and SIMS (see below) U–Pb dating are comparative techniques that require analysis of a reference material, which is as close as possible to the chemical composition and the structural state of the unknown (sample). It is analyzed under identical ablation conditions to the sample to determine the machine fractionation factor of any measured element concentration; this fractionation factor is then applied to the element ratios and concentrations of the unknowns. A series of analyses unknown ($\sim$10) is typically bracketed by analyses of a reference material ($\sim$2–4) to correct for elemental fractionation and monitor for machine drift. In addition, at least one secondary standard should be repeatedly analyzed during the same session in order to demonstrate the accuracy of the fractionation correction. This enables an estimate of the long-term excess variance of the laboratory that is required in the uncertainty propagation protocol (see below). A list of commonly used reference materials and their reference values is provided in Horstwood et al. (2016). Standards for LA-ICPMS and SIMS U–Pb dating should be homogenous in age, trace element composition, and have comparable trace element concentration and structural state (matrix match) as the unknowns (Košler et al. 2005). Failure to match the matrix of the unknown results in different ablation behavior (rate, stability, fractionation) and ultimately compromises the accuracy of the date (Klötzli et al. 2009). Therefore, a mineral of unknown age should be standardized using a reference material from the same mineral. Furthermore, different degrees of metamictization also impact on the matrix match between standards and unknowns and can be an important source of inaccuracy for zircon dates (as much as 5% inaccurate; Allen and Campbell 2012; Marillo-Sialer et al. 2014) and possibly for other minerals as well (e.g., titanite, allanite, columbo-tantalite).

Interlaboratory comparisons for LA-ICPMS and SIMS U–Pb dating have highlighted discrepancies of U–Pb ages for a series of standards measurements which is sometimes outside of the reported 2σ uncertainties (Košler et al. 2013). This is thought to reflect different data reduction strategies in different laboratories (e.g., Fisher et al. 2010) and uncertainty propagation protocols, that are not always thoroughly documented. This has triggered a community driven effort to establish standard data reduction workflow, uncertainty propagation protocols, and data reporting templates (Horstwood et al. 2016) that should be embraced by the LA-ICPMS community. New community-derived standards for LA-ICPMS dating suggest the use of the *x/y/z/w* notation for uncertainty reporting where: *x* refers to the analytical (or random) uncertainty, *y* includes the variability of standards measured in the same lab, *z* includes the systematic uncertainty of the primary standard isotopic composition (and of the common Pb correction if appropriate), and *w* includes the decay constant uncertainty (Horstwood et al. 2016; McLean et al. 2016). Comparing LA-ICPMS U–Pb data with data from other LA-ICPMS, SIMS or ID-TIMS laboratories should be done at the *z* uncertainty level, while comparison with geochronological data from other isotopic systems have to include decay constant uncertainties (Chiaradia et al. 2013). Raw data processing, visualization and uncertainty propagation protocols for LA-ICPMS U–Pb dating have been implemented in the freely available *ET_Redux* software (McLean et al. 2016) and allow more robust interlaboratory data comparison and collaborative science.

3.2 Secondary Ion Mass Spectrometry (SIMS)

Compared to LA-ICPMS, SIMS U–Pb analysis has greater spatial resolution and sensitivity, allowing for the analysis of microscopic rims or domains in zircon, monazite, xenotime or other minerals. SIMS analysis involves the ablation of sample with a high-energy O^- or O_2^- ion beam within a high vacuum chamber. A small fraction of the ablated material forms atomic ions or molecular ionic compounds that are subsequently accelerated into a mass spectrometer. Typical SIMS craters are 10–15 μm in diameter and 1–2 μm deep, therefore this technique has higher spatial resolution and is by far less destructive than LA-ICPMS and permit subsequent isotopic analysis (e.g., O, Hf–Lu) to be done on the same spot (slight repolishing would be required before SIMS analysis). Analysis is done directly from a thin section, polished grains mounted in epoxy resin, or from entire grains pressed into indium when analyzing U and Pb isotopes along a profile from the surface to the interior of a grain (depth profiling). The accuracy of the obtained result depends on extrinsic factors such as the position of standard and unknowns in the mount and the quality of the polishing. SIMS analysis of zircon typically yields U–Pb dates of 0.1–1% precision and accuracy (Fig. 5); it is the preferred method when analyzing complex minerals (e.g., thin metamorphic rims), very small grains (e.g., xenotime outgrowths on zircon; McNaughton et al. 1999) or valuable material.

Pb isotopic fractionation in SIMS is subordinate when compared to LA-ICPMS techniques. Therefore, $^{207}Pb/^{206}Pb$ dates can be calculated directly from counting statistics. In contrast, there is a significant difference in the relative sensitivity factors for Pb^+ and U^+ ions during SIMS analysis. The fractionation of the $^{206}Pb^+/^{238}U^+$ ratios is highly correlated with simultaneous changes in the $^{254}UO^+/^{238}U^+$ ratios which forms the basis of a functional relationship that enables the calibration of the $^{206}Pb/^{238}U$ dates. Although the $^{206}Pb^+/^{238}U^+$ versus $^{254}UO^+/^{238}U^+$ calibration is the most widely used, other combinations of $^{238}U^+$, $^{254}UO^+$ and $^{270}UO_2$ have proved successful. As in the case of LA-ICPMS, the SIMS $^{206}Pb/^{238}U$ calibration is carried out with reference to a matrix matched reference material (e.g., Black et al. 2004). This is quite straightforward for zircon and baddeleyite (ZrO_2), but more difficult for chemically and structurally more complex minerals (e.g., phosphates, complex silicates, oxides). In the latter cases, matrix correction procedures using a suite of reference materials accounting for the effect of highly variable amount of trace elements have been developed (e.g., Fletcher et al. 2004, 2010). Calibration biases are also introduced through different crystal orientation (Wingate and Compston 2000) or different degrees of structural damage from radioactive decay (White and Ireland 2012). It is highly recommended to analyze a reference zircon as unknown again to control the accuracy of the technique (validation or secondary standard; Schaltegger et al. 2015).

The common Pb correction is carried out via measurement of ^{204}Pb, ^{207}Pb or ^{208}Pb masses. The main challenge of SIMS analysis is the resolution of molecular interferences on the masses of interest (Ireland and Williams 2003), which requires careful consideration when analyzing phosphates or oxides.

No standard data treatment protocol exists for SIMS dates. In fact, the two types of equipment (SHRIMP from Australian Scientific Instruments and IMS 1280/90 from CAMECA) provide very differently structured data that require different data treatment software.

3.3 Isotope Dilution-Thermal Ionization Mass Spectrometry (ID-TIMS)

The U–Pb method that offers the highest precision and accuracy is Chemical Abrasion, Isotope Dilution, Thermal Ionization Mass Spectrometry (CA-ID-TIMS; Table 1, Fig. 5). This method involves the dissolution and analysis of entire zircon grains and other accessory minerals, and, hence, disregards any protracted growth history recorded in this grain. Zircon imaging prior to dating can be taken to increase the chances of analyzing a single-aged grain or grain population. The ID-TIMS community is organized as a part of the EARTHTIME consortium (Bowring et al. 2005), which is working together to improve precision and accuracy of U–Pb dating.

It is now standard to pre-treat zircons with the "chemical abrasion" procedure of Mattinson (2005). This process involves heating the zircon at 900 °C for 48 h, followed by partial dissolution in HF + HNO_3 at 180–210 °C for 12 to 18 h (Widmann et al. 2019). The heating re-establishes the zircon crystalline structure by annealing any radiation-related structural damage in slightly affected domains. The partial dissolution procedure then only removes domains with more severe structural damage and leaves a proportion of the original grain behind. The surviving zircon fragment is then considered to be perfectly crystalline and is used for isotope ratio analysis. Chemically abraded zircon grains are recognized to be more concordant and provide more reproducible U–Pb results. This treatment is not currently applied for SIMS or LA-ICPMS analysis techniques, but initial experiments have yielded positive results (Kryza et al. 2012; Crowley et al. 2014; von Quadt et al. 2014). The procedure has been tested on other accessory phases including baddeleyite (Rioux et al. 2010), but without clear evidence of improving concordance.

The dissolved grains are mixed with a (^{202}Pb–)^{205}Pb–^{233}U–^{235}U tracer solution (e.g., as provided by EARTHTIME; ET535 and ET2535; Condon et al. 2015; McLean et al. 2015), and the Pb and U isotopes isolated from other trace elements through chromatography. Isotopic compositions are most commonly measured as Pb^+ and UO_2^+ on a thermal ionization mass spectrometer from the same filament either by ion counting methods (using a secondary electron multiplier or a Daly-based photomultiplier device), or by a combination of ion counters and high-sensitivity, high-resistance Faraday collectors. Uranium may also be measured separately as U^+ by solution MC-ICP-MS utilizing a mixed ion counting—Faraday measurement setup, or as U^+ on a double or triple filament assembly in a TIMS.

An important part of high-precision, high-accuracy U–Pb geochronology is the correct treatment of all sources of uncertainty and their correct propagation into the final age. The ID-TIMS community has been adopting the *x/y/z* notation for uncertainty reporting (e.g., 35.639 ± 0.011/0.014/0.041 Ma) where: *x* is the random uncertainty (or analytical; including counting statistics, common Pb and Th-disequilibrium corrections), y includes the systematic uncertainty from tracer calibration and, z includes the decay constant uncertainty (Schoene et al. 2006; Schoene and Bowring 2006; McLean et al. 2011). Comparison of ID-TIMS U–Pb data with U–Pb data from SIMS or LA-ICPMS techniques should consider the y uncertainty level, while comparison with data from other isotopic systems (e.g., Re-Os, $^{40}Ar/^{39}Ar$) should include both decay constant and systematic uncertainties (z level). Final age precision is mainly defined by the ratio of radiogenic to common Pb (Pb^*/Pb_c), which is, in the case of zircon, a function mainly of procedural Pb blank. Total blank levels of < 0.5 pg of Pb are currently state-of-the-art.

The EARTHTIME community has generally accepted and adopted a software package consisting of *Tripoli* raw data statistics and *U–Pb_Redux* data treatment and visualization (Bowring et al. 2011; McLean et al. 2011).

4 Guidelines for Interpreting U–Pb Dates

4.1 Date and Age

Isotopic dating makes a distinction between a *date* and an *age*. The term 'date' refers to a number in time unit (usually Ga, Ma or ka) calculated from an age equation (Eqs. 1–5). The term 'apparent age' is sometimes used as a synonym for 'date'. A 'date' becomes an 'age' as soon as in can be interpreted in terms of a geological process (Schoene 2014). Both terms may be appropriate for single grain/spot or weighted mean data and may be accurate or inaccurate. This semantic distinction reflects the clear distinction that should be made between data and their interpretation, which is at the core of scientific rigor and integrity.

As discussed in the preceding sections, the interpretation of U–Pb dates is not straightforward, even for concordant data. It requires a close and quantitative control of the way how an analytical result has been produced, including the knowledge of sources of error and their correct propagation into the final result (metrology), a good characterization of the sample material, and finally a good knowledge of the geological context. The lack of considering these aspects may very well lead to over-interpretations and erroneous conclusions.

4.2 Geochronology Versus Thermochronology

All minerals used for U–Pb dating can be theoretically subjected to some degree of thermally activated volume diffusion of U and Pb. The measured date reflects the time elapsed since closure of the isotopic system. While geochronology corresponds to dating of a mineral that has crystallized, rapidly cooled or remained below it closure temperature, thermochronology deals with minerals that have crystallized and/or spent some time above their respective closure temperatures, or in the partial retention temperature window of their daughter nuclide. As discussed above (causes of discordance) partial resetting of the U–Pb system by diffusion is a possible source of discordance. While the effect of post crystallization diffusion can usually be neglected for zircon, monazite and most other minerals due to their high closure temperature for Pb (> 700 Cherniak and Watson 2001, 2003; Cherniak et al. 2004); Fig. 6), Pb diffusion in minerals such as titanite, rutile and apatite is more likely to occur and should carefully be evaluated before interpreting U–Pb dates as they might record the age of closure rather the age of crystallization. Ultimately, thermochronological U–Pb data on these minerals may be used to constrain the high-temperature (> 350 °C) thermal history of the studied geological object (Schoene and Bowring 2007; Kooijman et al. 2010; Blackburn et al. 2011; Cochrane et al. 2014). Nevertheless, it appears that most minerals used for U–Pb dating can be used as geochronometers, of which partial resetting of the U–Pb system is often controlled by the stability of the mineral phase itself or Pb-loss along fast diffusion pathways (cracks, metamict domains), rather by volume diffusion (Fig. 6).

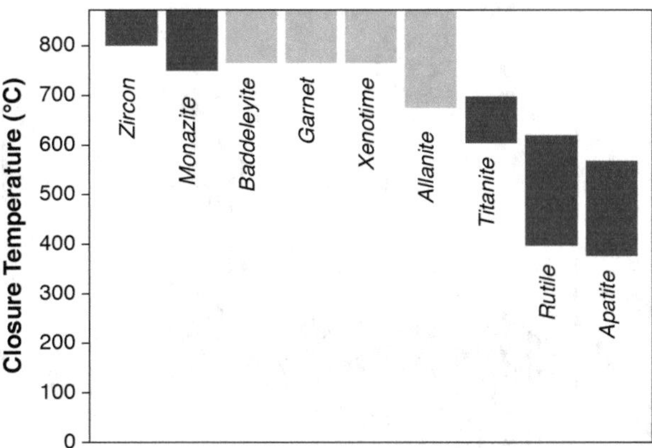

Fig. 6 Typical range of closure temperature for minerals used for U–Pb dating. Dark grey bars indicate robust closure estimates while light grey bars indicate approximate estimates. Modified from Chiaradia et al. (2014), with additional data for apatite (Cochrane et al. 2014), rutile (Vry and Baker 2006), baddeleyite (Heaman and LeCheminant 2001), garnet (Mezger et al. 1989), xenotime and allanite (Dahl 1997)

4.3 Precision and Weighted Mean

The weighted mean age is the most common representation of the age of a relatively short-lived geological event recorded at the scale of the sample (e.g., magma emplacement, hydrothermal fluid circulation) and is usually interpreted as the best age estimate. Weighted mean calculations are applied to a set of individual analyses in order to reduce the uncertainty of the population. It implicitly assumes that the data correspond to repeated analyses (samples) of the exact same value and that the uncertainties are only due to analytical scatter. In this case, the mean square of the weighted deviates (MSWD or reduced chi-squared) of a data population to the weighted mean should be around to 1. In turn, MSWD >> 1 would suggest excess scatter of the data given their respective uncertainties (i.e., they are unlikely to represent a single population), and values << 1 suggest that the reported uncertainties are larger than what would be expected from a single population. In detail, acceptable MSWD values actually depend on the number of points pooled together (Wendt and Carl 1991; Spencer et al. 2016). For example, values between 0.5 and 1.5 are acceptable for a population of 30 points (at 2σ).

However, the accuracy of weighted mean ages has been repeatedly questioned (Chiaradia et al. 2013, 2014; Schoene 2014). Indeed, the advent of high precision dating techniques (CA-ID-TIMS) has highlighted that data that might look statistically equivalent at the level of their uncertainties, can actually hide a spread of data that can only become apparent with more precise dating methods. An illustration of this is provided in Fig. 7 which shows LA-ICPMS and CA-ID-TIMS $^{206}Pb/^{238}U$ zircon dates from a porphyry intrusion from the Coroccohuayco porphyry-skarn deposit, Peru (Chelle-Michou et al. 2014). It is noteworthy that those grains analyzed by CA-ID-TIMS have previously been analyzed with LA-ICPMS (with 1 to 3 spots each) before being removed from the epoxy mount for further processing. Data points are plotted at the level of their analytical uncertainties and weighted mean dates include additional dispersion and standard/tracer calibration uncertainties (see caption of Fig. 7 for more details) so that they can be compared at their right level of uncertainties (i.e., neglecting only decay constant uncertainties). Both the LA-ICPMS (36.05 ± 0.25 Ma, n = 30, MSWD = 1.3) and CA-ID-TIMS (35.639 ± 0.014 Ma, n = 7, MSWD = 1.8) weighted means yield acceptable MSWDs (in agreement with their respective number of data points), thus suggesting they could correspond to statistically equivalent data populations, respectively. Independently from each other, these weighted dates would be

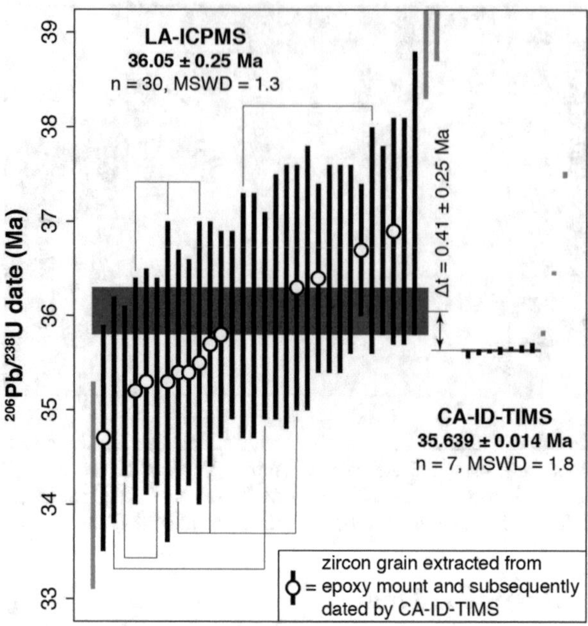

Fig. 7 Ranked LA-ICPMS and CA-ID-TIMS $^{206}Pb/^{238}U$ zircon dates and weighted means for the hornblende-biotite porphyry (sample 10CC51) from the Eocene Coroccohuayco porphyry-skarn deposit, Peru. Data from Chelle-Michou et al. (2014). Single spot/grain analyses are plotted at the level of their analytical uncertainties (2σ) and weighted mean dates include the analytical uncertainties and: (i) an additional excess variance obtained from repeated measurement of the secondary standard (91,500) and the systematic uncertainty in the standard mineral isotopic composition, for LA-ICPMS data; (ii) the systematic uncertainty related to the composition of the isotopic tracer, for CA-ID-TIMS data. Data bars in black are included in the calculation of the weighed mean date. Multiple LA-ICPMS dates from the same zircon grain are connected with thin lines

interpreted as the age of the porphyry intrusion at the Coroccohuayco deposit. However, Fig. 7 highlights that these ages do not overlap within uncertainties ($\Delta t = 0.41 \pm 0.25$ Ma), therefore indicating that at least one of them is inaccurate. In this case, the more precise single grain CA-ID-TIMS ages highlight more than 1 Ma of zircon crystallization in deep-seated crystal mushed (or proto-plutons) before their incorporation into felsic melts, ascent and emplacement of the porphyry intrusion at an upper crustal level (Chelle-Michou et al. 2014). These older zircon crystallization events cannot be resolved at the uncertainty level of LA-ICPMS dating for which data points pool together that are actually not part of the same population and therefore include data older than the emplacement age, resulting in a weighted mean age that is too old. While it is common practice in zircon CA-ID-TIMS dating to take the youngest point as best representative of the age of magma emplacement or eruption, this practice is not appropriate for in-situ or CA-free ID-TIMS dating techniques where the weighted mean date of the youngest cluster having an acceptable MSWD remains the best option, although it might sometimes be slightly inaccurate.

This example highlights the limitations of the weighted mean approach to complex and protracted natural processes. The statistical improvement in precision may be done at the cost the accuracy of the dated process. The calculated weighted mean date can be either too old (e.g., if grains crystallized from an earlier pulse of magma are included), too young (e.g., if several grains have suffered similar amounts of unrecognized Pb-loss) or just right by coincidence. In fact, the time resolution of geochronology is ultimately limited by the precision of single data points, rather than by the number of data that are pooled together to statistically reduce the age uncertainty.

4.4 Accuracy of Legacy U–Pb Data and Misinterpretation

Cases where the same rock has been dated several times using the same isotopic system and the same mineral are rare but necessary examples to put some perspective of the accuracy of legacy U–Pb data. Ore-related porphyry intrusions at the Miocene Bajo de la Alumbrera porphyry copper deposit have received much attention over the past decade. These rocks have been repeatedly dated by U–Pb zircon geochronology using different analytical methods (LA-ICPMS and CA-ID-TIMS) at different times (Harris et al. 2004, 2008; von Quadt et al. 2011; Buret et al. 2016). The early LA-ICPMS zircon dating survey of Harris et al. (2004, 2008) concluded that the deposit formed on a million-year time scale. However, subsequent high precision CA-ID-TIMS studies have decreased this duration by almost two orders of magnitude, to a maximum duration of 29 ka (Buret et al. 2016).

Available data for three porphyries are compiled Fig. 8 with their respective weighted means. Single LA-ICPMS date broadly range from 8.5 to 6.5 Ma while those obtained by CA-ID-TIMS are significantly less scattered between 8.2 and 7.1 Ma. Weighted mean dates can show as much as $\sim$1 Ma of age difference for the same porphyry between LA-IPCMS and CA-ID-TIMS which is far outside the reported analytical uncertainties (see P2 porphyry on Fig. 8). The same is true for high-precision CA-ID-TIMS data, which show differences up to $\sim$0.1 Ma in excess of the analytical uncertainty. Furthermore, these discrepancies persist even when systematic uncertainties are taken into account (i.e., 3% reproducibility for LA-ICPMS, calibration of the primary standard or of the tracer solution). Similar age discrepancies up to $\sim$0.8 Ma between LA-ICPMS and SIMS U–Pb zircon weighed mean ages have been noted by Ballard et al. (2001) on porphyries from the Eocene Chuquicamata Cu deposit, Chile.

It would be presumptuous to name the causes of these discrepancies without having the entire set of original technical and analytical data at our disposal. Nevertheless, we can make some conjectures. Potential causes may be: (1) that different populations of zircons grains or domains (within a single grain) where hand-picked and dated; (2) the use of inappropriate data reduction, common Pb correction, initial Th-correction and error propagation protocols; (3) a distinct difference in ablation rate between sample and standard zircon resulted in inaccurate correction for fractionation (for LA-ICPMS data); (4) inaccurate isotopic tracer calibration (for ID-TIMS data); and/or (5) unidentified concordia parallel Pb-loss (for the LA-ICPMS data).

In the case of Bajo de la Alumbrera, the most recent data by Buret et al. (2016) are deemed to be the most accurate (in addition of being the most precise) and tightly constrain the age of porphyry emplacement and zircon crystallization. This example illustrates the difficulty of dealing with legacy U–Pb data which might or might not be accurate. Obviously, there are published ages that are inaccurate, but they would remain unnoticed until new dating is done with state-of-the-art techniques. In particular, reporting of *x/y/z* (for ID-TIMS) and *x/y/z/w* (for LA-ICPMS) uncertainties and comparison of disparate U–Pb dates at the level of their *y* uncertainty should be systematic. Again, these potential biases should be carefully accounted for when interpreting short time differences on the order of the analytical uncertainty of single dates. This also highlight the need for thorough reporting of analytical and data handling procedures, or even, using common analytical procedures and data reduction platforms (Košler et al. 2013).

5 What Mineral Can We Date with the U–Pb System and What Does It Date?

As of today, a great number of minerals have been used for U–Pb dating, many of which in the context of mineral deposits. A non-exhaustive list of these minerals is provided in Table 2

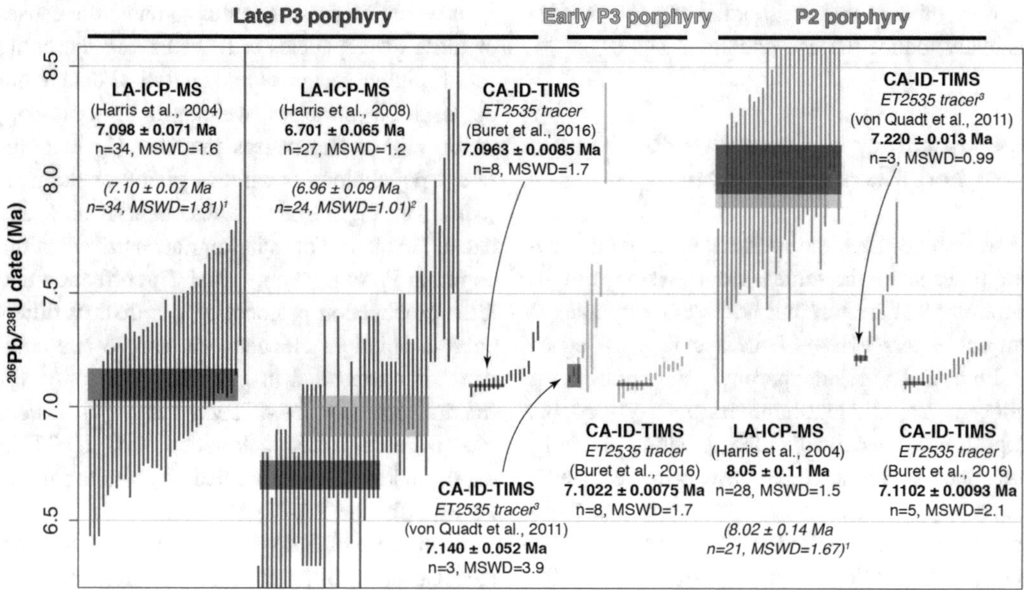

Fig. 8 Compilation of $^{206}Pb/^{238}U$ Th-corrected dates acquired with different methods for three porphyry intrusions at the Bajo de la Alumbrera porphyry copper deposit, Argentina. Data are from Harris et al. (2004), Harris et al. (2008), von Quadt et al. (2011), and Buret et al. (2016). The horizontal grey bands represent the weighted mean dates recalculated by us and include analytical uncertainties based on U–Pb dates from tables provided in the aforementioned publications. [1]weighted mean date reported in Harris et al. (2004). [2]weighted mean date reported in Harris et al. (2008). [3]tracer used in von Quadt et al. (2011) (written communication to the authors). All uncertainties are given at 2σ (95% confidence)

which presents their main characteristics and usefulness for dating ore deposits. It is noteworthy that this table only presents a selection of some useful minerals, but others might also be amenable to U–Pb dating. Furthermore, ongoing and future developments will likely improve our understanding of the U–Pb system in these and new mineral species while allowing better precision, accuracy and interpretation of the dates.

Ideal minerals for U–Pb dating should necessarily contain traces of U, and as little common (initial) Pb as possible. They should also have a low diffusivity for Pb so as to accurately record the radiogenic Pb ingrowth. Many minerals used for U–Pb dating are accessory minerals (zircon, baddeleyite, titanite, monazite, xenotime) but a handful of them are major rock forming minerals (calcite, garnet) or even ore minerals (cassiterite, columbo-tantalite, uraninite, wolframite) (see Table 2). This exceptional mineralogical diversity allows most types of ore deposit and ore forming processes to be dated directly or indirectly with the U–Pb method. However, in detail, all minerals do not provide equally precise, accurate and/or meaningful dates. In Table 2, we have classified the minerals in three categories depending on the average quality of the date that they can provide. Nevertheless, we stress that this classification should only be taken as a 'rule of thumb' and that each case would be different. For example, zircon might give very imprecise and discordant dates while xenotime from the same sample would return more precise and concordant dates (e.g., Cabral and Zeh 2015).

5.1 Low Common Pb, High U and Structurally Robust Minerals

The most dated mineral is arguably zircon. This is mainly due to its virtual ubiquity in the geological environment, its chemical and mechanical

Table 2 Minerals suitable for U–Pb dating in the context of mineral deposits (as of year 2017)

Mineral	Formula	Main types of mineral deposit where U–Pb dating can be done (non-exhaustive list)	Event dated	Principal limitations	Additional comments	Average quality of U–Pb dating[a]	Some references with application to mineral deposits
Allanite	$(Ca, Ce, La, Y)_2(Fe^{2+}, Fe^{3+}, Al)_3O(SiO_7)(SiO_4)(OH)$	Skarn; IOCG: Fault-related U (±REE)	Hydrothermal activity, metasomatism; magmatism	Low to moderate amount of common Pb; No matrix-matched standard available; High amount of excess ^{206}Pb (initial ^{230}Th excess)[b];	Possibility of Th–Pb dating	XX	Pal et al. (2011); Chen and Zhou (2014); Deng et al. (2014)
Apatite	$Ca_5(PO_4)_3(F, OH, Cl)$	IOCG; REE(±U, P) vein; Magmatic Ni–Cu-PGE(±Co) sulphide;	Cooling; hydrothermal activity	Low to high amount of common Pb; low U concentration; No matrix-matched international standard available (zircon or in-house standard; see Chew et al., 2011); Can be sensitive to initial ^{230}Th excess[b]	No metamictization; Possibility of Th–Pb dating	X	Romer (1996); Amelin et al. (1999); Gelcich et al. (2005); Stosch et al. (2010); Seo et al. (2015); Huston et al. (2016)
Baddeleyite	ZrO_2	Magmatic Ni–Cu-PGE(±Co) sulphide; Banded iron formation; Orogenic Au; Diamond-bearing kimberlite; Rare-metal carbonatite	Alkaline and mafic to ultramafic magmatism, hydrothermal activity	Crystal orientation affects $^{206}Pb/^{238}U$ ratios and dates measured with SIMS (Wingate and Compston, 2000);	Limited common Pb, no metamictization	XXX	Corfu and Lightfoot (1996); Schärer et al. (1997); Amelin et al. (1999); Wingate and Compston (2000); Müller et al. (2005); Li et al. (2005); Wu et al. (2011); Zhang et al. (2013); Bjärnborg et al. (2015); Wall and Scoates (2016)
Brannerite	$(U, Ca, Ce)(Ti, Fe)_2O_6$	Fault-related U(±REE-F-Ba-Th); Magmatic-hydrothermal/epithermal U (± Ni-Co-As-Mo-Pb-PGE-Au); Archean Au paleoplacer	Hydrothermal activity	Moderate to high amount of common Pb; easy resetting of the U–Pb system (Pb loss) with hydrothermal fluids; no matrix-matched standard available		X	Frei (1996); Zartman and Smith (2009); Oberthür et al. (2009); Bergen and Fayek (2012)
Calcite	$CaCO_3$	MVT Pb–Zn ± F	Hydrothermal activity, diagenesis	Moderate to high amount of common Pb; easy resetting of the U–Pb system (Pb and U	Date sometimes determined with the isochron method;	X	DeWolf and Halliday (1991); Brannon et al. (1996); Coveney et al. (2000);

(continued)

Table 2 (continued)

Mineral	Formula	Main types of mineral deposit where U–Pb dating can be done (non-exhaustive list)	Event dated	Principal limitations	Additional comments	Average quality of U–Pb dating[a]	Some references with application to mineral deposits
				mobility) with hydrothermal fluids; difficulty to interpret the event being dated; No international matrix-matched standard available (in-house standard)	Inverse discordance is not uncommon (U loss). Mostly Pb_c uncorrected $^{238}U/^{206}Pb$ dates		Grandia et al. (2000); Rasbury and Cole (2009); Burisch et al. (2017)
Cassiterite	SnO_2	Granite-related Sn(± Mo–W–Cu–Pb–Zn–Sb–Ag) greisen, skarn and lode; Supergene Sn	Hydrothermal activity, supergene alteration (?)	High amount of common Pb; No international matrix-matched standard available (in-house standard)	Date often determined with the isochron method	X	Gulson and Jones (1992); Yuan et al. (2011); Chen et al. (2014); Zhang et al. (2014); Li et al. (2016)
Colombo-tantalite	$(Mn, Fe^{2+})(Nb, Ta)_2O_6$	Rare-metal (± Sn–W) pegmatite, greisen and granite	Late magmatic stage, hydrothermal resetting	Low to moderate amount of common Pb; in-situ dating often standardized to zircon mineral, the use of Coltan-139 standard is suggested by Che et al. (2015); can be highly metamict	Inverse discordance is not uncommon (maybe related to inclusions); possible inclusions of uraninite; chemical abrasion is possible	XXX	Romer and Wright (1992); Romer and Smeds (1994); Romer and Smeds (1996); Romer et al. (1996); Romer and Smeds (1997); Glodny et al. (1998); Smith et al. (2004); Baumgartner et al. (2006); Dewaele et al. (2011); Melleton et al. (2012); Melcher et al. (2015); Che et al. (2015); Van Lichtervelde et al. (2016)
Garnet	$(Ca, Ce, La, Y)_2(Fe^{2+}, Fe^{3+}, Al)_3O(SiO_7)(SiO_4)(OH)$	Skarn; Metamorphosed deposit	Metasomatim, metamorphism	Moderate to high amount of common Pb; low U content; no matrix-matched standard available	Andradite garnet tend to have higher U content. Date sometimes determined with the isochron method	X	Mezger et al. (1989); Mueller et al. (1996); Glodny et al. (1998); Jung and Mezger (2003); Seman et al. (2017)
Perovskite	$CaTiO_3$	Diamond-bearing kimberlite; Rare-metal carbonatite	Alkaline and ultramafic magmatism	Moderate to high amount of common Pb; prone to Pb loss; in-situ dating often	Possibility of Th–Pb dating	X	Smith et al. (1989); Heaman (2003); Lehmann et al. (2010); Donnelly et al.

(continued)

Table 2 (continued)

Mineral	Formula	Main types of mineral deposit where U–Pb dating can be done (non-exhaustive list)	Event dated	Principal limitations	Additional comments	Average quality of U–Pb dating[a]	Some references with application to mineral deposits
				standardized to zircon mineral, perovskite standard described in Heaman (2009)			(2012); Zhang et al. (2013); Rao et al. (2013); Wu et al. (2013a, 2013b); Griffin et al. (2014); Heaman et al. (2015); Castillo-Oliver et al. (2016)
REE-Phosphate (Monazite and Xenotime)	(Ce, La, Th) PO_4	Rare-metal (± Sn–W) pegmatite, greisen and granite; Orogenic Au; Banded iron formation; Archean Au paleoplacer; Stratabound polymetallic (Co, Cu, Pb, Zn, Fe, Au, Ag, Bi, W, REE); Unconformity-related U; MVT Pb–Zn; IOCG; granite-related U–Mo; Cordilleran polymetallic	Hydrothermal activity, metamorphim, magmatism	High amount of excess ^{206}Pb (initial ^{230}Th excess)[b]; strong matrix effect due to trace elements needs to be taken in consideration for SIMS dating (e.g., Fletcher et al. 2010)	Limited common Pb. No metamictization. Possibility of Th–Pb dating	XXX	Glodny et al. (1998); Torrealday et al. (2000); Petersson et al. (2001); Pigois et al. (2003); Tallarico et al. (2004); Salier et al. (2004, 2005); Fletcher et al. (2004); Schaltegger et al. (2005); Vallini et al. (2006); Michael Meyer et al. (2006); Rasmussen et al. (2007a, b, 2008); Lobato et al. (2007); Mueller et al. (2007); Kempe et al. (2008); Vielreicher et al. (2010, 2015); Fletcher et al. (2010); Sarma et al. (2011); Muhling et al. (2012); Aleinikoff et al. (2012a,b); Mosoh Bambi et al. (2013); Moreto et al. (2014); Cabral and Zeh (2015); Zi et al. (2015); McKinney et al. (2015); Taylor et al. (2015); Catchpole et al. (2015); Huston et al. (2016); Van Lichtervelde et al. (2016)
	YPO_4			Moderate excess ^{206}Pb (intial ^{230}Th excess)[b]; strong matrix effect due to trace elements needs to be taken in consideration for SIMS dating (e.g., Fletcher et al. 2010)			

(continued)

Table 2 (continued)

Mineral	Formula	Main types of mineral deposit where U–Pb dating can be done (non-exhaustive list)	Event dated	Principal limitations	Additional comments	Average quality of U–Pb dating[a]	Some references with application to mineral deposits
Rutile	TiO_2	Metamorphic and magmatic Ti; Porphyry Cu–Au; Orogenic Au	Cooling; hydrothermal activity	Low U concentration in most cases, but rutiles from high-grade metamorphic rocks tend to have higher U contents (Meinhold 2010); moderate amount of common Pb		XX	de Ronde et al. (1992); Norcross et al. (2000); von Quadt et al. (2005); Kouzmanov et al. (2009); Morisset et al. (2009); Shi et al. (2012)
Titanite (Sphene)	$CaTiOSiO_4$	Skarn; IOCG; Orogenic Au; VMS	Late magmatic stage, hydrothermal activity; metasomatism; metamorphism; (cooling)	Low to moderate amount of common Pb; Titanites BLR-1 and MKED-1 proposed as matrix-matched standards (Aleinikoff et al., 2007; Spandler et al., 2016), common use of zircon or in-house standards	Possibility of Th–Pb dating	XX	Corfu and Muir (1989); Romer and Öhlander (1994); Romer et al. (1994); Eichhorn et al. (1995); Mueller et al. (1996, 2007); Norcross et al. (2000); Salier et al. (2004); Bucci et al. (2004); Wanhainen et al. (2005); De Haller et al. (2006); Skirrow et al. (2007); Chiaradia et al. (2008); Smith et al. (2009); Li et al. (2010); Dziggel et al. (2010); Chelle-Michou et al. (2015); Deng et al. (2015a,b); Seo et al. (2015); Fu et al. (2016); Poletti et al. (2016)
Uraninite	UO_2	U(±Au, REE,…) deposits; Stratiform polymetallic (Co, As, Pb, Zn, Cu, Mo, …) epithermal	Hydrothermal activity; Metamorphism	Prone to U and Pb mobility (loss and gain) and recrystallization; low to high amount of common Pb; no international matrix-matched standard available (in-house standard)	Chemical dating (EMP) is common	X	Hofmann and Eikenberg (1991); Hofmann and Knill (1996); Fayek et al. (2000); Polito et al. (2005); Alexandre et al. (2007); Ono and Fayek (2011); Carl et al. (2011); Philippe et al. (2011); Dieng et al. (2013); Decree et al. (2014); Luo et al. (2015); Skirrow et al. (2016)

(continued)

Table 2 (continued)

Mineral	Formula	Main types of mineral deposit where U–Pb dating can be done (non-exhaustive list)	Event dated	Principal limitations	Additional comments	Average quality of U–Pb dating[a]	Some references with application to mineral deposits
Zircon	$ZrSiO_4$	All type of intermediate to acidic magmatism, differentiate products of mafic to ultramafic magmatism, Porphyry-systems (incl. Epithermal, Skarn), VMS, …	Magmatic, hydrothermal, detrital provenance and deposition, metamorphic	Can present very complex zoning patterns with several age domains	Limited common Pb. Chemical abrasion is possible	XXX	Most of the references provided for the other minerals also present zircon U–Pb dating
Wolframite	$(Fe^{2+}, Mn) WO_4$	Granite-related W ± Sn ± Mo deposits, greisen, skarn, lodes and pegmatite	Hydrothermal activity	Low to moderate amount of common Pb; no matrix-matched standard available; Possible alteration of the mineral; Prone to host fluid and mineral inclusions		X	Romer and Lüders (2006); Pfaff et al. (2009); Lecumberri-Sanchez et al. (2014); Harlaux et al. (2017)
Other minerals: (urano) thorite, vesuvianite, bastnaesite, polycrase, coffinite, …		U deposits; REE deposits; …	Hydrothermal activity; Metamorphism; Supergene alteration	–	–	–	Romer (1992); Romer (1996); Rasmussen et al. (2008); Dill et al. (2010, 2013); Wu et al. (2011); Bergen and Fayek (2012); Cottle (2014); Downes et al. (2016)

[a]XXX: Low common Pb, high U and structurally robust minerals; XX: Moderate common Pb, low U and structurally robust minerals; X: Common Pb-rich, low U, structurally and/or chemically weak minerals. This classification should only be taken as a ‘rule of thumb’ as each case would be different

[b]Refer to Fig. 4 to assess the magnitude of initial ^{230}Th-disequilibrium of the expected age of the mineral

resistance in a range of extreme geological processes from the surface to the deep Earth crust and to the low diffusivity of U and Pb in its crystal lattice (Cherniak et al. 1997; Cherniak and Watson 2001, 2003; Harley and Kelly 2007). Importantly, zircon may contain tens to thousands of ppm of U (Hoskin and Schaltegger 2003) while essentially excluding initial Pb upon crystallization (Watson et al. 1997). This is mainly due to the large charge and ionic radius differences between Pb^{2+} (1.26 Å) and Zr^{4+} (0.84 Å) in eight-fold coordination in zircon. In fact, common Pb in zircon is often limited to small inclusions and to structurally damaged parts of the crystal which are readily removed with a chemical abrasion procedure while preserving the crystalline portion of the mineral (Mattinson 2005). The quality and ubiquity of this mineral has triggered most of the technical development of U–Pb geochronology including a wealth of international reference materials used for in-situ dating methods in all laboratories around the world.

Nevertheless, other minerals such as baddeleyite, columbite group minerals (columbo-tantalite), and rare earth element (REE)-phosphates (monazite and xenotime) present U enrichment and common Pb exclusion properties comparable to zircon. Despite their occurrence in the geological environment being more restricted than that of zircon, published data often show the same level of precision as for zircon, according to the analytical method used. Chemical abrasion techniques have been tested on these minerals but show contrasting behavior. In the case of monazite and baddeleyite, chemical abrasion has not shown any significant improvement in term of precision, reproducibility and concordance (Rioux et al. 2010; Peterman et al. 2012). This might be due to the fact that monazite and baddeleyite do not suffer metamictization (Seydoux-Guillaume et al. 2002, 2004; Trachenko, 2004). However, baddeleyite is suggested to become tetragonal at high ion radiation doses, a phase change that may facilitate radiogenic Pb mobility (Schaltegger and Davies 2017). Additionally, chemical abrasion has been successfully applied to columbo-tantalite minerals and improved the concordance of the data (Romer and Wright 1992). It is thought to remove small inclusions of Pb bearing minerals such as uraninite or secondary Nb- and Ta-bearing minerals (Romer et al. 1996).

5.2 Moderate Common Pb, Low U and Structurally Robust Minerals

Titanite, rutile and allanite represent very interesting properties for U–Pb dating. These accessory mineral species usually have low to moderate amounts of common Pb while being sufficiently enriched in U to allow precise dating in most cases. Analytical protocols and matrix-matched standards for in-situ dating have been developed and allow some labs to routinely date these mineral (Storey et al. 2006, 2007; Aleinikoff et al. 2007; Gregory et al. 2007; Luvizotto et al. 2009; Zack et al. 2011; Darling et al. 2012; Schmitt and Zack 2012; Smye et al. 2014). The use of titanite and especially rutile as geochronometers might be limited by their relatively lower closure temperature of the U–Pb system compared to zircon. Hydrothermal titanite (e.g., in skarn deposits) would crystallize near or just below its closure temperature allowing its use as a geochronometer (Chiaradia et al. 2008; Chelle-Michou et al. 2015), and helping to pinpoint antecrystic zircon growth (i.e., crystallized in earlier magma pulses and incorporated in a later pulse; Miller et al. 2007) in the skarn-forming magmatic intrusion. Rutile is involved in high temperature metamorphic reactions and can produce new zircon upon recrystallization at lower temperature and expulsion of Zr (e.g., Pape et al. 2016). Allanite may have exceedingly high Th/U ratios requiring a very careful approach for accurately correcting and interpreting initial ^{230}Th disequilibrium (Oberli et al. 2004).

5.3 Common Pb-Rich, Low U, Structurally and/or Chemically Weak Minerals

A wealth of other minerals can be used for U–Pb geochronology but tend (most of the time) to produce lower quality data than the minerals described above. This is mainly due to the high ratio of common to radiogenic lead in these mineral ($\gg$ 1 ppm) together with low U concentrations (< 10 ppm). This results in the chosen common Pb correction having a critical impact on the accuracy and precision of the dates. The best dates are usually obtained with the 3D isochron method or 238U/206Pb intercept ages of mixing lines (so-called "isochrons") in a Tera-Wasserburg concordia space from LA-ICP-MS dating (Schoene and Bowring 2006) and potential accompanied with the measurement of a cogenetic common Pb-rich phase (such as the magnetite-apatite geochronometer; Gelcich et al. 2005).

Furthermore, some species such as brannerite, calcite, uraninite, and, to a lesser extent, perovskite and wolframite are prone to resetting of the U–Pb system (Pb-loss), or even U mobility in the presence of hydrothermal fluids that may also promote dissolution/recrystallization of the mineral (e.g., Zartman and Smith 2009; Rasbury and Cole 2009; Ono and Fayek 2011; Bergen and Fayek 2012; Donnelly et al. 2012; Decree et al. 2014; Harlaux et al. 2017). This often results in markedly normally or inversely discordant common Pb-corrected data. Recent, advances in calcite U–Pb dating by LA-ICPMS and ID-TIMS make it possible to routinely achieve uncertainties on the order of 2–5% despite the high amount of common Pb (Li et al. 2014; Coogan et al. 2016; Roberts and Walker 2016; Burisch et al. 2017). Due to the ubiquity of calcite in vein, cement or replacement phase in mineral deposits, calcite U–Pb dating is expected to open to new opportunities for ore deposit research and to address the timing of crustal fluid flow through direct dating. Yet, the main difficulty of calcite dating is to correctly interpret the event being dated, or if unsure, allow for all reasonable possibilities (e.g., see the case of the Hamersley spherule beds, Australia; Woodhead et al. 1998; Rasmussen et al. 2005).

5.4 Choosing the Best Mineral for U–Pb Dating

The choice of the mineral targeted for U(–Th)–Pb dating should be dictated by the particular event or process of interest, cross-cutting and paragenetic information, and geochemical and/or structural data. Dating without consideration of the geological/petrographic context of the mineral will very likely lead to erroneous interpretation. One such example is the case of post-mineralization rhyodacite porphyry at the Corrocohuayco deposit, Peru. There, most zircon grains (11/13) from this post-mineralization porphyry were dated ~0.5 Ma older that the syn-mineralization porphyries it crosscuts (Chelle-Michou et al. 2014). This unambiguous field relationship shows that it could only be interpreted in the context of proto-pluton remelting, rather than as the age of magma emplacement.

Magmatism is arguably the most easily dated geological process. In the vast majority of cases zircon would be the mineral of choice. Even relatively mafic rocks can host zircon in the most differentiated 'melt pockets' (e.g. the Bushveld complex, South Africa; Zeh et al. 2015). In the cases where zircon is absent from the magmatic rock, usually in ultramafic, mafic or alkaline rocks, baddeleyite or perovskite present good alternatives. Finally, crust-derived granitoids often host zircon grains that are dominantly inherited from their source and minimally reflect new growth from the granitic liquid (e.g., Clemens 2003). In such cases, dating of monazite may be preferred. The main goal of dating these magmatic minerals is to constrain the age of magma emplacement in the crust or of volcanic eruption.

The increasing precision of zircon dates achievable with the CA-ID-TIMS method sheds new light on the long-lived history of magmatic systems. At the sample scale, more than 0.1 Ma of protracted zircon crystallization has been documented at a number of silicic systems, some of which are associated with porphyry copper mineralization (Schütte et al. 2010; Wotzlaw et al. 2013; Chelle-Michou et al. 2014; Barboni et al. 2015; Buret et al. 2017). When combined with complimentary geochemical data, zircon crystallization ages can provide valuable insights into the specific petrological processes responsible the transition from barren to ore-producing intrusions (Chelle-Michou et al. 2014; Tapster et al. 2016; Buret et al. 2016).

Despite its common usage in ore deposit research, the dating of magma intrusion only rarely dates the mineralization itself. In fact, this is only restricted to places where the ore minerals have crystallized under magmatic conditions such as the magmatic Ni–Cu–Cr(±Au ± PGE) deposits and possibly some magmatic REE deposits as well. If appropriate crosscutting relationship with the mineralization can be observed, dating magmatic intrusions can elegantly bracket the timing of ore deposition (e.g., von Quadt et al. 2011). In the case of porphyry, greisen, or volcanogenic massive sulfide (VMS) deposits the age of the ore-related intrusion or of the associated volcanics may often provide a good, if not excellent, approximation for age of the mineralization. Yet, this approach requires much caution as even in classical magmatic-hydrothermal deposits such as W–Sn granite deposits or porphyry Cu deposits, the mineralization can have been sourced by a hidden intrusion at depth while being hosted in a previously emplaced one (e.g., Schaltegger et al. 2005). However, for deposits where the relationship between ore formation and a particular magma intrusion is ambiguous (e.g., iron oxide copper–gold (IOCG) deposits, orogenic Au deposit, epithermal deposits, distal skarns) or even totally absent (e.g., Mississippi Valley-type (MVT) deposits) it is much more advantageous to determine directly the timing of hydrothermal fluid circulation and/or of ore deposition. The list of ore minerals suitable for U–Pb dating include cassiterite (for Sn deposits), wolframite (for W deposits) columbo-tantalite (in some rare-metal granite, greisen and pegmatite deposits), rutile (for Ti deposits), and minerals associated with U deposits (e.g., uraninite, brannerite). This restricts the types of ore that can dated with the U–Pb method. Alternatively, several gangue mineral species can be used to date hydrothermal fluid circulation, metasomatism and metamorphism. Their relevance for the genesis or reworking of the studied ore deposit is fundamentally linked to their position in the paragenetic sequence with respect to the ore minerals. REE-phosphates such as monazite and xenotime are common in a wide variety of hydrothermal systems ranging from granite-related rare metal deposits to MVT deposits (Table 2) and, if available, would be the ideal minerals to date hydrothermal processes. In few cases, hydrothermal zircons at skarn (Niiranen et al. 2007; Wan et al. 2012; Deng et al. 2015c), IOCG (Valley et al. 2009), orogenic Au (Kerrich and Kyser 1994; Pelleter et al. 2007), and alkaline/carbonatite magmatism related rare-metal deposits (Yang et al. 2013; Campbell et al. 2014) have been reported and can date hydrothermal activity and metasomatism. However, in the absence of these hydrothermal minerals (which is not uncommon), other minerals listed in Table 2 with non-negligible amounts of common Pb can be called on. Titanite or allanite can provide excellent dates for skarn (Chiaradia et al. 2008; Deng et al. 2014, 2015b; Chelle-Michou et al. 2015) and IOCG deposits (Skirrow et al. 2007; Smith et al. 2009; De Haller et al. 2011). Ore-stage calcite or apatite may sometimes represent the only minerals suitable for U–Pb dating at MVT deposits (Grandia et al. 2000) or some REE-P deposits (Huston et al. 2016). The ability of apatite to keep record of Cl, F, OH and SO_4^{2-} of the hydrothermal fluid (or magma) from which it crystallizes (Webster and Piccoli 2015; Harlov 2015) coupled with the possibility to date it with the U–Pb method (Chew and Spikings 2015) opens interesting opportunities to refine ore forming models. Finally, U–Pb minerals such as rutile, apatite and/or titanite can provide invaluable thermochronological information on the thermal evolution of the studied ore deposit during and after its genesis.

6 Case Studies of Applications of U–Pb Dating to Mineral Deposits

In the following section, we present two case studies that make very distinct use of U–Pb geochronology. The first one focuses on the Witwatersrand gold deposits, South Africa, and illustrates how geochronology based on several mineral species can be used to bracket the age of multiple geological events over long timescales (> 10 Ma). The second one discusses how rather than the absolute age, the duration of the mineralizing event at porphyry copper deposits can help understand the ore-forming processes and the main controls on the size (metal content) of the deposits. These two examples embody different timescales of reasoning, different precision, accuracy and spatial resolution requirements, and different uses of the geochronological data.

6.1 Input of Multi-mineral U–Pb Dating for Understanding Gold Deposition and Remobilization in the Witwatersrand Basin, South Africa

About 32% of all gold ever mined and about the same proportion of known gold resources comes from deposits hosted in the Witwatersrand Basin, South Africa (Frimmel and Hennigh 2015), a Mesoarchean detrital sedimentary basin deposited on the Kaapvaal Craton (Fig. 9). The genesis of this enormous accumulation of gold in the crust has triggered one of the "*greatest debate in the history of economic geology*" (see summary in Muntean et al. 2005). Proposed models for the deposition of gold range from a modified paleoplacer to a purely hydrothermal origin. These disparate views arises from contradicting observations that are selectively put forward to favor either model (Frimmel et al. 2005; Law and Phillips 2005; Muntean et al. 2005). In fact, probably none of these end-member models can account for all the geological, chemical and isotopic observations. The most recent models rather consider the very peculiar conditions that prevailed in the Mesoarchean atmosphere, hydrosphere and biosphere (Frimmel and Hennigh 2015; Heinrich 2015). At this time, redox reations mediated by microbial life could have triggered the synsedimentary precipitation of the large quantities of gold dissolved in acidic and reduced meteoric and shallow sea waters.

U–Pb geochronology has been instrumental in the understanding of the formation Witwatersrand goldfields. It first played an essential role in calibrating the depositional age of the sediments (Fig. 9). One of the most significant contribution comes from Armstrong et al. (1991) who dated zircons from volcanic rocks distributed along the sedimentary pile of the basin. They constrained the deposition of the Witwatersrand Supergroup to within a timeframe of ca. 360 Ma from 3074 to 2714 Ma. Subsequent studies have focused on detrital zircon and xenotime from the main formations present along the stratigraphic column and intimately associated with the gold-bearing reefs (England et al. 2001; Kositcin and Krapež 2004; Koglin et al. 2010). These have confirmed the previous depositional ages but provide additional insight in the source of the detritus that filled the basin, as well as secular changes in the catchment area of the basin over time. Results show that the source area of detritus has an increasing age-range of rocks undergoing erosion over time. Apart from the lowermost part of the Witwatersrand Supergroup (Orange Formation, West Rand Group) which has dates clustering around 3.21 Ga, zircon dates from the West Rand Group cluster around 3.06 Ga, with only few older and younger dates (Fig. 9). Furthermore, zircon dates from the Central Rand Group shows additional peaks at 2.96–2.92 Ga and 3.44–3.43 Ga with several intervening dates in between these main peaks. Dates of detrital xenotime are mostly within the 3.1–2.9 Ga range but also extend as low as 2.8 Ga (Fig. 9). Koglin et al. (2010) and Ruiz et al. (2006) further link the gold-rich sediments to the presence of the 3.06 Ga zircon age peak. When compared with

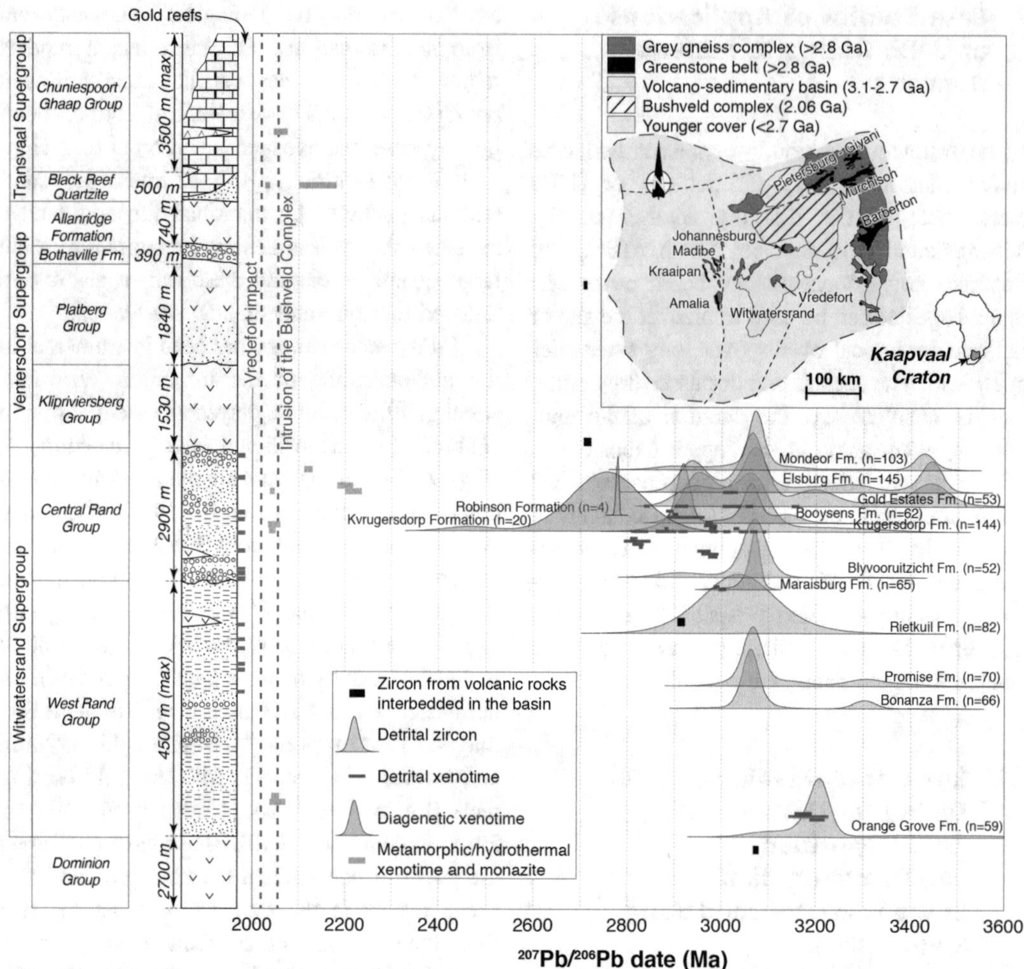

Fig. 9 Compilation of available U–Pb data from the Witwatersrand basin plotted against the stratigraphic position of the sample. Stratigraphic column of the Archean to early Proterozoic succession in South Africa from Muntean et al. (2005). Data from the Witwatersrand basin are from Armstrong et al. (1991), England et al. (2001), Kositcin et al. (2003) Kositcin and Krapež (2004), Rasmussen et al. (2007a) and Koglin et al. (2010). Ages of the intrusion of the Bushveld Complex and of the Vredefort impact are from Zeh et al. (2015) and Moser (1997), respectively. Kernel density estimates (KDE) where obtained using DensityPlotter (Vermeesch 2012). Selected data are 95–105% concordant and data points are plotted at the 2σ uncertainty level. Inset map of the Kaapvaal craton is modified from Poujol (2007)

outcropping Archean terrains of the Kaapvaal Craton, these zircons could have originated from the greenstone belts west of the Witwatersrand basin (Madibe and Kraaipan), rocks in the immediate proximity of the basin (e.g., Johannesburg and Vrefefort Dome) or from equivalent units located northwest of the basin that might be present below the post-Witwatersrand cover (Koglin et al. 2010). More distal candidates such as the Murchison and the Barberton greenstone belts have also been proposed (Ruiz et al. 2006; Koglin et al. 2010). This interpretation is also compatible with paleocurrent directions and isotopic data (Koglin et al. 2010).

A paleoplacer model requires that all of the gold deposited in the basin originated from the same eroding massifs that sourced the sediments. However, such gigantic quantities of gold are two orders of magnitude in excess of all the gold ever mined and discovered in the potential outcropping massifs that sourced the zircons. This observation has been a major argument against any sort of paleoplacer model (e.g., Phillips and Law 2000; Law and Phillips 2005; Frimmel and Hennigh 2015). The existence of a now vanished or buried, hypothetical massif as a source of this huge amount of gold would pose an equally important question about how this massif would have been exceptionally well endowed with gold.

An epigenetic (hydrothermal) origin of the gold is supported by several petrographic observations. Yet, cross-cutting relationship suggest that hydrothermal activity took place before deposition of the Platberg Group, that is, before ca. 2.7 Ga (e.g., Law and Phillips 2005; Meier et al. 2009). U–Pb dating of diagenetic xenotime have yielded a major peak between 2.78–2.72 Ga which could be related to a heating event and flood-basalt volcanism during the deposition of the Klipriviersberg Group, immediately following the deposition of the Witwatersrand Supergroup (Fig. 9; England et al. 2001; Kositcin et al. 2003). Although this timing for gold introduction would be consistent with temporal constrains, the association of gold with this 2.78–2.72 Ga xenotime has not been reported. Additionally, U–Pb dating of metamorphic-hydrothermal REE-phosphates (monazite and xenotime) paragenetically associated with some gold or unrelated to gold mostly records ages between 2.06 and 2.03 Ga throughout the stratigraphic succession from the Witwatersrand to the Transvaal Supergroups (Fig. 9; England et al. 2001; Kositcin et al. 2003; Rasmussen et al. 2007a). This age is consistent with the emplacement age 1of the Bushveld complex on the northern flank of the Witwatersrand Basin (Zeh et al. 2015) which most likely triggered fluid circulation, gold remobilization and peak greenschist metamorphic conditions in the basin (Rasmussen et al. 2007a).

While none of the available U–Pb data for the Witwatersrand basin (Fig. 9) can firmly date gold deposition, or conclusively explain how gold was deposited, they have provided the necessary temporal framework on which to challenge relative chronological data. They have brought significant arguments against each of the classical models invoked for the formation of this district (syngenetic vs epigenetic) while confirming that gold remobilization occurred long after the formation of the deposit and contributed to the emergence of new ore forming models (Frimmel and Hennigh 2015; Heinrich 2015). This example highlights the necessity to properly constrain each U–Pb date against paragenetic, cross-cutting and stratigraphic observations in order to draw meaningful conclusions. The Witwatersrand gold deposits result from a long-lived and multi-episodic geological history where U–Pb geochronology provided constraints on basin formation, sediment provenance, diagenesis and metamorphism. It is noteworthy that the different minerals that were dated (zircon, monazite, xenotime), individually record a limited portion of the multiple processes that shaped the Witwatersrand basin and proved to be highly complementary to each other. Unveiling this protracted history did not require particularly high-precision dating methods, as LA-ICPMS and SIMS instruments with high sample throughput (Table 1) proved very effective. Additionally, the very high spatial resolution achievable with a SIMS instrument was crucial in unlocking the U–Pb information in tiny xenotime and monazite crystals identified from thin sections.

6.2 Zircon U–Pb Insights into the Genesis on Porphyry Copper Deposits

Porphyry copper deposits (PCDs) typically form at convergent margins in association with subduction or post-subduction magmatism (e.g., Richards 2009). Metals and sulfur fixed in these deposits are thought to have been sourced from a cooling and degassing fluid-saturated magma body emplaced at shallow depths within the upper crust and transported to the site of deposition by magmatic-hydrothermal fluids (Hedenquist and Lowenstern 1994; Sillitoe 2010; Pettke et al. 2010; Simon and Ripley 2011; Richards

2011). Ultimately, very efficient fluid focusing and sulfide precipitation together with post-mineralization ore deposit preservation will favor the presence of economic porphyry deposits at erosion levels.

The USGS global database of PCDs shows that these deposits span more than four orders of magnitude in copper endowment (Singer et al. 2008; Fig. 10a). Yet, the specific factors that control the size of these deposits have remained speculative. Comparing 'standard' and 'giant' PCDs, Richards (2013) speculated that the formation of the largest deposits result from a combination of copper enrichment in the magma, the focusing of fluids in structural corridors and, long-lived hydrothermal activity may favor the formation of the largest deposits. Among these possible factors, the timescale of PCD formation may play a significant role in their size. Compiling geochronological data (U–Pb on zircon and Re–Os on molybdenite) from PCDs around the world, Chelle-Michou et al. (2017) and Chiaradia and Caricchi (2017) have highlighted a correlation between the duration of the mineralizing event and the total mass of copper deposited, suggesting an average copper deposition rate of about 40 Mt/Ma (Fig. 10b). This relationship probably reflects the mass balance requirement for a giant deposit to be sourced by a large body of magma, which is incrementally injected into the upper crust over long timescales (see Chelle-Michou et al. 2017).

Similar conclusions where reached by Caricchi et al. (2014) who suggested that magmatism associated with economic PCDs is distinguishable from background pluton-forming

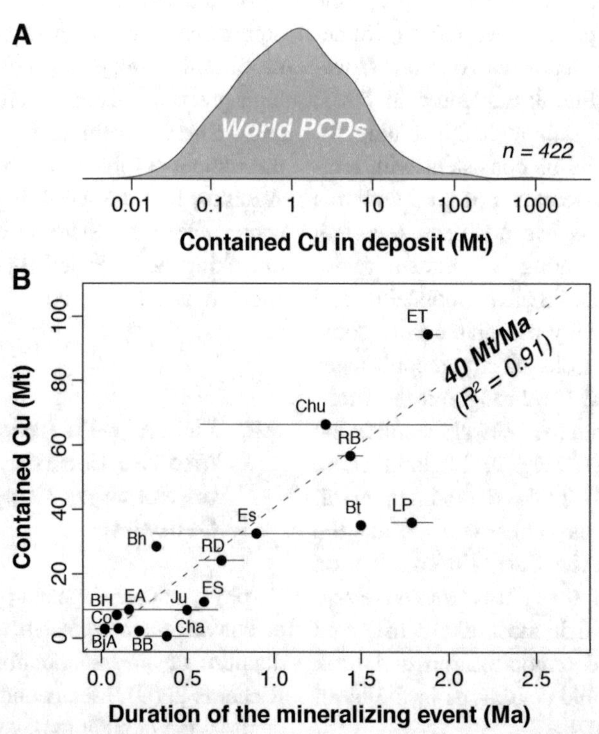

Fig. 10 **a** probability density distribution of Cu endowment in global porphyry copper deposits (PCDs). Data from Singer et al. (2008). **b** Correlation between the duration of the mineralizing event and the total amount of Cu deposited (adapted from Chiaradia and Caricchi, 2017). BH: Batu Hijau (Indonesia), BjA: Bajo de la Alumbrera (Argentina), Co: Coroccohuayco (Peru), EA: El Abra (Chile), BB: Boyongan-Bayugo (Philippines)), Bh: Bingham (US), Cha: Chaucha (Ecuador), Ju: Junin (Ecuador), ES: El Salvador (Chile), Es: Escondida (Chile), LP: Los Pelambres (Chile), Chu: Chuquicamata (Chile), RB: Rio Blanco (Chile), Bt: Butte (US), ET: El Teniente (Chile), RD: Reko Diq (Pakistan)

magmatism and large-eruption-forming magmatism by large magma volumes emplaced at average rate of magma injection ($\sim$0.001 km^3/yr). This conclusion was drawn through inverse thermal modelling of high-precision CA-ID-TIMS U–Pb zircon age distributions.

While geochronology on PCDs has been mostly used to determine the formation age of these deposits, high-precision geochronological data can now be used to elucidate the duration of the ore-forming process. Figure 10b shows that the duration of ore-formation may be a significant control on their size (i.e., metal endowment) and, by inference, on the specific processes responsible for their formation. In addition, high-precision geochronological data may be able to help test the validity of numerical models of PCD formation (e.g., Weiss et al. 2012), or directly as input data into numerical models aiming at quantifying the time-volume-flux-geochemistry relationships of the magmatism associated with PCD genesis (e.g., Caricchi et al. 2014; Chelle-Michou et al. 2017; Chiaradia and Caricchi 2017). These studies only start to unearth the great potential of high-precision CA-ID-TIMS U–Pb dating for PCD exploration, and can also significantly contribute to a better understanding of PCD magmatic ore-forming processes.

7 Concluding Remarks

Over the past two decades U–Pb geochronology has become an essential tool for the study of ore deposits. After a century of development, more than 16 minerals can now be dated with the U–Pb technique allowing its use for most types of ore deposits. U–Pb dating is most commonly used to provide the age of a particular geological event related to a studied deposit (e.g., magmatism, hydrothermal activity, sedimentation, metamorphism, ore deposition and remobilization), depending on the mineral(s) available for dating. The choice of the mineral(s) and of the analytical technique (LA-ICPMS, SIMS or ID-TIMS) used for dating mainly depends on the scientific questions that need to be answered and on the opportunities offered by the studied deposit. This point is perhaps one of the main limitations of the U–Pb dating of ore deposits. For example, MVT deposits rarely contain minerals suitable for U–Pb dating (potentially calcite, provided it has low initial Pb), in which case the use of other isotopic systems will be necessary (e.g., Rb–Sr on sphalerite, Re–Os on sulfides). In addition, as we have seen in the case of the Witwatersrand basin, the spatial resolution required for the analysis may sometimes critically guide the choice of the analytical method.

Recent advances that combine numerical modelling with U–Pb geochronology for porphyry copper deposits suggest that high-precision zircon U–Pb data may also be used as a window to better understand the magmatic aspect of the ore-forming process (Caricchi et al. 2014) and to unravel the fundamental controls on the size of the deposit (Chelle-Michou et al. 2017; Chiaradia and Caricchi, 2017). Comparable studies on other deposit types could potentially advance our understanding of ore-forming processes and may generate innovative tools for mineral exploration.

A further important development of U–Pb geochronology concerns its coupling with textural and geochemical data (e.g., trace elements, Lu–Hf isotopes, O isotopes) obtained on the same grain or on the same spot as the U–Pb data. This is commonly referred to as 'petrochronology' and allows temporal information relative to the evolution and/or the source of the liquid (a magma or an aqueous fluid) from which the mineral precipitated and potentially the rate of its evolution (e.g., Ballard et al. 2002; Smith et al. 2009; Valley et al. 2010; Pal et al. 2011; Rao et al. 2013; Yang et al. 2013; Griffin et al. 2014; Rezeau et al. 2016; Poletti et al. 2016; Gardiner et al. 2017). In particular, high-precision petrochronology on zircon and baddeleyite can provide unprecedented insights into the processes at play during magma evolution, the potential turning point leading to mineralization, or, the link between small intrusive bodies (dykes or stocks) or volcanic products and their larger deep-seated plutonic source (e.g.,Wotzlaw et al. 2013, 2015; Chelle-Michou et al. 2014; Tapster et al. 2016; Buret et al. 2016; 2017; Schaltegger and Davies, 2017).

The field of U–Pb geochronology is working towards a level of maturity whereby inter-laboratory reproducibility will be guaranteed in most labs around the world and where each date and its uncertainty can be fully traceable to SI units. This however, should not mask the high-level of competency and training required to certify the quality of the analysis, to maintain the lab at the best level (picogram levels of common Pb contaminations can be dramatic in a CA-ID-TIMS lab) and, very importantly, the interpretation of the dates into geologically relevant ages. As we have shown, there are numerous potential pitfalls that, if not carefully accounted for, can result in unsupported or even wrong conclusions.

The improving precision, accuracy and spatial resolution of analyses now achievable, challenges paradigms of ore-forming processes and will continue to contribute significant breakthroughs in ore deposit research and potentially also contribute to the development of new mineral exploration tools. The full added value of U–Pb geochronology will however only be assured through its coupling with geochemical data, high-quality field and petrographic observations and numerical modelling.

Acknowledgements We are grateful to the D. Huston for inviting us to write this chapter. Constructive reviews from Andrew Cross and Neal McNaughton significantly improved the clarity of the manuscript and were very much appreciated. C. Chelle-Michou acknowledges financial support from the European Commission and the Université Jean Monnet during the preparation of the manuscript. Both authors acknowledge the support of the Swiss National Science Foundation.

References

Aleinikoff JN, Wintsch RP, Tollo RP, Unruh DM, Fanning CM, Schmitz MD (2007) Ages and origins of rocks of the Killingworth dome, south-central Connecticut: implications for the tectonic evolution of southern New England. Am J Sci 307:63–118. https://doi.org/10.2475/01.2007.04

Aleinikoff JN, Hayes TS, Evans KV, Mazdab FK, Pillers RM, Fanning CM (2012a) SHRIMP U–Pb Ages of xenotime and monazite from the Spar Lake red bed-associated Cu–Ag deposit, western Montana: implications for ore genesis. Econ Geol 107:1251–1274. https://doi.org/10.2113/econgeo.107.6.1251

Aleinikoff JN, Slack JF, Lund K, Evans KV, Fanning CM, Mazdab FK, Wooden JL, Pillers RM (2012b) Constraints on the timing of Co-Cu ± Au mineralization in the Blackbird District, Idaho, using SHRIMP U–Pb ages of monazite and xenotime plus zircon ages of related Mesoproterozoic orthogneisses and metasedimentary rocks. Econ Geol 107:1143–1175. https://doi.org/10.2113/econgeo.107.6.1143

Alexandre P, Kyser K, Thomas D, Polito P, Marlat J (2007) Geochronology of unconformity-related uranium deposits in the Athabasca Basin, Saskatchewan, Canada and their integration in the evolution of the basin. Miner Depos 44:41–59. https://doi.org/10.1007/s00126-007-0153-3

Allen CM, Campbell IH (2012) Identification and elimination of a matrix-induced systematic error in LA–ICP–MS $^{206}Pb/^{238}U$ dating of zircon. Chem Geol 332–333:157–165. https://doi.org/10.1016/j.chemgeo.2012.09.038

Amelin Y, Li C, Naldrett AJ (1999) Geochronology of the Voisey's Bay intrusion, Labrador, Canada, by precise U–Pb dating of coexisting baddeleyite, zircon, and apatite. Lithos 47:33–51

Amelin Y, Zaitsev AN (2002) Precise geochronology of phoscorites and carbonatites: the critical role of U-series disequilibrium in age interpretations. Geochim Cosmochim Acta 66:2399–2419. https://doi.org/10.1016/S0016-7037(02)00831-1

Anczkiewicz R, Oberli F, Burg JP, Villa IM, Günther D, Meier M (2001) Timing of normal faulting along the Indus suture in Pakistan Himalaya and a case of major $^{231}Pa/^{235}U$ initial disequilibrium in zircon. Earth Planet Sci Lett 191:101–114. https://doi.org/10.1016/S0012-821X(01)00406-X

Arevalo R Jr (2014) Laser ablation ICP-MS and laser fluorination GS-MS. In: Treatise on geochemistry. Elsevier, pp 425–441

Arevalo R, Bellucci J, McDonough WF (2010) GGR Biennial Review: advances in laser ablation and solution ICP-MS from 2008 to 2009 with particular emphasis on sensitivity enhancements, mitigation of fractionation effects and exploration of new applications. Geostand Geoanal 34:327–341. https://doi.org/10.1111/j.1751-908X.2010.00934.x

Armstrong RA, Compston W, Retief EA, Williams IS, Welke HJ (1991) Zircon ion microprobe studies bearing on the age and evolution of the Witwatersrand triad. Precambrian Res 53:243–266. https://doi.org/10.1016/0301-9268(91)90074-K

Aston FW (1929) The mass-spectrum of uranium lead and the atomic weight of protactinium. Nature 123:313. https://doi.org/10.1038/123313a0

Ballard J, Palin M, Campbell I (2002) Relative oxidation states of magmas inferred from Ce(IV)/Ce(III) in zircon: application to porphyry copper deposits of northern Chile. Contrib Mineral Petrol 144:347–364. https://doi.org/10.1007/s00410-002-0402-5

Ballard JR, Palin JM, Williams IS, Campbell IH, Faunes A (2001) Two ages of porphyry intrusion resolved for the super-giant Chuquicamata copper deposit of northern Chile by ELA-ICP-MS and SHRIMP. Geology 29:383–386. https://doi.org/10.1130/0091-7613(2001)029%3c0383:TAOPIR%3e2.0.CO;2

Barboni M, Annen C, Schoene B (2015) Evaluating the construction and evolution of upper crustal magma reservoirs with coupled U/Pb zircon geochronology and thermal modeling: a case study from the Mt. Capanne pluton (Elba, Italy). Earth Planet Sci Lett 432:436–448. https://doi.org/10.1016/j.epsl.2015.09.043

Baumgartner R, Romer RL, Moritz R, Sallet R, Chiaradia M (2006) Columbite–tantalite-bearing granitic pegmatites from the Seridó belt, northeastern Brazil: genetic constraints from U–Pb dating and Pb isotopes. Can Mineral 44:69–86. https://doi.org/10.2113/gscanmin.44.1.69

Becquerel H (1896a) Sur les radiations invisibles émises par les corps phosphorescents. C R Acad Sci Paris 122:501–503

Becquerel H (1896b) Sur les radiations invisibles émises par les sels d'uranium. C R Acad Sci Paris 122:689–694

Bergen L, Fayek M (2012) Petrography and geochronology of the Pele Mountain quartz-pebble conglomerate uranium deposit, Elliot Lake District, Canada. Am Min 97:1274–1283. https://doi.org/10.2138/am.2012.4040

Beyer C, Berndt J, Tappe S, Klemme S (2013) Trace element partitioning between perovskite and kimberlite to carbonatite melt: new experimental constraints. Chem Geol 353:132–139. https://doi.org/10.1016/j.chemgeo.2012.03.025

Bjärnborg K, Scherstén A, Söderlund U, Maier D, W, (2015) Geochronology and geochemical evidence for a magmatic arc setting for the Ni–Cu mineralised 1.79 Ga Kleva gabbro–diorite intrusive complex, southeast Sweden. GFF 137:83–101. https://doi.org/10.1080/11035897.2015.1015265

Black LP, Kamo SL, Allen CM, Davis DW, Aleinikoff JN, Valley JW, Mundil R, Campbell IH, Korsch RJ, Williams IS, Foudoulis C (2004) Improved ^{206}Pb/^{238}U microprobe geochronology by the monitoring of a trace-element-related matrix effect; SHRIMP, ID–TIMS, ELA–ICP–MS and oxygen isotope documentation for a series of zircon standards. Chem Geol 205:115–140. https://doi.org/10.1016/j.chemgeo.2004.01.003

Blackburn T, Bowring SA, Schoene B, Mahan K, Dudas F (2011) U–Pb thermochronology: creating a temporal record of lithosphere thermal evolution. Contrib Mineral Petrol 162:479–500. https://doi.org/10.1007/s00410-011-0607-6

Boltwood BB (1907) Ultimate disintegration products of the radioactive elements; Part II, Disintegration products of uranium. Am J Sci Series 4 23:78–88. https://doi.org/10.2475/ajs.s4-23.134.78

Bowring SA, Erwin D, Parrish R, Renne P (2005) EARTHTIME: a community-based effort towards high-precision calibration of earth history. Geochim Cosmochim Acta Supplement 69:A316

Bowring JF, McLean NM, Bowring SA (2011) Engineering cyber infrastructure for U–Pb geochronology: Tripoli and U–Pb_Redux. Geochem Geophys Geosyst 12:Q0AA19. https://doi.org/10.1029/2010GC003479

Brandt S, Schenk V, Raith MM, Appel P, Gerdes A, Srikantappa C (2011) Late Neoproterozoic P-T evolution of HP-UHT Granulites from the Palni Hills (South India): new constraints from phase diagram modelling, LA-ICP-MS zircon dating and in-situ EMP monazite dating. J Petrol 52:1813–1856. https://doi.org/10.1093/petrology/egr032

Brannon JC, Cole SC, Podosek FA, Ragan VM, Coveney RM, Wallace MW, Bradley AJ (1996) Th–Pb and U–Pb Dating of ore-stage calcite and Paleozoic fluid flow. Science 271:491–493. https://doi.org/10.1126/science.271.5248.491

Bucci LA, McNaughton NJ, Fletcher IR, Groves DI, Kositcin N, Stein HJ, Hagemann SG (2004) Timing and duration of high-temperature gold mineralization and spatially associated granitoid magmatism at Chalice, Yilgarn Craton, Western Australia. Econ Geol 99:1123–1144. https://doi.org/10.2113/gsecongeo.99.6.1123

Buret Y, von Quadt A, Heinrich C, Selby D, Wälle M, Peytcheva I (2016) From a long-lived upper-crustal magma chamber to rapid porphyry copper emplacement: reading the geochemistry of zircon crystals at Bajo de la Alumbrera (NW Argentina). Earth Planet Sci Lett 450:120–131. https://doi.org/10.1016/j.epsl.2016.06.017

Buret Y, Wotzlaw J-F, Roozen S, Guillong M, von Quadt A, Heinrich CA (2017) Zircon petrochronological evidence for a plutonic-volcanic connection in porphyry copper deposits. Geology 45:623–626. https://doi.org/10.1130/G38994.1

Burisch M, Gerdes A, Walter BF, Neumann U, Fettel M, Markl G (2017) Methane and the origin of five-element veins: mineralogy, age, fluid inclusion chemistry and ore forming processes in the Odenwald, SW Germany. Ore Geol Rev 81:42–61

Cabral AR, Zeh A (2015) Detrital zircon without detritus: a result of 496-Ma-old fluid–rock interaction during the gold-lode formation of Passagem, Minas Gerais, Brazil. Lithos 212–215:415–427. https://doi.org/10.1016/j.lithos.2014.10.011

Campbell LS, Compston W, Sircombe KN, Wilkinson CC (2014) Zircon from the East Orebody of the Bayan Obo Fe–Nb–REE deposit, China, and SHRIMP ages for carbonatite-related magmatism and REE mineralization events. Contrib Mineral Petrol 168:1041. https://doi.org/10.1007/s00410-014-1041-3

Caricchi L, Simpson G, Schaltegger U (2014) Zircons reveal magma fluxes in the Earth's crust. Nature 511:457–461. https://doi.org/10.1038/nature13532

Carl C, von Pechmann E, Höhndorf A, Ruhrmann G (2011) Mineralogy and U/Pb, Pb/Pb, and Sm/Nd geochronology of the Key Lake uranium deposit, Athabasca Basin, Saskatchewan, Canada. Can J Earth Sci 29:879–895. https://doi.org/10.1139/e92-075

Carlson RW (2014) Thermal ionization mass spectrometry. In: Treatise on geochemistry. Elsevier, pp 337–354

Castillo-Oliver M, Galí S, Melgarejo JC, Griffin WL, Belousova E, Pearson NJ, Watangua M, O'Reilly SY (2016) Trace-element geochemistry and U–Pb dating of perovskite in kimberlites of the Lunda Norte province (NE Angola): Petrogenetic and tectonic implications. Chem Geol 426:118–134. https://doi.org/10.1016/j.chemgeo.2015.12.014

Catchpole H, Kouzmanov K, Bendezú A, Ovtcharova M, Spikings R, Stein H, Fontboté L (2015) Timing of porphyry (Cu–Mo) and base metal (Zn-Pb-Ag-Cu) mineralisation in a magmatic-hydrothermal system – Morococha district, Peru. Miner Depos 50:895–922. https://doi.org/10.1007/s00126-014-0564-x

Chakhmouradian AR, Reguir EP, Kamenetsky VS, Sharygin VV, Golovin AV (2013) Trace-element partitioning in perovskite: implications for the geochemistry of kimberlites and other mantle-derived undersaturated rocks. Chem Geol 353:112–131. https://doi.org/10.1016/j.chemgeo.2013.01.007

Che XD, Wu F-Y, Wang R-C, Gerdes A, Ji W-Q, Zhao Z-H, Yang J-H, Zhu Z-Y (2015) In situ U–Pb isotopic dating of columbite–tantalite by LA–ICP–MS. Ore Geol Rev 65:979–989. https://doi.org/10.1016/j.oregeorev.2014.07.008

Chelle-Michou C, Chiaradia M, Ovtcharova M, Ulianov A, Wotzlaw J-F (2014) Zircon petrochronology reveals the temporal link between porphyry systems and the magmatic evolution of their hidden plutonic roots (the Eocene Coroccohuayco deposit, Peru). Lithos 198–199:129–140. https://doi.org/10.1016/j.lithos.2014.03.017

Chelle-Michou C, Chiaradia M, Selby D, Ovtcharova M, Spikings RA (2015) High-resolution geochronology of the Coroccohuayco porphyry-skarn deposit, Peru: a rapid product of the Incaic Orogeny. Econ Geol 110:423–443. https://doi.org/10.2113/econgeo.110.2.423

Chelle-Michou C, Rottier B, Caricchi L, Simpson G (2017) Tempo of magma degassing and the genesis of porphyry copper deposits. Sci Rep 7:40566. https://doi.org/10.1038/srep40566

Chen WT, Zhou M-F (2014) Ages and compositions of primary and secondary allanite from the Lala Fe–Cu deposit, SW China: implications for multiple episodes of hydrothermal events. Contrib Mineral Petrol 168:1043. https://doi.org/10.1007/s00410-014-1043-1

Chen X-C, Hu R-Z, Bi X-W, Li H-M, Lan J-B, Zhao C-H, Zhu J-J (2014) Cassiterite LA-MC-ICP-MS U/Pb and muscovite $^{40}Ar/^{39}Ar$ dating of tin deposits in the Tengchong-Lianghe tin district, NW Yunnan, China. Miner Depos 49:843–860. https://doi.org/10.1007/s00126-014-0513-8

Cheng H, Lawrence Edwards R, Shen C-C, Polyak VJ, Asmerom Y, Woodhead J, Hellstrom J, Wang Y, Kong X, Spötl C, Wang X, Calvin Alexander E (2013) Improvements in ^{230}Th dating, ^{230}Th and ^{234}U half-life values, and U–Th isotopic measurements by multi-collector inductively coupled plasma mass spectrometry. Earth Planet Sci Lett 371:82–91. https://doi.org/10.1016/j.epsl.2013.04.006

Cherniak DJ, Watson EB (2001) Pb diffusion in zircon. Chem Geol 172:5–24. https://doi.org/10.1016/S0009-2541(00)00233-3

Cherniak DJ, Watson EB (2003) Diffusion in Zircon. Rev Mineral Geochem 53:113–143. https://doi.org/10.2113/0530113

Cherniak DJ, Hanchar JM, Watson EB (1997) Diffusion of tetravalent cations in zircon. Contrib Mineral Petrol 127:383–390. https://doi.org/10.1007/s004100050287

Cherniak DJ, Watson EB, Grove M, Harrison TM (2004) Pb diffusion in monazite: a combined RBS/SIMS study. Geochim Cosmochim Acta 68:829–840. https://doi.org/10.1016/j.gca.2003.07.012

Chew DM, Spikings RA (2015) Geochronology and thermochronology using apatite: time and temperature, lower crust to surface. Elements 11:189–194. https://doi.org/10.2113/gselements.11.3.189

Chew DM, Sylvester PJ, Tubrett MN (2011) U–Pb and Th–Pb dating of apatite by LA-ICPMS. Chem Geol 280:200–216. https://doi.org/10.1016/j.chemgeo.2010.11.010

Chiaradia M, Vallance J, Fontboté L, Stein H, Schaltegger U, Coder J, Richards J, Villeneuve M, Gendall I (2008) U–Pb, Re–Os, and $^{40}Ar/^{39}Ar$ geochronology of the Nambija Au-skarn and Pangui porphyry Cu deposits, Ecuador: implications for the Jurassic metallogenic belt of the Northern Andes. Miner Depos 44:371–387. https://doi.org/10.1007/s00126-008-0210-6

Chiaradia M, Caricchi L (2017) Stochastic modelling of deep magmatic controls on porphyry copper deposit endowment. Sci Rep 7:44523. https://doi.org/10.1038/srep44523

Chiaradia M, Schaltegger U, Spikings R, Wotzlaw JF, Ovtcharova M (2013) How accurately can we date the duration of magmatic-hydrothermal events in porphyry systems? An invited paper. Econ Geol 108:565–584. https://doi.org/10.2113/econgeo.108.4.565

Chiaradia M, Schaltegger U, Spikings RA (2014) Time scales of mineral systems—advances in understanding over the past decade. Soc Econ Geol Spec Publ 18:37–58

Clemens J (2003) S-type granitic magmas—petrogenetic issues, models and evidence. Earth Sci Rev 61:1–18. https://doi.org/10.1016/S0012-8252(02)00107-1

Cochrane R, Spikings RA, Chew DM, Wotzlaw J-F, Chiaradia M, Tyrrell S, Schaltegger U, van der Lelij R (2014) High temperature (> 350 °C) thermochronology and mechanisms of Pb loss in apatite. Geochim Cosmochim Acta 127:39–56. https://doi.org/10.1016/j.gca.2013.11.028

Condomines M, Gauthier P-J, Sigmarsson O (2003) Timescales of magma chamber processes and dating of young volcanic rocks. Rev Mineral Geochem 52:125–174. https://doi.org/10.2113/0520125

Condon DJ, Schoene B, McLean NM, Bowring SA, Parrish RR (2015) Metrology and traceability of U–Pb isotope dilution geochronology (EARTHTIME Tracer Calibration Part I). Geochim Cosmochim Acta 164:464–480. https://doi.org/10.1016/j.gca.2015.05.026

Coogan LA, Parrish RR, Roberts NMW (2016) Early hydrothermal carbon uptake by the upper oceanic crust: insight from in situ U–Pb dating. Geology 44:147–150. https://doi.org/10.1130/G37212.1

Corfu F (2013) A century of U–Pb geochronology: the long quest towards concordance. Geol Soc Am Bull 125:33–47. https://doi.org/10.1130/B30698.1

Corfu F, Lightfoot PC (1996) U–Pb geochronology of the sublayer environment, Sudbury igneous complex, Ontario. Econ Geol 91:1263–1269. https://doi.org/10.2113/gsecongeo.91.7.1263

Corfu F, Muir TL (1989) The Hemlo-Heron Bay greenstone belt and Hemlo Au-Mo deposit, Superior Province, Ontario, Canada 2. Timing of metamorphism, alteration and Au mineralization from titanite, rutile, and monazite U–Pb geochronology. Chem Geol 79:201–223. https://doi.org/10.1016/0168-9622(89)90030-4

Cottle JM (2014) In-situ U–Th/Pb geochronology of (urano)thorite. Am Min 99:1985–1995. https://doi.org/10.2138/am-2014-4920

Coveney RM, Ragan VM, Brannon JC (2000) Temporal benchmarks for modeling Phanerozoic flow of basinal brines and hydrocarbons in the southern Midcontinent based on radiometrically dated calcite. Geology 28:795–798. https://doi.org/10.1130/0091-7613(2000)028%3c0795:tbfmpf%3e2.3.co;2

Crowley Q, Heron K, Riggs N, Kamber B, Chew D, McConnell B, Benn K (2014) Chemical abrasion applied to LA-ICP-MS U–Pb zircon geochronology. Minerals 4:503–518

Curie P, Curie M, Bémont G (1898) Sur une nouvelle substance fortement radio-active contenue dans la pechblende. C R Acad Sci Paris 127:1215–1217

Curie P, Skolodowska Curie M (1898) Sur une substance nouvelle radioactive, contenue dans la pechblende. C R Acad Sci Paris 127:175–178

Dahl PS (1997) A crystal-chemical basis for Pb retention and fission-track annealing systematics in U-bearing minerals, with implications for geochronology. Earth Planet Sci Lett 150:277–290. https://doi.org/10.1016/S0012-821X(97)00108-8

Darling JR, Storey CD, Engi M (2012) Allanite U–Th–Pb geochronology by laser ablation ICPMS. Chem Geol 292–293:103–115

Davis DW, Krogh TE, Williams IS (2003) Historical development of zircon geochronology. Rev Mineral Geochem 53:145–181. https://doi.org/10.2113/0530145

De Haller A, Corfú F, Fontboté L, Schaltegger U, Barra F, Chiaradia M, Frank M, Alvarado JZ (2006) Geology, geochronology, and Hf and Pb isotope data of the Raúl-Condestable iron oxide-copper-gold deposit, central coast of Peru. Econ Geol 101:281–310. https://doi.org/10.2113/gsecongeo.101.2.281

De Haller A, Tarantola A, Mazurek M, Spangenberg J (2011) Fluid flow through the sedimentary cover in northern Switzerland recorded by calcite–celestite veins (Oftringen borehole, Olten). Swiss J Geosci 104:493–506. https://doi.org/10.1007/s00015-011-0085-x

de Ronde CEJ, Spooner ETC, de Wit MJ, Bray CJ (1992) Shear zone-related, Au quartz vein deposits in the Barberton greenstone belt, South Africa; field and petrographic characteristics, fluid properties, and light stable isotope geochemistry. Econ Geol 87:366–402. https://doi.org/10.2113/gsecongeo.87.2.366

Decree S, Deloule É, De Putter T, Dewaele S, Mees F, Baele J-M, Marignac C (2014) Dating of U-rich heterogenite: new insights into U deposit genesis and U cycling in the Katanga Copperbelt. Precambrian Res 241:17–28

Deng X-D, Li J-W, Wen G (2014) Dating iron skarn mineralization using hydrothermal allanite-(La) U–Th–Pb isotopes by laser ablation ICP-MS. Chem Geol 382:95–110

Deng X-D, Li J-W, Zhao X-F, Wang H-Q, Qi L (2015a) Re–Os and U–Pb geochronology of the Laochang Pb–Zn–Ag and concealed porphyry Mo mineralization along the Changning-Menglian suture, SW China: implications for ore genesis and porphyry Cu–Mo exploration. Miner Depos 51:237–248. https://doi.org/10.1007/s00126-015-0606-z

Deng X-D, Li J-W, Zhou M-F, Zhao X-F, Yan D-R (2015b) In-situ LA-ICPMS trace elements and U–Pb analysis of titanite from the Mesozoic Ruanjiawan W–Cu–Mo skarn deposit, Daye district, China. Ore Geol Rev 65:990–1004. https://doi.org/10.1016/j.oregeorev.2014.08.011

Deng XD, Li JW, Wen G (2015c) U–Pb geochronology of hydrothermal zircons from the Early Cretaceous iron skarn deposits in the Handan-Xingtai district, North China Craton. Econ Geol 110:2159–2180. https://doi.org/10.2113/econgeo.110.8.2159

Dewaele S, Henjes-Kunst F, Melcher F, Sitnikova M, Burgess R, Gerdes A, Fernandez MA, Clercq FD, Muchez P, Lehmann B (2011) Late Neoproterozoic overprinting of the cassiterite and columbite-tantalite bearing pegmatites of the Gatumba area, Rwanda (Central Africa). J Afr Earth Sci 61:10–26. https://doi.org/10.1016/j.jafrearsci.2011.04.004

DeWolf CP, Halliday AN (1991) U–Pb dating of a remagnetized Paleozoic limestone. Geophys Res Lett 18:1445–1448. https://doi.org/10.1029/91GL01278

Dieng S, Kyser K, Godin L (2013) Tectonic history of the North American shield recorded in uranium deposits in the Beaverlodge area, northern Saskatchewan, Canada. Precambrian Res 224:316–340

Dill HG, Gerdes A, Weber B (2010) Age and mineralogy of supergene uranium minerals — Tools to unravel geomorphological and palaeohydrological processes in granitic terrains (Bohemian Massif, SE Germany).

Geomorphology 117:44–65. https://doi.org/10.1016/j.geomorph.2009.11.005

Dill HG, Hansen BT, Weber B (2013) U/Pb age and origin of supergene uranophane-beta from the Borborema Pegmatite mineral province, Brazil. J South Am Earth Sci 45:160–165. https://doi.org/10.1016/j.jsames.2013.03.014

Donnelly CL, Griffin WL, Yang J-H, O'Reilly SY, Li Q-L, Pearson NJ, Li X-H (2012) In situ U–Pb dating and Sr–Nd isotopic analysis of perovskite: constraints on the age and petrogenesis of the Kuruman Kimberlite province, Kaapvaal Craton, South Africa. J Petrol 53:2497–2522. https://doi.org/10.1093/petrology/egs057

Downes PJ, Dunkley DJ, Fletcher IR, McNaughton NJ, Rasmussen B, Jaques AL, Verrall M, Sweetapple MT (2016) Zirconolite, zircon and monazite-(Ce) U–Th–Pb age constraints on the emplacement, deformation and alteration history of the Cummins Range carbonatite complex, Halls Creek Orogen, Kimberley region, Western Australia. Miner Petrol 110:199–222. https://doi.org/10.1007/s00710-015-0418-y

Drake H, Tullborg E-L, MacKenzie AB (2009) Detecting the near-surface redox front in crystalline bedrock using fracture mineral distribution, geochemistry and U-series disequilibrium. Appl Geochem 24:1023–1039. https://doi.org/10.1016/j.apgeochem.2009.03.004

Dziggel A, Poujol M, Otto A, Kisters AFM, Trieloff M, Schwarz WH, Meyer FM (2010) New U–Pb and $^{40}Ar/^{39}Ar$ ages from the northern margin of the Barberton greenstone belt, South Africa: implications for the formation of Mesoarchaean gold deposits. Precambrian Res 179:206–220. https://doi.org/10.1016/j.precamres.2010.03.006

Eichhorn R, Schärer U, Höll R (1995) Age and evolution of scheelite-hosting rocks in the Felbertal deposit (Eastern Alps): U–Pb geochronology of zircon and titanite. Contrib Mineral Petrol 119:377–386. https://doi.org/10.1007/BF00286936

England GL, Rasmussen B, McNaughton NJ, Fletcher IR, Groves DI, Krapež B (2001) SHRIMP U–Pb ages of diagenetic and hydrothermal xenotime from the Archaean Witwatersrand Supergroup of South Africa. Terra Nova 13:360–367. https://doi.org/10.1046/j.1365-3121.2001.00363.x

Fayek M, Harrison TM, Grove M, Coath CD (2000) A rapid in situ method for determining the ages of uranium oxide minerals: evolution of the Cigar Lake deposit, Athabasca Basin. Int Geol Rev 42:163–171

Fisher CM, Longerich HP, Jackson SE, Hanchar JM (2010) Data acquisition and calculation of U–Pb isotopic analyses using laser ablation (single collector) inductively coupled plasma mass spectrometry. J Anal at Spectrom 25:1905–1920. https://doi.org/10.1039/C004955G

Fletcher IR, McNaughton NJ, Aleinikoff JA, Rasmussen B, Kamo SL (2004) Improved calibration procedures and new standards for U–Pb and Th–Pb dating of Phanerozoic xenotime by ion microprobe. Chem Geol 209:295–314. https://doi.org/10.1016/j.chemgeo.2004.06.015

Fletcher IR, McNaughton NJ, Davis WJ, Rasmussen B (2010) Matrix effects and calibration limitations in ion probe U–Pb and Th–Pb dating of monazite. Chem Geol 270:31–44

Frei R (1996) The extent of inter-mineral isotope equilibrium: a systematic bulk U–Pb and Pb step leaching (PbSL) isotope study of individual minerals from the Tertiary granite of Jerissos (northern Greece). Eur J Mineral 8:1175–1190. https://doi.org/10.1127/ejm/8/5/1175

Frimmel HE, Hennigh Q (2015) First whiffs of atmospheric oxygen triggered onset of crustal gold cycle. Miner Depos 50:5–23. https://doi.org/10.1007/s00126-014-0574-8

Frimmel HE, Groves DI, Kirk J, Ruiz J, Chesley J, Minter WEL (2005) The formation and preservation of the Witwatersrand goldfields, the world's largest gold province. Econ Geol 100th Anniv:769–797

Froude DO, Ireland TR, Kinny PD, Williams IS, Compston W, Williams IR, Myers JS (1983) Ion microprobe identification of 4100–4200 Myr-old terrestrial zircons. Nature 304:616–618. https://doi.org/10.1038/304616a0

Fryer BJ, Jackson SE, Longerich HP (1993) The application of laser ablation microprobe-inductively coupled plasma-mass spectrometry (LAM-ICP-MS) to in situ (U)–Pb geochronology. Chem Geol 109:1–8. https://doi.org/10.1016/0009-2541(93)90058-Q

Fu Y, Sun X, Zhou H, Lin H, Yang T (2016) In-situ LA–ICP–MS U–Pb geochronology and trace elements analysis of polygenetic titanite from the giant Beiya gold–polymetallic deposit in Yunnan Province, Southwest China. Ore Geol Rev 77:43–56

Gardiner NJ, Hawkesworth CJ, Robb LJ, Whitehouse MJ, Roberts NMW, Kirkland CL, Evans NJ (2017) Contrasting granite metallogeny through the zircon record: a case study from Myanmar. Sci Rep 7:748. https://doi.org/10.1038/s41598-017-00832-2

Gehrels GE, Valencia VA, Ruiz J (2008) Enhanced precision, accuracy, efficiency, and spatial resolution of U–Pb ages by laser ablation–multicollector–inductively coupled plasma–mass spectrometry. Geochem Geophys Geosyst 9:Q03017. https://doi.org/10.1029/2007GC001805

Geisler T, Schaltegger U, Tomaschek F (2007) Re-equilibration of zircon in aqueous fluids and melts. Elements 3:43–50. https://doi.org/10.2113/gselements.3.1.43

Gelcich S, Davis DW, Spooner ETC (2005) Testing the apatite-magnetite geochronometer: U–Pb and $^{40}Ar/^{39}Ar$ geochronology of plutonic rocks, massive magnetite-apatite tabular bodies, and IOCG mineralization in northern Chile. Geochim Cosmochim Acta 69:3367–3384. https://doi.org/10.1016/j.gca.2004.12.020

Glodny J, Grauert B, Fiala J, Vejnar Z, Krohe A (1998) Metapegmatites in the western Bohemian massif: ages of crystallisation and metamorphic overprint, as constrained by U–Pb zircon, monazite, garnet, columbite and Rb–Sr muscovite data. Geol Rundsch 87:124–134. https://doi.org/10.1007/s005310050194

Grandia F, Asmerom Y, Getty S, Cardellach E, Canals A (2000) U–Pb dating of MVT ore-stage calcite: implications for fluid flow in a Mesozoic extensional basin from Iberian Peninsula. J Geochem Explor 69–70:377–380. https://doi.org/10.1016/S0375-6742(00)00030-3

Gregory CJ, Rubatto D, Allen CM, Williams IS, Hermann J, Ireland T (2007) Allanite microgeochronology: a LA-ICP-MS and SHRIMP U–Th–Pb study. Chem Geol 245:162–182. https://doi.org/10.1016/j.chemgeo.2007.07.029

Griffin WL, Batumike JM, Greau Y, Pearson NJ, Shee SR, O'Reilly SY (2014) Emplacement ages and sources of kimberlites and related rocks in southern Africa: U–Pb ages and Sr–Nd isotopes of groundmass perovskite. Contrib Mineral Petrol 168:1032. https://doi.org/10.1007/s00410-014-1032-4

Guillong M, Horn I, Günther D (2003) A comparison of 266 nm, 213 nm and 193 nm produced from a single solid state Nd:YAG laser for laser ablation ICP-MS. J Anal at Spectrom 18:1224–1230. https://doi.org/10.1039/B305434A

Gulson BL, Jones MT (1992) Cassiterite: potential for direct dating of mineral deposits and a precise age for the Bushveld Complex granites. Geology 20:355–358. https://doi.org/10.1130/0091-7613(1992)020%3c0355:CPFDDO%3e2.3.CO;2

Günther D, Frischknecht R, Heinrich CA, Kahlert H-J (1997) Capabilities of an argon fluoride 193 nm excimer laser for laser ablation inductively coupled plasma mass spectometry microanalysis of geological materials. J Anal at Spectrom 12:939–944. https://doi.org/10.1039/A701423F

Harlaux M, Romer RL, Mercadier J, Morlot C, Marignac C, Cuney M (2017) 40 Ma years of hydrothermal W mineralization during the Variscan orogenic evolution of the French Massif Central revealed by U–Pb dating of wolframite. Miner Depos 334:1–31. https://doi.org/10.1007/s00126-017-0721-0

Harley SL, Kelly NM (2007) Zircon tiny but timely. Elements 3:13–18. https://doi.org/10.2113/gselements.3.1.13

Harlov DE (2015) Apatite: a fingerprint for metasomatic processes. Elements 11:171–176. https://doi.org/10.2113/gselements.11.3.171

Harris AC, Allen CM, Bryan SE, Campbell IH, Holcombe RJ, Palin MJ (2004) ELA-ICP-MS U–Pb zircon geochronology of regional volcanism hosting the Bajo de la Alumbrera Cu–Au deposit: implications for porphyry-related mineralization. Miner Depos 39:46–67. https://doi.org/10.1007/s00126-003-0381-0

Harris AC, Dunlap WJ, Reiners PW, Allen CM, Cooke DR, White NC, Campbell IH, Golding SD (2008) Multimillion year thermal history of a porphyry copper deposit: application of U–Pb, $^{40}Ar/^{39}Ar$ and (U–Th)/He chronometers, Bajo de la Alumbrera copper-gold deposit, Argentina. Miner Depos 43:295–314. https://doi.org/10.1007/s00126-007-0151-5

Heaman L (2003) The timing of kimberlite magmatism in North America: implications for global kimberlite genesis and diamond exploration. Lithos 71:153–184

Heaman LM (2009) The application of U–Pb geochronology to mafic, ultramafic and alkaline rocks: an evaluation of three mineral standards. Chem Geol 261:43–52. https://doi.org/10.1016/j.chemgeo.2008.10.021

Heaman LM, LeCheminant AN (2001) Anomalous U–Pb systematics in mantle-derived baddeleyite xenocrysts from Île Bizard: evidence for high temperature radon diffusion? Chem Geol 172:77–93. https://doi.org/10.1016/S0009-2541(00)00237-0

Heaman LM, Pell J, Grütter HS, Creaser RA (2015) U–Pb geochronology and Sr/Nd isotope compositions of groundmass perovskite from the newly discovered Jurassic Chidliak kimberlite field, Baffin Island, Canada. Earth Planet Sci Lett 415:183–199. https://doi.org/10.1016/j.epsl.2014.12.056

Hedenquist JW, Lowenstern JB (1994) The role of magmas in the formation of hydrothermal ore deposits. Nature 370:519–527. https://doi.org/10.1038/370519a0

Heinrich CA (2015) Witwatersrand gold deposits formed by volcanic rain, anoxic rivers and Archaean life. Nature Geosci 8:206–209. https://doi.org/10.1038/ngeo2344

Hiess J, Condon DJ, McLean N, Noble SR (2012) $^{238}U/^{235}U$ systematics in terrestrial uranium-bearing minerals. Science 335:1610–1614. https://doi.org/10.1126/science.1215507

Hinthorne JR, Andersen CA, Conrad RL, Lovering JF (1979) Single-grain $^{207}Pb/^{206}Pb$ and U/Pb age determinations with a 10-μm spatial resolution using the ion microprobe mass analyzer (IMMA). Chem Geol 25:271–303. https://doi.org/10.1016/0009-2541(79)90061-5

Hinton RW, Long JVP (1979) High-resolution ion-microprobe measurement of lead isotopes: variations within single zircons from Lac Seul, northwestern Ontario. Earth Planet Sci Lett 45:309–325. https://doi.org/10.1016/0012-821X(79)90132-8

Hofmann B, Eikenberg J (1991) The Krunkelbach uranium deposit, Schwarzwald, Germany; correlation of radiometric ages (U–Pb, U–Xe–Kr, K–Ar, $^{230}Th–^{234}U$). Econ Geol 86:1031–1049. https://doi.org/10.2113/gsecongeo.86.5.1031

Hofmann BA, Knill MD (1996) Geochemistry and genesis of the Lengenbach Pb–Zn–As–Tl–Ba-mineralisation, Binn Valley, Switzerland. Miner Depos 31:319–339. https://doi.org/10.1007/BF02280795

Holden NE (1990) Total half-lives for selected nuclides. Pure Appl Chem 62:941–958. https://doi.org/10.1351/pac199062050941

Holmes A (1911) The association of lead with uranium in rock-minerals, and its application to the measurement of geological time. Proc R Soc Lond A 85:248–256. https://doi.org/10.1098/rspa.1911.0036

Holmes A (1913) The age of the Earth. Harper and Brothers, London and New York

Holmes A (1954) The oldest dated minerals of the Rhodesian Shield. Nature 173:612–614

Horn I, Rudnick RL, McDonough WF (2000) Precise elemental and isotope ratio determination by simultaneous solution nebulization and laser ablation-ICP-MS: application to U–Pb geochronology. Chem Geol 164:281–301

Horstwood MSA, Košler J, Gehrels G, Jackson SE, McLean NM, Paton C, Pearson NJ, Sircombe K, Sylvester P, Vermeesch P, Bowring JF, Condon DJ, Schoene B (2016) Community-derived standards for LA-ICP-MS U–(Th–)Pb geochronology - uncertainty propagation, age interpretation and data reporting. Geostand Geoanal Res 40:311–332. https://doi.org/10.1111/j.1751-908X.2016.00379.x

Hoskin PWO, Schaltegger U (2003) The composition of zircon and igneous and metamorphic petrogenesis. Rev Mineral Geochem 53:27–62. https://doi.org/10.2113/0530027

Huston DL, Maas R, Cross A, Hussey KJ, Mernagh TP, Fraser G, Champion DC (2016) The Nolans Bore rare-earth element-phosphorus-uranium mineral system: geology, origin and post-depositional modifications. Miner Depos 51:797–822. https://doi.org/10.1007/s00126-015-0631-y

Ireland TR (2014) Ion microscopes and microprobes. Treatise on geochemistry. Elsevier, Amsterdam, pp 385–409

Ireland TR, Williams IS (2003) Considerations in zircon geochronology by SIMS. Rev Mineral Geochem 53:215–241. https://doi.org/10.2113/0530215

Jaffey AH, Flynn KF, Glendenin LE, Bentley WC, Essling AM (1971) Precision measurement of half-lives and specific activities of ^{235}U and ^{238}U. Phys Rev C 4:1889–1906

Jung S, Mezger K (2003) U–Pb garnet chronometry in high-grade rocks – case studies from the central Damara orogen (Namibia) and implications for the interpretation of Sm–Nd garnet ages and the role of high U–Th inclusions. Contrib Mineral Petrol 146:382–396. https://doi.org/10.1007/s00410-003-0506-6

Kelly NM, Harley SL (2005) An integrated microtextural and chemical approach to zircon geochronology: refining the Archaean history of the Napier Complex, east Antarctica. Contrib Mineral Petrol 149:57–84. https://doi.org/10.1007/s00410-004-0635-6

Kempe U, Lehmann B, Wolf D, Rodionov N, Bombach K, Schwengfelder U, Dietrich A (2008) U–Pb SHRIMP geochronology of Th-poor, hydrothermal monazite: an example from the Llallagua tin-porphyry deposit, Bolivia. Geochim Cosmochim Acta 72:4352–4366. https://doi.org/10.1016/j.gca.2008.05.059

Kerrich R, Kyser TK (1994) 100 Ma timing paradox of Archean gold, Abitibi greenstone belt (Canada): new evidence from U–Pb and Pb–Pb evaporation ages of hydrothermal zircons. Geology 22:1131–1134. https://doi.org/10.1130/0091-7613(1994)022%3c1131:MTPOAG%3e2.3.CO;2

Klemme S, Meyer H-P (2003) Trace element partitioning between baddeleyite and carbonatite melt at high pressures and high temperatures. Chem Geol 199:233–242

Klemme S, Prowatke S, Hametner K, Günther D (2005) Partitioning of trace elements between rutile and silicate melts: implications for subduction zones. Geochim Cosmochim Acta 69:2361–2371. https://doi.org/10.1016/j.gca.2004.11.015

Klötzli U, Klötzli E, Günes Z, Košler J (2009) Accuracy of laser ablation U–Pb zircon dating: results from a test using five different reference zircons. Geostand Geoanal Res 33:5–15. https://doi.org/10.1111/j.1751-908X.2009.00921.x

Koglin N, Zeh A, Frimmel HE, Gerdes A (2010) New constraints on the auriferous Witwatersrand sediment provenance from combined detrital zircon U–Pb and Lu–Hf isotope data for the Eldorado Reef (Central Rand Group, South Africa). Precambrian Res 183:817–824

Kooijman E, Mezger K, Berndt J (2010) Constraints on the U–Pb systematics of metamorphic rutile from in situ LA-ICP-MS analysis. Earth Planet Sci Lett 293:321–330. https://doi.org/10.1016/j.epsl.2010.02.047

Kositcin N, Krapež B (2004) Relationship between detrital zircon age-spectra and the tectonic evolution of the Late Archaean Witwatersrand Basin, South Africa. Precambrian Res 129:141–168. https://doi.org/10.1016/j.precamres.2003.10.011

Kositcin N, McNaughton NJ, Griffin BJ, Fletcher IR, Groves DI, Rasmussen B (2003) Textural and geochemical discrimination between xenotime of different origin in the Archaean Witwatersrand Basin, South Africa. Geochim Cosmochim Acta 67:709–731. https://doi.org/10.1016/S0016-7037(02)01169-9

Košler J, Wiedenbeck M, Wirth R, Hovorka J, Sylvester P, Míková J (2005) Chemical and phase composition of particles produced by laser ablation of silicate glass and zircon—implications for elemental fractionation during ICP-MS analysis. J Anal at Spectrom 20:402–409. https://doi.org/10.1039/B416269B

Košler J, Sláma J, Belousova E, Corfú F, Gehrels GE, Gerdes A, Horstwood MSA, Sircombe KN, Sylvester PJ, Tiepolo M, Whitehouse MJ, Woodhead JD (2013) U–Pb detrital zircon analysis—results of an inter-laboratory comparison. Geostand Geoanal Res 37:243–259. https://doi.org/10.1111/j.1751-908X.2013.00245.x

Kouzmanov K, Moritz R, von Quadt A, Chiaradia M, Peytcheva I, Fontignie D, Ramboz C, Bogdanov K (2009) Late Cretaceous porphyry Cu and epithermal Cu–Au association in the southern Panagyurishte district, Bulgaria: the paired Vlaykov Vruh and Elshitsa deposits. Miner Depos 44:611–646. https://doi.org/10.1007/s00126-009-0239-1

Krogh TE (1973) A low-contamination method for hydrothermal decomposition of zircon and extraction of U and Pb for isotopic age determinations. Geochim Cosmochim Acta 37:485–494. https://doi.org/10.1016/0016-7037(73)90213-5

Krogh TE (1982) Improved accuracy of U–Pb zircon ages by the creation of more concordant systems using an air abrasion technique. Geochim Cosmochim Acta 46:637–649. https://doi.org/10.1016/0016-7037(82)90165-X

Krogh TE, Davis GL (1975) The production and preparation of ^{205}Pb for use as a tracer for isotope dilution analyses. Carnegie Institute of Washington, Yearbook 74:416–417

Kröner A, Kovach VP, Kozakov IK, Kirnozova T, Azimov P, Wong J, Geng HY (2015) Zircon ages and Nd–Hf isotopes in UHT granulites of the Ider complex: a cratonic terrane within the Central Asian Orogenic Belt in NW Mongolia. Gondwana Res 27:1392–1406. https://doi.org/10.1016/j.gr.2014.03.005

Kryza R, Crowley QG, Larionov A, Pin C, Oberc-Dziedzic T, Mochnacka K (2012) Chemical abrasion applied to SHRIMP zircon geochronology: an example from the Variscan Karkonosze Granite (Sudetes, SW Poland). Gondwana Res 21:757–767. https://doi.org/10.1016/j.gr.2011.07.007

Lancelot J, Vitrac A, Allegre CJ (1976) Uranium and lead isotopic dating with grain-by-grain zircon analysis: a study of complex geological history with a single rock. Earth Planet Sci Lett 29:357–366. https://doi.org/10.1016/0012-821X(76)90140-0

Larsen ES, Keevil NB, Harrison HC (1952) Method for determining the age of igneous rocks using the accessory minerals. Geol Soc Am Bull 63:1045–1052. https://doi.org/10.1130/0016-7606(1952)63[1045:MFDTAO]2.0.CO;2

Law JDM, Phillips GN (2005) Hydrothermal replacement model for Witwatersrand gold. Econ Geol 100th Anniv:799–811

Lecumberri-Sanchez P, Romer RL, Lüders V, Bodnar RJ (2014) Genetic relationship between silver–lead–zinc mineralization in the Wutong deposit, Guangxi Province and Mesozoic granitic magmatism in the Nanling belt, southeast China. Miner Depos 49:353–369. https://doi.org/10.1007/s00126-013-0494-z

Lehmann B, Burgess R, Frei D, Belyatsky B, Mainkar D, Rao NVC, Heaman LM (2010) Diamondiferous kimberlites in central India synchronous with Deccan flood basalts. Earth Planet Sci Lett 290:142–149. https://doi.org/10.1016/j.epsl.2009.12.014

Li XH, Su L, Chung SL, Li ZX, Liu Y, Song B, Liu DY (2005) Formation of the Jinchuan ultramafic intrusion and the world's third largest Ni–Cu sulfide deposit: associated with the ~825 Ma south China mantle plume? Geochem Geophys Geosyst 6:Q11004. https://doi.org/10.1029/2005GC001006

Li J-W, Deng X-D, Zhou M-F, Liu Y-S, Zhao X-F, Guo J-L (2010) Laser ablation ICP-MS titanite U–Th–Pb dating of hydrothermal ore deposits: a case study of the Tonglushan Cu–Fe–Au skarn deposit, SE Hubei Province, China. Chem Geol 270:56–67. https://doi.org/10.1016/j.chemgeo.2009.11.005

Li Q, Parrish RR, Horstwood MSA, McArthur JM (2014) U–Pb dating of cements in Mesozoic ammonites. Chem Geol 376:76–83. https://doi.org/10.1016/j.chemgeo.2014.03.020

Li C-Y, Zhang R-Q, Ding X, Ling M-X, Fan W-M, Sun W-D (2016) Dating cassiterite using laser ablation ICP-MS. Ore Geol Rev 72:313–322. https://doi.org/10.1016/j.oregeorev.2015.07.016

Lobato LM, Santos JOS, McNaughton NJ, Fletcher IR, Noce CM (2007) U–Pb SHRIMP monazite ages of the giant Morro Velho and Cuiabá gold deposits, Rio das Velhas greenstone belt, Quadrilátero Ferrífero, Minas Gerais, Brazil. Ore Geol Rev 32:674–680. https://doi.org/10.1016/j.oregeorev.2006.11.007

Ludwig KR (1980) Calculation of uncertainties of U–Pb isotope data. Earth Planet Sci Lett 46:212–220

Ludwig KR (1998) On the treatment of concordant uranium-lead ages. Geochim Cosmochim Acta 62:665–676. https://doi.org/10.1016/S0016-7037(98)00059-3

Ludwig KR (2012) Isoplot 3.75, a geochronological toolkit for Microsoft Excel. Berkeley Geochronology Center Special Publication 75

Ludwig KA, Shen C-C, Kelley DS, Cheng H, Edwards RL (2011) U–Th systematics and ^{230}Th ages of carbonate chimneys at the Lost City hydrothermal field. Geochim Cosmochim Acta 75:1869–1888. https://doi.org/10.1016/j.gca.2011.01.008

Luo J-C, Hu R-Z, Fayek M, Li C-S, Bi X-W, Abdu Y, Chen Y-W (2015) In-situ SIMS uraninite U–Pb dating and genesis of the Xianshi granite-hosted uranium deposit, South China. Ore Geol Rev 65:968–978. https://doi.org/10.1016/j.oregeorev.2014.06.016

Luvizotto GL, Zack T, Meyer HP, Ludwig T, Triebold S, Kronz A, Münker C, Stöckli DF, Prowatke S, Klemme S, Jacob DE, von Eynatten H (2009) Rutile crystals as potential trace element and isotope mineral standards for microanalysis. Chem Geol 261:346–369. https://doi.org/10.1016/j.chemgeo.2008.04.012

Marillo-Sialer E, Woodhead J, Hergt J, Greig A, Guillong M, Gleadow A, Evans N, Paton C (2014) The zircon "matrix effect": evidence for an ablation rate control on the accuracy of U–Pb age determinations by LA-ICP-MS. J Anal at Spectrom 29:981–989. https://doi.org/10.1039/C4JA00008K

Mattinson JM (1972) Preparation of hydrofluoric, hydrochloric, and nitric acids at ultralow lead levels. Anal Chem 44:1715–1716. https://doi.org/10.1021/ac60317a032

Mattinson JM (1994) A study of complex discordance in zircons using step-wise dissolution techniques. Contrib Mineral Petrol 116:117–129. https://doi.org/10.1007/BF00310694

Mattinson JM (2005) Zircon U–Pb chemical abrasion ("CA-TIMS") method: combined annealing and multi-step partial dissolution analysis for improved precision and accuracy of zircon ages. Chem Geol 220:47–66. https://doi.org/10.1016/j.chemgeo.2005.03.011

Mattinson JM (2013) Revolution and evolution: 100 years of U–Pb geochronology. Elements 9:53–57. https://doi.org/10.2113/gselements.9.1.53

McKinney ST, Cottle JM, Lederer GW (2015) Evaluating rare earth element (REE) mineralization mechanisms in Proterozoic gneiss, Music Valley, California. Geol Soc Am Bull 127:1135–1152. https://doi.org/10.1130/B31165.1

McLean NM, Bowring JF, Bowring SA (2011) An algorithm for U–Pb isotope dilution data reduction and uncertainty propagation. Geochem Geophys Geosyst 12:Q0AA18. https://doi.org/10.1029/2010GC003478

McLean NM, Condon DJ, Schoene B, Bowring SA (2015) Evaluating uncertainties in the calibration of isotopic reference materials and multi-element isotopic tracers (EARTHTIME Tracer Calibration Part II). Geochim Cosmochim Acta 164:481–501. https://doi.org/10.1016/j.gca.2015.02.040

McLean NM, Bowring JF, Gehrels G (2016) Algorithms and software for U–Pb geochronology by LA-ICPMS. Geochem Geophys Geosyst 17. https://doi.org/10.1002/2015GC006097

McNaughton NJ, Rasmussen B, Fletcher IR (1999) SHRIMP uranium-lead dating of diagenetic xenotime in siliciclastic sedimentary rocks. Science 285:78–80. https://doi.org/10.1126/science.285.5424.78

Meier DL, Heinrich CA, Watts MA (2009) Mafic dikes displacing Witwatersrand gold reefs: evidence against metamorphic-hydrothermal ore formation. Geology 37:607–610. https://doi.org/10.1130/G25657A.1

Meinhold G (2010) Rutile and its applications in earth sciences. Earth Sci Rev 102:1–28. https://doi.org/10.1016/j.earscirev.2010.06.001

Melcher F, Graupner T, Gäbler HE, Sitnikova M, Henjes-Kunst F, Oberthür T, Gerdes A, Dewaele S (2015) Tantalum–(niobium–tin) mineralisation in African pegmatites and rare metal granites: constraints from Ta–Nb oxide mineralogy, geochemistry and U–Pb geochronology. Ore Geol Rev 64:667–719. https://doi.org/10.1016/j.oregeorev.2013.09.003

Melleton J, Gloaguen E, Frei D, Novák M, Breiter K (2012) How are the emplacement of rare-element pegmatites, regional metamorphism and magmatism interrelated in the Moldanubian domain of the Variscan Bohemian Massif, Czech Republic? Can Mineral 50:1751–1773. https://doi.org/10.3749/canmin.50.6.1751

Mezger K, Hanson GN, Bohlen SR (1989) U–Pb systematics of garnet: dating the growth of garnet in the late Archean Pikwitonei granulite domain at Cauchon and Natawahunan lakes, Manitoba, Canada. Contrib Mineral Petrol 101:136–148. https://doi.org/10.1007/BF00375301

Michael Meyer F, Kolb J, Sakellaris GA, Gerdes A (2006) New ages from the Mauritanides Belt: recognition of Archean IOCG mineralization at Guelb Moghrein, Mauritania. Terra Nova 18:345–352. https://doi.org/10.1111/j.1365-3121.2006.00698.x

Michard-Vitrac A, Lancelot J, Allegre CJ, Moorbath S (1977) U–Pb ages on single zircons from the early Precambrian rocks of West Greenland and the Minnesota River Valley. Earth Planet Sci Lett 35:449–453. https://doi.org/10.1016/0012-821X(77)90077-2

Miller JS, Matzel JEP, Miller CF, Burgess SD, Miller RB (2007) Zircon growth and recycling during the assembly of large, composite arc plutons. J Volcanol Geotherm Res 167:282–299. https://doi.org/10.1016/j.jvolgeores.2007.04.019

Möller A, O'Brien PJ, Kennedy A, Kröner A (2003) Linking growth episodes of zircon and metamorphic textures to zircon chemistry: an example from the ultrahigh-temperature granulites of Rogaland (SW Norway). Geol Soc Spec Pub 220:65–81. https://doi.org/10.1144/GSL.SP.2003.220.01.04

Moreto CPN, Monteiro LVS, Xavier RP, Creaser RA, DuFrane SA, Melo GHC, da Silva MAD, Tassinari CCG, Sato K (2014) Timing of multiple hydrothermal events in the iron oxide–copper–gold deposits of the Southern Copper Belt, Carajás Province, Brazil. Miner Depos 50:517–546. https://doi.org/10.1007/s00126-014-0549-9

Morisset C-E, Scoates JS, Weis D, Friedman RM (2009) U–Pb and $^{40}Ar/^{39}Ar$ geochronology of the Saint-Urbain and Lac Allard (Havre-Saint-Pierre) anorthosites and their associated Fe–Ti oxide ores, Québec: evidence for emplacement and slow cooling during the collisional Ottawan Orogeny in the Grenville Province. Precambrian Res 174:95–116. https://doi.org/10.1016/j.precamres.2009.06.009

Moser DE (1997) Dating the shock wave and thermal imprint of the giant Vredefort impact, South Africa. Geology 25:7–10. https://doi.org/10.1130/0091-7613(1997)025%3c0007:DTSWAT%3e2.3.CO;2

Moser DE, Davis WJ, Reddy SM, Flemming RL, Hart RJ (2009) Zircon U–Pb strain chronometry reveals deep impact-triggered flow. Earth Planet Sci Lett 277:73–79. https://doi.org/10.1016/j.epsl.2008.09.036

Mosoh Bambi CK, Frimmel HE, Zeh A, Suh CE (2013) Age and origin of Pan-African granites and associated U-Mo mineralization at Ekomédion, southwestern Cameroon. J Afr Earth Sci 88:15–37. https://doi.org/10.1016/j.jafrearsci.2013.08.005

Mueller AG, Campbell IH, Schiotte L, Sevigny JH, Layer PW (1996) Constraints on the age of granitoid emplacement, metamorphism, gold mineralization, and subsequent cooling of the Archean greenstone terrane at Big Bell, Western Australia. Econ Geol 91:896–915. https://doi.org/10.2113/gsecongeo.91.5.896

Mueller AG, Hall GC, Nemchin AA, Stein HJ, Creaser RA, Mason DR (2007) Archean high-Mg monzodiorite–syenite, epidote skarn, and biotite–sericite gold lodes in the Granny Smith-Wallaby district, Australia: U–Pb and Re–Os chronometry of two intrusion-related hydrothermal systems. Miner Depos 43:337–362. https://doi.org/10.1007/s00126-007-0164-0

Muhling JR, Fletcher IR, Rasmussen B (2012) Dating fluid flow and Mississippi Valley type base-metal mineralization in the Paleoproterozoic Earaheedy Basin, Western Australia. Precambrian Res 212–213:75–90. https://doi.org/10.1016/j.precamres.2012.04.016

Müller SG, Krapež B, Barley ME, Fletcher IR (2005) Giant iron-ore deposits of the Hamersley province related to the breakup of Paleoproterozoic Australia: new insights from in situ SHRIMP dating of baddeleyite from mafic intrusions. Geology 33:577–580. https://doi.org/10.1130/G21482.1

Muntean JL, Frimmel HE, Phillips N, Law J, Myers R (2005) Controversies on the origin of world-class gold deposits, Part II: Witwatersrand gold deposits. Soc Econ Geol Newsl 60:7–12–19

Niiranen T, Poutiainen M, Mänttäri I (2007) Geology, geochemistry, fluid inclusion characteristics, and U–Pb age studies on iron oxide–Cu–Au deposits in the Kolari region, northern Finland. Ore Geol Rev 30:75–105. https://doi.org/10.1016/j.oregeorev.2005.11.002

Norcross C, Davis DW, Spooner E, Rust A (2000) U–Pb and Pb–Pb age constraints on Paleoproterozoic magmatism, deformation and gold mineralization in the Omai area, Guyana Shield. Precambrian Res 102:69–86. https://doi.org/10.1016/s0301-9268(99)00102-3

Oberli F, Meier M, Berger A, Rosenberg CL, Gieré R (2004) U–Th–Pb and $^{230}Th/^{238}U$ disequilibrium isotope systematics: precise accessory mineral chronology and melt evolution tracing in the Alpine Bergell intrusion. Geochim Cosmochim Acta 68:2543–2560. https://doi.org/10.1016/j.gca.2003.10.017

Oberthür T, Melcher F, Henjes-Kunst F, Gerdes A, Stein H, Zimmerman A, El Ghorfi M (2009) Hercynian age of the cobalt-nickel-arsenide-(gold) ores, Bou Azzer, Anti-Atlas, Morocco: Re–Os, Sm–Nd, and U–Pb age determinations. Econ Geol 104:1065–1079. https://doi.org/10.2113/econgeo.104.7.1065

Ono S, Fayek M (2011) Decoupling of O and Pb isotope systems of uraninite in the early Proterozoic conglomerates in the Elliot Lake district. Chem Geol 288:1–13. https://doi.org/10.1016/j.chemgeo.2010.03.015

Pal DC, Chaudhuri T, McFarlane C, Mukherjee A, Sarangi AK (2011) Mineral chemistry and in situ dating of allanite, and geochemistry of its host rocks in the Bagjata uranium mine, Singhbhum Shear Zone, India—implications for the chemical evolution of REE mineralization and mobilization. Econ Geol 106:1155–1171. https://doi.org/10.2113/econgeo.106.7.1155

Pape J, Mezger K, Robyr M (2016) A systematic evaluation of the Zr-in-rutile thermometer in ultra-high temperature (UHT) rocks. Contrib Mineral Petrol 171:44. https://doi.org/10.1007/s00410-016-1254-8

Paquette JL, Tiepolo M (2007) High resolution (5 μm) U–Th–Pb isotope dating of monazite with excimer laser ablation (ELA)-ICPMS. Chem Geol 240:222–237. https://doi.org/10.1016/j.chemgeo.2007.02.014

Parrish RR (1987) An improved micro-capsule for zircon dissolution in U–Pb geochronology. Chem Geol 66:99–102. https://doi.org/10.1016/0168-9622(87)90032-7

Parrish RR (1990) U–Pb dating of monazite and its application to geological problems. Can J Earth Sci 27:1431–1450. https://doi.org/10.1139/e90-152

Parrish RR, Noble SR (2003) Zircon U–Th–Pb geochronology by isotope dilution—thermal ionization mass spectrometry (ID-TIMS). Rev Mineral Geochem 53:183–213. https://doi.org/10.2113/0530183

Pelleter E, Cheilletz A, Gasquet D, Mouttaqi A, Annich M, El Hakour A, Deloule É, Féraud G (2007) Hydrothermal zircons: a tool for ion microprobe U–Pb dating of gold mineralization (Tamlalt–Menhouhou gold deposit—Morocco). Chem Geol 245:135–161. https://doi.org/10.1016/j.chemgeo.2007.07.026

Peterman EM, Mattinson JM, Hacker BR (2012) Multi-step TIMS and CA-TIMS monazite U–Pb geochronology. Chem Geol 312–313:58–73

Petersson J, Whitehouse MJ, Eliasson T (2001) Ion microprobe U–Pb dating of hydrothermal xenotime from an episyenite: evidence for rift-related reactivation. Chem Geol 175:703–712. https://doi.org/10.1016/S0009-2541(00)00338-7

Pettke T, Oberli F, Heinrich CA (2010) The magma and metal source of giant porphyry-type ore deposits, based on lead isotope microanalysis of individual fluid inclusions. Earth Planet Sci Lett 296:267–277. https://doi.org/10.1016/j.epsl.2010.05.007

Pfaff K, Romer RL, Markl G (2009) U–Pb ages of ferberite, chalcedony, agate, "U-mica" and pitchblende: constraints on the mineralization history of the Schwarzwald ore district. Eur J Mineral 21:817–836. https://doi.org/10.1127/0935-1221/2009/0021-1944

Philippe S, Lancelot JR, Clauer N, Pacquet A (2011) Formation and evolution of the Cigar Lake uranium deposit based on U–Pb and K–Ar isotope systematics. Can J Earth Sci 30:720–730. https://doi.org/10.1139/e93-058

Phillips GN, Law JDM (2000) Witwatersrand gold fields; geology, genesis, and exploration. Rev Econ Geol 13:439–500

Pigois J-P, Groves DI, Fletcher IR, McNaughton NJ, Snee LW (2003) Age constraints on Tarkwaian palaeoplacer and lode-gold formation in the Tarkwa-Damang district, SW Ghana. Miner Depos 38:695–714. https://doi.org/10.1007/s00126-003-0360-5

Poletti JE, Cottle JM, Hagen-Peter GA, Lackey JS (2016) Petrochronological constraints on the origin of the Mountain Pass ultrapotassic and carbonatite intrusive suite, California. J Petrol 57:1555–1598. https://doi.org/10.1093/petrology/egw050

Polito PA, Kyser TK, Thomas D, Marlatt J, Drever G (2005) Re-evaluation of the petrogenesis of the Proterozoic Jabiluka unconformity-related uranium deposit, Northern Territory, Australia. Miner Depos 40:257–288. https://doi.org/10.1007/s00126-005-0007-9

Porcelli D, Swarzenski PW (2003) The behavior of U- and Th-series nuclides in groundwater. Rev Mineral Geochem 52:317–361. https://doi.org/10.2113/0520317

Poujol M (2007) An overview of the Pre-Mesoarchean rocks of the Kaapvaal Craton, South Africa. Dev Precambrian Geol 15:453–463. https://doi.org/10.1016/S0166-2635(07)15051-9

Prowatke S, Klemme S (2005) Effect of melt composition on the partitioning of trace elements between titanite and silicate melt. Geochim Cosmochim Acta 69:695–709. https://doi.org/10.1016/j.gca.2004.06.037

Prowatke S, Klemme S (2006) Trace element partitioning between apatite and silicate melts. Geochim Cosmochim Acta 70:4513–4527. https://doi.org/10.1016/j.gca.2006.06.162

Rao NVC, Wu F-Y, Mitchell RH, Li Q-L, Lehmann B (2013) Mesoproterozoic U–Pb ages, trace element and Sr–Nd isotopic composition of perovskite from kimberlites of the eastern Dharwar craton, southern India: distinct mantle sources and a widespread 1.1 Ga tectonomagmatic event. Chem Geol 353:48–64. https://doi.org/10.1016/j.chemgeo.2012.04.023

Rasbury ET, Cole JM (2009) Directly dating geologic events: U–Pb dating of carbonates. Rev Geophys 47: RG3001. https://doi.org/10.1029/2007RG000246

Rasmussen B, Blake TS, Fletcher IR (2005) U–Pb zircon age constraints on the Hamersley spherule beds: evidence for a single 2.63 Ga Jeerinah-Carawine impact ejecta layer. Geology 33:725–728. https://doi.org/10.1130/G21616.1

Rasmussen B, Fletcher IR, Muhling JR, Mueller AG, Hall GC (2007a) Bushveld-aged fluid flow, peak metamorphism, and gold mobilization in the Witwatersrand basin, South Africa: constraints from in situ SHRIMP U–Pb dating of monazite and xenotime. Geology 35:931–934. https://doi.org/10.1130/G23588A.1

Rasmussen B, Fletcher IR, Muhling JR, Thorne WS, Broadbent GC (2007b) Prolonged history of episodic fluid flow in giant hematite ore bodies: evidence from in situ U–Pb geochronology of hydrothermal xenotime. Earth Planet Sci Lett 258:249–259. https://doi.org/10.1016/j.epsl.2007.03.033

Rasmussen B, Mueller AG, Fletcher IR (2008) Zirconolite and xenotime U–Pb age constraints on the emplacement of the Golden Mile Dolerite sill and gold mineralization at the Mt Charlotte mine, Eastern Goldfields Province, Yilgarn Craton, Western Australia. Contrib Mineral Petrol 157:559–572. https://doi.org/10.1007/s00410-008-0352-7

Rezeau H, Moritz R, Wotzlaw J-F, Tayan R, Melkonyan R, Ulianov A, Selby D, D'Abzac F-X, Stern RA (2016) Temporal and genetic link between incremental pluton assembly and pulsed porphyry Cu–Mo formation in accretionary orogens. Geology 44:627–630. https://doi.org/10.1130/G38088.1

Richards JP (2009) Postsubduction porphyry Cu–Au and epithermal Au deposits: products of remelting of subduction-modified lithosphere. Geology 37:247–250. https://doi.org/10.1130/G25451A.1

Richards JP (2011) Magmatic to hydrothermal metal fluxes in convergent and collided margins. Ore Geol Rev 40:1–26. https://doi.org/10.1016/j.oregeorev.2011.05.006

Richards JP (2013) Giant ore deposits formed by optimal alignments and combinations of geological processes. Nature Geosci 6:911–916. https://doi.org/10.1038/ngeo1920

Rioux M, Bowring S, Dudás F, Hanson R (2010) Characterizing the U–Pb systematics of baddeleyite through chemical abrasion: application of multi-step digestion methods to baddeleyite geochronology. Contrib Mineral Petrol 160:777–801. https://doi.org/10.1007/s00410-010-0507-1

Robert J, Miranda CF, Muxart R (1969) Mesure de la période du protactinium 231 par microcalorimétrie. Radiochim Acta 11:104–108. https://doi.org/10.1524/ract.1969.11.2.104

Roberts NMW, Walker RJ (2016) U–Pb geochronology of calcite-mineralized faults: absolute timing of rift-related fault events on the northeast Atlantic margin. Geology 44:531–534. https://doi.org/10.1130/G37868.1

Romer RL (1992) Vesuvianite—new tool for the U–Pb dating of skarn ore deposits. Miner Petrol 46:331–341

Romer RL (1996) U–Pb systematics of stilbite-bearing low-temperature mineral assemblages from the Malmberget iron ore, northern Sweden. Geochim Cosmochim Acta 60:1951–1961. https://doi.org/10.1016/0016-7037(96)00066-X

Romer RL, Lüders V (2006) Direct dating of hydrothermal W mineralization: U–Pb age for hübnerite ($MnWO_4$), Sweet Home Mine, Colorado. Geochim Cosmochim Acta 70:4725–4733. https://doi.org/10.1016/j.gca.2006.07.003

Romer RL, Öhlander B (1994) U–Pb age of the Yxsjöberg tungsten-skarn deposit, Sweden. GFF 116:161–166. https://doi.org/10.1080/11035899409546179

Romer RL, Smeds S-A (1994) Implications of U–Pb ages of columbite-tantalites from granitic pegmatites for the Palaeoproterozoic accretion of 1.90–1.85 Ga magmatic arcs to the Baltic Shield. Precambrian Res 67:141–158. https://doi.org/10.1016/0301-9268(94)90008-6

Romer RL, Smeds S-A (1996) U–Pb columbite ages of pegmatites from Sveconorwegian terranes in southwestern Sweden. Precambrian Res 76:15–30. https://doi.org/10.1016/0301-9268(95)00023-2

Romer RL, Smeds S-A (1997) U–Pb columbite chronology of post-kinematic Palaeoproterozoic pegmatites in Sweden. Precambrian Res 82:85–99. https://doi.org/10.1016/S0301-9268(96)00050-2

Romer RL, Wright JE (1992) U–Pb dating of columbites: a geochronologic tool to date magmatism and ore deposits. Geochim Cosmochim Acta 56:2137–2142. https://doi.org/10.1016/0016-7037(92)90337-I

Romer RL, Martinsson O, Perdahl JA (1994) Geochronology of the Kiruna iron ores and hydrothermal alterations. Econ Geol 89:1249–1261. https://doi.org/10.2113/gsecongeo.89.6.1249

Romer RL, Smeds SA, Černý P (1996) Crystal-chemical and genetic controls of U–Pb systematics of columbite-tantalite. Miner Petrol 57:243–260. https://doi.org/10.1007/BF01162361

Rubatto D, Hermann J (2007) Experimental zircon/melt and zircon/garnet trace element partitioning and implications for the geochronology of crustal rocks. Chem Geol 241:38–61. https://doi.org/10.1016/j.chemgeo.2007.01.027

Ruiz J, Valencia VA, Chesley JT, Kirk J, Gehrels G, Frimmel HE (2006) The source of gold for the Witwatersrand from Re–Os and U–Pb detrital zircon geochronology. Geochim Cosmochim Acta 70:A543

Rutherford E (1906) Radioactive transformations. Yale University Press, New Haven

Rutherford E (1929) Origin of actinium and age of the Earth. Nature 123:313–314

Salier BP, Groves DI, McNaughton NJ, Fletcher IR (2004) The world-class Wallaby gold deposit, Laverton, Western Australia: an orogenic-style overprint on a magmatic-hydrothermal magnetite-calcite alteration pipe? Miner Depos 39:473–494. https://doi.org/10.1007/s00126-004-0425-0

Salier BP, Groves DI, McNaughton NJ, Fletcher IR (2005) Geochronological and stable isotope evidence for widespread orogenic gold mineralization from a deep-seated fluid source at ca 2.65 Ga in the Laverton gold province. Western Australia. Econ Geol 100:1363–1388. https://doi.org/10.2113/gsecongeo.100.7.1363

Sarma DS, Fletcher IR, Rasmussen B, McNaughton NJ, Mohan MR, Groves DI (2011) Archaean gold mineralization synchronous with late cratonization of the Western Dharwar Craton, India: 2.52 Ga U–Pb ages of hydrothermal monazite and xenotime in gold deposits. Miner Depos 46:273–288. https://doi.org/10.1007/s00126-010-0326-3

Schaltegger U, Davies JHFL (2017) Petrochronology of zircon and baddeleyite in igneous rocks: reconstructing magmatic processes at high temporal resolution. Rev Mineral Geochem 83:297–328. https://doi.org/10.2138/rmg.2017.83.10

Schaltegger U, Pettke T, Audétat A, Reusser E, Heinrich CA (2005) Magmatic-to-hydrothermal crystallization in the W–Sn mineralized Mole Granite (NSW, Australia). Chem Geol 220:215–235. https://doi.org/10.1016/j.chemgeo.2005.02.018

Schaltegger U, Schmitt AK, Horstwood MSA (2015) U–Th–Pb zircon geochronology by ID-TIMS, SIMS, and laser ablation ICP-MS: recipes, interpretations, and opportunities. Chem Geol 402:89–110. https://doi.org/10.1016/j.chemgeo.2015.02.028

Schärer U (1984) The effect of initial ^{230}Th disequilibrium on young U–Pb ages: the Makalu case, Himalaya. Earth Planet Sci Lett 67:191–204. https://doi.org/10.1016/0012-821x(84)90114-6

Schärer U, Corfu F, Demaiffe D (1997) U–Pb and Lu–Hf isotopes in baddeleyite and zircon megacrysts from the Mbuji-Mayi kimberlite: constraints on the subcontinental mantle. Chem Geol 143:1–16. https://doi.org/10.1016/S0009-2541(97)00094-6

Schmitt AK (2007) Ion microprobe analysis of (^{231}Pa)/(^{235}U) and an appraisal of protactinium partitioning in igneous zircon. Am Mineral 92:691–694. https://doi.org/10.2138/am.2007.2449

Schmitt AK, Zack T (2012) High-sensitivity U–Pb rutile dating by secondary ion mass spectrometry (SIMS) with an O^{2+} primary beam. Chem Geol 332–333:65–73. https://doi.org/10.1016/j.chemgeo.2012.09.023

Schmitz MD, Bowring SA (2001) U–Pb zircon and titanite systematics of the Fish Canyon Tuff: an assessment of high-precision U–Pb geochronology and its application to young volcanic rocks. Geochim Cosmochim Acta 65:2571–2587

Schoene B (2014) U–Th–Pb geochronology. Treatise on geochemistry. Elsevier, Amsterdam, pp 341–378

Schoene B, Bowring SA (2006) U–Pb systematics of the McClure Mountain syenite: thermochronological constraints on the age of the ^{40}Ar/^{39}Ar standard MMhb. Contrib Mineral Petrol 151:615–630. https://doi.org/10.1007/s00410-006-0077-4

Schoene B, Bowring SA (2007) Determining accurate temperature–time paths from U–Pb thermochronology: an example from the Kaapvaal craton, southern Africa. Geochim Cosmochim Acta 71:165–185. https://doi.org/10.1016/j.gca.2006.08.029

Schoene B, Crowley JL, Condon DJ, Schmitz MD, Bowring SA (2006) Reassessing the uranium decay constants for geochronology using ID-TIMS U–Pb data. Geochim Cosmochim Acta 70:426–445. https://doi.org/10.1016/j.gca.2005.09.007

Schütte P, Chiaradia M, Beate B (2010) Geodynamic controls on Tertiary arc magmatism in Ecuador: constraints from U–Pb zircon geochronology of Oligocene-Miocene intrusions and regional age distribution trends. Tectonophysics 489:159–176. https://doi.org/10.1016/j.tecto.2010.04.015

Seman S, Stöckli DF, McLean NM (2017) U–Pb geochronology of grossular-andradite garnet. Chem Geol 460:106–116. https://doi.org/10.1016/j.chemgeo.2017.04.020

Seo J, Choi S-G, Kim DW, Park J-W, Oh CW (2015) A new genetic model for the Triassic Yangyang iron-oxide–apatite deposit, South Korea: constraints from in situ U–Pb and trace element analyses of accessory minerals. Ore Geol Rev 70:110–135. https://doi.org/10.1016/j.oregeorev.2015.04.009

Seydoux-Guillaume AM, Wirth R, Nasdala L, Gottschalk M, Montel JM, Heinrich W (2002) An XRD, TEM and Raman study of experimentally annealed natural monazite. Phys Chem Min 29:240–253. https://doi.org/10.1007/s00269-001-0232-4

Seydoux-Guillaume A-M, Wirth R, Deutsch A, Schärer U (2004) Microstructure of 24–1928 Ma concordant monazites; implications for geochronology and nuclear waste deposits. Geochim Cosmochim Acta

68:2517–2527. https://doi.org/10.1016/j.gca.2003.10.042

Shi G, Li X, Li Q, Chen Z, Deng J, Liu Y, Kang Z, Pang E, Xu Y, Jia X (2012) Ion microprobe U–Pb age and Zr-in-rutile thermometry of rutiles from the Daixian rutile deposit in the Hengshan Mountains, Shanxi Province, China. Econ Geol 107:525–535. https://doi.org/10.2113/econgeo.107.3.525

Sillitoe RH (2010) Porphyry copper systems. Econ Geol 105:3–41. https://doi.org/10.2113/gsecongeo.105.1.3

Simon AC, Ripley EM (2011) The role of magmatic sulfur in the formation of ore deposits. Rev Mineral Geochem 73:513–578. https://doi.org/10.2138/rmg.2011.73.16

Singer DA, Berger VI, Moring BC (2008) Porphyry copper deposits of the world: database and grade and tonnage models. US Geol Surv Open File Rep:2008–1155

Skirrow RG, Bastrakov EN, Barovich K, Fraser GL, Creaser RA, Fanning CM, Raymond OL, Davidson GJ (2007) Timing of iron oxide Cu–Au–(U) hydrothermal activity and Nd isotope constraints on metal sources in the Gawler Craton, South Australia. Econ Geol 102:1441–1470

Skirrow RG, Mercadier J, Armstrong R, Kuske T, Deloule É (2016) The Ranger uranium deposit, northern Australia: timing constraints, regional and ore-related alteration, and genetic implications for unconformity-related mineralisation. Ore Geol Rev 76:463–503. https://doi.org/10.1016/j.oregeorev.2015.09.001

Skolodowska Curie M (1898) Rayons émis par les composés de l'uranium et du thorium. C R Acad Sci 126:1101–1103

Smith CB, Allsopp HL, Garvie OG, Kramers JD, Jackson PFS, Clement CR (1989) Note on the U–Pb perovskite method for dating kimberlites: examples from the Wesselton and De Beers mines, South Africa, and Somerset Island, Canada. Chem Geol 79:137–145. https://doi.org/10.1016/0168-9622(89)90016-X

Smith SR, Foster GL, Romer RL, Tindle AG, Kelley SP, Noble SR, Horstwood M, Breaks FW (2004) U–Pb columbite-tantalite chronology of rare-element pegmatites using TIMS and laser ablation-multi collector-ICP-MS. Contrib Mineral Petrol 147:549–564. https://doi.org/10.1007/s00410-003-0538-y

Smith MP, Storey CD, Jeffries TE, Ryan C (2009) In situ U–Pb and trace element analysis of accessory minerals in the Kiruna district, Norrbotten, Sweden: new constraints on the timing and origin of mineralization. J Petrol 50:2063–2094. https://doi.org/10.1093/petrology/egp069

Smye AJ, Roberts NMW, Condon DJ, Horstwood MSA, Parrish RR (2014) Characterising the U–Th–Pb systematics of allanite by ID and LA-ICPMS: implications for geochronology. Geochim Cosmochim Acta 135:1–28. https://doi.org/10.1016/j.gca.2014.03.021

Spandler C, Hammerli J, Sha P, Hilbert-Wolf H, Hu Y, Roberts E, Schmitz M (2016) MKED1: a new titanite standard for in situ analysis of Sm–Nd isotopes and U–Pb geochronology. Chem Geol 425:110–126. https://doi.org/10.1016/j.chemgeo.2016.01.002

Spencer CJ, Kirkland CL, Taylor RJM (2016) Strategies towards statistically robust interpretations of in situ U–Pb zircon geochronology. Geosci Front 7:581–589. https://doi.org/10.1016/j.gsf.2015.11.006

Stacey JS, Kramers JD (1975) Approximation of terrestrial lead isotope evolution by a two-stage model. Earth Planet Sci Lett 26:207–221. https://doi.org/10.1016/0012-821X(75)90088-6

Stelten ME, Cooper KM, Vazquez JA, Calvert AT, Glessner JJG (2015) Mechanisms and timescales of generating eruptible rhyolitic magmas at Yellowstone Caldera from zircon and sanidine geochronology and geochemistry. J Petrol 56:1607–1642. https://doi.org/10.1093/petrology/egv047

Stepanov AS, Hermann J, Rubatto D, Rapp RP (2012) Experimental study of monazite/melt partitioning with implications for the REE, Th and U geochemistry of crustal rocks. Chem Geol 300:200–220. https://doi.org/10.1016/j.chemgeo.2012.01.007

Storey CD, Jeffries TE, Smith M (2006) Common lead-corrected laser ablation ICP–MS U–Pb systematics and geochronology of titanite. Chem Geol 227:37–52

Storey CD, Smith MP, Jeffries TE (2007) In situ LA-ICP-MS U–Pb dating of metavolcanics of Norrbotten, Sweden: records of extended geological histories in complex titanite grains. Chem Geol 240:163–181. https://doi.org/10.1016/j.chemgeo.2007.02.004

Stosch H-G, Romer RL, Daliran F, Rhede D (2010) Uranium–lead ages of apatite from iron oxide ores of the Bafq District, East-Central Iran. Miner Depos 46:9–21. https://doi.org/10.1007/s00126-010-0309-4

Tallarico FHB, McNaughton NJ, Groves DI, Fletcher IR, Figueiredo BR, Carvalho JB, Rego JL, Nunes AR (2004) Geological and SHRIMP II U–Pb constraints on the age and origin of the Breves Cu–Au-(W–Bi–Sn) deposit, Carajás, Brazil. Miner Depos 39:68–86. https://doi.org/10.1007/s00126-003-0383-y

Tapster S, Condon DJ, Naden J, Noble SR, Petterson MG, Roberts NMW, Saunders AD, Smith DJ (2016) Rapid thermal rejuvenation of high-crystallinity magma linked to porphyry copper deposit formation; evidence from the Koloula Porphyry Prospect, Solomon Islands. Earth Planet Sci Lett 442:206–217. https://doi.org/10.1016/j.epsl.2016.02.046

Taylor RD, Goldfarb RJ, Monecke T, Fletcher IR, Cosca MA, Kelly NM (2015) Application of U–Th–Pb phosphate geochronology to young orogenic gold deposits: new age constraints on the formation of the Grass Valley gold district, Sierra Nevada Foothills Province, California. Econ Geol 110:1313–1337. https://doi.org/10.2113/econgeo.110.5.1313

Tera F, Wasserburg GJ (1972a) U–Th–Pb systematics in lunar highland samples from the Luna 20 and Apollo 16 missions. Earth Planet Sci Lett 17:36–51

Tera F, Wasserburg GJ (1972b) U–Th–Pb systematics in three Apollo 14 basalts and the problem of initial Pb in lunar rocks. Earth Planet Sci Lett 14:281–304. https://doi.org/10.1016/0012-821X(72)90128-8

Tiepolo M, Oberti R, Vannucci R (2002) Trace-element incorporation in titanite: constraints from experimentally determined solid/liquid partition coefficients. Chem Geol 191:105–119. https://doi.org/10.1016/S0009-2541(02)00151-1

Tilton GR (1960) Volume diffusion as a mechanism for discordant lead ages. J Geophys Res: Solid Earth 65:2933–2945. https://doi.org/10.1029/JZ065i009p02933

Tilton GR, Patterson C, Brown H, Inghram M, Hayden R, Hess D, Larsen E (1955) Isotopic composition and distribution of lead, uranium, and thorium in a Precambrian granite. Geol Soc Am Bull 66:1131–1148. https://doi.org/10.1130/0016-7606(1955)66[1131:ICADOL]2.0.CO;2

Tilton GR, Davis GL, Wetherill GW, Aldrich LT (1957) Isotopic ages of zircon from granites and pegmatites. Eos T Am Geophys Un 38:360–371. https://doi.org/10.1029/TR038i003p00360

Torrealday HI, Hitzman MW, Stein HJ, Markley RJ, Armstrong R, Broughton D (2000) Re–Os and U–Pb dating of the vein-hosted mineralization at the Kansanshi copper deposit, northern Zambia. Econ Geol 95:1165–1170. https://doi.org/10.2113/gsecongeo.95.5.1165

Trachenko K (2004) Understanding resistance to amorphization by radiation damage. J Phys: Condens Matter 16:R1491–R1515. https://doi.org/10.1088/0953-8984/16/49/R03

Valley PM, Hanchar JM, Whitehouse MJ (2009) Direct dating of Fe oxide-(Cu–Au) mineralization by U/Pb zircon geochronology. Geology 37:223–226. https://doi.org/10.1130/G25439A.1

Valley PM, Fisher CM, Hanchar JM, Lam R, Tubrett M (2010) Hafnium isotopes in zircon: a tracer of fluid-rock interaction during magnetite–apatite ("Kiruna-type") mineralization. Chem Geol 275:208–220. https://doi.org/10.1016/j.chemgeo.2010.05.011

Vallini DA, Groves DI, McNaughton NJ, Fletcher IR (2006) Uraniferous diagenetic xenotime in northern Australia and its relationship to unconformity-associated uranium mineralisation. Miner Depos 42:51–64. https://doi.org/10.1007/s00126-005-0012-z

Van Lichtervelde M, Grand'Homme A, de Saint Blanquat M, Olivier P, Gerdes A, Paquette J-L, Melgarejo JC, Druguet E, Alfonso P (2016) U–Pb geochronology on zircon and columbite-group minerals of the Cap de Creus pegmatites, NE Spain. Miner Petrol:1–21.https://doi.org/10.1007/s00710-016-0455-1

Vermeesch P (2012) On the visualisation of detrital age distributions. Chem Geol 312–313:190–194. https://doi.org/10.1016/j.chemgeo.2012.04.021

Vermeesch P (2018) IsoplotR: a free and open toolbox for geochronology. Geosci Front 9:1479–1493. https://doi.org/10.1016/j.gsf.2018.04.001

Vielreicher NM, Groves DI, Snee LW, Fletcher IR, McNaughton NJ (2010) Broad synchroneity of three gold mineralization styles in the Kalgoorlie Gold Field: SHRIMP, U–Pb, and $^{40}Ar/^{39}Ar$ geochronological evidence. Econ Geol 105:187–227. https://doi.org/10.2113/gsecongeo.105.1.187

Vielreicher N, Groves D, McNaughton N, Fletcher I (2015) The timing of gold mineralization across the eastern Yilgarn craton using U–Pb geochronology of hydrothermal phosphate minerals. Miner Depos 50:391–428. https://doi.org/10.1007/s00126-015-0589-9

von Grosse A (1932) On the origin of the actinium series of radioactive elements. Phys Rev 42:565–570. https://doi.org/10.1103/PhysRev.42.565

von Quadt A, Moritz R, Peytcheva I, Heinrich CA (2005) 3: Geochronology and geodynamics of Late Cretaceous magmatism and Cu–Au mineralization in the Panagyurishte region of the Apuseni–Banat–Timok–Srednogorie belt, Bulgaria. Ore Geol Rev 27:95–126. https://doi.org/10.1016/j.oregeorev.2005.07.024

von Quadt A, Erni M, Martinek K, Moll M, Peytcheva I, Heinrich CA (2011) Zircon crystallization and the lifetimes of ore-forming magmatic-hydrothermal systems. Geology 39:731–734. https://doi.org/10.1130/G31966.1

von Quadt A, Gallhofer D, Guillong M, Peytcheva I, Waelle M, Sakata S (2014) U–Pb dating of CA/non-CA treated zircons obtained by LA-ICP-MS and CA-TIMS techniques: impact for their geological interpretation. J Anal at Spectrom 29:1618–1629. https://doi.org/10.1039/C4JA00102H

Vry JK, Baker JA (2006) LA-MC-ICPMS Pb–Pb dating of rutile from slowly cooled granulites: confirmation of the high closure temperature for Pb diffusion in rutile. Geochim Cosmochim Acta 70:1807–1820. https://doi.org/10.1016/j.gca.2005.12.006

Wall CJ, Scoates JS (2016) High-precision U–Pb zircon-baddeleyite dating of the J-M reef platinum group element deposit in the Stillwater Complex, Montana (USA). Econ Geol 111:771–782. https://doi.org/10.2113/econgeo.111.3.771

Wan B, Xiao W, Zhang L, Han C (2012) Iron mineralization associated with a major strike–slip shear zone: radiometric and oxygen isotope evidence from the Mengku deposit, NW China. Ore Geol Rev 44:136–147. https://doi.org/10.1016/j.oregeorev.2011.09.011

Wanhainen C, Billström K, Martinsson O, Stein H, Nordin R (2005) 160 Ma of magmatic/hydrothermal and metamorphic activity in the Gällivare area: Re–Os dating of molybdenite and U–Pb dating of titanite from the Aitik Cu–Au–Ag deposit, northern Sweden. Miner Depos 40:435–447

Watson EB, Chemiak DJ, Hanchar JM, Harrison TM, Wark DA (1997) The incorporation of Pb into zircon. Chem Geol 141:19–31. https://doi.org/10.1016/S0009-2541(97)00054-5

Webber GR, Hurley PM, Fairbairn HW (1956) Relative ages of eastern Massachusetts granites by total lead ratios in zircon. Am J Sci 254:574–583. https://doi.org/10.2475/ajs.254.9.574

Webster JD, Piccoli PM (2015) Magmatic apatite: a powerful, yet deceptive, mineral. Elements 11:177–182. https://doi.org/10.2113/gselements.11.3.177

Weis P, Driesner T, Heinrich CA (2012) Porphyry-copper ore shells form at stable pressure-temperature fronts within dynamic fluid plumes. Science 338:1613–1616. https://doi.org/10.1126/science.1225009

Wendt I (1984) A three-dimensional U–Pb discordia plane to evaluate samples with common lead of unknown isotopic composition. Chem Geol 46:1–12. https://doi.org/10.1016/0009-2541(84)90162-1

Wendt I, Carl C (1991) The statistical distribution of the mean squared weighted deviation. Chem Geol 86:275–285. https://doi.org/10.1016/0168-9622(91)90010-T

Wetherill GW (1956) Discordant uranium-lead ages, I. Eos T Am Geophys Un 37:320–326. https://doi.org/10.1029/TR037i003p00320

White LT, Ireland TR (2012) High-uranium matrix effect in zircon and its implications for SHRIMP U–Pb age determinations. Chem Geol 306–307:78–91. https://doi.org/10.1029/2011GC003726

Widmann P, Davies JHFL, Schaltegger U (2019) Calibrating chemical abrasion_ Its effects on zircon crystal structure, chemical composition and UPb age. Chem Geol 511:1–10. https://doi.org/10.1016/j.chemgeo.2019.02.026

Wingate MTD, Compston W (2000) Crystal orientation effects during ion microprobe U–Pb analysis of baddeleyite. Chem Geol 168:75–97. https://doi.org/10.1016/S0009-2541(00)00184-4

Woodhead JD, Hergt JM, Simonson BM (1998) Isotopic dating of an Archean bolide impact horizon, Hamersley basin, Western Australia. Geology 26:47–50. https://doi.org/10.1130/0091-7613(1998)026%3c0047:IDOAAB%3e2.3.CO;2

Wotzlaw JF, Schaltegger U, Frick DA, Dungan MA, Gerdes A, Günther D (2013) Tracking the evolution of large-volume silicic magma reservoirs from assembly to supereruption. Geology 41:867–870. https://doi.org/10.1130/G34366.1

Wotzlaw J-F, Bindeman IN, Watts KE, Schmitt AK, Caricchi L, Schaltegger U (2014) Linking rapid magma reservoir assembly and eruption trigger mechanisms at evolved Yellowstone-type supervolcanoes. Geology 42:807–810

Wotzlaw J-F, Bindeman IN, Stern RA, D'Abzac F-X, Schaltegger U (2015) Rapid heterogeneous assembly of multiple magma reservoirs prior to Yellowstone supereruptions. Sci Rep 5:14026. https://doi.org/10.1038/srep14026

Wu F-Y, Yang Y-H, Li Q-L, Mitchell RH, Dawson JB, Brandl G, Yuhara M (2011) In situ determination of U–Pb ages and Sr–Nd–Hf isotopic constraints on the petrogenesis of the Phalaborwa carbonatite Complex, South Africa. Lithos 127:309–322

Wu F-Y, Arzamastsev AA, Mitchell RH, Li Q-L, Sun J, Yang Y-H, Wang R-C (2013a) Emplacement age and Sr–Nd isotopic compositions of the Afrikanda alkaline ultramafic complex, Kola Peninsula, Russia. Chem Geol 353:210–229. https://doi.org/10.1016/j.chemgeo.2012.09.027

Wu F-Y, Mitchell RH, Li Q-L, Sun J, Liu C-Z, Yang Y-H (2013b) In situ U–Pb age determination and Sr–Nd isotopic analysis of perovskite from the Premier (Cullinan) kimberlite, South Africa. Chem Geol 353:83–95. https://doi.org/10.1016/j.chemgeo.2012.06.002

Yang W-B, Niu H-C, Shan Q, Sun W-D, Zhang H, Li N-B, Jiang Y-H, Yu X-Y (2013) Geochemistry of magmatic and hydrothermal zircon from the highly evolved Baerzhe alkaline granite: implications for Zr–REE–Nb mineralization. Miner Depos 49:451–470. https://doi.org/10.1007/s00126-013-0504-1

York D (1968) Least squares fitting of a straight line with correlated errors. Earth Planet Sci Lett 5:320–324. https://doi.org/10.1016/S0012-821X(68)80059-7

Yuan S, Peng J, Hao S, Li H, Geng J, Zhang D (2011) In situ LA-MC-ICP-MS and ID-TIMS U–Pb geochronology of cassiterite in the giant Furong tin deposit, Hunan Province, South China: new constraints on the timing of tin–polymetallic mineralization. Ore Geol Rev 43:235–242. https://doi.org/10.1016/j.oregeorev.2011.08.002

Zack T, Stockli DF, Luvizotto GL, Barth MG, Belousova E, Wolfe MR, Hinton RW (2011) In situ U–Pb rutile dating by LA-ICP-MS: ^{208}Pb correction and prospects for geological applications. Contrib Mineral Petrol 162:515–530. https://doi.org/10.1007/s00410-011-0609-4

Zartman RE, Smith JV (2009) Mineralogy and U–Th–Pb age of a uranium-bearing jasperoid vein, Sunshine Mine, Coeur d'Alene district, Idaho, USA. Chem Geol 261:185–195. https://doi.org/10.1016/j.chemgeo.2008.09.006

Zeh A, Ovtcharova M, Wilson AH, Schaltegger U (2015) The Bushveld Complex was emplaced and cooled in less than one million years—results of zirconology, and geotectonic implications. Earth Planet Sci Lett 418:103–114. https://doi.org/10.1016/j.epsl.2015.02.035

Zhang D, Zhang Z, Santosh M, Cheng Z, He H, Kang J (2013) Perovskite and baddeleyite from kimberlitic intrusions in the Tarim large igneous province signal the onset of an end-Carboniferous mantle plume. Earth Planet Sci Lett 361:238–248. https://doi.org/10.1016/j.epsl.2012.10.034

Zhang D, Peng J, Coulson IM, Hou L, Li S (2014) Cassiterite U–Pb and muscovite ^{40}Ar–^{39}Ar age constraints on the timing of mineralization in the Xuebaoding Sn–W–Be deposit, western China. Ore Geol Rev 62:315–322. https://doi.org/10.1016/j.oregeorev.2014.04.011

Zi J-W, Rasmussen B, Muhling JR, Fletcher IR, Thorne AM, Johnson SP, Cutten HN, Dunkley DJ, Korhonen FJ (2015) In situ U–Pb geochronology of xenotime and monazite from the Abra polymetallic deposit in the Capricorn Orogen, Australia: dating hydrothermal mineralization and fluid flow in a long-lived crustal structure. Precambrian Res 260:91–112. https://doi.org/10.1016/j.precamres.2015.01.010

The ^{187}Re-^{187}Os and ^{190}Pt-^{186}Os Radiogenic Isotope Systems: Techniques and Applications to Metallogenic Systems

Marc D. Norman

Abstract

Rhenium, Os, and Pt are redox sensitive elements that are concentrated in highly reducing environments such as those associated with black shales but mobile under more oxidizing conditions such as those associated with arc volcanism. They are chalcophile in many terrestrial ore-forming environments, and their isotopic systematics provide unique opportunities to date the formation of sulfide ore deposits and understand their petrogenesis. Fractional crystallization of magmatic sulfide ores generates primary variations in Re/Os and Pt/Os that allow mineral and whole rock isochron ages to be determined and discrimination of crustal and mantle sources based on initial Os isotopic compositions. Molybdenite is especially well suited for geochronology due to its high Re/Os and resistance to resetting. Rhenium concentrations in molybdenite tend to reflect the composition or provenance of the ore-forming fluids, with higher concentrations associated with more primitive sources or more oxidized fluids and lower concentrations with more evolved and/or reduced conditions, although local and regional factors also have a significant influence. Many studies have used pyrite for dating but its typically low Re concentration, variable initial Os isotopic composition (reflecting fluid mxing), and susceptibility to re-equilibration makes its use as a geochronometer problematic in many cases. Other sulfide minerals such as bornite and arsenopyrite have shown promise for Re–Os isotope geochronology but additional studies are needed to evaluate their broader applicability for dating of ore deposits. The isobaric beta decay of parent isotope ^{187}Re to ^{187}Os has restricted investigation of this system by microbeam techniques such as ion microprobe or laser ablation mass spectrometry, especially for geochronology. This requires either chemically processing the sample to separate the elements or novel techniques such as collision-cells that preferentially ionize the Re and Os during the analysis. Thermal ionization mass spectrometry (TIMS) and inductively-coupled plasma mass spectrometry (ICPMS) are the most widely applied techniques for Re-Pt-Os isotopic analyses. Specialized techniques for sample digestion to ensure redox equilibrium between Os in the sample and the isotopically enriched spikes used for isotope dilution measurements are typically required. This chapter briefly reviews development of the ^{187}Re-^{187}Os and ^{190}Pt-^{186}Os isotopic systems for earth science, physico-chemical controls

M. D. Norman (✉)
Research School of Earth Sciences, The Australian National University, Canberra, ACT 2601, Australia
e-mail: marc.norman@anu.edu.au

D. Huston and J. Gutzmer (eds.), *Isotopes in Economic Geology, Metallogenesis and Exploration*, Mineral Resource Reviews, https://doi.org/10.1007/978-3-031-27897-6_4

on their behavior in ore-forming environments, and applications to metallogenic systems.

1 Introduction

The ^{187}Re-^{187}Os isotopic system has a number of characteristics that distinguish it from other commonly applied radiogenic isotopic systems and make it especially useful for ore deposit research. The system is based on the long-lived β-decay of ^{187}Re to ^{187}Os. Both Re and Os tend to be chalcophile or siderophile when sulfide or metallic melts are present, leading to pronounced enrichments of these elements in magmatic sulfide deposits. During partial melting of the Earth's mantle, Os behaves compatibly and is retained in the depleted residue. In contrast, Re is moderately incompatible and is progressively extracted into the crust. This fractionation of Re/Os between the crust and depleted mantle has produced large variations in the proportion of ^{187}Os (expressed as $^{187}Os/^{188}Os$) in various reservoirs through time, with the crust evolving to highly radiogenic values (high Re/Os and $^{187}Os/^{188}Os$) and strongly depleted mantle evolving to less radiogenic compositions (Fig. 1).

Osmium isotopes are therefore sensitive tracers of the relative proportions of crustal and mantle inputs to ore systems, as well as being useful for geochronology. In addition, both elements are redox sensitive, being immobilized in organic-rich reducing environments such as those associated with black shales, but mobile in more oxidizing environments such as those associated with arc volcanism and near-surface magmatic outgassing. As a consequence of these diverse characteristics, Os isotopes have been applied to a wide variety of problems in earth science. However, both elements occur in very low concentrations in most common rock types (ppb or less), making the analysis and broader application of the Re–Os system a challenging task. This chapter briefly reviews aspects of the ^{187}Re-^{187}Os system that are especially relevant to ore deposits, including sulfide-related geochronology and source tracing. The ^{190}Pt-^{186}Os decay scheme is mentioned, although it has so far found relatively limited applications to ore systems (Sun et al. 2003d). The articles cited here are intended to be illustrative rather than a comprehensive review of the literature. Reviews of the Re–Os isotopic system of particular relevance to ore deposits have been presented by Shirey and Walker (1998), Lambert et al. (1999), Chesley (1999), Carlson (2005), Stein (2014), and Stein and Hannah (2015). Barnes and Ripley (2016) review the application of Re–Os isotopes to magmatic sulfide deposits, and considerable relevant information is provided by other chapters in that volume.

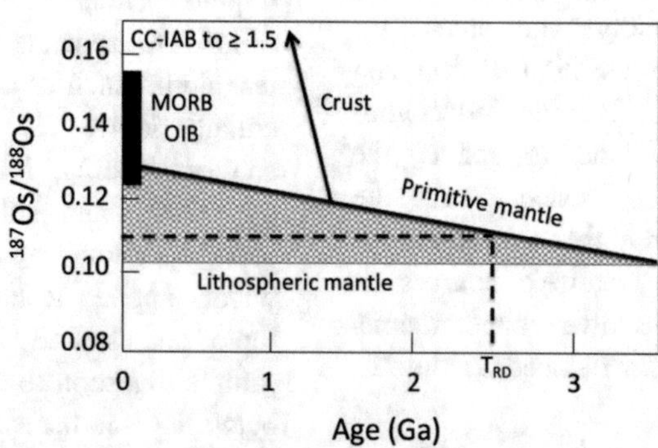

Fig. 1 Time versus $^{187}Os/^{188}Os$ ratios for various reservoirs in the Earth. The line labeled 'primitive mantle' is representative of much of the Earth's fertile convecting mantle through time. The black box shows representative compositions of modern mid-ocean ridge basalts (MORB) and ocean islnd basalts (OIB). Recently erupted island arc basalts (IAB) have $^{187}Os/^{188}Os$ compositions that range from MORB-like to highly radiogenic values. 'Crust' includes basaltic and continental crust. The stippled field is occupied primarily by strongly depleted lithospheric mantle. The dashed line illustrates Re depletion model ages (T_{RD})

2 Background

Osmium has seven naturally occurring isotopes (Table 1; Rosman and Taylor 1998), all of which can be considered stable (^{186}Os decays to ^{182}W with a half-life of 2×10^{15} years; Arblaster 2004). The minor isotopes ^{186}Os and ^{187}Os are radioactive decay products of ^{190}Pt and ^{187}Re, respectively, and these isotopic systems provide the basis for geochronology and source tracing as applied to ore deposits and related systems. Naldrett and Libby (1948) established the weak radioactivity of rhenium and showed that it was a beta decay of ^{187}Re to ^{187}Os. Hintenberger et al. (1954) showed that the osmium in a molybdenite was $\geq$ 99.5% radiogenic ^{187}Os, in contrast to the $\leq$ 2% in common Os, thereby establishing the ^{187}Re-^{187}Os system as a possible geochronometer. Subsequent studies showed that geologically plausible ages could be obtained from some molybdenites (Herr et al. 1968) but the methods then available were hindered by poor sensitivity, the resulting requirement for very large samples (1–10 g), and uncertainties in both the isotopic measurements and the ^{187}Re half-life. Isotope dilution measurements of Re concentrations in a large suite of Australian molybdenites (Riley 1967) demonstrated an extremely wide range of Re concentrations (0.25–1690 ppm), and neutron activation analyses of Australian molybdenites by Morgan et al. (1968) confirmed their high Re/non-radiogenic Os ratios and pointed to unusually low Re concentrations of molybdenites from Tasmania compared to those from mainland Australia.

Table 1 The naturally occurring isotopes of osmium

Isotope	Abundance (atomic %)
^{184}Os	0.02
^{186}Os	1.59
^{187}Os	1.96
^{188}Os	13.24
^{189}Os	16.15
^{190}Os	26.26
^{192}Os	40.78

The modern era for application of Re–Os isotopes to source tracing and ore deposit geochronology began with Allègre and Luck (1980) and Luck and Allègre (1982). Allègre and Luck (1980) established the Os isotopic evolution trend of the mantle based on compositions of samples of osmiridium with a range of assumed ages, and they showed that the much of the mantle apparently has evolved with a near-chondritic Re/Os since the origin of the Earth. This was surprising considering the complex history of mantle dynamics that was already known from Sr, Nd, and Pb isotopes, and it provided the basis for calculating mantle evolution model ages based on initial Os isotopic compositions. Luck and Allègre (1982) showed that geologically reasonable dates with useful precision could be obtained from molybdenite with a wide range of Re concentrations, although a few exceptions were noted.

3 The ^{187}Re-^{187}Os and ^{190}Pt-^{186}Os Isotope Systems

The ^{187}Re decay constant has been constrained by both direct counting and calibration against other geochronometers (reviewed by Selby et al. 2007a). Direct counting experiments have yielded uncertainties of several percent. The most widely used value in geosciences is 1.666×10^{-11} y^{-1}, based on a best-fit for iron meteorites with an assumed age of 4557.8 $\pm$ 0.4 Ma (Smoliar et al. 1996). Selby et al. (2007a) re-evaluated the λ^{187}Re based on a comparison of molybdenite ages against U–Pb ages of magmatic zircons that could reasonably be considered as coeval with the molybdenite. They proposed a value of $1.6668 \pm 0.0034 \times 10^{-11}$ y^{-1}, within uncertainty of the Smoliar et al. (1996) value. Begemann et al. (2001) also reviewed the half-lives for commonly applied isotopic geochronometers.

Macfarlane and Kohman (1961) established ^{190}Pt as an alpha emitter using enriched isotopes. Walker et al. (1997) demonstrated the potential of the ^{190}Pt-^{186}Os system for dating platinum-group mineralization through a study of ores from the Noril'sk region of the Siberian flood basalts. Begemann et al. (2001), Cook et al. (2004), and Tavares et al. (2006) reviewed the status of the ^{190}Pt decay constant and concluded that there are significant discrepancies between values obtained by direct counting and theoretical approaches compared to calibrations against other isotopic geochronometers such as ^{187}Re-^{187}Os and U–Pb, and that these discrepancies are much larger than the precision of $\sim$1–2% typically obtained by the latter. Begemann et al. (2001) recalculated the λ^{190}Pt proposed by Walker et al. (1997) to a value of 1.477×10^{-12} y^{-1} in order to account for a revised atomic percentage of ^{190}Pt. Cook et al. (2004) found that a value of 1.41×10^{-12} y^{-1} with an uncertainty of $\pm$ 1–2% provides the best match between ^{187}Re-^{187}Os and ^{190}Pt-^{186}Os isochron ages of magmatic iron meteorites. Additional work in this area is needed to refine λ^{190}Pt and account for the discrepancy between the geological versus experimental and theoretical calibrations.

4 Analytical Methods

Reisberg and Meisel (2002) and Meisel et al (2003) provide comprehensive reviews of analytical techniques used for Re–Os isotopic analysis. See also Walker and Fassett (1986) and Shirey and Walker (1998) for useful summaries of historical developments. Meisel and Horan (2016) provide a recent review of analytical methods for platinum-group elements more generally that includes information about Re, Os, and Pt. Here we provide a brief review of the development of analytical techniques for Re–Os isotopic analysis.

4.1 Instrumental Techniques

As the parent and daughter isotopes both have effectively the same mass, either chemical separations or differential ionization of the Re and Os are essential for geochronology. This isobaric relationship effectively rules out commonly available microbeam techniques such as ion microprobe or laser ablation ICPMS for in-situ dating of high Re/Os phases such as molybdenite. Those techniques have, however, been applied to source tracing of osmiridiums and other PGE-bearing minerals with low Re/Os ratios (Allègre and Luck 1980; Hart and Kinloch 1989; Hattori and Hart 1991; Meibom et al. 2002, 2004; Hirata et al. 1998; Pearson et al. 2002; Ahmed et al. 2006). Hogmalm et al. (2019) presented a novel technique using collision cells and ICP tandem mass spectrometry for the online separation of Re from Os during laser ablation analyses of molybdenite, but the method has so far not been widely applied to ore deposit geochronology.

The low concentrations and high ionization potential of both Re (7.9 eV) and Os (8.7 eV) presented particular challenges to the development of geochemically useful techniques. A recurring theme in the development of Os isotopic analysis is to utilize the volatile character of the compound OsO_4 (osmium tetroxide). The boiling point of OsO_4 is $\sim$130 °C but it can be distilled at much lower temperatures. It is also highly poisonous and reacts rapidly with organics such as those comprising various body parts of laboratory analysts. Early studies ionized OsO_4 by electron bombardment for introduction into a magnetic sector mass spectrometer (Nier 1937). Herr et al. (1968) applied this technique to measure ^{187}Os in molybdenites for Re–Os dating, with Os concentrations quantified by isotope dilution using a ^{190}Os spike and Re concentrations measured by neutron activation (e.g., Morgan 1965). These pioneering techniques provided reasonable precision ($\sim$3% for ^{187}Os/^{192}Os quoted by Nier 1937) but they required large sample sizes (e.g., 1–10 g) and ppm quantities of Os for the measurement.

A significant advance was the development of techniques for Re–Os isotopic analysis by secondary ionization mass spectrometry (SIMS or ion microprobe), either in situ for phases such as osmiridium or following chemical separations for whole rock analysis (Luck et al. 1980; Luck and Allègre 1983). They obtained precision on the isotope ratios of $\leq$ 1% relative and substantially reduced the amount of Os needed for analysis. This was an important development because it revitalized interest in Re–Os as a geochemically important isotopic system. Luck and Allègre (1982) applied a modified version of this technique to determine ^{187}Re/^{187}Os ages of a suite of molybdenite samples. Building on this success and a variety of technological developments, several innovative techniques for Os isotopic analysis were trialled. These included laser ionization mass spectrometry (Englert and Herpers 1980; Simons 1983), resonance ionization mass spectrometry (Walker and Fassett 1986; Blum et al. 1990), inductively-coupled plasma mass spectrometry (Russ et al. 1987; Dickin et al. 1988; Richardson et al. 1989) and accelerator mass spectrometry (Fehn et al. 1986; Sie et al. 2002). Each of these instruments had certain advantages, but all of them became obsolete after development of the negative-ion thermal ionization technique (N-TIMS) described below. Notable here, however, is the concept of introducing OsO_4 gas directly into an inductively-coupled plasma mass spectrometer (ICP-MS) as described by Russ et al. (1987). That step has been revived for some applications to ore deposit geochronology, as also discussed below.

Thermal ionization mass spectrometry is based on the formation of ions of the elements or oxides during desorption from a hot filament, followed by focussing of the ions into a beam by electrostatic lenses and separation of the isotopes according to their mass/charge by an electromagnet or mass analyzer. The ion current for each mass is then measured by some type of detector, typically either an electron multiplier or a Faraday cup. Isotope ratios can be measured using a single collector by cycling the magnet to focus each mass in the detector, or all masses of interest can be collected simultaneously using a multi-collector array. Multi-collection provides better precision and faster analysis times, and is considered as state-of-the-art for high-precision isotope ratio analysis. Conventional isotope ratio analyses of elements like Sr, Nd, and Pb are based on creating positive ions but the high ionization potential of Os makes this impractical. In contrast, Os readily forms negative ions such as OsO_3^- at relatively low T (800–900 °C) when loaded on the filament with a Ba or La activator (Völkening et al. 1991; Creaser et al. 1991; Walczyk et al. 1991). Analytical precisions of $< 0.01\%$ on isotope ratios can be obtained from very small amounts (few ng) of Os by N-TIMS using multiple collection. A potential disadvantage to N-TIMS is the need to obtain highly purified aliquots of Os and Re for analysis. Reisberg and Meisel (2002) provide additional advice and details regarding analytical conditions.

An alternative approach is ionization via an inductively-coupled plasma (ICP). Argon plasmas have a first ionization potential of 15.8 eV compared to 8.7 eV for Os and 7.9 eV for Re, so Re and Os will be efficiently ionized. For isotopic analysis, ICP ionization sources are coupled with either a quadrupole mass analyser or a magnetic sector mass spectrometer. Sample introduction can be by solution aspiration via a nebulizer, a dry gas stream produced by laser ablation, or vapor phase sample introduction. Osmium is prone to memory effects during solution aspiration and chemical procedures have been developed to offset this (Schoenberg et al. 2000; Nowell et al. 2008a). In contrast, vapour phase introduction of OsO_4, also known as sparging, produces better intensities and fewer memory effects as well as providing an efficient separation of ^{187}Os from ^{187}Re without additional chemistry (Russ et al. 1987; Richardson et al. 1989; Gregoire 1990). Hassler et al. (2000) further developed the use of sparging for Os isotopic analysis by single-collector magnetic ICP-MS and obtained a precision similar to that of Russ et al. (1987).

A significant technological advance was the commercial availability of multi-collector magnetic sector ICPMS instruments (MC-ICPMS) as these provided the efficient ionization of an ICP with higher measurement precision provided by multiple Faraday cup collector arrays. Schoenberg et al. (2000), Norman et al. (2002), and Nozaki et al. (2012) describe sparging experiments for Os isotopic analysis using MC-ICPMS. Because the ICP effectively decomposes the oxide molecules, these measurements are made on the nominal masses of the element, rather than the oxides as for N-TIMS. While N-TIMS remains the method of choice for low-level Os isotopic analysis, MC-ICPMS can be useful for certain applications. For example, molybdenite samples from a variety of settings have been dated using a combination of sparging and solution aspiration for isotope dilution determinations of ^{187}Os and Re concentrations, respectively (Norman et al. 2004a; Armistead et al. 2017; Kemp et al. 2020; Wells et al. 2021). Magnetic sector ICP-MS instruments can be also be configured with multiple electron multiplier array detectors for measurement of less intense beams (Birck et al. 2016; Zhu et al. 2019).

Microbeam techniques such as electron microprobe, laser ablation ICPMS, and secondary ionization mass spectrometry (SIMS) have been used to examine the spatial distribution of Re and Os within individual grains and have found that many grains are remarkably heterogeneous (Zaccarini et al. 2014; Barra et al. 2017; Hnatyshin et al. 2020; Wells et al. 2021).

4.2 Chemical Procedures

As both Re and Os are typically trace constituents of geological materials (osmiridium and rheniite being notable exceptions), chemical methods for concentrating these elements from appropriate volumes of samples are usually necessary for the isotopic analysis. This requires recipes for dissolving or digesting the sample without losing volatile Os compounds or incompletely attacking potentially important phases like spinels. Because OsO_4 is volatile at low T, samples cannot be simply dissolved on a hotplate using oxidizing acid solutions such as HF-HNO_3, as commonly done for Rb–Sr or Sm–Nd isotopes. For geochronology and age-corrections, it is necessary to ensure complete equilibration between the enriched isotope spikes and natural Re and Os from the sample; this is especially challenging for Os due to its multiple oxidation states and different behaviours when analysed in oxidized versus reduced forms.

The most widely applied method of sample digestion for Re–Os isotopic measurements uses Carius tubes and mixed HNO_3–HCl acids in proportions 2:1 (inverse *aqua regia*). Carius tubes are thick-walled borosilicate glass tubes, sealed at both ends by a glassblower (Shirey and Walker 1995). When heated to temperatures of 240–260 °C, they generate large internal pressures and conditions that ensure spike-sample equilibration through complete oxidation of the Os and Re. An advantage of this method is that the Os (as OsO_4) can be sparged or distilled directly from the resulting solution. A disadvantage is that sample size typically is limited compared to other methods of decomposition such as fire assay. For example, < 0.5 g aliquots of sulfides are commonly used (e.g., Selby et al. 2009; Morelli et al. 2010; Hnaytshin et al. 2020) due in part to practical limits on the volume of acid required to fully oxidize the spike and sample (Frei et al. 1998), whereas 1–3 g of silicate samples can be digested by this technique (Shirey and Walker 1995; Meisel et al. 2001a). A variant of the technique is the high-pressure asher, which achieves even higher T and internal pressures resulting in faster and more complete sample dissolution (Meisel et al. 2001b). For most samples encountered in ore deposit studies, however, conventional Carius tube digestion seems adequate.

Rhenium-Os geochronology and isotope tracing can also be applied to sediment-hosted deposits and hydrocarbon systems through analysis of black shales and associated organics (e.g., reviews by Stein and Hannah 2015; Zeng et al. 2014; Gao et al. 2020), but these kinds of studies require specialized techniques to disentangle hydrogenous components associated with the organic phases from mixtures with detrital components. Selby and Creaser (2003) showed that the hydrogenous Re and Os can be extracted selectively from black shales using a carius tube digestion with CrO_3–H_2SO_4 instead of reverse aqua regia, and procedures for dating petroleum using extracted organic fractions were developed by Selby et al. (2007b), Sen and Peucker-Erinbrink (2014) and Georgiev et al. (2016). Economic applications of these techniques have been primarily to hydrocarbon systems (e.g., Cumming et al. 2014; Liu et al. 2019; DiMarzio et al. 2018; Meng et al. 2021; Li et al. 2021; Georgiev et al. 2021) although they appear to have considerable potential for application to sediment-hosted metallogenic systems as well, especially those associated with organic matter (e.g., Saintilan et al. 2019).

Two other approaches to sample digestion that have been applied to ore deposit studies are alkali fusion (Morgan and Walker 1989; Markey et al. 1998) and NiS fire assay (Hoffmann et al. 1978; Fehn et al. 1986; Ravizza and Pyle 1997; Savard et al. 2010). While the alkali fusion method can produce accurate and precise results, it is a complex, multi-step procedure that can be prone to incomplete spike-sample equilibration, incomplete dissolution of the sample, and higher procedural blanks compared to Carius tube digestion. In addition, the resulting fusion cake must be dissolved for separation of the Os and Re, and the procedure for this does not appear to be suitable for direct analysis. The NiS fire assay can cope with large samples (> 10 g), thereby overcoming nugget effects and sample heterogeneity, but yields and procedural blanks can be variable. Although Re can be poorly recovered by NiS fire assay, good results for isotope dilution measurements for Re by this method have been reported (Ravizza et al. 1991; Park et al. 2013). Maintaining a reducing environment during the fusion appears to be a key step to obtaining useful Re data by this method (Reisberg and Meisel 2002).

For N-TIMS analysis, it is necessary to obtain pure aliquots of Re and Os. Purification of the Os is a moderately complex process involving solvent extractions and microdistillation, the latter specifically designed for Os purification utilising the volatile nature of OsO_4 (Birck et al. 1997; Morgan et al. 1991; Cohen and Waters 1996; Shirey and Walker 1998; Reisberg and Meisel 2002). In this respect, sparging the OsO_4 directly into an ICP mass spectrometer is a more efficient approach as it provides a chemical separation of Os from the Re and matrix elements and an isotopic analysis in a single step, provided that the samples are appropriate. Purification of the Re is more straightforward as it is not volatile under ambient conditions and can be separated from matrix elements using conventional anion exchange chromatography.

4.3 Data Handling

Thermal ionization and ICP mass spectrometers rarely measure exactly the correct isotope ratio for any element. Thermal ionization is based on evaporation of a finite mass of the element of interest, which tends to produce a Raleigh distillation effect as light masses evaporate preferentially to heavier masses. This often causes the measured isotopic composition to change with time during a run, or to be offset from the actual values. The physics behind this type of process is reasonably well understood and can be corrected for by assuming a nominal ratio of two stable isotopes in the analyte and some form of the distillation law, typically exponential (Habfast 1983, 1998; Andreasen and Sharma 2009). For Os isotopic analysis by N-TIMS, the fractionation correction uses the oxide equivalent masses of either ^{192}Os/^{188}Os or, less commonly, ^{190}Os/^{192}Os. This method also requires corrections for oxygen isotopic compositions. In contrast to TIMS, ICPMS isotopic analysis usually relies on consumption of a constant composition sample, either as a solution or bulk extraction of a homogeneous phase as for sparging. In this case, the composition of the sample does not change during the analysis. However, ICP mass spectrometers typically measure isotopic compositions that are offset from the actual value by factors of $\sim 1\%$ and that vary as a regular function of mass (Albarede et al. 2004). The physical processes that are responsible for this mass bias are less well understood (e.g., Andrén et al. 2004), but corrections assuming fractionation laws similar to those used for TIMS appear to be adequate for most applications.

If the sample contains a sufficient amount of common Os, then the fractionation or mass bias corrections can be made based on the isotopes intrinsic to the sample. In this case, concentrations can be determined by spiking the sample with a single enriched isotope such as ^{190}Os for a conventional isotope dilution measurement. Alternatively, for samples with negligible amounts of initial Os, such as molybdenite, the intrinsic Os in the sample will be essentially all radiogenic and therefore monoisotopic (^{187}Os). In this case, the sample can be spiked with a well-calibrated solution of common Os or with a specially prepared double spike containing, for example, ^{188}Os and ^{190}Os in known proportions. Either approach allows both mass bias corrections and concentrations to be determined by isotope dilution. The advantage of the double spike method is that it allows a monitor of any initial Os in the sample using other masses, which can be useful if the sample is young or contains low concentrations of Re (Markey et al. 2003). A potential complication is mass-independent isotopic fractionation during MC-ICPMS analysis. Zhu et al. (2018) measured a series of Os standard solutions and suggested biases both positive and negative of up to $\sim$0.5% on ^{187}Os relative to ^{188}Os and ^{190}Os. However, such effects have not been found in other studies (Nanne et al. 2017) and the topic requires further investigation with particular attention to potential interferences (Birck et al. 2016). Fractionation or mass bias corrections for Re are more difficult because it has only 2 isotopes (mass 185 and 187). In this case, the corrections can be applied by normalization to a reference solution run under the same conditions as the samples (also known as "standard-sample bracketing"), or by spiking the sample with another element of known isotopic composition and using that element for the fractionation or mass bias corrections. For example, Miller et al. (2009) used W for mass bias corrections of Re isotopes analysed by MC-ICPMS.

Modern geochemical studies typically report Os isotopic data as ^{187}Os/^{188}Os and ^{186}Os/^{188}Os ratios as these reflect the time-integrated parent/daughter ratios. The ^{187}Os/^{188}Os ratios, either measured or age-corrected to initial values, are often reported as the quantity γ^{187}Os (gamma), which is the percent deviation of the data from the chondritic evolution curve. In contrast, ^{186}Os/^{188}Os ratios are reported as ε^{186}Os (epsilon), or deviations from chondritic in parts per 10,000. As discussed below, this is useful because the Earth's mantle appears to have evolved with near-chondritic Re/Os and Pt/Os since at least 3.8 Ga (Bennett et al. 2002). A source of potential confusion was the reporting of Os isotopic data as ^{187}Os/^{186}Os ratios in some of the early literature. This practice was discontinued once high-precision measurements demonstrated variations in ^{186}Os due to decay of ^{190}Pt.

5 Geochemical and Mineral Systems: Controls on Re–Os Behaviour in Magmatic and Hydrothermal Environments

5.1 Silicate Mineral and Oxide Partitioning

The partitioning behavior of Re, Os and other PGE during mantle melting is controlled primarily by the presence or absence of sulfide in the source and to a lesser extent by partitioning into other silicate and oxides phases (Brenan et al. 2016). Rhenium becomes increasingly incompatible under more oxidizing conditions, such as those applicable to volcanic arcs, due to the larger fractions of sulphate relative to sulfide, and Re^{6+} relative to Re^{4+} (Brenan et al. 2003; Fonseca et al. 2007; Mallmann and O'Neill 2007). Osmium is only sparingly soluble in basaltic melts under all conditions, with partitioning into alloys and spinel contributing to its retention in the mantle even after sulfide is exhausted. A long-standing observation is the high concentrations of Os, Ir, and other refractory PGE and low concentrations of Re in Cr-spinels relative to silicates (Page and Talkigton 1984; Zhou et al. 1998; Park et al. 2012; Lorand and Luget 2016). This appears to reflect a combination of mineral-melt partitioning, and preferential inclusion of platinum-group minerals such as laurite-erlichmanite (RuS_2–OsS_2) into the spinel, possibly due to small-scale redox gradients at the crystal-melt interface during melting (Brenan et al. 2016; Lorand and Luget 2016; O'Driscoll and Gonzalez-Jimenez 2016). As Cr-spinel is resistant against hydrothermal alteration, it has been used to recover the initial Os isotopic compositions of serpentinites, greenstones, and metamorphosed ultramafic rocks where the isotopic composition of the whole rock has been compromised (Bennett et al. 2002; Standish et al. 2002; Walker et al 2002).

Osmium is largely incompatible in silicate minerals (Brenan et al. 2016), but it follows other compatible elements such as MgO, Ni, and Cr during basalt fractionation (Hart and Ravizza 1996). This leads to very low concentrations ($\leq$ 0.01 ppb) in moderately evolved basalts, making them prone to compositional modification by crustal contamination or alteration (Lassiter and Luhr 2001; Gannoun et al. 2007, 2016; Pitcher et al 2009). As noted above, the compatibility of Os seems to reflect a physicochemical association with Cr-spinel, although the exact mechanism responsible for this association remains somewhat obscure (Lorand and Luget 2016; Gannoun et al. 2016).

As an incompatible element, Re concentrations tend to increase during fractional crystallization of mafic and intermediate magmas (Righter et al. 2008; Pitcher et al. 2009; Li 2014). It can, however, be lost from these magmas as a volatile species during near-surface degassing (Lassiter 2003; Sun et al. 2003a, b, c, 2004; Norman et al. 2004b; Righter et al. 2008; Pitcher et al. 2009). Failure to recognize Re loss due to outgassing could lead to incorrect inferences regarding source composition and sulfide saturation (Hauri and Hart 1997; Bennett et al. 2000). In more evolved magmas, Re can be compatible in FeTi oxides, probably due to substitution of Re^{4+} for Ti^{4+} (Righter et al. 1998; Mallmann and O'Neill 2007; Li 2014; Park et al. 2013). Righter et al. (1998) estimated magnetite/melt partition coefficients of 20–50 in the absence of sulfide, plausibly accounting for the systematic decrease in Re contents of evolved arc magmas (Li 2014). Crystallization of magnetite can also promote fluid exsolution (Sun et al. 2004) or the saturation of Cu-Au sulfides from oxidized arc magmas through conversion of sulfate in the melt to sulfide via the reaction: $SO_4^{2-} + 8Fe^{2+}O = S^{2-} + 8Fe^{3+}O_{1.5}$, a process that has been referred to as the 'magnetite crisis' (Jenner et al. 2010).

5.2 Sulfide Melt—Silicate Melt Partitioning

The siderophile behaviour of Re, Os, and Pt is clearly demonstrated by partitioning experiments that yield liquid metal/liquid silicate partition coefficients of 10^9 to 10^{12} at 1200–1400 °C and fO_2 = IW (regression of Brenan et al. 2016). Temperature and redox (fO_2) appear to be the primary parameters that control metal-silicate partitioning, with silicate melt composition as a secondary control and pressure not very important. Metal-silicate partitioning is applicable primarily to planetary scale processes such as core formation.

In crustal magmatic environments, sulfide melt/silicate melt immiscibility and solid sulfide/liquid sulfide partitioning is more relevant. The partitioning of Re, Os, Pt, and other platinum-group elements (PGE) is complicated, however, because fS_2, fO_2, and silicate melt composition all exert significant controls on the oxidation states of the elements as well as the composition and stability of the sulfide phase. For example, Re and Os become increasingly soluble in silicate melts with increasing fO_2, consistent with predominantly Re^{6+} and Os^{4+} species in most terrestrial magmas (Brenan et al. 2016). In contrast, a variety of Pt species may be present in silicate melts, including Pt^0, Pt^{2+}, and Pt^{4+} with increasing fO_2 (Brenan et al. 2016). The solubility of Pt appears to decrease with increasing SiO_2 in the melt (Borisov and Danyushevsky 2011; Brenan et al. 2016) but similar experiments have not been done for Re and Os.

Experimental measurements of sulfide melt/silicate melt partition coefficients ($D_{SUL/SIL}$) for Re have demonstrated a strong influence on relative fO_2 and fS_2 (Fonseca et al. 2007; Brenan 2008; Brenan et al. 2016). In relatively reducing and sulphur-rich magmas, possibly relevant to MORB, Re is chalcophile with $D_{SUL/SIL}$ on the order of $\sim 10^2$–10^3, whereas in more oxidizing conditions it appears to be lithophile with $D_{SUL/SIL}$ <<1. Brenan et al. (2016) describe this as an exchange reaction between oxide and sulfide species:

$$Re_xO_{y(silicate)} + z/2 \times S_2 = Re_xS_{z(sulfide)} + y/2 \times O_2.$$

in which the fO_2 and fS_2 (expressed as $0.5\log fS_2 - 0.5\log fO_2$) controls the metal/sulfur composition of the sulfide melt. Kiseeva and Wood (2013) proposed an alternate approach that makes use of an exchange reaction involving the element of interest and Fe in the sulfide or silicate. Their formulation does not have an explicit dependence on fO_2 or fS_2.

Experimental determinations of $D_{SUL/SIL}$ for Os, Pt, and other PGE have been plagued by concerns over nuggets and other experimental effects, with 'best guess' estimates of $\sim 10^4$–10^7 (see discussion in Brenan et al. 2016; Barnes and Ripley 2016). Similar $D_{SUL/SIL}$ values have been inferred from the Re and Os compositions of sulfide globules in MORB and submarine OIB (Roy-Barman et al. 1998). These $D_{SUL/SIL}$ values imply the potential for substantial depletion of Re relative to Os and Pt during melting and/or crystallization when sulfide is present, consistent with the compositions of basaltic magmas and mantle xenoliths as described in a subsequent section. However, uncertainties over bulk sulfide contents, redox states, and possible sulfide mobility in the mantle (Gaetani and Grove 1999; Rehkämper et al. 1999) have complicated attempts to model the magmatic behaviour of these elements precisely.

5.3 Sulfide Mineral-Melt Partitioning

Once a sulfide melt forms, fractional crystallization of that melt can produce large variations in the relative abundances of Re, Pt, and Os, which can be exploited for geochronology. On crystallization of a homogeneous sulfide liquid, the first phase to form is a Fe-rich monosulfide solid solution (MSS) that is typically enriched in Os, Re, Ir, Ru, and Rh. Further crystallization of the residual melt produces a Cu-rich intermediate solid solution (ISS) that becomes enriched in Pt, Pd, and other incompatible chalcophile elements. Experimentally determined values of $D_{MSS/sulfide\ melt}$ show that Os is more compatible than Re in the MSS (D_{Os} ~5–20, D_{Re} ~2–10) whereas Pt is incompatible (D_{Pt}~0.03–0.2) (Brenan et al. 2016). This implies that MSS cumulates will have relatively high Re and Os concentrations but only moderate fractionations of Re/Os and very low Pt/Os ratios.

Fractional crystallization of sulfide melts has been recognized in Ni–Cu–PGE deposits associated with the Sudbury Igneous Complex (SIC; Naldrett et al. 1982, 1999), basaltic intrusions in the Noril'sk district (Czamanske et al. 1992), and some komatiite-related ores (Barnes et al. 1997). Smaller scale examples of these process have also been observed in sulfide globules from MORB and mantle xenoliths (Brenan et al. 2016; Barnes and Ripley 2016). Sulfide ores in the Noril'sk district are mineralogically and chemically complex with broad ranges and strong spatial zonation in the absolute and relative abundances of Cu, Ni, PGE and related mineral assemblages (Czamanske et al. 1992; Stekhin 1994; Torgashin 1994; Distler 1994; Naldrett et al. 1994; Zientek et al. 1994). This compositional diversity is reflected in the Re–Os-Pt isotopic data, with, for example, $^{187}Re/^{188}Os$ ratios of the ores (S > 30 wt%) ranging from ~4.3 to 930 (Walker et al. 1994; Malitch and Latypov 2011).

While the isotopic data provide information about ages and sources of magmatic sulfide deposits, constraints on the geochemical evolution of sulfide melts can be obtained from the concentrations of Re and Os, with the Noril'sk district ores being especially suitable for this due to their wide compositional variations. As shown in Fig. 2, Re/Os increases whereas concentrations of both Re and Os decrease by 2–3 orders of magnitude with increasing Cu content (data from Walker et al. 1994). Following the approach of Naldrett et al. (1994), these compositional variations were modelled as fractional crystallization of an MSS phase from an evolving

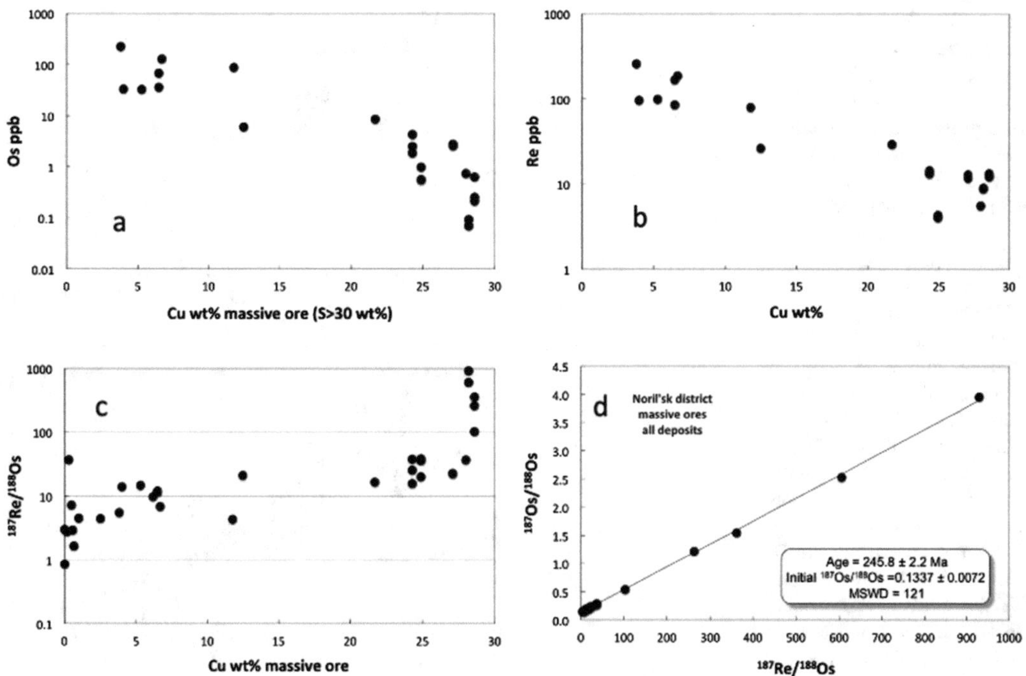

Fig. 2 Copper (wt%) in Noril'sk district massive ores versus **a** Os concentrations, **b** Re concentrations, and **c** ^{187}Re/^{188}Os ratios. **d** ^{187}Re/^{188}Os versus ^{187}Os/^{188}Os isochron for Noril'sk district massive ores. Data from Walker et al. (1994)

sulfide melt (Fig. 3). In this model, the ores represent predominantly cumulus MSS with variable proportions of trapped sulfide melt. The models presented in Fig. 3 assume initial melt compositions of 5 wt% for Cu, 500 ppb for Re, and 250 ppb for Os, and $D_{MSS/sulfide\ melt} = 0.1$ for Cu, 3.5 for Re, and 4 for Os. The values of initial Cu concentration and D_{Cu} are from Naldrett et al. (1994). For Re and Os, the values provide empirical fits to the data; however, these values for Os are similar to those inferred by Naldrett et al. (1994) for the geochemically similar element Ir, and the D_{Re} and D_{Os} values are similar to values measured experimentally by Brenan (2002) and those used by Lambert et al. (1999) to model the evolution of magmatic sulfide deposits. Overall, the observed variations of Re and Os in the Noril'sk ores are broadly consistent with previous conclusions based on other PGE that the massive ores are predominantly MSS cumulates with variable proportions of trapped sulfide melt (Naldrett et al. 1994; Zientek et al. 1994). The strong increase in Re/Os at constant Cu concentration of ~30 wt% probably reflects crystallization of an intermediate solid solution (ISS) phase from the highly fractionated sulfide liquid as this would increase D_{Cu} dramatically and prevent further enrichments of Cu in the melt (Naldrett et al. 1994).

5.4 Diffusion and Closure Temperatures

Limited data for diffusion and closure temperatures are available for Re and Os in geological environments and ore-forming systems. Brenan et al. (2000) conducted a study of Os diffusion in pyrite and pyrrhotite. They calculated closure temperatures of ~300–400 °C for 10–1000 micron grains of pyrrhotite, and 'core retention times' of < 0.5 Ma for a 500 micron grain of pyrrhotite. This implies that pyrrhotite will tend to equilibrate rapidly during relatively mild thermal events, and so may be unsuitable for Re–Os dating. In contrast, pyrite appears to be more

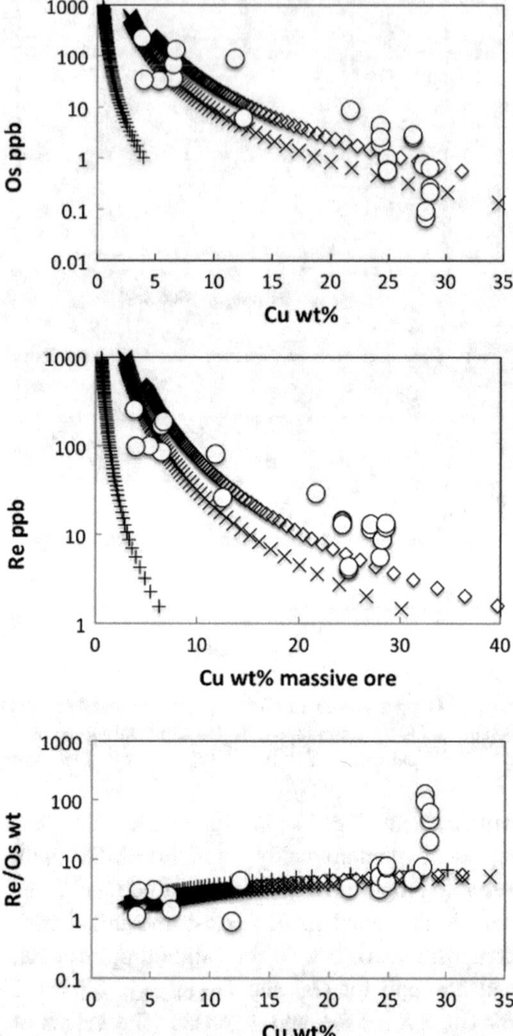

Fig. 3 Rhenium-osmium compositions of Norilsk massive ores modelled as fractional crystallization of an MSS phase from an evolving sulfide melt. Model trends show the composition of the MSS cumulate (+), the complementary sulfide residual liquid (open diamonds), and a mixture of MSS plus 50% trapped residual sulfide liquid (X). Markers indicate 1% increments of residual melt remaining for F = 0.12–0.99. Initial contents of the sulfide melt are assumed to be 5 wt% Cu, 500 ppb Re, and 250 ppb Os. MSS-sulfide liquid distribution coefficients are D (Cu) = 0.1, D(Re) = 3.5, and D(Os) = 4 (after Naldret et al. 1994). Noril'sk data (open circles) from Walker et al. (1994)

robust. Brenan et al. (2000) observed Os uptake by pyrite only during new growth with no evidence of diffusion at temperatures of 400–600 ° C, and they calculate a minimum 'core retention time' of > 10 Ma for Os in pyrite. Suzuki et al. (1996) estimated a closure temperatures for molybdenite of ~500 °C, similar to that of Rb–Sr and considerably higher than that of K–Ar in granitic systems.

Stein et al. (1998) showed that the primary crystallization ages of molybdenite associated with 2.7 Ga intrusions in the Fennoscandian Shield were preserved during subsequent thermal metamorphism and deformation. In that study, pyrite analyses recorded a younger age, reflecting post-metamorphic recrystallization. Bingen and Stein (2003) noted that molybdenite that formed during granulite-facies metamorphism at 973 Ma was not affected by subsequent contact metamorphism associated with anorthosite intrusion at 930 Ma, implying closure at conditions of 4.7 kb and 710 °C. Stein et al. (2003) reached a similar conclusion regarding the immunity of molybdenite to subsequent metamorphism. As emphasized by Stein (2014), however, chemical exchange may be more important than a thermal threshold for Re–Os isotopic closure. When it is present, molybdenite typically contains orders of magnitude more Re and Os than other phases, so while there would be a large chemical gradient, diffusive equilibration would be impaired by the relative stability of Os in molybdenite compared to many other sulfides (Stein et al. 2003; Takahashi et al. 2007). In some situations, minerals with low Os concentrations can acquire ^{187}Os by diffusion from coexisting molybdenite; examples of unrealistically old ages of chalcopyrite attributed to this process are described by Stein et al. (2003).

5.5 Transport and Deposition of Re and Os in Volcanogenic Hydrothermal Systems

Rhenium is readily transported in chloride-rich solutions at high fO_2 and neutral to acidic pH (Xiong and Wood 1999, 2001; Xiong et al. 2006). Its solubility increases with ionic strength of the solution, consistent with a chloride complex. In contrast, Re precipitates from sulfide-rich solutions such that mixing of oxidised, saline solutions with reduced sulfur should be an effective mechanism for depositing rhenium in sulfides, either in solid solution or incorporated as nanoparticles. A prediction would be that the Re contents of molybdenite (MoS_2) and related phases should be related to the fO_2 and salinity of the transporting fluid. Few studies have addressed the stability of Os in hydrothermal solutions, in part due to lack of thermodynamic data. In contrast to Re, Os-chloride complexes appear to be stable only at very low pH ($\leq$ 3) with oxygen-bearing, neutral, and sulfur-bearing species stable under relatively oxidizing to more reducing conditions, respectively (Mountain and Wood 1988). In addition, Os-bearing alloys and sulfides can form during hydrothermal alteration (e.g., 350–800 °C; Petrou and Economou-Eliopoulos 2009), sequestering the Os and further restricting its mobility in these environments.

Numerous Re–Os isotopic studies of active or recent ocean-floor hydrothermal systems have demonstrated mixing between seawater and crustal components, with seawater typically dominating the Os isotopic composition of the sulfides (Ravizza et al. 1996; Brügmann et al. 1998; Cave et al. 2003; Zeng et al. 2014). Osmium and Re concentrations also reflect mixing between oxidised seawater and reduced hydrothermal fluid, as well as variations in redox conditions within the system, creating variations in ^{187}Re/^{188}Os that can be useful for geochronology (Zeng et al. 2014). However, variations in initial ^{187}Os/^{188}Os related to mixing of seawater and primary hydrothermal fluids as well as post-depositional redistribution of Os and loss of Re during sulfide oxidation (Ravizza et al. 2001) would tend to introduce scatter and degrade any isochron relationship (Zeng et al. 2014). Nonetheless, Re–Os isochron ages have been obtained from a number of ancient volcanogenic massive sulfide (VMS) deposits, including the Iberian Pyrite Belt (Mathur et al. 1999; Munhá et al. 2005), the Urals VMS deposit (Gannoun et al. 2003; Tessalina et al 2008a), the Iimori Besshi-type Cu deposit (Nozaki et al. 2010), the Hitachi VMS deposit (Nozaki et al. 2014), and the Gacun Ag-rich VMS deposit (Hou et al. 2003) among others. At the Iimori and Hitachi deposits, the Re–Os system was not affected by later regional metamorphism and allowed the primary depositional age of the deposit to be established (Nozaki et al. 2010).

Both Re and Os can also be mobile as volatile species in magmatic environments depending on conditions. Osmium appears to be volatile primarily under oxidizing conditions and at high temperatures (> 1000 °C) (Wood 1987; Fleet and Wu 1993) whereas Re volatility is enhanced by the presence of Cl $\pm$ S as well as high fO_2 (MacKenzie and Canil 2006; Johnson and Canil 2011). A number of studies have documented Re and Os volatility in natural environments. Tessalina et al. (2008b) found high concentrations of both Re and Os in magmatic gasses collected at Kudryavy volcano, and they determined a Re–Os isochron age of 79 $\pm$ 11 years based on replicate analyses of the mineral rheniite (ReS_2). Rhenium loss by magmatic outgassing has also been demonstrated for arc volcanics (Sun et al. 2003a; Righter et al. 2008) and subaerial ocean island basalts (Lassiter 2003; Norman et al. 2004b).

5.6 Subduction Recycling

The efficiency of Os and Re recycling through subduction zones, which has implications for sources of metals in ore deposits associated with volcanic arcs, remains a topic of some debate. Brandon et al. (1996) and McInnes et al. (1999) provided evidence for incorporation of crustal Os into subduction-modified mantle xenoliths, but the genetic relationship between those xenoliths, arc basalts, and associated ore systems such as Lihir is unclear. Oxidizing conditions clearly favors mobilization of metals from the mantle to the crust, but the suggestion that slab-derived brines transport Os (Brandon et al. 1996; Borg et al. 2000) is not strongly supported by the experimental data on Os complexes (previous section). Alternatively, stabilization of Os in the mantle wedge might contribute to the recycling of crustal Os into the source regions of OIB (Borg et al. 2000; Suzuki et al. 2002).

Many arc basalts have radiogenic Os isotopic compositions but also very low Os concentrations, which makes them prone to modification by even small degrees of crustal interaction. Alves et al. (2002) recognized a broad negative correlation between Os concentrations and Re/Os ratios for a global set of arc lavas that parallels similar trends in MORB and OIB. They also found systematically higher ^{187}Os/^{188}Os in the lavas with individual arcs forming discrete linear arrays. This implies a mixing process between a less radiogenic component similar to the MORB source and crustal components that vary from arc to arc, but the authors were not able to distinguish between subduction contamination of the mantle versus magmatic assimilation. Subsequent studies have tended to favour high-level crustal contamination during evolution of arc magmas as the primary process responsible for their radiogenic Os isotopic compositions rather than a subduction modified mantle source (Hart et al. 2002; Woodhead and Brauns 2004; Turner et al. 2009; Bezard et al. 2015). In this case, the Os isotopic compositions of ore deposits associated with subduction environments may reflect crustal processes more than source characteristics.

6 Isotopic Dating and Source Tracing Using the ^{187}Re-^{187}Os and ^{190}Pt-^{186}Os Isotopic Systems

6.1 Isotopic Dating

As mentioned earlier, the ^{187}Re-^{187}Os and ^{190}Pt-^{186}Os systems are unique among commonly used radiogenic isotope decay schemes because the parent and daughter isotopes of both systems are chalcophile or siderophile when sulfides or metal are present. This provides an opportunity for dating the formation of ore minerals directly rather than relying on associated phases such as zircons or micas, whose relationship to the mineralisation might be unclear or affected by later events. Like most long-lived radiogenic isotope systems, ^{187}Re-^{187}Os and ^{190}Pt-^{186}Os can be used for geochronology and source tracing. Conventional isochrons are based on the correlation of parent/daughter isotopes versus radiogenic isotopic compositions with age, e.g., ^{187}Re/^{188}Os versus ^{187}Os/^{188}Os. The slope of this correlation is proportional to the time since the system closed and the y-intercept provides the initial ^{187}Os/^{188}Os of the system. Fundamentally, any process that geochemically fractionates Re or Pt from Os can be dated using this approach if cogenetic samples with a range of parent/daughter ratios are available or using individual phases with high Re/Os or Pt/Os provided the initial Os isotopic is known or can be assumed. Such processes could include sulfide immiscibility, fractional crystallization of either the sulfide melt or the silicate melt, or the in-situ crystallization of phases, typically sulfides, with high parent/daughter ratios. Because the half-life of ^{187}Re is relatively fast (similar to ^{87}Rb), measurable differences in ^{187}Os/^{188}Os can develop quickly depending on the Re/Os. For example, a basalt with a ^{187}Re/^{188}Os of 100, which would be representative of modern MORB or OIB, would develop measurable differences in ^{187}Os/^{188}Os relative to its initial composition within $\sim 10^4$ years, assuming an analytical precision of 0.01%. An impressive example of the

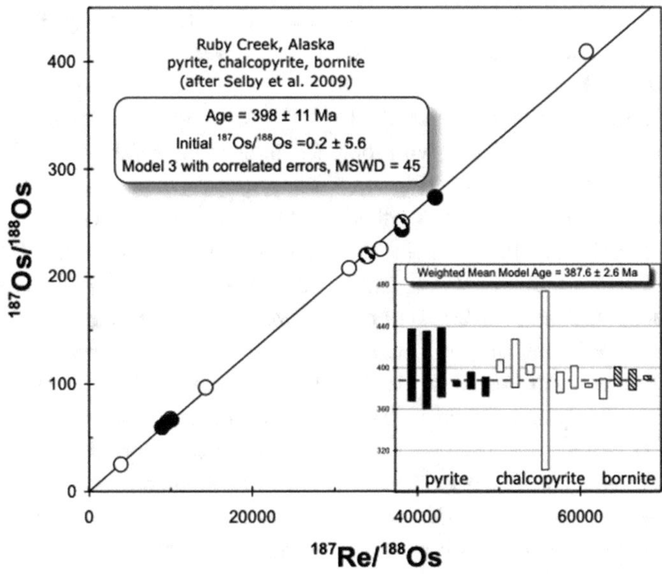

Fig. 4 A Re–Os isochron based on pyrite (black filled symbols), chalcopyrite (open symbols), and bornite (diagonal filled symbols) for the Ruby Creek, Alaska, deposit. Reproduced with permission from Selby et al. (2009); Copyright 2009 Society of Economic Geologists. The insert shows model ages obtained from each analysis. Size of the symbols on the isochron is fixed and not proportional to analytical uncertainties; these are indicated by the length of the bars shown for the model ages

ability to date young events is the absolute age of 79 ± 11 years based on fumarolic rheniite (ReS_2) by Tessalina et al. (2008b). Current uncertainty on the ^{187}Re decay constant is ~0.2% (Selby et al. 2007a), similar to that for ^{238}U (Villa et al. 2016).

Isochrons can be based on either 'whole rock' samples of ore (e.g., Foster et al. 1996; Mathur et al. 1999; McInnes et al. 2008; Nozaki et al. 2010) or separates of minerals such as pyrite (Stein et al. 2000; Barra et al. 2003; Gao et al. 2020; Hnatyshin et al. 2020), arsenopyrite (Morelli et al. 2005, 2010), and chalcopyrite (Zhimin and Yali 2013) (Fig. 4). Selby et al. (2009) reported highly precise multi-mineral isochron ages based on pyrite, chalcopyrite, and bornite separates but inclusions of cogenetic minor phases such as renierite ($(Cu, Zn)_{11}(Ge, As)_2Fe_4S_{16}$) or germanite ($Cu_{26}Fe_4Ge_4S_{32}$) in those samples might be the actual hosts for the Re. Pyrrhotite appears to be susceptible to disturbance and does not appear to be generally suitable for Re–Os dating (Frei et al. 1998; Brenan et al. 2000; Morelli et al. 2010).

Single-mineral ages can also be determined for phases such as molybdenite that form with very high Re/Os ratios, analogous to U–Pb dating of zircon or Rb–Sr dating of biotite. As discussed in Sect. 7.2 of this chapter, molybdenite (MoS_2) is especially suitable for Re–Os dating because of its typically high Re concentrations (often 10's-1000's of ppm), negligible initial Os, and resistance to subsequent resetting (Suzuki et al. 1993; Stein et al. 2001; Chiaradia et al. 2013). In this case, the age is given directly by the measured $^{187}Re/^{187}Os$ ratio assuming negligible initial Os (i.e., y-intercept = 0). In contrast, most other common sulfides such those mentioned above typically contain about a thousand times less Re, often in the low ppb range, but their initial Re/Os ratios are often still sufficiently high that they are suitable for dating. Stein et al. (2000) referred to this type of sulfide as Low Level Highly Radiogenic (LLHR) and suggested that plots of ^{187}Re vs ^{187}Os concentrations (rather than parent/daughter ratios as for conventional isochrons) are more suitable for determining their ages so as to avoid introducing correlated errors based on poorly determined initial Os compositions, although this may not always be necessary (e.g., Barra et al. 2003; Selby et al. 2009; Hnatyshin et al. 2020). Ideally, the age calculated from this type of isochron is consistent with the ^{187}Re/^{187}Os model age of each phase (Fig. 4). Sulfide mineral separates can be obtained either by handpicking coarsely crushed ore in an attempt to maximize the variation in Re contents (e.g., Barra et al. 2003; Majzlan et al. 2022) or by a combination of magnetic and density separations, minimizing contact with metal to avoid contamination (e.g., Hnatyshin et al. 2020; Saintilan et al. 2020).

The ^{187}Re-^{187}Os system has also been used to date titanomagnetite (Morgan et al. 2000; Lambert et al. 2000; Mathur et al. 2002; Zhou et al. 2005; Huang et al. 2014), deposition of black shales (Ravizza and Turekian 1989; Creaser et al. 2002; Hannah et al. 2004; Selby and Creaser 2005a), formation of bitumen and petroleum (Selby and Creaser 2003, 2005b; Selby et al. 2007b; Georgiev et al. 2016), eruption and alteration of komatiites and related ores (Foster et al. 1996; Gangopadhyay and Walker 2003; Gangopadhyay et al. 2005; Puchtel et al. 2004), gold deposits (Kirk et al. 2002; McInnes et al. 2008; Schaefer et al. 2010), sulfide inclusions in diamonds (Harvey et al. 2016), graphite (Toma et al. 2022), and a variety of minor phases including arsenides and sulfarsenides (e.g., Saintilan et al. 2017; Majzlan et al. 2022).

The ^{190}Pt-^{186}Os system has not been as widely applied for geochronology due to the low relative abundance and long half-life of the parent isotope. Consequently, variations of ^{186}Os/^{188}Os are generally much less than those of ^{187}Os/^{188}Os, which results in relatively large uncertainties on ages calculated from ^{190}Pt-^{186}Os isochrons compared to those obtained using ^{187}Re-^{187}Os (e.g., Puchtel et al. 2004). Nonetheless, it has been applied with some success to dating of mineralization in the Bushveld Complex (Coggon et al. 2012) and alluvial grains that are otherwise difficult to date (Nowell et al. 2008b; Coggan et al. 2011).

6.2 Source Tracing and Re–Os Model Ages

The Os isotopic composition of a source reflects its long-term Re/Os and Pt/Os. At the global scale, the fundamental process that fractionates Re from Os is crust formation. Continental crust has a highly radiogenic ^{187}Os/^{188}Os isotopic composition reflecting its high time-integrated ^{187}Re/^{188}Os (Esser and Turekian 1993, Saal et al. 1998; Peucker-Ehrenbrink and Jahn 2001; Fig. 1). However, the behavior of Re, Os, and Pt in the mantle differs from other commonly applied isotopic systems based on incompatible lithophile elements such Rb–Sr, Lu–Hf, and Sm–Nd in that they are not strongly fractionated by small to moderate degrees of mantle melting. As a consequence, fertile to moderately depleted mantle (i.e., MORB-source) has an Os isotopic composition consistent with long-term evolution at near-chondritic ratios of Re/Os and Pt/Os (Meisel et al. 2001a; Gannoun et al. 2007), and this seems to have been the case for at least the last 3.8 billion years (Bennett et al. 2002). Komatiite-hosted magmatic sulfide ores often have ^{187}Os/^{188}Os isotopic compositions consistent with this type of mantle source, precluding assimilation of significant amounts of older continental crust (Lambert et al. 1998, 1999). Assuming chondritic Re–Os and Pt–Os isotopic evolution for the mantle, therefore, provides a useful baseline for evaluating the relative contributions of crustal and mantle inputs to an ore system.

Model ages relative to a reference composition such as the mantle evolution curve can be calculated. Two types of model ages have been widely applied. One of these assumes that the Re and Os in the sample are both primary and the measured ^{187}Os/^{188}Os is corrected for in-situ decay until the isotopic composition intersects that of the mantle evolution curve. This "age" is referred to as T_{MA} and is analogous to the T_{DM} ages often calculated from Sm–Nd isotopic data. It can be applied to any material regardless of its Re/Os ratio but requires the assumptions of a normal mantle source and a single Re/Os fractionation event. A second type of model age assumes that the sample contains no Re (i.e., Re/Os = 0) such that the measured Os isotopic composition provides the mantle value directly, and therefore the time of Re depletion, presumably in a melting event. This Re-depletion model age (T_{RD}; Fig. 1) is suitable only for materials with low Re/Os ratios, such as highly depleted mantle xenoliths or chromite. Both of these model ages (T_{MA}, T_{RD}) assume that the sample was derived directly from the mantle, and they are therefore most commonly applied to materials such as mantle xenoliths and basalts. Neither type of model age is especially useful for most ore-forming systems, which often contain a crustal component and may have had multi-stage histories. Crustal model ages have not been widely applied, in part because the crust is highly radiogenic and appears to be heterogeneous in its Os isotopic composition (Johnson et al. 1996).

In contrast to the continental crust and fertile upper mantle, the subcontinental lithospheric mantle, as sampled by mantle xenoliths, preserves evidence of ancient melt depletion events in their low Re/Os ratios, subchondritic $^{187}Os/^{188}Os$, and correlations of these parameters with geochemical indicators of melt extraction such as Al_2O_3 abundances (Meisel et al. 2001a; Carlson 2005; Aulbach et al. 2016; Harvey et al. 2016). A common application of Os model ages is constraining the timing of ancient melt depletion events (Shirey and Walker 1998; Carlson 2005; Rudnick and Walker 2009; Dijkstra et al. 2016; Harvey et al. 2016). Osmium isotopic evidence for ancient depletion events has also been found in abyssal peridotites (Brandon et al. 2000; Harvey et al. 2006) and peridotite xenoliths from arc (Parkinson et al. 1998) and mantle plume settings (Bizimis et al. 2007). Kimberlites, lamproites, and diamonds are often linked to such sources (Lambert et al. 1995; Carlson et al. 1996; Graham et al. 1999; Aulbach et al. 2016; Harvey et al. 2016).

While such data are generally considered to have broad geochronological significance, it should be emphasized that these are model ages analogous to those obtained from common Pb isotopic compositions of galena or the depleted mantle ages (T_{DM}) calculated from $^{143}Nd/^{144}Nd$ isotopic compositions, rather than crystallization ages. The significance of model ages depends on the applicability of the model and the time-integrated history of parent/daughter isotopic evolution in the measured materials and their source regions. In this case, Os isotopic model ages can be affected by a variety of processes such as melt-rock reactions, refertilization, metasomatism, metamorphism, Re volatility, or intrinsic isotopic heterogeneity in the source (Burton et al. 2000; Griffin et al. 2004; Gangopadhyay et al. 2005; Rudnick and Walker 2009; Kochergina et al. 2016; Aulbach et al. 2016).

Volcanogenic base-metal deposits such as porphyries and manto-type deposits often show clear evidence for involvement of older crust in the elevated initial $^{187}Os/^{188}Os$ isotopic compositions of the ores (Ruiz and Mathur 1999; Mathur et al. 2000a, 2005; Barra et al. 2003). However, the relative contributions of crustal and mantle sources remain contentious and variable signatures may reflect the tectonic and/or thermal evolution of the system (Freydier et al. 1997; Mathur et al. 2000b; McBride et al. 2001; Zimmerman et al. 2014; Saintilan et al. 2021). Gregory et al. (2008) concluded that the elevated $^{187}Os/^{188}Os$ isotopic compositions of Cu ore from the sediment-hosted Mt. Isa deposit reflects extraction of the Cu from older basalts that had developed radiogenic isotopic compositions through a combination of in-situ decay (high Re/Os) and crustal contamination or alteration during their emplacement. In contrast, the isochrons obtained by McInnes et al. (2008) for gold and copper–gold mineralization in the Proterozoic Tanami and Tennant Creek districts of the Northern Territory, Australia, have mantle-like initial $^{187}Os/^{188}Os$ isotopic compositions despite their emplacement into radiogenic crustal settings.

Interestingly, few ore deposits show clear isotopic evidence for involvement of ancient, Re-depleted mantle. While a hydrated and metasomatised peridotite might be an attractive source for the 'boninitic' (high Si, Mg) characterstics of some layered intrusion magmas (Lambert et al. 1994; Carlson 2005), the Os isotopic signature expected for an ancient, highly depleted mantle is typically absent in these intrusions. An exception is the Proterozoic Ipuera-Medrado chromite deposits in Brazil (Marques et al. 2003). Stratigraphically lower chromitites have the unusual combination of unradiogenic Os (Re-depleted) and unradiogenic Nd (LREE-enriched) isotopic compositions, which appears to be a unique signature of the subcontinental lithospheric mantle. As for many other layered intrusions, the Nd-Os isotopic compositions at Ipuera-Medrado change upsection in a way that is consistent with increasing degrees of crustal contamination.

7 Applications to Metallogenic Systems

7.1 Magmatic Sulfide Deposits

Osmium isotopic studies of magmatic sulfide deposits, which are significant resources for PGE, Ni, Co, Cr, and other metals, have provided definitive information about the timing of ore formation and the sources of ore metals. Examples include komatiite-associated NiS deposits such as those in Western Australia and the Canadian Shield (Walker et al. 1988; Foster et al. 1996; Lambert et al. 1998, 1999; Lahaye et al. 2001; Hulbert et al. 2005), layered mafic intrusions such as the Bushveld (Hart and Kinloch 1989; Schoenberg et al. 1999; Reisberg et al. 2011; Coggon et al. 2012), Stillwater (Martin 1989; Marcantonio et al. 1993; Lambert et al. 1994; Horan et al. 2001), Voisey's Bay (Lambert et al. 2000; Hannah and Stein 2002), Muskox (Day et al. 2008), Great Dyke (Schoenberg et al. 2003), Portimo (Andersen et al. 2006), and Duluth complexes (Ripley et al. 1998, 2008; Williams et al. 2010), and Ni–Cu–PGE ores associated with flood basalts such as those in the Noril'sk region of Siberia (Walker et al. 1994) and the Xinjie layered intrusion in the Emeishan large igneous province of China (Zhong et al. 2011).

Komatiite-associated NiS deposits have yielded geologically reasonable Re–Os isochron ages although precision is often degraded by Re mobility related to subsequent deformation and alteration. Perhaps the most significant contribution that Re–Os isotope geochemistry has made to understanding these deposits is the clear definition of a mantle signature in the Os isotopic compositions. Although some komatiite flows may have experienced crustal assimilation (Lahaye et al. 2001), the data clearly show that the Ni and PGE (including Os) in these ores were derived predominantly from the mantle. This has allowed a detailed evaluation of volcanological processes and sources associated with emplacement of these magmatic systems, in particular by comparison with mass-independent variations in Δ^{33}S (Lambert et al. 1999; Bekker et al. 2009; Fiorentini et al. 2012).

In contrast to komatiite-hosted ores, Os isotopic compositions of mineralization in layered mafic intrusions show that mixing of crustal and mantle components is a ubiquitous process. The proportions of crust and mantle components vary widely (Shirey and Walker 1998), and the mechanism(s) by which the enriched (crustal) component becomes incorporated is not always clear. Assimilation of country rocks, magma mixing, mantle metasomatism, and overprinting by late hydrothermal fluids have all been suggested as possible mechanisms to explain the Os isotopic heterogeneity observed within many layered intrusions. In the Bushveld complex, initial ^{187}Os/^{188}Os ratios increase with stratigraphic height through the Critical Zone, reaching a maximum at the level of the Merensky Reef before decreasing sharply again within a few meters (Fig. 5; Reisberg et al. 2011, and references therein). In that scenario, ore grades in the Merensky Reef may have been enhanced by transient assimilation of PGE-rich black shales (Reisberg et al. 2011). Sproule et al. (2002) showed how Nd and Os isotopes can be decoupled depending on the timing of sulfide saturation relative to crustal assimilation. The magma-fluid-country rock interactions documented at the Duluth Complex (Ripley et al. 2001; Williams et al. 2010) illustrate how complex these

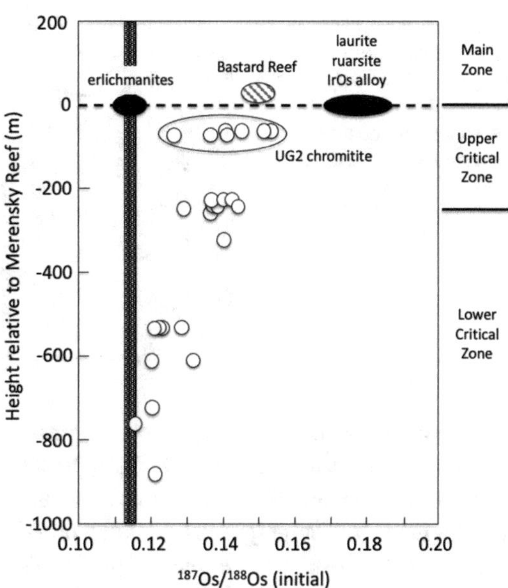

Fig. 5 Initial Os isotope stratigraphy of the Bushveld complex, after Reisberg et al. (2011). Filled black symbols are samples from the Merensky Reef. The vertical hashed bar shows the mantle composition at 2.054 Ga. Crustal host rocks would have had $^{187}Os/^{188}Os$ composition of ~0.5 at that time. Increasing initial $^{187}Os/^{188}Os$ ratios with stratigraphic height indicates progressively greater contributions of crust in the magma

processes can be. An implication of these studies would be that binary mixing or assimilation-fractional crystallization (AFC) calculations are unlikely to capture the physico-chemical reality of these processes. Late-stage mobility of Re due to hydrothermal activity is commonly inferred (Marcantonio et al. 1994; Andersen et al. 2006; Schoenberg et al. 2003; Day et al. 2008).

The Sudbury Igneous Complex provides another example of the use of Re–Os isotopes to define the timing of ore formation and metal sources. Walker et al. (1991) and Morgan et al. (2002) showed that the Sudbury ores formed contemporaneously with crystallization of the igneous complex at 1850 Ma, followed by minor resetting at ~1770–1780 Ma. However, the limited range of $^{187}Re/^{187}Os$ and scatter introduced by either small-scale heterogeneities in initial $^{187}Os/^{188}Os$ isotopic compositions or minor post-crystallization disturbance limited the precision on some of the isochrons (e.g., 1825 ± 340 Ma, MSWD = 228 for Falconbridge and 1835 ± 70 Ma, MSWD = 45 for McCreedy West; Morgan et al. 2002). Notable in these data are the much higher Pt and lower Re and Os concentrations, and higher $^{187}Re/^{188}Os$ and $^{190}Pt/^{188}Os$ ratios in the Deep Copper Zone (DCZ) of the Strathcona mine compared to those from Falconbridge, McCreedy West, and the other Strathcona ores. This is consistent with an origin of the DCZ ores predominantly as residual sulfide melt and the Falconbridge, McCreedy West and other Strathcona ores predominantly as MSS-dominated cumulates. Morgan et al. (2002) also report ^{190}Pt-^{186}Os data, which show that the DCZ ores have highly radiogenic $^{186}Os/^{188}Os$ isotopic compositions as expected from their high Pt/Os ratios. However, the low Os concentrations in these ores and apparent variability in initial $^{186}Os/^{188}Os$ isotopic compositions precluded a precise age determination based on this system. Although the initial $^{187}Os/^{188}Os$ isotopic compositions of Sudbury ores are heterogeneous they are all totally dominated by radiogenic crust, possibly reflecting variable mixing of the crustal target rocks during formation of the SIC by impact melting (Morgan et al. 2002). The predominantly crustal composition of the SIC contrasts with that of many other layered intrusions such as Bushveld, Stillwater, Muskox, Rum, and Xinjie, which have Os isotopic compositions indicating mixtures of crustal and mantle components (Reisberg et al. 2011; Marcantonio et al. 1993; Lambert et al. 1994; Day et al. 2008; O'Driscoll et al. 2009; Zhong et al. 2011).

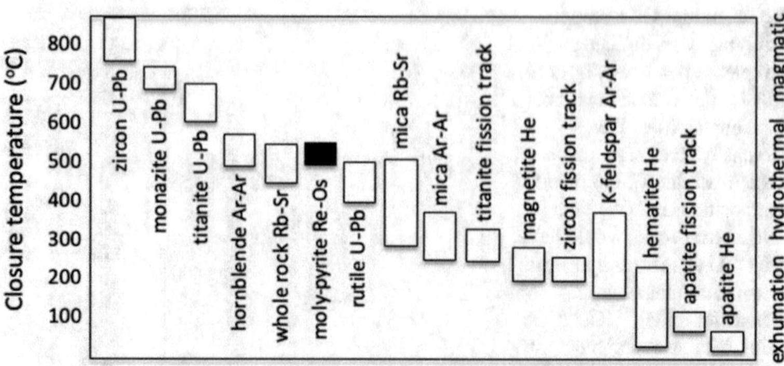

Fig. 6 Diagram illustrating approximate closure temperatures for various mineral-isotope systems relevant to ore deposit geochronology. Modified after Chiaradia et al. (2013)

A somewhat different style of magmatic sulfide deposit is associated with flood basalts such as those in the Noril'sk region of Siberia and the Emeishan large igneous province of SW China. At Noril'sk, sulfide ores yield Re–Os isochron ages of ~245–248 Ma (Fig. 2), somewhat younger than the canonical eruption age of the Siberian flood basalts (250–252 Ma; Renne 1995; Kamo et al. 2003). However, uncertainties on the isochrons are large (±~ 2–10%) and the exact age obtained depends on which samples are included in the regressions. As for Sudbury, excess scatter in the data is indicated by the large MSWD's of these isochrons (e.g., MSWD = 45–288 for the four isochrons presented by Malitch and Latypov 2011), possibly reflecting local variations in the $^{187}Os/^{188}Os$ of the ores at the time they formed. Alternatively, the Re–Os isotopic systematics of the ores might have been disturbed by subsequent events such as the minor silicic magmatism that followed emplacement of the flood basalts (Kamo et al. 2003; Malitch et al. 2010). In contrast to Sudbury, the initial $^{187}Os/^{188}Os$ isotopic compositions of Noril'sk district ores indicate a slightly enriched mantle source, consistent with the formation of the Siberian flood basalt from a mantle plume. The Emeishan large igneous province appears to be part of a broad episode of flood basalt eruption in the Late Permian with the Xinjie mafic–ultramafic layered intrusion yielding a ^{187}Re-^{187}Os isochrons age of 262 ± 27 Ma and an initial $^{187}Os/^{188}Os$ consistent with a moderately enriched mantle plume source (Zhong et al. 2011).

7.2 Molybdenite

Molybdenite (MoS_2) has been widely used for Re–Os geochronology because it often contains 10's–100's of ppm of Re and essentially no initial Os (Stein et al. 2001; Stein 2014). Precipitation of molybdenite from hydrothermal solutions is favored by decreasing fO_2, salinity, and T, and high fS_2 (Cao 1989; Ulrich and Mavrogenes 2008). It has a high closure temperature and is robust against subsequent thermal events and metamorphism (Fig. 6). Because Mo partitions into condensed fluids relative to vapour, primary molybdenite will tend to occur closer to the source of hydrothermal fluids compared to more vapour-mobile elements such as Au and Cu (Zajacz et al. 2017). Although a few examples of molybdenite with compositionally distinct cores and rims have been reported (Stein et al. 2004; Aleinikoff et al. 2012), this is rare. Rather, multiple episodes of molybdenite crystallization tend to form as discrete crystal populations such that age distributions can sometimes be at least partially resolved through analysis of multiple aliquots of different grain size fractions (Wilson et al. 2007; Aleinikoff et al. 2012). Variations in Re contents of molybdenite can also contribute to recognition of multiple mineralization events (Selby and Creaser 2001; Barra et al. 2005; Wilson et al. 2007; Wells et al. 2021). Controls on Re concentrations in molybdenite are discussed further in the next section.

While the reliability of Re–Os molybdenite dating is now well established, care is needed when sampling the molybdenite because small-scale decoupling of ^{187}Re/^{187}Os within individual crystals has been inferred from apparent variations in the spatial distribution of ^{185}Re/$\Sigma m187$, where $\Sigma m187$ = the net signal intensity measured on mass 187, comprising ^{187}Re plus radiogenic ^{187}Os, measured by laser ablation ICPMS (Stein et al. 2003; Selby and Creaser 2004; Košler et al. 2003). Such decoupling might be due to intra-crystal migration of ^{187}Os as a consequence of the misfit of Os in the crystal structure of molybdenite whereas Re substitutes for Mo as a limited solid solution of ReS_2 in MoS_2 (Takahashi et al. 2007; Drabek et al. 2010). In contrast, imaging of the spatial distribution of Re and Os in molybdenite using nano-SIMS (Barra et al. 2017) and a laser ablation study using a MC-ICPMS equipped with a collision cell to separate Re from Os during the analysis (Hogmalm et al. 2019) did not observe such decoupling, although both of those studies have been criticised as inconclusive (Zimmerman et al. 2022). Clearly, further work is necessary to address this question.

Molybdenite can also form as a primary magmatic phase, often as small inclusions in quartz phenocrysts (Audétat et al. 2011; Sun et al. 2014). It seems to occur more frequently in within-plate or rift-related rhyolites compared to arc rhyolites, probably due to the higher fO_2 and resulting higher proportion of sulphate versus sulfide in the latter. Audétat et al. (2011) developed a thermodynamic model that estimates magmatic fO_2 and fS_2 provided that the melt is saturated in molybdenite + pyrrhotite, and the Mo concentration and T can be constrained. Alternatively, fS_2 can be calculated if fO_2 can be constrained independently, for example, through FeTi-oxide thermobarometry. Molybdenum solubility in silicic melts is enhanced by oxidizing conditions and depolymerized (e.g., strongly peralkaline) compositions (Sun et al. 2014).

7.3 Controls on Re Concentrations in Molybdenite

Bulk rhenium concentrations in molybdenite display a broad correlation with the lithologic association and style of the deposit (Barton et al. 2020). Terada et al. (1971) showed that rhenium contents of molybdenite generally decrease in the sequence: volcanic sublimate, porphyry copper deposits, contact-metasomatic (skarn) deposits, disseminated greisen or stockwork deposits, and quartz veins, which suggests a broad gradient with temperature and/or source of the metals. An extreme example of high-T molybdenites are those associated with fumaroles at Kudryavy volcano in the Kurile volcanic arc, which have ~ 1.5 wt% Re (Tessalina et al. 2008b). Although rheniite (ReS2) has also been found at this volcano, the Re content of this molybdenite appears to be primary and reflects the high concentrations of Re and other volatile metals in the volcanic gases (Tessalina et al. 2008b). Rhenium volatility during magmatic outgassing appears to be a common process that has been documented at a variety of settings that include ocean island and island arc volcanoes (Sun et al. 2003a; Lassiter 2003; Norman et al. 2004b; Righter et al. 2008).

Porphyry deposits often carry molybdenite with high Re concentrations although this appears to vary with the specific style of mineralization (Berzina et al. 2005; Barton et al. 2020). In general, molybdenite associated with Cu and Au porphyries tend to have systematically higher Re contents (100's to 1000's of ppm) than that in Mo-dominated systems (10's to 100's of ppm), possibly due to dilution of the Re by the large volume of molybdenite in the latter (Stein et al. 2001; Stein 2014). Molybdenite in deposits associated with compositionally intermediate granitic systems often contain 10's to 100's of ppm Re whereas those related to highly evolved felsic rocks, and especially Sn-W deposits, tend to have low Re contents (<1–10 ppm) (e.g., Morgan et al. 1968; Selby et al. 2007a; Barton et al. 2020). It has been suggested that the general trend of higher Re in molybdenite associated with more mafic systems to lower Re in more evolved systems reflects a transition from mantle- to crustally-dominated systems (e.g., Stein et al. 2001), but it may also be due to co-factors such as redox and/or sulfidation state, with more oxidized, higher sulfidation systems producing molybdenite with higher Re contents (Berzina et al. 2005; Barton et al. 2020).

In addition to these broad lithologic associations and physical–chemical controls, there may also be a degree of underlying geological influence on the composition of molybdenite in some deposits. For example, some provinces appear to produce molybdenite with systematically lower concentrations (<20 to sub-ppm; Morgan et al. 1968; Stein et al. 2001) possibly due to derivation of the system by metamorphic dehydration melting of mid-crustal rocks, a process that should be reflected in other isotopic and compositional characteristics (Stein 2006; Voudouris et al. 2013). In contrast, Voudouris et al. (2013) suggested that the regional distribution of molybdenite with high to very high Re contents in northeastern Greece may indicate a Re-enriched source, perhaps related to its tectonic history.

7.4 Case Study: Crystallization and Uplift of the Boyongan and Bayugo Cu–Au Porphyry Deposit

While Re–Os isotopic studies of molybdenite are typically aimed at establishing the age of mineralisation, the study by Braxton et al. (2012) provides an example of the added value that molybdenite dating can provide when placed in a broader geological and geochronological context. Braxton et al. (2012) studied the Pleistocene age Boyongan and Bayugo Cu–Au porphyry deposits, which are located in the Surigao district of the Phillipines, using multiple radiogenic isotope systems. The Boyongan complex is characterized by an exceptionally thick (600 m) oxidation profile.

Porphyry magmatism occurred between 2.09 and 2.31 Ma based on U–Pb isotopic ages of zircons from the earliest and latest intrusive phases of the complex. Rhenium-Os ages of 2.115 ± 0.008 and 2.120 ± 0.007 Ma on late molybdenite are consistent with a close genetic link between igneous intrusion and the mineralization, while K–Ar ages of 2.09–2.12 Ma on hydrothermal illite inferred to post-date the molybdenite shows that the system cooled rapidly following the final stage of porphyry emplacement and mineralization. In contrast, (U–Th)/He apatite thermochronology provided ages between 1.49 and 1.98 Ma, with ages decreasing systematically with increasing depth into the intrusive complex. Those cooling ages provide constraints on the uplift and exhumation of the complex. Finally, the timing of late extension and sedimentation was established from ^{40}Ar-^{39}Ar ages of post-mineralization volcanics and ^{14}C dating of wood and plant material, which indicate burial began at ~ 0.6 Ma and continued to ~ 1.6 ka. These relationships are summarized in Fig. 7.

The geochronological data on the Boyongan-Bayugo complex imply porphyry mineralization at ~ 2.1 Ma was followed by rapid uplift at a rate of 1–5 km/Ma, with exposure and intense

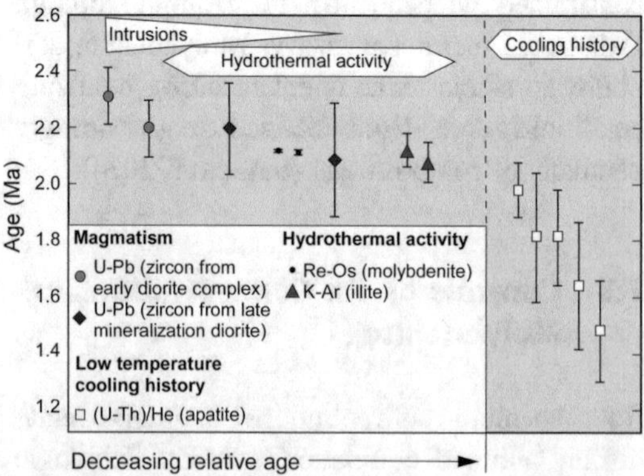

Fig. 7 Diagram illustrating the sequence of igneous, hydrothermal, and cooling ages at the Boyongan and Bayugo porprhyry deposits, Philippines (after Braxton et al. 2012)

supergene alteration occurring predominantly between ~0.6 and 1.5 Ma. These uplift, alteration, and burial rates are considerably greater than the global averages but similar to other sites with analogous tectonic and climatic conditions. Braxton et al. (2012) interpret the uplift and subsequent burial history of the complex to reflect a transition from strongly compressional tectonism related to collision of the Philippine arc with the Eurasian plate to the present-day transtensional setting in the mid-Pleistocene, highlighting the dynamic tectonism of the Philippine Mobile Belt and related geological environments.

8 Conclusions

The ^{187}Re-^{187}Os isotope system is well suited for direct dating of sulfide minerals. Molybdenite is especially valuable for ore deposit geochronology because of its relatively high Re concentrations, high initial ^{187}Re/^{187}Os, high closure T, and resistance to resetting, but it is not always present; many other sulfide minerals have been dated successfully. Initial ^{187}Os/^{188}Os isotopic compositions of magmatic sulfide deposits have established a predominantly mantle provenance for the PGE in these systems, whereas the ore-forming fluids responsible for hydrothermal deposits carry a strong signature of crustal Os. The ^{190}Pt-^{186}Os system has so far been useful only for a limited set of applications, mainly in the dating of alluvial nuggets with high Pt/Os. Future developments can expect to see increasing applications to the dating of other mineral systems, and a more integrated approach to place mineralisation ages within a broader geotectonic context and the duration of ore-forming systems. Recent efforts to measure stable isotope variations of Re (Miller et al. 2009), Os (Nanne et al. 2017), and Pt (Creech et al. 2014) may find future applications to ore deposits. Further studies of Os isotopes coupled with stable isotopic compositions of S and O will continue to provide new insights into the sources of mineralising fluids and conditions of mineralization.

Acknowledgements Reviews by Svetlana Tessalina and Christopher Lawley improved the chapter and are appreciated. Editorial handling and assistance from David Huston is greatly appreciated.

References

Ahmed AH, Hanghøj K, Kelemen PB, Hart SR, Arai S (2006) Osmium isotope systematics of the Proterozoic and Phanerozoic ophiolitic chromitites: in situ ion probe analysis of primary Os-rich PGM. Earth Planet Sci Lett 245:777–791

Albarede F, Telouk P, Blichert-Toft J, Boyet M, Agranier A, Nelson B (2004) Precise and accurate isotopic measurements using multiple-collector ICPMS. Geochim Cosmochim Acta 68:2725–2744

Aleinikoff JN, Creaser RA, Lowers HA, Magee CW Jr, Grauch RI (2012) Multiple age components in individual molybdenite grains. Chem Geol 300:55–60

Allègre CJ, Luck JM (1980) Osmium isotopes as petrogenetic and geological tracers. Earth Planet Sci Lett 48:148–154

Alves S, Schiano P, Capmas F, Allègre CJ (2002) Osmium isotope binary mixing arrays in arc volcanism. Earth Planet Sci Lett 198:355–369

Andersen JCØ, Thalhammer OA, Schoenberg R (2006) Platinum-group element and Re–Os isotope variations of the high-grade Kilvenjärvi platinum-group element deposit, Portimo layered igneous complex, Finland. Econ Geol 101:159–177

Andreasen R, Sharma M (2009) Fractionation and mixing in a thermal ionization mass spectrometer source: implications and limitations for high-precision Nd isotope analyses. Int J Mass Spectrom 285:49–57

Andrén H, Rodushkin I, Stenberg A, Malinovsky D, Baxter DC (2004) Sources of mass bias and isotope ratio variation in multi-collector ICP-MS: optimization of instrumental parameters based on experimental observations. J Anal at Spectrom 19:1217–1224

Arblaster JW (2004) The discoverers of the osmium isotopes. Platin Met Rev 48:173–179

Armistead SE, Skirrow RG, Fraser GL, Huston DL, Champion D, Norman MD (2017) Gold and intrusion-related Mo-W mineral systems in the southern Thomson Orogen, New South Wales. Geosci Austr Rec 2017/005

Audétat A, Dolejš D, Lowenstern JB (2011) Molybdenite saturation in silicic magmas: occurrence and petrological implications. J Petrol 52:891–904

Aulbach S, Mungall JE, Pearson DG (2016) Distribution and processing of highly siderophile elements in cratonic mantle lithosphere. Rev Mineral Geochem 81:239–304

Barnes SJ, Ripley EM (2016) Highly siderophile and strongly chalcophile elements in magmatic ore systems. Rev Mineral Geochem 81:725–774

Barnes SJ, Makovicky E, Makovicky M, Rose-Hansen J, Karup-Moller S (1997) Partition coefficients for Ni, Cu, Pd, Pt, Rh, and Ir between monosulfide solid solution and sulfide liquid and the formation of compositionally zoned Ni–Cu sulfide bodies by fractional crystallization of sulfide liquid. Can J Earth Sci 34:366–374

Barra F, Ruiz J, Mathur R, Titley S (2003) A Re–Os study of sulfide minerals from the Bagdad porphyry Cu–Mo deposit, northern Arizona, USA. Miner Depos 38:585–596

Barra F, Ruiz J, Valencia VA, Ochoa-Landín L, Chesley JT, Zurcher L (2005) Laramide porphyry Cu-Mo mineralization in northern Mexico: age constraints from Re-Os geochronology in molybdenite. Econ Geol 100:1605–1616

Barra F, Deditius A, Reich M, Kilburn MR, Guagliardo P, Roberts MP (2017) Dissecting the Re-Os molybdenite geochronometer. Sci Rep 7:1–7

Barton IF, Rathkopf CA, Barton MD (2020) Rhenium in molybdenite: a database approach to identifying geochemical controls on the distribution of a critical element. Min Metall Explor 37:21–37

Bekker A, Barley ME, Fiorentini ML, Rouxel OJ, Rumble D, Beresford SW (2009) Atmospheric sulfur in Archean komatiite-hosted nickel deposits. Science 326:1086–1089

Begemann F, Ludwig K, Lugmair G, Min K, Nyquist L, Patchett P, Renne P, Shih C-Y, Villa IM, Walker R (2001) Call for an improved set of decay constants for geochronological use. Geochim Cosmochim Acta 65:111–121

Bennett VC, Norman MD, Garcia MO (2000) Rhenium and platinum group element abundances correlated with mantle source components in Hawaiian picrites: sulfides in the plume. Earth Planet Sci Lett 183:513–526

Bennett VC, Nutman AP, Esat TM (2002) Constraints on mantle evolution from $^{187}Os/^{188}Os$ isotopic compositions of Archean ultramafic rocks from southern West Greenland (3.8 Ga) and Western Australia (3.46 Ga). Geochim Cosmochim Acta 66:2615–2630

Berzina AN, Sotnikov VI, Economou-Eliopoulos M, Eliopoulos DG (2005) Distribution of rhenium in molybdenite from porphyry Cu–Mo and Mo–Cu deposits of Russia (Siberia) and Mongolia. Ore Geol Rev 26:91–113

Bezard R, Schaefer BF, Turner S, Davidson JP, Selby D (2015) Lower crustal assimilation in oceanic arcs: insights from an osmium isotopic study of the Lesser Antilles. Geochim Cosmochim Acta 150:330–344

Bingen B, Stein H (2003) Molybdenite Re–Os dating of biotite dehydration melting in the Rogaland high-temperature granulites, S Norway. Earth Planet Sci Lett 208:181–195

Birck JL, Barman MR, Capmas F (1997) Re-Os isotopic measurements at the femtomole level in natural samples. Geostand Newsl 21:19–27

Birck JL, Mounier L, Lloyd N, Schwieters JB (2016) Low level Os isotopic measurements using multiple ion counting. Thermoscientific Application Note 30355. https://assets.thermofisher.com/TFS-Assets/CMD/Application-Notes/AN-30355-ICP-MS-Multiple-Ion-Counting-AN30355-EN.pdf

Bizimis M, Griselin M, Lassiter JC, Salters VJ, Sen G (2007) Ancient recycled mantle lithosphere in the Hawaiian plume: osmium–hafnium isotopic evidence from peridotite mantle xenoliths. Earth Planet Sci Lett 257:259–273

Blum JD, Pellin MJ, Calaway WF, Young CE, Gruen DM, Hutcheon ID, Wasserburg GJ (1990) Insitu measurement of osmium concentrations in iron meteorites by resonance ionization of sputtered atoms. Geochim Cosmochim Acta 54:875–881

Borg LE, Brandon AD, Clynne MA, Walker RJ (2000) Re–Os isotopic systematics of primitive lavas from the Lassen region of the Cascade arc, California. Earth Planet Sci Lett 177:301–317

Borisov A, Danyushevsky L (2011) The effect of silica contents on Pd, Pt and Rh solubilities in silicate melts: an experimental study. Eur J Mineral 23:355–367

Brandon AD, Creaser RA, Shirey SB, Carlson RW (1996) Osmium recycling in subduction zones. Science 272:861

Brandon AD, Snow JE, Walker RJ, Morgan JW, Mock TD (2000) ^{190}Pt–^{186}Os and ^{187}Re–^{187}Os systematics of abyssal peridotites. Earth Planet Sci Lett 177:319–335

Braxton DP, Cooke DR, Dunlap J, Norman M, Reiners P, Stein H, Waters P (2012) From crucible to graben in 2.3 Ma: a high-resolution geochronological study of porphyry life cycles, Boyongan-Bayugo copper-gold deposits, Philippines. Geology 40:471–474

Brenan JM (2002) Re–Os fractionation in magmatic sulfide melt by monosulfide solid solution. Earth Planet Sci Lett 199:257–268

Brenan JM (2008) Re–Os fractionation by sulfide melt–silicate melt partitioning: a new spin. Chem Geol 248:140–165

Brenan JM, Cherniak DJ, Rose LA (2000) Diffusion of osmium in pyrrhotite and pyrite: implications for closure of the Re–Os isotopic system. Earth Planet Sci Lett 180:399–413

Brenan JM, McDonough WF, Dalpe C (2003) Experimental constraints on the partitioning of rhenium and some platinum-group elements between olivine and silicate melt. Earth Planet Sci Lett 212:135–150

Brenan JM, Bennett NR, Zajacz Z (2016) Experimental results on fractionation of the highly siderophile elements (HSE) at variable pressures and temperatures during planetary and magmatic differentiation. Rev Mineral Geochem 81:1–87

Brügmann GE, Birck JL, Herzig PM, Hofmann AW (1998) Os isotopic composition and Os and Re distribution in the active mound of the TAG hydrothermal system, Mid-Atlantic Ridge. Proc Ocean Drill Program Sci Results 158:91–100

Burton KW, Schiano P, Birck JL, Allegre CJ, Rehkämper M, Halliday AN, Dawson JB (2000) The distribution and behaviour of rhenium and osmium amongst mantle minerals and the age of the lithospheric mantle beneath Tanzania. Earth Planet Sci Lett 183:93–106

Cao X (1989) Solubility of molybdenite and the transport of molybdenum in hydrothermal solutions. Unpublished PhD thesis, Iowa State University, http://lib.dr.iastate.edu/rtd/9026/

Carlson RW (2005) Application of the Pt–Re–Os isotopic systems to mantle geochemistry and geochronology. Lithos 82:249–272

Carlson RW, Esperanca S, Svisero DP (1996) Chemical and Os isotopic study of Cretaceous potassic rocks from southern Brazil. Contrib Mineral Petrol 125:393–405

Cave RR, Ravizza GE, German CR, Thomson J, Nesbitt RW (2003) Deposition of osmium and other platinum-group elements beneath the ultramafic-hosted Rainbow hydrothermal plume. Earth Planet Sci Lett 210:65–79

Chesley JT (1999) Integrative geochronology of ore deposits: new insights into the duration and timing of hydrothermal circulation. Rev Econ Geol 12:115–141

Chiaradia M, Schaltegger U, Spikings R, Wotzlaw JF, Ovtcharova M (2013) How accurately can we date the duration of magmatic-hydrothermal events in porphyry systems?—an invited paper. Econ Geol 108:565–584

Coggon JA, Nowell GM, Pearson DG, Parman SW (2011) Application of the ^{190}Pt-^{186}Os isotope system to dating platinum mineralization and ophiolite formation: an example from the Meratus mountains, Borneo. Econ Geol 106:93–117

Coggon JA, Nowell GM, Pearson DG, Oberthür T, Lorand JP, Melcher F, Parman SW (2012) The ^{190}Pt–^{186}Os decay system applied to dating platinum-group element mineralization of the Bushveld Complex, South Africa. Chem Geol 302:48–60

Cohen AS, Waters FG (1996) Separation of osmium from geological materials by solvent extraction for analysis by thermal ionisation mass spectrometry. Anal Chim Acta 332:269–275

Cook DL, Walker RJ, Horan MF, Wasson JT, Morgan JW (2004) Pt-Re-Os systematics of group IIAB and IIIAB iron meteorites. Geochim Cosmochim Acta 68:1413–1431

Cumming VM, Selby D, Lillis PG, Lewan MD (2014) Re–Os geochronology and Os isotope fingerprinting of petroleum sourced from a Type I lacustrine kerogen: insights from the natural Green River petroleum system in the Uinta Basin and hydrous pyrolysis experiments. Geochim Cosmochim Acta 138:32–56

Creaser RA, Papanastassiou DA, Wasserburg GJ (1991) Negative thermal ion mass spectrometry of osmium, rhenium and iridium. Geochim Cosmochim Acta 55:397–401

Creaser RA, Sannigrahi P, Chacko T, Selby D (2002) Further evaluation of the Re–Os geochronometer in organic-rich sedimentary rocks: a test of hydrocarbon maturation effects in the Exshaw Formation, Western Canada Sedimentary Basin. Geochim Cosmochim Acta 66:3441–3452

Creech JB, Baker JA, Handler MR, Bizzarro M (2014) Platinum stable isotope analysis of geological standard reference materials by double-spike MC-ICPMS. Chem Geol 363:293–300

Czamanske GK, Kunilov VE, Zientek ML, Cabri LJ, Likhachev AP, Calk LC, Oscarson RL (1992) A proton microprobe study of magmatic sulfide ores from the Noril'sk-Talnakh District, Siberia. Can Mineral 30:249–287

Day JM, Pearson DG, Hulbert LJ (2008) Rhenium–osmium isotope and platinum-group element constraints on the origin and evolution of the 1–27 Ga Muskox layered intrusion. J Petrol 49:1255–1295

Dickin AP, McNutt RH, McAndrew JI (1988) Osmium isotope analysis by inductively coupled plasma mass spectrometry. J Anal at Spectrom 3:337–342

Dijkstra AH, Dale CW, Oberthür T, Nowell GM, Pearson DG (2016) Osmium isotope compositions of detrital Os-rich alloys from the Rhine River provide evidence for a global late Mesoproterozoic mantle depletion event. Earth Planet Sci Lett 452:115–122

DiMarzio JM, Georgiev SV, Stein HJ, Hannah JL (2018) Residency of rhenium and osmium in a heavy crude oil. Geochim Cosmochim Acta 220:180–200

Distler VV (1994) Platinum mineralization of the Noril'sk deposits. Ontario Geol Survey Spec Publ 5:243–260

Drábek M, Rieder M, Böhmová V (2010) The Re–Mo–S system: new data on phase relations between 400 and 1200 C. Eur J Mineral 22:479–484

Englert P, Herpers U (1980) Isotopic anomalies of osmium from different deposits determined by the laser microprobe mass analyzer lamma. Inorg Nucl Chem Lett 16:37–43

Esser BK, Turekian KK (1993) The osmium isotopic composition of the continental crust. Geochim Cosmochim Acta 57:3093–3104

Fehn U, Teng R, Elmore D, Kubik PW (1986) Isotopic composition of osmium in terrestrial samples determined by accelerator mass spectrometry. Nature 323:707–710

Fiorentini M, Beresford S, Barley M, Duuring P, Bekker A, Rosengren N, Cas R, Hronsky J (2012) District to camp controls on the genesis of komatiite-hosted nickel sulfide deposits, Agnew-Wiluna greenstone belt, Western Australia: insights from the multiple sulfur isotopes. Econ Geol 107:781–796

Fleet ME, Wu TW (1993) Volatile transport of platinum-group elements in sulfide-chloride assemblages at 1000 C. Geochim Cosmochim Acta 57:3519–3531

Fonseca RO, Mallmann G, O'Neill HSC, Campbell IH (2007) How chalcophile is rhenium? An experimental study of the solubility of Re in sulfide mattes. Earth Planet Sci Lett 260:537–548

Foster JG, Lambert DD, Frick LR, Maas R (1996) Re–Os isotopic evidence for genesis of Archaean nickel ores from uncontaminated komatiites. Nature 382:703

Frei R, Nägler TF, Schönberg R, Kramers JD (1998) Re–Os, Sm–Nd, U–Pb, and stepwise lead leaching isotope systematics in shear-zone hosted gold mineralization: genetic tracing and age constraints of crustal hydrothermal activity. Geochim Cosmochim Acta 62:1925–1936

Freydier C, Ruiz J, Chesley J, McCandless T, Munizaga F (1997) Re–Os isotope systematics of sulfides from felsic igneous rocks: application to base metal porphyry mineralization in Chile. Geology 25:775–778

Gaetani GA, Grove TL (1999) Wetting of mantle olivine by sulfide melt: implications for Re/Os ratios in mantle peridotite and late-stage core formation. Earth Planet Sci Lett 169:147–163

Gangopadhyay A, Walker RJ (2003) Re–Os systematics of the ca. 2.7-Ga komatiites from Alexo, Ontario. Canada. Chem Geol 196:147–162

Gangopadhyay A, Sproule RA, Walker RJ, Lesher CM (2005) Re–Os systematics of komatiites and komatiitic basalts at Dundonald Beach, Ontario, Canada: Evidence for a complex alteration history and implications of a late-Archean chondritic mantle source. Geochim Cosmochim Acta 69:5087–5098

Gannoun A, Tessalina S, Bourdon B, Orgeval JJ, Birck JL, Allègre CJ (2003) Re–Os isotopic constraints on the genesis and evolution of the Dergamish and Ivanovka Cu (Co, Au) massive sulphide deposits, south Urals, Russia. Chem Geol 196:193–207

Gannoun A, Burton KW, Parkinson IJ, Alard O, Schiano P, Thomas LE (2007) The scale and origin of the osmium isotope variations in mid-ocean ridge basalts. Earth Planet Sci Lett 259:541–556

Gannoun A, Burton KW, Day JM, Harvey J, Schiano P, Parkinson I (2016) Highly siderophile element and Os isotope systematics of volcanic rocks at divergent and convergent plate boundaries and in intraplate settings. Rev Mineral Geochem 81:651–724

Gao B, Zhang L, Jin X, Li Z, Li W (2020) Rhenium-Osmium isotope systematics of an early mesoproterozoic SEDEX polymetallic pyrite deposit in the North China Craton: implications for geological significance and the marine osmium isotopic record. Ore Geol Rev 117:103331

Georgiev SV, Stein HJ, Hannah JL, Galimberti R, Nali M, Yang G, Zimmerman A (2016) Re–Os dating of maltenes and asphaltenes within single samples of crude oil. Geochim Cosmochim Acta 179:53–75

Georgiev SV, Stein HJ, Hannah JL, di Primio R (2021) Timing and origin of multiple petroleum charges in the Solveig oil field, Norwegian North Sea: a rhenium-osmium isotopic study. AAPG Bull 105:109–134

Gregoire DC (1990) Sample introduction techniques for the determination of osmium isotope ratios by inductively coupled plasma mass spectrometry. Anal Chem 62:141–146

Graham S, Lambert DD, Shee SR, Smith CB, Reeves S (1999) Re-Os isotopic evidence for Archean lithospheric mantle beneath the Kimberley block, Western Australia. Geology 27:431–434

Gregory MJ, Schaefer BF, Keays RR, Wilde AR (2008) Rhenium–osmium systematics of the Mount Isa copper orebody and the Eastern Creek Volcanics, Queensland, Australia: implications for ore genesis. Miner Depos 43:553–573

Griffin WL, Graham S, O'Reilly SY, Pearson NJ (2004) Lithosphere evolution beneath the Kaapvaal Craton: Re–Os systematics of sulfides in mantle-derived peridotites. Chem Geol 208:89–118

Habfast K (1983) Fractionation in the thermal ionization source. Int J Mass Spectrom Ion Phy 51:165–189

Habfast K (1998) Fractionation correction and multiple collectors in thermal ionization isotope ratio mass spectrometry. Int J Mass Spectrom 176:133–148

Hannah JL, Stein HJ (2002) Re–Os model for the origin of sulfide deposits in anorthosite-associated intrusive complexes. Econ Geol 97:371–383

Hannah JL, Bekker A, Stein HJ, Markey RJ, Holland HD (2004) Primitive Os and 2316 Ma age for marine shale: implications for Paleoproterozoic glacial events and the rise of atmospheric oxygen. Earth Planet Sci Lett 225:43–52

Hart SR, Kinloch ED (1989) Osmium isotope systematics in Witwatersrand and Bushveld ore deposits. Econ Geol 84:1651–1655

Hart SR, Ravizza GE (1996) Os partitioning between phases in lherzolite and basalt. Geophys Mon Ser 95:123–134

Hart GL, Johnson CM, Shirey SB, Clynne MA (2002) Osmium isotope constraints on lower crustal recycling and pluton preservation at Lassen Volcanic Center, CA. Earth Planet Sci Lett 199(3):269–285

Harvey J, Gannoun A, Burton KW, Rogers NW, Alard O, Parkinson IJ (2006) Ancient melt extraction from the oceanic upper mantle revealed by Re–Os isotopes in abyssal peridotites from the Mid-Atlantic ridge. Earth Planet Sci Lett 244:606–621

Harvey J, Warren JM, Shirey SB (2016) Mantle sulfides and their role in Re–Os and Pb isotope geochronology. Rev Mineral Geochem 81:579–649

Hassler DR, Peucker-Ehrenbrink B, Ravizza GE (2000) Rapid determination of Os isotopic composition by sparging OsO_4 into a magnetic-sector ICP-MS. Chem Geol 166:1–14

Hattori K, Hart SR (1991) Osmium-isotope ratios of platinum-group minerals associated with ultramafic intrusions: Os-isotopic evolution of the oceanic mantle. Earth Planet Sci Lett 107:499–514

Hauri EH, Hart SR (1997) Rhenium abundances and systematics in oceanic basalts. Chem Geol 139:185–205

Herr W, Woelfe R, Eberhardt P, Kopp E (1968) Development and recent applications of the Re/Os dating method. Radioactive dating and methods of low-level counting. International Atomic Energy Agency, Vienna, pp 499–508

Hintenberger H, Herr W, Voshage H (1954) Radiogenic osmium from rhenium-containing molybdenite. Phys Rev 95:1690

Hirata T, Hattori M, Tanaka T (1998) In-situ osmium isotope ratio analyses of iridosmines by laser ablation–multiple collector–inductively coupled plasma mass spectrometry. Chem Geol 144:269–280

Hnatyshin D, Creaser RA, Meffre S, Stern RA, Wilkinson JJ, Turner EC (2020) Understanding the micro-scale spatial distribution and mineralogical residency of Re in pyrite: examples from carbonate-hosted Zn–Pb ores and implications for pyrite Re–Os geochronology. Chem Geol 533:119427

Hoffman EL, Naldrett AJ, Van Loon JC, Hancock RGV, Manson A (1978) The determination of all the platinum group elements and gold in rocks and ore by neutron activation analysis after preconcentration by a nickel sulfide fire-assay technique on large samples. Anal Chimica Acta 102:157–166

Hogmalm KJ, Dahlgren I, Fridolfsson I, Zack T (2019) First in situ Re–Os dating of molybdenite by LA-ICP-MS/MS. Miner Depos 54:821–828

Horan MF, Morgan JW, Walker RJ, Cooper RW (2001) Re-Os isotopic constraints on magma mixing in the peridotite zone of the Stillwater Complex, Montana, USA. Contrib Mineral Petrol 141:446–457

Hou Z, Wang S, Du A, Qu X, Sun W (2003) Re–Os dating of sulfides from the volcanogenic massive sulfide deposit at Gacun, Southwestern China. Resource Geol 53:305–310

Huang X, Qi L, Wang Y, Liu Y (2014) Re–Os dating of magnetite from the Shaquanzi Fe–Cu deposit, eastern Tianshan, NW China. Sci China Earth Sci 57:267–277

Hulbert LJ, Hamilton MA, Horan MF, Scoates RFJ (2005) U–Pb zircon and Re–Os isotope geochronology of mineralized ultramafic intrusions and associated nickel ores from the Thompson Nickel Belt, Manitoba, Canada. Econ Geol 100:29–41

Jenner FE, O'Neill HSC, Arculus RJ, Mavrogenes JA (2010) The magnetite crisis in the evolution of arc-related magmas and the initial concentration of Au, Ag and Cu. J Petrol 51:2445–2464

Johnson CM, Shirey SB, Barovich KM (1996) New approaches to crustal evolution studies and the origin of granitic rocks: what can the Lu–Hf and Re–Os isotope systems tell us? Geol Soc America Spec Pap 315:339–352

Johnson A, Canil D (2011) The degassing behavior of Au, Tl, As, Pb, Re, Cd and Bi from silicate liquids: experiments and applications. Geochim Cosmochim Acta 75:1773–1784

Kamo SL, Czamanske GK, Amelin Y, Fedorenko VA, Davis DW, Trofimov VR (2003) Rapid eruption of Siberian flood-volcanic rocks and evidence for coincidence with the Permian-Triassic boundary and mass extinction at 251 Ma. Earth Planet Sci Lett 214:75–91

Kemp AIS, Blevin PL, Norman MD (2020) A SIMS U-Pb (zircon) and Re-Os (molybdenite) isotope study of the early Paleozoic Macquarie arc, southeastern Australia: implications for the tectono-magmatic evolution of the paleo-Pacific Gondwana margin. Gondwana Res 82:73–96

Kiseeva ES, Wood BJ (2013) A simple model for chalcophile element partitioning between sulfide and silicate liquids with geochemical applications. Earth Planet Sci Lett 383:68–81

Kirk J, Ruiz J, Chesley J, Walshe J, England G (2002) A major Archean, gold-and crust-forming event in the Kaapvaal Craton, South Africa. Science 297:1856–1858

Kochergina YV, Ackerman L, Erban V, Matusiak-Małek M, Puziewicz J, Halodová P, Špaček P, Trubač J, Magna T (2016) Rhenium–osmium isotopes in pervasively metasomatized mantle xenoliths from the Bohemian Massif and implications for the reliability of Os model ages. Chem Geol 430:90–107

Košler J, Simonetti A, Sylvester PJ, Cox RA, Tubrett MN, Wilton DH (2003) Laser ablation ICP–MS measurements of Re/Os in molybdenite and implications for Re–Os geochronology. Can Mineral 41:307–320

Lahaye Y, Barnes SJ, Frick LR, Lambert DD (2001) Re–Os isotopic study of komatiitic volcanism and magmatic sulfide formation in the southern Abitibi greenstone belt, Ontario, Canada. Can Mineral 39:473–490

Lambert DD, Walker RJ, Morgan JW, Shirey SB, Carlson RW, Zientek ML, Lipin BR, Koski MS, Cooper RL (1994) Re–Os and Sm–Nd isotope geochemistry of the Stillwater Complex, Montana: implications for the petrogenesis of the JM Reef. J Petrol 35:1717–1753

Lambert DD, Shirey SB, Bergman SC (1995) Proterozoic lithospheric mantle source for the Prairie Creek lamproites: Re–Os and Sm–Nd isotopic evidence. Geology 23:273–276

Lambert DD, Foster JG, Frick LR, Hoatson DM, Purvis AC (1998) Application of the Re–Os isotopic system to the study of Precambrian magmatic sulfide deposits of Western Australia. Austr J Earth Sci 45:265–284

Lambert DD, Foster JG, Frick LR, Ripley EM (1999) Re–Os isotope geochemistry of magmatic sulfide ore systems. Rev Econ Geol 12:29–58

Lambert DD, Frick LR, Foster JG, Li C, Naldrett AJ (2000) Re–Os isotope systematics of the Voisey's Bay Ni–Cu–Co magmatic sulfide system, Labrador, Canada: II. implications for parental magma chemistry, ore genesis, and metal redistribution. Econ Geol 95:867–888

Lassiter JC (2003) Rhenium volatility in subaerial lavas: constraints from subaerial and submarine portions of the HSDP-2 Mauna Kea drillcore. Earth Planet Sci Lett 214:311–325

Lassiter JC, Luhr JF (2001) Osmium abundance and isotope variations in mafic Mexican volcanic rocks: evidence for crustal contamination and constraints on the geochemical behavior of osmium during partial melting and fractional crystallization. Geochem Geophys Geosys 2: 2000GC000116

Li Y (2014) Comparative geochemistry of rhenium in oxidized arc magmas and MORB and rhenium partitioning during magmatic differentiation. Chem Geol 386:101–114

Li SJ, Wang XC, Wilde SA, Chu Z, Li C, He S, Liu K, Ma X, Zhang Y (2021) Revisiting rhenium-osmium isotopic investigations of petroleum systems: from geochemical behaviours to geological interpretations. J Earth Sci 32:1226–1249

Liu J, Selby D, Zhou H, Pujol M (2019) Further evaluation of the Re–Os systematics of crude oil: implications for Re–Os geochronology of petroleum systems. Chem Geol 513:1–22

Lorand JP, Luguet A (2016) Chalcophile and siderophile elements in mantle rocks: trace elements controlled by trace minerals. Rev Mineral Geochem 81:441–488

Luck JM, Allègre CJ (1982) The study of molybdenites through the ^{187}Re-^{187}Os chronometer. Earth Planet Sci Lett 61:291–296

Luck JM, Allègre CJ (1983) ^{187}Re–^{187}Os systematics in meteorites and cosmochemical consequences. Nature 302:130–132

Luck JM, Birck JL, Allègre CJ (1980) ^{187}Re–^{187}Os systematics in meteorites: early chronology of the Solar System and age of the Galaxy. Nature 283:256–259

Macfarlane RD, Kohman TP (1961) Natural alpha radioactivity in medium-heavy elements. Phys Rev 121:1758

MacKenzie JM, Canil D (2006) Experimental constraints on the mobility of rhenium in silicate liquids. Geochim Cosmochim Acta 70:5236–5245

Majzlan J, Mikuš T, Kiefer S, Creaser RA (2022) Rhenium-osmium geochronology of gersdorffite and skutterudite-pararammelsbergite links nickel–cobalt mineralization to the opening of the incipient Meliata Ocean (Western Carpathians, Slovakia). Miner Depos 57:621–629

Malitch KN, Latypov RM (2011) Re–Os and S isotope constraints on timing and source heterogeneity of PGE–Cu–Ni sulfide ores: a case study at the Talnakh ore junction, Noril'sk Province, Russia. Can Mineral 49:1653–1677

Malitch KN, Belousova EA, Griffin WL, Badanina IY, Pearson NJ, Presnyakov SL, Tuganova EV (2010) Magmatic evolution of the ultramafic–mafic Kharaelakh intrusion (Siberian Craton, Russia): insights from trace-element, U–Pb and Hf-isotope data on zircon. Contrib Mineral Petrol 159:753–768

Mallmann G, O'Neill HSC (2007) The effect of oxygen fugacity on the partitioning of Re between crystals and silicate melt during mantle melting. Geochim Cosmochim Acta 71:2837–2857

Marcantonio F, Zindler A, Reisberg L, Mathez EA (1993) Re–Os isotopic systematics in chromitites from the Stillwater Complex, Montana, USA. Geochim Cosmochim Acta 57:4029–4037

Marcantonio F, Reisberg L, Zindler A, Wyman D, Hulbert L (1994) An isotopic study of the Ni–Cu–PGE-rich Wellgreen intrusion of the Wrangellia Terrane: evidence for hydrothermal mobilization of rhenium and osmium. Geochim Cosmochim Acta 58:1007–1018

Markey R, Stein H, Morgan J (1998) Highly precise Re–Os dating for molybdenite using alkaline fusion and NTIMS. Talanta 45:935–946

Markey R, Hannah JL, Morgan JW, Stein HJ (2003) A double spike for osmium analysis of highly radiogenic samples. Chem Geol 200:395–406

Marques JC, Ferreira Filho CF, Carlson RW, Pimentel MM (2003) Re–Os and Sm–Nd isotope and trace element constraints on the origin of the chromite deposit of the Ipueira-Medrado Sill, Bahia, Brazil. J Petrol 44:659–678

Martin CE (1989) Re-Os isotopic investigation of the Stillwater Complex, Montana. Earth Planet Sci Lett 93:336–344

Mathur R, Ruiz J, Tornos F (1999) Age and sources of the ore at Tharsis and Rio Tinto, Iberian Pyrite Belt, from Re–Os isotopes. Miner Depos 34:790–793

Mathur R, Ruiz J, Titley S, Gibbins S, Margotomo W (2000a) Different crustal sources for Au-rich and Au-poor ores of the Grasberg Cu–Au porphyry deposit. Earth Planet Sci Lett 183:7–14

Mathur R, Ruiz J, Munizaga F (2000b) Relationship between copper tonnage of Chilean base-metal porphyry deposits and Os isotope ratios. Geology 28:555–558

Mathur R, Marschik R, Ruiz J, Munizaga F, Leveille RA, Martin W (2002) Age of mineralization of the Candelaria Fe oxide Cu–Au deposit and the origin of the Chilean iron belt, based on Re–Os isotopes. Econ Geol 97:59–71

Mathur R, Titley S, Ruiz J, Gibbins S, Friehauf K (2005) A Re–Os isotope study of sedimentary rocks and copper–gold ores from the Ertsberg District, West Papua, Indonesia. Ore Geol Rev 26:207–226

McBride JS, McInnes BIA, Keays RR (2001) COMMENT: Relationship between copper tonnage of Chilean base-metal porphyry deposits and Os isotope ratios. Geology 29:467–468

McInnes BI, McBride JS, Evans NJ, Lambert DD, Andrew AS (1999) Osmium isotope constraints on ore metal recycling in subduction zones. Science 286:512–516

McInnes BIA, Keays RR, Lambert DD, Hellstrom J, Allwood JS (2008) Re–Os geochronology and isotope systematics of the Tanami, Tennant Creek and Olympic Dam Cu–Au deposits. Austr J Earth Sci 55:967–981

Meibom A, Sleep NH, Chamberlain CP, Coleman RG, Frei R, Hren MT, Wooden JL (2002) Re–Os isotopic evidence for long-lived heterogeneity and equilibration processes in the Earth's upper mantle. Nature 419:705–708

Meibom A, Frei R, Sleep NH (2004) Osmium isotopic compositions of Os-rich platinum group element alloys from the Klamath and Siskiyou Mountains. J Geophys Res Solid Earth 109:B02203

Meisel T, Walker RJ, Irving AJ, Lorand JP (2001a) Osmium isotopic compositions of mantle xenoliths: a global perspective. Geochim Cosmochim Acta 65:1311–1323

Meisel T, Moser J, Fellner N, Wegscheider W, Schoenberg R (2001b) Simplified method for the determination of Ru, Pd, Re, Os, Ir and Pt in chromitites and other geological materials by isotope dilution ICP-MS and acid digestion. Analyst 126:322–328

Meisel T, Reisberg L, Moser J, Carignan J, Melcher F, Brügmann G (2003) Re–Os systematics of UB-N, a serpentinized peridotite reference material. Chem Geol 201:161–179

Meisel T, Horan MF (2016) Analytical methods for the highly siderophile elements. Rev Mineral Geochem 81:89–106

Meng QA, Wang X, Huo QL, Dong ZL, Li Z, Tessalina SG, Ware BD, McInnes BI, Wang XL, Liu T, Zhang L (2021) Rhenium–osmium (Re–Os) geochronology of crude oil from lacustrine source rocks of the Hailar Basin, NE China. Pet Sci 18:1–9

Miller CA, Peucker-Ehrenbrink B, Ball L (2009) Precise determination of rhenium isotope composition by multi-collector inductively-coupled plasma mass spectrometry. J Anal at Spectrom 24:1069–1078

Morelli RM, Creaser RA, Selby D, Kontak DJ, Horne RJ (2005) Rhenium-Osmium geochronology of arsenopyrite in Meguma Group gold deposits, Meguma Terrane, Nova Scotia, Canada: evidence for multiple gold-mineralizing events. Econ Geol 100:1229–1242

Morelli RM, Bell CC, Creaser RA, Simonetti A (2010) Constraints on the genesis of gold mineralization at the Homestake Gold Deposit, Black Hills, South Dakota from rhenium–osmium sulfide geochronology. Miner Depos 45:461–480

Morgan JW (1965) The simultaneous determination of rhenium and osmium in rocks by neutron activation analysis. Anal Chim Acta 32:8–16

Morgan JW, Walker RJ (1989) Isotopic determinations of rhenium and osmium in meteorites by using fusion, distillation and ion-exchange separations. Anal Chim Acta 222:291–300

Morgan JW, Lovering JF, Ford RJ (1968) Rhenium and non-radiogenic osmium in Australian molybdenites and other sulfide minerals by neutron activation analysis. J Geol Soc Austr 15:189–194

Morgan JW, Golightly DW, Dorrzapf AF (1991) Methods for the separation of rhenium, osmium and molybdenum applicable to isotope geochemistry. Talanta 38:259–265

Morgan JW, Stein HJ, Hannah JL, Markey RJ, Wiszniewska J (2000) Re–Os study of Fe–Ti–V oxide and Fe–Cu–Ni sulfide deposits, Suwałki Anorthosite Massif, northeast Poland. Miner Depos 35:391–401

Morgan JW, Walker RJ, Horan MF, Beary ES, Naldrett AJ (2002) ^{190}Pt-^{186}Os and ^{187}Re-^{187}Os systematics of the Sudbury igneous complex, Ontario. Geochim Cosmochim Acta 66:273–290

Mountain BW, Wood SA (1988) Solubility and transport of platinum-group elements in hydrothermal solutions: thermodynamic and physical chemical constraints. Geo-Platinum 87. Springer, Amsterdam, pp 57–82

Munhá J, Relvas JMRS, Barriga FJAS, Conceição P, Jorge RCGS, Mathur R, Ruiz J, Tassinari CCG (2005) Osmium isotope systematics in the Iberian Pyrite Belt. In: Mineral deposit research: meeting the global challenge, Springer, Berlin-Heidelberg, pp 663–666

Naldrett SN, Libby WF (1948) Natural radioactivity of rhenium. Phys Rev 73:487

Naldrett AJ, Innes DG, Sowa J, Gorton MP (1982) Compositional variations within and between five Sudbury ore deposits. Econ Geol 77:1519–1534

Naldrett AJ, Asif M, Gorbachev NS, Kunilov VE, Fedorenko VA, Lightfoot PC (1994) The composition of the Ni–Cu ores of the Noril'sk region. In The Sudbury-Noril'sk Symposium. Ontario Geol Surv Spec Publ 5:357–371

Naldrett AJ, Asif M, Scandl E, Searcy T, Morrison GG, Binney WP, Moore C (1999) Platinum-group elements in the Sudbury ores; significance with respect to the origin of different ore zones and to the exploration for footwall orebodies. Econ Geol 94:185–210

Nanne JA, Millet MA, Burton KW, Dale CW, Nowell GM, Williams HM (2017) High precision osmium stable isotope measurements by double spike MC-ICP-MS and N-TIMS. J Anal at Spectrom 32:749–765

Nier AO (1937) The isotopic constitution of osmium. Phys Rev 52:885

Norman M, Bennett V, McCulloch M, Kinsley L (2002) Osmium isotopic compositions by vapor phase sample introduction using a multi-collector ICP-MS. J Anal at Spectrom 17:1394–1397

Norman M, Bennett V, Blevin P, McCulloch M (2004a) New Re–Os ages of molybdenite from granite-related deposits of eastern Australia using an improved multi-collector ICP-MS Technique. Geol Soc Austr Abstr 74:129–132

Norman MD, Garcia MO, Bennett VC (2004b) Rhenium and chalcophile elements in basaltic glasses from Ko'olau and Moloka'i volcanoes: Magmatic outgassing and composition of the Hawaiian plume. Geochim Cosmochim Acta 68:3761–3777

Nowell GM, Luguet A, Pearson DG, Horstwood MSA (2008a) Precise and accurate ^{186}Os/^{188}Os and ^{187}Os/^{188}Os measurements by multi-collector plasma ionisation mass spectrometry (MC-ICP-MS) part I: solution analyses. Chem Geol 248:363–393

Nowell GM, Pearson DG, Parman SW, Luguet A, Hanski E (2008b) Precise and accurate ^{186}Os/^{188}Os and ^{187}Os/^{188}Os measurements by multi-collector plasma ionisation mass spectrometry, part II: laser ablation and its application to single-grain Pt–Os and Re–Os geochronology. Chem Geol 248:394–426

Nozaki T, Kato Y, Suzuki K (2010) Re–Os geochronology of the Iimori Besshi-type massive sulfide deposit in the Sanbagawa metamorphic belt, Japan. Geochim Cosmochim Acta 74:4322–4331

Nozaki T, Suzuki K, Ravizza G, Kimura JI, Chang Q (2012) A method for rapid determination of Re and Os isotope compositions using ID-MC-ICP-MS combined with the sparging method. Geostand Geoanal Res 36:131–148

Nozaki T, Kato Y, Suzuki K (2014) Re–Os geochronology of the Hitachi volcanogenic massive sulfide deposit: the oldest ore deposit in Japan. Econ Geol 109:2023–2034

O'Driscoll B, Day JM, Daly JS, Walker RJ, McDonough WF (2009) Rhenium–osmium isotopes and platinum-group elements in the Rum Layered Suite, Scotland: Implications for Cr-spinel seam formation and the composition of the Iceland mantle anomaly. Earth Planet Sci Lett 286:41–51

O'Driscoll B, González-Jiménez JM (2016) Petrogenesis of the platinum-group minerals. Rev Mineral Geochem 81:489–578

Page NJ, Talkington RW (1984) Palladium, platinum, rhodium, ruthenium and iridium in peridotites and chromitites from ophiolite complexes in Newfoundland. Can Mineral 22:137–149

Park JW, Campbell IH, Eggins SM (2012) Enrichment of Rh, Ru, Ir and Os in Cr spinels from oxidized magmas: evidence from the Ambae volcano, Vanuatu. Geochim Cosmochim Acta 78:28–50

Park JW, Campbell IH, Ickert RB, Allen CM (2013) Chalcophile element geochemistry of the Boggy Plain zoned pluton, southeastern Australia: a S-saturated barren compositionally diverse magmatic system. Contrib Mineral Petrol 165:217–236

Parkinson IJ, Hawkesworth CJ, Cohen AS (1998) Ancient mantle in a modern arc: osmium isotopes in Izu-Bonin-Mariana forearc peridotites. Science 281:2011–2013

Pearson NJ, Alard O, Griffin WL, Jackson SE, O'Reilly SY (2002) In situ measurement of Re–Os isotopes in mantle sulfides by laser ablation multicollector-inductively coupled plasma mass spectrometry: analytical methods and preliminary results. Geochim Cosmochim Acta 66:1037–1050

Petrou AL, Economou-Eliopoulos M (2009) Platinum-group mineral formation: evidence of an interchange process from the entropy of activation values. Geochim Cosmochim Acta 73:5635–5645

Peucker-Ehrenbrink B, Jahn BM (2001) Rhenium-osmium isotope systematics and platinum group element concentrations: loess and the upper continental crust. Geochem Geophys Geosys 2: 2001GC000172

Pitcher L, Helz RT, Walker RJ, Piccoli P (2009) Fractionation of the platinum-group elements and Re during crystallization of basalt in Kilauea Iki Lava Lake, Hawaii. Chem Geol 260:196–210

Puchtel IS, Brandon AD, Humayun M (2004) Precise Pt–Re–Os isotope systematics of the mantle from 2.7-Ga komatiites. Earth Planet Sci Lett 224:157–174

Ravizza G, Pyle D (1997) PGE and Os isotopic analyses of single sample aliquots with NiS fire assay preconcentration. Chem Geol 141:251–268

Ravizza G, Turekian KK (1989) Application of the 187Re-187Os system to black shale geochronometry. Geochim Cosmochim Acta 53:3257–3262

Ravizza G, Turekian KK, Hay BJ (1991) The geochemistry of rhenium and osmium in recent sediments from the Black Sea. Geochim Cosmochim Acta 55:3741–3752

Ravizza G, Martin CE, German CR, Thompson G (1996) Os isotopes as tracers in seafloor hydrothermal systems: Metalliferous deposits from the TAG hydrothermal area, 26 N Mid-Atlantic Ridge. Earth Planet Sci Lett 138:105–119

Ravizza G, Blusztajn J, Prichard HM (2001) Re–Os systematics and platinum-group element distribution in metalliferous sediments from the Troodos ophiolite. Earth Planet Sci Lett 188:369–381

Rehkämper M, Halliday AN, Alt J, Fitton JG, Zipfel J, Takazawa E (1999) Non-chondritic platinum-group element ratios in oceanic mantle lithosphere: petrogenetic signature of melt percolation? Earth Planet Sci Lett 172:65–81

Reisberg L, Meisel T (2002) The Re–Os isotopic system: a review of analytical techniques. Geostand Geoanal Res 26:249–267

Reisberg L, Tredoux M, Harris C, Coftier A, Chaumba J (2011) Re and Os distribution and Os isotope composition of the Platreef at the Sandsloot-Mogolakwena mine, Bushveld complex, South Africa. Chem Geol 281:352–363

Renne PR (1995) Excess ^{40}Ar in biotite and hornblende from the Noril'sk 1 intrusion, Siberia: implications for the age of the Siberian Traps. Earth Planet Sci Lett 131:165–176

Richardson JM, Dickin AP, McNutt RH, McAndrew JI, Beneteau SB (1989) Analysis of a rhenium-osmium solid-solution spike by inductively coupled plasma mass spectrometry. J Anal at Spectrom 4:465–471

Righter K, Chesley JT, Geist D, Ruiz J (1998) Behavior of Re during magma fractionation: an example from Volcan Alcedo, Galapagos. J Petrol 39:785–795

Righter K, Chesley JT, Caiazza CM, Gibson EK, Ruiz J (2008) Re and Os concentrations in arc basalts: the roles of volatility and source region fO_2 variations. Geochim Cosmochim Acta 72:926–947

Riley GH (1967) Rhenium concentration in Australian molybdenites by stable isotope dilution. Geochim Cosmochim Acta 31:1489–1497

Ripley EM, Lambert DD, Frick LR (1998) Re–Os, Sm–Nd, and Pb isotopic constraints on mantle and crustal contributions to magmatic sulfide mineralization in the Duluth Complex. Geochim Cosmochim Acta 62:3349–3365

Ripley EM, Park YR, Lambert DD, Frick LR (2001) Re-Os isotopic variations in carbonaceous pelites hosting the Duluth Complex: implications for metamorphic and metasomatic processes associated with mafic magma chambers. Geochim Cosmochim Acta 65:2965–2978

Ripley EM, Shafer P, Li C, Hauck SA (2008) Re–Os and O isotopic variations in magnetite from the contact zone of the Duluth Complex and the Biwabik Iron Formation, northeastern Minnesota. Chem Geol 249:213–226

Rosman KJR, Taylor PDP (1998) Isotopic compositions of the elements 1997 (Technical Report). Pure Appl Chem 70:217–235

Roy-Barman M, Wasserburg GJ, Papanastassiou DA, Chaussidon M (1998) Osmium isotopic compositions and Re–Os concentrations in sulfide globules from basaltic glasses. Earth Planet Sci Lett 154:331–347

Rudnick RL, Walker RJ (2009) Interpreting ages from Re–Os isotopes in peridotites. Lithos 112:1083–1095

Ruiz J, Mathur R (1999) Metallogenesis in continental margins: Re–Os evidence from porphyry copper deposits in Chile. Rev Econ Geol 12:59–72

Russ GP III, Bazan JM, Date AR (1987) Osmium isotopic ratio measurements by inductively coupled plasma source mass spectrometry. Anal Chem 59:984–989

Saal AE, Rudnick RL, Ravizza GE, Hart SR (1998) Re–Os isotope evidence for the composition, formation and age of the lower continental crust. Nature 393: 58–61

Saintilan NJ, Creaser RA, Bookstrom AA (2017) Re–Os systematics and geochemistry of cobaltite (CoAsS) in the Idaho Cobalt belt, Belt-Purcell Basin, USA: evidence for middle Mesoproterozoic sediment-hosted Co–Cu sulfide mineralization with Grenvillian and Cretaceous remobilization. Ore Geol Rev 86:509–525

Saintilan NJ, Spangenberg JE, Chiaradia M, Chelle-Michou C, Stephens MB, Fontboté L (2019) Petroleum as source and carrier of metals in epigenetic sediment-hosted mineralization. Sci Rep 9:1–7

Saintilan NJ, Selby D, Hughes JW, Schlatter D, Kolb J, Boyce A (2020) Mineral separation protocol for accurate and precise rhenium-osmium (Re–Os) geochronology and sulphur isotope composition of individual sulphide species. Methodsx 7:100944

Saintilan NJ, Sproson AD, Selby D, Rottier B, Casanova V, Creaser RA, Kouzmanov K, Fontboté L, Piecha M, Gereke M, Zambito JJ IV (2021) Osmium isotopic constraints on sulphide formation in the epithermal environment of magmatic-hydrothermal mineral deposits. Chem Geol 564:120053

Savard D, Barnes SJ, Meisel T (2010) Comparison between nickel-sulfur fire assay Te co-precipitation and isotope dilution with high-pressure asher acid digestion for the determination of platinum-group elements, rhenium and gold. Geostand Geoanal Res 34:281–291

Schaefer BF, Pearson DG, Rogers NW, Barnicoat AC (2010) Re–Os isotope and PGE constraints on the timing and origin of gold mineralisation in the Witwatersrand Basin. Chem Geol 276:88–94

Schoenberg R, Kruger FJ, Nägler TF, Meisel T, Kramers JD (1999) PGE enrichment in chromitite layers and the Merensky Reef of the western Bushveld Complex; a Re–Os and Rb–Sr isotope study. Earth Planet Sci Lett 172:49–64

Schoenberg R, Nägler TF, Kramers JD (2000) Precise Os isotope ratio and Re–Os isotope dilution measurements down to the picogram level using multicollector inductively coupled plasma mass spectrometry. Int J Mass Spectrom 197:85–94

Schoenberg R, Nägler TF, Gnos E, Kramers JD, Kamber BS (2003) The source of the Great Dyke, Zimbabwe, and its tectonic significance: evidence from Re–Os isotopes. J Geol 111:565–578

Selby D, Creaser RA (2001) Late and mid-Cretaceous mineralization in the northern Canadian Cordillera: constraints from Re–Os molybdenite dates. Econ Geol 96:1461–1467

Selby D, Creaser RA (2003) Re–Os geochronology of organic rich sediments: an evaluation of organic matter analysis methods. Chem Geol 200:225–240

Selby D, Creaser RA (2004) Macroscale NTIMS and microscale LA-MC-ICP-MS Re–Os isotopic analysis of molybdenite: testing spatial restrictions for reliable Re–Os age determinations, and implications for the decoupling of Re and Os within molybdenite. Geochim Cosmochim Acta 68:3897–3908

Selby D, Creaser RA (2005a) Direct radiometric dating of the Devonian-Mississippian time-scale boundary using the Re–Os black shale geochronometer. Geology 33:545–548

Selby D, Creaser RA (2005b) Direct radiometric dating of hydrocarbon deposits using rhenium-osmium isotopes. Science 308:1293–1295

Selby D, Creaser RA, Stein HJ, Markey RJ, Hannah JL (2007a) Assessment of the ^{187}Re decay constant by cross calibration of Re–Os molybdenite and U–Pb zircon chronometers in magmatic ore systems. Geochim Cosmochim Acta 71:1999–2013

Selby D, Creaser RA, Fowler MG (2007b) Re–Os elemental and isotopic systematics in crude oils. Geochim Cosmochim Acta 71:378–386

Selby D, Kelley KD, Hitzman MW, Zieg J (2009) Re-Os sulfide (bornite, chalcopyrite, and pyrite) systematics of the carbonate-hosted copper deposits at Ruby Creek, southern Brooks Range, Alaska. Econ Geol 104:437–444

Sen IS, Peucker-Ehrenbrink B (2014) Determination of osmium concentrations and ^{187}Os/^{188}Os of crude oils and source rocks by coupling high-pressure, high-temperature digestion with sparging OsO_4 into a multicollector inductively coupled plasma mass spectrometer. Anal Chem 86:2982–2988

Shirey SB, Walker RJ (1995) Carius tube digestion for low-blank rhenium-osmium analysis. Anal Chem 67:2136–2141

Shirey SB, Walker RJ (1998) The Re–Os isotope system in cosmochemistry and high-temperature geochemistry. Annu Rev Earth Planet Sci 26:423–500

Sie SH, Niklaus TR, Sims DA, Bruhn F, Suter G, Cripps G (2002) AUSTRALIS: a new tool for the study of isotopic systems and geochronology in mineral systems. Austr J Earth Sci 49:601–611

Simons DS (1983) Isotopic analysis with the laser microprobe mass analyzer. Int J Mass Spectrom Ion Process 55:15–30

Smoliar MI, Walker RJ, Morgan JW (1996) Re–Os ages of group IIA, IIIA, IVA, and IVB iron meteorites. Science 271:1099

Sproule RA, Lambert DD, Hoatson DM (2002) Decoupling of the Sm–Nd and Re–Os isotopic systems in sulfide-saturated magmas in the Halls Creek Orogen, Western Australia. J Petrol 43:375–402

Standish JJ, Hart SR, Blusztajn J, Dick HJB, Lee KL (2002) Abyssal peridotite osmium isotopic compositions from Cr-spinel. Geochem Geophys Geosys 3:1–24

Stein HJ (2006) Low-rhenium molybdenite by metamorphism in northern Sweden: recognition, genesis, and global implications. Lithos 87:300–327

Stein HJ (2014) Dating and tracing the history of ore formation. Treatise Geochem 13:87–118

Stein H, Hannah J (2015) Rhenium–osmium geochronology: sulfides, shales, oils, and mantle. Encyclopedia of scientific dating methods. Springer, Dordrecht, pp 707–723

Stein HJ, Sundblad K, Markey RJ, Morgan JW, Motuza G (1998) Re–Os ages for Archean molybdenite and pyrite, Kuittila-Kivisuo, Finland and Proterozoic molybdenite, Kabeliai, Lithuania: testing the chronometer in a metamorphic and metasomatic setting. Miner Depos 33:329–345

Stein HJ, Morgan JW, Scherstén A (2000) Re–Os dating of low-level highly radiogenic (LLHR) sulfides: The Harnäs gold deposit, southwest Sweden, records continental-scale tectonic events. Econ Geol 95:1657–1671

Stein HJ, Markey RJ, Morgan JW, Hannah JL, Scherstén A (2001) The remarkable Re–Os chronometer in molybdenite: how and why it works. Terra Nova 13:479–486

Stein H, Scherstén A, Hannah J, Markey R (2003) Subgrain-scale decoupling of Re and ^{187}Os and assessment of laser ablation ICP-MS spot dating in molybdenite. Geochim Cosmochim Acta 67:3673–3686

Stein HJ, Hannah JL, Zimmerman A, Markey RJ, Sarkar SC, Pal AB (2004) A 2.5 Ga porphyry Cu–Mo–Au deposit at Malanjkhand, central India: implications for Late Archean continental assembly. Precambrian Res 134:189–226

Stekhin AI (1994) Mineralogical and geochemical characteristics of the Cu–Ni ores of the Oktyabrsky and Talnakh deposits. Ontario Geol Survey Spec Publ 5:217–230

Sun W, Bennett VC, Eggins SM, Kamenetsky VS, Arculus RJ (2003a) Enhanced mantle-to-crust rhenium transfer in undegassed arc magmas. Nature 422:294–297

Sun W, Bennett VC, Eggins SM, Arculus RJ, Perfit MR (2003b) Rhenium systematics in submarine MORB and back-arc basin glasses: laser ablation ICP-MS results. Chem Geol 196:259–281

Sun W, Arculus RJ, Bennett VC, Eggins SM, Binns RA (2003c) Evidence for rhenium enrichment in the mantle wedge from submarine arc–like volcanic glasses (Papua New Guinea). Geology 31:845–848

Sun W, Xie Z, Chen J, Zhang X, Chai Z, Du A, Zhao J, Zhang C, Zhou T (2003d) Os–Os dating of copper and molybdenum deposits along the middle and lower reaches of the Yangtze River, China. Econ Geol 98:175–180

Sun W, Bennett VC, Kamenetsky VS (2004) The mechanism of Re enrichment in arc magmas: evidence from Lau Basin basaltic glasses and primitive melt inclusions. Earth Planet Sci Lett 222:101–114

Sun W, Audétat A, Dolejš D (2014) Solubility of molybdenite in hydrous granitic melts at 800 °C, 100–200 MPa. Geochim Cosmochim Acta 131:393–401

Suzuki K, Shimizu H, Masuda A (1993) Reliable Re–Os age for molybdenite. Geochim Cosmochim Acta 57:1625–1628

Suzuki K, Shimizu H, Masuda A (1996) Re–Os dating of molybdenites from ore deposits in Japan: implication for the closure temperature of the Re–Os system for molybdenite and the cooling history of molybdenum ore deposits. Geochim Cosmochim Acta 60:3151–3159

Suzuki K, Aizawa Y, Tatsumi Y (2002) Osmium and rhenium transport during serpentinite dehydration at 1.0 GPa. Frontier Research on Earth Evolution, Tokyo, Mapan Marine Scince and Technology Centre 1:107–110

Takahashi Y, Uruga T, Suzuki K, Tanida H, Terada Y, Hattori KH (2007) An atomic level study of rhenium and radiogenic osmium in molybdenite. Geochim Cosmochim Acta 71:5180–5190

Tavares OAP, Terranova ML, Medeiros EL (2006) New evaluation of alpha decay half-life of 190 Pt isotope for the Pt–Os dating system. Nucl Instrum Methods Phys Res Sect b: Beam Interact Mater at 243:256–260

Terada K, Osaki S, Ishihara S, Kiba T (1971) Distribution of rhenium in molybdenites from Japan. Geochem J 4:123–141

Tessalina SG, Bourdon B, Maslennikov VV, Orgeval JJ, Birck JL, Gannoun A, Capmas F, Allègre CJ (2008a) Osmium isotope distribution within the Palaeozoic Alexandrinka seafloor hydrothermal system in the Southern Urals, Russia. Ore Geol Rev 33:70–80

Tessalina SG, Yudovskaya MA, Chaplygin IV, Birck JL, Capmas F (2008b) Sources of unique rhenium enrichment in fumaroles and sulfides at Kudryavy volcano. Geochim Cosmochim Acta 72:889–909

Toma J, Creaser RA, Card C, Stern RA, Chacko T, Steele-MacInnis M (2022) Re–Os systematics and chronology of graphite. Geochim Cosmochim Acta 323:164–182

Torgashin AS (1994) Geology of the massive and copper ores of the western part of the Oktyabr'sky deposit. Ontario Geol Survey Spec Publ 5:231–242

Turner S, Handler M, Bindeman I, Suzuki K (2009) New insights into the origin of O–Hf–Os isotope signatures in arc lavas from Tonga-Kermadec. Chem Geol 266:187–193

Ulrich T, Mavrogenes J (2008) An experimental study of the solubility of molybdenum in H_2O and KCl–H_2O solutions from 500 °C to 800 °C, and 150 to 300 MPa. Geochim Cosmochim Acta 72:2316–2330

Villa IM, Bonardi ML, De Bièvre P, Holden NE, Renne PR (2016) IUPAC-IUGS status report on the half-lives of ^{238}U, ^{235}U and ^{234}U. Geochim Cosmochim Acta 172:387–392

Völkening J, Walczyk T, Heumann KG (1991) Osmium isotope ratio determinations by negative thermal ionization mass spectrometry. Int J Mass Spectrom Ion Process 105:147–159

Voudouris P, Melfos V, Spry PG, Bindi L, Moritz R, Ortelli M, Kartal T (2013) Extremely Re-rich molybdenite from porphyry Cu–Mo–Au prospects in northeastern Greece: mode of occurrence, causes of enrichment, and implications for gold exploration. Minerals 3:165–191

Walczyk T, Hebeda EH, Heumann KG (1991) Osmium isotope ratio measurements by negative thermal ionization mass spectrometry (NTI-MS). Fresenius' J Anal Chem 341:537–541

Walker RJ, Fassett JD (1986) Isotopic measurement of subananogram quantities of rhenium and osmium by resonance ionization mass spectrometry. Anal Chem 58:2923–2927

Walker RJ, Shirey SB, Stecher O (1988) Comparative Re–Os, Sm–Nd and Rb–Sr isotope and trace element systematics for Archean komatiite flows from Munro Township, Abitibi Belt, Ontario. Earth Planet Sci Lett 87:1–12

Walker RJ, Morgan JW, Naldrett AJ, Li C, Fassett JD (1991) Re–Os isotope systematics of Ni–Cu sulfide ores, Sudbury Igneous Complex, Ontario: evidence for a major crustal component. Earth Planet Sci Lett 105:416–429

Walker RJ, Morgan JW, Horan MF, Czamanske GK, Krogstad EJ, Fedorenko VA, Kunilov VE (1994) Re–Os isotopic evidence for an enriched-mantle source for the Noril'sk-type, ore-bearing intrusions, Siberia. Geochim Cosmochim Acta 58:4179–4197

Walker RJ, Morgan JW, Beary ES, Smoliar MI, Czamanske GK, Horan MF (1997) Applications of the ^{190}Pt-^{186}Os isotope system to geochemistry and cosmochemistry. Geochim Cosmochim Acta 61:4799–4807

Walker RJ, Prichard HM, Ishiwatari A, Pimentel M (2002) The osmium isotopic composition of convecting upper mantle deduced from ophiolite chromites. Geochim Cosmochim Acta 66:329–345

Wells TJ, Cooke DR, Baker MJ, Zhang L, Meffre S, Steadman J, Norman MD, Hoye JL (2021) Geology and geochronology of the two-thirty prospect, Northparkes district, NSW. Aust J Earth Sci 68:659–683

Williams CD, Ripley EM, Li C (2010) Variations in Os isotope ratios of pyrrhotite as a result of water–rock and magma–rock interaction: constraints from Virginia Formation-Duluth Complex contact zones. Geochim Cosmochim Acta 74:4772–4792

Wilson AJ, Cooke DR, Stein HJ, Fanning CM, Holliday JR, Tedder IJ (2007) U–Pb and Re–Os geochronologic evidence for two alkalic porphyry ore-forming events in the Cadia district, New South Wales, Australia. Econ Geol 102:3–26

Wood SA (1987) Thermodynamic calculations of the volatility of the platinum group elements (PGE): the PGE content of fluids at magmatic temperatures. Geochim Cosmochim Acta 51:3041–3050

Woodhead J, Brauns M (2004) Current limitations to the understanding of Re–Os behaviour in subduction systems, with an example from New Britain. Earth Planet Sci Lett 221:309–323

Xiong Y, Wood SA (1999) Experimental determination of the solubility of ReO_2 and the dominant oxidation state of rhenium in hydrothermal solutions. Chem Geol 158:245–256

Xiong Y, Wood SA (2001) Hydrothermal transport and deposition of rhenium under subcritical conditions (up to 200 °C) in light of experimental studies. Econ Geol 96:1429–1444

Xiong Y, Wood S, Kruszewski J (2006) Hydrothermal transport and deposition of rhenium under subcritical conditions revisited. Econ Geol 101:471–478

Zaccarini F, Garuti G, Fiorentini ML, Locmelis M, Kollegger P, Thalhammer OA (2014) Mineralogical hosts of platinum group elements (PGE) and rhenium in the magmatic Ni–Fe–Cu sulfide deposits of the Ivrea Verbano Zone (Italy): an electron microprobe study. Neues Jahrb Mineral Abh 191:169–187

Zajacz Z, Candela PA, Piccoli PM (2017) The partitioning of Cu, Au and Mo between liquid and vapor at magmatic temperatures and its implications for the genesis of magmatic-hydrothermal ore deposits. Geochim Cosmochim Acta 207:81–101

Zeng Z, Chen S, Selby D, Yin X, Wang X (2014) Rhenium–osmium abundance and isotopic compositions of massive sulfides from modern deep-sea hydrothermal systems: implications for vent associated ore forming processes. Earth Planet Sci Lett 396:223–234

Zhimin Z, Yali S (2013) Direct Re–Os dating of chalcopyrite from the Lala IOCG deposit in the Kangdian Copper Belt, China. Econ Geol 108:871–882

Zhong H, Qi L, Hu RZ, Zhou MF, Gou TZ, Zhu WG, Liu BG, Chu ZY (2011) Rhenium–osmium isotope and platinum-group elements in the Xinjie layered intrusion, SW China: implications for source mantle composition, mantle evolution, PGE fractionation and mineralization. Geochim Cosmochim Acta 75:1621–1641

Zhou MF, Sun M, Keays RR, Kerrich RW (1998) Controls on platinum-group elemental distributions of podiform chromitites: a case study of high-Cr and high-Al chromitites from Chinese orogenic belts. Geochim Cosmochim Acta 62:677–688

Zhou J, Jiang S, Wang X, Yang J, Zhang M (2005) Re-Os isochron age of Fankeng basalts from Fujian of SE China and its geological significance. Geochem J 39:497–502

Zhu Z, Meija J, Tong S, Zheng A, Zhou L, Yang L (2018) Determination of the isotopic composition of osmium using MC-ICPMS. Anal Chem 90:9281–9288

Zhu LY, Liu YS, Jiang SY, Lin J (2019) An improved in situ technique for the analysis of the Os isotope ratio in sulfides using laser ablation-multiple ion counter inductively coupled plasma mass spectrometry. J Anal at Spectrom 34:1546–1552

Zientek ML, Likhachev AP, Kunilov VE, Barnes SJ, Meier AL, Carlson RR, Briggs PH, Fries TL, Adrian BM, Lightfoot PC, Naldrett AJ (1994) Cumulus processes and the composition of magmatic ore deposits: examples from the Talnakh district, Russia. Ontario Geol Surv Spec 5:373–392

Zimmerman A, Stein HJ, Morgan JW, Markey RJ, Watanabe Y (2014) Re–Os geochronology of the El Salvador porphyry Cu–Mo deposit, Chile: tracking analytical improvements in accuracy and precision over the past decade. Geochim Cosmochim Acta 131:13–32

Zimmerman A, Yang G, Stein HJ, Hanna JL (2022) A critical review of molybdenite ^{187}Re parent-^{187}Os daughter intra-crystalline decoupling in light of recent in situ micro-scale observations. Geostand Geoanal Res 46:761–772. https://doi.org/10.1111/ggr.12448

Applications of Neodymium Isotopes to Ore Deposits and Metallogenic Terranes; Using Regional Isotopic Maps and the Mineral Systems Concept

David C. Champion and David L. Huston

Abstract

Although radiogenic isotopes historically have been used in ore genesis studies for age dating and as tracers, here we document the use of regional- and continental-scale Sm–Nd isotope data and derived isotopic maps to assist with metallogenic interpretation, including the identification of metallogenic terranes. For the Sm–Nd system, calculated Nd model ages, which are time independent, are of most value for small-scale isotopic maps. Typically, one- or two-stage depleted mantle model ages (T_{DM}, T_{2DM}) are used to infer age when the isotope characteristics of the rock were in isotopic equilibrium with a modelled (mantle) reservoir. An additional advantage is that Nd model ages provide, with a number of assumptions, an estimate of the approximate age of continental crust in a region. Regional- and continental-scale Nd model age maps, constructed from rocks such as granites, which effectively sample the middle to lower crust, therefore, provide a proxy to constrain the nature of the crust within a region. They are of increasing use in metallogenic analysis, especially when combined with a mineral systems approach, which recognizes that mineral deposits are the result of geological processes, at a scales from the ore shoot to the craton. These maps can be used empirically and/or predictively to identify and target large parts of mineral systems that may be indicative, or form part of, metallogenic terranes. Examples presented here include observed spatial relationships between mineral provinces and isotopic domains; the identification of old and/or thick cratonic blocks; determination of tectonic regimes favorable for mineralization; identification of isotopically juvenile zones that may indicate rifts or primitive arcs; recognition of crustal breaks that define metallogenic terrane boundaries or delineate fluid pathways; and, as baseline maps. Of course, any analysis of Sm–Nd and similar isotopic maps are predicated on integration with geological, geochemical and geophysical information data. In the future, research in this area should focus on the spatial and temporal evolution of the whole lithosphere at the province- to global-scales to more effectively targeting mineral exploration. This must involve integration of radiogenic isotopic data with other data, in particular, geophysical data, which has the advantage of being able to directly image the crust and lithosphere and being of a more continuous nature as compared to invariably incomplete isotopic data sets.

D. C. Champion (✉) · D. L. Huston
Geoscience Australia, GPO Box 378, Canberra, ACT 2601, Australia
e-mail: David.Champion@ga.gov.au

D. Huston and J. Gutzmer (eds.), *Isotopes in Economic Geology, Metallogenesis and Exploration*, Mineral Resource Reviews, https://doi.org/10.1007/978-3-031-27897-6_5

1 Introduction

Samarium-Neodymium (Sm–Nd) isotopes have commonly been used in economic geology studies, not just for geochronology, but also as tracers that provide constraints on mineralizing and geological processes. As a tracer Sm–Nd data constrain the age and source of mineralization, fluid pathways, metal sources and mineralizing processes. There is extensive literature regarding such use of radiogenic isotopes; review papers include Tosdal et al. (1999), Lambert et al. (1999), and Ruiz and Mathur 1999). Recently, the use of the isotopes, such as Sm–Nd, has extended to regional-scale metallogenic analysis, importantly utilizing isotopic data from both mineralized and unmineralized rocks (e.g., Champion 2013; Champion and Huston 2016). This builds on recognition that regional-scale isotopic data can assist with identification of metallogenic terranes (Zartman 1974; Farmer and DePaolo 1984; Wooden et al. 1998). This approach has benefitted from the large quantity of isotopic data now available and graphical software, which have enabled the rapid and repeatable generation of isotopic maps at regional to continental scales. Examples for the Sm–Nd system include studies by Cassidy et al. (2002), Champion and Cassidy (2007, 2008), Champion (2013), Huston et al. (2014), Mole et al. (2013), and Wu et al. (2021). Champion's (2013) Australia map was the first Sm–Nd isotopic map produced for an entire continent, undertaken at the Australian continent-scale.

Following pioneering studies such as Zartman (1974), the usefulness of isotopic distribution patterns for metallogenesis has become increasingly recognized. Although initially driven by empiricism (Zartman 1974; Wooden et al. 1998; Cassidy and Champion 2004; Huston et al. 2005, 2014), these studies are now strongly influenced by the mineral systems concept (Wyborn et al. 1994), which recognizes that mineral deposits, although small, result from geological processes that occur, and can be mapped, at larger scales (Fig. 1: e.g., McCuaig et al. 2010; Hronsky et al. 2012; Huston et al. 2016). It is the larger camp- to continent-scale scales that regional Sm–Nd isotopic maps have most use, for example to infer continent- to province-scale lithospheric structures and architecture. This places both known deposits in context, but also allows targeting of undiscovered deposits by constraining geodynamic settings, fluid and magma pathways, and energy and metal sources.

This contribution, therefore, details the Sm–Nd isotopic system, which has been use for over 40 years (Farmer and DePaolo 1983, 1984; Bennett and DePaolo 1987), and for which large regional isotopic datasets are available (e.g., Fraser et al. 2020). It discusses the uses and implications Sm–Nd isotopes and isotopic maps to different mineral systems, focusing on the craton- to district-scale. Examples include identifying lithospheric boundaries that commonly control sites of mineralization, determining fertile metallogenic provinces using isotopic characteristics, and establishing rock types essential to mineralization by their isotopic signature. We present general principles of Sm–Nd system; identify time-independent isotopic variables; and show how these variables can be used to generate isotopic maps useful to metallogenic studies. Much of the discussion and usage presented here applies to other isotopic systems, especially Lu–Hf (Osei et al. 2021; Waltenberg 2023) and U–Th–Pb (Champion and Huston 2016; Huston and Champion 2023).

2 The Samarium-Neodymium Isotopic System

As there are many reviews written on isotope systematics (e.g., Faure 1977; Dickin 1995) in general and on the Sm–Nd system in particular (e.g., DePaolo 1988; Champion and Huston 2016), only a brief introduction is provided here.

The equation describing the evolution of the Sm–Nd system through time can be written as follows:

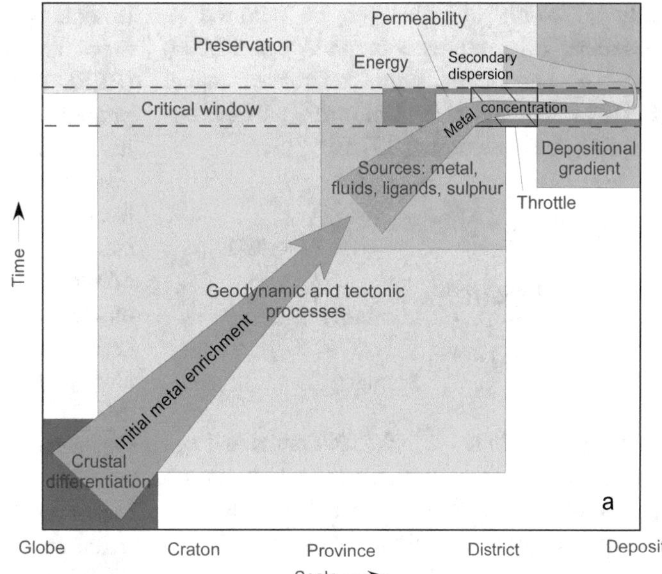

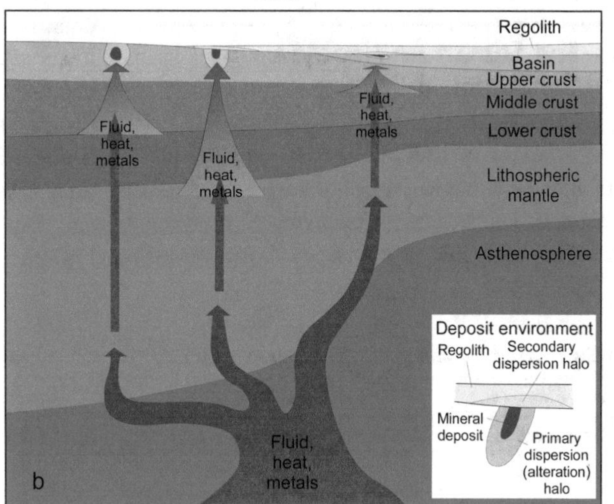

Fig. 1 Mineral system model and examples. **a** Mineral system model illustrating the range of processes required to produce a mineral deposit. Such processes operate at a range of spatial and temporal scales (modified from Huston et al. 2016). **b** Cartoon illustrating potential processes that may be important for a specific mineral deposit. Note the large range of scales relative to the scale of the deposit (modified extensively from Australian Academy of Science 2012)

$$\left({}^{143}\mathrm{Nd}/{}^{144}\mathrm{Nd}\right)_{(\mathrm{now})} = \left({}^{143}\mathrm{Nd}/{}^{144}\mathrm{Nd}\right)_{(\mathrm{initial})} + \left({}^{147}\mathrm{Sm}/{}^{144}\mathrm{Nd}\right)_{(\mathrm{now})} * \left(e^{\lambda t} - 1\right) \quad (1)$$

where: λ = the decay constant for ^{147}Sm to ^{143}Nd (6.54×10^{-12} yr^{-1}: Dickin 1995), (now) = abundance of the isotope as measured in the present day, (initial) = abundance of the isotope at time t in the past, and t = the age of the last isotopic disturbance rock in years. The isotopic disturbance age can record the magmatic, metamorphic, or mineralization age of the rock.

For tracer applications, often used in metallogenic studies, the initial parent/daughter ratio (initial ratio), which is commonly more useful, can be determined as follows:

$$\left({}^{143}\mathrm{Nd}/{}^{144}\mathrm{Nd}\right)_{(\mathrm{initial})} = \left({}^{143}\mathrm{Nd}/{}^{144}\mathrm{Nd}\right)_{(\mathrm{now})} - \left({}^{147}\mathrm{Sm}/{}^{144}\mathrm{Nd}\right)_{(\mathrm{now})} * \left(e^{\lambda t} - 1\right) \quad (2)$$

Initial $^{143}Nd/^{144}Nd$ ratios can be reported as deviations (ε_{Nd}, reported in parts per 10,000) from a chondritic earth reference model (CHUR = Chondritic Uniform Reservoir: DePaolo and Wasserburg 1976):

$$\varepsilon_{Nd} = 10000 * \left[\left(^{143}Nd/^{144}Nd \right)_{Sample(T)} - \left(^{143}Nd/^{144}Nd \right)_{CHUR(T)} \right] / \left(^{143}Nd/^{144}Nd \right)_{CHUR(T)} \quad (3)$$

Although both initial $^{143}Nd/^{144}Nd$ ratios and ε_{Nd} values are used, the latter has the advantage on ε_{Nd}—time plots where depleted mantle and crustal reservoirs have distinct signatures.

3 The Samarium-Neodymium Geochemical System

The behavior of isotopic systems and the interpretation of isotopic data, is mostly a function of the relative geochemical behavior of the parent and the daughter isotopes. For the Sm–Nd isotopic system, both parent and daughter isotopes are lanthanides, or rare earth elements (REE), that all have very similar geochemical properties. Consequently, Sm and Nd have similar, predictable, geochemical behavior in most geological processes, resulting in very minor fractionation and minimal variation in Sm/Nd for common crustal rocks (Table 1). The fractionation that does occur is mostly related to lanthanide contraction, whereby lanthanides with higher atomic number have progressively smaller atomic radii. Hence, Nd is more incompatible than Sm, and Nd is preferentially concentrated in the melt. As a result, processes such as partial melting and fractional crystallization result in higher levels of the light REE and lower Sm/Nd ratios in the more differentiated end-members (DePaolo 1988). As a consequence, the Earth's crust is enriched in the light REE and also has lower Sm/Nd than complementary depleted mantle reservoirs. Differences in Sm/Nd of mantle and crustal reservoirs result in divergent isotopic behaviors in these two reservoirs with time (DePaolo and Wasserburg 1976; Bennett and DePaolo 1987; DePaolo 1988; Table 1; Fig. 2).

Table 1 Average Nd and Sm concentrations (in parts-per-million) and Sm/Nd ratios of mantle and crustal reservoirs

Reservoir	Nd	Sm	Sm/Nd
Upper crust	27	4.7	0.17
Middle crust	25	4.6	0.18
Lower crust	11	2.8	0.26
Bulk crust	20	3.9	0.20
OIB	38.5	10.0	0.26
MORB	9.8	3.25	0.26
E-MORB	9.0	2.6	0.29
N-MORB	7.3	2.63	0.36
Depleted mantle	0.71	0.27	0.38
Silicate earth	1.25	0.41	0.33
Bulk earth	0.84	0.27	0.32
Chondrite	0.474	0.154	0.32

Sources Continental crust reservoirs from Rudnick and Gao (2003), mid-ocean ridge basalt (MORB) from Arevalo and McDonough (2010); ocean island basalt (OIB), enriched- and depleted-MORB (E-MORB and N-MORB) from Sun and McDonough (1989); Depleted mantle from Salters and Stracke (2004); bulk earth from McDonough (2003), silicate earth from McDonough and Sun (1995), C1 carbonaceous chondrite from Palme and Jones (2003)

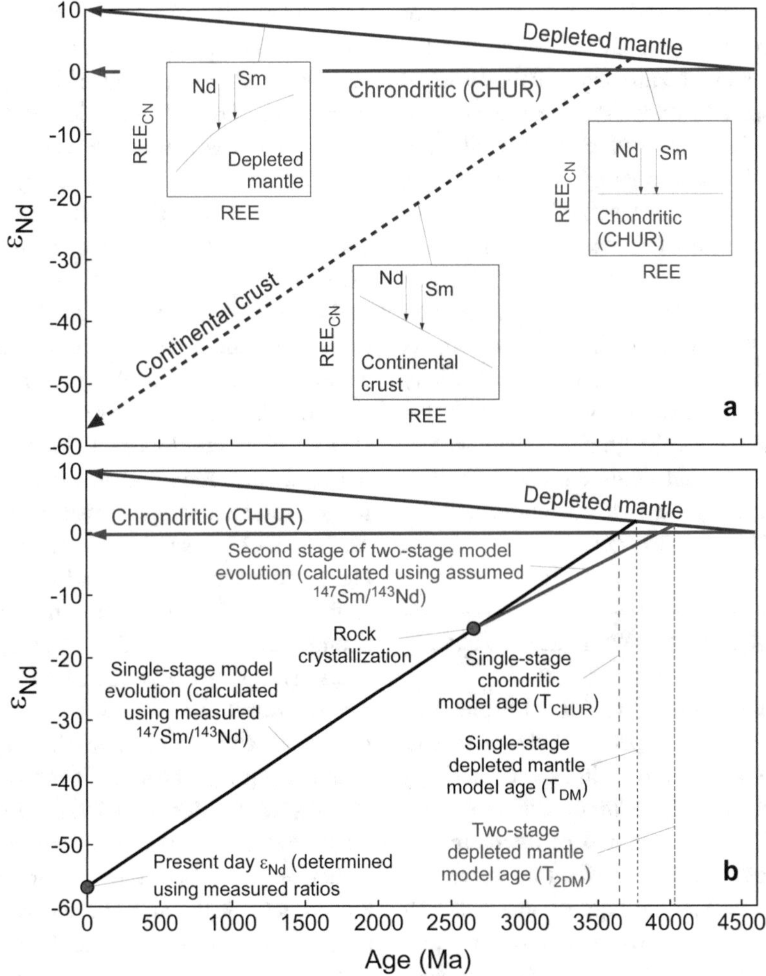

Fig. 2 Epsilon Nd (ε_{Nd}) through time and model age plots (modified after Champion 2013). **a** Time-integrated behaviour of ε_{Nd} in continental crustal reservoirs versus the complementary depleted mantle reservoir. Schematic REE plots illustrate the idealised change in Sm/Nd ratio (normalised to chondrite) between the reservoirs. CHUR equals Chondritic Uniform Reservoir. **b** Nd model ages. Single stage Nd model ages (T_{CHUR}, T_{DM}) assume no fractionation of Sm/Nd (= measured ratio). The intersection of the sample evolution curve with the mantle evolution curve (chondritic (CHUR) or depleted mantle (DM)), defines the model age. Two-stage model ages (T_{2DM}) assume a change in Sm/Nd at some point in the crustal history of the protolith. For felsic magmatic rocks this is typically the magmatic age. For ages older than the magmatic age a model $^{147}Sm/^{144}Nd$ ratio is used, typically that of average continental crust. To calculate ε_{Nd}, values of $^{147}Sm/^{144}Nd$ and $^{143}Nd/^{144}Nd$ have to be assigned for the DM and CHUR reservoirs. DM values used here are 0.2136 and 0.513163, respectively

The long half-life of ^{147}Sm (10^6 Gyr) and the geochemistry of the lanthanides make the Sm–Nd system ideal for documenting crustal development. Because of the long half-life, significant changes in the Nd isotopic signature require very long time frames, making the Sm–Nd system very useful to identify ancient crust. The isotopic effect of addition of isotopically juvenile material, through the incorporation of a juvenile mantle component in granitic magmas, depends mostly on the volume of juvenile material added. As REEs are enriched in granitic rocks relative to mantle material, the Nd isotopic signature of granites is often insensitive to addition of significant amounts of such mantle material. For example, Kirkland et al. (2013) show that the isotopic signatures of Proterozoic REE-enriched granites in the Musgrave Province of central Australia have been insensitive to the addition of as much as 85% mantle input. While an extreme case, this behavior means that the Sm–Nd system can be used to effectively 'see' through many crustal processes and can provide information on the nature of the source of the rocks in question (DePaolo 1988). For voluminous crustal rocks such as granites this provides a potentially powerful proxy in constraining the nature of the various crustal blocks the granites occur within, i.e. in effect broadly mapping the crust and thus timing of crustal growth as demonstrated by Bennett and DePaolo (1987).

4 Model Ages and Residence Ages

Although initial ratios and related measures like ε_{Nd} are useful when comparing rocks of similar ages, the time factor implicit in Eqs. (2) and (3) makes comparison of rocks of different ages problematic. This problem can be overcome by calculating model ages, ages when the measured rock was last in isotopic equilibrium with the modelled reservoir from which it was extracted. Model ages, if calculated from the same model, are the simplest way to spatially compare isotopic data from rocks of different ages in the Sm–Nd system. Model ages determined from the Sm–Nd systems (Nd model ages) approximate the average age of continental crust in a province (McCulloch and Wasserburg 1978; DePaolo 1981, 1988; Farmer and DePaolo 1983, 1984; Liew and McCulloch 1985; Bennett and DePaolo 1987; McCulloch 1987; Fig. 2) due to the differing time-integrated behavior of Nd isotopic signatures in mantle and crustal reservoirs (Fig. 2), and 'minimal' changes of Sm/Nd during many crustal processes. Hence, Nd model ages estimate the time when a rock was separated from its modelled source, typically depleted mantle. This approach is most useful for magmatic rocks, but has been used for all rock types (McCulloch and Wasserburg 1978).

Although historically model ages were calculated assuming a chondritic mantle (T_{CHUR}), after recognition that depleted upper mantle is the dominant reservoir from which crust is extracted, most model ages now are calculated assuming depleted mantle as the source (T_{DM}: DePaolo 1981, 1988). Several models have been developed to model the evolution of depleted mantle. These include a model that assumes an increasingly depleted mantle (DePaolo 1981), and a linear depletion model, used herein, which assumes linear depletion from $\varepsilon_{Nd} = 0$ at ~4.56 Ga to + 10 today (Fig. 2). McCulloch (1987) also developed a linear model with assumed depletion commencing at 2.75 Ga. Recent research suggests that mantle older than ca. 3.5–3.7 Ga may have been chondritic (e.g., Hiess and Bennett 2016), and so a model similar to that of McCulloch (1987) but with depletion commencing at ca. 3.5–3.7 Ga may be more valid. This is a controversial area, however, of active research. It has been largely based on Lu–Hf studies, and it is apparent that there may have been decoupling of Sm–Nd from Lu–Hf isotopic systems (e.g., Vervoort 2011; Vervoort et al. 2019, 2020), with depletion commencing much earlier for Sm–Nd. This apparent decoupling may reflect disturbance of the Sm–Nd system (e.g., Hammerli and Kemp 2021). Until this matter is resolved, we still use the linear depletion model back to 4.5 Ga as our preferred model for the depleted mantle growth curve and for calculating Nd model ages. The linear depletion model results in older calculated model ages. Given that in isotopic mapping it is the relative spatial changes in model ages that are of more concern than absolute ages, the choice of model is not expected to have a significant effect on results and conclusions derived from isotopic mapping.

Model ages can also be determined using single stage or multi-stage (mostly two-stage) models. Single stage models (Fig. 2) infer that the main change in Sm/Nd occurred during mantle extraction and that crustal processes (e.g., fractional crystallization, contamination, alteration) have not significantly modified this ratio. Although geochemically unrealistic (see variations in Sm/Nd of crustal reservoirs in Table 1), single stage models were a feature of many early studies and did provide useful results (e.g., Bennett and DePaolo 1987). As single stage model ages (T_{DM}) become increasingly prone to error with increasing Sm/Nd when sample evolution curves become sub-parallel with mantle evolution curves, model ages should not be calculated for $^{147}Sm/^{144}Nd$ ratios over 1.4–1.5. This is generally not a problem as most felsic igneous rocks have $^{147}Sm/^{144}Nd$ ratios between 0.09 and 0.12 (e.g., Sun et al. 1995).

Two-stage model ages (denoted as T_{2DM}) are commonly used for felsic igneous rocks (Liew and McCulloch 1985) to correct for changes in Sm/Nd ratios due to partial melting, fractional crystallization, magma mixing, contamination or hydrothermal alteration. Two-stage models use the measured $^{147}Sm/^{144}Nd$ ratio to model evolution back in time to the independently-determined magmatic age of the rock. From the magmatic age to the modelled extraction age, an assumed $^{147}Sm/^{144}Nd$ ratio is used to calculate the sample evolution (Fig. 2). In this review, a $^{147}Sm/^{144}Nd$ ratio of 0.11, which is based on the average upper continental Sm and Nd concentrations reported by Rudnick and Gao (2003), was used. For a given sample or study, T_{2DM} may be younger or older than T_{DM}, depending on the assumed age and the measured and assumed $^{147}Sm/^{144}Nd$ ratios used. Two-stage model ages provide more consistent ages on a regional basis (Champion and Cassidy 2008), and allow ages to be determined for samples with high measured $^{147}Sm/^{144}Nd$.

Residence ages (T_{Res}), which model age differences between the modelled crustal extraction from the mantle and melting and crystallization of the sample, can also be used as a variable in isotopic maps, and are calculated using the equation:

$$T_{Res} = T_{2DM} - T_{magmatic} \quad (4)$$

Residence ages provide an indication of the time between when the granite source was extracted and when the source was reworked (melted) to produce the granite. Granite formed by the reworking of largely juvenile (i.e., young) crust will have younger residence ages than granites formed from older crust. Combining T_{2DM} and T_{Res} with geochronological data such as magmatic and inherited zircon ages provides constraints on the age of crust in a region, the age of crustal reworking and melting, and on the protolith of individual granite units.

It must be stressed that model and residence ages are just expression of a model and that absolute ages differ between models. As discussed by many workers (DePaolo 1988; Kemp et al. 2009; Vervoort and Kemp 2016) processes that produce the isotopic signature of a rock are complex, and Sm–Nd model age calculations contain a number of assumptions, as discussed above and outlined in detail by Champion and Huston (2016) and Vervoort and Kemp (2016). As an example, models ages for felsic igneous rocks, if taken at face value, imply that the source of the rock in question was homogeneous and formed in a single event, assumptions that are rarely true. Moreover, even when these criteria are met, a model age is still only an approximation with many uncertainties. There is abundant evidence of complex sources as the mixed role of crustal, juvenile and assimilated components has been demonstrated by many studies (Kemp et al. 2007, 2009; Fisher et al. 2017). In such complex open systems the calculated Nd model ages are best interpreted as average ages of all preceding crustal growth events. Importantly, as discussed by Champion and Huston (2016) it is the regional variations in model ages that are most applicable and not the absolute ages. These relative variations highlight regional changes in mantle flux, crustal growth and geodynamic environments. Moreover, use of isotopic data in a regional manner, especially in regions with a significant range of ages, tends to smooth out and downplay these 'secondary' effects. As shown by Champion and Huston (2016) this is evident when the map scale is taken into account. Small-scale images, covering a large region (1000+ km), exhibit significantly less scatter of data (less noise) than larger scale images (100's of km). This needs to be

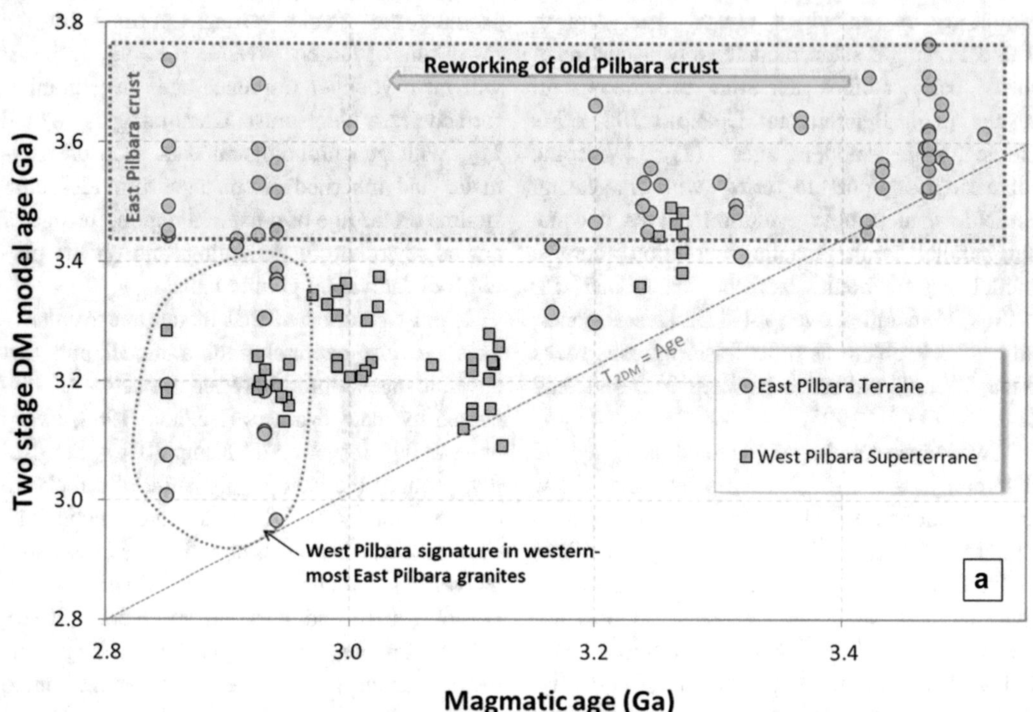
Reworking of old Pilbara crust
East Pilbara crust
Two stage DM model age (Ga)
T2DM = Age
East Pilbara Terrane
West Pilbara Superterrane
West Pilbara signature in western-most East Pilbara granites
a
Magmatic age (Ga)

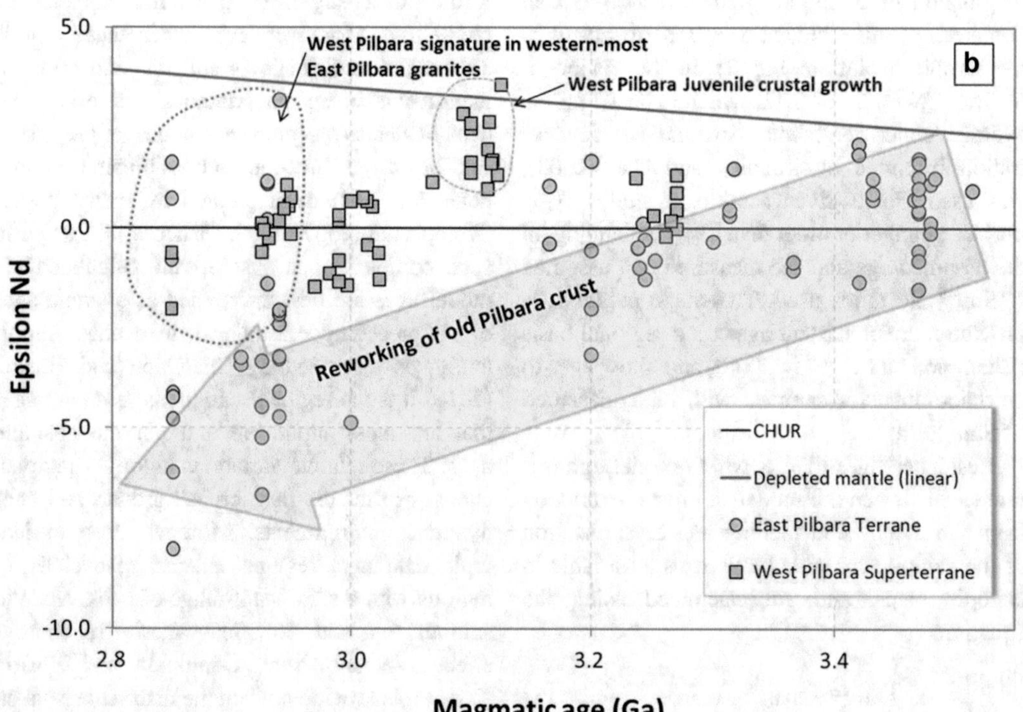
West Pilbara signature in western-most East Pilbara granites
West Pilbara Juvenile crustal growth
b
Epsilon Nd
Reworking of old Pilbara crust
CHUR
Depleted mantle (linear)
East Pilbara Terrane
West Pilbara Superterrane
Magmatic age (Ga)

◀ **Fig. 3** Two-stage depleted mantle model ages (**a**) and ε_{Nd} (**b**) versus magmatic ages for the Pilbara Craton, Western Australia (modified from Champion and Huston 2016). Isotopic data points are coloured by geological province: East Pilbara Terrane versus the Karratha, Regal and Sholl terranes of the West Pilbara Superterrane (see Fig. 4). The East Pilbara Terrane shows decreasing ε_{Nd} and approximately constant T_{2DM} with decreasing magmatic age (Type 1 behaviour in Table 2), consistent with reworking of old crust with minimal juvenile involvement. The exceptions to this are 2.95–2.85 Ga magmatism in the western part of the East Pilbara Terrane, which include a more significant juvenile component (younger T_{2DM}, more positive ε_{Nd}), probably related to crustal growth of the West Pilbara Superterrane (Champion and Smithies 2000; Smithies et al. 2005). Data shown spatially in Fig. 4. Data sources as reported in Champion (2013). Modified after Champion and Huston (2016)

considered when using isotopic maps, particularly when using large scale maps. For metallogenic analysis it is the continent to regional scale isotopic maps that are of most use.

5 Mineral Systems and Spatial and Temporal Variations in Sm–Nd Isotopic Signatures

The advantage of a mineral systems approach is that increased knowledge of the 4D (space and time) evolution of a geological terrane (i.e. regions with mutual tectonic histories), effectively allows for a better understanding of the metallogeny of that terrane, and by extension, more effective mineral exploration (McCuaig et al. 2010; Huston et al. 2016). This includes gaining a greater knowledge of the lithosphere of a terrane, its architecture, and its evolution in space and time (Fig. 1b). Radiogenic isotopes are very applicable for this purpose for geochronology and, as tracers, are able to constraint the nature, composition and evolution of the lithosphere. Examples of this approach, using Sm–Nd isotopic data coupled with other geochemical and geochronological data sets from felsic igneous rocks to help constrain crustal evolution, include the western United States (e.g., Zartman 1974; Farmer and DePaolo 1983, 1984; Bennett and DePaolo 1987; Wooden et al. 1998), and more recently Australia (Champion and Huston 2016). As granites are mainly derived from the lower and middle continental crust and may have volumes of X0 to X00 km^3, these rocks can constrain the evolution of crustal growth. Studies of granites and related rocks provide indirect constraints on the origin of the crustal domains that host these rocks (Farmer and DePaolo 1983, 1984).

As indicated above, most information from isotopic data can be obtained by assessing spatial and secular changes of isotopic signatures (Fig. 3). Temporal variation can be included by time slices (e.g., Mole et al. 2014; Champion and Huston 2016; Osei et al. 2021), or by using time-independent isotopic variables such as model ages. Documenting geographic changes in isotopic parameters relies on data classification. Rapid and objective visualization of regional isotopic data, such as Nd model ages, is best produced using computer-assisted imaging, including interpolated images, where data are grouped and displayed as classes (Fig. 4). Figure 4 shows regional changes in Nd model ages that correspond well to patterns deduced from the isotope data alone (Fig. 3) and constrains crustal growth in the Pilbara Craton. The results of interpolation will depend to some extent on the interpolation process used and the technique used to bin interpolated results (i.e., identify intervals using equal intervals, percentiles, natural breaks etc.), both of which will affect the produced image. After considering interpolation techniques (Slocum et al. 2009), Champion and Huston (2016) concluded that natural neighbor classification using natural breaks in data values worked best for Sm–Nd model age images, and produced results akin to those produced by manual contouring.

Absolute values of isotopic parameters such as model ages are less useful than geographical and/or secular variations. Champion and Huston (2016) illustrated interpretation of such maps, as well as their advantages and disadvantages. In addition to problems such as data smoothing/averaging, and areas of no data and artefacts of interpolation, a consistent difficulty concerns the ambiguity of the isotopic data itself with a variety of possible interpretations. These images must be used with the original data (ε_{Nd}-time and T_{2DM}-time plots), to fully understand the isotopic signatures and complexities within each block, and with other geological and geophysical data, including data and images from other isotopic systems (U–Th–Pb: Huston and Champion 2023; Lu–Hf: Osei et al. 2021).

6 Using Radiogenic Isotope Maps and Mineral Systems: Specific Examples

Although radiogenic isotopes have been used in many ways during metallogenic studies, it is the continent to regional scale isotopic maps that are of interest here. Maps at these scales can be used to identify large-scale components and processes from a variety of mineral systems (Champion and Huston, 2016):

- Determining relationships between mineral systems and isotopic domains. Examples include: orogenic gold, volcanic-hosted massive sulfide (VHMS) base-metal and komatiite-associated nickel deposits (KANS) in the Yilgarn Craton, Western Australia (Cassidy and Champion 2004; Cassidy et al. 2005; Huston et al. 2005; Mole et al. 2013, 2014; Champion and Huston 2016; Osei et al. 2021); porphyry Cu and Mo deposits in the western United States (Zartman 1974) and in

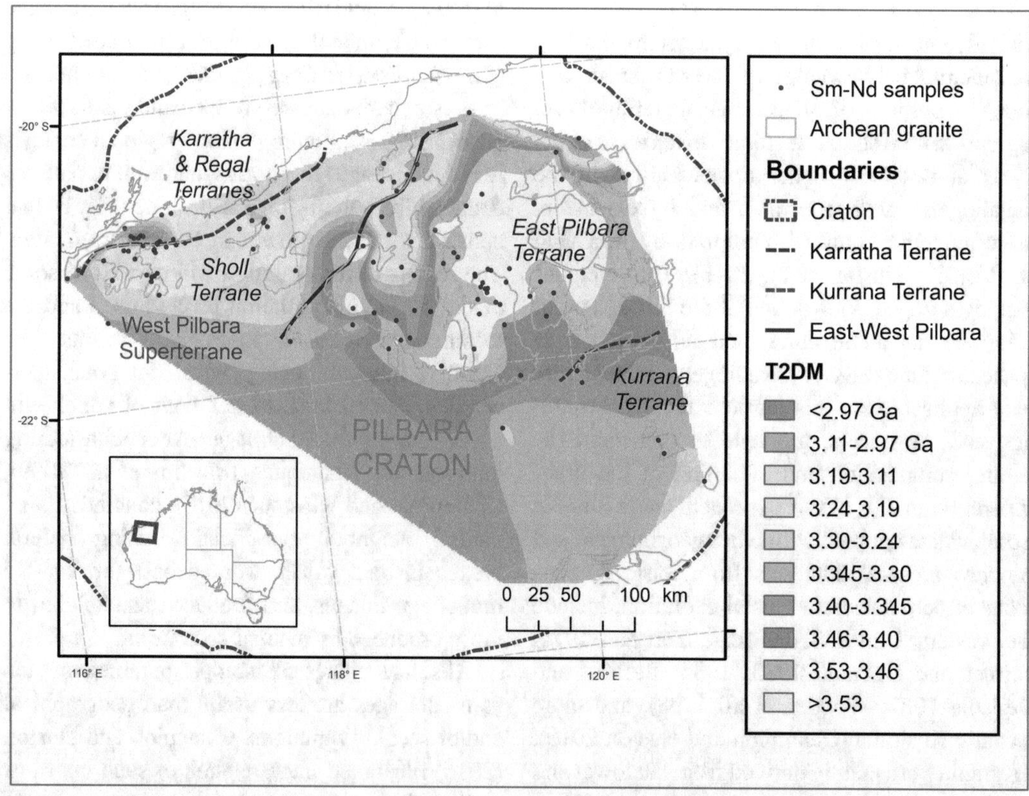

Fig. 4 Gridded Nd two-stage depleted mantle model age (T_{2DM}) map for the Pilbara Craton, Western Australia. Isotopic data used to create the grid are also shown (data and data sources as reported in Champion 2013). Also shown are: the boundary between the Karratha & Regal Terranes and the Sholl Terrane (all three comprise the West Pilbara Superterrane); the boundary between the West Pilbara Superterrane and East Pilbara Terrane and the boundary between the East Pilbara Terrane and the Kurrana Terrane (nomenclature follows van Kranendonk et al. 2007). Grid colours in areas with no samples are based on interpolation and may have no relationship with underlying deeper crust. Modified after Champion and Huston (2016)

the Central Asian Orogenic Belt (Wu et al. 2021); and Proterozoic IOCG belts in Australia (Skirrow 2013). These empirical correlations can be extracted, tested and applied as predictive tools. Osei et al. (2021) provide a good example of extracting and testing correlations between isotopic signatures and mineral deposits for the Yilgarn Craton. Similarly, Huston et al. (2014) show how correlations observed initially in the Yilgarn Craton were applicable for Archean and Proterozoic VHMS deposits, in general.

- Identifying cratonic blocks that may have greater metal endowment and have other favorable metallogenic characteristics, (for example, regions of thicker lithosphere that focus mantle melts around their margins (Begg et al. 2010; Mole et al. 2014). Recently, Hoggard et al. (2020) demonstrated that margins of thicker lithosphere coincide with the location of the world's sediment-hosted metal deposits, and suggested a role for long-lived lithospheric edge stability. Such old stable margins should be evident in isotopic maps and Huston et al. (2020), using Pb isotopes, show that such a relationship does exist. Old crustal blocks and their underlying mantle lithosphere may also have been the foci of (repeated) mantle melts that provided metal sources for later reworking (e.g., Groves et al. 2010). Similarly, Wu et al. (2021) showed a relationship between crustal thickness and isotopic signature in the Central Asian Orogenic Belt and that the relative distribution of porphyry Cu–Au and porphyry Cu–Mo deposits was related to both isotopic signature and crustal thickness, with porphyry Cu–Au deposits in the more isotopically juvenile thin crust zones.
- Establishing old continental margins, particularly accretionary margins that are favorable sites for porphyry copper and related deposits, especially where accompanied by juvenile isotopic signatures (Champion 2013; Wu et al. 2021). This also includes the ability to identify and map continental fragments within such accretionary orogens. For example, Wu et al. (2021) were able to use Sm–Nd isotopic mapping to map the extent of cratons, and locations of microcontinents, and juvenile crustal blocks. Kemp et al. (2009, 2020) and Champion and Huston (2016) identified interpreted island arc fragments in eastern Australia based on juvenile isotopic signatures. These fragments may be important for later mineral systems as accretionary orogens are can involve lithospheric metasomatism and provide metal sources for later reworking (Groves et al. 2010).
- Identifying juvenile zones indicative of extension and/or rifting and associated deposits, such as VHMS (Huston et al. 2014) or deposits formed in primitive arc crust (e.g., porphyry Cu–Au: Champion and Huston 2016; Wu et al. 2021).
- Recognizing crustal breaks that represent major faults and sutures, or were pathways for fluids and magmas (Wooden et al. 1998). Such breaks also may delineate boundaries of metallogenic terranes (VHMS deposits, Huston et al. 2014; porphyry deposits, Zartman 1974).
- Constructing baseline maps that determine spatial or temporal zones with high magmatic, especially mantle, flux and which have been overprinted by younger crustal growth.

Many of the above act in combination and repeatedly and are discussed below to highlight steps involved proceeding from conceptual mineral systems to exploration targeting (McCuaig et al. 2010).

6.1 Regional Samarium–Neodymium Isotopic Signatures in the Yilgarn Craton and Their Relationship to Nickel, Copper–Zinc–Lead and Gold Deposits

In the Yilgarn Craton, features of the regional Sm–Nd signature of granites of the Yilgarn Craton correspond with a range of mineralization styles, including KANS, VHMS and orogenic gold deposits (Fig. 5; Cassidy et al. 2002, 2005;

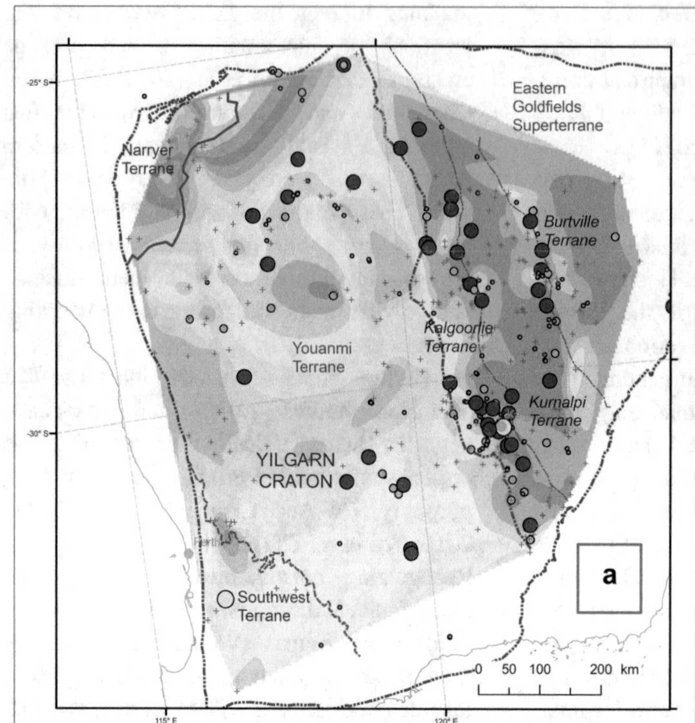
Eastern
Goldfields
Superterrane
Narryer
Terrane
Burtville
Terrane
Kalgoorlie
Terrane
Youanmi
Terrane
Kurnalpi
Terrane
YILGARN
CRATON
Southwest
Terrane
a
0 50 100 200 km
25° S
30° S
115° E
120° E

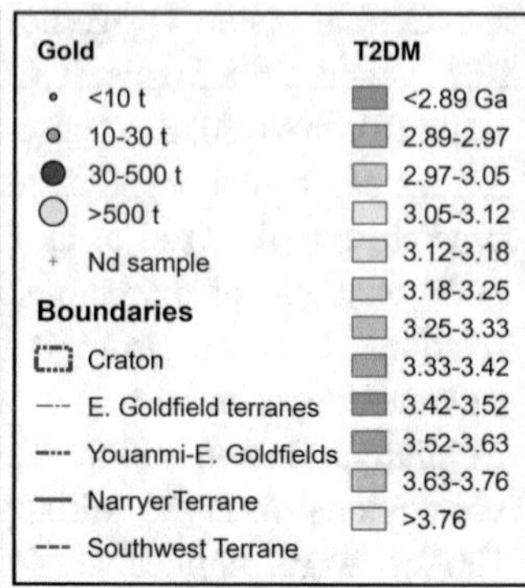
Gold
<10 t
10-30 t
30-500 t
>500 t
Nd sample
Boundaries
Craton
E. Goldfield terranes
Youanmi-E. Goldfields
NarryerTerrane
Southwest Terrane
T2DM
<2.89 Ga
2.89-2.97
2.97-3.05
3.05-3.12
3.12-3.18
3.18-3.25
3.25-3.33
3.33-3.42
3.42-3.52
3.52-3.63
3.63-3.76
>3.76

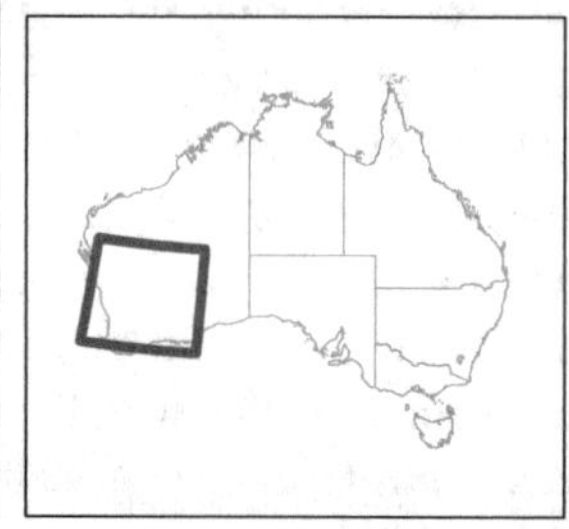

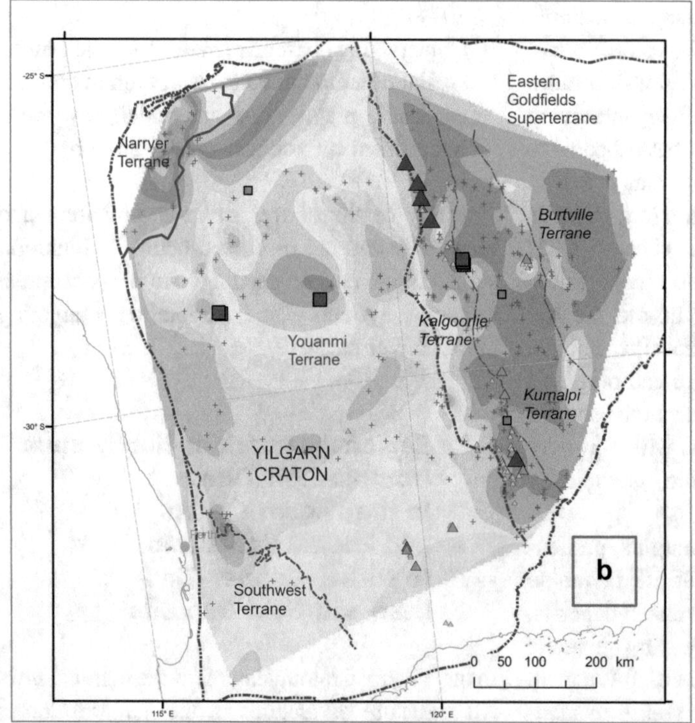
Eastern
Goldfields
Superterrane
Narryer
Terrane
Burtville
Terrane
Kalgoorlie
Terrane
Youanmi
Terrane
Kurnalpi
Terrane
YILGARN
CRATON
Southwest
Terrane
b
0 50 100 200 km
25° S
30° S
115° E
120° E

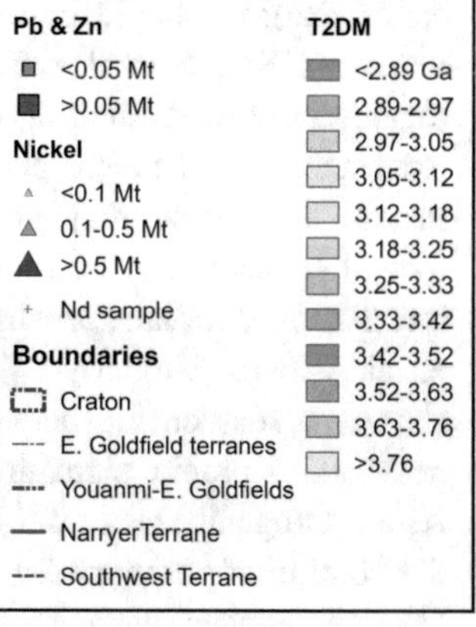
Pb & Zn
<0.05 Mt
>0.05 Mt
Nickel
<0.1 Mt
0.1-0.5 Mt
>0.5 Mt
Nd sample
Boundaries
Craton
E. Goldfield terranes
Youanmi-E. Goldfields
NarryerTerrane
Southwest Terrane
T2DM
<2.89 Ga
2.89-2.97
2.97-3.05
3.05-3.12
3.12-3.18
3.18-3.25
3.25-3.33
3.33-3.42
3.42-3.52
3.52-3.63
3.63-3.76
>3.76

◀ **Fig. 5** **a** Location of gold, and **b** nickel (nickel sulphide) and volcanic hosted massive sulphide (Pb–Zn) deposits, by size, in the Yilgarn Craton, superimposed over gridded two-stage depleted mantle model age (T_{2DM}) map. Image constructed from 305 data points. Data and data sources as reported in Champion (2013), from which the figure was modified. Grid colours in areas with no samples are purely based on interpolation and may have no relationship to the underlying crust. Mineral deposit locations are from Geoscience Australia (2022). Yilgarn Craton terrane boundaries from Cassidy et al. (2006). WAE = Western Australian Element, CAE = Central Australian Element. Element boundaries as in Huston et al. (2012). Refer to Osei et al. (2021) for more updated images of the same region. Modified after Champion and Huston (2016)

Huston et al. 2005, 2014; Champion and Cassidy 2008). Osei et al. (2021), using both Sm–Nd and Lu–Hf isotopes with a more rigorous statistical approach, showed that KANS deposits are spatially associated with terranes with pre-existing evolved crust, as identified by Nd T_{2DM} model ages (Cassidy et al. 2005; Fig. 5). Osei et al. (2021) refined this relationship and showed that these deposits were located along the edge of, and not within, older cratonic blocks. This relationship holds despite variations in komatiite type and age. Cassidy et al. (2005) inferred a relationship between orogenic gold deposits and older crustal terranes, again updated by Osei et al. (2021) to show a more complex relationship although still located along edges of older crust. There is also a global antithetic relationship between KANS and VHMS deposits (Groves and Batt 1984; Cassidy et al. 2005). Huston et al. (2005, 2014) and Osei et al. (2021) showed that VHMS deposits in the Yilgarn Craton are associated with regions with isotopically primitive crust (Nd model ages close to magmatic ages, i.e. with young residence (T_{Res}) ages). The latter are age-independent, and readily identify young (juvenile) versus older crustal domains. In regions where felsic magmatic ages have small ranges, T_{Res} maps are very similar to T_{2DM} maps. Either works well and T_{Res} maps of the Yilgarn Craton also identify juvenile zones and their spatial association with VHMS deposits.

Reasons for the Yilgarn isotope-deposit correspondence are not well understood, particularly the linkage of nickel and gold deposits to isotopically more evolved crust. In part, this reflects uncertainty over the interpretation of the crustal isotopic signature, i.e., is it a direct signature of the mineral system or simply is it a proxy for some other important feature. Subsequent work (Huston et al. 2014; Osei et al. 2021) inferred both, with VHMS deposits reflecting the former and KANS deposits the latter.

Reasons for the association of gold endowment with isotopic domains of intermediate T_{2DM} are more enigmatic, and several mechanisms have been proposed (Cassidy et al. 2005; Osei et al. 2021). Blewett et al. (2010), Czarnota et al. (2010) and McCuaig et al. (2010) argued that the Nd model age map identified lithospheric-scale architecture adjacent to paleocraton margins. This is not dissimilar to models, such as those of Goldfarb and Santosh (2014), which suggested deep (possibly slab-related) ore-forming fluids in the Phanerozoic gold deposits of Jiaodong Province, China, were channeled into the upper crust along continental-scale fault systems. Such faults occur proximal to old craton margins, readily identifiable by isotope systems such as Sm–Nd. As outlined by Osei et al. (2021) these lithospheric structures may also be controlling location of magmas, such as sanukitoids and lamprophyres, providing a possible link between location of gold deposits and the magmas and magma pathways that may have carried the gold into the crust (e.g., Beakhouse 2007). A recent example of this possible sanukitoid association are the recently discovered Hemi gold deposit in the Pilbara Craton, Australia (De Grey Mining Ltd 2022). This, and associated deposits, occur closely associated with a belt of ca. 2.945 Ga sanukitoid intrusions in the Mallina Basin, southwest of Port Hedland, in an isotopically more juvenile zone, that approximates the Sholl Terrane but also extends to within the westernmost part of the neighboring older East Pilbara Terrane (Smithies and Champion 2000; Figs. 3, 4).

6.1.1 Isotopic Domains and Komatiite-Associated Nickel Deposits: Control By Lithospheric Architecture?

The distribution of KANS deposits in Yilgarn crustal domains has been assessed by many authors. Barnes and Fiorentini (2010a, b, 2012) found that in the Yilgarn, nickel endowment in the Yilgarn was concentrated in the Kalgoorlie Terrane, suggesting that this concentration resulted from factors that enabled high and prolonged fluxes of komatiitic magmas into the Kalgoorlie Terrane, following the craton-margin model of Begg et al. (2010; Fig. 6). This model suggests that komatiitic melts from upwelling mantle plumes were channeled away from regions of thicker lithosphere to areas of thinner lithosphere with accompanying decompression melting. The Yilgarn Nd isotopic map (Champion and Cassidy 2007, 2008; Mole et al. 2013; Osei et al. 2021) is consistent with this model with as it clearly identifies a break between two lithospheric blocks that represents an old continental margin (Begg et al. 2010) and/or marginal basin (Krapez et al. 2000). Mole et al. (2014) used Lu–Hf data from inherited zircons, along with their U–Pb ages, to show that similar architecture may have also controlled older komatiitic magmatic events and associated KANS deposits.

Crustal isotopic maps define lithospheric architecture. Begg et al. (2010) summarized several mechanisms whereby lithospheric architecture, particularly thinner lithosphere along craton margins, would be more favorable for KANS and related deposits. These mechanisms include the preferred flow of mantle plumes and accompanying partial melts into such regions, and passage of such melts into the crust along large-scale fault systems. Isotopic systems identify favorable regions at a craton- to district-scale, whereas other factors, such as the availability of sulfur, control mineralization at the district-scale (Fiorentini et al. 2012).

Samarium–Nd isotopic maps (e.g., Fig. 5) clearly identify old continental margins, and can be used in exploration targeting as proxies to recognize such architecture. For example, the Nd isotopic map of the Pilbara (Fig. 4) identifies old cratonic margins, and identification of old margins is important for a number of other mineral systems.

There are other possible explanations for the correlation between the Nd isotopic data and nickel mineralization. McCuaig et al. (2010) and Fiorentini et al. (2012) highlighted other proxies, such as the location of felsic volcanic rocks and VHMS deposits, to identify extensional zones, which, like old continental margins, are another

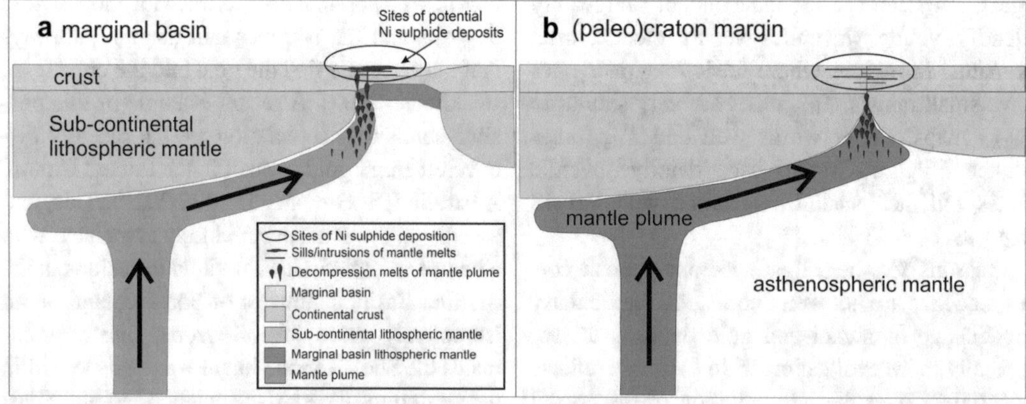

Fig. 6 The generalised model of Begg et al. (2010) illustrates preferential flow of impinging mantle plume towards, and subsequent localisation of nickel sulfide deposits within, thinner lithosphere along the margins of thick (> 150 km) lithospheric blocks, e.g. with intervening marginal basin (**a**) or along an old paleo-margin (**b**). Either model may be applicable for the Yilgarn Craton (Fig. 5). Modified after Begg et al. (2010)

active pathway for komatiite lava. This is consistent with Osei et al. (2021) who demonstrate that Ni–Cu–PGE mineralization is located along the margins of older crust. Juvenile isotopic zones also allow recognition of major extensional zones that are evident within the Yilgarn Craton, especially in the Eastern Goldfields Superterrane. These juvenile isotopic zones are bordered by more isotopically evolved zones, and it could be inferred that KANS mineral systems are closely spatially associated with second order extensional zones. It is probable that both mechanisms were operative in the Yilgarn Craton, albeit at different scales.

6.1.2 Isotopic Domains and Volcanic-Hosted Massive Sulfide Deposits

Huston et al. (2005, 2014), and subsequently Osei et al. (2021), showed an empirical relationship between the distribution of VHMS deposits in the Yilgarn Craton and more juvenile crust, as identified by younger T_{2DM} (Fig. 7). Huston et al. (2014) also showed a similar relationship with Pb isotopes in galena. They used the close correspondence between the Nd isotopes in the granites and Pb isotope signatures in ore minerals to suggest that the juvenile Nd signature reflected a critical aspect of the VHMS mineral system. More, importantly, they suggested this signature could be used predictively for VHMS deposits in exploration targeting. Huston et al. (2014) investigated this further and demonstrated a strong relationship between endowment and isotopic signature, at least for Archean (and, possibly, Proterozoic) VHMS provinces around the world. The association of VHMS deposits with juvenile isotopic zones strongly suggests a link between juvenile crustal growth and VHMS mineralization. Neodymium and Pb data in highly endowed domains indicate limited interaction with pre-existing crust. Huston et al. (2014) argued that isotope data indicate a favorable setting for VHMS deposits in extensional zones that are characterized by high-temperature juvenile magmas and extensive structuring.

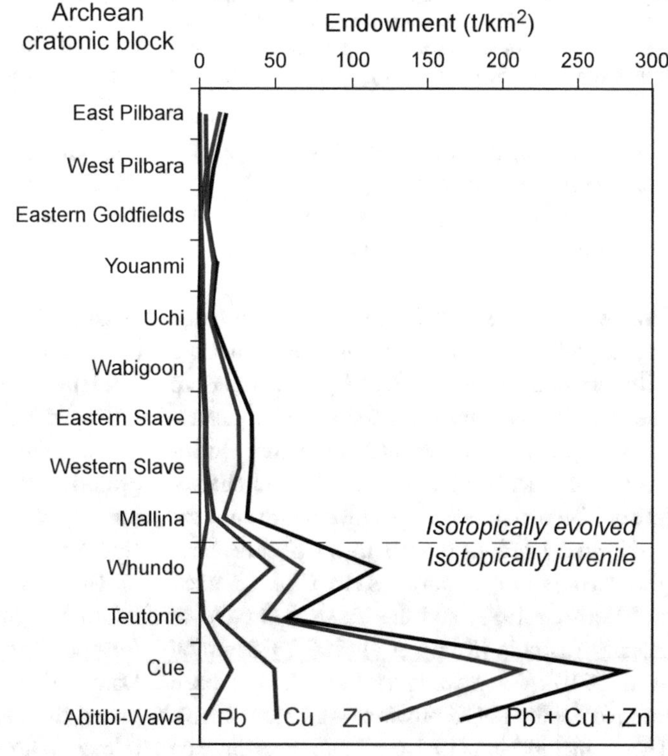

Fig. 7 Plot of Cu, Zn, Pb and combined Cu–Pb–Zn metal endowment (in tonnes per km^2) for volcanic-hosted massive sulfide deposits in Archean cratonic blocks in Canada and Australia, highlighting the much greater endowment in blocks with a juvenile isotopic signature (Whundo, Teutonic, Cue and Abitib-Wawa). Endowment figures from Table 3 of Huston et al. (2014), which were updated from Franklin et al. (2005). Modified after Champion and Huston (2016)

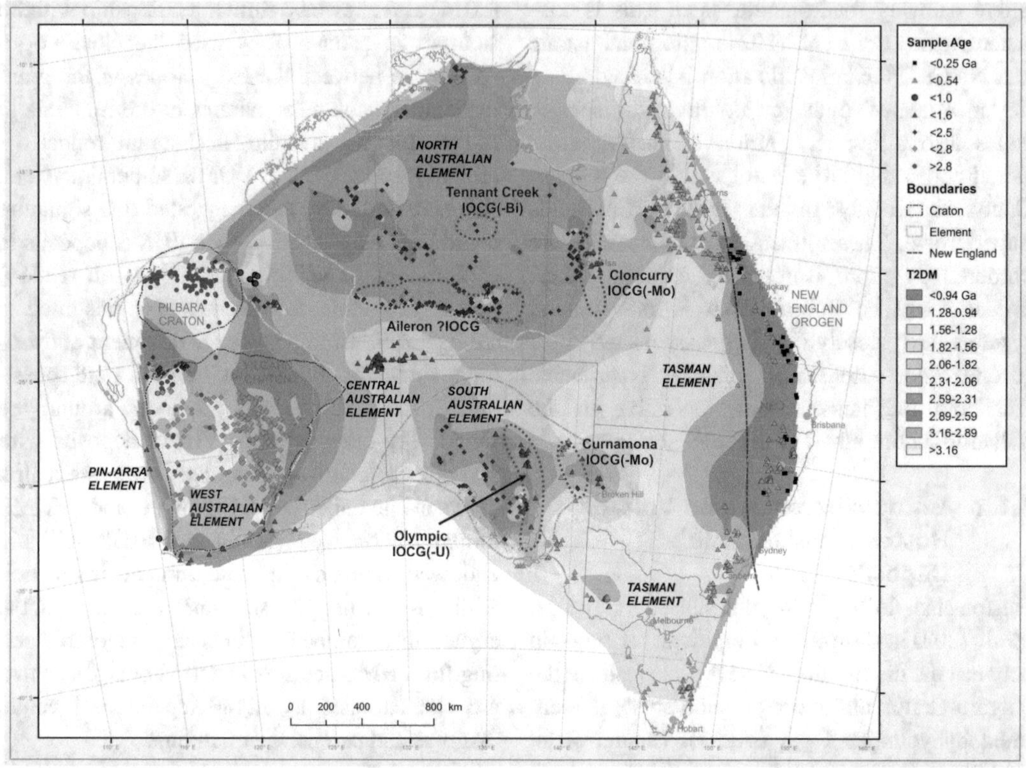

Fig. 8 Plot of iron oxide copper gold (IOCG) provinces overlain over gridded two-stage depleted mantle model age (T_{2DM}) map of Australia (Champion 2013). Updated from Skirrow (2013). As noted by Skirrow (2013), there is a good correlation between these mineral provinces and apparent gradients (breaks) in the regional T_{2DM} map. Refer to Champion (2013) for Sm–Nd data sources. Element nomenclature follows Huston et al. (2012)

6.2 Iron Oxide–Copper–Gold Deposits and Isotopic Gradients: Mapping Old Continental Margins?

Skirrow (2013) showed that many iron-oxide copper–gold (IOCG) provinces are associated with isotopic gradients (Fig. 8), a relationship observed at continental and regional scales. These gradients may record an earlier accretionary continental margin. The isotopic characteristics become increasingly juvenile toward the continental margin, as seen in continental margins such as in the southwestern USA (Bennett and DePaolo 1987) and the Tasman Orogen of eastern Australia (Kemp et al. 2007; Champion et al. 2010). An accretionary margin has been inferred for the IOCG provinces (Skirrow 2013). These include at 1.85 Ga and possibly ca 1.95 Ga for the Mount Isa region (Korsch et al. 2011; McDonald et al. 1997; Gregory et al. 2008), ages of ca. 1880–1650 Ma for the southern Arunta region (Zhao and McCulloch 1995; Scrimgeour et al. 2005), and 1.85 Ga and possibly 2.5 Ga for the Gawler Craton at (Swain et al. 2005; Payne et al. 2009; Korsch et al. 2011;). As noted by Champion and Huston (2016), the isotopic data for the southern part of the Northern Territory becomes increasingly primitive southward, consistent with a relatively long-lived or episodic accretionary margin (Fig. 9).

The association with old margins has been inferred by others (e.g., Skirrow 2010), based on seismic reflection data for the Olympic Dam IOCG deposit. Groves et al. (2010) interpreted that large IOCG deposits were located in intracratonic settings but close to lithospheric

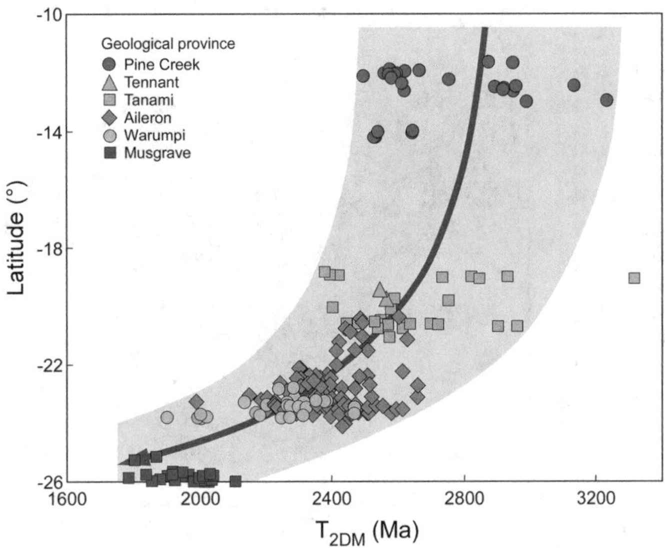

Fig. 9 Two-stage depleted mantle model (T_{2DM}) ages versus latitude for felsic igneous rocks of the Northern Territory (modified after Champion and Huston 2016). Note the general decrease in minimum and max T_{2DM} with decreasing latitude, highlighted by gray area and black arrow. This is particularly evident south of latitude 20 degrees. Figure modified from Champion (2013); refer to that reference for Sm–Nd data and data sources

boundaries indicated by regional isotopic patterns. Groves et al. (2010) also suggested that larger IOCGs formed shortly after (within 100–200 Myr) supercontinent formation, indicating that age data can assist in identifying potential IOCG provinces. This relationship, however, in part reflects classification of IOCG deposits. Inferred IOCG deposits at Tennant Creek, Northern Territory, for example, which Groves et al. (2010) interpreted as high grade Au (+Cu) deposits, not true IOCGs, were generated during formation of the Nuna supercontinent (Huston et al. 2012).

Given the association of IOCG deposits with craton margins/sutures, Groves et al. (2010) postulated that subduction-related metasomatism of lithospheric mantle may have been an important element in the IOCG mineral system. Partial melts from the mantle lithosphere transported copper, gold and volatiles into the crust, although local country rocks may have supplied uranium (Skirrow et al. 2007). These lithospheric melts would have been oxidized (e.g., Rowe et al. 2009) with enhanced capacity to carry copper and gold (e.g., Jégo et al. 2010; Loucks 2014). The presence of igneous rocks derived from metasomatized lithosphere combined with regions of pronounced gradients in isotopic maps may highlight zones with IOCG mineral potential. Melts from metasomatized lithosphere would not necessarily have juvenile isotopic signatures as the timing of metasomatism and the presence of sediments in the subduction component responsible for metasomatism may produce evolved isotopic signatures.

For example, Tasmanian Jurassic dolerites in Tasmania have evolved signatures ($\varepsilon_{Nd} \sim -6$) even though these rocks were interpreted as being derived from subduction-related metasomatized mantle lithosphere (Hergt et al. 1989). More recently, Lu et al. (2013) have highlighted isotopically evolved mafic magmas associated with porphyry Cu mineralization in the Western Yangtze Craton, China, which they suggested were derived from ancient metasomatized lithospheric mantle.

Johnson and McCulloch (1995) and Skirrow et al. (2007) suggested, based on correlations between Sm–Nd isotope signatures and Cu contents, that juvenile mantle melts sourced copper

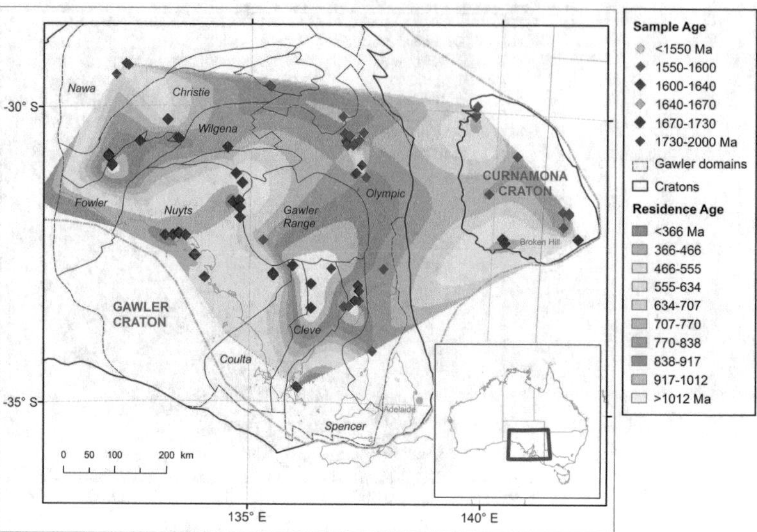

Fig. 10 Crustal residence age map for the Gawler Craton, based on T_{Res} of ca. 2000–1550 Ma granites from that craton. Outlines of Gawler and Curnamona Cratons and the geological domains of the Gawler Craton are from Ferris et al. (2002). Refer to Champion (2013) for relevant data and data sources. Modified after Champion and Huston (2016)

at the giant Olympic Dam deposit, although no compelling evidence exists that these are lithosphere melts. Arndt (2013) points out that large volumes of melt are unlikely to be generated from such sources. This does not, however, change the apparent relationship between Nd (and Pb) isotopic signatures and IOCG belts as pointed out by Skirrow (2013; Fig. 8).

Isotopic data can also be visualized with maps of crustal residence (T_{Res}). This approach can be applied to the Gawler Craton where widespread granites have similar or slightly older ages (ca. 1860–1570 Ma) than the accepted age of IOCG mineralization (ca. 1600–1570 Ma; Skirrow et al. 2007). Hence, the crustal residence map (Fig. 10) provides an image of the characteristics of Gawler crust during IOCG mineralization. Figure 10 shows that isotopic data support the presence of continental margins on both sides of the Gawler Craton. This relationship is consistent with geological relationships indicative of a convergent margin setting on the southwestern part of the Gawler Craton at ca. 1630–1610 Ma and ca. 1700 Ma (Swain et al. 2008; Ferris and Schwartz 2004; Payne et al. 2010). This raises the question of why IOCG mineralization is localized on the eastern and northeastern margins of the Gawler Craton and not elsewhere in the craton, and suggests additional controls on the location of mineralization. For example, mineralization may in part reflect focusing of mantle-derived melts into zones with thinner lithosphere. Skirrow (2010) suggested a model for the Olympic Dam Province in which delamination resulted in mafic and ultramafic magmatism that was focused into, the delaminated region, which was ultimately responsible for mineralization. There is perhaps evidence for such a mantle-derived magmatic flux in the Sm–Nd isotopic data, best seen through specific time slice isotopic plots.

6.2.1 Refining Iron Oxide–Copper-Gold Search Space—Isotopic Time Slice Diagrams

Time slice Nd model age maps effectively provide an isotopic snap shot of the crust at the time of any specific geological event, such as IOCG mineralization. Such plots can be simply constructed. For example, for 1800 Ma the model age is simply calculated as follows:

$$T_{2DM(1800\ Ma)} = T_{2DM} - 1800;\ \text{for all}\ T_{2DM} > 1800\ Ma \qquad (5)$$

In provinces with a limited age range for magmatism, this will produce similar images to T_{Res} maps. In provinces with a large range of

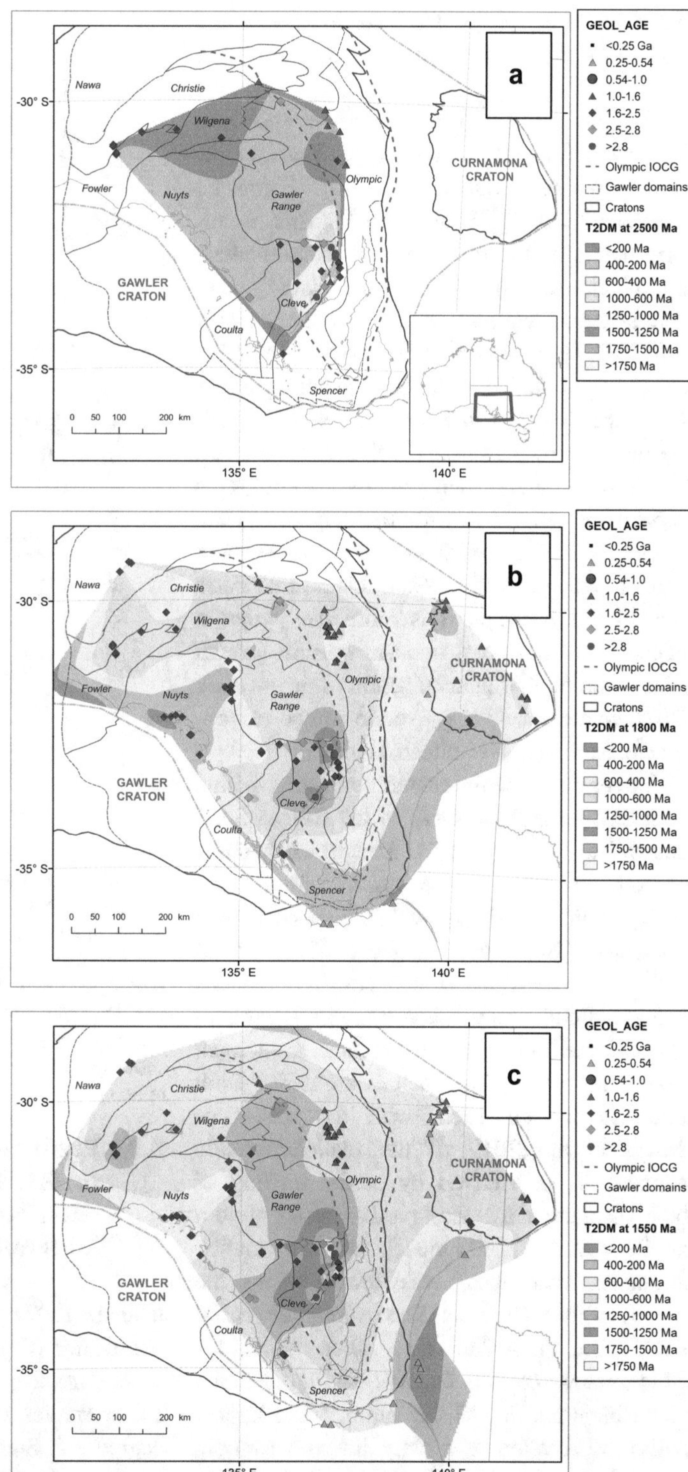

Fig. 11 Model age time slice figures (A: at 2500 Ma; B: at 1800 Ma; C: at 1550 Ma) for the Gawler and Curnamona Cratons of the South Australian Element, Australia. The Olympic IOCG field (named after the Olympic Dam IOCG deposit; from Skirrow 2013) is denoted by orange dashed line in A and C. Note the persistent more juvenile isotopic embayment located around the Olympic Dam IOCG deposit (near the word 'Olympic' on **a**, **b** and **c**). Refer to Champion (2013) for Sm–Nd data and data sources. Modified after Champion and Huston (2016)

magmatic ages, maps for pseudo-time slices based on Nd model ages can be produced (Fig. 11).

The Gawler Craton is particularly amenable to this approach as felsic magmatism spanned over 1500 Myr (Reid and Hand 2012). Figure 11 illustrates 2500, 1800 and 1550 Ma time slices, corresponding to the Archean, intermediate (pre-mineralization) and immediate post-mineralization craton structure. The Olympic Dam IOCG province straddles the margin of the 2500 Ma and older crust, and in the younger time slices, the isotopic data are zoned along this margin, with an old crustal block in the south and the juvenile embayment in the north. This embayment coincides with the Olympic Dam deposit (Fig. 11c). Although this zone could represent a long-lived juvenile feature that dates back to the Archean, it is more likely that it reflects juvenile post-Archean crustal growth close to the time of mineralization (Skirrow et al. 2007), consistent with a juvenile source of copper (Johnson and McCulloch 1995).

Further evidence for contemporaneous mantle input lies in the felsic rocks of the Hiltaba Suite magmatism (ca. 1595–1570 Ma) which become increasingly more A-type toward the deposit (Budd 2006), consistent with a greater mantle input, towards the deposit (Budd 2006). Metal could be derived from contemporaneous mafic/ultramafic magmatism through magmatic-hydrothermal processes or leached from such pre-existing rocks (Campbell et al. 1998). Both processes are consistent with the model of Groves et al. (2010) if the mafic/ultramafic magmatism was derived by partial melting of the lithosphere and at least partly consistent with the Begg et al. (2010) model, in which mafic–ultramafic magmatism was focused into the region around Olympic Dam along a crustal break. The 1550 Ma time slice (Fig. 11c) approximates crustal features during Hiltaba magmatism, and IOCG formation. More importantly, the location of a more juvenile isotopic zone along the isotopically defined paleo-continental suture may be highlighting a more favorable exploration target.

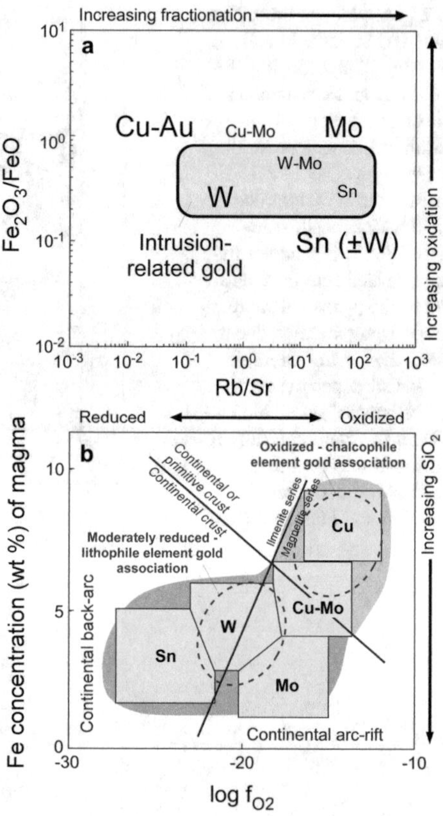

Fig. 12 **a** Rb/Sr ratio versus Fe_2O_3/FeO ratio plot modified after Blevin et al. (1996). The plot illustrates the relationship between the degree of oxidation and compositional evolution of the magma (based on whole-rock compositions) and the dominant commodities in related mineralisation. Intrusion related gold deposit (IRGD) field from Blevin (2004). **b** Plot of oxygen fugacity versus total Fe concentration in the magma, also showing tectonic setting. Plot modified after Thompson et al. (1999) and Lang et al. (2000)

6.3 Predictive Analysis Based on Radiogenic Signatures: Granite-Related Mineralization

Ishihara (1977, 1981) first recognized that the redox state of granite magmas appeared to have some control over the associated metallogenesis. These studies were expanded to show that the degree of chemical evolution of the granite magma also controls metallogenesis (Blevin and Chappell 1992; Blevin et al. 1996; Thompson et al. 1999; Lang et al. 2000; Blevin 2004;

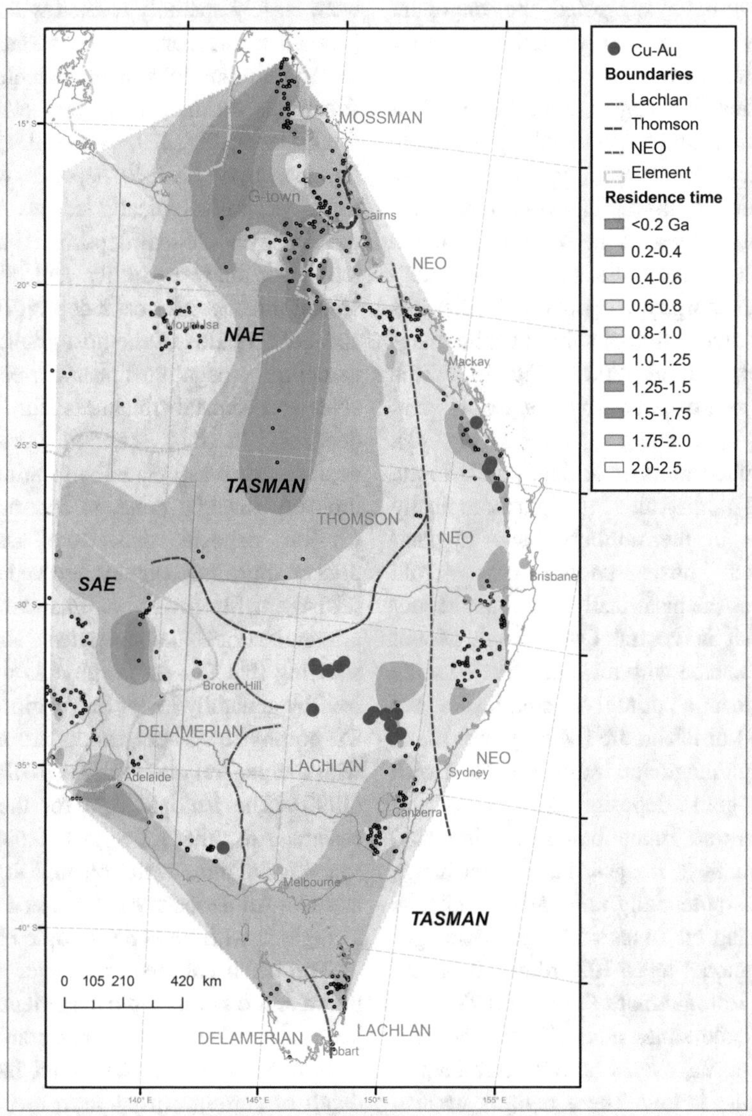

Fig. 13 Location of Cu–Au deposits superimposed on the gridded Nd residence age map for the Tasman Element and surrounding regions (South (SAE) and North (NAE) Australian elements). Copper–gold deposits (from Geoscience Australia's Australian Mines Atlas, downloadable via their portal https://portal.ga.gov.au/persona/minesatlas) are shown as blue circles. Also shown are the Delamerian, Lachlan, Thomson and Mossman orogens of the Tasman Element. There is a good correlation between many copper–gold deposits and zones with young residence ages, notably in the central Lachlan Orogen and the central New England Orogen. Copper–gold deposits outside of these two zones are almost exclusively not magmatic-related. Location of Nd samples used to create the grid are shown as black circles. Data and data sources are given in Champion (2013), after which the figure is modified. Boundaries of Delamerian, Lachlan, Thomson, Mossman and New England (NEO) orogens modified from Stewart et al. (2013)

Fig. 12). Thompson et al. (1999) showed that rare metal (Sn, Mo and W) deposits are associated with continental material in backarc or post/non-arc extensional environments, whereas copper ± gold deposits are related to more primitive crust arcs, consistent with the oxidized nature of arc rocks (Parkinson and Arculus 1999). More recently, Richards (2011) and

Loucks (2014) have suggested the important additional role of water in arc-related magmas for copper mineralization.

These relationships suggest that Sm–Nd (and other) isotopic data can delineate prospective zones for porphyry copper and copper–gold deposits, by identifying regions with more juvenile isotopic signatures. In the Macquarie Arc in New South Wales, Australia, latest Ordovician–earliest Silurian porphyry copper–gold deposits are associated with isotopically juvenile magmatism (Cooke et al. 2007; Fig. 13). This domain is well-mapped by regional- and national-scale T_{2DM} and T_{Res} images (Fig. 13), despite the limited number of data points in the data set. The T_{Res} map also highlight isotopically juvenile areas in the northern New England Orogen, which hosts known copper–gold deposits such as the historically important Mount Morgan deposit in central Queensland. Mount Morgan is associated with rocks interpreted to be related to either a primitive continental arc (Morand 1993) or island arc (Murray and Blake 2005). Although magmatic rocks associated with these copper–gold deposits are isotopically primitive, they are much better discriminated using T_{Res} than T_{2DM} images. The former image better identifies potentially mineralized units, as they are focusing on zones with magmatic ages very close to model ages. Regardless, it is evident that both isotopic maps (T_{2DM} and T_{Res}) are indicating juvenile zones amenable to porphyry and related deposits, even where the numbers of analyzed samples is low. These regions are also identifiable in other isotopic figures, such as ε_{Nd} versus time (see Champion 2013).

Recently, Wu et al. (2021) demonstrated similar relationships between Nd isotopic signatures and maps and the many porphyry Cu–Au and Cu–Mo deposits in the Central Asian Orogenic Belt (CAOB). Using an extensive Sm–Nd isotopic dataset, they were able to map locations and extents of cratons, as well as microcontinents and juvenile crustal blocks. Wu et al. (2021) were able to show that the porphyry deposits were largely spatially located within isotopically juvenile crust, marginal to cratons or microcontinents, similar to that documented for eastern Australia. Further, consistent with Thompson et al. (1999; Fig. 12), Wu et al. (2021) showed that porphyry Cu–Mo deposits were generally found within isotopically less juvenile crust than the porphyry Cu–Au deposits. Wu et al. (2021) took their analysis further, and, after estimating crustal thickness for each deposit (based on Sr/Y and La/Yb ratios of the host granites), were able to define a negative linear correlation between εNd and crustal thickness for the porphyry deposits. In this scheme, porphyry Cu–Au deposits were associated with thinner, more isotopically juvenile crust, in contrast to porphyry Cu–Mo deposits that were associated with thicker more isotopically evolved crust. Loucks (2014), in his review of Phanerozoic porphyry mineralization, demonstrated similar results, showing that Cu–Au porphyries were characterized by generally lower Sr/Y ratios than Au-poor Cu porphyries. These results are also consistent with the general observations of Thompson et al. (1999). The isotopic data for the CAOB, and eastern Australia show that exploration search space for regions with potential for porphyry Cu and Cu–Au deposits can be effectively narrowed to regions with juvenile isotopic characteristics.

The usual caveats, of course, apply. Neodymium and other isotopic data identify potentially juvenile terranes, but do not convey any information on additional important factors, such as depth of current crustal exposure, the degree of preservation of the porphyry and related deposits, the rate of uplift, the degree of metamorphism and other factors. Porphyry copper–gold deposits are also common in more mature continental arcs although the delineation of these on the basis of their isotopic signature is more problematical. Further, as shown by Lu et al. (2013), porphyry Cu-related magmas, even along suture zones, need not have juvenile isotopic signatures, especially if ancient metasomatized lithosphere is involved in the genesis of such magmas.

Table 2 Secular variations in T_{2DM} and ε_{Nd} in felsic magmatism and possible interpretations

Type	Variations with decreasing age	Interpretation	Examples
1	T_{2DM} approximately constant (especially maximum values) ε_{Nd} decreasing	Largely reworking of pre-existing crust Any juvenile input is cryptic	South-western United States (Bennett and DePaolo 1987); East Pilbara Terrane, Western Australia (Champion 2013)
2	T_{2DM} decreasing ε_{Nd} ~ constant	Involvement of both pre-existing crust and juvenile material (either reworking of young crust and/or direct mantle input).	Pine Creek-Tennant Creek-Tanami-Aileron provinces, Northern Territory (Champion 2013)
3	T_{2DM} decreasing (markedly) ε_{Nd} increasing	Significant juvenile input (often related to long-lived accretionary orogens). Pre-existing older crust thinned or largely absent. May be associated with a significant sedimentary component.	Tasman Orogen (e.g. Kemp et al 2007); Western United States (Farmer and DePaolo 1983)

The three identified types can be regarded as end-members, with gradations between all three. Note that a range of isotopic signatures (e.g. ε_{Nd} and T_{2DM}) are indicators for multiple age sources and/or components

6.4 Isotopic Mapping and Accretionary Tectonic Settings

Isotopic maps can be integrated with other features that characterize mineral systems, including tectonic settings. An example is identification of accretionary orogens that commonly host specific mineral systems such as those that form porphyry copper and associated deposits (Hronsky et al. 2012). Accretionary margins commonly are characterized by increasingly juvenile isotopic signatures outward and with time as the margin evolves. This gradient can be useful for regional exploration targeting as seen in data for the western United States (Farmer and DePaolo 1983; Bennett and DePaolo 1987), eastern Australia (Kemp et al. 2009; Champion et al. 2010; Fig. 8) and in the southern half of Northern Territory (Fig. 9). Collins et al. (2011) suggested that different types of orogenic systems can be discriminated. For example, dominantly accretionary (or Internal: circum-Pacific) versus dominantly collisional (or External: southern Europe) can be recognized on the basis of isotopic signatures. Although we agree with this concept in general, Champion (2013) indicate that the isotopic signatures for 'Internal' orogens are not unique (Table 2).

Isotopic maps at various scales can be used to characterize larger scale components that are indicative of metallogenic provinces, for many mineral systems. As discussed by McCuaig et al. (2010), exploration models are predicated on the use of many geological, geochemical and geophysical datasets over a range of scales. Isotopic maps, including those based on Sm–Nd data are one of many layers to be integrated with other data sets. One example of such integration is the use of isotopic data in conjunction with crustal boundaries interpreted from geophysical data (e.g., Korsch and Doublier 2016).

7 Future Developments and In-Situ Analysis of Radiogenic Isotopes

The recent development of in-situ methods of analysis of radiogenic isotopes have led to a number of innovative approaches applicable to mineral system studies. These include in-situ Sm–Nd analysis of accessory minerals, such as apatite and other REE-rich accessory phases (Fisher et al. 2017, 2020; Hammerli and Kemp 2021), which enables assessment of inherited components within grains and allows characterization of temporal changes in the isotopic characteristics of these inherited components, particularly where independent age data are available from U–Pb analyses (Fisher et al. 2017). Much of this work is in its infancy and so regional data sets are not currently available. Good examples of what such data may eventually provide can be illustrated using the Lu–Hf isotopic system which behaves comparably to the Sm–Nd system (Vervoort and Patchett 1996; Chauvel et al. 2008). Examples of regional isotopic maps using Lu–Hf data for zircon investigating mineralization include Mole et al. (2014), Wang et al. (2016) and Osei et al. (2021).

Mole et al. (2014) determined Lu–Hf signatures and U–Pb ages of inherited and xenocrystic domains from zircons extracted from Yilgarn. Based on these data, they produced a series of temporal Lu–Hf snapshots (at 3050–2820 Ma, 2820–2720 Ma and 2720–2600 Ma). These snapshots indicated the Yilgarn Craton formed from several Archean micro-continents. These snapshots enabled documentation of the relationship between ca. 2.9 Ga and ca. 2.7 Ga komatiite deposits and paleo-crustal structures. The ca. 2.7 Ga craton margin of Mole et al. (2014) matched the one determined from Sm–Nd whole rock data (Champion and Cassidy 2007, 2008). Mole et al. (2014), however, identified an additional crustal block in the southwest part of the Yilgarn.

Techniques developed by Mole et al. (2014) are an advance on temporal images compiled from isotope model ages alone (Fig. 11), as the former can exclude complexities due to juvenile input related to younger crustal growth, magma mixing and other processes. The approach is not without problems though, in particular, the presence of inherited components in zircons derived from sedimentary rocks. For such zircons, the isotopic information they carry may not represent the crust in the general area of the host intrusive, but may have been brought in from other provinces. This is less problematic in Archean provinces, such as the Yilgarn and Pilbara cratons, as sedimentary rocks are not a large component of the geology (Champion 2013). This contrasts with younger terranes where sedimentary rocks are more abundant (e.g., eastern Australia: Kemp et al. 2009), particularly when transport distances of detrital zircons are considered (Sircombe 1999). The possible influence of sedimentary components, however, is also problematic for whole rock Nd maps (Champion, 2013).

The approach of Mole et al. (2014) provides a good example of what may also be achievable with in-situ microanalysis of accessory minerals for Sm–Nd. More importantly, as outlined by Fisher et al. (2017) and Hammerli and Kemp (2021), the combined use of in-situ Lu–Hf, Sm–Nd analysis and other isotopic systems offers a powerful way forward in deciphering isotopic signatures of components within granitic and other rocks, and, hence, making more robust isotopic maps. A number of studies (Mole et al. 2013, 2014; Osei et al. 2021) have already made use of in-situ Lu–Hf analysis combined with whole rock Sm–Nd data. On this note, given the comparable behavior and model age calculations for both Sm–Nd and Lu–Hf, there is no reason why hybrid maps using both systems comined cannot be produced (e.g., Fraser et al. 2020).

8 Links with Complementary Isotopic Systems and Geophysical Data Sets

Radiogenic isotopic studies have significant potential for understanding and constraining mineralization, not just as metal source tracers, but also, as demonstrated here, indicators of favorable geological setting, lithospheric architecture, and geodynamic environment. This contribution concentrated largely on isotopic maps based on Sm–Nd data from felsic magmatic rocks. Clearly, more can be done. Obvious first steps are to better link and integrate different isotopic systems. The U–Th–Pb system, using isotopes from ore minerals (Huston et al. 2020), provides one example. Champion and Huston (2016) discussed aspects of linking these two systems and showed that often, though not universally, the two systems convey similar information: regions with old or young Nd model ages correlate with regions of high or low μ

($^{238}U/^{204}Pb$), respectively. A good example of this Nd–Pb correlation is provided by Huston et al. (2014) for VHMS deposits in the Yilgarn Craton. The link between regional Nd isotopic data from granitic rocks and regional Pb isotopic data from mineralization is perhaps not surprising for granite-related mineralization, but is unexpected for non-magmatic mineral systems, for example shale-hosted Zn–Pb deposits. This linkage between the isotopic systems needs to be investigated in a more robust manner. In particular, there should also be investigation of regions where regional patterns from one system, e.g., U–Th–Pb in mineralized samples, differ from those from Sm–Nd (or Lu–Hf) in granites, as is evident, for example, in parts of the Pilbara Craton (see Champion and Huston 2016).

Further work can also extend the isotopic mapping approach here to other isotopic systems and sample media. As mineral systems commonly involve the entire lithosphere as well as the upper asthenosphere (Huston et al. 2016), the entire lithosphere, including the mantle component, should be mapped as shown by Begg et al. (2009) for Africa. Therefore, future isotope mapping should focus on characterizing the entire lithosphere at a range of scales, applying multiple isotopic systems to different rock types and minerals, and integrating the results with geological, geochemical and geophysical data.

Major recent advances have been made on geophysical mapping of the lithosphere, using data sets including passive/active seismic, magnetotellurics and other methods. Such geophysical data sets are available at many scales, including the global-scale for some datasets (Kennett and Salmon 2012; Kennett et al. 2013; Priestley and McKenzie 2013; Hoggard et al. 2020). In contrast to isotopic data sets, geophysical data are commonly collected in systematic grids. Significant additional interpretative power could be achieved in combining the two data types. This benefit works in both directions as geophysical data are a measure of the present-day situation and do not provide time resolution. Isotopic data, where integrated, will help to provide this temporal framework. As demonstrated by Begg et al. (2009), full use of geophysical data depends on integration with other data sets, including geochronological, geochemical and petrological data, including isotopic and data (O'Reilly and Griffin 2006; Griffin et al. 2013).

One isotopic system that has not seen systematic use in isotopic mapping is the Re–Os system. This system potentially can provide better constraints on growth and nature of the lithospheric mantle than isotopic systems such as Sm–Nd, Lu–Hf and U–Th–Pb (Shirey and Walker 1998; Carlson 2005). This potential is important as it provides a link between the lithospheric mantle and the crust which is more effectively characterized by the Sm–Nd and U–Th–Pb systems. Gregory et al. (2008), for example, linked copper mineralization at Mount Isa (Queensland, Australia) to metasomatized lithospheric mantle lithosphere possibly produced during subduction associated with a paleo-margin similar to that imaged by the Sm–Nd data. Model ages (including both mantle model ages and model depletion ages; T_{RD}) for mantle lithosphere extraction using the Re–Os system can be determined on lithospheric mantle samples (mantle nodules and the like: Pearson et al. 1995; O'Reilly and Griffin 2006). As the availability of these samples is limited, any future Re–Os mapping project must depend on proxies such as mantle-derived magmas (c.f., Gregory et al. 2008).

Mapping lithospheric mantle is important to mineral systems that involve, at some point in their evolution, metasomatism of the lithospheric mantle. Metasomatized lithospheric mantle may be an important source of ore metals or volatiles for many mineral styles, for example orthomagmatic nickel-copper-platinum group elements, IOCG, porphyry copper–gold, and thorium-REE deposits (Zhang et al. 2008; Groves et al. 2010; Hronsky et al. 2012; Griffin et al. 2013). Zhang et al. (2008) argued that formation of deposits related to flood basalts involved mantle plumes that interacted with metasomatized lithosphere and were emplaced within or marginal to crustal blocks of either Archean and/or Paleoproterozoic age. Griffin et al. (2013) extended this interpretation to involve reworking (melting) of (subduction-related) metasomatized mantle lithosphere for many other mineral systems. The role of metasomatized mantle in mineral systems, however, is controversial, and Arndt (2013) presented counter-arguments against this involvement.

If metasomatized lithosphere is important for mineral systems related to large igneous provinces and subduction, then recognition of zones of metasomatized mantle by isotopic and other data (e.g., magnetotellurics) is valuable for area selection (e.g., Hronsky et al. 2012). Even if mantle metasomatism is not an important process (Arndt 2013), isotopic maps still provide information on other processes. As discussed earlier, the topography of the lithosphere-asthenosphere boundary may be just as important for KANS and other orthomagmatic mineral systems (Begg et al. 2010). Hence, although the root causes of association with mineral systems with specific lithospheric architectures varies according to metallogenic model, datasets and images generated from Sm–Nd and other isotopic systems can identify this architecture and, thereby, fertile metallogenic terranes.

9 Conclusions

Although Sm–Nd isotope data have historically been used in metallogenic studies to stablish age and metal source, these data, presented as regional to continental isotope map, assist identification of metallogenic terranes favorable for many mineral systems. This approach is aided by large, spatially (and in some cases temporally) constrained data packages that are easily downloadable (Fraser et al. 2020). This in combination with available 2-D interpolation software, have allowed repeatable generation of isotopic maps, using various isotopic variables, at regional to continental scales, allowing for metallogenic interpretation over similarly large regions. Interpretations of these isotopic maps is enhanced by a mineral systems approach in which mineral deposits are recognized as products of geological processes that occur over a wide range of scales.

The use of isotopic data and maps strongly depends on the relative geochemical behavior of parent and daughter isotopes used and the rock types and minerals analyzed. For the Sm–Nd system detailed here, the largest fractionation is seen between depleted mantle and continental crust. This means that the Sm–Nd system can be used to effectively 'see' through many crustal processes to provide information on the source of granitic rocks. This includes model age calculations, which, with several assumptions, approximate of the average age of the crust. For large volume rocks like granites, the model age is an excellent proxy that constrains the character of crustal blocks that host the granite. This capability is enhanced by regional to continental scale maps constructed with Sm–Nd data for these granitic rocks. These maps can be used to identify larger-scale portions of mineral systems indicative of metallogenic tracts, for many mineral systems. Examples of uses include:

- demonstration of associations if specific mineral systems and characteristic isotopic domains;
- recognition of old and/or thick cratonic blocks;
- identification of tectonic regimes favorable for mineralization;
- mapping of isotopically juvenile zones indicative of zones of rifts or primitive arcs;
- identification of crustal breaks and possible sutures that localize mineral systems for various reasons; and
- producing baseline maps that assist in identifying regions/periods characterized by magmatic, especially mantle, input.

As exploration models currently are built with many different types of geological, geochemical and geophysical data over many scales, isotopic maps are just another layer for integration. Future work should focus on the spatial and temporal evolution of the entire lithosphere, including the lithospheric mantle, to assist in more effective mineral exploration. This must involve integration of radiogenic isotope data with other data, particularly geophysical data, which has the advantage of being able to directly image the lithosphere and being of a more continuous nature that often decidedly lumpy isotopic data sets.

Acknowledgements This contribution has benefited from many discussions with Drs Geoff Fraser, Michael Doublier, Roger Skirrow and Kathryn Waltenberg (Geoscience Australia) and Drs Hugh Smithies and Yong-Jun Lu (Geological Survey of Western Australia), and helpful and constructive reviews by journal reviewers Dr Steffen Hagemann and an anonymous reviewer, and editor Dr Jens Getzmer. It is published with permission of the Chief Executive Officer of Geoscience Australia.

References

Arevalo R Jr, McDonough WF (2010) Chemical variations and regional diversity observed in MORB. Chem Geol 271:70–85

Arndt N (2013) The lithospheric mantle plays no active role in the formation of orthomagmatic ore deposits. Econ Geol 108:1953–1970

Australian Academy of Science (2012) Searching the deep earth—a vision for exploration geoscience in Australia. Australian Academy of Science, Canberra, p 42

Barnes SJ (2010a) Fiorentini ML (2010a) Komatiite-hosted nickel sulphide deposits: what's so special about the Kalgoorlie Terrane. Geol Surv West Austr Rec 18:281–283

Barnes SJ (2010b) Fiorentini ML (2010b) Comparative lithogeochemistry of komatiites in the Eastern Goldfields Superterrane and the Abitibi Greenstone Belt, and implications for distribution of nickel sulphide deposits. Geol Surv West Austr Rec 20:23–26

Barnes SJ, Fiorentini ML (2012) Komatiite magmas and Ni sulfide deposits: a comparison of variably endowed Archean terranes. Econ Geol 107:755–780

Beakhouse GP (2007) Gold, granite and Late Archean tectonics: a Superior province perspective. Geosci Austr Rec 2007(14):191–195

Begg GC, Griffin WL, Natapov LM, O'Reilly SY, Grand S, O'Neill CJ, Poudjom Djomani Y, Deen T, Bowden P (2009) The lithospheric architecture of Africa: seismic tomography, mantle petrology and tectonic evolution. Geosphere 5:23–50

Begg GC, Hronsky JAM, Arndt NT, Griffin WL, O'Reilly SY, Hayward N (2010) Lithospheric, cratonic, and geodynamic setting of Ni–Cu–PGE sulfide deposits. Econ Geol 105:1057–1070

Bennett VC, DePaolo DJ (1987) Proterozoic crustal history of the western United States as determined by neodymium isotopic mapping. Bull Geol Soc America 99:674–685

Blevin PL (2004) Redox and compositional parameters for interpreting the granitoid metallogeny of eastern Australia: implications for gold-rich ore systems. Res Geol 54:241–252

Blevin PL, Chappell BW (1992) The role of magma sources, oxidation states and fractionation in determining the granitoid metallogeny of eastern Australia. Trans Roy Soc Edinburgh Earth Sci 83:305–316

Blevin PL, Chappell BW, Allen CM (1996) Intrusive metallogenic provinces in eastern Australia based on granite source and composition. Trans Roy Soc Edinburgh Earth Sci 87:281–290

Blewett RS, Henson PA, Roy IG, Champion DC, Cassidy KF (2010) Scale integrated architecture of a world-class gold mineral system: the Archaean eastern Yilgarn Craton, Western Australia. Precambrian Res 183:230–250

Budd AR (2006) The Tarcoola goldfield of the central Gawler gold province, and the Hiltaba association granites, Gawler craton, South Australia. Unpublished PhD thesis, Australian National University, p 507

Campbell IH, Compston DM, Richards JP, Johnson JP, Kent AJR (1998) Review of the application of isotopic studies to the genesis of Cu–Au mineralisation at Olympic Dam and Au mineralisation at Porgera, the Tennant Creek district and Yilgarn Craton. Austr J Earth Sci 45:201–218

Carlson RW (2005) Application of the Pt–Re–Os isotopic systems to mantle geochemistry and geochronology. Lithos 82:249–272

Cassidy KF, Champion DC (2004) Crustal evolution of the Yilgarn Craton from Nd isotopes and granite geochronology: implications for metallogeny. Univ West Australia Pub 33:317–320

Cassidy KF, Champion DC, McNaughton NJ, Fletcher IR, Whitaker AJ, Bastrakova IV, Budd AR (2002) Characterization and metallogenic significance of Archaean granitoids of the Yilgarn Craton, Western Australia. Australian Mineral and Industry Research Association Project P482/MERIWA Project M281, final report, p 514

Cassidy KF, Champion DC, Huston DL (2005) Crustal evolution constraints on the metallogeny of the Yilgarn Craton. In: Mao J, Bierlein FP (eds) Mineral deposit research: meeting the global challenge. Springer, Berlin/Heidelberg, pp 901–904

Cassidy KF, Champion DC, Krapež B, Barley ME, Brown SJA, Blewett RS, Groenewald PB, Tyler IM (2006) A revised geological framework for the Yilgarn Craton. Geol Surv West Austr Rep 2006/8

Champion DC (2013) Neodymium depleted mantle model age map of Australia: explanatory notes and user guide. Geosci Austr Rec 2013/044. https://doi.org/10.11636/Record.2013.044

Champion DC, Cassidy KF (2007) An overview of the Yilgarn Craton and its crustal evolution. Geosci Austr Rec 14:8–13

Champion DC, Cassidy KF (2008) Using geochemistry and isotopic signatures of granites to aid mineral systems studies: an example from the Yilgarn Craton. Geosci Austr Rec 09:7–16

Champion DC, Huston DL (2016) Radiogenic isotopes, ore deposits and metallogenic terranes: novel approaches based on regional isotopic maps and the mineral systems concept. Ore Geol Rev 76:229–256

Champion DC, Smithies RH (2000) The geochemistry of the Yule Granitoid Complex, East Pilbara Granite–

Greenstone Terrane; evidence for early felsic crust. Geol Surv West Austr 1999–2000 Ann Rev 42–48

Champion DC, Bultitude RJ, Blevin PL (2010) Geochemistry and isotope systematics of Carboniferous to Triassic felsic magmatism in northeastern Australia—putting the New England Orogen in its place. In: Buckman S, Blevin PL (eds) New England Orogen 2010. Proceedings of a conference held at the University of New England, Armidale, New South Wales, Australia, November 2010. University of New England, Armidale, pp 112–118

Chauvel C, Lewin E, Carpentier M, Arndt NT, Marini J-C (2008) Role of recycled oceanic basalt and sediment in generating the Hf–Nd mantle array. Nat Geosc 1:64–67. https://doi.org/10.1038/ngeo.2007.51

Collins WJ, Belousova EA, Kemp AIS, Murphy JB (2011) Two contrasting Phanerozoic orogenic systems revealed by hafnium isotope data. Nat Geosc 4:333–337. https://doi.org/10.1038/NGEO1127

Cooke DR, Wilson AJ, House MJ, Wolfe RC, Walshe JL, Lickfold V, Crawford AJ (2007) Alkalic porphyry Au–Cu and associated mineral deposits of the Ordovician to Early Silurian Macquarie Arc, New South Wales. Austr J Earth Sci 55:445–463

Czarnota K, Champion DC, Cassidy KF, Goscombe B, Blewett RS, Henson PA, Groenewald PB (2010) Late Archaean geodynamic processes: how the Eastern Goldfields Superterrane evolved in time and space. Precambrian Res 183:175–202

De Grey Mining Limited (2022) 6.8Moz Hemi Maiden Mineral Resource drives Mallina Gold Project. Unpublished De Grey Mining Limited announcement to the Australian Stecurities Exchange, 23 June 2021. https://degreymining.com.au/asx-releases/

DePaolo DJ (1981) Neodymium isotopes in the Colorado Front Range and crust–mantle evolution in the Proterozoic. Nature 291:193–196

DePaolo DJ (1988) Neodymium isotope geochemistry: an introduction. Springer-Verlag, Berlin, p 187

DePaolo DJ, Wasserburg GJ (1976) Nd isotopic variations and petrogenetic models. Geophys Res Lett 3:249–252

Dickin AP (1995) Radiogenic isotope geology. Cambridge University Press, Cambridge, p 490

Farmer GL, DePaolo DJ (1983) Origin of Mesozoic and Tertiary granite in the western United States and implications for pre-Mesozoic crustal structure I. Nd and Sr isotopic studies in the geocline of the northern Great Basin. J Geophy Res 88:3379–3401

Farmer GL, DePaolo DJ (1984) Origin of Mesozoic and Tertiary granite in the western United States and implications for pre-Mesozoic crustal structure II. Nd and Sr isotopic studies of unmineralised and Cu- and Mo-mineralized granite in the Precambrian craton. J Geophy Res 89:10141–10160

Faure G (1977) Principles of isotope geology, 2nd edn. Wiley, New York, p 589

Ferris GM, Schwarz MP, Heithersay P (2002) The geological framework, distribution and controls of Fe-oxide Cu–Au mineralisation in the Gawler Craton, South Australia. Part I—geological and tectonic framework. In: Porter TM (ed) Hydrothermal iron oxide copper-gold and related deposits: a global perspective, vol 2. PGC Publishing, Adelaide, pp 9–31

Ferris GM, Schwarz M (2004) Definition of the Tunkillia Suite, western Gawler Craton. MESA J 34:32–41

Fiorentini ML, Beresford SW, Barley ME, Duuring P, Bekker A, Rosengren N, Cas R, Hronsky J (2012) District to camp controls on the genesis of komatiite-hosted nickel sulfide deposits, Agnew-Wiluna greenstone belt, Western Australia: insights from the multiple sulfur isotopes. Econ Geol 107:781–796

Fisher CM, Hanchar JM, Miller CF, Phillips S, Vervoort JD, Whitehouse MJ (2017) Combining Nd isotopes in monazite and Hf isotopes in zircon to understand complex open-system processes in granitic magmas. Geology 45:267–270. https://doi.org/10.1130/G38458.1

Franklin JM, Gibson HL, Jonasson IR, Galley AG (2005) Volcanogenic massive sulfide deposits. Econ Geol 100th Anniv 523–560

Fraser GL, Waltenberg K, Jones SL, Champion DC, Huston D.L, Lewis CJ, Bodorkos S, Forster M, Vasegh D, Ware B, Tessalina S (2020) An isotopic atlas of Australia. In: Czarnota K, Roach I, Abbott S, Haynes M, Kositcin N, Ray A, Slatter E (eds) Exploring for the Future: Extended Abstracts. Geoscience Australia, Canberra. https://doi.org/10.11636/133772

Geoscience Australia (2022) Australian Mines Atlas. Geoscience Australia, Canberra. https://portal.ga.gov.au/persona/minesatlas

Goldfarb R, Santosh M (2014) The dilemma of the Jiaodong gold deposits: are they unique? Geosc Front 5:139–153. https://doi.org/10.1016/j.gsf.2013.11.001

Gregory MJ, Schaefer BF, Keays RR, Wilde AR (2008) Rhenium–osmium systematics of the Mount Isa copper orebody and the Eastern Creek Volcanics, Queensland, Australia: implications for ore genesis. Mineral Deposita 43:553–573

Griffin WL, Begg GC, O'Reilly SY (2013) Continental-root control on the genesis of magmatic ore deposits. Nat Geosci 6:905–910. https://doi.org/10.1038/NGEO1954

Groves DI, Batt WD (1984) Spatial and temporal variations of Archaean metallogenic associations in terms of evolution of granitoid-greenstone terrains with particular emphasis on the Western Australian shield. In: Kroner A, Hanson GN, Goodwin AM (eds) Archaean geochemistry. Springer-Verlag, Berlin, pp 73–98

Groves DI, Bierlein FP, Meinert LD, Hitzman MW (2010) Iron oxide copper-gold (IOCG) deposits through Earth history: implications for origin, lithospheric setting, and distinction from other epigenetic iron oxide deposits. Econ Geol 105:641–654

Hammerli J, Kemp AIS (2021) Combined Hf and Nd isotope microanalysis of co-existing zircon and REE-rich accessory minerals: high resolution insights into crustal processes. Chem Geol 581(2021):120393. https://doi.org/10.1016/j.chemgeo.2021.120393

Hiess J, Bennett VC (2016) Chondritic Lu/Hf in the early crust–mantle system as recorded by zircon populations from the oldest Eoarchean rocks of Yilgarn Craton, West Australia and Enderby Land, Antarctica. Chem Geol 427:125–143. https://doi.org/10.1016/j.chemgeo.2016.02.011

Hergt JM, Chappell BW, McCulloch MT, McDougall I, Chivas AR (1989) Geochemical and isotopic constraints on the origin of the Jurassic dolerites of Tasmania. J Petrol 30:841–883

Hoggard MJ, Czarnota K, Richards FD, Huston DL, Jaques AL, Ghelichkhan S (2020) Global distribution of sediment-hosted metals controlled by craton edge stability. Nat Geosc 13:504–510. https://doi.org/10.1038/s41561-020-0593-2

Hronsky JMA, Groves DI, Loucks RR, Begg GC (2012) A unified model for gold mineralisation in accretionary orogens and implications for regional-scale exploration targeting methods. Mineral Deposita 47:339–358

Huston DL, Champion DC (2023) Applications of lead isotopes to ore geology, metallogenesis and exploration. In: Huston DL, Gutzmer J (eds), Isotopes in economic geology, metallogensis and exploration, Springer, Berlin, this volume

Huston DL, Champion DC, Cassidy KF (2005) Tectonic controls on the endowment of Archean cratons in VHMS deposits: evidence from Pb and Nd isotopes. In: Mao J, Bierlein FP (eds) Mineral deposit research: meeting the global challenge. Springer, Berlin/Heidelberg, pp 15–18

Huston DL, Blewett RS, Champion DC (2012) Australia through time: a summary of its tectonic and metallogenic evolution. Episodes 35:23–43

Huston DL, Champion DC, Cassidy KF (2014) Tectonic controls on the endowment of Neoarchean cratons in volcanic-hosted massive sulfide deposits: evidence from lead and neodymium isotopes. Econ Geol 109:11–26. https://doi.org/10.2113/econgeo.109.1.11

Huston DL, Mernagh TP, Hagemann SG, Doublier M, Fiorentini M, Champion DC, Jaques AL, Czarnota K, Cayley R, Skirrow R, Bastrakov E (2016) Tectono-metallogenic systems—the place of mineral systems within tectonic evolution. Ore Geol Rev 76:168–210. https://doi.org/10.1016/j.oregeorev.2015.09.005

Huston DL, Champion DC, Czarnota K, Hutchens M, Hoggard MJ, Ware B, Richards F, Gibson GM, Carr G, Tessalina S (2020) Lithospheric-scale controls on zinc–lead–silver deposits of the North Australian Zinc Belt: evidence from isotopic and geophysical data. In: Czarnota K, Roach I, Abbott S, Haynes M, Kositcin N, Ray A, Slatter E (eds) Exploring for the Future: Extended Abstracts. Geoscience Australia, Canberra. https://doi.org/10.11636/134276

Ishihara S (1977) The magnetite-series and ilmenite-series granitic rocks. Mining Geol 27:293–305

Ishihara S (1981) The granitoid series and mineralization. Econ Geol 75th Anniv 458–484

Jégo S, Pichavant M, Mavrogenes JA (2010) Controls on gold solubility in arc magmas: an experimental study at 1000 °C and 4 kbar. Geochim Cosmochim Acta 74:2165–2189

Johnson JP, McCulloch MT (1995) Sources of mineralising fluids for the Olympic Dam deposit (South Australia): Sm–Nd isotopic constraints. Chem Geol 121:177–199

Kemp AIS, Hawkesworth CJ, Foster GL, Paterson BA, Woodhead JD, Hergt JM, Gray CM, Whitehouse MJ (2007) Magmatic and crustal differentiation history of granitic rocks from Hf–O isotopes in zircon. Science 315:980–983

Kemp AIS, Hawkesworth CJ, Collins WJ, Gray CM, Blevin PL, EIMF (2009) Isotopic evidence for rapid continental growth in an extensional accretionary orogen: the Tasmanides, eastern Australia. Earth Planet Sci Lett 284:455–466

Kemp AIS, Blevin PL, Norman MD (2020) A SIMS U–Pb (zircon) and Re–Os (molybdenite) isotope study of the early Paleozoic Macquarie Arc, southeastern Australia: implications for the tectono-magmatic evolution of the paleo-Pacific Gondwana margin. Gondwana Res 82:73–96. https://doi.org/10.1016/j.gr.2019.12.015

Kennett BLN, Salmon M (2012) AuSREM: Australian seismological reference model. Austr J Earth Sci 59:1091–1103. https://doi.org/10.1080/08120099.2012.736406

Kennett BLN, Fichtner A, Fishwick S, Yoshizawa K (2013) Australian Seismological Reference Model (AuSREM): mantle component. Geophys J Int 192:871–887. https://doi.org/10.1093/gji/ggs065

Kirkland CL, Smithies RH, Woodhouse AJ, Howard HM, Wingate MTD, Belousova EA, Cliff JB, Murphy RC, Spaggiari CV (2013) Constraints and deception in the isotopic record; the crustal evolution of the west Musgrave Province, central Australia. Gondwana Res 23:759–781. https://doi.org/10.1016/j.gr.2012.06.001

Korsch RJ, Doublier MP (2016) Major crustal boundaries of Australia and their significance in mineral systems targeting. Ore Geol Rev 76:211–228. https://doi.org/10.1016/j.oregeorev.2015.05.010

Korsch RJ, Kositcin N, Champion DC (2011) Australian island arcs through time: geodynamic implications for the Archean and Proterozoic. Gondwana Res 19:716–734

Krapez B, Brown SJA, Hand J, Barley ME, Cas RAF (2000) Age constraints on recycled crustal and supracrustal sources of Archaean metasedimentary sequences, Eastern Goldfields Province, Western Australia: evidence from SHRIMP zircon dating. Tectonophysics 322:89–133

Lambert DD, Foster JG, Frick LR, Ripey EM (1999) Re–Os isotope geochemistry of magmatic sulfide ore systems. Rev Econ Geol 12:29–58

Lang JR, Baker T, Hart CJR, Mortensen JK (2000) An exploration model for intrusion-related gold systems. Soc Econ Geol Newsl 40:1–15

Liew TC, McCulloch MT (1985) Genesis of granitoid batholiths of Peninsular Malaysia and implications for models of crustal evolution: evidence from a Nd–Sr

isotopic and U–Pb, zircon study. Geochim Cosmochim Acta 49:587–600
Loucks RR (2014) Distinctive composition of copper-ore-forming arc magmas. Austr J Earth Sci 61:5–16. https://doi.org/10.1080/08120099.2013.865676
Lu Y-J, Kerrich R, Kemp AIS, McCuaig TC, Hou Z-Q, Hart CJR, Li Z-X, Cawood PA, Bagas L, Yang Z-M, Cliff J, Belousova EA, Jourdan F, Evans NJ (2013) Intracontinental Eocene-Oligocene porphyry Cu mineral systems of Yunnan, western Yangtze Craton, China: compositional characteristics, sources, and implications for continental collision metallogeny. Econ Geol 108:1541–1576. https://doi.org/10.2113/econgeo.108.7.1541
McCuaig TC, Beresford S, Hronsky J (2010) Translating the mineral systems approach into an effective exploration targeting system. Ore Geol Rev 38:128–138
McCulloch MT (1987) Sm–Nd isotopic constraints on the evolution of Precambrian crust in the Australian continent. Am Geophys Union Geodyn Ser 17
McCulloch MT, Wasserburg GJ (1978) Sm–Nd and Rb–Sr chronology of continental crust formation. Science 200:1003–1011
McDonald GD, Collerson KD, Kinny PD (1997) Late Archean and Early Proteozoic crustal evolution of the Mount Isa Block, northwest Queensland, Australia. Geology 25:1095–1098
McDonough WF (2003) Compositional model for the Earth's core. Treatise Geochem 2:547–568
McDonough WF, Sun S-S (1995) Composition of the Earth. Chem Geol 120:223–253
Mole DR, Fiorentini ML, Cassidy KF, Kirkland CL, Thebaud N, McCuaig TC, Doublier MP, Duuring ML, Romano SS, Maas R, Belousova EA, Barnes SJ, Miller J (2013) Crustal evolution, intra-cratonic architecture and the metallogeny of an Archaean craton. Geol Soc London Spec Pub 393. https://doi.org/10.1144/SP393.8
Mole DR, Fiorentini ML, Thebaud N, Cassidy KF, McCuaig TC, Kirkland CL, Romano SS, Doublier MP, Belousova EA, Barnes SJ, Miller J (2014) Archean komatiite volcanism controlled by the evolution of early continents. Proc Nat Acad Sci 111:10083–10088. https://doi.org/10.1073/pnas.1400273111
Morand VJ (1993) Stratigraphy and tectonic setting of the Calliope Volcanic Assemblage, Rockhampton area, Queensland. Austr J Earth Sci 40:15–30
Murray CG, Blake PR (2005) Geochemical discrimination of tectonic setting for Devonian basalts of the Yarrol Province of the New England Orogen, central coastal Queensland: an empirical approach. Austr J Earth Sci 52:993–1034
O'Reilly SY, Griffin WL (2006) Imaging global chemical and thermal heterogeneity in the subcontinental lithospheric mantle with garnets and xenoliths: geophysical implications. Tectonophysics 416:289–309
Osei KP, Kirkland CL, Mole DR (2021) Nd and Hf isoscapes of the Yilgarn Craton, Western Australia and implications for its mineral systems. Gondwana Res 92:253–265. https://doi.org/10.1016/j.gr.2020.12.027
Palme H, Jones A (2003) Solar system abundances of the elements. Treatise Geochem 1:41–61
Parkinson IJ, Arculus RJ (1999) The redox state of subduction zones: insights from arc-peridotites. Chem Geol 160:409–423
Payne JL, Hand M, Barovich KM, Reid A, Evans DAD (2009) Correlations and reconstruction models for the 2500–1500 Ma evolution of the Mawson Continent. Geol Soc London Spec Pub 323:319–355
Payne JL, Ferris G, Barovich KM, Hand M (2010) Pitfalls of classifying ancient magmatic suites with tectonic discrimination diagrams: an example from the Paleoproterozoic Tunkillia Suite, southern Australia. Precambrian Res 177:227–240
Pearson DG, Carlson RW, Shirey SB, Boyd FR, Nixon PH (1995) Stabilisation of Archaean lithospheric mantle: a Re–Os isotope study of peridotite xenoliths from the Kaapvaal craton. Earth Planet Sci Lett 134:341–357
Priestley K, McKenzie DP (2013) The relationship between shear wave velocity, temperature, attenuation and viscosity in the shallow part of the mantle. Earth Planet Sci Lett 381:78–91
Reid AJ, Hand M (2012) Mesoarchean to Mesoproterozoic evolution of the southern Gawler Craton, South Australia. Episodes 35:216–225
Richards JP (2011) High Sr/Y arc magmas and porphyry Cu ± Mo ± Au deposits: just add water. Econ Geol 106:1075–1081. https://doi.org/10.2113/econgeo.106.7.1075
Rowe MC, Kent AJR, Nielsen RL (2009) Subduction Influence on oxygen fugacity and trace and volatile elements in basalts across the Cascade Volcanic Arc. J Petrol 50:61–91
Rudnick RL, Gao S (2003) Composition of the continental crust. In: Treatise on geochemistry, vol 3, pp 1–65
Ruiz J, Mathur R (1999) Metallogenesis in continental margins: Re–Os evidence from porphyry copper deposits in Chile. Rev Econ Geol 12:59–72
Salters VJM, Stracke A (2004) Composition of the depleted mantle. Geochem Geophys Geosyst 5: Q05B07. https://doi.org/10.1029/2003GC000597
Scrimgeour IR, Kinny PD, Close DF, Edgoose CJ (2005) High-T granulites and polymetamorphism in the southern Arunta Region, central Australia: evidence for a 1.64 Ga accretional event. Precambrian Res 142:1–27
Shirey SB, Walker RJ (1998) The Re–Os system in cosmochemistry and high-temperature geochemistry. Ann Rev Earth Planet Sci 26:423–500
Sircombe KN (1999) Tracing provenance through the isotope ages of littoral and sedimentary detrital zircon, eastern Australia. Sediment Geol 124:47–67
Skirrow RG (2010) "Hematite-group" IOCG ± U ore systems: tectonic settings, hydrothermal characteristics, and Cu–Au and U mineralizing processes. Geol Assoc Can Short Course Notes 20:39–58

Skirrow RG (2013) Australia's iron oxide Cu–Au provinces: world-class opportunities. Presentation given at Prospectors and Developers Association of Canada (PDAC) Convention, Toronto, Canada, 3–6 March 2013

Skirrow RG, Bastrakov EN, Barovich K, Fraser GL, Creaser RA, Fanning CM, Raymond OL, Davidson GJ (2007) Timing of iron oxide Cu–Au–(U) hydrothermal activity and Nd isotope constraints on metal sources in the Gawler Craton, South Australia. Econ Geol 102:1441–1470

Slocum TA, McMaster RB, Kessler FC, Howard HH (2009) Thematic cartography and geovisualization, 3rd edn. Pearson Prentice Hall, New Jersey, p 561

Smithies RH, Champion DC (2000) The Archean high-Mg diorite suite: links to tonalite-trondhjemite-granodiorite magmatism and implications for Early Archean crustal Growth. J Petrol 41:1653–1671

Smithies R, Champion DC, Van Kranendonk MJ, Howard HM, Hickman AH (2005) Modern-style subduction processes in the Mesoarchaean: geochemical evidence from the 3.12 Ga Whundo intra-oceanic arc. Earth Planet Sci Lett 231:221–237

Stewart AJ, Raymond OL, Totterdell JM, Zhang W, Gallagher R (2013) Australian Geological Provinces, 2013.01 edition, scale 1:2500000. Geoscience Australia, Canberra, Australia

Sun S-S, McDonough WF (1989) Chemical and isotopic systematics of oceanic basalts: implications for mantle composition and processes. Geol Soc London Spec Pub 42:313–345

Sun S-S, Warren RG, Shaw RD (1995) Nd isotope study of granites from the Arunta Inlier, central Australia: constraints on geological models and limitation of the method. Precambrian Res 71:301–314

Swain G, Woodhouse A, Hand M, Barovich K, Schwarz M, Fanning CM (2005) Provenance and tectonic development of the late Archaean Gawler Craton, Australia; U–Pb zircon, geochemical and Sm–Nd isotopic implications. Precambrian Res 141:106–136

Swain G, Barovich K, Hand M, Ferris G, Schwarz M (2008) Petrogenesis of the St Peter Suite, southern Australia: arc magmatism and Proterozoic crustal growth of the South Australian Craton. Precambrian Res 166:283–296

Thompson JFH, Sillitoe RH, Baker T, Lang JR, Mortensen JK (1999) Intrusion-related gold deposits associated with tungsten-tin provinces. Mineral Deposita 34:323–334

Tosdal RM, Wooden JL, Bouse RM (1999) Pb isotopes, ore deposits and metallogenic terranes. Rev Econ Geol 12:1–28

Van Kranendonk MJ, Smithies RH, Hickman AH, Champion DC (2007) Paleoarchean development of a continental nucleus: the East Pilbara Terrane of the Pilbara Craton, Western Australia. Dev Precambrian Geol 15:307–337

Vervoort J (2011) Evolution of the depleted mantle—revisited. AGU Fall Meeting Abstracts, p 06. https://ui.adsabs.harvard.edu/abs/2011AGUFM.V43E..06V

Vervoort JD, Kemp AIS (2016) Clarifying the zircon Hf isotope record of crust–mantle evolution. Chem Geol 425:65–75

Vervoort JD, Patchett PJ (1996) Behavior of hafnium and neodymium isotopes in the crust: constraints from Precambrian crustally derived granites. Geochim Cosmochim Acta 60:3717–3733. https://doi.org/10.1016/0016-7037(96)00201-3

Vervoort JD, Fisher CM, Kemp AI, Salerno RA (2019) Resolving the conflicting Hf and Nd isotope records of early Earth crust-mantle evolution. AGU Fall Meeting Abstracts. abstract #V43D-0121. https://ui.adsabs.harvard.edu/abs/2019AGUFM.V43D0121V

Vervoort J, Fisher C, Salerno R (2020) Resolving the Hf–Nd paradox of early Earth crust-mantle evolution. EGU General Assembly 2020, Online, 4–8 May 2020, EGU2020-13269. https://doi.org/10.5194/egusphere-egu2020-13269

Waltenberg K (2023) Application of Lu–Hf isotopes to ore geology, metallogenesis and exploration. In: Huston DL, Gutzmer J (eds), Isotopes in economic geology, metallogensis and exploration, Springer, Berlin

Wang C, Bagas L, Lu Y, Santosh M, Du B, McCuaig TC (2016) Terrane boundary and spatio-temporal distribution of ore deposits in the Sanjiang Tethyan Orogen: Insights from zircon Hf-isotopic mapping. Earth-Sci Rev 156:39–65. https://doi.org/10.1016/j.earscirev.2016.02.008

Wooden JL, Kistler RW, Tosdal RM (1998) Pb isotopic mapping of crustal structure in the northern Great Basin and relationships to Au deposit trends. US Geol Surv Open-File Rep 98–338:20–33

Wu C, Chen H, Lu Y (2021) Crustal structure control on porphyry copper systems in accretionary orogens: insights from Nd isotopic mapping in the Central Asian Orogenic Belt. Mineral Deposita 57:631–641

Wyborn LAI, Heinrich CA, Jaques AL (1994) Australian Proterozoic mineral systems: essential ingredients and mappable criteria. Austr Inst Mining Metall Pub Ser 5 (94):109–115

Zartman RE (1974) Lead isotopic provinces in the Cordillera of the western United States and their geologic significance. Econ Geol 69:792–805

Zhang M, O'Reilly SA, Wang K-L, Hronsky J, Griffin WL (2008) Flood basalts and metallogeny: the lithospheric mantle connection. Earth-Sci Rev 86:145–174

Zhao J, McCulloch MT (1995) Geochemical and Nd isotopic systematics of granites from the Arunta Inlier, central Australia: implications for Proterozoic crustal evolution. Precambrian Res 71:265–299

Applications of Lead Isotopes to Ore Geology, Metallogenesis and Exploration

David L. Huston and David C. Champion

Abstract

Although lead isotopes are most commonly used to date geological events, including mineralizing events, they also can provide information on many aspects of metallogeny and can be directly used in mineral exploration. Lead isotope data are generally reported as ratios of radiogenic isotopes normalized to the non-radiogenic isotope ^{204}Pb (e.g. ^{206}Pb/^{204}Pb, ^{207}Pb/^{204}Pb and ^{208}Pb/^{204}Pb). These ratios can be used in exploration to characterize the style of mineralization, metal (i.e. Pb) source and as vectors to ore. When combined with lead isotope evolution models, the data can be used to indicate the age and tectonic environment of mineralization. The raw ratios and evolution models enable calculation of derived parameters such as μ (^{238}U/^{204}Pb), κ (^{232}Th/^{238}U) and ω (^{232}Th/^{204}Pb), which provide more information about tectonic setting and can be contoured to identify crustal boundaries and metallogenic provinces. In some cases, tectonic boundaries, mapped using gradients in μ and other derived parameters, are fundamental controls on the distribution of certain deposit types in space and time. Moreover, crustal character, as determined by lead and other radiogenic isotopes (e.g. Nd) can be an indicator of province fertility for many deposit types. The development of cost effective analytical techniques and the assembly of large geo-located datasets for lead and other isotope data has enabled significant advances in understanding the genesis and localization of many deposit type, particularly when the isotopic data are integrated with other independent datasets such as potential field, magnetotelluric, passive seismic and geochemical data.

1 Introduction

Radiogenic isotopes have had an important use in the geological sciences as chronometers, providing constraints on the ages of rocks and geological events, including mineralizing events (see also Chelle-Michou and Schaltegger 2023; Chiaradia 2023). Another important, though less widespread, use of radiogenic isotopes is as a tracer of geological processes, including mineralizing processes. The use of radiogenic isotopes as tracers takes advantage not only of radioactive decay between parent and daughter isotopes, but also geochemical partitioning between the parent and daughter elements: the isotopic evolution of a given volume of rock is not only governed by radiogenic decay, but also by geochemical

D. L. Huston (✉) · D. C. Champion
Geoscience Australia, GPO Box 378, Canberra, ACT 2601, Australia
e-mail: David.Huston@ga.gov.au

D. Huston and J. Gutzmer (eds.), *Isotopes in Economic Geology, Metallogenesis and Exploration*, Mineral Resource Reviews, https://doi.org/10.1007/978-3-031-27897-6_6

processes within the rock volume and between it and other rock volumes. In this chapter we review how spatial and temporal patterns in lead isotope data have been used in metallogenic and tectonic studies. Similar reviews for the Sm–Nd and Lu–Hf isotopic systems are provided in other chapters of this volume (Champion and Huston (2023) for Sm–Nd, Waltenberg (2023) for Lu–Hf). These three reviews illustrate that large scale variations in the spatial and temporal characteristics of radiogenic isotopes can be used to identify tectonic and metallogenic processes at the global to province scale—information important for regional target area selection in exploration. At the smaller scale, isotopic signatures and patterns may be used more directly for local target definition.

2 The Uranium–Thorium-Lead Isotopic and Geochemical System

The lead isotope composition of a particular mineral, rock or geological reservoir is the sum of the isotopic composition of lead present in the geological entity (mineral, rock or reservoir) when it formed and the isotopic composition of lead formed subsequently by radioactive decay of uranium and thorium within the entity. ^{238}U decays through a complex reaction chain to ^{206}Pb; similarly ^{235}U decays to ^{207}Pb and ^{232}Th decays to ^{208}Pb. The decay constants for the chains are 0.155125×10^{-9} yr^{-1}, 0.98485×10^{-9} yr^{-1}, and 0.049475×10^{-9} yr^{-1}, which correspond to half-lives of 4.468 Gyr, 0.704 Gyr and 14.05 Gyr, respectively (LeRoux and Glendenin 1963; Jaffey et al. 1971).

For the Earth as a whole, the isotopic composition at formation is generally taken as that of Canyon Diablo meteorite (Tatsumoto et al. 1973) at 4.54 Ga, with radioactive lead growth dependent on time, the decay constants of ^{238}U, ^{235}U and ^{232}Th, and chondritic abundances of uranium, thorium and lead at 4.54 Ga. The relative abundances of uranium, thorium and lead can be expressed as $^{238}U/^{204}Pb$ (μ), $^{232}Th/^{238}U$ (κ) and $^{232}Th/^{204}Pb$ (ω). Ancient values of these ratios can be calculated using measured lead isotope ratios from lead isotope evolution models. Although uranium, thorium and lead are all concentrated in the crust relative to the mantle (Wedepohl and Hartmann 1994; Palme and O'Neill 2003), the relative fractionation of these elements differs (Fig. 1), implying that the isotopic evolution of a reservoir is dependent upon the geochemical processes through which the reservoir formed. Figure 2a shows the effect on lead isotopic growth of a hypothetical geochemical differentiation event in which an isotopically homogeneous reservoir produced two separate reservoirs. As shown in the figure, the isotopic characteristics of the two reservoirs diverge, with the reservoir with higher μ producing ever higher $^{206}Pb/^{204}Pb$ and $^{207}Pb/^{204}Pb$ ratios over time relative to the second lower μ reservoir. The concentrations of lead isotopes that grow radiogenically (i.e. ^{206}Pb, ^{207}Pb and ^{208}Pb) are generally normalized to the concentration of ^{204}Pb, which does not grow radiogenically, to allow comparison.

The above scenario reflects crust formation. Because uranium partitions more strongly into the crust than lead (Fig. 1), the upper trajectory in Fig. 2a, which is associated with higher μ, reflects the crustal reservoir, whereas the lower trajectory reflects the residual mantle reservoir. The scenario suggests that an important control of lead isotope evolution is crust formation, a process that, on modern Earth, occurs mostly along plate margins. As U/Pb, Th/Pb and Th/U also vary within the crust (Fig. 1), lead isotopic ratios not only reflect crust formation, but also geochemical processes that occur within the crust.

3 Lead Isotope Evolution Models, Model Ages, and μ, κ and ω Values

Once an isotopic reservoir has been identified, growth of lead within the reservoir can be modelled using lead isotope evolution models. These mathematical models simulate radiogenic isotope growth within the closed reservoir as a function of geological time. Holmes (1946) and

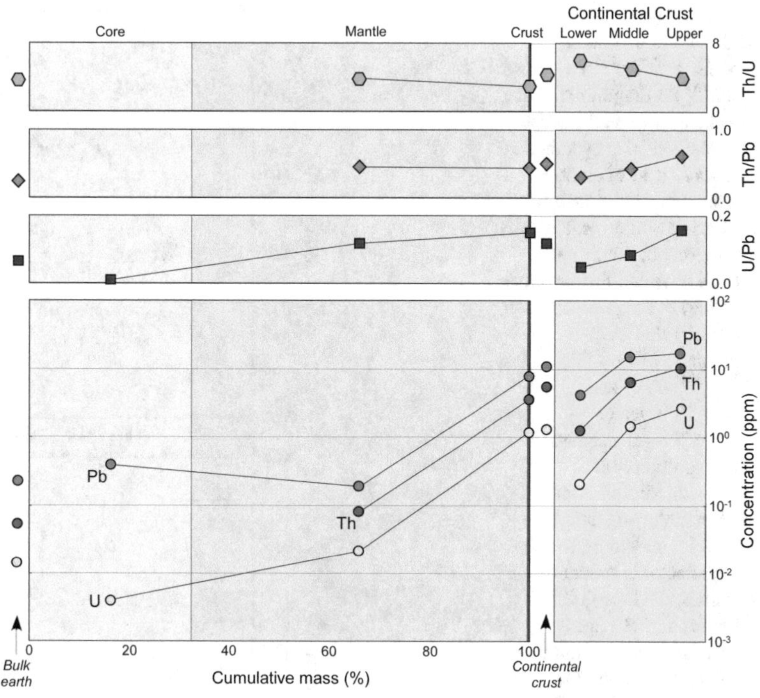

Fig. 1 Variations in the concentrations of uranium, thorium and lead and in U/Pb, Th/U and Th/Pb ratios between major terrestrial reservoirs (element concentrations for the major reservoirs are from McDonough 2003; Palme and O'Neill 2003; Rudnick and Gao 2003). The abscissa on the main diagram indicates the depth range that the various reservoirs occupy

Houtermans (1946) independently proposed a model whereby the lead isotope characteristics of galena (and other Pb-rich and U-poor minerals) reflects lead growth in a reservoir until time t, at which time the lead was extracted from the reservoir and was crystalized into galena. Because the galena contains virtually no uranium the uranogenic isotopic ratios are frozen in and closely reflect "initial" ratios at the time of crystallization. Using the measured ratios of the galena and the isotopic ratio of the reservoir when it formed, a two-point isochron can be determined and the time since formation of the reservoir can be estimated. Holmes (1946) and Houtermans (1946) treated the Earth (or solar nebula) as this reservoir and the age of the Earth as time when the reservoir formed. Hence, the age t, as determined from the isochron is the age since formation of the reservoir, in this case the age of the Earth. Subsequently it has been shown that the Holmes-Houtermans model produces erroneous ages, and other models for reservoir lead evolution have been developed to allow estimation of ages from galena isotopic ratios.

These models are quite varied and make different assumptions. In "single-stage" models the reservoir evolution is modelled using initial lead isotopic ratios, and initial μ, κ and ω values. The isotopic evolution as a function of time is then calculated simply using these initial values and the decay constants. A single-stage model was used in Fig. 2a until the point at which chemical differentiation occurred. Most single-stage models are geologically unreasonable except at the broadest scale as they do not consider later geological/geochemical events during which uranium and thorium are fractionated relative to lead.

Maltese and Mezger (2020) recently proposed a more geologically reasonable single stage model for the evolution of bulk silicate Earth (i.e. the crust and mantle). This model was developed to resolve the so-called "future lead paradox" (Sinha and Tilton 1973) in which major crust and mantle lead reservoirs are more radiogenic that expected assuming lead growth from a chondritic initial reservoir. It infers that a relatively small chondritic impactor, Theia, collided with a volatile-rich proto-Earth at ca 4.50 Ga

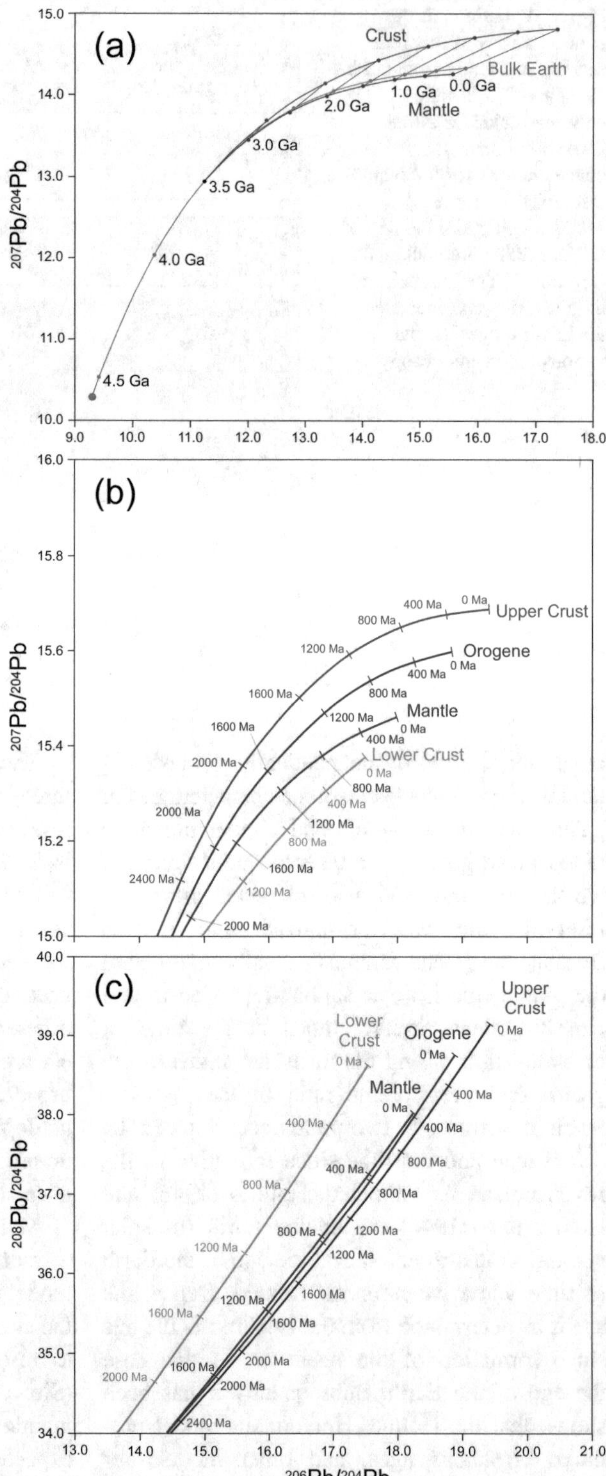

Fig. 2 Diagrams showing lead isotope evolution models. **a** $^{206}Pb/^{204}Pb$ versus $^{207}Pb/^{204}Pb$ diagram showing the changes in lead isotope composition in a hypothetical two stage lead evolution model. The model assumes lead growth in an isotopically homogeneous Earth until 3.5 Ga at which time a (crust-forming) event produces two reservoirs that evolve separately. $^{206}Pb/^{204}Pb$ versus $^{207}Pb/^{204}Pb$ (**b**) and $^{206}Pb/^{204}Pb$ versus $^{208}Pb/^{204}Pb$ (**c**) diagrams showing evolution of uranogenic and thorogenic lead in major terrestrial reservoirs according to the plumbotectonic model of Zartman and Doe (1981). Parts (**a**) and (**b**) are reproduced with permission from Zartman and Doe (1981); Copyright 1981 Elsevier

($\sim$70 Myr after formation of the solar system). This collision not only resulted in the formation of the moon, but the formation of a bulk silicate Earth reservoir that was modelled by the Maltese and Mezger (2020) model. This model, a significant departure from previous two stage models (see below), explains the isotopic evolution of bulk silicate Earth relatively well.

Two-stage models begin like single-stage models, but at some point during the evolution, a geochemical or geological event is assumed to cause fractionation between uranium, thorium and lead, and two isotopic reservoirs form with different isotopic evolution trajectories. At the point where geochemical differentiation of Earth's crust and mantle occurred (at 3.5 Ga), the evolution model shown in Fig. 2a becomes a two-stage model. Interaction or exchange between the two reservoirs can be modelled as mixing between to the two evolution curves that formed after differentiation by varying μ (i.e. using an intermediate value of μ). The most commonly used global lead isotope evolution model, that of Stacey and Kramers (1975), is a two-stage model, which models the evolution of "terrestrial" lead after geochemical differentiation. Another global model (Cumming and Richards 1975) assumes that fractionation between uranium, thorium and lead occurs continuously and models lead growth assuming U/Pb and Th/U change continuously through time. Although these models approximate the evolution of lead at the global scale reasonably well, the geological processes that form reservoirs are more complex than these relatively simple models.

Doe, Zartman and co-workers (Doe and Zartman 1979; Zartman and Doe 1981; Zartman and Haines 1988) developed a more complex model, in which the evolution of four major isotopic reservoirs was modelled assuming periodic events when uranium, thorium and lead were exchanged between reservoirs. During each event, the fourth reservoir, the so-called orogene reservoir, was created by recycling and extraction from the other three reservoirs—mantle, lower crust and upper crust. The model was based upon the concepts of modern plate tectonics—hence the name 'plumbotectonics'—although Zartman and Haines (1988) indicated that the modelling would remain valid for geodynamic processes other than plate tectonics if such processes involved a similar magnitude of reservoir interaction.

The earlier versions of plumbotectonics assumed an orogene event every 400 Myr (Zartman and Doe 1981), with the mantle input decreasing through time. The later versions (e.g. Zartman and Haines 1988) increased the frequency of events and the complexity of the models, although the overall trends were similar. It must be noted that the term orogene, as used by Doe, Zartman and co-workers, is not a volume of rock that has undergone orogenesis, but rather quantifies the global distribution of crust formation (Zartman and Haines 1988).

Figure 2b, c illustrate the evolution of the four major reservoirs based on plumbotectonics version II (Zartman and Doe 1981). The plumbotectonic model replicates modern isotopic characteristics of the four main reservoirs well, and the general evolution of each reservoir matches natural secular variations reasonably well. The plumbotectonic model highlights important differences in the four major reservoirs, particularly the lower crust in which uranogenic lead growth has been retarded relative to other reservoirs (although thorogenic lead growth is similar to other reservoirs). As a consequence the plumbotectonic evolution curves have been widely used to infer reservoir sources for many mineral deposits, as discussed below. Doe and Zartman (1979) assessed the correspondence of a range of (mainly young) deposits to the plumbotectonic model. Like other general models, model ages can be quite inaccurate, but with a few notable exceptions (e.g. some Mississippi Valley-type deposits) the modelled source of lead in the deposits is consistent with the geological environment inferred for ore formation.

Kramers and Tolstikhin (1997) used a similar concept of isotopic and elemental exchange between isotopic reservoirs to model lead isotope evolution. Their approach differed from that of Doe and co-workers in that it involved nine—instead of four—different reservoirs, including the Earth's core, and used crustal growth rates as an input into the model. Like Doe and co-workers, Kramers and Tolstikhin (1997) were able to model present-day crustal and mantle reservoirs reasonably well.

Due to inaccuracies of global lead isotope evolution models, some workers (Thorpe et al. 1992; Thorpe 1999; Carr et al. 1995; Sun et al. 1996) have produced evolution models that model lead growth at the province scale. These models can very accurately and precisely model isotopic growth, but direct comparison of these models or extending the models outside of their calibrated reservoirs are problematic. Hence, for global or large-scale comparisons, use of a global model (e.g. Stacey and Kramers 1975; Cumming and Richards 1975; Zartman and Doe 1981) is more effective.

As mentioned above, lead isotope model ages assume that no radiogenic growth has occurred since crystallization of the analysed mineral—that is the concentrations of uranium (and Th) are so low that the additional radiogenic lead produced since crystallization is negligible. Other minerals also meet the criteria of being Pb-rich and U- and Th-poor. The most common of these minerals is potassium feldspar. In many cases measured ratios of potassium feldspars are virtually identical to the initial ratios inherited from crystallization (see discussion regarding analytical techniques below). Less common lead selenide (clausthalite), telluride (altaite) and sulfosalt (beaverite, bournonite and cosalite) minerals also can be analyzed (Thorpe 2008) and retain initial ratios.

The discussion above assumes that the mineral of interest has been a closed system since crystallization; if the system has been open (i.e. due to recrystallization, alteration or a second mineralizing event), the isotopic ratios will reflect either the ratios at the time when the system opened or a mixture between lead produced during initial crystallization and lead introduced during the resetting event.

4 Analytical Methods

For metallogenic studies and exploration, the most common methods of lead isotope analysis are thermal ionization mass spectrometry (TIMS) or multi-collector inductively-coupled plasma mass spectrometry (MC-ICP-MS) on solutions derived by the dissolution of mineral separates and rock powders. Over the last two decades, the development of MC-ICP-MS has seen a revolution in lead isotopic analysis, producing high quality analyses for a much lower cost in comparison with TIMS. This is largely because MC-ICP-MS analyses are more rapid and less labour-intensive than TIMS analyses.

For both TIMS and MC-ICP-MS analyses, samples are dissolved using a range of chemical attacks and then elutriated using ionic distillation columns to remove matrix elements. For conventional TIMS analysis, which was the most common analytical method until the early-mid 1990s; lead from the elutriant was loaded onto a rhenium filament in a silica gel-phosphoric acid suspension (Cameron et al. 1969), in some cases after addition of a ^{202}Pb or ^{207}Pb spike.[1] A current is passed through the filament and the lead is ionised, accelerated using an electrical potential gradient, forming a beam which is split by mass using and electromagnet, and then analyzed using Faraday cup detectors. Mass fractionation during analysis is then corrected by normalization using fractionation factors determined from standards. External analytical precisions (2σ) for conventional TIMS analysis are typically 0.05% for ^{206}Pb/^{204}Pb, and 0.1% for ^{207}Pb/^{204}Pb and ^{208}Pb/^{204}Pb (cf. Carr et al. 1995).

As precisions of conventional analyses are insufficient to resolve many geological problems, new methods were developed over the last 20–30 years to improve precision. The first of these, which came into widespread use in the mid-1990s, was TIMS analysis in which spikes are added to an aliquot of the elutriant prior to

[1] Spikes are solutions containing known abundances and ratios of selected lead, thallium and/or uranium isotopes that are added to the sample or sample solution to quantify mass fractionation during mass spectrometry.

loading onto the filament. The spiked and unspiked aliquots are then loaded and analyzed as described above, with the unspiked analyses corrected for mass fractionation using the spiked analyses and mathematical equations developed by Dodson (1963). Double-spike TIMS analyses involve spikes containing two isotopes (^{202}Pb-^{205}Pb: Todt et al. 1993; ^{204}Pb-^{207}Pb: Woodhead et al. 1995), whereas triple-spike analyses, which are even more precise, involve spikes containing three isotopes (^{204}Pb-^{206}Pb-^{207}Pb: Galer and Abouchami 1998). Typical external analytical precisions (2σ) for double and triple spike analyses are 0.008–0.034% for $^{206}Pb/^{204}Pb$, 0.010–0.047% for $^{207}Pb/^{204}Pb$ and 0.012–0.057% for $^{208}Pb/^{204}Pb$ (based on repeated analysis of SRM981 from a range of labs: Thirlwall 2000). The extra time and expense of double spike analyses (two mass spectrometer runs are required for each sample) is justified for many geological studies by the significant improvement in precision. For more detailed descriptions of TIMS analytical methods the reader is referred to the above references as well as Tuttas and Habfest (1982).

For MC-ICP-MS analysis, an aliquot of doped tracer solution containing thallium isotopes (^{203}Tl and ^{205}Tl: Hirata 1996; Belshaw et al. 1998; Rehkämper and Halliday 1998; Woodhead 2002) of known concentrations and ratios is added to the sample during dissolution. The resulting solution is purified by ionic distillation and the elutriant is then aspirated into the ionization chamber of the MC-ICP-MS and analysed. Instrumental mass fractionation is corrected using mass fractionation factors determined from the doped isotopes added during dissolution; this produces more precise analyses than conventional TIMS analyses but less precise analyses than double or triple spike TIMS analyses. Typical external analytical precisions (2σ) for MC-ICP-MS analyses are 0.018–0.053% for $^{206}Pb/^{204}Pb$, 0.030–0.047% for $^{207}Pb/^{204}Pb$ and 0.012–0.057% for $^{208}Pb/^{204}Pb$ (Thirlwall 2000; R Maas, 2021, pers. comm.). More details of the methods used by MC-ICP-MS analysis are described in the references above as well as Albarède et al. (2004) and Baker et al. (2004).

Because potassium feldspars are common rock forming minerals that strongly concentrate lead relative to uranium and thorium, these minerals can provide information on the initial isotopic composition of their host rocks, such as granites. However, as potassium feldspars are readily altered or recrystallized, they can exhibit open system behaviour. To minimize these effects and exclude the contribution of radiogenic lead, potassium feldspar and, less commonly, clinopyroxene are analysed with using a sequential acid leach Typically, the first leach uses a relatively weak acid or combinations of acids (HCl-HNO_3) followed by leaches using stronger acids (concentrated HF-HNO_3) (e.g. Carr et al. 1995). The earlier leaches removes labile, radiogenic lead, whereas the later leaches dissolved common lead held in the mineral lattice and are taken as the closest approximation of the initial ratios that characterize the mineral.

Material for isotopic analysis can also be extracted directly from samples using either of two microanalytical techniques—secondary ion mass spectrometry (SIMS) and multi-collector laser-ablation inductively-coupled-plasma mass spectrometry (MC-LA-ICP-MS). In both cases the analytical spots are generally below 100 μm in diameter, with SIMS analyses, in some cases, less than 10 μm in width. Methods involving SIMS, which includes the sensitive high-resolution ion microprobe (SHRIMP), have most application in U–Pb geochronology and are described by Chelle-Michou and Schaltegger (2023). Gigon et al. (2020) describe a study that has documented micro-scale variations in lead isotopes from SIMS analyses galena, as discussed in the section on future developments below. Methods involving MC-LA-ICP-MS can be less precise and require larger spot sizes (Zametzer et al. 2022), although these disadvantages are set against more rapid analysis, lower costs and the ability to analyze a greater range of sample types. For instance, Pettke et al. (2010) have analysed lead isotopes in fluid inclusions and used these data to infer the source of lead in the Bingham Canyon porphyry copper district (see below). Typical external precisions (2σ) for MC-LA-ICP-MS analyses are 0.10–

0.18% for $^{206}Pb/^{204}Pb$, 0.084–0.17% for $^{207}Pb/^{204}Pb$ and 0.085–0.16% for $^{208}Pb/^{204}Pb$ (based on repeated analysis of internal feldspar standard "Albany": Zametzer et al. 2022).

5 Application of Lead Isotopes to Metallogenic Studies and Exploration

Over the last few decades, lead isotope characteristics have found many uses and potential uses in both academic metallogenic studies and practical exploration. The most common academic use has been identifying metal source regions using isotopes as a fingerprinting tool, but more recently spatial variations in isotopic properties have been used to identify and map the extent, character, endowment and origin of tectonic and metallogenic provinces. For practical exploration, district-scale variations in lead isotope ratios may serve as direct vectors to ore, lead isotope ratios can be used to discriminate different styles of mineralization in some provinces, and lead isotope characteristics can be indicative of fertility for some deposit types and, in some cases, size potential. Examples of all of these uses are presented below. These studies can use both analyses of mineral separates and whole rocks, although interpretation of analyses from Pb-poor samples (both mineral separates and whole rock) can be problematic as determining initial ratios, which are essential in many studies, requires information on the age and the concentrations of lead, uranium and thorium of the analysed sample. Moreover, open system behaviour is more difficult to determine in such analyses. Nonetheless, some of the case studies below use whole rock samples in addition to mineral separates for analyses.

5.1 Determining Lead Sources

Unlike most elements for which isotopic data is collected in ore genesis studies, lead is commonly extracted as an ore metal from many deposits for which lead isotope data are collected. Hence, no assumptions are required to link the isotopic system with the ore metal assemblage in these deposits. Metallic stable isotope systems such as iron (Troll et al. 2019; Lobato et al. 2023), copper (Li et al. 2010; Wilkinson 2023) and zinc (John et al. 2008; Mathur and Zhao 2023) also have this advantage. On the other hand, relating the Nd-Sm, Re–Os, Hf–Lu radiogenic systems and most light stable isotope systems to ore minerals and assemblages, is often not straightforward. Lead is commonly present, but not extracted, in other deposit types as a trace or minor element in the ore assemblage or in ore-related fluid inclusions. In these instances, lead isotopes can also be used as a tracer to determine metal sources, although assumptions are required regarding the relationship between lead and the minerals/metals of economic interest. Below we give a few examples of studies where lead is recovered and where lead is present, but not recovered. It must be stressed that the literature on this topic is extensive, so only a select number of studies are included in this review. Russell and Farquhar (1960), Cannon et al. (1961), Doe and Stacey (1974), Doe and Zartman (1979) and Gulson (1986) summarize other examples.

5.1.1 Studies of Deposits Where Lead is an Ore Metal

Although lead is recovered from many deposit types, the most relevant sources of lead are shale-hosted Zn–Pb–Ag, Mississippi Valley-type Zn–Pb (MVT; including Irish-type) and volcanic-hosted massive sulfide (VHMS) deposits. The origin of lead (and by inference Zn) in these deposit types has been historically contentious, with hypotheses involving crustal or mantle, and magmatic-hydrothermal or leached country rock sources advocated by different authors over time. As examples of the use of lead isotopes to resolve these questions, the studies of Brevart et al. (1982), Fehn et al. (1983), Relvas et al. (2001), Leach et al. (2005) and Vaasjoki and Gulson (1986) are summarized below. These examples were chosen to cover all of the major types of lead deposits and to illustrate different approaches to determine lead sources using lead isotopes.

The Massif Central in south-central France contains a relatively large number of small Zn–Pb deposits and occurrences including stratiform shale-hosted deposits, epigenetic carbonate-hosted and cross-cutting vein deposits. These deposits are hosted by Phanerozoic metasedimentary rocks that have been intruded by Hercynian granites. Brevart et al. (1982) used analyses of galena from these deposits to identify two discrete populations, a less radiogenic population (population 1) dominated by carbonate-hosted, stratiform deposits in Paleozoic sedimentary successions, and a more radiogenic population (population 2) dominated by vein deposits but also including the stratiform, sandstone-hosted Largentiére deposit (Fig. 3a). The second, more radiogenic population overlaps with the isotopic composition of Hercynian granites as determined from K-feldspar analyses (Vitrac et al. 1981; Brevart et al. 1982; Fig. 3a). These results suggest two mineralizing events, a syngenetic or early diagenetic event, and a second event associated with Hercynian magmatism and deformation. During the latter event, lead may have been sourced either via magmatic-hydrothermal processes associated with intrusion of the granites or, alternatively, through leaching of the granites subsequent to their intrusion and crystallization.

As a consequence of systematic studies by Japanese and international researchers in the late 1970s and early 1980s, Zn–Pb–Cu–Ag-Au deposits in the Miocene Hokuroku district of Japan are among the best understood deposits in the world (Ohmoto and Skinner, 1983). This research has been key to the understanding of ore forming processes of VHMS deposits. As part of this research program, Fehn et al. (1983) undertook a systematic lead isotope study of deposits (using mostly galena analyses, with subordinate whole-rock ore analyses) in this district to determine lead source(s). The authors found a consistent pattern in individual deposits in which samples of yellow (Cu-rich) ores are systematically less radiogenic than black (Zn–Pb-rich) ores (Fig. 3b). Fehn et al. (1983) interpreted these results and data from potential source rocks to indicate that the lead in the Zn–Pb-rich ores was mainly sourced from the host volcanic succession (or from related intrusive rocks), whereas there was a significant component of basement lead in the Cu-rich ores. The difference in sources was interpreted to be the result of deeper penetration of the ore fluids into the basin as the hydrothermal system waxed from lower temperature Zn–Pb-rich mineralization to higher temperature Cu-rich mineralization (Fehn et al. 1983).

The early Carboniferous Neves-Corvo deposit in Portugal is unusual for VHMS deposits in that tin, along with copper, zinc and lead, are present in the ores, raising questions as to the source of the anomalous tin. Lead isotope studies by Relvas et al. (2001), supported by results from the Sm–Nd and Rb–Sr isotopic systems, suggest the presence of two types of lead at Neves Corvo. The $^{206}Pb/^{204}Pb$ versus $^{207}Pb/^{204}Pb$ plot in Fig. 3c identifies two trends in the isotopic data, a steeper trend that corresponds with other VHMS deposits in the Iberian Pyrite Belt, and a shallower trend characterized by cassiterite analyses. The apparent age associated with the cassiterite trend is unrealistic ($\sim$934 Ma). The host rocks to the deposit have a late Strunian age (354.8–354.0 Ma) based on faunal assemblages (Oliveira et al. 2004), and a sulfide-cassiterite Rb–Sr errorchron and a pyrite Re-Os isochron yielded ages of 347 $\pm$ 25 and 354 $\pm$ 29 Ma, respectively (Relvas et al. 2001; Munhá et al. 2005). Relvas et al. (2001) suggested that the cassiterite Pb–Pb trend reflected the mixing of a much more radiogenic lead source into the mineralizing system possibly from a magmatic source, although such a source is not exposed in the region.

In their synthesis of sediment-hosted Zn–Pb deposits, Leach et al. (2005) documented and compared the lead isotope characteristics of both shale-hosted and MVT Zn–Pb deposits around the world, leading to several important conclusions relating to the origin of lead in these

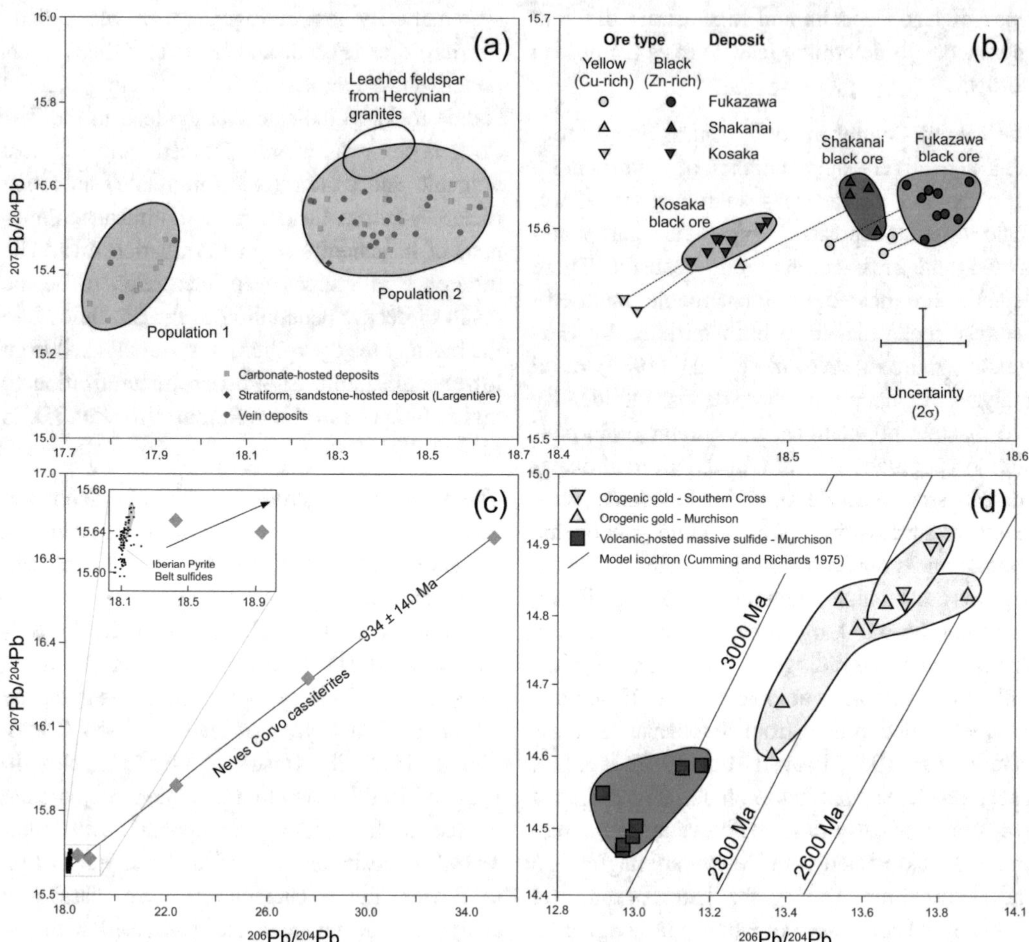

Fig. 3 $^{206}Pb/^{204}Pb$ versus $^{207}Pb/^{204}Pb$ diagrams showing **a** the lead isotope composition of Pb-rich minerals from mineral deposits and occurrences from the Massif Central, France (Brevart et al. 1982) in comparison with the initial isotopic composition of local Hercynian granites (Vitrac et al. 1981; a composite of diagrams in Brevart et al. 1982); **b** the lead isotope composition of Pb-rich minerals from the Hokuroku district of Japan illustrating differences in the lead isotope characteristics of yellow (Cu-rich) and black (Zn-rich) (modified after Fehn et al. 1983); **c** variations in the lead isotope composition of cassiterite from the Neves Corvo deposit in comparison to the lead isotope composition of sulfides from the Neves Corvo and other Iberian Pyrite Belt massive sulfide deposits (modified after Relvas et al. 2001; Iberian Pyrite Belt sulfide data from Marcoux 1998); **d** the lead isotope compositions of galena and altaite from volcanic-hosted massive sulfide deposits and orogenic gold deposits in the Youanmi Terrane overlain on isochrons calculated from the Cumming and Richards (1975) lead evolution model (modified after Browning et al. 1987)

deposits. The most important conclusion was that the lead in both deposit types was derived almost entirely from crustal sources, with little or no mantle input. The authors also noted several important differences between these two major types of sediment-hosted Zn–Pb deposits. They found that most of the large range in lead isotope ratios observed in sediment-hosted deposits is accounted for by MVT deposits. Phanerozoic MVT deposits commonly have future model ages no matter the model used to calculate the age.

Nier (1938) originally recognized the anomalously radiogenic character of many MVT ores, later termed Joplin-type, or J-type, lead by Houtermans (1953) after the Joplin mine in the

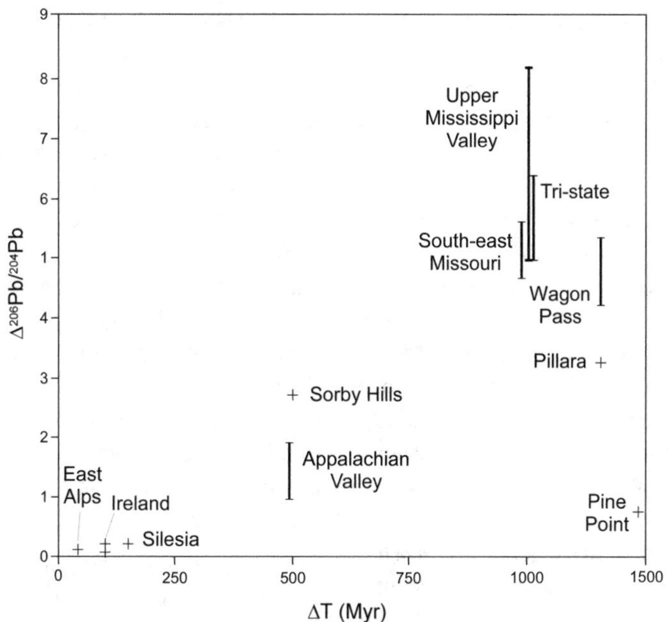

Fig. 4 Relationship of excess radiogenic lead ($\Delta^{206}Pb/^{204}Pb$ = difference between measured $^{206}Pb/^{204}Pb$ and average lead calculated using the Cumming and Richards (1975) model) to age difference between crystalline basement and host rocks (ΔT = age of basement—age of host unit). Note the increasing radiogeneity with increasing ΔT for all deposits except Pine Point. Reproduced with permission from Vaasjoki and Gulson (1986); Copyright 1986 Soceity of Economic Geologists. Note also change in scale at 1000 Myr

Tri-State MVT district in the United States. Cannon et al. (1961) found that whereas many North American MVT ores had J-type lead, Many European MVT deposits did not; Vaasjoki and Gulson (1986) noted that Australian MVT deposits had J-type lead. In addition to the radiogenic character of many MVT deposits, Vaasjoki and Gulson (1986) two types of MVT deposits with respect to their lead isotope characteristics: (1) deposits with isotopically homogeneous lead, and (2) deposits with highly heterogeneous lead that define linear arrays on $^{206}Pb/^{204}Pb$ versus $^{207}Pb/^{204}Pb$ diagrams. Vaasjoki and Gulson (1986) found that the former type of deposits occurs in basins in which the basement is only slightly older than the basin, whereas the latter deposits occur in basins that are significantly younger (to 1300 Myr) than the underlying basement (Fig. 4).

Mechanisms to explain these arrays and their highly radiogenic character are problematic. Cannon et al. (1961) suggested either (1) incremental accumulation of radiogenic lead over a protracted period, or (2) mixing of radiogenic lead with common lead during a short period. The key for both processes is the separation of radiogenic lead from common lead. Chiaradia and Fontboté (2003) demonstrated that radiogenic lead is preferentially leached from rock powders during weak acid attack, whereas common lead is preferentially leached during stronger acid attacks; a similar response has been noted in K-feldspar (cf. Carr et al. 1995). These results should also apply to lead source rocks (Ströbele et al. 2012), hence, radiogenic lead is more easily leached than common lead.

The combination of the Chiaradia and Fontboté (2003) and Ströbele et al. (2012) results may account for the two types of MVT deposits (Vaasjoki and Gulson 1986). Extraction of highly radiogenic lead from old, crystalline basement (i.e. granite and high-grade metamorphic rocks) during non-pervasive alteration of variable intensity would lead to highly heterogeneous lead that characterize type (2) deposits of Vaasjoki and Gulson (1986). If the alteration was more intense and pervasive and affected low-grade metasedimentary and mafic rocks, common lead would be extracted, leading to the homogeneous lead characteristic of type (1) MVT deposits.

5.1.2 Studies of Deposits Where Lead is a Minor or Trace Element

Lead is present as an anomalous minor to trace element in many types of mineral deposits, occurring as galena or substituted into other ore and gangue minerals. As discussed above, although lead isotope data can be used to infer the sources of lead in these deposits, caution must be exercised in inferring the source of ore metals from the lead isotope data. Below case studies of orogenic gold deposits in Western Australia, the Bingham Canyon porphyry copper deposit in Utah (United States) and the Dahu Au-Mo deposit in China are used to illustrate the potential use and pitfalls of lead isotopes to determine metal source when lead is not a major ore component.

The Yilgarn Craton in Western Australia is one of two major Archean orogenic gold provinces, the other being the Abitibi Subprovince in eastern Canada. Resources and production from the Yilgarn Craton total approximately 8 kt contained gold (Robert et al. 2005). Given the controversy at the time regarding the timing of gold mineralization (i.e. syngenetic versus epigenetic) Browning et al. (1987) undertook lead isotope analyses mostly of galena from orogenic gold, VHMS and komatiite-associated Ni-Cu-PGE deposits from the Murchison and Southern Cross Provinces (Youanmi Terrane in current terminology) and Eastern Goldfields Province (Superterrane) to determine relative and absolute ages of mineralization. In the Youanmi Terrane, Browning et al. (1987) estimated model ages of between 3050 and 2970 Ma for VHMS deposits and mostly between 2865 and 2755 Ma for orogenic gold deposits using the evolution model of Cumming and Richards (1975) (Fig. 3d). Although these model ages are 100–200 Myr older than currently accepted ages for these deposits ($\sim$2950 Ma for VHMS deposits; Wang et al. 1998) and 2670–2620 Ma for orogenic gold deposits (Robert et al. 2005)), the difference in model ages supported the interpretation that the gold deposits had an epigenetic rather than syngenetic origin. Results from the Eastern Goldfields Superterrane were ambiguous: the isotopic characteristics of the lode gold were not sufficiently different from the VHMS deposits to confidently ascribe a younger, epigenetic origin to the lode gold deposits. The Eastern Goldfields data suggested that there was significant variability in μ, which Browning et al. (1987) ascribed to the presence of old crust in parts of the Eastern Goldfields Superterrane, an inference highlighted by Oversby (1975) and later lead isotope mapping.

In the first study of its type, Pettke et al. (2010) used high precision MC-LA-ICP-MS analysis to determine the lead isotopic composition of well-characterized fluid inclusions from the Bingham Canyon porphyry copper deposit. This study differs from others presented here in that the lead isotope analyses are of the ore fluids. The fluid inclusions analysed, from ore-related quartz, were interpreted as pseudosecondary and were located on planes that contained ore minerals such as molybdenite. These fluid inclusions, which can contain hundreds to thousands ppm Pb, were preferred to analysis of Pb-bearing minerals in the deposit as the latter were considered by Pettke et al. (2010) to be more susceptible to post-depositional isotopic disturbances. The fluid inclusion analyses define a tight group that plots at the more radiogenic end of an array in $^{208}Pb/^{206}Pb$ versus $^{207}Pb/^{206}Pb$ space defined by fluid inclusion, K-feldspar and galena data (Fig. 5a). This array trends toward less radiogenic present day values of MORB-source mantle. In $^{206}Pb/^{204}Pb$ versus $^{207}Pb/^{204}Pb$ space, the fluid inclusion data also define a tight array, that lies between the trend of sub-continental lithospheric mantle defined using data from regional (relative to Bingham Canyon) metasomatized mantle xenoliths and a tight grouping of galena analyses (Fig. 5b). In combination with Monte-Carlo simulations, Pettke et al. (2010) interpreted the data to indicate that

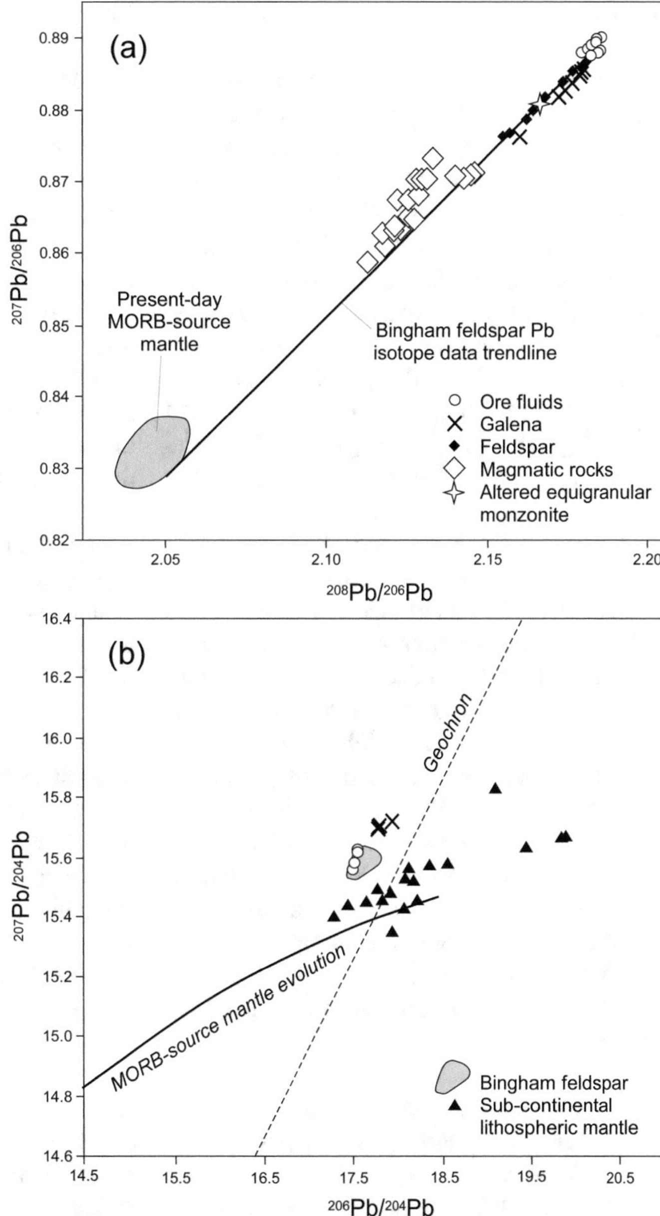

Fig. 5 ^{208}Pb/^{206}Pb versus ^{207}Pb/^{206}Pb (**a**) and ^{206}Pb/^{204}Pb versus ^{208}Pb/^{204}Pb (**b**) diagrams showing the lead isotopic composition of fluid inclusions, galena and magmatic K-feldspar from the Bingham Canyon deposit and associated magmatic rocks (modified after Pettke et al. 2010). The field showing the composition of present-day MORB-source mantle (A) and MORB-source mantle evolution (B) are from Kramers and Tolstikhin (1997)

the Bingham Canyon magmas and mineralization were derived through the melting of sub-continental lithospheric mantle that was metasomatized during the Paleoproterozoic during north-dipping subduction below the Wyoming Craton. They further suggested that such ancient metasomatized mantle may be the key to formation of major porphyry Cu–Mo and molybdenum deposits in the western United States. The importance of pre-existing metasomatized mantle is increasingly being recognized as a key in forming porphyry copper (Richards 2009) and other deposit types around the world.

The study of Ni et al. (2012) on the Dahu deposit in China illustrates an evolving trend to use multiple radiogenic isotope systems, in this case U–Th–Pb, Rb–Sr and Sm–Nd, to determine metal and fluid sources. The Dahu Au–Mo deposit is a structurally controlled quartz vein deposit hosted by migmatite and biotite-plagioclase gneiss of the Archean to Paleoproterozoic Taihua Supergroup in the Qinling Orogen of central China. SHRIMP U–Pb analysis of hydrothermal monazite intergrown with molybdenite indicated an age of 218 ± 5 Ma (Li et al. 2011) for the main mineralizing event; an isotopic disturbance towards 125 Ma is also indicated by the data. Initial lead isotope ratios (and $^{87}Sr/^{86}Sr_{218\ Ma}$ and $\varepsilon_{Nd,\ 218\ Ma}$) from the Dahu deposit differ from those of the host Taihua Supergroup (Fig. 6), suggesting that the lead (and Sr and Nd: Fig. 6a), and possibly the gold and molybdenum, were sourced, at least in part, exogenously. The lead isotope data from the ores define a gross trend from more radiogenic host rock towards a less radiogenic source (hypothetical fluid in diagram), particularly on the $^{206}Pb/^{204}Pb$ versus $^{207}Pb/^{204}Pb$ diagram (Fig. 6 b). Based on the three radiogenic isotope systems used, Ni et al. (2012) argued that the lead, strontium and neodymium—and by inference the ore-forming fluid, Au and Mo—were derived from either depleted mantle or a refractory, subducted oceanic slab. The isotopic composition of the ores indicate mixing of this fluid with a wall rock reservoir represented by the Taihua Supergroup.

5.1.3 Determining Metal Sources with Lead Isotopes—Buyer Beware

Like other isotopic systems and elements used as tracers of geochemical and geological processes a number of factors must be considered in assessing metal sources as determined using lead isotopes. These include the potential that lead isotope ratios can change after mineralization, the possibility that the sources of the lead and other ore metals might be different, and the ambiguity of resolving sources.

For robust interpretation of lead isotope systematics in mineral systems, the signature of the isotopic system must be that at the time of mineralization. As discussed above, lead can be introduced into a mineralized sample by either the addition of common lead during a subsequent mineralizing event or by post-mineralization ingrowth of radiogenic lead from decay of uranium and thorium. When interpreting the lead isotope data as a source tracer, both processes must be considered and, if possible, corrected.

In deposits where lead is recovered, lead isotopic characteristics directly reflect the provenance of an ore metal, but in deposits in which lead is not recovered, an implicit assumption that the source of lead is similar to that of the ore metal (e.g. Cu, Au or Mo in the cases discussed above) is made. As documented by Browning et al. (1987) and McNaughton et al. (1993), and highlighted by Huston et al. (2014), the lead isotopic signatures of orogenic gold deposits in the Eastern Goldfields Superterrane are very provincial, suggesting local crustal sources of lead. Although it is tempting to infer a crustal source for gold based on the lead isotope data, it is possible that the lead and gold had different sources and a crustal gold source cannot be definitively inferred.

The data of Ni et al. (2012) highlight the ambiguity of using lead isotopes to identify metal sources. Notwithstanding the question as to whether the source(s) of gold, molybdenum and lead in the Dahu deposit were the same, the origin of the lead is somewhat ambiguous. As shown in Fig. 6b, the lead isotopic composition of the ore fluid lies between the Zartman and Doe (1981) lower crust and mantle reservoirs, consistent with the possibility that the ore fluids sourced a significant component of their lead from the lower crust in addition to the preferred interpretation of Ni et al. (2012) of a mantle or ocean slab source. Perhaps a more important result to be taken from this type of study is what reservoirs can be excluded as metal sources. In the case of the Dahu deposit, both the upper crust and orogene reservoirs can be excluded as potential sources of lead (and, possibly, Mo and Au), which places important constraints on genetic models.

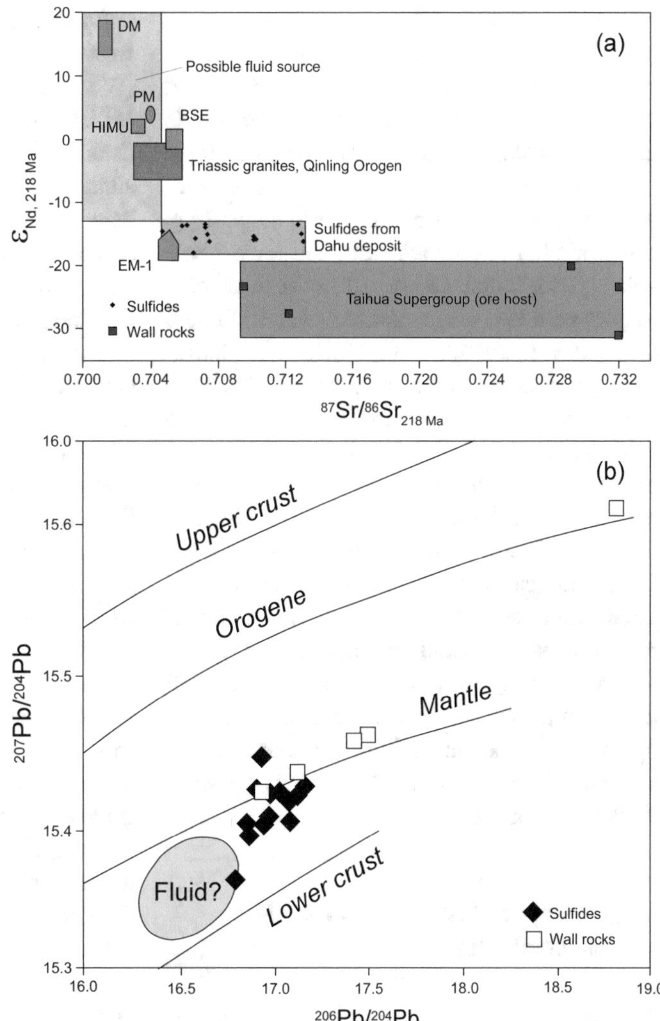

Fig. 6 Diagrams showing isotopic characteristics of the Dahu deposit, China: (**a**) $^{87}Sr/^{86}Sr_{218\ Ma}$ versus $\varepsilon_{Nd,\ 218\ Ma}$ diagram, and (**b**) $^{206}Pb/^{204}Pb$ versus $^{207}Pb/^{204}Pb$ diagram (modified after Ni et al. 2012). In both diagrams, isotopic values have been corrected to the time of ore formation (218 Ma); the isotopic compositions of the ores and wall rocks are shown as fields, with individual analyses shown as symbols. Hypothesized ore fluids, prior to mixing with the wall rock reservoir at the site of deposition, are also shown as fields. Part (**b**) also includes plumbotectonic evolution curves of Zartman and Doe (1981)

5.2 Model Ages, μ and the Age of Mineralization

Lead isotope evolution models can be used for a given lead isotope analysis to determine the model age, μ, κ and ω. It must be stressed that these parameters are dependent upon the model used—model ages, μ κ and ω determined using, for example, the Stacey and Kramers (1975) model differ from those determined using the Cumming and Richards (1975) model. Model ages must not be considered equivalent to geochronological ages—there are many processes that affect lead evolution models and, hence model ages. In fact spatial variations in the difference between model ages and independently-established mineralizing ages can be used to infer these processes.

Of parameters produced from evolution models, the model age and μ are most relevant to metallogenic studies. Although the use of locally calibrated lead evolution models can produce model ages that are accurate and precise (e.g. Thorpe et al. 1992; Thorpe 1999; Carr et al. 1995; Sun et al. 1996), a significant proportion (~20% based on the application of the Thorpe (1999) model to volcanic-hosted massive sulfide deposits in the Abitibi Subprovince in Canada) of the model ages can be inaccurate; global models are even less accurate. Model ages should be considered indicative only, and followed up using more reliable techniques to determine the age of mineralization. Model ages are inaccurate for a number of reasons, for example application of an inappropriate model. Another possible reason for inaccurate ages is the use of isotopic ratios that are not initial ratios. Analysis of samples with low U/Pb and Th/Pb ratios (e.g. galena or other Pb-rich minerals, and some whole rocks) yield initial ratios. As a general rule, samples with lead concentrations over 1000 ppm yield initial ratios unless they also have unusually high concentrations of U and/or Th. Leachate analysis of potassium feldspars can also yield initial ratios, but analyses of Pb-poor samples (e.g. most whole rocks and many pyrite) need to be corrected for post-crystallization ingrowth of lead from uranium and thorium decay. This can only happen if the age of lead introduction and U/Pb and Th/Pb are known. If there has been a second lead introduction event, measured ratios, even of Pb-rich material, will not be initial ratios. When undertaking lead isotope studies, initial ratios should be sought, and it is preferable to have multiple analyses of a deposit or rock unit.

5.3 Isotopic Mapping Using Lead Isotope Data

The development of GIS spatial data analysis packages and the availability of large sets of radiogenic analyses has enabled isotopic mapping at regional to continental scales, and maps thus produced have implication not only to metallogenesis, but also to tectonics. Even before the development of GIS analysis, however, many workers had recognized that isotopic data could map tectonic boundaries. For example, Rb–Sr data had been used to identify and map Proterozoic margins both in the Western Cordillera of North America (Armstrong et al. 1977; Kistler and Peterman 1978; Burchfiel and Davis 1981) and in eastern Australia (Webb and McDougall 1968) many decades ago. Bennett and DePaolo (1987) mapped the extent of Proterozoic crust in the Western Cordillera using neodymium isotopes, and Zartman (1974) defined broad lead isotope provinces in the western United States. Wooden and co-workers (Wooden and Aleinikoff 1987; Wooden and Miller 1990; Wooden and DeWitt 1991) used detailed variations in lead isotope ratios in Arizona and California to identify a boundary zone between the Proterozoic provinces that in a broad sense corresponded to boundaries defined by neodymium isotope data (Fig. 7). In addition, Chiaradia et al. (2006) identified distinctive lead isotope provinces and boundaries in the Altaid Orogen of central Asia, and Tessalina et al. (2016) defined systematic decreases in μ from VHMS deposits from west to east across the southern Urals orogenic zone. In the latter study, the higher μ zones are associated with a back-arc zone, whereas the lower μ zones are associated with an island arc. Tosdal et al. (1999) summarized many of the results of earlier studies, illustrating the potential of lead isotope mapping in tectonic and metallogenic studies.

These studies, however, did not produce maps showing variations in lead isotope ratios and derived parameters, but rather produced either generalized maps showing province boundaries (e.g. Figure 7a), general statements about the location of provinces or broad transects (e.g. Chiaradia et al. 2006). Although results illustrated in this manner produce useful trends, contouring (or mapping) the results on map images can show subtleties not evident using other techniques. Some of the earliest studies that mapped spatial variations in lead isotope parameters include Duebendorfer et al. (2006) and Huston et al. (2005). Duebendorfer et al. (2006) showed that in detail the boundary zone between the Central Arizona and Mojave lead isotope (sub)provinces, originally defined by Wooden and Miller (1990) is relatively complex with a number of discrete zones of more juvenile crust (Fig. 7b) interpreted as rift basins. As shown by Duebendorfer et al. (2006), mapping of derived lead isotope parameters (in this case the ΔJ (delta Jerome) parameter; see caption to Fig. 7 for definition) can identify subtle variations in patterns not shown in the more generalized approach. Moreover, these more subtle variations may have metallogenic and

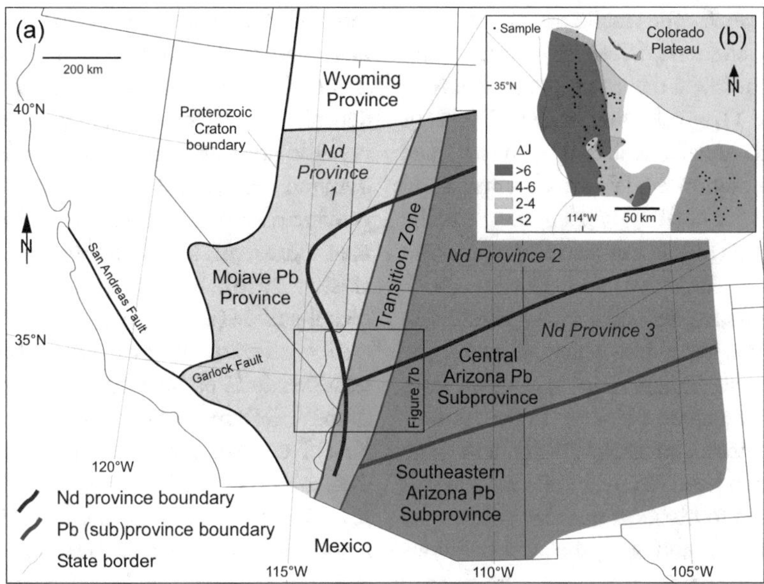

Fig. 7 Map (**a**) showing lead isotope and neodymium isotope provinces of the western United States (modified after Wooden and DeWitt 1991; Nd provinces from Bennett and DePaolo 1987). The colours in (**a**) indicate the extent of lead isotope provinces defined by Wooden and deWitt (1991). The heavy brown lines indicate boundaries between neodymium isotope provinces as defined by Bennett and DePaolo (1987); the heavy blue line define the boundary between Central Arizona and Southern Arizona lead isotope subprovinces (Wooden and deWitt 1991). The inset (**b**) shows spatial variations in ΔJ (Delta Jerome) in the Central Arizona and Mohave lead isotope provinces from (mostly) northwestern Arizona (modified after Duebendorfer et al. 2006). The ΔJ parameter was defined by Wooden and DeWitt (1991) as follows: "In order to emphasize the differences in $^{207}Pb/^{204}Pb$ among these samples a normalization technique is used. A model 1.70 Ga $^{207}Pb/^{204}Pb$—$^{206}Pb/^{204}Pb$ isochron was calculated that passes through the galena lead isotopic data from the United Verde mine at Jerome ($^{206}Pb/^{204}Pb$ = 15.725, $^{207}Pb/^{204}Pb$ = 15.270, $^{208}Pb/^{204}Pb$ = 35.344). For any given $^{206}Pb/^{204}Pb$ measured for a sample a $^{207}Pb/^{204}Pb$ ratios can be calculated that falls on this model isochron. The $^{207}Pb/^{204}Pb$ measured from the sample is compared to the calculated value by subtracting the model value from the sample value and multiplying by 100. This derived number is referred to as the Delta Jerome value."

exploration significance, as discussed below. It should be noted, however, that the details mapped out by studies such as Duebendorfer et al. (2006) require a relatively high density of data. It is important to stress that data density has a marked effect on patterns produced by isotopic mapping; these patterns are unreliable in areas of low data density.

Below we describe a number of recent studies in which maps showing variations in parameters derived from lead isotope data have been produced at the province and continental scales. These include studies by Huston et al. (2005, 2014) in the Archean Yilgarn Craton in Western Australia and Abitibi-Wawa Subprovince in Canada, Blichert-Toft et al. (2016) in Europe, and Huston et al. (2016, 2017) in the Tasman Element in eastern Australia. Champion and Huston (2016, 2023) present additional examples of lead (and Nd) isotope mapping.

5.3.1 Lead Isotope Mapping of the Eastern Goldfields Superterrane, Western Australia and Abitibi-Wawa Subprovince, Canada

Huston et al. (2005, 2014) used lead isotope data from Neoarchean orogenic gold and VHMS deposits in the Eastern Goldfields Superterrane and the Abitibi-Wawa Subprovince in Canada to map spatial variations in μ using the Abitibi-Wawa lead isotope evolution model (Thorpe et al.

1992; Thorpe 1999). The results from the Eastern Goldfields Superterrane define an internal, north trending zone of low μ that corresponds to a zone of low T_{2DM} (T_{2DM} is the modelled age of extraction from depleted mantle using a two stage neodymium evolution model (see Champion and Huston (2023) for details) mapped using granite neodymium data (Fig. 8; see also Champion and Cassidy 2008; Champion and Huston 2016, 2023). The data also defined a major gradient in T_{2DM} across the Ida Fault, which separates the Eastern Goldfields Superterrane to the east from the Youanmi Terrane to the west. In the Abitibi-Wawa Subprovince (not shown) the data indicate an east–west trend marked by an internal lower μ zone that grades to higher μ margins both to the north and south. Importantly, the Abitibi-Wawa Subprovince is characterized by significantly lower μ (7.69–7.96) than the Eastern Goldfields Superterrane (8.05–9.05), with only the north-trending low-μ zone (8.05–8.15) in the latter terrane approaching the values observed in the Abitibi-Wawa Subprovince. Huston et al. (2014) interpreted these low μ zones as extensional zones with a greater mantle input, and found that most VHMS deposits were localized within the low-μ zones, commonly along gradients. They found that Archean and Proterozoic VHMS deposits were preferentially associated with juvenile crust as determined from lead and neodymium isotope data, with juvenile zones and provinces having significantly higher endowment than more evolved zones (Fig. 9: Champion and Huston 2016). In contrast, komatiite-associated nickel-sulfide (KANS) deposits are more strongly associated with more evolved crust. Although this relationship is particularly well developed in the Eastern Goldfields Superterrane (Fig. 8), where KANS deposits are common, limited data from the Abitibi-Wawa Subprovince suggests that the few KANS deposits in this juvenile terrane are associated with more evolved signatures. Barrie et al. (1999) found that KANS ores in the Abitibi-Wawa Subprovince have Abitibi-Wawa model μ values of 7.94 (Alexo deposit: Tilton 1983) and 8.28 ± 0.12 (Marbridge deposit: Deloule et al. 1989), much higher than the range in local VHMS ores.

More recently, Zametzer et al. (2022) undertook MC-LA-ICP-MS analysis of potassium

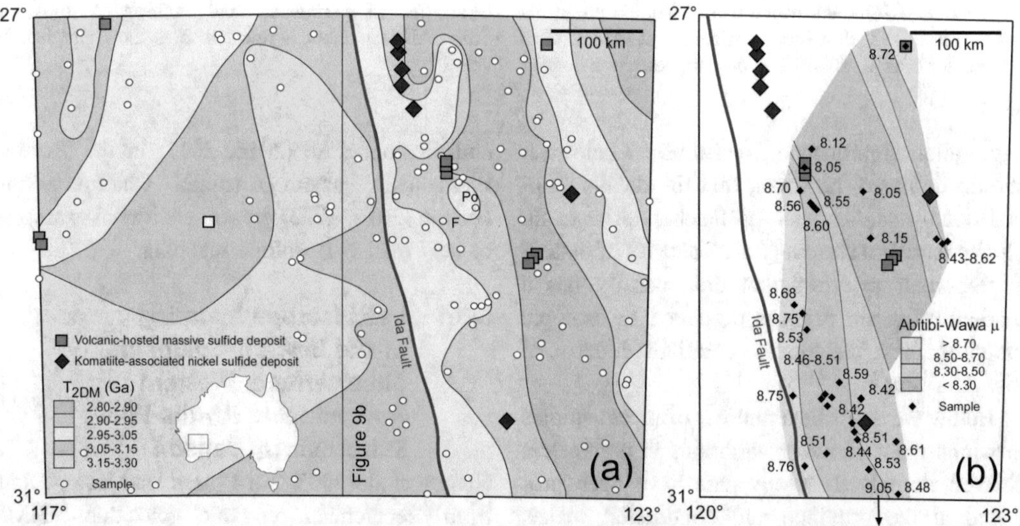

Fig. 8 Variations in T_{2DM} (**a**: from granite analyses) and μ (**b**: from volcanic-hosted massive sulfide and orogenic gold deposits) in the Eastern Goldfields Superterrane (modified after Huston et al. 2014). The location and deposit types of analyses (mostly from Browning et al. 1987) are shown as different symbols. The location of major komatiite-associated nickel sulfide deposits are also shown; data from these deposits were not used in determining variations in μ

feldspar from granites along a transect that crossed the boundary between the Eastern Goldfields Superterrane and the Youanmi Terrane. This study used the same granite samples used in the original neodymium T_{2DM} study (Champion and Cassidy 2008), and demonstrated that like the T_{2DM} data, parameters derived from lead isotope data, such as μ and ω, also varied across the Ida Fault, with granites from the Youanmi Terrane characterized by more evolved (higher μ) lead isotope signatures relative to the Eastern Goldfields Superterrane. The Zametzer et al. (2022) study suggested that spatial variations seen in neodymium and lead data are genetically linked.

5.3.2 Lead Isotope Mapping of Europe

Blichert-Toft et al. (2016) used ore lead isotope analyses from a range of deposits in Europe to determine model ages and calculate μ and $^{232}Th/^{204}Pb$ values using the Stacey and Kramers (1975) model. The calculated model ages matched the major tectonic cycles in Europe (Svecofennian, Pan-African, Variscan, Caledonian and Alpine: Fig. 10) and were consistent with the ages of major metallogenic epochs in Europe. Maps showing μ and ω ($^{232}Th/^{204}Pb$) values of individual analyses by Blichert-Toft et al. (2016) show systematic patterns, particularly when compared with tectonic subdivisions of Europe. Figure 11 uses the data of Blichert-Toft et al. (2016) to calculated a μ map similar to ones presented for the Eastern Goldfield Superterrane and Abitibi-Wawa Subprovince (see above) and the Tasman Element (see below).

Overall, southern Europe is characterized by higher μ (Fig. 11: Blichert-Toft et al. 2016) than northern Europe, but there are zones within this broad subdivision that correspond to tectonic boundaries and metallogenic domains. In northern Europe, the Caledonian Orogen is characterized by low μ. In the northern British Isles, the boundary between the Caledonian Orogen and Avalonia is marked by a μ gradient with which carbonate-hosted Zn-Pb deposits of the Irish Midlands are strongly associated (Hollis et al.

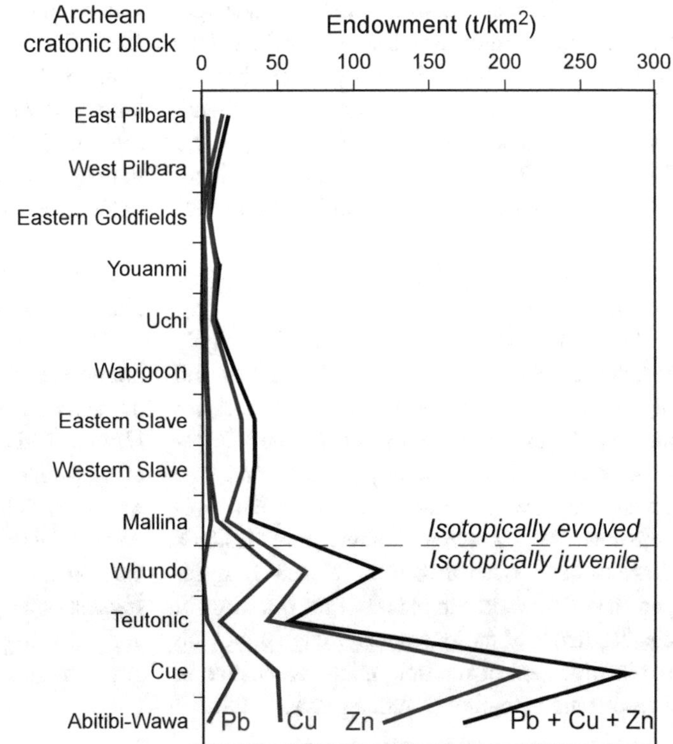

Fig. 9 Diagram showing the relationship of the endowment of total base metal (Zn + Pb + Cu) from volcanic-hosted massive sulfide deposits in selected Archean terranes classified according to crustal character (juvenile or evolved) based on lead and neodymium isotope data (modified after Champion and Huston 2016)

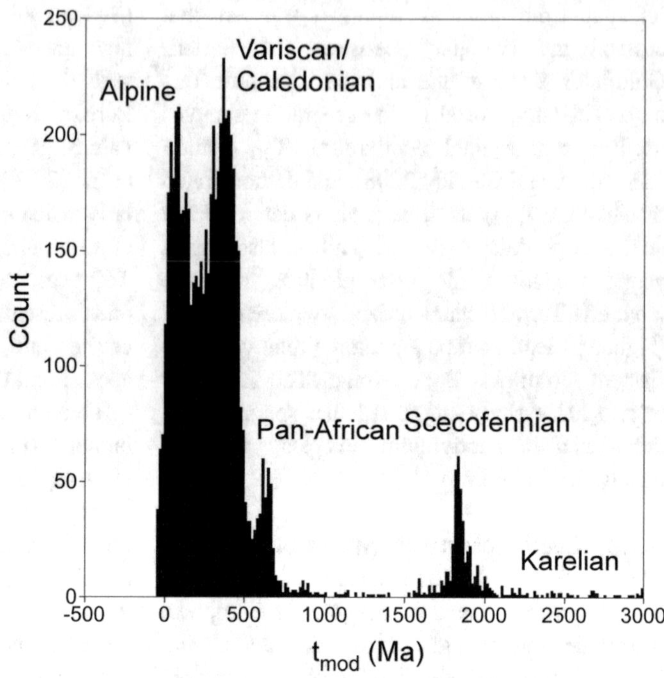

Fig. 10 Histogram of lead model ages (using Stacey and Kramers 1975 model) from Europe showing correspondence with major tectonic cycles (modified after Blichert-Toft et al. 2016)

2019). Avalonia is characterized by moderately high μ values of 9.6–9.9 (Fig. 11). Moderately low values of μ (mostly 9.5–9.6) characterize the northern Variscan Orogen, but in southern France, this tectonic province is dominated by high μ values of 9.7–9.9. This gradient broadly corresponds to the northern boundary of Gondwanan terranes within the Variscan Orogen (c.f. Plant et al. 2005). The Alpine Orogen is dominated by values of 9.7–9.9, though with discrete low-μ zones. μ decreases strongly along the suture between the Alpine and Variscan orogens in central Europe.

In northern Europe, Svecofennia has highly variable μ with discrete zones of both low and high μ. Gradients to lower μ mark the western and northwestern margins, with the Trans-Scandinavian granite-porphyry belt and the Caledonian Orogen, respectively. Within Svecofennia, the Skellfteå, Kaitele and Bothnia domains have lower μ than the Ulmeå. Bergslagen and Tavastia domains. The differing μ characteristics of these terranes is consistent with the involvement of multiple microcontentents in Svecofennia assembly (Lahtinen et al. 2005).

Like Archean and, possibly, Paleoproterozoic provinces globally, Paleoproterozoic and Phanerozoic VHMS districts and provinces in Europe are associated with low μ zones. The Skellefte, Norwegian Caledonide, Pontide and Troodos VHMS provinces, are located in zones with μ < 9.6 (mostly < 9.5: Fig. 11). Even the Iberian Pyrite Belt has μ values lower than domains to the south and west. Hence, the association of VHMS deposits with juvenile (low μ) crustal domains, as noted in the Archean (Huston et al. 2014), is also seen in European Phanerozoic domains.

The relationships of μ gradients to tectonic boundaries and the μ characteristics to tectonic domains suggest that lead isotope evolution in Europe (and elsewhere) reflects, and can be used to map tectonic domains. Moreover, modelling of the evolution of lead in these domains may provide insights into the processes that produced the domains (Blichert-Toft et al. 2016). The data suggest that, metallogenic provinces can be defined using lead isotope characteristics and/or gradients (c.f. Hollis et al. 2019).

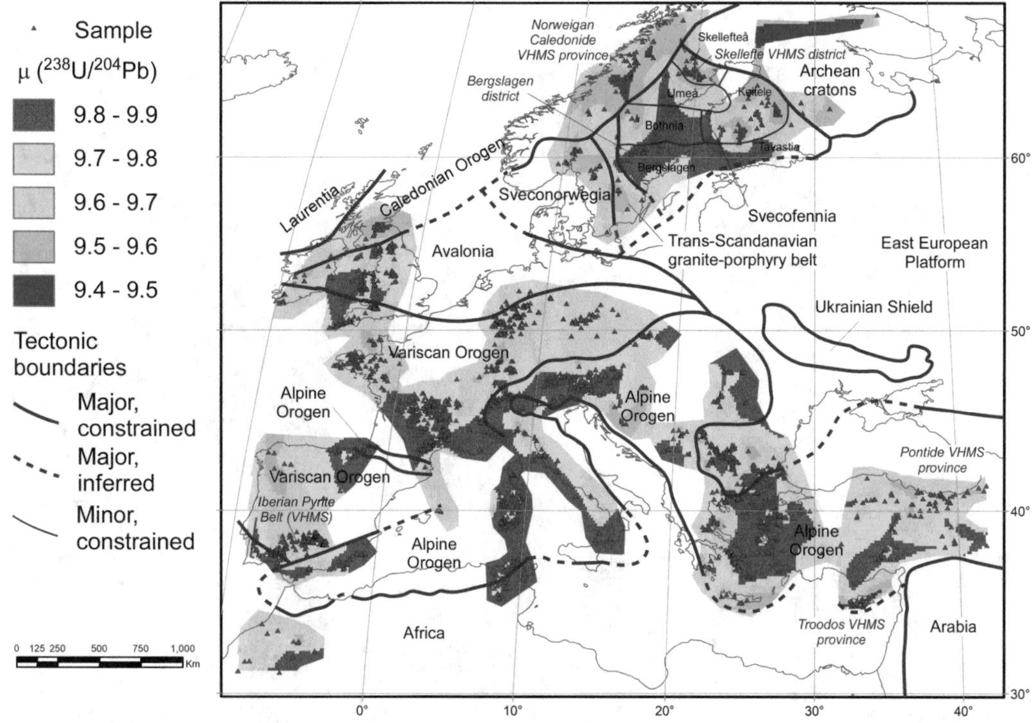

Fig. 11 Variations in μ, as determined mostly from the analyses of Pb-rich mineralized samples, overlain by European tectonic boundaries (lead isotope data used to construct image from Blichert-Toft et al. 2016; tectonic boundaries from Plant et al. 2005; Lahtinen et al. 2005; Richards 2015; Faure and Farrière 2022). Significant volcanic hosted massive sulfide districts are also indicated

5.3.3 Lead Isotope Mapping of the Tasman Element, Eastern Australia

Huston et al. (2016, 2017) mapped the largely Phanerozoic Tasman Element of eastern Australia using a range of parameters derived from lead isotope analyses of Pb-rich samples from mineral deposits and occurrences. This study followed an earlier study of the same area by Carr et al. (1995) that presented much of the data used and an evolution model to account for the isotopic variability. Figure 12a shows that variations in μ reflect province boundaries as well as mineral provinces. For example, a strong gradient in μ occurs along much of the boundary between the Eastern and Central Lachlan. The high-μ Central Lachlan hosts tin deposits of the Wagga tin province, whereas the low-μ zone in the central part of the East Lachlan corresponds very closely with the Macquarie porphyry Cu-Au province. Huston et al. (2017) discuss other patterns present in the μ map of the Tasman Element, and the broad similarity in this map to the T_{2DM} map constructed using neodymium isotope data from granites (Fig. 12c).

Lead isotope variations in the Tasman Element cannot be fully explained by the Carr et al. (1995) evolution model, which was developed for the Lachlan Orogen and models the spread in isotopic data as the resulting of the interaction of two lead sources that evolved through time, Lachlan crust and Lachlan mantle. Figure 13a compares the isotopic characteristics of Cambrian (520–490 Ma) deposits and Silurian (435–420 Ma) deposits from the southern Tasman element with the evolution of the two lead sources proposed by Carr et al. (1995). For the Silurian deposits that the model was designed to explain, it works well, with independently-determined ages of mineralization similar to

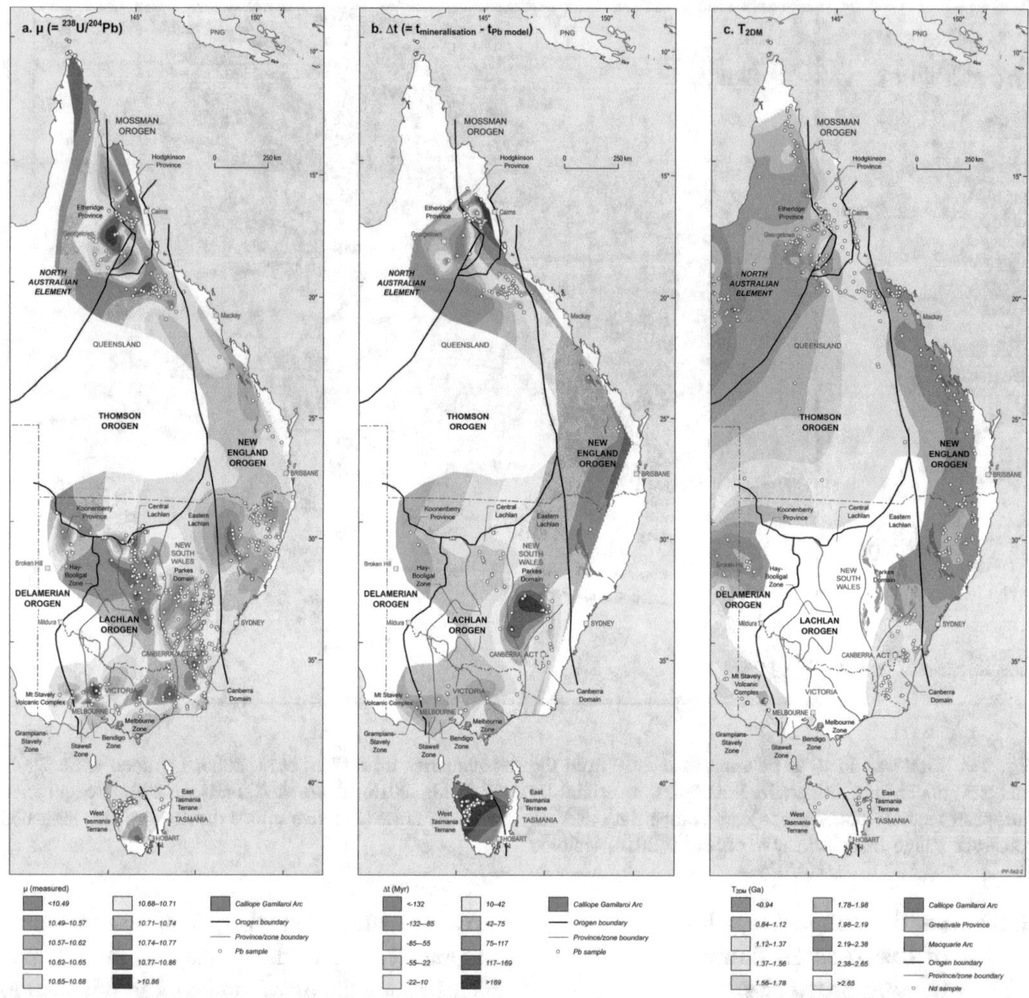

Fig. 12 Diagrams showing variations in µ ($^{238}U/^{204}Pb$) (a), Δt (b) and T_{2DM} (c) in the Tasman Element and surrounds. Parts (**a**) and (**b**) were determined using the Cumming and Richards (1975) lead evolution model. Modified from diagrams in Champion (2013) and Huston et al. (2017). Province and zone boundaries are from Glen (2013)

model ages. When expanded outside of the Lachlan Orogen, the model breaks down: the Lachlan crustal curve models the age of Cambrian deposits (Kanmantoo, Eclipse and Cymbric Vale) to the west of the Lachan Orogen well, but model ages of Cambrian deposits in the Mount Read Volcanics, western Tasmania (Rosebery, Que River, Hellyer, Mount Lyell and Henty) are ~200 Myr too young. Some Cambrian deposits along the western edge of the Tasman Element (Mount Ararat and Ponto) yield model ages more than 200 Myr too old using the Lachlan mantle model (Fig. 13a).

These discrepancies can be resolved by expanding the Carr et al. (1995) model so that the Lachlan crust and Lachlan mantle sources are considered parts of two lead isotopic systems. If the µ of the Lachlan mantle curve is increased to values of 10.2, the resulting evolution curve yields model ages that closely match the ages of western Tasmania deposits. Similarly, if the µ of the Lachlan crustal curve is decreased to 13.0,

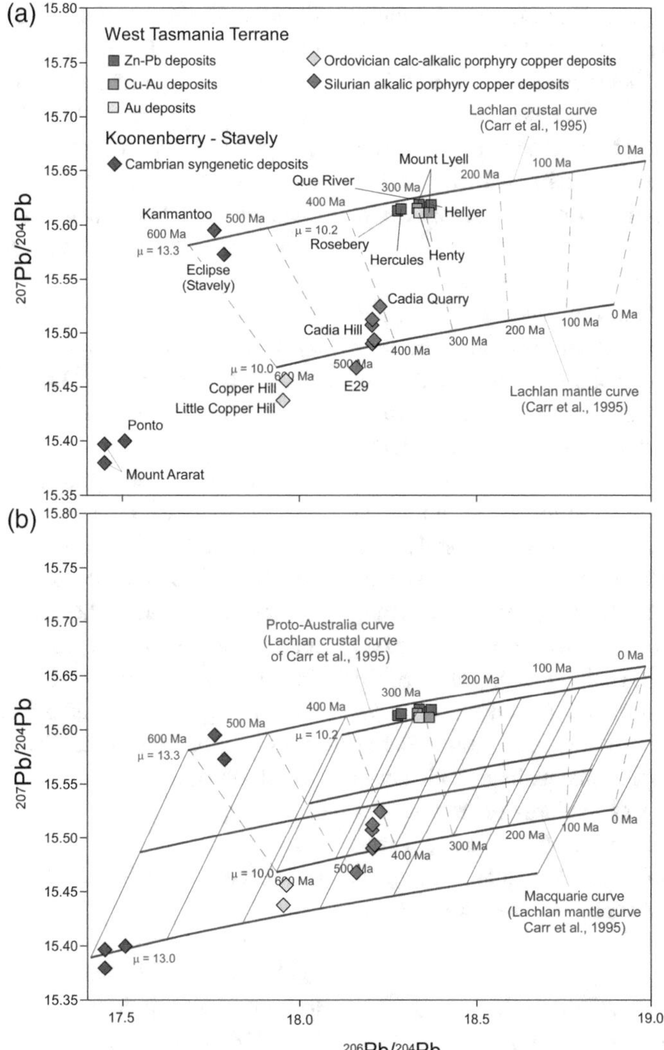

Fig. 13 Comparison of lead isotope data from selected deposits from the Tasman Element with lead isotope evolution curves: **a** Cambrian volcanic-hosted massive sulfide and Silurian porphyry-related deposits with the Lachlan crustal and Lachlan mantle curves of Carr et al. (1995); and **b** Cambrian volcanic-hosted massive sulfide and related deposits with a four component isotopic evolution model consisting of the two component proto-Pacific system (Lachlan mantle curve of Carr et al. (1995) and contemporaneous crustal curve) and a two component proto-Australia system (Lachlan crustal curve of Carr et al. (1995) and contemporaneous mantle curve). For both the proto-Pacific and the proto-Australian systems, the second curve was calculated by changing the μ associated with original Carr et al. (1995) curve. The dashed lines are mixing lines between the Lachlan crustal and Lachlan mantle curves

the resulting evolution curve produces ages that match the ages of western Tasman Element deposits (Fig. 13b). Based on these relationships, Huston et al. (2016, 2017) suggested that the formation of the Tasman Element involved the interaction of two isotopic systems that include both crustal and mantle components: one that characterized a proto-Pacific plate that was consumed during largely west-dipping subduction, and a second that characterized pre-existing proto-Australian crust in the over-riding plate. The vast majority of the variability in lead isotopes in the Tasman Element can be effectively explained by mixing between these four sources (Fig. 13b). Geologically, the West Tasmania Terrane forms part of the Selwyn Block (or Vandieland), which is considered an exotic block accreted onto Australia (Cayley 2011), consistent with the model of two separate isotopic systems.

Figure 12b illustrates variations in the parameter Δt, the difference between the age and model age of mineralisation. As the Cumming and Richards (1975) model used to calculate Δt has an evolution similar to that of the Lachlan crustal curve of Carr et al. (1995). The negative to near zero values in the western mainland part of the Tasman Element indicate a dominant proto-Australia influence, whereas strongly positive Δt values, particularly in western Tasmania and the New England Orogen, indicate a strong proto-Pacific influence. Huston et al. (2017) interpreted the low-μ, positive Δt signature of the Macquarie Cu-Au province to reflect the influence of volatiles from the subducting Pacific plate. As both isotopic systems have both low-μ and high-μ components, patterns in Δt are independent of variations in μ: some low-μ zones have positive proto-Pacific Δt signature whereas others have a near-zero to negative proto-Australia Δt signature.

5.3.4 Potential Problems with Lead Isotope Mapping—Some Traps for Young Players

A significant observation from this review is the variety of parameters that have been used to map or model variations in lead isotopes. Many of these parameters (e.g. ΔJ and $\Delta^{207}Pb/^{204}Pb$) are locally defined and, although they work in the local areas in which they were defined, do not have universal application. In addition, in many cases isotope mapping uses parameters derived from local lead evolution models and not more universal models such as Stacey and Kramers (1975) or Cumming and Richards (1975). The use of local models makes direct comparison between maps difficult. It should be stressed, however, that although absolute values of parameters derived from different evolution models differ, the observed patterns should be similar.

Like other geochemical data, the contouring of lead isotope data is strongly influenced by data density. In areas of low data density or no data, commonly used contouring packages can produce artefacts, hence patterns in areas of low data density should regarded critically, and areas with no data masked.

6 Applications of Lead Isotope Data to Exploration

Although not widely recognized as a direct tool, lead isotope data can and has been used in mineral exploration, from the province to the deposit scale. At the province scale, lead isotopes can be used assess the fertility of a particular region for different styles of mineralization. At the district to deposit scales, lead isotope data have many uses including assessing potential deposit type and potential size and defining vectors to ore.

6.1 Inferring Metallogenic Fertility

As discussed above, at the province scale, lead (and similarly Nd) isotopes can indicate fertility for some deposit types. This is best shown for VHMS deposits, where the most fertile provinces occur within isotopically juvenile terranes. In contrast, more fertile provinces for KANS deposits are isotopically more evolved. Although this was first documented for Archean provinces, emerging data suggest the relationship of fertile VHMS provinces with juvenile terranes also applies to younger provinces (see above).

At the large scale, Hollis et al. (2019) and Huston et al. (2020) have documented the association of major Irish-style and shale-hosted massive sulfide deposits, respectively with gradients in μ (and also κ and ω), in the Irish Midlands, the North Australian Zinc Belts and the Northern Cordillera of North America. These gradients are interpreted to mark cratonic edges and are also visible in geophysical datasets (Huston et al. 2020).

6.2 Deposit Type and Size

Since the pioneering work of Cannon et al. (1961), many workers have suggested that lead isotope characteristics can indicate the type and, potentially, the size of mineral deposits/ occurrences at an early stage of exploration. Many mineral provinces are characterized by the presence of different types of deposits, and determining deposit type is important to develop and apply exploration models, although limited geological data can make this difficult. As lead isotope data reflect both the source of the lead and the age of mineralization, they can provide an early indication of the origin of mineralization. Below we give examples of the use of lead isotope data to categorize mineral occurrences and gossans at an early stage of exploration. Then we discuss suggestions that lead isotope characteristics can indicate size potential.

6.2.1 Fingerprinting Deposit Types, a Case Study from Western Tasmania

An example of lead isotope fingerprinting during an exploration program was described by Gulson et al. (1987) in western Tasmania, Australia. Western Tasmania, one of the most richly mineralized provinces in Australia, has a range of mineral deposit types, including Cambrian VHMS and related deposits, Ordovician Irish-style or MVT deposits and Devonian granite-related tin deposits. Gulson et al. (1987) showed that major deposits of these groups have different lead isotope signatures (Fig. 14). These signatures were then used to classify and rank mineral occurrences from a range of sample media (steam sediments, soils, gossans/exposure, drill chips and drill core). Another example where lead isotope data have been used to determine the origin of deposits is in the Coeur d'Alene region in Idaho and Montana, United States, where lead isotope data distinguish Mesoproterozoic Cordilleran-type veins deposits from vein deposits associated with Tertiary intrusions (Zartman and Stacey 1971).

6.2.2 Assessing the Origin of Gossans

An important task during early exploration is determining the origin of surficial ironstones—are they true (i.e. related to mineralisation) or false gossans? Based on data from stratiform (VHMS and shale-hosted) Zn–Pb deposits, Gulson and Mizon (1979) found that true gossans usually had very uniform lead isotope characteristics that plotted close to locally determined growth curves and gave model ages consistent with expected ages. In contrast, false gossans were characterized by variable ratios that yielded model ages significantly younger than the expected ages. Like the deposit fingerprinting described above, lead isotope data have potential for early-stage ranking of surficial gossans, with resulting savings in time and money.

6.2.3 Inferring Potential Size of Deposits

One of the earliest suggested uses of lead isotope data in exploration was the inference that the data could be used to indicate deposit size (e.g. Cannon et al. 1961). Early workers (Cannon et al., 1961; Stacey et al. 1967, Stacey and Nkomo 1968; Zartman and Stacey 1971; Doe 1978; Stacey and Hedlund 1983) found that in intrusion-centred hydrothermal systems, the larger deposits near the causative intrusion tended to have less radiogenic and more homogeneous isotopic signatures than peripheral deposits. The larger deposits formed in the centres of these systems may be dominated by homogeneous, less radiogenic lead associated with the intrusion, with the smaller, peripheral deposits incorporating a larger component of more radiogenic lead from the wall rocks.

Another characteristic that has been used to infer deposit size is the homogeneity of the isotopic data: large deposits have more homogeneous lead isotope ratios than small deposits or occurrences. This relationship has been suggested for a range of deposit types, including orogenic gold (Browning et al. 1987), MVT (Foley et al. 1981) and shale-hosted massive

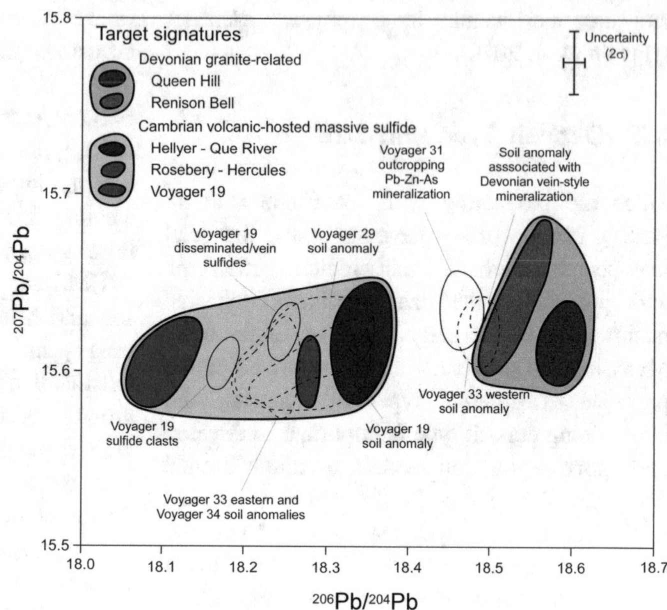

Fig. 14 $^{206}Pb/^{204}Pb$ versus $^{207}Pb/^{204}Pb$ diagram comparing the fields of mineral prospects from the Elliott Bay area with major mineral deposits from western Tasmania, Australia (compiled from Gulson et al. 1987). For this study, which targeted Cambrian VHMS deposits, prospects overlapping with major Cambrian deposits were given priority over prospects overlapping major Devonian deposits

sulfide (Carr et al. 2001; Large et al. 2005) deposits. The homogeneous isotopic character of the large deposits may be the consequence of large hydrothermal systems that homogenize lead isotope signatures, whereas the heterogeneous isotopic character of the small deposits may indicate the predominance of smaller hydrothermal systems tapping different metal sources.

In a critical assessment of the utility of lead isotopic data as indications of deposit size, Gulson (1986) cites several examples of small deposits characterized by uniform lead isotope data. Moreover, in the Zeehan district (Tasmania, Autralia), the largest deposit, at the core of the system, is characterized by a more radiogenic isotopic signature (Huston et al. 2017). Hence, the relationships between deposit size and lead isotope characteristics do not appear to be consistent and have limited exploration utility in greenfields provinces, although they may have utility in brownfield provinces where lead isotope characteristics are better established.

6.3 Vectors to Ore

In addition to discriminating deposit type, lead isotope ratios have also been used as vectors to ore from a range of sample media. For example, in the Snow Lake district, Manitoba, Canada, Bell and Franklin (1993) found that the least radiogenic lead isotope ratios from the fine-grained fractions of till samples were located down ice from the Chisel Lake deposit, suggesting that lead isotope data from tills could be used as vectors in glaciated terranes. In addition, lead isotope ratios may provide vectors toward ore in ground and surface waters in mineralized provinces, as described below. It must be stressed that use of lead isotopes as vectors to ores and as discriminators of deposit type is strongly provincial: the signatures differ according to mineral province and deposit type, so local orientation studies are required for use.

6.3.1 Lead Isotopes in Surface and Groundwater, Restigouche Deposit, New Brunswick, Canada

Leybourne et al. (2009) undertook a combined study documenting the variations in lead isotope compositions of ore samples from the Ordovician Restigouche VHMS deposit in the Bathurst district, New Brunswick. They found that the isotopic composition of ores from this deposit were similar to those of other VHMS deposits in the district but different from volcanic and

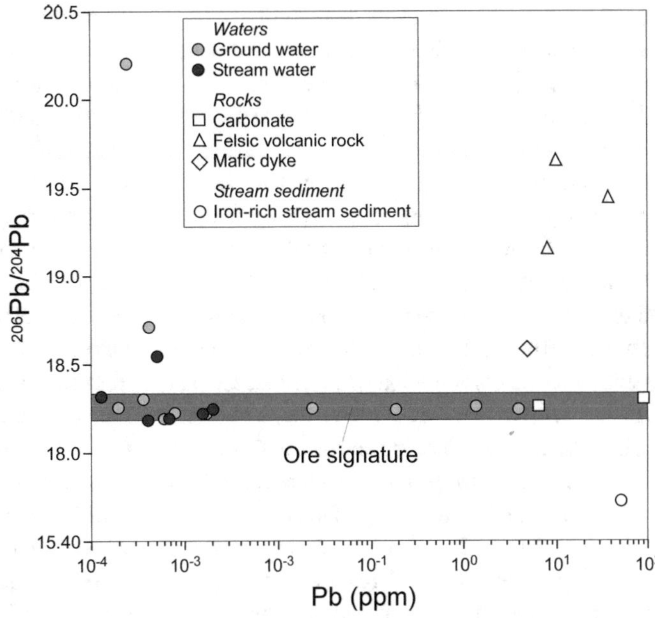

Fig. 15 $^{206}Pb/^{204}Pb$ versus lead concentration (ppm) diagram comparing the isotopic composition of surface and ground waters from in and around the Restigouche deposit, New Brunswick, Canada (modified after Leybourne et al. 2009)

sedimentary host rocks to the deposit, which were significantly more radiogenic (Fig. 15). They also found that surface and ground water sampled proximal to the deposit had an ore signature, but distal groundwater had more radiogenic lead isotope compositions due to weathering of host volcanic and sedimentary rocks (Fig. 15). Similarly, in the Broken Hill district, New South Wales, de Caritat et al. (2005) found that in the vicinity of deposits groundwater lead isotope compositions reflected the composition of the deposit; moreover, the groundwater data could be used to identify the style of mineralisation.

6.4 Summary and Potential Traps for Young Players

As described above, lead isotopes have a large range of potential uses in exploration, including determining fertility of mineral provinces, identifying crustal boundaries that controls major deposits, fingerprinting deposit (and gossan) types and as direct vectors to ore. Another suggested use is prediction of potential deposit size. Although the criteria to determine deposit size have not lived up to their early potential (e.g. Gulson 1986), it does appear that heterogeneous lead isotope signature is indicative of a small deposit for many (though not all) deposit types (i.e. shale-hosted massive sulfide, VHMS and orogenic gold deposits, for instance, but not MVT deposits).

The use of lead isotopes in exploration at the district to prospect scale is largely empirical, and site or district specific. Although particular patterns might be expected in district and prospect-scale datasets, it is important that orientation studies are undertaken to validate the utility of lead isotopes prior to undertaking major data collection. Like other uses of lead isotopes, it is preferable to determine initial ratios, if possible, when using lead isotopes directly in exploration.

7 Conclusions and Future Developments

As with many other scientific fields, the field of lead isotopes has changed dramatically over the last two decades. Using new analytical techniques such as MC-ICP-MS and MC-LA-ICP-MS, the cost of analyses has dropped, the precision has improved, and types of materials analysed has diversified. These developments, combined with the increasing availability of large datasets and new methods of data representation and analysis (e.g. GIS) has allowed significant advances in the use of lead and other radiogenic isotopes in metallogenesis and exploration. In particular, development of regional lead isotope evolution models has improved the accuracy of model ages, micro-scale analytical techniques and the combination of lead isotope data with other radiogenic and stable isotope systems can better constrain the sources of ore and related metals, and the combination of large datasets and GIS has allowed construction of maps showing variations in lead isotope ratios and derived parameters such as μ and Δt that have both metallogenic and tectonic significance.

In exploration, lead isotopes have many and varied uses. Regional lead isotope (and Nd isotope) mapping can identify zones with higher potential for certain types of deposits. For example, in Archean and Proterozoic terranes, VHMS deposits tend to be more abundant in juvenile crust, which can be mapped using lead (and neodymium) isotopes. In addition, lead isotope characteristics can distinguish between different types of mineralization, for example VHMS versus granite-related tin in western Tasmania or fertile versus barren gossans. There is some suggestion that isotopically homogeneous deposits are larger than isotopically heterogeneous deposits. Finally, lead isotopes can be used as empirical vectors to ore both at the camp to deposit scale, using a range of media ranging from galena in minor occurrences, through groundwater and tills.

The recent developments in analytical techniques and GIS open up a range of applications of lead isotopes to metallogenic, ore and tectonic studies. In particular, high precision laser ablation lead isotope analysis of fluid inclusions (Pettke et al. 2010) offer the opportunity of direct sampling of ore fluids, and, if the fluid inclusions are well constrained, direct insights into the behaviour of lead during fluid flow and ore deposition. The opportunity for microscale sampling of ore minerals presents similar opportunities to those seen with the development of micro-analytical capabilities for other isotopic systems such as oxygen, sulfur and boron (Sharp 1990; Eldridge et al. 1987; Foster et al. 2018; Troll et al. 2019). For example, Gigon et al. (2020) documented apparent micro-scale variations in lead isotopes in galena from the HYC (McArthur River) deposit in the Northern Territory, Australia from SIMS analysis. They interpreted these variations as mixing between two lead sources, but "bulk" high-precision double spike galena analyses did not show the same trends (e.g. Carr et al. 2001); reasons for this dichotomy in results have not been resolved. As spatially resolved micro-analyses of lead isotopes in a range of media become more widely available, new insights will emerge.

However, the opportunities for advancement extend well beyond microanalytical studies. In the regional context isotopic mapping of thorogenic lead and combination of the U–Th–Pb isotopic system with other radiogenic isotope systems such as Sm–Nd and Re–Os has potential to provide new insights into the role of different parts of the crust and mantle in metallogenesis and tectonics. For example, the combination of uranogenic and thorogenic lead may constrain the role of the lower to middle crust in metallogenesis. Radiogenic isotope studies from the continent- to fluid inclusion-scales offer the potential to understand province to ore-shoot controls on ore formation, data that will assist in developing genetic and exploration models, both for area selection and vectoring.

Acknowledgements This contribution has benefited over the years from discussion with a large range of people, including colleagues with whom we have closely worked, and some of the authors of studies cited by this review. In particularly we would like to thank Graham Carr, Michael Doublier and the late Shen-Su Sun for discussion about lead and neodymium isotope systematics over the years. This contribution benefited from reviews by Kathryn Waltenberg and two anonymous reviewers and additional comments by editor Jens Gutzmer, and is published with permission of the Chief Executive Officer of Geoscience Australia.

References

Albarède F, Telouk P, Blichert-Toft J, Boyet M, Agranier A, Nelson B (2004) Precise and accurate isotopic measurements using multiple-collector ICPMS. Geochim Cosmochim Acta 68:2725–2744

Armstrong RL, Taubeneck WH, Hales PO (1977) Rb–Sr and K-Ar geochronometry of Mesozoic granitic rocks and their Sr isotopic composition, Oregon, Washington, and Idaho. Geol Soc Am Bull 88:397–411

Baker J, Peate D, Waight T, Meyzen C (2004) Pb isotopic analysis of standards and samples using a ^{207}Pb-^{204}Pb double spike and thallium to correct for mass bias with a double-focusing MC-ICP-MS. Chem Geol 211:275–303

Barrie CT, Cousens BL, Hannington MD, Bleeker W, Gibson HL (1999) Lead and neodymium isotope systematics of the Kidd Creek mine stratigraphic sequence and ore, Abitibi Subprovince, Canada. Econ Geol Mon 10:497–510

Bell K, Franklin JM (1993) Application of lead isotopes to mineral exploration in glaciated terrains. Geology 21:1143–1146

Belshaw NS, Freedman PA, O'Nions RK, Frank M, Guo Y (1998) A new variable dispersion double-focussing plasma mass spectrometer with performance illustrated for Pb isotopes. Int J Mass Spectrom 181:51–58

Bennett VC, DePaolo DJ (1987) Proterozoic crustal history of the western United States as determined by neodymium isotopic mapping. Geol Soc Am Bull 99:674–685

Blichert-Toft J, Delile H, Lee C-T, Stos-Gale Z, Billström K, Anderson T, Hannu H, Albarède F (2016) Large-scale tectonic cycles in Europe revealed by distinct Pb isotope provinces. Geochem Geophy Geosyst 17:3854–3864

Brevart O, Dupré B, Allegre CJ (1982) Metallogenic provinces and the remobilization process studied by lead isotopes; lead-zinc ore deposits from the southern Massif Central, France. Econ Geol 77:564–575

Browning P, Groves DI, Blockley JG, Rosman KJR (1987) Lead isotope constraints on the age and source of gold mineralization in the Archean Yilgarn Block, Western Australia. Econ Geol 82:971–986

Burchfiel BC, Davis GA (1981) Triassic and Jurassic tectonic evolution of the Klamath Mountains-Sierra Nevada geologic terrane. In: Ernst WG (ed) The geotectonic development of California (Rubey Volume 1). Prentice-Hall, Englewood Cliffs, pp 50–70

Cameron AE, Smith DH, Walker RL (1969) Mass spectrometry of nanogram-size samples of lead. Anal Chem 41:525–526

Cannon RS Jr, Pierce AP, Antweiler JC, Buck KL (1961) The data of lead isotope geology related to problems of ore genesis. Econ Geol 56:1–38

Carr GR, Dean JA, Suppel DW, Heithersay PS (1995) Precise lead isotope fingerprinting of hydrothermal activity associated with Ordovician to Carboniferous metallogenic events in the Lachlan fold belt of New South Wales. Econ Geol 90:1467–1505

Carr GR, Denton GJ, Korsch MJ, Parr J, Andrew AS, Whitford D, Wyborn LAI, Sun S-S (2001) User friendly isotope technologies in mineral exploration: northern Australia Proterozoic basins. Australian Mineral Industry Research Association Final Report P480

Cayley RA (2011) Exotic crustal block accretion to the eastern Gondwanaland margin in the Late Cambrian—Tasmania, the Selwyn Block, and implications for the Cambrian-Silurian evolution of the Ross, Delamerian, and Lachlan orogens. Gondwana Res 19:628–649

Champion DC (2013) Neodymium depleted mantle model age map of Australia: explanatory notes and user guide. Geosci Austr Rec 2013/44

Champion DC, Cassidy KF (2008) Using geochemistry and isotopic signatures of granites to aid mineral systems studies: an example from the Yilgarn Craton. Geoscience Australia Record 09:7–16

Champion DC, Huston DL (2016) Radiogenic isotopes, ore deposits and metallogenic terranes: novel approaches based on regional isotopic maps and the mineral systems concept. Ore Geol Rev 76:229–256

Champion DC, Huston DL (2023) Applications of Nd isotopes to ore deposits and metallogenic terranes; using regional isotopic maps and the mineral systems concept. In: Huston DL, Gutzmer J (eds), Isotopes in economic geology, metallogensis and exploration, Springer, Berlin, this volume

Chelle-Michou C, Schaltegger U (2023) U-Pb dating of mineral deposits: from age constrains to ore-forming processes. In: Huston DL, Gutzmer J (eds), Isotopes in economic geology, metallogensis and exploration, Springer, Berlin, this volume

Chiaradia M (2023) Radiometric dating applied to ore deposits: theory and methods. In: Huston DL, Gutzmer J (eds), Isotopes in economic geology, metallogensis and exploration, Springer, Berlin, this volume

Chiaradia M, Fontboté L (2003) Separate lead isotope analyses of leachate and residue rock fractions: implications for metal source tracing in ore deposit studies. Mineral Deposita 38:185–195

Chiaradia M, Konopelko D, Seltmann R, Cliff RA (2006) Lead isotope variations across terrane boundaries of the Tien Shan and Chinese Altay. Mineral Deposita 41:411–428

Cumming GL, Richards JR (1975) Ore lead isotope ratios in a continuously changing Earth. Earth Planet Sci Lett 28:155–171

de Caritat P, Kirste D, Carr G, McCulloch M (2005) Groundwater in the Broken Hill region, Australia: recognising interaction with bedrock and mineralisation using S, Sr and Pb isotopes. Appl Geochem 20:767–787

Deloule E, Gariepy C, Dupre B (1989) Metallogenesis of the Abitibi greenstone belt of Canada: a contribution from the analyses of trace lead in sulfide minerals. Can J Earth Sci 26:2429–2540

Dodson MH (1963) A theoretical study of the use of internal standards for precise isotopic analysis by the surface ionization technique: part I. General first-order algebraic solutions. J Sci Instrum 40:289–295

Doe BR, Stacey JS (1974) The application of lead isotopes to the problems of ore genesis and ore prospect evaluation: a review. Econ Geol 69:757–776

DoeBR (1978) The applicaton of lead isotopes to mineral prospect evaluation of Cretaceous-Tertiary magmatothermal ore deposits in the western United States. In: Watterson JR, Theobald PK (eds) Proceedings of the 7thInternational Geochemical Exploration Symposium, Golden, Colorado. pp 227–232

Doe BR, Zartman RE (1979) Plumbotectonics, the Phanerozoic. In: Barnes HL (ed) Geochemistry of hydrothermal ore deposits, 2nd edn. Wiley, New York, pp 22–70

Duebendorfer EM, Chamberlain KR, Fry B (2006) Mojave-Yavapai boundary zone, southwestern United States: a rifting model for the formation of an isotopically mixed crustal boundary zone. Geology 34:681–684

Eldridge CS, Compston W, Williams IS, Walshe JL, Both RA (1987) In situ microanalysis for $^{32}S^{34}S$ ratios using the ion microprobe SHRIMP. Intl J Mass Spectr Ion Proc 76:65–83

Faure M, Farriere J (2022) Reconstructing the Variscan terranes in the Alpine basement: facts and arguments for an Alpidic orocline. Geosciences 12:65. https://doi.org/10.3390/geosciences12020065

Fehn U, Doe BR, Delevaux MH (1983) The distribution of lead isotopes and the origin of Kuroko ore deposits in the Hokuroku district, Japan. Econ Geol Mon 5:488–506

Foley NK, Sinha AK, Craig JR (1981) Isotopic composition of lead in the Austinville-Ivanhoe Pb–Zn district, Virginia. Econ Geol 76:2012–2017

Foster GL, Marschall HR, Palmer MR (2018) Boron isotope analysis of geological materials. Adv Isot Geochem 7:13–31

Galer SJG, Abouchami W (1998) Practical application of lead triple spiking for correction of instrumental mass discrimination. Mineral Mag 62A:491–492

Gigon J, Deloule E, Mercadier J, Huston DL, Richard A, Annesley AR, Wygralak AS, Skirrow RG, Mernagh TP, Masterman K (2020) Tracing metal sources for the giant McArthur River Zn–Pb deposit (Australia) using lead isotopes. Geology 48:478–482

Glen RA (2013) Refining accretionary orogen models for the Tasmanides of eastern Australia. Aust J Earth Sci 60:315–370

Gulson BL (1986) Lead isotopes in mineral exploration. Dev Econ Geol 23, 245

Gulson BL, Mizon KJ (1979) Lead isotopes as a tool for gossan assessment in base metal exploration. J Geochem Explor 11:299–320

Gulson BL, Large RR, Porritt PM (1987) Base metal exploration of the Mount Read Volcanics, western Tasmania; Pt. III, application of lead isotopes at Elliott Bay. Econ Geol 82:308–327

Hirata T (1996) Lead isotopic analyses of NIST standard reference materials using multiple collector inductively coupled plasma mass spectrometry coupled with a modified external correction method for mass discrimination effect. Analyst 121:1407–1411

Hollis SP, Doran AL, Menuge JF, Daly JS, Güven J, Piercey SJ, Cooper M, Turney O, Unitt R (2019) Mapping Pb isotope variations across Ireland: from terrane delineation to deposit-scale fluid flow. In: Proceedings of the 15th SGA Biennial Meeting, 27–30 August 2019, Glasgow, Scotland, pp 1196–1199

Holmes A (1946) An estimate of the age of the earth. Nature 157:680–684

Houtermans FG (1946) Die Isotopen-Haufigkeiten im naturlichen Blei und dasAalter des Urans. Naturwissenschaften 33:185–187

Houtermans FG (1953) Determination of the age of the earth from isotopic composition of meteoritic lead. Nuovo Cimento 10:1623–1633

Huston DL, Champion DC, Cassidy KF (2005) Tectonic controls on the endowment of Archean cratons in VHMS deposits: evidence from Pb and Nd isotopes. Mineral deposit research: meeting the global challenge. Springer, Berlin Heidelberg, pp 15–18

Huston DL, Champion DC, Cassidy KF (2014) Tectonic controls on the endowment of Neoarchean cratons in volcanic-hosted massive sulfide deposits: evidence from lead and neodymium isotopes. Econ Geol 109:11–26

Huston DL, Champion DC, Mernagh TP, Downes PM, Jones P, Carr G, Forster D, David V (2016) Metallogenesis and geodynamics of the Lachlan Orogen: new (and old) insights from spatial and temporal variations in lead isotopes. Ore Geol Rev 76:257–267

Huston DL, Champion DC, Morrison G, Maas R, Thorne JP, Carr G, Beams S, Bottrill R, Chang Z-S, Dhnaram C, Downes PM, Forster DB, Gemmell JB, Lisitsin V, McNeill A, Vicary M (2017) Spatial variations in lead isotopes, Tasman Element, eastern Australia. Geoscience Australia Record 2017/09

Huston DL, Champion DC, Czarnota K, Hutchens M, Hoggard M, Ware B, Richards F, Tessalina S, Gibson GM, Carr G (2020) Lithospheric-scale controls on zinc-leas-silver deposits of the North Australian Zinc Belt: evidence from isotopic and geophysical data. Exploring for the Future: extended abstracts, Geoscience Australia, Canberra, https://doi.org/10.11636/135130

Jaffey AH, Flynn KF, Glendinin LE, Bentley WC, Essling AM (1971) Precision measurement of half-lives and specific activities of ^{235}U and ^{238}U. Phys Rev C 4:1889

John SG, Rouxel OJ, Craddock PR, Engwall AM, Boyle EA (2008) Zinc stable isotopes in seafloor hydrothermal vent fluids and chimneys. Earth Planet Sci Lett 269:17–28

Kistler RW, Peterman ZE (1978) Reconstruction of crustal blocks of California on the basis of initial strontium isotopic composition of Mesozoic granitic rocks. US Geol Surv Prof Paper 1071, 17

Kramers JD, Tolstikhin IN (1997) Two terrestrial lead isotope paradoxes, forward transport modelling, core formation and the history of the continental crust. Chem Geol 139:75–110

Lahtinen R, Korja A, Nironen M (2005) Paleoproterozoic tectonic evolution. In: Lehtinen M, Nurmi PA, Rämö OT (eds) Precambrian geology of Finland—key to the evoluton of the Fennoscandian Shield. Elsevier BV, Amsterdam, pp 481–532

Large RR, Bull SW, McGoldrick PJ, Walters W, Derrick GM, Carr GR (2005) Stratiform and strata-bound Zn–Pb–Ag deposits in Proterozoic sedimentary basins, northern Australia. Econ Geol 100th Anniv 931–963

Leach DL, Sangster DF, Kelley KD, Large RR, Garven G, Allen CR, Gutzmer J, Walters S (2005) Sediment-hosted Pb–Zn deposits: a global perspective. Econ Geol 100 Anniv 561–607

LeRoux LJ, Glendenin LL (1963) Half-life of thorium-232. In: Proceedings of national conference on nuclear energy, Pretoria, South Africa, p 83

Leybourne MI, Cousens BL, Goodfellow WD (2009) Lead isotopes in ground and surface waters: fingerprinting heavy metal sources in mineral exploration. Geochem Explor Env Anal 9:115–123

Li W, Jackson SE, Pearson NJ, Graham S (2010) Copper isotopic zonation in the Northparkes porphyry Cu–Au deposit, SE Australia. Geochim Cosmochim Acta 74:4078–4096

Li N, Chen YJ, Fletcher IR, Zeng QT (2011) Triassic mineralization with Cretaceous overprint in the Dahu Au–Mo deposit, Xiaoqinling gold province: constraints from SHRIMP monazite U–Th–Pb geochronology. Gondwana Res 20:543–552

Lobato LM, Figueiredo e Silva RC, Angerer T, Mendes, M, Hagemann S, Halversn GP (2023) Fe isotopes applied to BIF-hosted iron ore deposits. In: Huston DL, Gutzmer J (eds), Isotopes in economic geology, metallogensis and exploration, Springer, Berlin, this volume

Maltese A, Mezger K (2020) The Pb isotope evolution of Bulk Silicate Earth: constraints from its accretion and early differentiation history. Geochim Cosmochim Acta 271:179–193

Marcoux E (1998) Lead isotope systematics of the giant massive sulfide deposits in the Iberian Pyrite Belt. Mineral Deposita 33:45–58

Mathur R, Zhao Y (2023) Copper isotopes used in mineral exploration. In: Huston DL, Gutzmer J (eds) Isotopes in economic geology, metallogensis and exploration, Springer, Berlin, this volume

McDonough WF (2003) Compositional model for the earth's core. Treatise Geochem 2:547–568

McNaughton NJ, Groves DI, Witt WK (1993) The source of lead in Archaean lode gold deposits of the Menzies-Kalgoorlie-Kambalda region, Yilgarn Block, Western Australia. Mineral Deposita 28:495–502

Munhá J, Relvas JMRS, Barriga FJAS, Conceição P, Jorge RCGS, Mathur R, Ruiz J, Tassinari CCG (2005) Osmium isotope systematics in the Iberian Pyrite Belt. Mineral deposit research: meeting the global challenge. Springer, Berlin, pp 663–666

Ni ZY, Chen YJ, Li N, Zhang H (2012) Pb–Sr–Nd isotope constraints on the fluid source of the Dahu Au–Mo deposit in Qinling Orogen, central China, and implication for Triassic tectonic setting. Ore Geol Rev 46:60–67

Nier AO (1938) Variations in the relative abundances of common lead from various sources. J Am Chem Soc 60:1571–1576

Ohmoto H, Skinner BJ (eds) (1983) The Kuroko and related volcanogenic massive sulfide deposits. Econ Geol Mon 5

Oliveira JT, Pereira Z, Carvalho P, Pacheco N, Korn D (2004) Stratigraphy of the tectonically imbricated lithological succession of the Neves Corvo mine area, Iberian Pyrite Belt, Portugal. Mineral Deposita 39:422–436

Oversby VM (1975) Lead isotopic systematics and ages of Archaean acid intrusives in the Kalgoorlie-Norseman area. Western Australia: Geochim Cosmochim Acta 39:1107–1125

Palme H, O'Neill HSC (2003) Cosmochemical estimates of mantle composition. Treatise Geochem 2:1–38

Pettke T, Oberli F, Heinrich CA (2010) The magma and metal source of giant porphyry-type ore deposits, based on lead isotope microanalysis of individual fluid inclusions. Earth Planet Sci Lett 296:267–277

Plant JA, Whittaker A, Demetriades A, De Vuvi B, Lexa J (2005) The geological and tectonic framework of Europe. In: Salminen R (ed) FOREGS geochemical atlas of Europe, Part I: backgraound information, methodology and maps. Geological Survey of Finland, Espoo

Rehkämper M, Halliday AN (1998) Accuracy and long-term reproducibility of lead isotopic measurements by multiple- collector inductively coupled plasma mass spectrometry using an external method for correction of mass discrimination. Int J Mass Spectrom 181:123–133

Relvas JMRS, Tassinari CCG, Munhá J, Barriga FJAS (2001) Multiple sources for ore-forming fluids in the Neves-Corvo VHMS deposit of the Iberian pyrite belt (Portugal): strontium, neodymium and lead isotope evidence. Mineral Deposita 36:416–427

Richards JP (2009) Postsubduction porphyry Cu–Au and epithermal Au deposits: products of remelting of subduction-modified lithosphere. Geology 37:247–250

Richards JP (2015) Tectonic, magmatic, and metallogenic evolution of the Tethyan orogen: from subduction to collision. Ore Geol Rev 70:323–345

Robert F, Poulsen KH, Cassidy KF, Hodgson CJ (2005) Gold metallogeny of the Superior and Yilgarn Cratons. Econ Geol 100th Anniv 1001–1033

Rudnick RL, Gao S (2003) Composition of the continental crust. Treatise Geochem 3:1–64

Russell RD, Farquhar RM (1960) Lead isotopes in geology. Interscience, New York, p 243

Sharp ZD (1990) A laser-based microanalytical method for the in situ determination of oxygen isotope ratios of silicates and oxides. Geochim Cosmochim Acta 54:1353–1357

Sinha AK, Tilton GR (1973) Isotopic evolution of common lead. Geochim Cosmochim Acta 37:1823–1849

Stacey JS, Hedlund DC (1983) Lead-isotopic compositions of diverse igneous rocks and ore deposits from southwestern New Mexico and their implications for early Proterozoic crustal evolution in the western United States. Geol Soc Am Bull 94:43–57

Stacey JS, Kramers JD (1975) Approximation of terrestrial lead isotope evolution by a two stage model. Earth Planet Sci Lett 26:207–221

Stacey JS, Moore WJ, Rubright RD (1967) Precision measurements of lead isotope ratios: preliminary analyses from the U.S. mine, Bingham Canyon. Utah. Earth Planet Sci Lett 2:489–499

Stacey ZRE, Nkomo IT (1968) A lead isotope study of galenas and selected feldspars from mining districts in Utah. Econ Geol 63:796–814

Ströbele F, Staude S, Pfaff K, Premo WR, Hildebrandt LH, Baumann A, Pernicka E, Markl G (2012) Pb isotope constraints on fluid flow and mineralization processes in SW Germany. N Jb Miner Abh 189:287–309

Sun S-S, Carr GR, Page RW (1996) A continued effort to improve lead-isotope model ages. AGSO Res Newsl 24:19–20

Tatsumoto M, Knight RJ, Allégre CJ (1973) Time differences in the formation of meteorites as determined from the ratios of lead-207 to lead-206. Science 180:1279

Tessalina SG, Herrington RJ, Taylor RN, Sundblad K, Maslennikov VV, Orgeval JJ (2016) Lead isotopic systematics of massive sulphide deposits in the Urals: applications for geodynamic setting and metal sources. Ore Geol Rev 72:22–36

Thirlwall MF (2000) Inter-laboratory and other errors in Pb isotope analyses investigated using a $^{207}Pb–^{204}Pb$ double spike. Chem Geol 163:299–322

Thorpe RI (1999) The Pb isotope linear array for volcanogenic massive sulfides deposits of the Abitibi and Wawa Subprovinces, Canadian Shield. Econ Geol Mon 10:555–576

Thorpe R (2008) Release of lead isotope data in 4 databases: Canadian, western Superior, foreign, and whole rock and feldspar. Geol Survey Canada Open File 5664

Thorpe RI, Hickman AH, Davis DW, Mortensen JK, Trendall AF (1992) Constraints to models for Archean lead evolution from precise zircon U–Pb geochronology for the Marble Bar Region, Pilbara Craton, Western Australia. In: Glover JE, Ho SE (eds) The Archaean; terrains, processes and metallogeny; proceedings volume for the Third International Archaean Symposium. Geology Department (Key Centre) & University Extension, The University of Western Australia Publication vol 22, pp 395–407

Tilton GR (1983) Evolution of the depleted mantle: the lead perspective. Geochim Cosmochim Acta 47:1191–1197

Todt W, Cliff RA, Hanser A, Hofmann AW (1993) Re-calibration of NBS lead standards using a ^{202}Pb + ^{205}Pb double spike. Terra Nova 5, 396, Terra Abstracts supplement 1

Tosdal RM, Wooden JL, Bouse RM (1999) Pb isotopes, ore deposits and metallogenic terranes. Rev Econ Geol 12:1–28

Troll V, Weis F, Jonsson E, Andersson U, Majidi S, Hogdahl K, Harris C, Millet M-A, Chinnasamy S, Kooijman E, Nilsson K (2019) Global Fe–O isotope correlation reveals magmatic origin of Kiruna-type apatite-iron-oxide ores. Nature Commun 10:1712. https://doi.org/10.1038/s41467-019-09244-4

Tuttas D, Habfast K (1982) High precision lead isotope ratio measurements. Finnigan MAT Appl Rep 51

Vaasjoki M, Gulson BL (1986) Carbonate-hosted base metal deposits; lead isotope data bearing on their genesis and exploration. Econ Geol 81:156–172

Vitrac AM, Albarede F, Allègre CJ (1981) Lead isotopic composition of Hercynian granitic K-feldspars constrains continental genesis. Nature 291:460–464

Wang Q, Schiøtte L, Campbell IH (1998) Geochronology of supracrustal rocks from the Golden Grove area, Murchison Province, Yilgarn Craton, Western Australia. Aust J Earth Sci 45:571–577

Waltenberg K (2023) Application of Lu–Hf isotopes to ore geology, metallogenesis and exploration. In: Huston DL, Gutzmer J (eds) Isotopes in economic geology, metallogensis and exploration, Springer, Berlin, this volume

Webb AW, McDougall I (1968) The geochronology of the igneous rocks of eastern Queensland. J Geol Soc Aust 15:313–346

Wedepohl KH, Hartmann G (1994) The composition of the primitive upper earth's mantle. In: Meyer HOA, Leonardos OH (eds) Kimberlites, related rocks and mantle xenoliths. Companhia de Pesquisa de Recursos Minerais, Rio de Janeiro, pp 486–495

Wilkinson JJ (2023) The potential of Zn isotopes in the science and exploration of ore deposits. In: Huston DL, Gutzmer J (eds) Isotopes in economic geology, metallogensis and exploration, Springer, Berlin, this volume

Wooden JL, Aleinikoff JN (1987) Pb isotopic evidence for Early Proterozoic crustal evolution in the southwestern US. Geol Soc Am Abstr Programs 19(7):897

Wooden JL, DeWitt E (1991) Pb isotopic evidence for the boundary between the early Proterozoic Mojave and central Arizona crustal provinces in western Arizona. Arizona Geol Soc Dig 19:27–50

Wooden JL, Miller DM (1990) Chronologic and isotopic framework for Early Proterozoic crustal evolution in the eastern Mojave Desert region, SE California. J Geophys Res: Solid Earth 95(B12):20133–20146

Woodhead JD (2002) A simple method for obtaining highly accurate Pb isotope data by MC-ICP-MS. J Anal at Spectrom 17:1381–1385

Woodhead JD, Volker F, McCulloch MT (1995) Routine lead isotope determination using a lead-207-lead-204 double spike: a long-term assessment of analytical precision and accuracy. Analyst 120:35–39

Zametzer A, Kirkland CL, Hartnady MIH, Barham M, Champion DC, Bodorkos S, Smithies RH, Johnson SP (2022) Applications of Pb isotopes in granite K-feldspar and Pb evolution in the Yilgarn Craton. Geochim Cosmochim Acta 320:279–303. https://doi.org/10.1016/j.gca.2021.11.029

Zartman RE (1974) Lead isotope provinces in the Cordilleran of the western United States and their geological significance. Econ Geol 69:792–805

Zartman RE, Doe BR (1981) Plumbotectonics—the model. Tectonophysics 75:135–162

Zartman RE, Haines SM (1988) The plumbotectonic model for Pb isotopic systematics among major terrestrial reservoirs—a case for bi-directional transport. Geochim Cosmochim Acta 52:1327–1339

Zartman RE, Stacey JS (1971) Lead isotopes and mineralization ages in Belt Supergroup rocks, northwestern Montana and northern Idaho. Econ Geol 66:849–860

Application of the Lu–Hf Isotopic System to Ore Geology, Metallogenesis and Mineral Exploration

Kathryn Waltenberg

Abstract

The Lu-Hf isotopic system, much like the Sm-Nd isotopic system, can be used to understand crustal evolution and growth. Crustal differentiation processes yield reservoirs with differing initial Lu/Hf values, and radioactive decay of ^{176}Lu results in diverging ^{176}Hf/^{177}Hf between reservoirs over time. This chapter outlines the fundamentals of the Lu-Hf isotopic system, and provides several case studies outlining the utility of this system to mineral exploration and understanding formation processes of ore deposits. The current, rapid, evolution of this field of isotope science means that breadth of applications of the Lu-Hf system are increasing, especially in situations where high-precision, detailed analyses are required.

1 Introduction

The primary application of the Lu-Hf isotopic system is to trace crustal evolution and growth through time. Its direct applications to regional mineral systems include, but are not restricted to, mineralisation styles directly related to crustal growth via addition of juvenile mantle material rich in compatible elements (e.g. most Ni and platinum-group element deposits, and kimberlites), and those developed at large-scale tectonic interfaces between isotopically disparate crustal blocks (e.g. porphyry Cu-Au and orogenic Au deposits). The Lu-Hf system can be used to isotopically map and identify prospective crustal blocks and boundaries, to provide a combined geochemical-geophysical view of the lithosphere, and to act as a 'paleo-geophysical' tool (Hartnady et al. 2018) to reconstruct crustal and lithospheric changes through time. When integrated with geological, geochemical, structural and geophysical datasets, the Lu-Hf isotope system can be a powerful tool for refining mineral systems modelling (McCuaig et al. 2010) and narrowing the search space for a range of mineralisation styles.

The Lu-Hf isotopic system can be used to track crustal growth and evolution through time in a similar fashion to the Sm-Nd isotopic system (Champion and Huston 2023). One of the most important advantages of the Lu-Hf system is the ability to measure Lu-Hf in individual zircons. This is possible because the combination of high Hf (1–2 wt%) and low Lu (ppm-ppb) in zircon ensures that the preserved (present-day) Hf isotopic composition is close to initial (crystallisation) values; it requires little correction for ingrowth of radiogenic (post-crystallisation) Hf. It also means that Lu-Hf measurements can be securely linked to U-Pb measurements directly

K. Waltenberg (✉)
Geoscience Australia, GPO Box 378, Canberra, ACT 2601, Australia
e-mail: Kathryn.Waltenberg@ga.gov.au

D. Huston and J. Gutzmer (eds.), *Isotopes in Economic Geology, Metallogenesis and Exploration*, Mineral Resource Reviews, https://doi.org/10.1007/978-3-031-27897-6_7

constraining the age of the zircon. Measuring Lu-Hf on zircon also allows direct integration with other zircon measurements beyond U-Pb—such as oxygen isotopes and trace elements—allowing zircon to be 'fingerprinted' in multiple systems.

In this chapter, the fundamentals of the Lu-Hf system are outlined to provide the basis for application to exploration and metallogenic investigations. This includes better integration into exploration programs, whether by designing new data collection campaigns and their interpretation, or evaluating the quality and impact of existing datasets. Lutetium-Hf isotope data are useful in early planning stages to identify the search space for province selection and, depending on the commodity of interest and geological setting, may also be valuable at the district scale.

2 The Lutetium-Hafnium Isotopic and Geochemical System

The Lu-Hf isotopic system is based on the beta decay of ^{176}Lu to ^{176}Hf, with a half-life of about 37 billion years (Söderlund et al. 2004). Lutetium and Hf are fractionated differently during crust formation processes (Vervoort 2014). During mantle melting, Hf is more incompatible than Lu, so fractionation processes enrich the crust in Hf relative to Lu, and different crustal reservoirs (e.g., continental vs oceanic) take on different Lu-Hf isotopic characteristics. As ^{176}Lu decays, reservoirs with higher Lu/Hf will have a correspondingly high radiogenic ^{176}Hf abundance relative to the non-radiogenic and stable ^{177}Hf isotope (Eq. 1; Fig. 1).

There is widespread agreement amongst the Hf-isotope community on a value of 1.867×10^{-11} yr^{-1} for the ^{176}Lu decay constant (Scherer et al. 2001; Söderlund et al. 2004; Thrane et al. 2010). However, older published literature may use an alternate decay constant (e.g., Sguigna et al. 1982; Bizzarro et al. 2003). When compiling data from different sources, each author's choice of decay constant should be confirmed. This is particularly important in older terranes, or for high-Lu samples, where more ^{176}Hf will have accumulated from ^{176}Lu decay.

The specific equation for the evolution of the Hf isotope composition is:

$$\left(\frac{^{176}Hf}{^{177}Hf}\right)_{now} = \left(\frac{^{176}Hf}{^{177}Hf}\right)_0 + \frac{^{176}Lu}{^{177}Hf}\left(e^{\lambda t}+1\right) \quad (1)$$

where ^{176}Hf/^{177}Hf and ^{176}Lu/^{177}Hf are the present-day (measured) values, (^{176}Hf/^{177}Hf)$_0$ is the initial Hf isotope composition at time t in the past, and λ is the decay constant of ^{176}Lu.

Crustal differentiation processes yield reservoirs with differing initial Lu/Hf values, and the decay of ^{176}Lu will result in diverging ^{176}Hf/^{177}Hf between reservoirs over time. However, the long half-life of ^{176}Lu means that isotopic variations between reservoirs are generally very small, and distinguishing between them requires high analytical precision and accuracy.

2.1 Epsilon Hf (εHf)

The isotope ^{176}Hf continues to accumulate in all reservoirs over time, so the same ^{176}Hf/^{177}Hf value can have very different interpretations at different geologic times and in different geologic settings. The εHf parameter normalizes ^{176}Hf/^{177}Hf values against that of the chondritic uniform reservoir (CHUR), which serves as an estimate of the bulk Earth ^{176}Hf/^{177}Hf. This system is easier to work with than the isotopic ratios, because differences in ^{176}Hf/^{177}Hf are geologically significant at the fourth decimal place. The equation for calculating εHf at a time (t) is:

$$\varepsilon \mathrm{Hf_t} = 10^4 \times \left[\left(^{176}\mathrm{Hf}/^{177}\mathrm{Hf}\right)_{\mathrm{sample,t}} / \left(^{176}\mathrm{Hf}/^{177}\mathrm{Hf}\right)_{\mathrm{CHUR,t}} - 1\right] \quad (2)$$

2.2 Model Ages

As with the Sm-Nd system, model ages of crustal reservoirs can be calculated from Lu-Hf data in

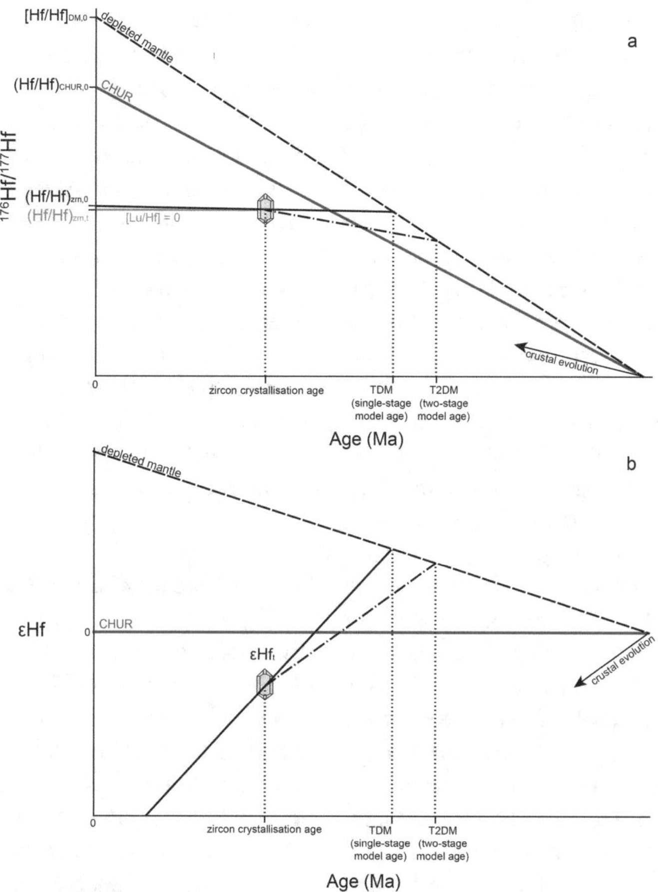

Fig. 1 (a) $^{176}Hf/^{177}Hf$ and (b) epsilon-Hf (εHf) plots showing main principles of the Lu-Hf isotope system as applied to measurements in zircon. To calculate the $^{176}Hf/^{177}Hf$ (a) or εHf (b) at the time of crystallisation (εHf_t), the measured $^{176}Hf/^{177}Hf$ of a zircon $(Hf/Hf)_{zrn,0}$ is corrected for the ingrowth of radiogenic ^{176}Hf since crystallisation based on the $^{176}Lu/^{177}Hf$ in the zircon. To calculate a single-stage model age (TDM), this trend is extrapolated until it intercepts the depleted mantle (DM) curve. A two-stage model age (T2DM) uses an estimate for the $^{176}Lu/^{177}Hf$ composition of the source to extrapolate from the crystallisation event until the depleted mantle intercept. CHUR, an estimate of bulk earth composition continues to accumulate ^{176}Hf as a function of ^{176}Lu abundance. On an εHf plot (b), CHUR is defined as zero throughout Earth history. $[Hf/Hf]_{CHUR,0}$ is the $^{176}Hf/^{177}Hf$ of CHUR at present day. $[Hf/Hf]_{DM,0}$ is the $^{176}Hf/^{177}Hf$ of the depleted mantle at present day (t=0).

order to estimate the time since extraction from the mantle (i.e. crust formation events). There are several assumptions that must be made to deduce these model ages, and so it is important not to confuse them with accurate, precise determinations of geologically-meaningful rock-forming event ages such as igneous crystallisation U-Pb or Ar/Ar ages. Model ages are estimates of average mantle extraction age, and do not directly date geologic events, so they should primarily be used as petrogenetic tools for comparing rock suites (Spencer et al. 2020).

The simplest type of model age is a single-stage model age, which assumes the sample was extracted directly from the mantle and has evolved as a closed system ever since, and that the Lu/Hf value of the analysed material reflects bulk-rock values. This is calculated by using the measured $^{176}Lu/^{177}Hf$ and $^{176}Hf/^{177}Hf$ to track backwards in time until the depleted mantle model curve is intercepted (Fig. 1). In this case, it is not necessary to know the crystallisation age of the rock. There are significant problems with this method, particularly when measuring Lu-Hf in zircon because zircon fractionates Lu and Hf extremely strongly, leading to a Lu/Hf in zircon that can be orders of magnitude lower than the protolith from which the zircon formed. The result is that Hf single-stage model ages calculated from zircons are always minimum estimates of the model age of extraction of associated crust from the mantle.

A better approximation of the time of extraction from the mantle (crustal residence age) is derived from a two-stage model age, in which the measured isotopic ratios are traced back in time the same way as the single-stage model age to the time of zircon crystallization, at which point an estimated Lu/Hf for the protolith or source region is used to calculate the intersection point with the depleted mantle curve (Fig. 1). This gives a more reasonable model age estimate, particularly when using measured values from zircons, but it does also entail more assumptions.

Firstly, the age of the rock (or analysed mineral grain) must be well-constrained. This is generally straightforward for zircon-bearing rocks, as the U-Pb age is usually determined on the same zircons, either beforehand or simultaneously with the Lu-Hf isotopic measurements.

Secondly, it is necessary to assign an initial Lu/Hf value to the precursor source material(s). Given that these sources are commonly destroyed or substantially modified during the crustal differentiation process responsible for forming the existing rock or minerals, it can be difficult to appropriately characterise their Lu/Hf, particularly when these sources are mixed, or at depth. Bulk ^{176}Lu/^{177}Hf is most often assigned the value of 0.015 (Griffin et al. 2002), estimated from average continental crust, though other researchers use other values based on other evidence or models for a different source, such as mafic or felsic crust (Amelin et al. 1999; Kemp et al. 2006; Vervoort and Kemp 2016).

Thirdly, there are multiple models for depleted mantle composition through time, and other suggested source regions, such as that similar to island arc crust (Dhuime et al. 2011; Vervoort and Kemp 2016).

Because of the large differences between different methods used to estimate model mantle evolution curves, the reader should be careful to understand which method the author has reported and why, particularly when compiling data from multiple sources. Figure 1 demonstrates the potential differences between the two types of model ages. The two-stage model age is so strongly preferred in recent literature that the ascribed notation for this model age is often TDM which can easily be confused with the single-stage model age notation.

2.3 Hf Isotope Evolution Models

The isotopic composition of Earth reservoirs evolves as more ^{176}Hf is accumulated through decay of ^{176}Lu (Fig. 1). This means that material extracted from the mantle reservoir at different times has different isotopic compositions. It is these differences which form the basis of the Lu-Hf system for characterising the nature of the crust and mantle lithosphere.

When a melt from a depleted-mantle source is extracted and emplaced at shallower levels, this material has an isotopic composition that lies on the depleted mantle curve, and is considered 'radiogenic' due to its high proportion of radiogenic ^{176}Hf relative to non-radiogenic ^{177}Hf. Some authors also describe these compositions as 'juvenile'. In contrast, where melt is extracted from crustally-derived material, the corresponding terminology is 'unradiogenic' and 'evolved'.

Figure 2 shows a synthesis of various ways that geological processes influence ^{176}Hf/^{177}Hf and eHf. Crust-mantle differentiation begins in the early Earth, marked on the figure by the initial divergence of the CHUR and depleted mantle curves. At position t_1, partial melting in the depleted mantle generates new crust. If there is no more mantle-derived material added to the crust after this point (i.e., crust is stable or internally reworked), the extracted crustal

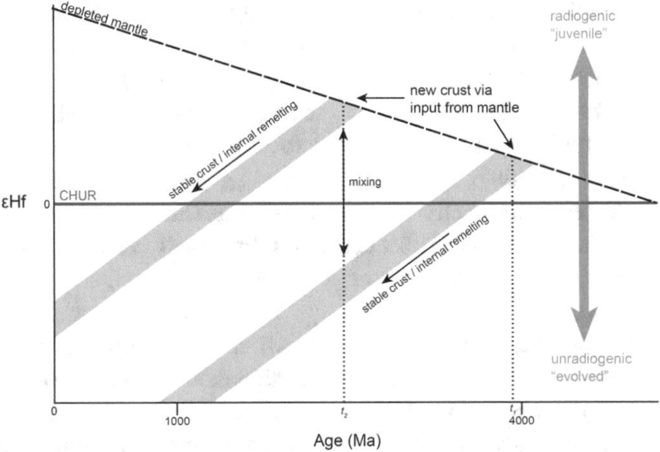

Fig. 2 Epsilon-Hf (εHf) plot showing generalised geological processes that influence Hf-isotope signatures. Material derived directly from a depleted mantle source will fall along the depleted mantle curve at the time of extraction. As this material resides in the crust, its Hf composition will become unradiogenic relative to depleted mantle, and remelting of these crustal sources will have an εHf signature that lies along the same trajectory indicated by shaded grey bars. Mixing between multiple sources extracted from the mantle at different times leads to intermediate εHf between the various sources, with exact signature a complex function of the proportions and Hf content of the disparate sources.

material simply accumulates ^{176}Hf proportional to its ^{176}Lu abundance. If there is a second partial melting event in the depleted mantle at t_2, any mixing between crust from the t_1 and t_2 events will have εHf intermediate between the two isotopic compositions.

Confidence in interpreting these types of events stems from the fact that the Lu-Hf system is robust, particularly in zircon (Cherniak and Watson 2003). This is because zircons themselves are generally robust: hard, refractory and resistant to alteration, thus able to remain intact during sedimentary, metamorphic and even some igneous processes. By comparison, the bulk-rock Lu-Hf isotope system is more susceptible to weathering- or alteration-related isotopic disturbance.

Not all influences on εHf come from geological sources. Analytical and data reduction artifacts can introduce apparent trends which do not reflect a geological process. Vervoort and Kemp (2016) and Spencer et al. (2020) provide a detailed synopses of many of the potential pitfalls and issues in collecting Lu-Hf isotopic data, as well as advice on analytical and interpretive approaches to avoid these problems.

3 Analytical Methods

The Lu-Hf isotopic system reflects similar geological processes to the Sm-Nd system, but has historically been more difficult to measure (Vervoort 2014). Bulk-rock Lu-Hf systematics can be interpreted in largely the same way as Sm-Nd systematics (Kinny and Maas 2003; Vervoort 2014).

The low abundance of Lu and Hf in Earth materials, the very small differences in isotopic ratios that mark geologically significant variation, and the large isobaric interferences on the isotopes of interest, pose difficulties for precise isotopic measurements (Halliday et al. 1995). Thermal ionisation mass spectrometry (TIMS) was the first technique used to collect high-precision bulk-rock Lu-Hf data (Patchett and Tatsumoto 1980; Patchett et al. 1982; Corfu and Noble 1992), but the high ionisation potential of Hf makes these measurements difficult (Vervoort 2014). The high Hf content of zircon makes analysis of mineral separates easier than whole-rock analysis (Kinny and Maas 2003). As inductively-coupled plasma mass spectrometry

(ICP-MS), and in particular multi-collector (MC) capability (Walder and Freedman 1992; Halliday et al. 1995; Blichert-Toft and Albarède 1997), and laser ablation (LA) functionality (Thirlwall and Walder 1995; Woodhead et al. 2004) has become more accessible, use of the Lu-Hf system has increased dramatically.

Lu-Hf data are now most routinely collected by laser ablation microanalysis (Kinny and Maas 2003), which allows close links to the age information necessary for appropriately determining time-corrected parameters (Amelin et al. 1999, 2001; Gerdes and Zeh 2006). Microanalysis also permits greater understanding of how multiple components in a single sample contribute to the isotopic signature, and therefore the history of the sample (Woodhead et al. 2004; Mole et al. 2014; Kirkland et al. 2015). Data collection is very rapid, however sample preparation involves mineral separation and preparing grain mounts in addition to the crushing steps of whole-rock analysis, and requires further mineral characterisation techniques such as cathodoluminescence imaging (Kinny and Maas 2003).

LA-MC-ICP-MS analysis has enabled rapid (on the order of 1–2 min per analysis; Thirlwall and Walder 1995; Woodhead et al. 2004) collection of data from individual minerals while sampling very small volumes of material (usually below 100 μm diameter ablation pit and excavating 0.5–1.0 μm per second; Kinny and Maas 2003). By targeting individual mineral grains, it is possible to gain extra insights from 'unmixing' different components of a rock. Several of the case studies presented in this chapter take advantage of this ability.

Laser ablation analyses are increasingly benefiting from split-stream technology, whereby material ablated from the target mineral can be 'split', and directed to multiple mass-spectrometer systems for simultaneous measurement. This permits truly synchronous acquisition of Lu-Hf using a multicollector mass spectrometer while using a second mass spectrometer to measure U-Pb, and trace elements (Woodhead et al. 2004; Yuan et al. 2008). This is beneficial not only because of the increased richness of data but also allows real-time assessment of Lu-Hf analysis quality. Oxygen isotopic analysis is also commonly paired with Hf and U-Pb analysis, and provides information on alteration and crustal recycling processes (Valley 2003; Kemp et al. 2007; Harrison et al. 2008). These oxygen analyses are usually done by ion-probe before laser analysis.

The technical aspects of Lu-Hf data collection (such as sensitivity improvements, down-hole fractionation and resolution of triple-interferences at mass 176) have been thoroughly reviewed by other authors (e.g., Kinny and Maas 2003; Fisher et al. 2014; Vervoort and Kemp 2016; Spencer et al. 2020). These reviews provide useful discussion of the power and pitfalls of Lu-Hf analyses, and also provide further criteria for determining data quality and for aiding interpretation of a diverse range of data trends—not all of which are geologically meaningful. Other papers cover the specifics of Hf analysis of zircons by LA-ICP-MS (Amelin et al. 2000; Woodhead et al. 2004; Xie et al. 2008; Yuan et al. 2008; Gerdes and Zeh 2009).

4 Geological Applications of the Lu-Hf Isotope System

Analysis of magmatic rocks or minerals can be used to characterise crustal blocks through time (Mole et al. 2014; Cross et al. 2018; Waltenberg et al. 2018) and to identify subsurface source regions (Flowerdew et al. 2009; Dolgopolova et al. 2013). The meaning and origin of geophysical and paleothermal anomalies can also be investigated (Hartnady et al. 2018; Siegel et al. 2018; Waltenberg et al. 2018). The Lu-Hf isotope system can be used to understand the origins of ore-bearing fluids in hydrothermal systems (Westhues et al. 2017). Further, Lu-Hf isotopic analysis can determine the metal transport pathways, and mechanisms of mineralisation caused by magmatic processes (Murgulov et al. 2008; Mole et al. 2014; Kirkland et al. 2015) and refine mineral systems models (Hou et al. 2015; Kirkland et al. 2015).

Analysis of metamorphic zircons can reveal information about the source of fluids involved in the metamorphic process. If the metamorphic domains have a different Hf signature, this may indicate an exotic source for metamorphic fluids, including the possibility of mantle-derived fluids. If the signature is the same as earlier magmatic domains, the source of metamorphic material is more likely to be proximal (e.g. Wu et al. 2009; Kirkland et al. 2015).

One of the strengths of the Lu-Hf system as applied to mineral separates is the ability to retrieve fossil isotopic information even when characteristics of the original rock have been lost to weathering and erosion. Lutetium-Hf studies of detrital minerals (e.g., zircon) allow investigation of the sources of detritus and characterized minerals can be used as indicators to detect possible nearby intrusions, for example in zircons derived from kimberlites in nearby creek catchments (Griffin et al. 2000; Belousova et al. 2001; Batumike et al. 2009). Analysis of detrital material in a catchment can be used to survey the entire provenance spectra, and narrow down targeting of regions. Isotopic 'fingerprints' can be used to match sediments with a library of magmatic data to determine provenance, and correlate sedimentary basins (Doe et al. 2013). This can be expanded by also collecting O-isotopes and split-stream U-Pb and trace elements on the same zircons to characterise sources in more detail (e.g., Belousova et al. 2002; Griffin et al. 2006; Lu et al. 2016; Purdy et al. 2016).

5 Target Materials and Minerals

The dominant method of Hf-isotope determination is analysis of zircons by MC-ICPMS methods. Zircon substitutes Hf in the Zr site in its structure due to identical charge and similar ionic radius between the two elements—most terrestrial zircons have 0.5–2.0 wt% HfO_2 (Speer and Cooper 1982), making zircon the primary host of Hf in most rocks (Kinny and Maas 2003). Zircon also excludes Lu, making it ideal for Lu-Hf isotopic studies (Amelin et al. 1999). Because it incorporates Hf so effectively, the original $^{176}Hf/^{177}Hf$ at the time of zircon formation is well-preserved within zircon crystals, and only a small correction is required for ingrown ^{176}Hf (Kinny and Maas 2003). This means that zircons preserve 'fossil' isotopic markers and are used to determine Hf isotope composition at the time of zircon growth. Zircon is informative for characterising a range of rock-forming processes including magmatic, metamorphic and detrital processes (Kinny and Maas 2003; Vervoort et al. 1996). Individual zircon grains are commonly complexly zoned and can robustly retain inherited cores and magmatic and metamorphic overgrowths, which can provide constraints on the growth of the zircon, and host rocks—one of the primary reasons why zircon is such a popular mineral for both geochronology and isotopic studies.

Rutile LA-ICPMS analytical methods have been developed (Luvizotto et al. 2009; Ewing et al. 2011) for igneous, metamorphic and detrital applications (e.g., Choukroun et al. 2005; Zack et al. 2011; Ewing et al. 2014). Ewing et al. (2011) documented an example of zircon and rutile Hf data from the same rock recording information on different parts of metamorphic P-T-t history and suggested that these two minerals give complementary information about metamorphism. Ewing et al. (2014) demonstrated that rutile preserves $^{176}Hf/^{177}Hf$ even through dissolution-reprecipitation events during metamorphism, provided rutile is not replaced by another mineral such as ilmenite. These authors proposed that rutile may be especially useful in UHT metamorphic rocks where metamorphic zircon is limited, and in other zircon-poor lithologies.

Likewise, baddeleyite analysis by LA-MC-ICPMS methods have been developed (Xie et al. 2008; D'Abzac et al. 2016) and these techniques have been applied to sediment-hosted baddeleyite grains (e.g., Schärer et al. 1997; Bodet and Schärer 2000), and igneous applications (Söderlund et al. 2005, 2006). Like zircon, baddeleyite contains very little Lu in its structure (Söderlund et al. 2005), so the correction for ^{176}Hf ingrowth since crystallisation is similarly small.

Baddeleyite rarely preserves inheritance and populations in magmatic rocks are predominantly cogenetic, so there may be benefit in analysing baddeleyite rather than zircon in silica-undersaturated samples with complex histories to retrieve unambiguously magmatic Lu-Hf signatures from the rock.

This chapter is focused on the use of Lu-Hf as an isotopic tracer, however minerals that incorporate the parent (^{176}Lu) isotope can be used as geochronometers (e.g., apatite, garnet, lawsonite; Vervoort 2014). Vervoort (2014) present an overview of the geochronological applications of the Lu-Hf system.

6 Application of the Lu-Hf Isotope System to Mineral Deposit Research and Exploration

The nature and configuration of crustal blocks is critically important to understand where and when mineral deposits form. The primary value of Lu-Hf isotopic data in mineral exploration is currently in focussing the search space, and guide the extension of exploration out from regions of known mineralisation. The structures of the deep crust and lithosphere are major controls on Lu-Hf isotopic signatures, and these are also strong controls on the occurrence of many mineral deposits. Geophysical techniques can be used to image these sub-surface regions in their present-day configuration, but an isotopic approach means that it is possible to map crustal blocks and lithospheric architecture *through time* (Champion and Huston 2016, 2023).

Several studies have demonstrated a strong link between isotopic signatures and lithospheric thickness in young terranes (e.g., Nash et al. 2006; Zhu et al. 2011; Yang et al. 2014; Hou et al. 2015) but isotopic information is especially valuable in older regions (e.g., Australia, Canada), where the environments and crustal configurations prevailing during mineral deposit formation differ significantly from those of the modern Earth. Isotope systems such as Sm-Nd and Lu-Hf can be used as paleo-geophysical tools (Hartnady et al. 2018), to reconstruct the Earth's dynamic history and place mineralisation in a contemporaneous context. The isotopic signature of pre-existing materials is preserved even though the pre-existing material may have been lost or is inaccessible.

Direct application of the Lu-Hf system to deposit-scale problems is an emerging field and there are not many publicly available case studies. However, larger-scale studies have value to explorers, in much the same way as regional geophysical datasets provide context. Regional studies are useful because they provide 'baseline' data against which camp-scale results can be compared. For example, identification of unusually radiogenic εHf in a region of non-radiogenic, old crust may indicate mantle-derived input in that area, and thus increased potential for particular types of mineralisation. Investigating temporal variations in isotopic character may indicate periods of mantle input, and enable the explorer to narrow the time window of interest and constrain timing of mineralisation. The key is to keep in mind the mineral systems approach (McCuaig et al. 2010), what components are required for the mineral deposit style of interest, and how these might be expressed isotopically.

The first three of the following case studies demonstrate some applications of the Lu-Hf isotope system in understanding metallogenesis, and emphasise the close links between Lu-Hf isotopes and lithospheric thicknesses and boundaries, both of which strongly influence mineral occurrence and endowment. The fourth case study demonstrates the power of the Lu-Hf isotope system applied to detrital minerals to act as pathfinders for kimberlite-hosted mineral deposits.

6.1 Lithospheric Controls on Mineralisation Style in the Lhasa Terrane

The Lu-Hf isotope system assists in understanding regional trends in lithospheric character and the resulting mineral potential for magmatic-associated systems. Hou et al. (2015) compiled new and published LA-ICP-MS zircon Hf and other geochemical data to investigate why magmatic-hydrothermal ore deposits occur in specific tectonic environments. These authors produced Hf model-age maps for the Lhasa Terrane of the Himalayan–Tibetan Orogen which demonstrate a correlation between the occurrence of several styles of mineralisation formed during the Jurassic-Miocene, and the Hf-signature of mineralisation-related igneous rocks.

The Hf model-age maps vary between the three distinct E-W trending crustal blocks—a central Proterozoic microcontinent with bounding Phanerozoic blocks to the north and south (Fig. 3). The Hf data also suggested the existence of two concealed N-S trending lithospheric-scale faults that cross-cut all three blocks. The authors argued that the configuration of these blocks and faults exert a first-order control on the formation of various magmatic-hydrothermal mineral deposits across the Lhasa Terrane.

Hou et al. (2015) interpreted that many of the deposit types in the region form in proximity to terrane margins and lithosphere-scale faults (as expressed by regions with the highest Hf-isotope gradients) because these regions of structural weakness promote transport of mantle-derived material to provide the metalliferous and heat/energy inputs needed to form ore deposits. Mantle-derived magma ascending through lithospheric faults sourced metal for skarn Fe-Cu ore deposits, whereas away from terrane boundaries the lack of structural conduits limits transport of mantle-derived material—so the potential for skarn Fe-Cu deposits is low.

Where mineral deposits occur away from boundaries and major faults, the authors also demonstrated that the major mineralisation style varies based on the age and character of the crust. Porphyry Cu systems are associated with relatively young crust (T2DM < 1200 Ma), but Mo-dominated porphyry systems occur in older crust (T2DM > 1200 Ma). Fe-Cu skarns are associated with young crust, however Fe-only skarns are more strongly correlated with older crust.

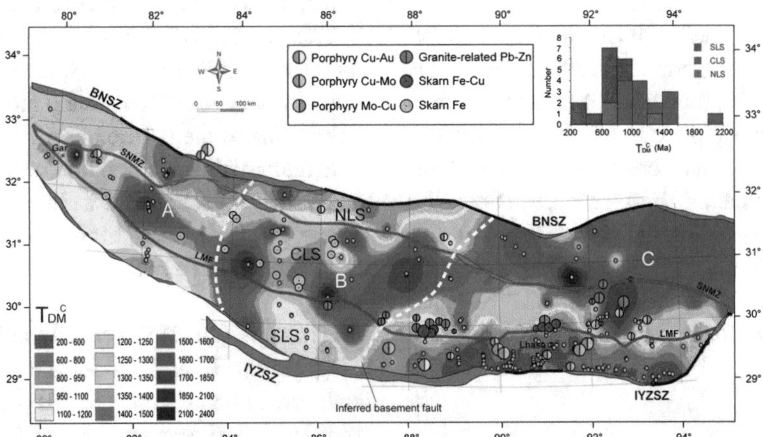

Fig. 3 Two-stage model-age map of the Lhasa terrane. Reproduced with permission from Hou et al. (2015); Copyright 2015 Society of Economic Geologists. The maps show the close spatial relationship between ore deposits and sub-terranes and boundaries mapped out by Lu-Hf isotopes. A central Proterozoic microcontinent (blue) is bounded to the north and south by two Phanerozoic blocks (red-yellow), and the Hf data (n=4762 analyses) are also suggestive of the existence of two concealed N-S trending lithospheric-scale faults (white dashed lines) crosscutting the blocks. The location of mineral deposits is closely linked to blocks of specific isotopic character, or to isotopic boundaries that may indicate regions of lithospheric weakness. A, B and C denote the author's delineations of three domains, based on spatial differences in Hf model age. NLS = northern Lhasa subterrane, CLS = central Lhasa subterrane, SLS = southern Lhasa subterrane, BNSZ = Bangong-Nujiang suture zone, IYZSZ = Indus-Yarlung-Tsangpo suture zone, SNMZ = Shiquan River-Nam Tso Mélange Zone, LMF = Luobadui-Milashan Fault. Red lines denote Miocene normal fault systems; white dashed lines denote inferred basement faults.

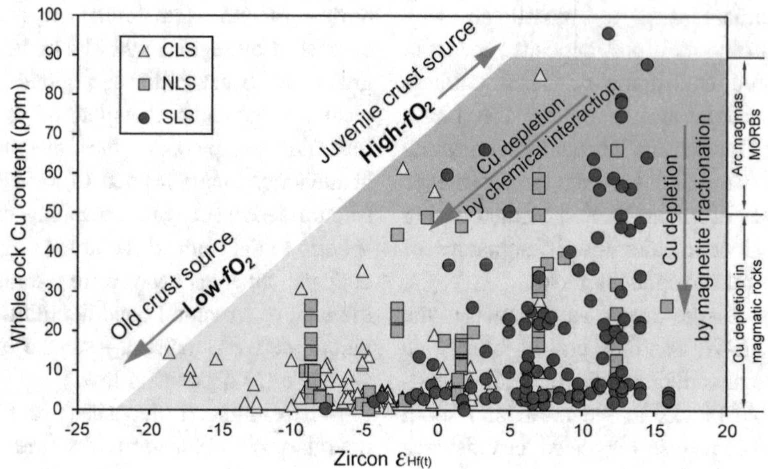

Fig. 4 Whole-rock copper content vs zircon εHf for igneous rocks of the Lhasa Terrane. Reproduced with permission from Hou et al. (2015); Copyright 2015 Society of Economic Geologists. Juvenile rocks tend to contain the highest Cu-content, demonstrating a link between positive εHf values and increased copper mineralisation potential.

Granite-related Pb-Zn deposits are associated with older crustal blocks. In addition to the spatial correlation, the authors also demonstrated that many rocks in the region with high zircon εHf are rich in Cu (Fig. 4).

Paired Lu-Hf and U-Pb age data from zircons enables further temporal constraints to be applied to tectonic reconstructions and crustal evolution models, including periods of crustal reworking and mantle input into crust. Integration of this isotopic information into reconstructions has enabled better understanding of the processes that formed the Lhasa Terrane (Zhu et al. 2011; Hou et al. 2015) and the development of a geodynamic model for its mineralisation (Hou et al. 2015).

6.2 Cautions and Considerations

Hou et al. (2015) showed that the Lu-Hf isotope system can be used as a mapping tool to identify crustal blocks of different character and origins and that boundaries between these blocks tend to be significant hosts of mineralisation. This study showed the power of high data density (4762 analyses total) to understand subtle differences across terranes. This example also demonstrates that there is no single 'golden' Hf-signature to look for—mineralisation can occur in old or young crust, but different styles tend to occur in regions of specific crustal character. The relative differences in Hf isotope signatures are as important as the absolute isotopic values—isotopic gradients can indicate the most prospective regions due to their correlation with lithosphere-scale faults acting as a conduit for mantle-derived material. The background isotopic signature of a terrane will be important information to consider when conducting a more localised or camp-scale study.

Hou et al. (2015) illustrated the Hf-signatures using both εHf and model-age (TDM) maps; however, mapping using the εHf parameter is not considered best-practise for comparing isotopic signatures between samples of different age, because εHf values are not time-independent (Fig. 1). The depleted mantle source continues to increase in $^{176}Hf/^{177}Hf$, which means that the same absolute εHf value can have different implications at different times. In cases where there is significant age variation in the analysed rocks, a more robust approach is to use T2DM (Champion and Huston 2016).

7 Mapping Lithospheric Evolution Through Time and Implications for Ni Mineralisation

The previous Mesozoic-Cenozoic example illustrated the strong relationship between Lu-Hf isotopes, lithospheric structure and transport of metals into the crust. Isotopic characterisation also provides unique insight into lithospheric processes. There is a strong link between isotopic mapping and geophysical imaging techniques if the lithospheric configuration at the time of igneous emplacement is preserved. In older or dynamic terranes where current-day configurations might be very different to those of the past, isotopes can preserve signatures of configurations through time, including pre-, syn- or post-mineralization. This ability to reconstruct past configurations can help guide mineral exploration and targeting (Champion and Huston 2016; Collins et al. 2011; Mole et al. 2014).

Mole et al. (2014) conducted a zircon LA-ICP-MS Lu-Hf isotopic study in the Eastern Goldfields Province in Western Australia to understand the isotopic architecture and evolution of the lithosphere for the purposes of understanding controls on komatiite volcanism and associated Ni occurrences. Hafnium isotope data were used to generate a series of isotopic time-slices across the region. Isotopic variations in both space and time were interpreted to reflect gross lithospheric architecture. The researchers found that two older crustal blocks and one younger block existed in the time interval 3050–2820 Ma, and the configuration of these blocks controlled the location of komatiite emplacement in the Forrestania and Lake Johnston greenstone belts at 2.9 Ga (Fig. 5). The 2820–2720 Ma time slice provides evidence for the formation of the Eastern Goldfields crustal block, along with two other new blocks and the three previously existing blocks. This time period is interpreted to be associated with significant crustal reworking and minimal komatiite emplacement. The third time slice (2720–2600 Ma) identified by Mole et al. (2014) coincides with the timing of the most voluminous komatiite emplacement (c. 2.7 Ga) and is characterised by the presence of only two major crustal blocks—the Eastern Goldfields and the West Yilgarn. In both periods

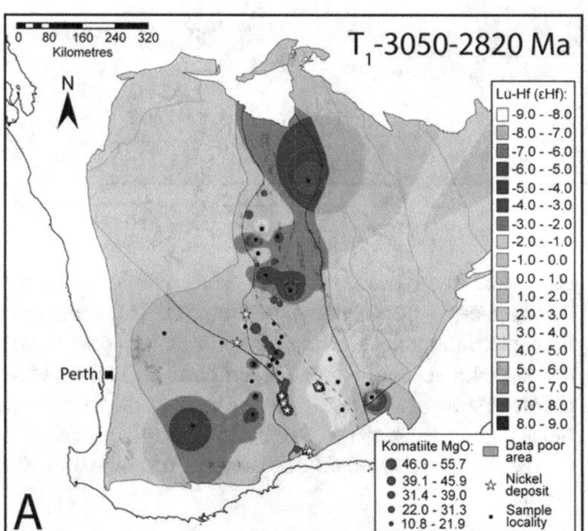

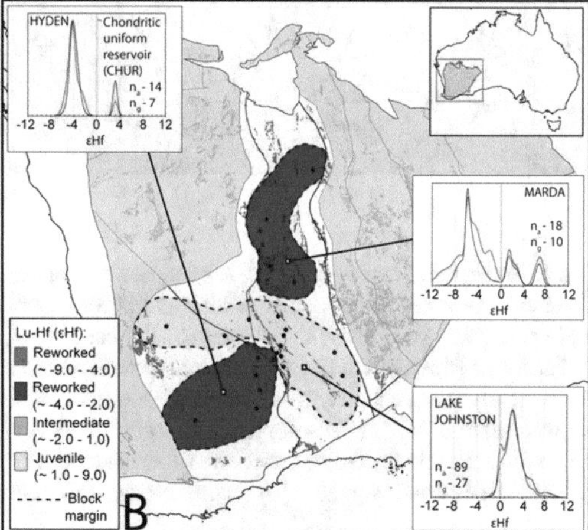

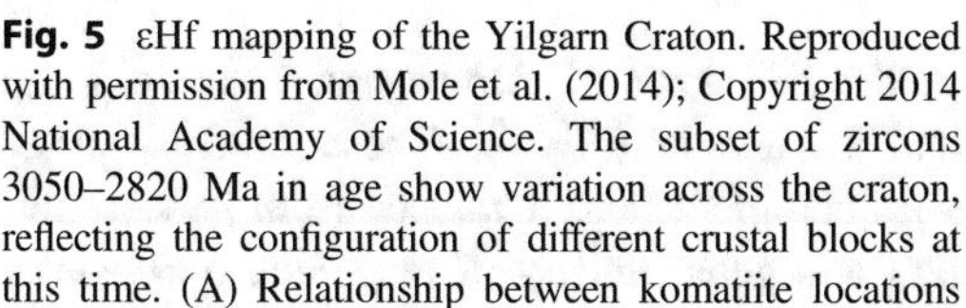

Fig. 5 εHf mapping of the Yilgarn Craton. Reproduced with permission from Mole et al. (2014); Copyright 2014 National Academy of Science. The subset of zircons 3050–2820 Ma in age show variation across the craton, reflecting the configuration of different crustal blocks at this time. (A) Relationship between komatiite locations and radiogenic (juvenile) εHf. (B) Interpreted configuration of the crustal blocks at this time; a juvenile east-west trending block separates two blocks of reworked crust. The probability density plots for each region are shown as insets.

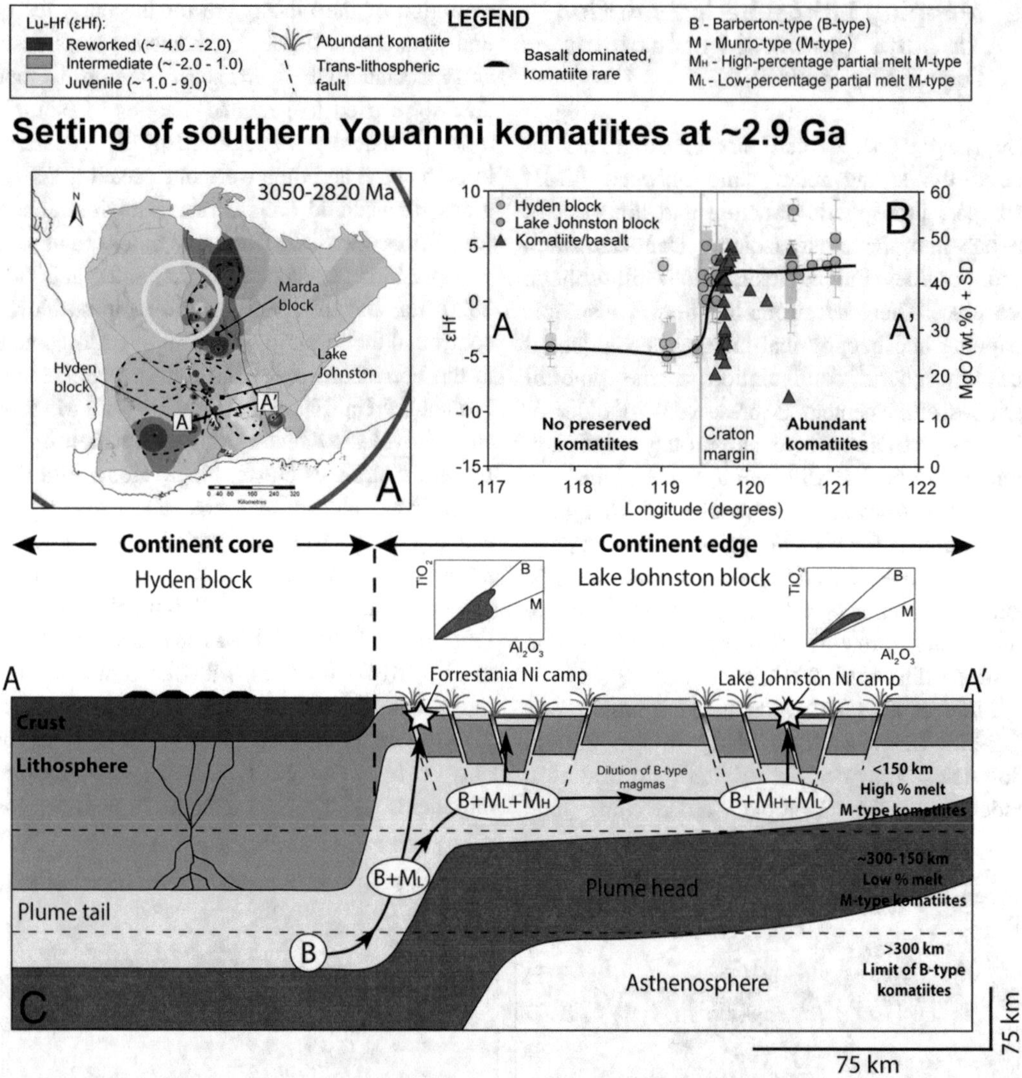

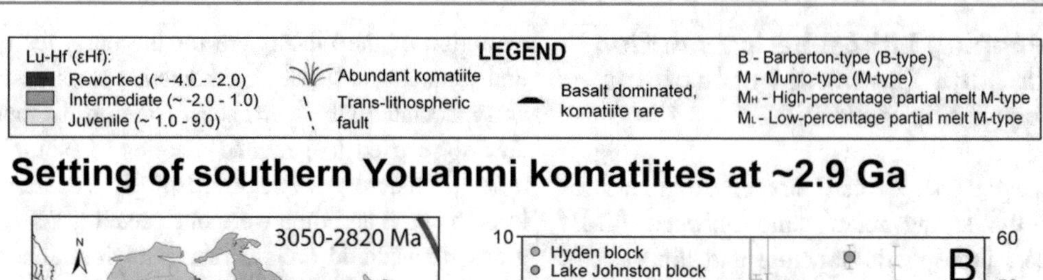

Fig. 6 Interpreted cross-section at 2.9 Ga in the southern Youanmi Terrane. Reproduced with permission from Mole et al. (2014); Copyright 2014 National Academy of Science. (A) Measured εHf values plotted for analyses in the 3050–2820 Ma age range, as per Figure 5 and location of cross section A-A'. (B) Isotopic cross-section between the Hyden Block and the Lake Johnston Block. The average Hf-isotope signature is significantly different across the two blocks and the interface is interpreted as a craton margin between the two blocks. (C) Interpreted lithospheric architecture at c. 2.9 Ga as interpreted from the isotopic results. Komatiite eruption is facilitated by plume-related extension at the interface of the two crustal blocks, and metals are preferentially deposited in the crust in the Lake Johnson Block where the lithosphere is thinner.

of komatiite generation (2.9 Ga and 2.7 Ga), the volcanism is constrained to radiogenic (juvenile) crustal blocks and margins.

Mole et al. (2014) mapped the changing lithospheric configuration through time, and produced geodynamic models for the generation of the Ni camps of Forrestania, Lake Johnston and Kambalda (Fig. 6). These models, along with complementary evidence from geochronology and geochemistry of magmatic rocks, infer the presence of thin lithosphere based on the radiogenic isotopic signatures, and propose that these regions of thinner lithosphere facilitated ascent of mantle-derived komatiitic magma metals into the upper crust.

7.1 Cautions and Considerations

Mole et al. (2014) combined U-Pb and Lu-Hf isotopes to map ancient craton boundaries through time, and track the progressive development of Archean crust. The earliest cratonic configurations no longer exist, but the isotopic signatures and mineralisation associated with these structures remain intact. The power of isotopes to look back through time is amplified by the zircon LA-ICP-MS method, as used in this study. A range of different zircon growth periods can be targeted and used to get a time-series of events—from a single sample. Mole et al. (2014) remarked upon the abundance of inherited zircons available in the analysed samples, and this allowed isotopic extraction of not only the magmatic population (the magmatic data considered in isolation is similar to whole-rock Sm-Nd data in the region), but also the isotopic signatures of earlier magmatic events as preserved in the inherited zircon population. This approach was especially valuable here, because direct exposure of granites older than c. 2800 Ma are rare in this region (Mole et al. 2014). There is an implicit assumption that the inherited material is from a relatively local source to the emplaced igneous body, and not transported laterally before incorporation into the magma.

Compromises must be made to reduce the complexity of multi-spot analytical data to enable meaningful spatial portrayal. The choice of this simplification (e.g., means vs medians, εHf vs T2DM) can have strong impacts on subsequent interpretations of the data. Mole et al. (2014) used median values from each population, but have provided a range of other portrayals in their supplementary material. These plots demonstrate that the overall trends remain the same for a range of data reduction techniques, but it is an important reminder that decisions on data portrayal have the potential to impact isotopic interpretations.

8 Metamorphism and Gold Mineralisation in the Tropicana Zone

Kirkland et al. (2015) examined potential models for gold deposit formation in the regolith-covered Archean Tropicana Zone, in the Albany-Fraser Orogen on the eastern margin of the Yilgarn Craton, Australia. They applied zircon LA-ICP-MS Lu-Hf analyses, zircon SIMS U-Pb and pyrite TIMS Re-Os geochronology, and whole-rock geochemistry to contextualise the gold mineralisation that occurs within the Tropicana Zone, and to draw comparisons with the adjacent, well-endowed Eastern Goldfields province.

As a component of the Albany-Fraser Orogen, the high-grade Tropicana Zone preserves a complex Proterozoic thermal history. The rocks investigated by Kirkland et al. (2015) are predominantly granulite-facies gneissic rocks, with sanukitoids as the protolith. Kirkland et al. (2015) used micro-analytical U-Pb and Lu-Hf analytical techniques to extract information from metamorphic, magmatic and inherited regions in zircons from these rocks to characterise fractionation events, and also to compare with other lithospheric blocks.

Kirkland et al. (2015) found that there are strong similarities in Hf-isotope character between the Tropicana Zone and the Eastern Goldfields Terrane. The authors interpret the Tropicana Zone was originally deep Archean crust that was structurally emplaced in its current configuration at or before a 1780–1760 Ma thermal overprinting event.

Timing of gold mineralisation in the Tropicana Zone (c. 2520 Ma and c 2100 Ma; Doyle et al. 2015) is much younger than that in the Eastern Goldfields (2660–2630 Ma; Vielreicher et al. 2015). Kirkland et al. (2015) pointed out that there is a known association between Archean sanukitoids and gold mineralisation, and argued that the previous configuration of the Tropicana Zone enabled the remobilisation of metalliferous (Au-bearing) fluids derived from sanukitoids at depth, and subsequent concentration of those fluids into fracture systems.

8.1 Cautions and Considerations

This study demonstrated the power of microbeam isotopic analysis on complex rocks; an advantage that zircon Lu-Hf has over more traditional whole-rock Sm-Nd analysis. By targeting zones within zircon grains, it is possible to disentangle the influence of multiple components of a rock and better understand the processes that generated the rocks and associated mineralisation. As well as magmatic zircons, metamorphic rims and inherited cores were also analysed, which enabled tracking of source fluids and magmas through different geological events, even high-grade metamorphism. This wealth of information is particularly valuable in regions where samples are sparse or difficult to collect, such as regolith-covered regions away from outcropping basement rocks.

This study highlighted the importance of CL imaging to delineate different growth zones in complex zircon grains. Without appropriate characterisation of the internal structure of the zircon grains, it would be impossible to distinguish the different growth phases, and so the isotopic data would be a geologically meaningless mixture of multiple growth phases.

The metamorphic rims in this study have a Lu-Hf signature that is more akin to recycled crust than juvenile crust—which is interpreted to mean that no new mantle-derived material was introduced to the system during the metamorphism that generated the metamorphic zircon rims. However, the authors noted that the Lu-Hf isotopic values in zircon rims appears to be derived more strongly from resorbed inherited zircon cores than primary magmatic zircon. This leaves open the possibility of a mantle contribution that was not preserved in the metamorphic zircon rims.

9 Kimberlite Pathfinding for Diamond Exploration

Schärer et al. (1997) and Batumike et al. (2009) applied ID-TIMS U-Pb, Hf and trace elemental analysis on mineral separates from modern drainage systems to characterise crustal and magmatic evolution in the central Congo-Kasai Craton of central Africa, and to investigate sources of alluvial diamonds in the region. It had previously been assumed that alluvial diamonds were derived solely from the Angolan kimberlite field which intruded Cretaceous sandstone, imposing a maximum age of c. 120 Ma on these occurrences (Batumike et al. 2009). However, some alluvial diamonds in the region contain natural irradiation features characteristic of ancient diamonds (Shmakov 2008), and at least some kimberlites have interacted with older crust, as evidenced by a diamond-hosted zircon inclusion that yielded a U-Pb age of 628 ± 12 Ma (Kinny and Meyer 1994).

To better understand the age, origin and mantle source characteristics of kimberlitic diamonds, Schärer et al. (1997) performed isotope-dilution U-Pb dating and Lu-Hf analysis on zircon and baddeleyite megacrysts with high-pressure formation characteristics (indicative of crystallisation at great depth) from the Mbuji-Mayi kimberlite in the Democratic Republic of Congo. The researchers derived ages of both zircon and baddeleyite of c. 70 Ma. The εHf values of both minerals were strongly positive: c. + 8 for zircon and + 5 to + 10 for baddeleyite, in line with a moderately to strongly depleted mantle source.

Batumike et al. (2009) characterised zircons from the sediments from the Luebo region based on the trace-element classification system of Belousova et al. (2002). This sediment-sampling approach allowed them to gather isotopic and trace-element information from any rock units that eroded heavy minerals into the catchment. Zircons were filtered by εHf and trace element composition to constrain the protolith composition, potentially identifying kimberlite occurrences that had not yet been discovered (Fig. 7). The Hf-isotopes from samples which were classified as kimberlite-derived had relatively positive εHf, consistent with mantle-derived material interacting with late Archean lithosphere during magma ascent. The zircon U-Pb ages comprise three groups: late Archean, Neoproterozoic and Cretaceous. The authors proposed three distinct episodes of diamondiferous kimberlite magmatism in the region despite previous assumptions that diamonds in the region were all sourced in a single time period. This increases the range of geology which may be host to kimberlitic intrusions from post-120 Ma into the Archean.

9.1 Cautions and Considerations

Hf-isotope data in combination with geochronology and trace-element geochemistry (all of which can be performed on the same zircons) can be used to understand kimberlite genesis and improve pathfinding in diamond exploration. Drainages sample heavy minerals from across a catchment and so provide an efficient method for capturing information on a range of nearby lithologies from a single sediment. This approach can provide first-order information in under-explored regions, including from lithologies that may not be accessible to surface sampling due to vegetative or alluvial cover. In this application, where contextual information on the source protolith is limited, Hf isotope data is most useful when accompanied by complementary geochronological and geochemical datasets.

10 Discussion, Conclusions, Future Developments

The Lu-Hf isotope system is now used in a well-developed and robust metholology for characterising magma and fluid sources, particularly crust-mantle interactions. Many studies using the Lu-Hf isotopic system investigate large-scale lithospheric processes, but there is plenty of opportunity within this framework for more focused investigations to aid mineral exploration.

The four case studies highlight the utility of Lu-Hf isotope studies to increase the understanding of lithospheric processes that control how and where mineral deposits form. This enables greater understanding of which crustal blocks and boundaries may have this highest potential for particular ore deposits (Hou et al. 2015). The Lu-Hf isotopic system preserves syn-mineralization information even if crustal configurations have been subsequently re-arranged, because zircons, even inherited ones, preserve Hf

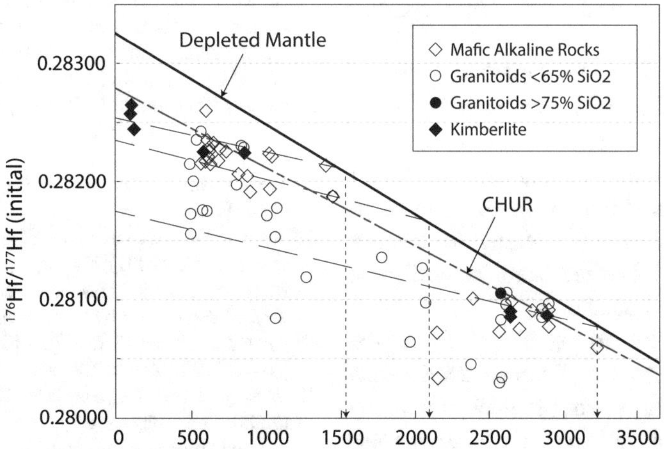

Fig. 7 ^{176}Hf/^{177}Hf data and interpreted rock-types from trace element data. Reproduced with permission from Batumike et al. (2009); Copyright 2009 Elsevier. Three kimberlite-related age groups (filled diamonds) were interpreted from sediment-hosted zircons in central Africa, as identified by zircon trace-element data. Associated εHf data shows the radiogenic (juvenile) nature of these zircons relative to the bulk sedimentary zircon load (all other symbols), providing evidence for three distinct periods of kimberlite emplacement rather than a single event as previously thought.

signatures from the time of their formation (Batumike et al. 2009; Kirkland et al. 2015; Mole et al. 2014; Schärer et al. 1997). The case studies demonstrate the wide applicability of this isotopic system across a range of commodities, including Cu-Mo-Fe-Pb-Zn (Hou et al. 2015), Ni (Mole et al. 2014), Au (Kirkland et al. 2015) and diamonds (Batumike et al. 2009; Schärer et al. 1997).

The Lu-Hf isotope system is best used as part of a toolkit that includes other geochemical and isotopic systems such as U-Pb, Sm-Nd, O-isotopes, and trace elements. The information from the Lu-Hf isotope system complements that from geophysical, structural and petrological investigations and can improve understanding of how geological systems and settings have changed through time.

Instrumentation, data quality criteria, interpretations and ideas continue to evolve rapidly in this dynamic field of research. Instrumental improvements are enabling a growing richness of data and improving the precision and spatial resolution of data, as well as enabling linked acquisition of additional elemental and isotopic data. With the increased ease and speed of data acquisition comes a need for improved data reduction and statistical treatment, and data storage systems to deal with large datasets easily.

Similarities between the Lu-Hf and Sm-Nd systems may see integrated datasets in the future, which leverage large existing Sm-Nd datasets and combine them with rapidly growing Lu-Hf data holdings. This will allow more comprehensive coverage and make it easier to identify regions of interest—be they isotopically distinct regions or isotopic boundaries.

Although there are large overlaps between the Sm-Nd and Lu-Hf systems, there are applications in which the Lu-Hf system provides new information. For example, it is possible to use the Lu-Hf isotopic system to date certain minerals that yield only poor ages with Sm-Nd or none at all (e.g. garnet, phosphate and carbonate minerals—see above). The big advantage of the Lu-Hf isotopic system is that Hf is hosted by zircon, which is both dateable via U-Pb, and sufficiently durable that it can be preserved during petrogenetic processes.

Acknowledgements Thank you to David Huston for inviting me to contribute, and Geoff Fraser and Simon Bodorkos (Geoscience Australia) for helpful comments on an early draft. Klaus Mezger and an anonymous reviewer are thanked for their thorough reviews, which significantly improved the manuscript. This chapter is published with the permission of the Chief Executive Officer, Geoscience Australia.

References

Amelin Y, Stern R, Wiechert U, Davis D, Lee D, Halliday A, Pidgeon RT (2001) Combined U-Pb, trace element, oxygen, Zr, and Lu-Hf isotopic systematics of 3.90–4.25 Ga detrital zircons from Jack Hills, Western Australia; progress report. Geosci Austr Rec 2001/037:33–34 https://doi.org/10.26186/5ca57944057bf

Amelin Y, Lee DC, Halliday AN (2000) Early-middle Archaean crustal evolution deduced from Lu-Hf and U-Pb isotopic studies of single zircon grains. Geochim Cosmochim Acta 64:4205–4225. https://doi.org/10.1016/S0016-7037(00)00493-2

Amelin Y, Lee DC, Halliday AN, Pidgeon RT (1999) Nature of the Earth's earliest crust from hafnium isotopes in single detrital zircons. Nature 399:252–255. https://doi.org/10.1038/20426

Batumike JM, Griffin WL, O'Reilly SY, Belousova EA, Pawlitschek M (2009) Crustal evolution in the central Congo-Kasai Craton, Luebo, D.R. Congo: insights from zircon U-Pb ages. Hf-Isotope and Trace-Element Data. Precambrian Res 170:107–115. https://doi.org/10.1016/j.precamres.2008.12.001

Belousova E, Griffin WL, O'Reilly SY, Fisher N (2002) Igneous zircon: trace element composition as an indicator of source rock type. Contrib Mineral Petrol 143:602–622. https://doi.org/10.1007/s00410-002-0364-7

Belousova EA, Griffin WL, Shee SR, Jackson SE, O'Reilly SY (2001) Two age populations of zircons from the Timber Creek kimberlites, Northern Territory, as determined by laser ablation ICP MS analysis. Austr J Earth Sci 48:757–765. https://doi.org/10.1046/j.1440-0952.2001.485894.x

Bizzarro M, Baker JA, Haack H, Ulfbeck D, Rosing M (2003) Early history of Earth's crust-mantle system inferred from hafnium isotopes in chondrites. Nature 421:931–933. https://doi.org/10.1038/nature01421

Blichert-Toft J, Albarède F (1997) The Lu-Hf isotope geochemistry of chondrites and the evolution of the mantle-crust system. Earth Planet Sci Lett 148:243–258. https://doi.org/10.1016/S0012-821X(97)00040-X

Bodet F, Schärer U (2000) Evolution of the SE-Asian continent from U-Pb and Hf isotopes in single grains of zircon and baddeleyite from large rivers. Geochim Cosmochim Acta 64:2067–2091. https://doi.org/10.1016/S0016-7037(00)00352-5

Champion DC, Huston DL (2016) Radiogenic isotopes, ore deposits and metallogenic terranes: novel approaches based on regional isotopic maps and the mineral systems concept. Ore Geol Rev 76:229–256. https://doi.org/10.1016/j.oregeorev.2015.09.025

Champion DC, Huston DL (2023) Applications of Nd isotopes to ore deposits and metallogenic terranes; using regional isotopic maps and the mineral systems concept. In: Huston DL, Gutzmer J (eds) Isotopes in economic geology, metallogenesis and exploration. Springer, Berlin, this volume.

Cherniak DJ, Watson EB (2003) Diffusion in Zircon. Rev Mineral Geochem 53:113–143. https://doi.org/10.2113/0530113

Choukroun M, O'Reilly SY, Griffin WL, Pearson NJ, Dawson JB (2005) Hf isotopes of MARID (mica-amphibole-rutile-ilmenite-diopside) rutile trace metasomatic processes in the lithospheric mantle. Geology 33:45–48. https://doi.org/10.1130/G21084.1

Collins WJ, Belousova EA, Kemp AI, Murphy JB (2011) Two contrasting Phanerozoic orogenic systems revealed by hafnium isotope data. Nat Geosci 4:333–337. https://doi.org/10.1038/ngeo1127

Corfu F, Noble SR (1992) Genesis of the southern Abitibi greenstone belt, Superior Province, Canada: evidence from zircon Hf isotope analyses using a single filament technique. Geochim Cosmochim Acta 56:2081–2097. https://doi.org/10.1016/0016-7037(92)90331-C

Cross AJ, Purdy DJ, Champion DC, Brown DD, Siégel C, Armstrong RA (2018) Insights into the evolution of the Thomson Orogen from geochronology, geochemistry, and zircon isotopic studies of magmatic rocks. Austr J Earth Sci 65:987–1008. https://doi.org/10.1080/08120099.2018.1515791

D'Abzac F-X, Davies JHFL, Wotzlaw J-F, Schaltegger U (2016) Hf isotope analysis of small zircon and baddeleyite grains by conventional multi collector-inductively coupled plasma-mass spectrometry. Chem Geol 433:12–23. https://doi.org/10.1016/j.chemgeo.2016.03.025

Dhuime B, Hawkesworth C, Cawood P (2011) When continents formed. Science 331:154–155. https://doi.org/10.1126/science.1201245

Doe MF, Jones JV III, Karlstrom KE, Dixon B, Gehrels G, Pecha M (2013) Using detrital zircon ages and Hf isotopes to identify 1.48–1.45 Ga sedimentary basins and fingerprint sources of exotic 1.6–1.5 Ga grains in southwestern Laurentia. Precambrian Res 231:409–421. https://doi.org/10.1016/j.precamres.2013.03.002

Dolgopolova A, Seltmann R, Armstrong R, Belousova E, Pankhurst RJ, Kavalieris I (2013) Sr–Nd–Pb–Hf isotope systematics of the Hugo Dummett Cu–Au porphyry deposit (Oyu Tolgoi, Mongolia). Lithos 164–167:47–64. https://doi.org/10.1016/j.lithos.2012.11.017

Doyle MG, Rasmussen B, Fletcher IR, Muhling JR, Foster J, Large RR, Meffre S, Mathur R, McNaughton NJ, Phillips D (2015) Geochronological constraints on the Tropicana gold deposit and Albany-Fraser Orogen, Western Australia. Econ Geol 110:355–386. https://doi.org/10.2113/econgeo.110.2.355

Ewing TA, Rubatto D, Eggins SM, Hermann J (2011) In situ measurement of hafnium isotopes in rutile by LA–MC-ICPMS: protocol and applications. Chem Geol 281:72–82. https://doi.org/10.1016/j.chemgeo.2010.11.029

Ewing TA, Rubatto D, Hermann J (2014) Hafnium isotopes and Zr/Hf of rutile and zircon from lower crustal metapelites (Ivrea–Verbano Zone, Italy): implications for chemical differentiation of the crust. Earth Planet Sci Lett 389:106–118. https://doi.org/10.1016/j.epsl.2013.12.029

Fisher CM, Vervoort JD, Hanchar JM (2014) Guidelines for reporting zircon Hf isotopic data by LA-MC-ICPMS and potential pitfalls in the interpretation of these data. Chem Geol 363:125–133. https://doi.org/10.1016/j.chemgeo.2013.10.019

Flowerdew MJ, Chew DM, Daly JS, Millar IL (2009) Hidden Archaean and Palaeoproterozoic crust in NW Ireland? evidence from zircon Hf isotopic data from granitoid intrusions. Geol Mag 146:903–916. https://doi.org/10.1017/S0016756809990227

Gerdes A, Zeh A (2006) Combined U-Pb and Hf isotope LA-(MC-)ICP-MS analyses of detrital zircons: comparison with SHRIMP and new constraints for the provenance and age of an Armorican metasediment in Central Germany. Earth Planet Sci Lett 249:47–61. https://doi.org/10.1016/j.epsl.2006.06.039

Gerdes A, Zeh A (2009) Zircon formation versus zircon alteration—new insights from combined U-Pb and Lu–Hf in-situ LA-ICP-MS analyses, and consequences for the interpretation of Archean zircon from the Central Zone of the Limpopo Belt. Chem Geol 261:230–243. https://doi.org/10.1016/j.chemgeo.2008.03.005

Griffin WL, Pearson NJ, Belousova E, Jackson SE, van Achterbergh E, O'Reilly SY, Shee SR (2000) The Hf isotope composition of cratonic mantle: LAM-MC-ICPMS analysis of zircon megacrysts in kimberlites. Geochim Cosmochim Acta 64:133–147. https://doi.org/10.1016/S0016-7037(99)00343-9

Griffin WL, Wang X, Jackson SE, Pearson NJ, O'Reilly SY, Xu X, Zhou X (2002) Zircon chemistry and magma mixing, SE China: in-situ analysis of Hf isotopes, Tonglu and Pingtan igneous complexes. Lithos 61:237–269. https://doi.org/10.1016/S0024-4937(02)00082-8

Griffin WL, Belousova EA, Walters SG, O'Reilly SY (2006) Archaean and proterozoic crustal evolution in the Eastern succession of the Mt Isa district, Australia: U-Pb and Hf-isotope studies of detrital zircons. Austr J Earth Sci 53:125–149. https://doi.org/10.1080/08120090500434591

Halliday AN, Lee D-C, Christensen JN, Walder AJ, Freedman PA, Jones CE, Hall CM, Yi W, Teagle D (1995) Recent developments in inductively coupled plasma magnetic sector multiple collector mass spectrometry. Int J Mass Spectrom Ion Process 146–147:21–33. https://doi.org/10.1016/0168-1176(95)04200-5

Harrison TM, Schmitt AK, McCulloch MT, Lovera OM (2008) Early ($\geq$ 4.5 Ga) formation of terrestrial crust: Lu–Hf, $\delta^{18}O$, and Ti thermometry results for Hadean zircons. Earth Planet Sci Lett 268:476–486. https://doi.org/10.1016/j.epsl.2008.02.011

Hartnady MIH, Kirkland CL, Dutch R, Bodorkos S (2018) Jagodzinski E (2018) Zircon Hf isotopic signatures of the Coompana Province in South Australia. Geol Surv S Austr Rep Book 00028:132–144

Hou Z, Duan L, Lu Y, Zheng Y, Zhu D, Yang Z, Yang Z, Wang B, Pei Y, Zhao Z, McCuaig TC (2015) Lithospheric architecture of the Lhasa Terrane and its control on ore deposits in the Himalayan-Tibetan Orogen. Econ Geol 110:1541–1575. https://doi.org/10.2113/econgeo.110.6.1541

Kemp AIS, Hawkesworth CJ, Paterson BA, Kinny PD (2006) Episodic growth of the Gondwana supercontinent from hafnium and oxygen isotopes in zircon. Nature 439:580–583. https://doi.org/10.1038/nature04505

Kemp AIS, Hawkesworth CJ, Foster GL, Paterson BA, Woodhead JD, Hergt JM, Gray CM, Whitehouse MJ (2007) Magmatic and crustal differentiation history of granitic rocks from Hf-O isotopes in zircon. Science 315:980–983. https://doi.org/10.1126/science.1136154

Kinny PD, Maas R (2003) Lu–Hf and Sm–Nd isotope systems in zircon. Rev Mineral Geochem 53:327–341. https://doi.org/10.2113/0530327

Kinny PD, Meyer HOA (1994) Zircon from the mantle: a new way to date old diamonds. J Geol 102:475–481. https://doi.org/10.1086/629687

Kirkland CL, Spaggiari CV, Smithies RH, Wingate MTD, Belousova EA, Gréau Y, Sweetapple MT, Watkins R, Tessalina S, Creaser R (2015) The affinity of Archean crust on the Yilgarn—Albany–Fraser Orogen boundary: implications for gold mineralisation in the Tropicana Zone. Precambrian Res 266:260–281. https://doi.org/10.1016/j.precamres.2015.05.023

Lu Y-J, Loucks RR, Fiorentini M, McCuaig TC, Evans NJ, Yang Z-M, Hou Z-Q, Kirkland CL, Parra-Avila LA, Kobussen A (2016) Zircon compositions as a pathfinder for porphyry Cu $\pm$ Mo $\pm$ Au deposits. Soc Econ Geol Spec Publ 19:329–347. https://doi.org/10.5382/SP.19.13

Luvizotto GL, Zack T, Meyer HP, Ludwig T, Triebold S, Kronz A, Münker C, Stockli DF, Prowatke S, Klemme S, Jacob DE, von Eynatten H (2009) Rutile crystals as potential trace element and isotope mineral standards for microanalysis. Chem Geol 261:346–369. https://doi.org/10.1016/j.chemgeo.2008.04.012

McCuaig TC, Beresford S, Hronsky J (2010) Translating the mineral systems approach into an effective exploration targeting system. Ore Geol Rev 38:128–138. https://doi.org/10.1016/j.oregeorev.2010.05.008

Mole DR, Fiorentini ML, Thebaud N, Cassidy KF, McCuaig TC, Kirkland CL, Romano SS, Doublier MP, Belousova EA, Barnes SJ (2014) Archean komatiite volcanism controlled by the evolution of early continents. Proc Natl Acad Sci 111:10083–10088. https://doi.org/10.1073/pnas.1400273111

Murgulov V, O'Reilly SY, Griffin WL, Blevin PL (2008) Magma sources and gold mineralisation in the Mount Leyshon and Tuckers Igneous Complexes, Queensland, Australia: U-Pb and Hf isotope evidence. Lithos 101:281–307. https://doi.org/10.1016/j.lithos.2007.07.014

Nash BP, Perkins ME, Christensen JN, Lee D-C, Halliday AN (2006) The Yellowstone hotspot in space and time: Nd and Hf isotopes in silicic magmas. Earth Planet Sci Lett 247:143–156. https://doi.org/10.1016/j.epsl.2006.04.030

Patchett PJ, Kouvo O, Hedge CE, Tatsumoto M (1982) Evolution of continental crust and mantle heterogeneity: evidence from Hf isotopes. Contrib Mineral Petrol 78:279–297. https://doi.org/10.1007/BF00398923

Patchett PJ, Tatsumoto M (1980) Lu–Hf total-rock isochron for the eucrite meteorites. Nature 288:571–574. https://doi.org/10.1038/288571a0

Purdy DJ, Cross AJ, Brown DD, Carr PA, Armstrong RA (2016) New constraints on the origin and evolution of the Thomson Orogen and links with central Australia from isotopic studies of detrital zircons. Gondwana Res 39:41–56. https://doi.org/10.1016/j.gr.2016.06.010

Schärer U, Corfu F, Demaiffe D (1997) U-Pb and Lu–Hf isotopes in baddeleyite and zircon megacrysts from the Mbuji-Mayi kimberlite: constraints on the subcontinental mantle. Chem Geol 143:1–16. https://doi.org/10.1016/S0009-2541(97)00094-6

Scherer E, Münker C, Mezger K (2001) Calibration of the lutetium-hafnium clock. Science 293:683–687. https://doi.org/10.1126/science.1061372

Sguigna AP, Larabee AJ, Waddington JC (1982) The half-life of ^{176}Lu by a γ–γ coincidence measurement. Can J Phys 60:361–364. https://doi.org/10.1139/p82-049

Shmakov I (2008) Mesozoic palaeochannels as diamond source of alluvial placers. In: 33rd International Geological Congress, abstract volume. Norwegian Academy of Science and Letters, Oslo, pp 5177

Siegel C, Bryan SE, Allen CM, Purdy DJ, Cross AJ, Uysal IT, Gust DA (2018) Crustal and thermal structure of the Thomson Orogen: constraints from the geochemistry, zircon U-Pb age, and Hf and O isotopes of subsurface granitic rocks. Austr J Earth Sci 65:967–986. https://doi.org/10.1080/08120099.2018.1447998

Söderlund U, Elming S-Å, Ernst RE, Schissel D (2006) The Central Scandinavian Dolerite Group—protracted hotspot activity or back-arc magmatism?: constraints from U-Pb baddeleyite geochronology and Hf isotopic data. Precambrian Res 150:136–152. https://doi.org/10.1016/j.precamres.2006.07.004

Söderlund U, Isachsen CE, Bylund G, Heaman LM, Patchett PJ, Vervoort JD, Andersson UB (2005) U-Pb baddeleyite ages and Hf, Nd isotope chemistry constraining repeated mafic magmatism in the Fennoscandian Shield from 1.6 to 0.9 Ga. Contrib Mineral Petrol 150:174–194. https://doi.org/10.1007/s00410-005-0011-1

Söderlund U, Patchett PJ, Vervoort JD, Isachsen CE (2004) The ^{176}Lu decay constant determined by Lu–Hf and U-Pb isotope systematics of Precambrian mafic intrusions. Earth Planet Sci Lett 219:311–324. https://doi.org/10.1016/S0012-821X(04)00012-3

Speer JA, Cooper BJ (1982) Crystal structure of synthetic hafnon, HfSiO4, comparison with zircon and the actinide orthosilicates. Ame Mineral 67:804–808

Spencer CJ, Kirkland CL, Roberts NMW, Evans NJ, Liebmann J (2020) Strategies towards robust interpretations of in situ zircon Lu–Hf isotope analyses. Geosci Front 11:843–853. https://doi.org/10.1016/j.gsf.2019.09.004

Thirlwall MF, Walder AJ (1995) In situ hafnium isotope ratio analysis of zircon by inductively coupled plasma multiple collector mass spectrometry. Chem Geol 122:241–247. https://doi.org/10.1016/0009-2541(95)00003-5

Thrane K, Connelly JN, Bizzarro M, Meyer BS, The L-S (2010) Origin of excess ^{176}Hf in meteorites. Astrophys J 717:861–867. https://doi.org/10.1088/0004-637X/717/2/861

Valley JW (2003) Oxygen isotopes in zircon. Rev Mineral Geochem 53:343–385. https://doi.org/10.2113/0530343

Vervoort J (2014) Lu-Hf dating: the Lu-Hf isotope system. In: Rink WJ, Thompson J (eds) Encyclopedia of scientific dating methods. Springer Netherlands, Dordrecht, pp 1–20. https://doi.org/10.1007/978-94-007-6326-5_46-1

Vervoort JD, Kemp AIS (2016) Clarifying the zircon Hf isotope record of crust–mantle evolution. Chem Geol 425:65–75. https://doi.org/10.1016/j.chemgeo.2016.01.023

Vervoort JD, Patchett PJ, Gehrels GE, Nutman AP (1996) Constraints on early Earth differentiation from hafnium and neodymium isotopes. Nature 379:624–627. https://doi.org/10.1038/379624a0

Vielreicher N, Groves D, McNaughton N, Fletcher I (2015) The timing of gold mineralization across the eastern Yilgarn craton using U-Pb geochronology of hydrothermal phosphate minerals. Miner Depos 50:391–428. https://doi.org/10.1007/s00126-015-0589-9

Walder AJ, Freedman PA (1992) Isotopic ratio measurement using a double focusing magnetic sector mass analyser with an inductively coupled plasma as an ion source. J Anal at Spectrom 7:571–575. https://doi.org/10.1039/JA9920700571

Waltenberg K, Bodorkos S, Armstrong R, Fu B (2018) Mid- to lower-crustal architecture of the northern Lachlan and southern Thomson orogens: evidence from O-Hf isotopes. Austr J Earth Sci 65:1009–1034. https://doi.org/10.1080/08120099.2018.1463928

Westhues A, Hanchar JM, LeMessurier MJ, Whitehouse MJ (2017) Evidence for hydrothermal alteration and source regions for the Kiruna iron oxide–apatite ore (northern Sweden) from zircon Hf and O isotopes. Geology 45:571–574. https://doi.org/10.1130/G38894.1

Woodhead JD, Hergt J, Shelley M, Eggins S, Kemp R (2004) Zircon Hf-isotope analysis with an excimer laser, depth profiling, ablation of complex geometries, and concomitant age estimation. Chem Geol 209:121–135. https://doi.org/10.1016/j.chemgeo.2004.04.026

Wu Y-B, Gao S, Zhang H-F, Yang S-H, Liu X-C, Jiao W-F, Liu Y-S, Yuan H-L, Gong H-J, He M-C (2009) U-Pb age, trace-element, and Hf-isotope compositions of zircon in a quartz vein from eclogite in the western Dabie Mountains: constraints on fluid flow during early exhumation of ultrahigh-pressure rocks. Am Mineral 94:303–312. https://doi.org/10.2138/am.2009.3042

Xie L, Zhang Y, Zhang H, Sun J, Wu F (2008) In situ simultaneous determination of trace elements, U-Pb and Lu-Hf isotopes in zircon and baddeleyite. Chin Sci Bull 53:1565–1573. https://doi.org/10.1007/s11434-008-0086-y

Yang Q, Santosh M, Shen J, Li S (2014) Juvenile vs. recycled crust in NE China: Zircon U-Pb geochronology, Hf isotope and an integrated model for Mesozoic gold mineralization in the Jiaodong Peninsula. Gondwana Res 25:1445–1468. https://doi.org/10.1016/j.gr.2013.06.003

Yuan H-L, Gao S, Dai M-N, Zong C-L, Günther D, Fontaine GH, Liu X-M, Diwu C (2008) Simultaneous determinations of U-Pb age, Hf isotopes and trace element compositions of zircon by excimer laser-ablation quadrupole and multiple-collector ICP-MS. Chem Geol 247:100–118. https://doi.org/10.1016/j.chemgeo.2007.10.003

Zack T, Stockli DF, Luvizotto GL, Barth MG, Belousova E, Wolfe MR, Hinton RW (2011) In situ U-Pb rutile dating by LA-ICP-MS: ^{208}Pb correction and prospects for geological applications. Contrib Mineral Petrol 162:515–530. https://doi.org/10.1007/s00410-011-0609-4

Zhu D-C, Zhao Z-D, Niu Y, Mo X-X, Chung S-L, Hou Z-Q, Wang L-Q, Wu F-Y (2011) The Lhasa Terrane: record of a microcontinent and its histories of drift and growth. Earth Planet Sci Lett 301:241–255. https://doi.org/10.1016/j.epsl.2010.11.005

Light Stable Isotopes (H, B, C, O and S) in Ore Studies—Methods, Theory, Applications and Uncertainties

David L. Huston, Robert B. Trumbull, Georges Beaudoin, and Trevor Ireland

Abstract

Variations in the abundances of light stable isotopes, particularly those of hydrogen, boron, carbon, oxygen and sulfur, were essential in developing mineralization models. The data provide constraints on sources of hydrothermal fluids, carbon, boron and sulfur, track interaction of these fluids with the rocks at both the deposit and district scales, and establish processes of ore deposition. In providing such constraints, isotopic data have been integral in developing genetic models for porphyry-epithermal, volcanic-hosted massive sulfide, orogenic gold, sediment-hosted base metal and banded-iron formation-hosted iron ore systems, as discussed here and in other chapters in this book. After providing conventions, definitions and standards used to present stable isotope data, this chapter summarizes analytical methods, both bulk and in situ, discusses processes that fractionate stable isotopes, documents the isotopic characteristics of major fluid and rock reservoirs, and then shows how stable isotope data have been used to better understand ore-forming processes and to provide vectors to ore. Analytical procedures, initially developed in the 1940s for carbon–oxygen analysis of bulk samples of carbonate minerals, have developed so that, for most stable isotopic systems, spots as small as a few tens of μm are routinely analyzed. This precision provides the paragenetic and spatial resolution necessary to answer previously unresolvable genetic questions (and create new questions). Stable isotope fractionation reflects geological and geochemical processes important in ore formation, including: (1) phase changes such as boiling, (2) water–rock interaction, (3) cooling, (4) fluid mixing, (5) devolatilization, and (6) redox reactions, including SO_2 disproportionation caused by the cooling of magmatic-hydrothermal fluids and photolytic dissociation in the atmosphere. These processes commonly produce gradients in isotopic data, both in time and in space. These gradients, commonly mappable in space, provide not only evidence of process but also exploration vectors. Stable isotope data can be used to estimate the conditions of alteration or

D. L. Huston (✉)
Geoscience Australia, GPO Box 378, Canberra, ACT 2601, Australia
e-mail: David.Huston@ga.gov.au

R. B. Trumbull
Helmholtz Centre Potsdam, GFZ German Research Centre for Geosciences, Telegrafenberg B125, Potsdam 14473, Germany

G. Beaudoin
Département de Géologie et de Génie Géologique, Université Laval, 1065, Avenue de La Médecine, Québec, QC G1V 0A6, Canada

T. Ireland
School of Earth and Environmental Sciences, University of Queensland, St Lucia, QLD 4072, Australia

D. Huston and J. Gutzmer (eds.), *Isotopes in Economic Geology, Metallogenesis and Exploration*, Mineral Resource Reviews, https://doi.org/10.1007/978-3-031-27897-6_8

mineralization when data for coexisting minerals are available. These estimates use experimentally- or theoretically-determined fractionation equations to estimate temperatures of mineral formation. If the temperature is known from isotopic or other data (e.g., fluid inclusion data or chemical geothermometers), the isotopic composition of the hydrothermal fluid components can be estimated. If fluid inclusion homogenization and compositional data exist, the pressure and depth of mineralization can be estimated. One of the most common uses of stable isotope data has been to determine, or more correctly delimit, fluid and sulfur sources. Estimates of the isotopic compositions of hydrothermal fluids, in most cases, do not define unequivocal sources, but, rather, eliminate sources. As an example, the field of magmatic fluids largely overlap that of metamorphic fluids in δ^{18}O-δD space, but are significantly different to the fields of meteoric waters and seawater. As such, a meteoric or seawater origin for a fluid source may be resolvable, but a magmatic source cannot be resolved from a metamorphic source. Similarly, although $\delta^{34}S \sim 0‰$ is consistent with a magmatic-hydrothermal sulfur source, the signature can also be produced by leaching of an igneous source. Recent analytical and conceptual advances have enabled gathering of new types of isotopic data and application of these data to resolve new problems in mineral deposit genesis and geosciences in general. Recent developments such as rapid isotopic analysis of geological materials or clumped isotopes will continue to increase the utility of stable isotope data in mineral deposit genesis and metallogeny, and, importantly, for mineral exploration.

1 Introduction

In the early 1900s isotopes, which are now recognized as atoms of a chemical element that differ in the number of neutrons, were discovered independently by studying radioactive decay series (Soddy 1913) and measuring the deflection of neon ions through a magnetic field (Thomson 1913). In the case of radioactive isotopes, different isotopes were originally described as different elements, however, both of these investigations concluded that there were varieties of chemical elements, which Soddy (1913) termed isotopes. It was not until the discovery of the charge neutral neutron by Chadwick (1932) that it was recognized that isotopes were the consequence of the number of neutrons in the nucleus. Isotope geochemistry then grew from the recognition that the vast majority of elements have more than one isotope (e.g. Nier 1937) and that the abundances could be affected by geological processes.

In a seminal paper, Urey (1947) estimated the temperature-dependent fractionation of oxygen isotopes between $CaCO_3$ and water and proposed that this fractionation could be used as a geothermometer. Subsequently Epstein et al. (1953) calibrated the geothermometer by growing molluscs at different temperatures, initiating the field of stable isotope geochemistry.

One of the first stable isotope studies on ore deposits was by Engel et al. (1958a) on changes in the carbon and oxygen isotope composition of limestone in the Leadville district (Colorado, USA) due to hydrothermal alteration. This initiated a blossoming of stable isotope research on mineral deposits, and by the late 1960s and 1970s, stable isotope geochemistry, particularly the use of hydrogen, boron, oxygen, carbon and sulfur isotopes, had become one of the mainstays of ore genesis research, providing methods by which temperatures of mineral deposition could be estimated, sources of ore fluids, sulfur and some metals could be identified, and chemical reactions in ore forming systems could be tracked. Furthermore, as discussed by Barker et al. (2013) and others, stable isotopes have great potential as exploration vectors to many types of mineral deposits. As an example, variations in whole rock oxygen isotope values were one of the vectors that led to the discovery of the 45 West volcanic-hosted massive sulfide deposit in Queensland, Australia (Miller et al. 2001).

Contributions in this section illustrate how light stable isotopes have been used in the study of and exploration volcanic-hosted massive sulfide (Huston et al. 2023), sediment-hosted base metal (Williams 2023), iron ore (Hagemann et al. 2023), and orogenic gold (Quesnel et al. 2023) systems. The use of stable isotopes in ore studies has been reviewed extensively, including in the second and third editions of *Geochemistry of Hydrothermal Ore Deposits* (Barnes 1979, 1997; Taylor 1979, 1997; Ohmoto and Rye 1979; Ohmoto and Goldhaber 1997). Other important overall reviews include those by Ohmoto (1986), Kerrich (1987), Taylor (1987), Seal (2006), and Shanks (2014), and there have been many reviews for specific deposit types. In addition, a number of more general reviews of stable isotope geochemistry have been published, including Valley et al. (1986), Kyser (1987), Valley and Cole (2001) and Hoefs (1997, 2021). This chapter provides context for the more detailed discussions of stable isotope geochemistry provided for individual mineral systems in later chapters.

2 Fundamentals of Light Stable Isotope Geochemistry

Light stable isotope geochemistry is concerned with variations in the relative abundance of stable isotopes of light elements, including hydrogen, helium, boron, carbon, oxygen, nitrogen, silicon, sulfur and chlorine. These elements share a number of characteristics: (1) low mass number, (2) large relative mass differences (difference in isotope mass relative to mass number) between the minor (generally the heavier) and the abundant isotope, and (3) sufficient abundance (>0.01%) of the minor isotope to allow precise measurements of the isotope ratio (Table 1). Of the above elements, only five—hydrogen, boron, carbon, oxygen and sulfur—are now used routinely in studies of mineral deposits and this review will concentrate on these elements. Helium (Simmons et al. 1987), nitrogen (Jia and Kerrich 1999), silicon (Zhou et al. 2007) and chlorine (Eastoe and Guilbert 1992) have also been used in mineral deposit studies, but not extensively, and are not discussed further.

3 Conventions, Definitions and Standards

Since the inception of stable isotope geochemistry as a separate field of geochemistry in the middle part of the last century, a set of conventions have been developed so that isotope data are reported systematically throughout the world. These conventions include definitions of notations used to report isotope data and standards to which isotope data are related.

The absolute isotope ratio, R, which is defined as the ratio of the number of atoms of the heavy isotope to that of the light isotope, is the fundamental parameter measured. However, differences in absolute ratios between two samples can be measured more precisely than absolute ratios. Consequently, the δ-value was introduced, which is a measure of the difference in absolute isotope ratios between the measured sample and a standard, to report the relative deviation of stable isotope abundances:

$$\begin{aligned}\delta_X(\text{‰}) &= 1000(R_X - R_{STD})/R_{STD} \\ &= 1000(R_X/R_{STD} - 1) \qquad (1)\end{aligned}$$

where R_X is the isotope ratio of an unknown sample and R_{STD} is the isotope ratio of a standard (McKinney et al. 1950). Because of the magnitude of most stable isotope variations in terrestrial materials, it is convenient to report the δ-value in per mil (‰) (Coplen 2011).

To facilitate interlaboratory comparisons, a set of international standards has been established for all geologically important light stable isotopes (Coplen et al. 1983). For hydrogen and oxygen, the internationally accepted reference standard is V-SMOW (Vienna[1] Standard Mean Oceanic Water; identical to the original SMOW); for boron the reference standard is NIST 951 (boric acid); for carbon the accepted reference standard is V-PDB (Vienna Peedee Belemnite; V-PDB is used as a reference standard for oxygen in some studies); and for sulfur the reference standard is V-CDT (Vienna Canyon Diablo

[1] Vienna denotes standard defined by the International Atomic Energy Agency based in Vienna.

Table 1 Characteristics of light stable isotope systems commonly used in mineral system studies and mineral exploration

Isotope system	Abundances of stable isotopes[1]	Isotope ratios of international standards
Hydrogen	^{1}H: 99.9885% ^{2}H (D or deuterium): 0.0115% ^{3}H (tritium) is a man-made radioactive isotope that does not naturally exist	V-SMOW[2] $^{2}H/^{1}H$: 155.75 (±0.08) × 10^{-6}
Boron	^{10}B: 19.9% ^{11}B: 80.1%	NIST 951[3] $^{11}B/^{10}B$: 4.04362 (±0.00137)
Carbon	^{12}C: 98.93% ^{13}C: 1.07% ^{14}C is a radioactive isotope with a short half-life that is produced by the interaction of cosmic rays with ^{14}N	V-PDB[4] $^{13}C/^{12}C$: 11,180.2 (±2.8) × 10^{-6}
Oxygen	^{16}O: 99.757% ^{17}O: 0.038% ^{18}O: 0.205%	V-SMOW[2] $^{18}O/^{16}O$: 2005.20 (±0.45) × 10^{-6} $^{17}O/^{16}O$: 379.9 (±0.8) × 10^{-6} V-PDB[4] $^{18}O/^{16}O$: 2067.2 × 10^{-6} $^{17}O/^{16}O$: 386.0 × 10^{-6}
Sulfur	^{32}S: 94.93% ^{33}S: 0.76% ^{34}S: 4.29% ^{36}S: 0.02%	V-CDT[5] $^{34}S/^{32}S$: 44,150.9 (±11.7) × 10^{-6} $^{33}S/^{32}S$: 7877.29 × 10^{-6}

Data sources: [1]Rosman and Taylor (1999), [2,4]Werner and Brand (2001), [3]Catanzaro et al. (1970) and [5]Ding et al. (2001)

Troilite). Table 1 gives the absolute abundance ratios (on an atomic basis) for these reference standards. Positive δ values imply that the heavy isotope (e.g. D (^{2}H), ^{11}B, ^{13}C, ^{18}O or ^{34}S) is enriched in the sample relative to the reference standard.

3.1 Isotope Systems

Table 1 summarizes the characteristics of the five major light stable isotope systems used in mineral systems research and exploration. Most of these systems (H, B and C) are characterized by two stable isotopes, but the oxygen and sulfur systems contain three and four isotopes, respectively. In addition, the hydrogen and carbon systems contain radiogenic isotopes that are produced by man-made nuclear reactions (^{3}H or tritium and ^{14}C: Health Physics Society 2011) or naturally by interaction of cosmic rays with the upper atmosphere (^{14}C and ^{3}H: Korff and Danforth 1939; Health Physics Society 2011). These radiogenic isotopes are generally not used for ore genesis studies, but have utility in the study of young geological and archeological systems.

4 Fractionation of Stable Isotopes

Like other chemical components, isotopes can be involved in chemical reactions that change their abundances between two or more chemical substances. As an example, galena and sphalerite exchange sulfur isotopes according to the following reaction:

$$Zn^{34}S + Pb^{32}S = Zn^{32}S + Pb^{34}S \quad (2)$$

Like other chemical reactions, the equilibrium constant (K) can be expressed as follows:

$$\begin{aligned} K &= \alpha = (a_{Zn^{32}S}a_{Pb^{34}S})/(a_{Zn^{34}S}a_{Pb^{34}S}) \\ &= (a_{Pb^{34}S}/a_{Pb^{32}S})/(a_{Zn^{34}S}/a_{Zn^{32}S}) \quad (3) \end{aligned}$$

where K and α are temperature dependent, and a_N indicates chemical activity of substance N. Because isotopes behave nearly ideally, the above activity ratios are essentially identical to isotope ratios, and the above equation reduces to the isotope fractionation α:

$$\alpha = R_{PbS}/R_{ZnS} \tag{4}$$

The difference in isotope composition between two minerals is commonly expressed as $\Delta_{A\text{-}B}$, which is defined as δ_A - δ_B. Using the approximation that 1000ln X ≈ X (valid only if X ~ 1.00) and the definitions of R and δ, $\alpha_{A\text{-}B}$ is related to $\Delta_{A\text{-}B}$ as follows:

$$1000\ln\alpha_{A-B} \approx \Delta_{A-B} \tag{5}$$

For $\Delta_{A\text{-}B}$ less than 10‰, this approximation is correct to within 0.25‰. For $|\Delta_{A\text{-}B}|$ greater than 10‰, the approximation becomes less accurate and isotope fractionation should be determined using the exact relationship between α and δ:

$$\begin{aligned}\alpha_{A-B} &= (1+\delta_A/1000)/(1+\delta_B/1000)\\ &= (1000+\delta_A)/(1000+\delta_B)\end{aligned} \tag{6}$$

For isotope systems in which fractionation is relatively small (e.g. oxygen and boron), the approximation is mostly valid, for but for the hydrogen, carbon and sulfur systems, in which fractionations can reach 100‰, the exact relationship should be used.

The fractionation of isotopes has been studied experimentally and theoretically for many minerals (c.f. Taylor 1979; Ohmoto and Rye 1979; Kieffer 1982; Clayton and Kieffer 1991; Kowalski and Wunder 2018, and many others). These data form the basis from which isotope data can be interpreted in a geologically meaningful manner. Readers are referred to Beaudoin and Therrien (2009) and the related web-based and macOS/iOS/padOS/Android AlphDelta isotope calculator (http://alphadelta.ggl.ulaval.ca) for an up-to-date compilation of these data for hydrogen, carbon, oxygen and sulfur and a tool for fractionation and equilibrium temperature isotope calculations.

4.1 Multiple Isotope Measurements

This convention so far considers mass-dependent fractionation, which can be expressed from the ratio of two isotopes of an element. This is sufficient for analysis and consideration of δD, $\delta^{11}B$, $\delta^{13}C$, and $\delta^{18}O$. While oxygen has another minor isotope ^{17}O, and its abundance is important in atmospheric chemistry, its use for ore studies has just begun (Peters et al. 2020), and multiple oxygen isotope analysis is not considered further here. However, for sulfur isotopes the abundances of the other minor isotopes are increasingly being used to examine potential processes affecting sulfide formation (one of the earliest examples of this is the study of Hulston and Thode 1965), and so some consideration of multiple isotope measurements is required.

Sulfur has four stable isotopes ^{32}S, ^{33}S, ^{34}S, and ^{36}S. As described above, the two-isotope ratio $^{34}S/^{32}S$, and hence $\delta^{34}S$, is generally used to consider an isotope fractionation related to mass dependent fractionation. Based on $\delta^{34}S$, a prediction can therefore be made concerning the magnitude of $\delta^{33}S$ and $\delta^{36}S$ in any sample. This can be considered in terms of relative mass difference from the major isotope. If the isotope fractionation is 1.0‰ for $\delta^{34}S$, then the prediction will be that $\delta^{33}S$ will be about 0.5‰ [0.5 being obtained from (33–32)/(34–32)], and $\delta^{36}S$ will be about 2.0‰ [2 being (36–32)/(34–32)]. This approximation can be refined using exact nuclidic masses, and also by changing the mass fractionation law that relates $\delta^{33}S$ and $\delta^{36}S$ to $\delta^{34}S$ (e.g. linear, power, exponential).

Another isotope parameter (Δ) can then be determined for the difference in isotope abundance measured, versus that predicted based on $\delta^{34}S$. A common formalism for $\Delta^{33}S$ and $\Delta^{36}S$ can be expressed as[2]:

$$\Delta^{33}S = \delta^{33}S_X - 0.515\delta^{34}S_X \tag{7}$$

and

[2] These definitions of $\Delta^{33}S$ and $\Delta^{36}S$ differ from that of $\Delta^{34}S$, which indicates the difference in isotopic composition between two minerals, that is $\Delta^{34}S_{A\text{-}B} = \delta^{34}S_A - \delta^{34}S_B$.

$$\Delta^{36}S = \delta^{36}S_X - 1.90\delta^{36}S_X \quad (8)$$

where the values 0.515 and 1.90 represent the mass differences based on the exact nuclidic masses, and a linear mass-dependent relationship is assumed. For mass dependent fractionation, $\Delta^{33}S$ and $\Delta^{36}S$ will be 0 ‰; for non-mass-dependent fractionation, or, mass independent fractionation, $\Delta^{33}S$ and $\Delta^{33}S$ can be non-zero. It should be noted that the normalized abundances of $\Delta^{33}S$ and $\Delta^{36}S$ are based on the concurrent measurements of $\Delta^{34}S$, with $\Delta^{33}S$ and $\Delta^{36}S$ and so the precision of $\Delta^{33}S$ and $\Delta^{36}S$ is not a function of the external reproducibility of $\Delta^{34}S$ in a sample suite.

5 Analytical Methods

Since the first description of analytical techniques for carbon and oxygen isotope analyses of carbonate minerals by McCrea (1950), analytical methods have evolved to the point where for most stable isotope systems, procedures exist that allow the analysis of less than a few pictograms (pg: 10^{-12} g) of sample for micro-analytical methods. As the methods for stable isotope analysis is a wide field and a number of volumes have reviewed the analytical techniques (e.g. de Groot 2004, 2009; Foster et al. 2018), the description below is brief.

In a broad sense, analytical methods for stable isotopes can be grouped into bulk methods that generally produce a gas that is analysed by a gas-source isotope ratio mass spectrometer, and microanalytical methods that generally extract material from the sample using either a laser or an ion beam followed by analysis using gas-sourced mass spectrometry, inductively coupled plasma mass spectrometry (ICP-MS) or secondary ion mass spectrometry (SIMS). Bulk methods, which were developed in the 1950s and 1960s, are still the mainstay for analyses in many laboratories. In contrast, microanalytical methods began to be developed in the 1980s and 1990s, and are only now becoming widespread as the number of instruments has increased and the relative costs of analyses have decreased.

5.1 Bulk Analytical Methods

Most bulk stable isotope analytical methods involve the conversion of solid minerals into gases that can be analysed using gas-source mass spectrometry. Carbonate minerals are typically reacted with phosphoric acid at various temperatures (depending on the mineral) to produce CO_2, which is then analysed for $\delta^{13}C$ and $\delta^{18}O$ (McCrea 1950). Typical external uncertainties (2σ)[3] for this method are 0.10‰ for $\delta^{13}C$ and 0.16‰ for $\delta^{18}O$.

Sulfide minerals are typically reacted with excess CuO at 800–1000 °C, with the sulfide converted to SO_2 gas, which is analysed for $\delta^{34}S$ (Grinenko 1962; Robinson and Kusakabe 1975). Alternatively, sulfide minerals can be reacted with BrF_5 or ClF_3 to produce SF_6 gas, which is then analysed (Puchelt et al. 1971). This latter method has an important advantage over the SO_2 method in that fluorine has only one isotope (versus three for oxygen), removing uncertainties in analysis due to non-S isotope variations. As a consequence, the fluorination method is used in multiple sulfur isotope studies (Rumble et al. 1993). External uncertainties (2σ) for the SO_2 method are typically 0.2–0.3‰ for $\delta^{34}S$, and uncertainties for the SF_6 method are typically 0.2‰. For $\Delta^{33}S$ and $\Delta^{36}S$, uncertainties of the order of 0.04 ‰ and 0.4 ‰ (2σ) reflect the relative abundances of ^{33}S and ^{36}S (Farquhar et al. 2007).

As the reagents that have traditionally been used to fluorinate sulfide (and other) minerals (BrF_5, ClF_3 and F_2) are hazardous and difficult to handle, Ueno et al. (2015) developed a rapid technique for fluorination of Ag_2S using solid CoF_3 as the fluorination agent. Following reaction at 590 °C using a Curie-point pyrolyzer, the resulting SF_6 gas was purified using a combination of cryogenic cleaning and gas

[3] To be consistent with radiogenic isotope systems, uncertainties cited here are external 2σ values. "External" refers to uncertainties determined from repeat analyses of individual samples or standards; internal uncertainties, which are usually less than external uncertainties, refer to uncertainties due to counting statistics on individual analyses.

chromoatogarphy. Details are provided by Ueno et al. (2015) and Caruso et al. (2022). External (2σ) uncertainties are 0.7‰ for $\delta^{34}S$, 0.01‰ for $\Delta^{33}S$ and 0.2‰ for $\Delta^{36}S$.

Prior to analysis, sulfate minerals are typically converted to Ag_2S using two methods. In the first method, described by Rafter (1957), the sulfate minerals are first dissolved, then precipitated as barite after the addition of $BaCl_2$ to the solution. The barite is reduced with graphite at 900–1050 °C to produce BaS and CO_2, then the BaS is dissolved and AgCl added to precipitated Ag_2S. Alternatively, the sulfate minerals are boiled either in an HI-H_3PO_2-HCl solution (Thode et al. 1961) or in Kiba reagent (Sasaki et al. 1979) to convert sulfur into H_2S, which is then precipitated as Ag_2S with the addition of AgCl. The Ag_2S produced in both methods is analysed using either the CuO reaction or fluorination methods described above for sulfide analysis. The graphite-reduction method has the advantage that $\delta^{18}O$ can also be determined from the CO_2 produced.

Oxygen is extracted from rocks and minerals by fluorination of the rock with either BrF_5 or ClF_3 at temperatures of 450–690 °C, depending on the mineral being analysed. The heating is done either with a heating element surrounding the reaction vessels or with lasers. This produces O_2 gas, which, historically, has been converted to CO_2 gas by Pt-calatysed reaction with graphite (Clayton and Mayeda 1963), although many laboratories currently use O_2 directly. The CO_2 or O_2 gas is then analysed. This method typically produces external uncertainties (2σ) in $\delta^{18}O$ of 0.2–0.4‰.

Hydrogen isotope analysis of whole rocks, minerals and fluid inclusions involves a fairly complicated procedure to produce H_2 gas suitable for mass spectrometry. First the sample is heated to 150–200 °C under vacuum for several hours to remove adsorbed water. After the adsorbed water has been removed, whole rock and mineral samples are heated to a temperature of 900 °C using either a resistance furnace or an induction oven to release water, which is collected on a liquid N_2 trap. To analyse fluid inclusions, after removal of adsorbed water, (generally) quartz chips are heated to 800 °C, which decrepitates the fluid inclusions, and water and other gases are collected using liquid N_2. After cryogenic purification, the collected water is then reduced to form H_2 gas using either heated uranium (Bigeleisen et al. 1952) or zinc (Vennemann and O'Neil 1993), with the H_2 gas then being analysed. This method typically produces external uncertainties (2σ) in δD of 4‰. It must be stressed that fluid inclusion decrepitation results in the collection of the mixing of all types of inclusions, which can introduce significant uncertainties in interpreting the resulting data.

Production of gases for bulk isotope analyses is always conducted on vacuum lines with all conversions *in vacuo*, and the resulting gases cleaned of impurities cryogenically. In most methods gas conversion is undertaken at high temperatures and involves the use of furnaces. In addition, preparation of SF_6 and O_2 by fluorination involves strong, and dangerous, oxidants. Hence, preparation of all gases for gas-sourced mass spectrometry has a degree of hazard, which some (though not all) micro-analytical methods eliminate.

In addition to the preparation methods described above, many laboratories use elemental analyzers to combust solids and produce gases such as CO_2 and SO_2, which are separated automatically using gas chromatography and analyzed by gas-source mass spectrometry (Pichlmayer and Blochberger 1988; Giesemann et al. 1994). With the removal of the need for cryogenic cleaning of the analyzed gases, this development has simplified analytical methods and made them much more rapid.

Bulk analytical methods for boron isotopes differ from those used for other light stable isotopes in that analysis is undertaken using either thermal ionization mass spectrometry (TIMS) or inductively-coupled plasma mass spectrometry (ICP-MS) rather than gas-source mass spectrometry. As a consequence, preparation for analysis involves solution chemistry and purification, whereby care is needed to avoid partial loss of volatile boron and resulting isotope fractionation. Details of sample preparation are described in detail by Sah and Brown (1997), Aggerwal and

You (2016) and Foster et al. (2018). Historically, boron isotopes were generally analysed using TIMS, but the development of ICP-MS analysis in the late 1990s has enabled much easier and more rapid analysis. TIMS analyses are done using positive (PTIMS) or negative ions (NTIMS), the pros and cons of which are described in the specialized literature cited. TIMS analysis yields a typical uncertainty (2σ) of 0.4–0.6‰, whereas ICP-MS analysis has an (2σ) uncertainty of 0.2–0.3‰ (c.f. Foster et al. 2013).

5.2 Micro-analytical Methods

The development of micro-analytical tools for analyzing stable isotopes began in the late 1980s and continues today, with both the spatial resolution and precision of the analyses improving. Micro-analytical methods have used two different techniques to extract and analyse samples: laser ablation followed by gas-source isotope ratio mass spectrometry or inductively-coupled-plasma mass spectrometry (ICP-MS, including triple quadrapole ICP-MS—ICP-QQQ-MS), and secondary-ion mass spectrometry (SIMS) utilising a focused ion beam of oxygen or cesium. The two techniques have both advantages and disadvantages. Laser-based systems tend to provide faster and less expensive analyses, but the analyses can have lower spatial and analytical precision and are relatively destructive. In contrast, secondary-ion mass spectrometry is slower and more expensive, but it can have better spatial and analytical precision, greater sensitivity, and is much less destructive, which can be important if multiple analyses are to be carried out of the same points.

5.2.1 Laser Heating and Laser Ablation

Lasers can be used in two distinct ways for isotope analysis – either for heating a small and localized sample, or for direct ablation. Isotope analysis by laser heating gas source isotope ratio mass spectrometry involves laser heating of a sample with an oxidant (O_2, or F_2) in a reaction chamber prior to purification of the product gas (O_2, SO_2 or SF_6: Elsenheimer and Valley 1992; Sharp 1990; Beaudoin and Taylor 1994) using cryogenic or chromatographic techniques, and measurement of the stable isotope composition by gas-source mass spectrometry. The advantage of the laser for heating is that localized heating of the sample can be effected without involving the container (crucible); hence blanks can be limited allowing smaller samples to be analysed. One of the first uses of laser heating for microanalysis was to determine $\delta^{18}O$ and $\delta^{34}S$ from silicate and sulfide minerals. Small crystals or small aliquots of powders are laser-heated using typically a Nd: YAG or CO_2 laser in a F_2 or BrF_5 (or ClF_3) atmosphere for oxygen analysis, or additionally with an O_2 atmosphere for sulfur analysis (Crowe et al. 1990; Kelley and Fallick 1990; Sharp 1990; Elsenheimer and Valley 1992; Akagi et al. 1993; Beaudoin and Taylor 1994).

Laser ablation, which can occur in both reactive and inert atmospheres, typically produces a crater 50–200 μm in diameter, depending on the mineral being ablated and laser power and focus. The depth of the pit is typically of a similar dimension. Reaction of the ablated mineral with a reactive atmosphere produces SO_2 (sulfide mineral in O_2 atmosphere), SF_6 (sulfide mineral in F_2, BrF_5 atmosphere) or O_2 (silicate or oxide mineral in BrF_5 atmosphere). In some systems the product gas is then cleaned offline and, in the case of oxygen, converted into CO_2 before analysis. In other systems, the product gas is automatically cleaned cryogenically or by gas chromatography and then directly introduced into the mass spectrometer for analysis. Laser ablation and a reactive atmosphere produces a reproducible, but mineral- and laboratory-dependent, fractionation effect that has to be corrected to produce final results. The typical external (2σ) errors associated with laser ablation are typically 0.2–0.4‰ for $\delta^{34}S$ and 0.2–0.6‰ for $\delta^{18}O$.

Laser ablation in an inert atmosphere has supplanted that in a reactive atmosphere. In this method ablation products are introduced directly into a (multi-collector) inductively coupled plasma mass spectrometer with a carrier gas (commonly He and/or Ar) (Mason et al. 2006; Bendall et al. 2006). These methods typically produce external errors (2σ) of 0.3–0.6‰.

Boron analysis using laser-ablation ICPMS has advantages of greater speed than SIMS and more flexibility in that it places lower demands on surface quality and is less strictly dependent on matrix-matched reference materials (but see Mikova et al., 2014). Studies report a similar level of external uncertainty as for SIMS (1‰ at 2σ). The main disadvantage, however, is the larger spot size and deeper penetration depth of the laser compared to the ion beam. In many cases a larger spot size is required as a greater volume of material is required for analysis if the boron concentrations are low.

5.2.2 Secondary Ion Mass Spectrometry

SIMS uses a focused ion beam, e.g. O^- or Cs^+, to ablate the sample. The interaction of the primary ion beam (10–20 keV) with the target leads to secondary ionization. The probability of ionization is element specific, with metals producing positive secondary ions with O^- primary ion beam, and non-metals being ionized to negative secondary ions with a Cs^+ ion beam. The chemistry of the ion beam also aids in ionization with electronegative oxygen producing strong electropositive element secondary ion beams and electropositive cesium producing strong electronegative element secondary ion beams (Ireland 1995). As such, SIMS differs from LA in that an ion beam is produced directly from the target. This makes it a sensitive technique, but also strongly matrix dependent and so mineral standards of close composition to the target minerals are required.

SIMS was initially developed in geochemistry in an attempt to measure trace element abundances that were inaccessible by electron beam techniques in the 1960s. The capability to examine radiogenic isotope systems was not developed until the SHRIMP ion microprobes when the combined issues of transmission (sensitivity) and high mass resolution were addressed. While stable isotope analysis does not require high sensitivity or high mass resolution, the low transmission of early instruments led to variable measured isotope compositions simply because of differences in ion paths through the instruments (Shimizu and Hart 1982). SIMS has also been applied to measuring stable isotopes including those of hydrogen, boron, carbon, oxygen, and sulfur. As such, early analyses suffered from reproducibility at a level of greater than one permil for oxygen and sulfur isotope systems (Valley and Graham 1991; Eldridge et al. 1987). Subsequent developments have led to much higher reproducibility with external uncertainties (2σ) of 0.3‰ for $\delta^{18}O$ analyses (e.g. Kita et al. 2009) and 0.3‰ for $\delta^{34}S$ analyses (e.g. Paterson et al. 1997; Kita et al. 2010). Moreover, recent developments have allowed analysis of all isotope ratios in both the oxygen (e.g. $\delta^{17}O$ and $\delta^{18}O$) and sulfur ($\delta^{34}S$, $\delta^{33}S$ and $\delta^{36}S$) isotope systems with similar uncertainties (Ireland et al. 2014).

A key aspect in stable isotope analysis is the development of multiple collection whereby all isotopes are measured at the same time thereby cancelling the uncertainty associated with the primary ion beam noise. This is now the standard methodology for large sector ion microprobes. For $\delta^{18}O$ measurements, sufficient $^{18}O^-$ is produced for analysis of both $^{18}O^-$ and $^{16}O^-$ in Faraday cup multiple collection mode. Similarly, for sulfur, ^{32}S, ^{33}S, and ^{34}S can be measured by Faraday cups. However, the low abundance ^{36}S cannot be measured with the Faraday cups using high-ohmic resistors because of the Johnson noise associated with the circuitry. Measurements of ^{36}S can use either an ion counter, or Faraday cup electrometers using charge mode (accumulation of charge across a capacitor). Precision for in situ analysis of $\Delta^{33}S$ can be better than 0.1 ‰ while $\Delta^{36}S$ measurements by electron multiplier are around 1 ‰ (2σ) (Whitehouse 2013) and around 0.4 ‰ (2σ) by charge mode (Ireland et al. 2014).

While the large sector ion microprobes are used to pursue highest precision stable isotope analyses, the nanoscale-SIMS offers high spatial resolution analysis. A typical "spot" for high precision stable isotope analysis is typically 10–30 μm, which allows sufficient material to be sampled to give sufficient ions for the required counting precisions. However, the purpose of the nanoscale-SIMS is to analyse targets at the

100 nm scale. At this sampling scale, only major isotopes can be measured to high precision.

Boron analysis by magnet-sector SIMS is commonly done with instruments set for mono-collection and mass-switching between ^{11} and ^{10}B, but multi-collection SIMS is routinely used in some installations of the large-geometry instruments. The external uncertainty obtained by mono-collection SIMS is typically between 1 and 2‰ (2σ), whereas multi-collection of the two isotopes can reduce this uncertainty by about half. Those values refer to "high-concentration" samples like tourmaline, which can contain percent levels of B. For low-concentration samples, often with less than 20 ppm B, the analytical precision may be limited by counting statistics. For example, large-geometry SIMS instruments have achieved precision of about 3‰ for samples with 1 ppm B (Foster et al. 2018).

An important, and for some applications limiting, issue for all SIMS analysis is the need for matrix-matched reference materials that are homogeneous on the micron scale or better. Such are available and internationally distributed for only a limited number of geologically relevant materials for stable isotope analysis. These include glass, tourmaline, carbonates, silicates/oxides and sulfides; these are primarily standards for mass-dependent fractionation. Further development is required for minor isotope abundances (e.g. for $\Delta^{33}S$ and $\Delta^{36}S$), boron isotope measurements at low levels, as well as a wider range of minerals relevant to ore deposits.

6 Processes That Fractionate Light Stable Isotopes

Fundamentally, fractionation of light stable isotopes is the result of small differences in translational, rotational and vibrational motions of bonds within the respective phases. Although it is beyond the scope of this review (see Kieffer 1982 for more details), bonds involving heavier isotopes are marginally stronger than bonds involving lighter isotopes of the same element. This effect is strongly dependent upon the relative mass difference (RMD = (AM_H - AM_L)/AM_L, where AM_H = atomic mass of heavy isotope and AM_L = atomic mass of light isotope) between isotope pairs: the effect is much stronger between hydrogen (^1H) and deuterium (^2H) (RMD ~ ((2–1)/1) = 1) than between isotopes of heavier isotopes (e.g. ^{34}S and ^{32}S; RMD ~ (34–32)/32 = 0.0625). Hence, isotope fractionation in the hydrogen system is much greater than that in the boron, carbon, oxygen and sulfur systems, in the absence of kinematic and biologic processes (C, S), and fractionations in light isotope systems are greater than fractionations in stable metal systems (e.g. Fe, Cu and Zn: Lobato et al. 2023; Mathur and Zhao 2023; Wilkinson 2023).

All geochemical processes involve some isotope fractionation, although for heavy elements (e.g. Pb, U), this effect is generally not measurable. The amount of fractionation depends not only on relative mass differences, but also on the type of geochemical reaction and the temperature at which the reaction occurs. For most stable isotope systems, the magnitude of isotope fractionation decreases with increasing temperature (Fig. 1), with mass dependent fractionation functions (except hydrogen) generally having the form:

$$1000\ln\alpha_{A-B}(\Delta_{A-B}) = A/T^2 \times 10^6 + B/T \times 10^3 + C \quad (9)$$

where T is temperature in Kelvin and A, B and C are experimentally- or theoretically-determined constants. This temperature dependence of isotope fractionation between minerals can be used to estimate paleotemperatures of mineral deposition (see below).

Most geochemical reactions can be divided into two general groups: those reactions that do not involve electron exchange, and those that do (redox reactions). In general, redox reactions involve much larger isotope fractionations. For example, $\Delta^{34}S$ at 300 °C for the reaction between sphalerite and aqueous H_2S, which is not a redox reaction, is 1.5‰, whereas $\Delta^{34}S$ at 300 °C for the reaction between anhydrite and aqueous H_2S, which is a redox reaction, is 22.0‰ (calculated using Alpha-Delta (Beaudoin

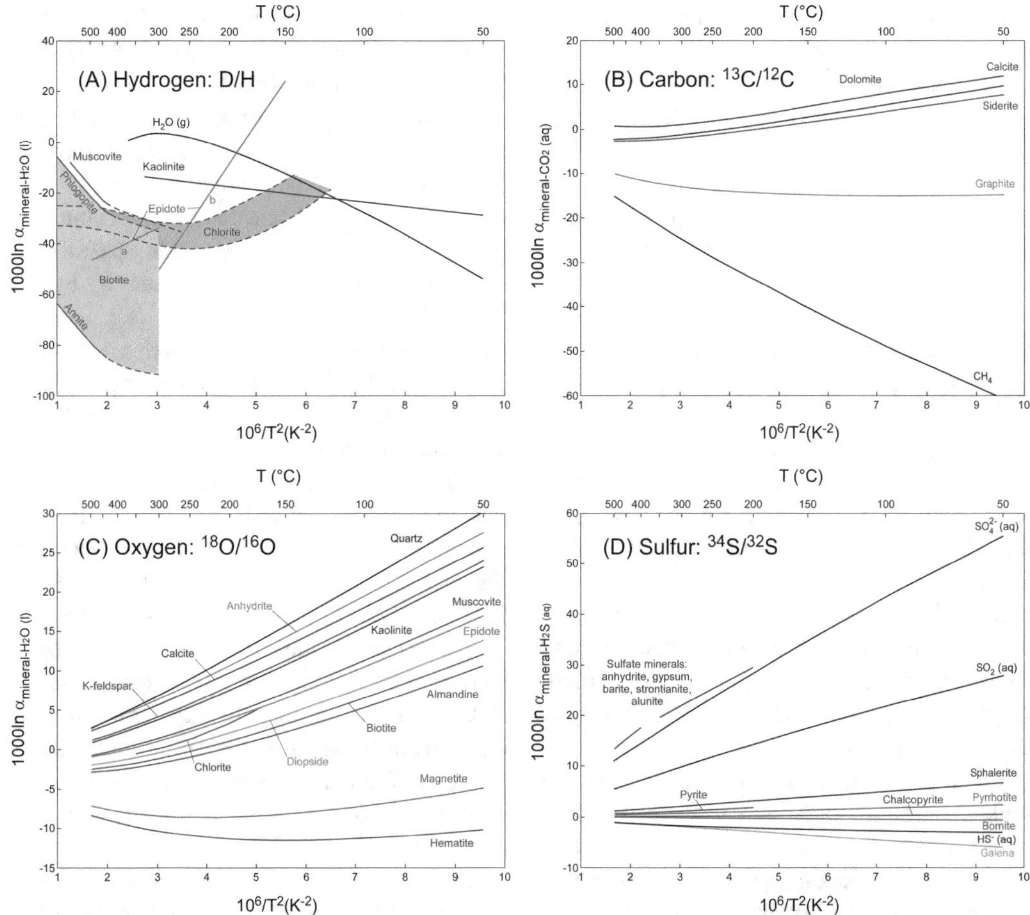

Fig. 1 Temperature versus 1000 ln α diagrams for the (A) hydrogen (1000 ln $\alpha_{mineral\text{-}H2O\ (l)}$), (B) carbon (1000 ln $\alpha_{mineral\text{-}CO2\ (aq)}$), (C) oxygen (1000 ln $\alpha_{mineral\text{-}H2O\ (l)}$) and (D) sulfur (1000 ln $\alpha_{mineral\text{-}H2S\ (aq)}$) isotope systems. For some systems, fractionation curves are also shown for dissolved species. Updated from Taylor (1979) and Ohmoto and Rye (1979). For hydrogen, isotope data of Taylor (1979: biotite and muscovite), Chacko et al. (1999: epidote curve a (300–600 °C)), Graham et al. (1980: epidote curve b (150–300 °C)), Gilg and Sheppard (1996: kaolinite) and Horita and Wesolowski (1994: H_2O vapor) were used. For carbon, isotope data of Ohmoto and Rye (1979) were used. For oxygen, isotope data of Cole and Ripley (1998: chlorite), Sharp et al. (2016: quartz), Zheng (1993a: almandine, diopside and K-feldspar), Zheng (1993b: biotite, epidote, kaolinite and muscovite), Zheng (1999: anhydrite and calcite), and Zheng and Simon (1991: hematite and magnetite) were used. For sulfur, isotope data of Eldridge et al. (2016: SO_2 (aq) and SO_4^{2-} (aq)), Li and Liu (2006: bornite, chalcopyrite, galena, pyrrhotite andsphalerite), and Ohmoto and Rye (1979: sulfate minerals (anhydrite, gypsum, barite, strontianite and alunite) and pyrite), were used

and Therrien 2009); most fractionation factors discussed below are calculated similarly). Moreover, redox reactions include some reactions, such as disproportionation, and photolytic and biogenic reactions, that are characterized by distinctive isotope effects, and at low temperatures (<250–300 °C), kinetic effects can strongly effect isotope fractionation.

6.1 Reactions Not Involving Reduction or Oxidation

Reactions that do not involve electron exchange are many and varied, but the most important reactions in mineral systems (and geological systems in general) involve the transition between liquid and solid. In some systems (e.g.

involving H and O) the transition between liquid and gas is also important.

The simplest reactions involve a physical phase change without a compositional change. Examples include boiling of pure water, condensation of steam or the recrystallization of aragonite to form calcite. Although these reactions involve changes in physical parameters such as density, the chemical composition of the reactants and products do not change. However, the isotope composition does change: $\Delta^{18}O_{liquid-vapour}$ for boiling or evaporation of pure H_2O is 2.5‰ at 200 °C.

Although phase changes are important for some mineralising processes (e.g. boiling in some epithermal systems), by far the most important processes in mineral systems involve dissolution or melting in, or precipitation of minerals from, aqueous fluids or magmas. Processes whereby ore metals and other components are extracted from source rocks into a transporting media generally involve either melting of rocks to form a metal-bearing magma or water–rock reaction to form a metal-bearing aqueous fluid. In contrast, processes whereby metal is extracted from the melts or hydrothermal fluids to form mineral deposits generally involve precipitation or crystallization of ore and gangue minerals from these fluids. Isotope data can be very useful in tracing these processes. A third type of chemical reaction involves changing the speciation and complexing of components within hydrothermal fluids. This type of reaction may not be observable in rock chemistry, but in some cases, it can result in isotope fractionation. The last group of processes involves the formation of two immiscible fluids from an original fluid. Examples include the exsolution of an immiscible sulfide melt from a parent silicate magma or the exsolution of an aqueous magmatic-hydrothermal fluid during crystallization of a silicate magma.

As the exsolution of a magmatic-hydrothermal fluid from a crystallising magma occurs at magmatic temperature (700–1000 °C: Burnham 1979), and oxygen isotope fractionation at these temperatures is small, the inferred oxygen isotope composition of the magmatic-hydrothermal fluid (5.5–10.0‰: Taylor 1979) overlaps with that of igneous rocks (5.5–11.0‰: Taylor and Sheppard 1986).

The reaction of hydrothermal fluids with rocks causes significant changes to the mineralogy and chemical and isotope composition of the rock and of the fluid, as a function of temperature and atomic water/rock ratios. This process of hydrothermal alteration is dominated by hydrolysis. Many alteration minerals (e.g. sericite, chlorite, epidote and others), which are enriched in H_2O or OH^- relative to the rock-forming minerals they have replaced, are the product of hydrolytic reactions in which water has been added to the rock. Tourmaline is a special case in this regard because it is not only a hydrous mineral but it is commonly the only mineral sink for boron in alteration assemblages. If the fluid/rock ratio is high (as is generally the case for pervasive alteration assemblages), the isotope composition of the altered rock is determined by that of the altering fluid. In these cases, the isotope composition of the altering fluid can be estimated from the isotope composition of the altered rock, and, in some cases, the source of this fluid can be inferred.

The last non-redox process that affects isotope characteristics in geological systems is mixing between components with differing isotope compositions that does not involve redox reaction. Mixing can result in significant gradients in isotope ratios, which can be used to identify and map this process (see below). This process commonly is an important mechanism of ore deposition, with important applications for mineral exploration.

6.2 Redox Reactions

Many elements involved in chemical reactions in geological systems exist in multiple valence states. In mineral systems the most important multivalent elements are Fe (Fe^0, Fe^{2+} and Fe^{3+}), S (S^{2-}, S^0, S^{4+} and S^{6+}) and C (C^{4-}, C^0, C^{4+}). Many of the reaction types discussed above can involve changes in valency and if so, they are considered redox reactions. An important characteristic of most redox reactions is that while one element is reduced (i.e. gained electrons) during the reaction, another must be oxidized (i.e. lost electrons). Redox reactions generally involve strong isotope fractionation, because of the equilibrium between the reduced species forming lower bond energy compounds (e.g. H_2S, CH_4) with higher bond energy in oxidized compounds (e.g. SO_4, CO_2). Hence, large variations in isotope ratios can be an indicator of redox reactions, particularly those involving S, C and Fe (the importance of redox in Fe isotope fractionation is discussed by Lobato et al. 2023).

Figure 2a illustrates the variations in $\Delta^{34}S_{H2S\text{-}fluid}$ as a function of pH and f_{O2} (i.e. redox) at a temperature of 250 °C. Although current fractionation data indicate no significant gradients are present as a function of pH, there is a strong gradient in $\Delta^{34}S_{H2S\text{-}fluid}$ just below the boundary between the hematite and pyrite stability fields (c.f., Ohmoto 1972). This gradient reflects the relative abundance of reduced S (H_2S and HS^-) and oxidized S (H_2SO_4, HSO_4^- and SO_4^{2-}) species, which is a reflection of the oxidation state of the system (i.e. $\Sigma SO_4/\Sigma H_2S$, f_{O2} and f_{H2}). Hence, if a large gradient in $\delta^{34}S$ exists either in space or time (i.e. paragenesis), this gradient could indicate the presence of redox reactions involving S species, particularly if accompanied by changes in Fe-S–O mineralogy.

Ohmoto (1972) indicated that carbon isotopes can experience shifts as a consequence of redox reactions involving graphite and carbonate minerals. Figure 2b shows variations in $\Delta^{13}C_{H2CO3\text{-}fluid}$ as a function of f_{O2} and pH at 250 °C. There is a strong gradient in $\Delta^{13}C_{H2CO3\text{-}fluid}$ that corresponds to the conversion of the C^{4-} to C^{4+} in the lower part of the pyrite field. Hence gradients in $\delta^{13}C$ may indicate redox processes, although other geological processes (e.g. mixing) can also produce gradients.

Although Fig. 2 shows variations at 250 °C at specific conditions (details in caption), the

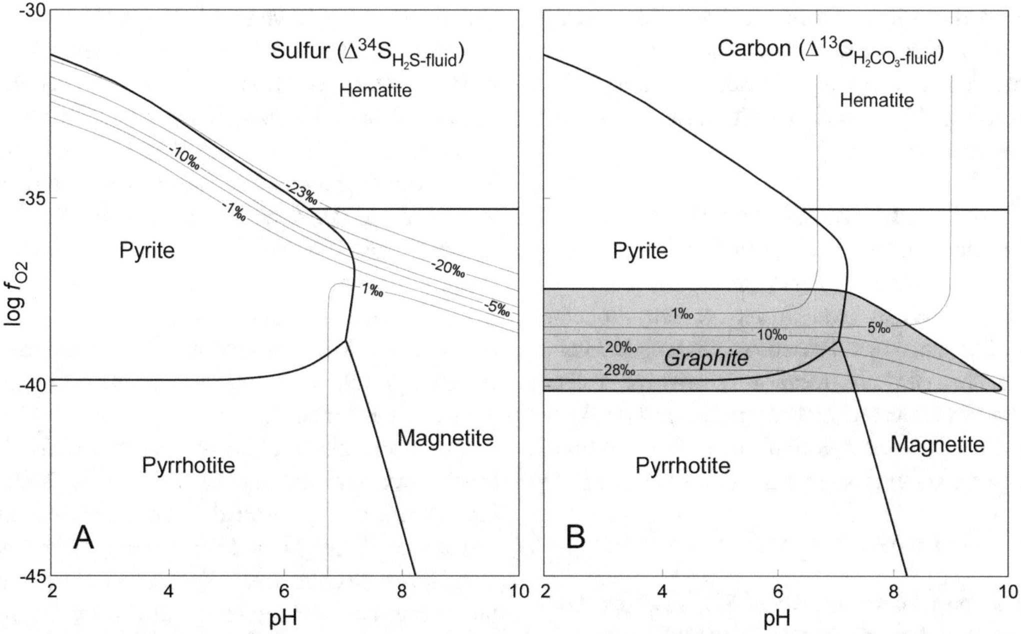

Fig. 2 Variations in (A) $\Delta^{34}S_{H2S\text{-}fluid}$ and (B) $\Delta^{13}C_{H2CO3\text{-}fluid}$ as a function of redox (f_{O2}) and pH at 250 °C. These diagrams are updated from the original diagrams in Ohmoto (1972) using updated thermodynamic and isotope fractionation data (Ohmoto and Rye 1979; Eldridge et al. 2016). The diagrams were calculated for a fluid with 5% dissolved NaCl, $10^{-2.5}$ m ΣS ($\sim$320 ppm S) and 1 m ΣC ($\sim$120 ppm C) taking into account activity coefficients and ion pairing

morphology of the diagram is retained for most hydrothermal conditions. For example, the gradient in $\Delta^{34}S_{H2S\text{-}fluid}$ near the upper margin of the pyrite field is retained at higher and lower temperature. Increasing the total sulfur content of the fluid expands the pyrite and, to a lesser extent, the pyrrhotite fields, but the sulfur isotope gradient is retained. Similarly, the gradient in $\Delta^{13}C_{H2CO3\text{-}fluid}$ also occurs near the bottom of the pyrite field and overlaps the graphite field under most hydrothermal temperatures.

Mixing of fluids with different redox characteristics can induce redox reactions and isotope fractionation. A good example of this is the mixing of relatively reduced H_2S-rich vapor generated by boiling of epithermal fluids into oxidized groundwater. As the gaseous H_2S is dissolved into groundwater, it is oxidized, producing sulfuric acid, which extensively alters rocks to aluminous assemblages (Simmons et al. 2005). In addition to the common hydrolytic, precipitation and mixing reactions, there are several special types of redox reactions that can have distinctive isotope effects, including disproportionation, photolytic and biologically-mediated reactions. In addition, many redox reactions are strongly affected by the speed, or kinetics, of reaction, which decreases as temperature decreases.

6.2.1 Disproportionation Reactions

Disproportionation reactions involve a reactant with intermediate valency producing products of higher valency and lower valency together. Unlike other redox reactions, disproportionation reactions do not necessarily involve valency changes to other chemical components during the reaction. The best example of such a reaction in mineral systems is the disproportionation of SO_2:

$$4SO_2 + 4H_2O = H_2S + 3H_2SO_4 \quad (10)$$

The most important role of SO_2 disproportionation in mineral systems is during the evolution of magmatic-hydrothermal fluids from moderately oxidized magmas. This reaction, which causes aluminous alteration due to the production of sulfuric acid, can produce distinctive isotope effects: in many porphyry systems in which this process is important, $\Delta^{34}S_{sulfate\text{-}sulfide} \sim 5\text{–}20‰$ (mostly 7–18‰: Rye 2005; Seal 2006). Hence, isotope fractionation can be indicative that this (and other) processes have occurred in a mineral system.

6.2.2 Mass-Independent Fractionation and Photolytic SO_2 Dissociation

For most reactions involving elements with three or more isotopes, fractionation between isotopes is "mass-dependent', that is the fractionation between isotopes is dependent on relative mass differences. In most cases $\delta^{17}O = 0.525\text{–}0.528 \times \delta^{18}O$ (Young et al. 2002; Pack and Herwartz 2014; Sharp et al. 2016), and $\delta^{33}S = 0.515 \times \delta^{34}S$ (Fig. 3a; Hulston and Thode 1965; Johnston 2011). Clayton et al. (1973, 1977), however, found that for O isotope fractionation within carbonaceous chondritic meteorites, the mass-dependent relationship did not hold. Subsequent studies have established this for anhydrous minerals from carbonaceous chondrites as well, where $\delta^{17}O = 0.941 \times \delta^{18}O - 4.00$ (Clayton 2003). Although it is beyond the scope of this contribution,[4] these meteoritic effects identified "mass-independent" isotope fractionation (MIF) for the first time (Ireland 2012). Subsequent studies found that the formation of ozone (Heidenreich and Thiemens 1986; Mauersberger 1987) and the photolytic dissociation of gaseous SO_2 (Farquhar et al. 2001) also involved MIF although it remains to be seen whether the MIF associated with ozone fractionation is related in any way to the ^{16}O variability in meteorites.

Photolytic dissociation of SO_2 and other S-bearing gases in the ancient (i.e. pre 2400 Ma) atmosphere was controlled by two factors, the abundance of the gas, and the shielding effects of other gas molecules (e.g. O_2 and O_3) present in the atmosphere (Farquhar et al. 2000; Lyons 2007; Ueno et al. 2009). The most important factor is likely shielding by O_2 and O_3, which

[4] See Clayton 2003 for more details and processes that cause fractionation in meteorites.

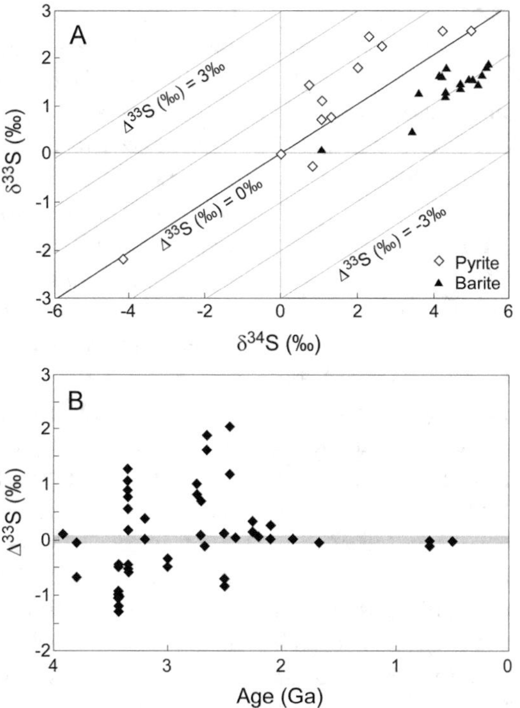

Fig. 3 Diagrams (modified after Farquhar et al. 2000) showing mass-independent fractionation in the ^{34}S-^{33}S-^{32}S isotope system. (A) $\delta^{34}S$ versus $\delta^{33}S$ diagram showing mass-dependent fractionation (heavy line with $\Delta^{33}S$ value of 0.0‰), mass-independent fractionation reflected as light lines with variable $\Delta^{33}S$ values, and analyses of syn-depositional sulfide and sulfate minerals from > 3.0 Ga sedimentary rocks. (B) Variations in $\Delta^{33}S$ of syn-sedimentary sulfur-bearing minerals with geological time. For the purpose this diagram syn-sedimentary minerals includes both syn-depositional and early diagenetic minerals. The gray bar in B encompasses the mean ± one standard deviations of 73 younger (<2.09 Ga) samples. The diagrams shown are the first documenting mass independent fractionation of sulfur isotopes from syn-sedimentary minerals; the data available have expanded by orders of magnitude since the initial study of Farquhar et al. (2000)

absorbs ultraviolet wavelengths of light thought to induce photolytic dissociation of SO_2 (Farquhar et al. 2001). As the concentration of free O_2 (and O_3) gas in the atmosphere prior to the Great Oxidation Event at ~ 2.42 Ga is thought to have been less than 1 ppm (Catling 2014), there would have been little shielding and photolytic dissociation of SO_2 would have been pronounced (Farquhar et al. 2001). In contrast, the increased shielding of O_2 (and O_3) and/or a decreased concentration of SO_2 in an oxygenated atmosphere after the Great Oxidation Event would have caused photolytic dissociation of atmospheric SO_2, and the mass-independent effects of this reaction, to cease.

Mass-independent fractionation in the S isotope system is measured by the deviation from mass-dependent fractionation. For ^{33}S, this deviation is measured by $\Delta^{33}S$ (Eq. 7). As the products (and isotope characteristics) of photolytic SO_2 dissociations in the atmosphere can be "rained out" into the hydrosphere, sedimentary and diagenetic S-bearing minerals may inherit the photolytic isotope signature. As shown in Fig. 3b, prior to the Great Oxidation Event, $\Delta^{33}S$ of syn-depositional or diagenetic S-bearing minerals is highly variable, with sulfide minerals (mostly pyrite) having mostly positive $\Delta^{33}S$ values and sulfate minerals having negative $\Delta^{33}S$ values (cf. Farquhar et al. 2000). This mass-independent fractionation effect is powerful as it can be used to identify S that was present in the atmosphere and hydrosphere during the Archean and trace it through subsequent geological cycles including tectonic and mineral systems, because it is indelible in younger mass-fractionation processes (LaFlamme et al. 2018a). Examples of how this technique has been used to identify sulfur sources and infer tectonic processes related to ore formation are discussed below.

6.2.3 Kinetic Effects, Thermochemical Sulfate Reduction and Biochemical Sulfate Reduction

Whereas most geochemical reactions occur very rapidly, redox reactions, involving the exchange of electrons, require significant time to occur, particularly at low temperatures (< 200 °C). Ohmoto and Lasaga (1982) were among the first to determine experimentally the rates of redox reactions between aqueous sulfate and aqueous sulfide. They found that at the pH conditions of most hydrothermal systems (4–7) thermochemical sulfate reduction to sulfide involved an intermediate thiosulfate species, and the rate at which thiosulfate formed was the rate-limiting step. Based upon this model they suggested that thermochemical sulfate reduction is an important hydrothermal process only at temperatures above 200 °C.

Subsequent studies have indicated that thermochemical sulfate reduction can be important at temperatures as low as 100–140 °C (Goldhaber and Orr 1995; Machel 2001; Thom and Anderson 2008), particularly when catalysed by the presence of H_2S, certain reactive metals (e.g. Ni, Co, Mn, Cu, Fe, Mg) and organic molecules (Machel 2001; Meshoulan et al. 2016). The isotope effects of thermochemical sulfate reduction depend not only on temperature, but also upon the availability of sulfate. In systems with excess sulfate the kinetic $\Delta^{34}S_{sulfate-sulfide}$ increases with decreasing temperature, from ~ 10‰ at 200 °C to ~ 20‰ at 100 °C (Machel et al. 1995; Meshoulan et al. 2016; and references therein). However, if the supply of sulfate is limited, or if the rate of sulfate supply is less than that of sulfate reduction, all sulfate is reduced to sulfide and $\delta^{34}S$ of the resulting sulfide (either as H_2S in sour gas or as sulfide minerals) approximates the $\delta^{34}S$ value of the original sulfate (Machel et al. 1995). Hence, the absence of significant fractionation between reactant sulfate and product sulfide may indicate limited availability of sulfate in a mineral system.

At lower temperatures (< 100 °C), biochemical sulfate reduction can be an important process to produce H_2S (Goldhaber and Orr 1995; Machel 2001). Due to its importance in a range of ecological and geological processes, including diagenesis, ore formation and sour gas formation, bacterial sulfate reduction has been extensively studied over the last three decades, and the geochemical and isotope effects of this process are well known. At present, sixty genera and over 220 species of sulfur-reducing bacteria are known (Barton and Fauque 2009), and kinetic sulfur isotope fractionation factors have been determined for 32 of these species, covering a large range of environments from freshwater and marine muds, salt lakes, arctic sediments, and hydrothermal environments. Detmers et al. (2001) reported experimental kinetic $\Delta^{34}S_{sulfate-sulfide}$ values of 2.0–42.0‰, and sulfate reduction rates of 0.9–434 fmole/cell/day.[5] These workers noted no correlation between $\Delta^{34}S_{sulfate-sulfide}$ and reduction rates, but found that the largest fractionations were associated with bacteria that completely oxidized electron-donor carbon to CO_2, whereas bacteria that only partially oxidize the carbon (to acetate) produce lower fractionation factors. The four largest fractionation factors (28.5–42.0‰) were produced by bacteria isolated from marine muds (Detmers et al. 2001). When both thermochemical and biochemical sulfate reduction processes are considered, the most extreme fractionations are associated with biochemical sulfate reduction, which Machel et al. (1995) highlighted as one that can be used to distinguish thermochemical (10–20‰) from biochemical sulfate reduction (15–60‰).

In addition, biochemical sulfate reduction appears to have mass-independent fractionation effects, as shown in Fig. 4. This fractionation, which has $\Delta^{36}S/\Delta^{33}S$ ~ -7, was interpreted by Johnston (2011) to indicate biogenic sulfate reduction. This contrasts with the mass-

[5] 1 fmole (femtomole) = 10^{-15} mol.

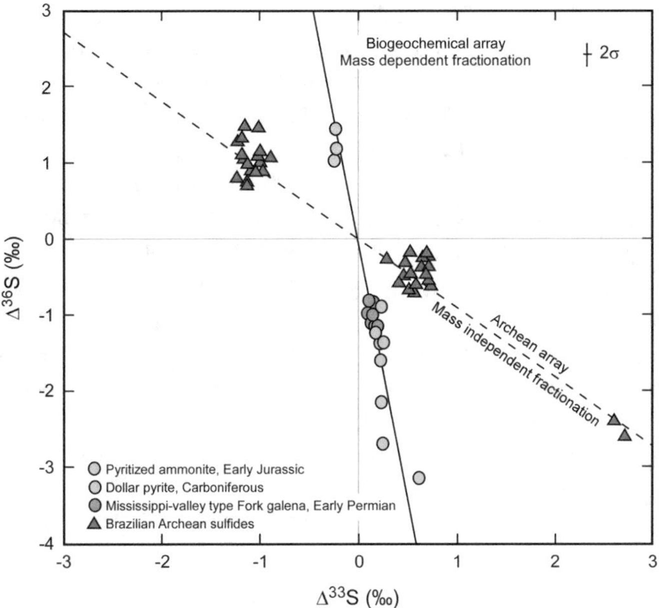

Fig. 4 S isotopic fractionations as measured in situ on the ion microprobe SHRIMP SI. $\Delta^{33}S$ and $\Delta^{36}S$ are the residuals in $^{33}S/^{32}S$ and $^{36}S/^{32}S$ after correction for mass-dependent fractionation (as expressed by $^{34}S/^{32}S$). Of particular significance is the resolution of the Biogeochemical Array (slope -7) from the Archean array (slope -1) which both can have subpermil anomalies in ^{33}S. Previously, the presence of $\Delta^{33}S$ anomalies alone has been used as a life signal, but in the Proterozoic with potential for Archean inheritance, a complete 4-isotope sulfur analysis is required to understand the origin of the sulfur

independent fractionation signal associated with Archean photolytic dissociation which has $\Delta^{36}S/\Delta^{33}S \sim -1$. Moreover, biogenic mass-independent fractionation is present in Phanerozoic rocks.

Biochemical (bacterial) reduction can also produce significant fractionation in the carbon isotope system. Reduction of acetate and CO_2 to form CH_4 produces significant, but variable fractionation ($\Delta^{13}C_{CO2\text{-}CH4}$) of up to 95‰ (Whiticar 1999), with typical fractionation between 25‰ and 60‰ (Conrad 2005). Kennedy et al. (2010) found that carbon associated with bacteriogenic iron oxides was depleted in ^{13}C by up to 22‰, with the greater depletion present in samples more highly enriched in organic carbon. Like biochemical sulfate reduction, the fractionation appears to be highly dependent upon the environment and it is likely to be also dependent upon the species of micro-organism involved.

6.3 Open- and Closed System Fractionation Between Isotope Reservoirs

Many geological processes, particularly devolatilization and degassing, involve the interaction of two distinct geochemical reservoirs. Isotope fractionation between these reservoirs can be modelled using two end-member processes: closed-system, or batch, fractionation, and open-system, or Rayleigh, fractionation. In closed-system fractionation, the two reservoirs are continuously in isotope equilibrium, whereas in open-system fractionation, aliquots of one reservoir are removed and isolated and do not interact with the other reservoir. Batch, or closed-system fractionation (Nabalek et al. 1984) can be modelled according to the relation:

$$\delta_{B,f} - \delta_{B,i} = -(1 - F)1000 \ln \alpha_{A-B} \quad (11)$$

Rayleigh, or open-system, fractionation (Broecker and Oversby 1971) can be modelled according to:

$$\delta_{B,f} - \delta_{B,i} = 1000\left(F^{(\alpha_{A-B}-1)} - 1\right) \quad (12)$$

where $\delta_{B,f}$ is the final isotope composition of reservoir B, $\delta_{B,i}$ is the initial isotope composition of reservoir B, and F is the fraction of reservoir B remaining.

Taylor (1986a) provided a good illustration of the differences between closed- and open-system hydrogen isotope fractionation during the degassing of a magmatic vapor from a magma (Fig. 5). In closed systems (curves a and b) where the vapor remains in contact and in isotope equilibrium with the melt, δD of both reservoirs decrease uniformly as the vapor evolves from the melt. If all hydrogen is extracted into the vapor, the final δD of the vapor is identical to that of the original magma.

In contrast, open-system fractionation (curves c and d) involves continuous removal and isolation of aliqots of evolved vapor from the magma, with the aliquots equilibrating with the magma only prior to removal. As a consequence, the δD characteristics of the magma and evolved vapor decrease dramatically as the amount of H_2O remaining in the magma (F) decreases (Taylor 1986a). Individual aliquots of vapor produced during open-system magma degassing can have dramatically lighter

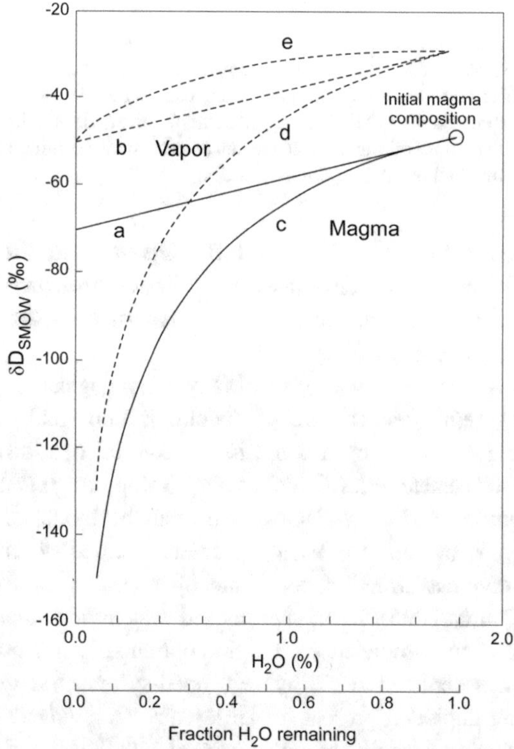

Fig. 5 Hydrogen isotope shifts in magmatic vapor and magma in hypothetical models of closed-system (batch) degassing (curves a and b), and open-system (Rayleigh) degassing (curves c and d; curve e indicates the accumulated magmatic vapor). The solid lines indicate modelled composition of the residual magma, and the dashed lines indicated with modelled composition of the evolved vapor. The initial magma had δD = -50‰ and $[H_2O]$ = 1.9%; a value for 1000 ln $\alpha_{vapor\text{-}melt}$ of 20‰ was used. Reproduced with permission from Taylor (1986a); Copyright 1986 Mineralogical Society of America

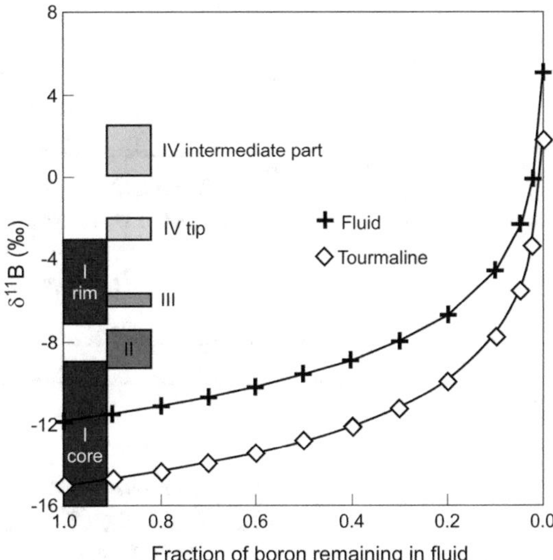

Fig. 6 Rayleigh fractionation model to explain boron isotope zoning in tourmaline from the Teremkyn gold deposit, Darasun district, eastern Russia (modified after Baksheev et al. 2015). The initial boron (12‰) is interpreted to be granite-derived (i.e., stage I core) while the composition of later tourmaline (stage I rim and stages II to IV) reflects Rayleigh fractionation driven by tourmaline crystallization (Baksheev et al. 2015)

δD than fluids produced during closed-system degassing, particularly as degassing proceeds towards completion. However, the accumulated vapor produced by open-system degassing has a much more moderate evolution (curve e), not dissimilar to that produced by closed-system degassing.

Open- vs. closed-system fractionation has also been demonstrated for the boron isotope system. In this case, the fluid-vapor fractionation is generally insignificant (only about 1 ‰ at 400° C: Kowalski and Wunder 2018) but fluid-mineral fractionation is strong owing to the contrasting coordination environment of the boron ion in aqueous fluid versus solid phases. Furthermore, in the case of tourmaline or other B-rich minerals, crystallization can strongly deplete the fluid reservoir in boron, so that Rayleigh fractionation is a common process. This was demonstrated in controlled experiments by Marschall et al. (2009) and Rayleigh fractionation has been invoked to explain systematic core-rim isotope variations in zoned tourmaline from hydrothermal ore deposits in several studies (e.g., Baksheev et al. 2015, Fig. 6).

7 Light Stable Isotope Characteristics of Major Geological Reservoirs

Ore fluids in mineral systems have a number of different sources and they carry metals, sulfur, carbon, boron, chlorine and other components that have been sourced from many different geochemical reservoirs. Stable isotope studies are one of very few tools to determine these sources as the reservoirs commonly (although not always) have distinctive stable isotope signatures. Figure 7 and Table 2 summarizes the isotope characteristics of a range of natural reservoirs. Figure 8a illustrates the characteristics of different crustal fluids on a $\delta^{18}O$ versus δD diagram, and Fig. 8b shows a similar plot of $\delta^{11}B$ versus δD used by Adlakha et al. (2017) to distinguish fluid sources of the McArthur River U deposit. Huston and Champion (2023), Lobato et al. (2023), Mathur and Zhao (2023) and Wilkinson (2023) discuss how radiogenic lead isotopes and heavy stable metal isotopes can be used to infer metal source.

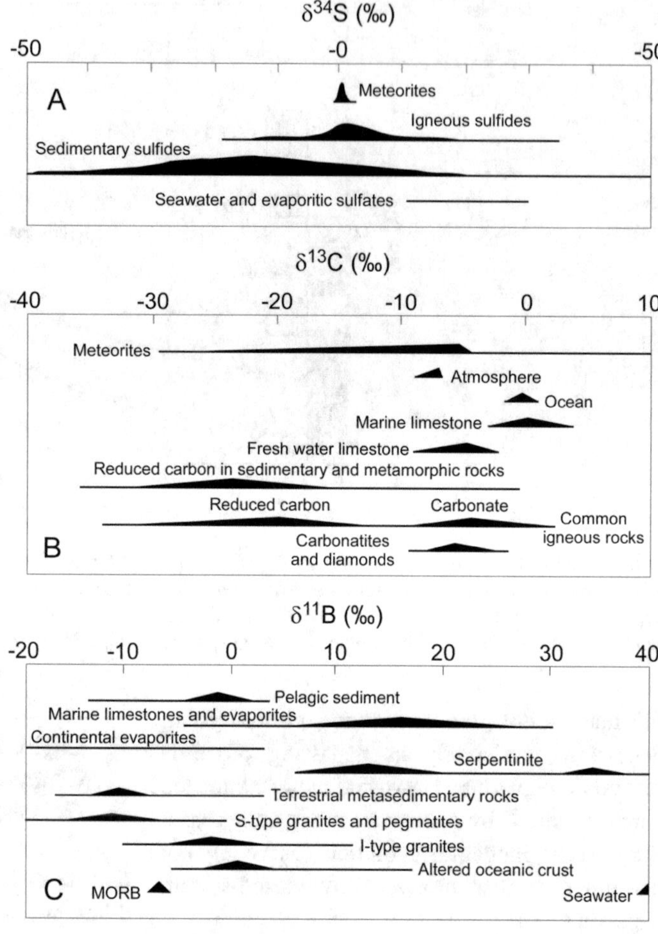

Fig. 7 Isotope characteristics of major reservoirs for (A) sulfur and (B) carbon reservoirs (modified after Ohmoto and Rye 1979), and (C) boron (derived from Marschall and Jiang 2011, deHoog and Savov 2018, Trumbull and Slack 2018)

Seawater is one of the major Earth reservoirs and it plays a key role in metallogenetic models for a range of stratiform ore deposit types. Modern ocean water has H- and O-isotope compositions identical to V-SMOW, and it is thought to have been relatively uniform through time, with ranges in δD and $\delta^{18}O$ of -25‰ to 0‰ and -3‰ to 0‰, respectively, since the Archean (Sheppard 1986). Seawater also has a uniform and distinctly heavy boron isotope composition ($\delta^{11}B \sim 40$ ‰) compared to all other Earth reservoirs (Table 2), which allows to distinguish marine and terrestrial provenance of boron in ore fluids and the minerals derived from them (e.g., Xavier et al. 2008). The H- and O-isotope composition of seawater is also the fundamental basis of the meteoric water line, an important reference on plots of H- and O-isotope compositions (Fig. 8a). The meteoric water line (Craig 1961) is defined by:

$$\delta D = 8\delta^{18}O + 10 \tag{13}$$

The isotope composition of meteoric water is governed by that of seawater and the temperatures of seawater evaporation and precipitation from clouds, both of which are related to latitude and altitude. Lower values of δD and $\delta^{18}O$ are characteristic of higher latitude and altitude (Sheppard 1986).

Traditionally, formation waters in sedimentary basins were thought to be seawater or brine entrained in sediments at the time of deposition (i.e. connate waters), but subsequent work has demonstrated that many (or most) formation waters are meteoric waters that infiltrated into

Table 2 Range in stable isotope values of the mantle and common upper-crustal sedimentary and igneous rocks

Reservoir	δD	$\delta^{18}O$	$\delta^{11}B$	$\delta^{13}C$	$\delta^{34}S$
Upper mantle	−80‰ to -50‰ for upper mantle; −46‰ to −32‰ for subduction-modified mantle	5.7 ± 0.3‰; 6.0–6.7‰ for subduction-modified mantle	7.1 ± 0.9 ‰ for MORB-source upper mantle	~ -5‰	0–1‰ (based on mantle xenolith data; overlaps with meteoritic data)
	Upper crustal sedimentary rocks				
Psammitic rocks	—	8–15‰	—	—	−70‰ to 70‰
Pelitic rocks	−90‰ to −40‰	13–20‰	−13 to 4 ‰ for pelagic clays, continental metapelites mostly less than -10 ‰	−30‰ to −15‰ (reduced carbon)	
Limestone and chert	—	25–35‰	−5 to 26 ‰ for marine carbonates, mostly above 5; −12 to 8 for chert	0 ± 5‰	
	Upper crustal igneous rocks				
Felsic rocks	−130‰ to −40‰ (this range is interpreted to be the result of Rayleigh degasing, with highest δD values associated with the least degased samples, as indicated by whole-rock H_2O content)	Mostly 6–10‰	−2 ± 5 ‰, rarely below -8 ‰ for I-type rocks; -11 ± 4 ‰, rarely above -8 ‰ for S-type felsic rocks	−35‰ to −10‰ (reduced carbon; −10‰ to 3‰ (oxidized carbon)	Mostly -5‰ to 10‰, but total range from −20‰ to 25‰
Mafic rocks	−120‰ to 0‰ (total range); −85‰ to −75‰ (primary magmatic composition of oceanic basalt)	4.9–8.0‰	−7.1 ± 0.9 ‰ for fresh MORB, −2 to 17 ‰ for altered basalt, gabbro in ophiolites, −3 to −12 ‰ for fresh OIB, higher values if seawater altered	−29.5‰ to −3.5‰ (total range); −11‰ to −3.5‰ for carbon extracted above 600 °C (the most likely magmatic range; most data cluster near -5‰,)	−3.0‰ to 2.5‰ (tholeiites); 2.5‰–6.0‰ (alkali basalts)

Data from Ohmoto and Rye (1979), Kyser (1986), Taylor and Sheppard (1986), Hoefs (1997) and Marschall and Foster (2018). In most cases metamorphosed rocks retain the isotope characteristics of their precursors (see text)

sedimentary basins subsequent to deposition and diagenesis (Kharaka and Hanor 2003). Evidence for this interpretation stems in part from isotope arrays in $\delta^{18}O$ versus δD plots that trend toward the meteoric water line and are consistent with the composition of meteoric fluids at likely recharge zones (see Kharaka and Hanor 2003 for discussion). On the other hand, some basins contain formation waters that are interpreted as true connate waters. As these connate waters were originally derived from seawater, their isotope signature should reflect that of seawater at the time of sedimentation (Sheppard 1986). Figure 8a indicates that connate waters, including evaporative brines, are characterized by δD and $\delta^{18}O$ values somewhat lower and higher, respectively, than modern seawater. These data from Gulf coast basins in the United States

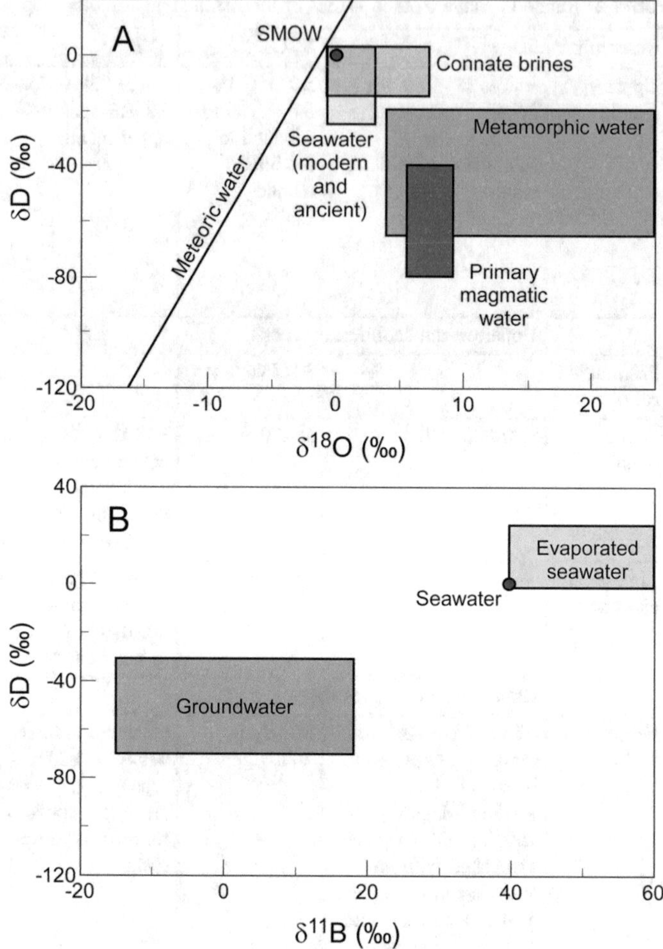

Fig. 8 Variations in stable isotope characteristics of fluids present at the Earth's surface or in upper crustal rocks. A. $\delta^{18}O$-δD diagram showing isotope characteristics of common crutal fluids. Data from Taylor (1979) and Kharaka and Hanor (2003). B. Variations in $\delta^{11}B$ (modified after Adlakha et al. 2017)

(Kharaka and Hanor 2003, and references therein), indicate a slight shift from oceanic water, possibly due to equilibration with the host rocks. Although limited, the data suggest that connate fluids may have a distinctive δD - $\delta^{18}O$ signature relative to meteoric water and other crustal fluids. Marine-derived connate fluids should also have a distinctive $\delta^{11}B$ signature relative to crustal fluids because of the uniquely high $\delta^{11}B$ value of seawater. This was used with success in studies of B-isotopes in tourmaline from U-deposits of the Athabasca Basin by Mercadier et al. (2012) and Adlakha et al. (2017) as shown on Fig. 8b.

The stable isotope composition of the upper mantle and mafic igneous rocks derived from it are thought to be relatively homogeneous, with $\delta^{13}C \sim -5‰$, $\delta^{18}O \sim 5.7 \pm 0.3‰$, $\delta^{34}S \sim 0$–1‰ (with $\Delta^{33}S$, $\Delta^{36}S \approx 0$) and $\delta^{11}B \sim 7 \pm 1$ ‰ (Table 2). The exception to this is where such rocks are altered on the seafloor or, in the case of arc settings, where the upper mantle is hydrated from slab-released fluids. The homogeneity of mantle-derived rocks contrasts with crustally-derived rocks, which are more variable (Fig. 7) due to the wider range of source components involved and to the number of chemical processes (metamorphism, melting, alteration and weathering) which cause isotope fractionation, particularly in the upper crust, atmosphere and hydrosphere. For example, siliciclastic sedimentary rocks have higher $\delta^{18}O$ values than igneous rocks from which they were derived due to fractionation during low temperature erosion and chemical weathering. Sedimentary rocks also have a wide range in $\delta^{13}C$ and $\delta^{34}S$ values (Table 2) due to biological processes and low-temperature, disequilibrium fractionation as discussed above, and their $\delta^{11}B$ values depend on their marine or terrestrial provenance (Table 2). The O-isotope discrimination of sedimentary versus igneous (juvenile, arc-related) sources of granites is well established (Taylor and Sheppard, 1986). Trumbull and Slack (2018) showed that the $\delta^{11}B$ composition of granites and related volcanic rocks also reflects the S- and I- type source dichotomy (see Table 2). The relatively heavy boron in I-type, continental arc magmas ($\delta^{11}B$ from -10‰ to 12‰, average -2‰), relates ultimately to seawater alteration of the sub-arc magma source, whereas the lighter boron in S-types ($\delta^{11}B$ from -22‰ to 0‰, average ~ -11‰) is due to continental (meta)sedimentary rocks in the source.

In cases where magmas were derived by partial melting of hydrothermally altered rocks (Muehlenbachs et al. 1974; Taylor 1986b), the $\delta^{18}O$ signature shifts, commonly to lower values as seen in igneous rocks from Iceland (Muehlenbachs et al. 1974) and the Yellowstone complex in Wyoming, USA (Friedman et al. 1974; Hildreth et al. 1984). Schmitt and Simon (2004) attributed the same effect of hydrothermal alteration of the magma source to explain $\delta^{11}B$ variations in volcanic rocks from the Long Valley caldera.

7.1 The Effects of Metamorphism, Metasomatism and Devolatilization

Metamorphism of a rock can change its stable isotope characteristics through two broad processes, devolatilization or the influx of exogeneous fluids. Devolatilization is the generic process whereby volatile components of a rock are lost during prograde metamorphism and includes dehydration, decarbonation and desulfidation. In some mineral systems, orogenic gold systems, for example, metamorphic fluids produced by these processes are thought to be ore fluids (e.g. Phillips and Groves 1983).

The most important devolatilization reaction is dehydration. According to Wedepohl (1969a) and Engel and Engel (1958b), the average water content of pelitic rocks decreases up metamorphic grade from 5.0% in shales, to 2.96% in phyllites and slates, to 2.41% in mica schists, to 2.02% in sillimanite gneisses, and to 0.62% in transitional amphibolite-granulite facies semi-pelitic rocks. On an atomic basis, metamorphic dehydration will remove at most 10% of oxygen, but up to 100% of the hydrogen if the process goes to completion. As a consequence, the effect on isotope composition is minor for oxygen (at most an increase in $\delta^{18}O$ of 0.6‰: Valley 1986), but can be significant for hydrogen (decreases in δD of up to 16‰ or 40‰, depending upon the process of devolatilization assumed: Valley 1986). As the water content in sandstone and igneous rocks is much lower than in pelitic rocks (Wedepohl 1969a,b), the effect of metamorphism on oxygen and hydrogen isotope characteristics of these rocks is much less. Hence, for most siliciclastic and igneous rocks, $\delta^{18}O$ of the metamorphosed rock is similar that of the protolith, but δD can significantly different.

Boron in clastic sedimentary rocks is hosted mainly in clays and micas (Trumbull and Slack 2018) and the breakdown of these hydrous minerals during metamorphism generally releases boron to the metamorphic fluid (Moran et al., 1992). At the same time, the B-isotope fractionation between sheet silicates and hydrous fluid causes a progressive lowering of $\delta^{11}B$ values in the higher-grade rocks and the opposite effect for the fluid phase. The situation changes if tourmaline forms during metamorphism because it has a high thermal stability and preserves the boron content in the rock, and its B-isotope composition, during prograde metamorphism (Trumbull and Slack 2018).

Decarbonation reactions can cause significant changes in both $\delta^{13}C$ and $\delta^{18}O$. Metamorphism and devolatilization of marly carbonate rocks commonly produces calc-silicate rocks containing wollastonite (after marly limestone) or diopside (after marly dolostone). Both of these reactions produce CO_2 as a product, which then escapes. Valley (1986) summarized the isotope effects of contact metamorphism on carbonate rocks, indicating isotope shifts of up to 14‰ in $\delta^{13}C$ and 22‰ in $\delta^{18}O$, although these shifts were spatially restricted to within 3 km of the causative granite. These shifts in skarns and marbles related to igneous intrusion require that the magmatic-hydrothermal fluids present also experienced significant isotope shifts during decarbonation reactions.

During regional metamorphism, shifts in $\delta^{13}C$ and $\delta^{18}O$ are likely to be restricted. Valley (1986) showed that steep isotope gradients are present at the contacts between compositionally and isotopically distinct units, whereas within these units the isotope signature is uniform. Based on this he suggested that regional metamorphism does not affect the isotope characteristics of the protolith unless significant, channelized fluid flow occurred. However, under situations involving the influx of isotopically distinct exogeneous fluids, isotope shifts in δD, $\delta^{13}C$ and $\delta^{18}O$ can occur. In contrast to the C, O and H isotope systems, the variation of rock $\delta^{11}B$ values during prograde metamorphism can vary strongly depending on the specific mineral assemblage and reactions involved (Trumbull and Slack 2018 and references therein).

Studies by Ferry (1981), Pitcairn et al. (2010), Zhong et al. (2015) and Finch and Tomkins (2017) have indicated that as temperatures increase during regional and contact metamorphism, pre-existing pyrite converts to pyrrhotite according to one or more of the following (or similar) reactions:

$$2FeS_2 \rightarrow 2FeS + S_2 \quad (14)$$

$$2FeS_2 + 2H_2O + C \rightarrow 2FeS + 2H_2S + CO_2 \quad (15)$$

Or

$$2FeS_2 + 2FeO_{rock} \rightarrow 4FeS + O_2 \quad (16)$$

The first two reactions are desulfidation reactions in that sulfur is lost from the rock, but the last reaction does not involve sulfur loss. The above studies indicate that the metamorphic conversion of pyrite to pyrrhotite occurs over a temperature range of 200 °C to 550 °C. Graphical analysis of whole rock geochemical data by Ferry (1981) suggested that reactions similar to (15) are the most likely reactions for pyrite to pyrrhotite conversion and require fluid influx from metamorphic dehydration reactions or from external sources.

The isotope effects of pyrite breakdown during prograde metamorphism are not well studied, with the best example being a laser ablation study by Alirezaei and Cameron (2001) who found no statistically significant fractionation between pyrite and pyrrhotite or between metamorphic zones. They interpreted these results to indicate that the conversion of pyrite to pyrrhotite involved uptake of iron from wall rocks (i.e. reaction 16) rather than sulfur loss. However, at temperatures of 300–450 °C, conditions where the conversion of pyrite to pyrrhotite is greatest (Ferry 1981), $\Delta^{34}S_{pyrite\text{-}H2S}$, $\Delta^{34}S_{pyrrhotite\text{-}H2S}$ and $\Delta^{34}S_{pyrite\text{-}pyrrhotite}$ are 0.8–1.2‰, 0.2–0.3‰ and 0.6–0.9‰, respectively. This suggests that $\delta^{34}S$ variations caused by metamorphic desulfidation are likely to be small, probably not recognizable given the large variations in $\delta^{34}S$ that characterize sedimentary pyrite. Moreover, the isotope composition of sulfur in metamorphic fluids is likely to be similar to that of the rocks that evolve the fluid.

Because primary metamorphic fluids cannot be sampled at or near the Earth's surface, their isotope composition has been inferred using the isotope, mineralogical and chemical composition of metamorphic rocks with estimates of fractionation factors at the temperature range characteristic of metamorphism. Due to the large range in the isotope composition of the protoliths and the range of metamorphic temperatures, the inferred isotope field of metamorphic fluids is large (Fig. 8). Despite this, metamorphic fluids have distinctive isotope characteristics relative to meteoric water, seawater and connate brines, although they overlap with the field of magmatic fluids.

The isotope composition of magmatic fluids also has been estimated from the isotope and chemical composition of igneous rocks and minerals, combined with crystallization temperatures and isotope fractionation factors (e.g. Taylor 1979; Sheppard 1986). The resulting field is large, although not as large as the metamorphic fluid field, with which it largely overlaps. Unlike metamorphic fluids, magmatic fluids have a distinctive isotope signature relative to other fluids such as seawater, basinal brines and meteoric waters present in the upper part of Earth's crust. It must be stressed that the isotope fields shown in Fig. 8 are for primary fluids. Reaction of these fluids with rock has the potential to shift the isotope composition of the fluid, particularly under rock-dominated conditions.

8 Applications of Light Stable Isotopes to Ore Genesis Studies and Exploration

As discussed above, light stable isotopes fractionate during a range of geological processes, hence variations in stable isotope ratios can be used to track these processes. Moreover, isotope data can provide information as to the temperatures and pressures at which mineral systems evolve, and the sources of fluids. The following section discusses how stable isotope data can be used to constrain these processes and conditions of ore formation. More importantly, it also discusses the limitations and ambiguities of some of these constraints.

8.1 Geothermometry and Geobarometry

Temperature, redox state, salinity and pH of the ore fluid largely determine what metals can be transported and how these metals can be deposited. These parameters can be determined from mineral assemblages, but also from fluid inclusion and stable isotope data. Due to temperature-dependent fractionation of stable isotopes between minerals (Fig. 1), the temperature of deposition can be estimated from $\Delta^{34}S$, $\Delta^{18}O$ and $\Delta^{13}C$ data. However, use of these parameters for geothermometry makes several implicit assumptions: (1) the minerals precipitated at the same time from an isotopically homogeneous fluid, (2) the fractionation between minerals has not been affected by later geological events, and (3) isotope fractionation between minerals occurred under equilibrium conditions. The first two assumptions can be largely (but not entirely) assessed by paragenetic studies, and at higher temperatures (>300 °C), disequilibrium becomes less important in most systems (see below).

Another factor that must be considered in assessing isotope thermometers is the purity of mineral separates analysed. Cross-contamination of mineral separates results in temperature estimates that are too high. For example, a hypothetical fluid with $\delta^{18}O$ of 8.0‰ will precipitate quartz and magnetite with $\delta^{18}O$ values of 17.4‰ and -1.5‰ at 250 °C, respectively, giving a $\Delta^{18}O_{quartz\text{-}magnetite}$ = 18.9‰. If the two separates are cross-contaminated by 10% (based on atomic oxygen), the impure quartz and magnetite separates will have with $\delta^{18}O$ values of 15.5‰ and 0.4‰, respectively, giving a $\Delta^{18}O_{quartz\text{-}magnetite}$ = 15.1‰, which corresponds to a temperature of 265 °C, which is 15 °C higher than the hypothetical conditions. This effect becomes stronger at higher temperature and with greater cross contamination. Hence, care must be taken to ensure sample purity, and temperatures estimated from isotope mineral pairs should be considered maximum temperatures. In-situ techniques like SIMS can solve many of the impurity issues, but at the cost of a larger analytical uncertainty, which equates to a larger temperature error.

If three or more minerals have been deposited concurrently from an isotopically homogeneous fluid, determining ΔX values for mineral pairs can be used to assess the reliability of the isotope thermometry. Determining δX of three coexisting minerals yields three separate ΔX and temperature estimates, and if all three temperature estimates are within error, greater confidence can be placed in the overall temperature estimate. If the temperatures differ, the mineral assemblage may not have formed concurrently under equilibrium conditions, or one or more minerals have been affected by post-depositional disturbance, or the mineral separates are not pure.

The uncertainties in stable isotope geothermometry depend on the relative uncertainties in determining δX. If the (2σ) uncertainties associated with δX for each mineral pair are similar, the (2σ) uncertainty in ΔX is $\sqrt{2} \times 2\sigma_{\delta X}$; if the uncertainties in δX are dissimilar, the uncertainty is ΔX is estimated by adding the individual uncertainties in δX in quadrature. This uncertainty in ΔX is then used to estimate the uncertainty in temperature. Note that the temperature uncertainty is asymmetric, with the lower-temperature uncertainty smaller than the higher-temperature uncertainty. This is because the intensity of fractionation in most stable isotope systems decreases with increasing temperature.

Once a temperature estimate has been made, it can be used to constrain other characteristics of the mineral system. Possibly the most important of these is the isotope characteristics of the ore fluid, which is calculated from temperature-dependent fluid-mineral fractionation factors. If sufficient data are available, estimates of δD_{fluid}, $\delta^{11}B_{fluid}$, $\delta^{13}C_{fluid}$, $\delta^{18}O_{fluid}$ and $\delta^{34}S_{fluid}$ can be made, which is very useful information for determining fluid sources. Furthermore, isotope temperature determinations can be combined with fluid inclusion data to estimate the pressure and, thereby, crustal depth at which mineralization occurred. The pressure can be estimated using the following equation:

$$P_t = P_h + (T_t - T_h)(\Delta P/\Delta T) \qquad (17)$$

where P_t is the trapping (geological) pressure of the fluid inclusion, P_h is the homogenization pressure (determined from fluid composition and T_h), T_t is the trapping (geological) temperature of the fluids (determined from isotopic data as described above or from another independent geothermometer), T_h is the fluid inclusion homogenization temperature, and $\Delta P/\Delta T$ is the slope of the isochore. Examples where this methodology has been applied include Huston et al. (1993) Honlet et al. (2018), and many others.

8.2 Tracking Fluid Sources

Stable isotopes are a powerful tool to determine the source of water in ore fluids, and, to a lesser extent the sources of boron, sulfur and carbon within them. As discussed earlier, many fluids present in the upper crust have quite distinctive isotope characteristics, which enables inference of the type of fluid from which the ore fluids were derived (see Fig. 8). An example is the strong contrast between the $\delta^{11}B$ values of seawater, marine carbonates and evaporitic rocks of marine origin on the one hand, and their equivalents from terrestrial, continental settings on the other (see Fig. 7). Thus the values of $\delta^{11}B$, $\delta^{34}S$ and $\delta^{13}C$ can be indicative of source (Fig. 7, Table 2), but because there is considerable overlap between potential reservoirs, in many (most) cases, the isotope signature of the ore and gangue minerals is not definitive of a particular source.

An important exception to this generalization for the sulfur isotope system is the use of multiple sulfur isotope ratios to identify mass-independent fractionation. The recognition of photolytic mass-independent fractionation of sulfur isotopes prior to the Great Oxidation Event (Farquhar et al. 2000, 2001) has opened a totally new method to identify the relative importance of sulfur that has or has not interacted with the atmosphere or hydrosphere, particularly in Archean and earliest Paleoproterozoic mineral systems. This tool is particularly useful for systems in which the source of sulfur is contested, for example volcanic-hosted massive sulfide mineral systems (Huston et al. 2023), orogenic gold deposits (LaFlamme et al. 2018b) and orthomagmatic Ni-Cu-PGE deposits. In the latter, one mechanism that has been proposed to produce an immiscible Ni-Cu-PGE sulfide melt from the parental mafic/ultramafic magmas is sulfur saturation of the magma by ingestion of sulfide-rich wall rocks (Lesher and Campbell 1993). As many of these deposits formed in the Neoarchean, the distinctive signature of mass-independent sulfur isotope fractionation can distinguish between mantle and crustal and mantle sulfur.[6] Bekker et al. (2009), Ding et al. (2012) and Fiorentini et al. (2012) used $\Delta^{33}S$, which indicates the presence ($|\Delta^{33}S| > 0.0‰$) or absence ($\Delta^{33}S \sim 0.0‰$) of crustal sulfur, to indicate that sulfur contamination from the crust was an important process in some, but not all, orthomagmatic Ni-Cu-PGE deposits, and that the presence of mass-independent fractionation may be a signature of the larger deposits (Fiorentini et al. 2012).

8.3 Tracking Geochemical Processes

One of the more powerful uses of stable isotopes in mineral systems studies is identifying and then tracking geochemical processes like redox reactions and phase separation (boiling, magma degassing). As discussed above, many geochemical processes cause significant, and, in some cases, diagnostic, isotope shifts in rocks and/or minerals, or cause spatial or temporal isotope gradients. These effects are most easily observed where there are strong isotope contrasts between the reactants and the products of the geochemical reaction. For example, in volcanic-hosted massive sulfide systems, evolved seawater, the main ore fluid, has hydrogen and oxygen isotope characteristics that contrast with the rocks and other potential fluids that it interacts with. As a consequence, hydrolytic alteration zones developed in these systems can be easily mapped using whole-rock $\delta^{18}O$ data, and the involvement of other fluids (e.g. magmatic-hydrothermal) can be assessed (Huston et al. 2011, 2023; and references therein).

One of the earliest uses of oxygen and hydrogen isotopes in alteration studies was determining the interaction of magmatic-hydrothermal ore fluids and meteoric water with host rocks in porphyry copper mineral systems. Sheppard et al. (1969, 1971) showed that hydrothermal minerals from core potassic alteration zones in these deposits formed from magmatic-hydrothermal fluids, whereas clay minerals in peripheral argillic alteration zones formed from heated, overprinting meteoric water. These isotope data were one of the key datasets in establishing the porphyry copper genetic model in the 1960s and 1970s, and the results of Sheppard et al. (1969, 1971) have been confirmed by studies of porphyry systems of different ages around the world.

Boron isotope studies of ore systems have focussed on tourmaline because this mineral is common, resists alteration, preserves zoning and contains percent levels of boron, but other minerals may become important (e.g. white mica, see "future directions"). One of the first uses of B-isotopes to understand ore fluid sources was the study of massive sulfide deposits associated with tourmalinite by Palmer and Slack (1989) and the method grew rapidly since the advent of in-situ studies using SIMS or LA-ICPMS (see Slack and Trumbull 2011; Trumbull et al. 2020). Common applications have been to constrain the fluid source or to define mixing of multiple sources and their temporal relationships. For example, Xavier et al. (2008) found distinctly high $\delta^{11}B_{tourmaline}$ values (> 10‰) that indicate a marine origin for high-salinity ore fluids in Brazilian IOCG deposits. Zoning and replacement textures combined with in-situ B-isotope ratios tracked fluid evolution and mixing (Pal et al. 2010; Baksheev et al. 2015; Lambert-Smith et al. 2016). In granite-related Sn-W deposits and pegmatites, tourmaline B-isotope studies have been used to recognize the magmatic-hydrothermal transition by its effect on B-isotope partitioning (e.g. Drivenes et al. 2015; Siegel et al. 2016).

Rotherham et al. (1998) interpreted large variations in $\delta^{34}S_{sulfide}$ to be indicative that reduction of an originally highly oxidized ore fluid by interaction with ironstone was the depositional mechanism for the Starra (now known as Selwyn) copper–gold deposit in northwest Queensland, Australia. Large (1975) and Huston et al. (1993) interpreted zonation in $\delta^{34}S_{sulfide}$ data from ironstone-hosted deposits in the Tennant Creek district in a similar manner. Large variations in $\delta^{34}S$ data can be indicative of

[6] In this case, defined as sulfur that has (crustal) and has not (mantle) interacted with the atmosphere prior to the Great Oxidation Event at ~ 2420 Ma.

redox reactions and spatial and temporal gradients can be used to track and map the progression of these reactions.

Pyrite- and sulfate-bearing advanced argillic alteration assemblages form by two processes in the porphyry-epithermal mineral system: (1) disproportionation of magmatic SO_2 to form aqueous H_2S and sulfate, and (2) condensation of hydrothermal H_2S into oxidized groundwater followed by oxidation of the H_2S to form sulfate. Both of these processes produce sulfuric acid, which reacts with rock to produce pyrophyllite and/or kandite (a complex mixture of members of the kaolinite-nacrite-dickite family), which characterize advanced argillic assemblages. These two processes, which occur in different parts of the porphyry-epithermal system, can produce mineralogically similar alteration assemblages that are difficult to distinguish. Comparison of sulfur isotope data from sulfide and sulfate minerals is one method of distinguishing between them (Rye et al. 1992). Rye (2005) found that because SO_2 disproportionation is a high temperature process, it produces equilibrium isotope fractionation, whereas oxidation of H_2S in groundwater, a low temperature process, generally produces disequilibrium isotope fractionation between sulfide and sulfate minerals. These characteristics, combined with other mineralogical data can be used as criteria to distinguish between SO_2 disproportionation and the oxidation of aqueous H_2S in groundwater as mechanisms to form advanced argillic alteration assemblages.

In summary, variations in stable isotope ratios can be used to identify and track geochemical processes in mineral systems. It must be stressed, however, that stable-isotope signatures can be affected by many geochemical processes, and these signatures are rarely diagnostic of one process alone. As an example, the presence of a $\delta^{34}S_{sulfide} \sim 0‰$ is commonly taken to indicate that the sulfur was derived (directly) from a magma. However, this signature can be produced by the fortuitous mixing of two non-magmatic sources, or by leaching of volcanic rocks, which would have a similar signature. Hence, when interpreting isotope data (and geochemical data in general) all processes that produce an isotope signature must be considered valid, and not just the favoured process the signature "proves". It is rare that an isotope signature is sufficiently unique to eliminate all but one process as its cause. The non-uniqueness of isotope signatures can, in some cases, be overcome by the use of multiple isotope systems, including both stable and radiogenic isotopes (see future directions).

8.4 Exploration and Discovery

Although stable isotopes have proved to be extremely useful in understanding fluid sources and processes that form ore deposits, discoveries of deposits using isotope data are uncommon. Two examples that we are aware of are the discovery of a new skarn lens in the Kamioka district in Japan using $\delta^{18}O$ data of carbonate minerals (Naito et al. 1995), and the discovery of the West 45 lens at the Thalanga volcanic-hosted massive sulfide deposit in Queensland, Australia using whole-rock $\delta^{18}O$ data (Miller et al. 2001). Despite well-documented and consistent zonation of isotope values for a number of deposit types (see above and later papers in this volume), isotope methods for exploration have not been taken up by industry to any significant extent. Barker et al. (2013) attribute the lack of uptake by industry to: (1) the costs of analyses, (2) the requirement of specialist analytical labs, and (3) the slow turnaround time for analyses. They noted that for isotopes to be taken up by industry, the quantity of analyses on a project would need to increase from the tens or hundreds typical for ore genesis studies to thousands for exploration. Barker et al. (2013) reported a study in the Carlin district, Nevada, USA in which they used large numbers of analyses to show decrease in $\delta^{18}O$ of carbonate minerals towards the Screamer gold deposit. Because of the large variability of the data at the drill hole scale, the statistical robustness of large data sets was required to document this zonation. Hence, uptake of stable isotope variations as an exploration tool will require a reduction in the costs and time required for analyses, and a greater number of laboratories able to provide them.

9 Future Directions

As with other science fields, the techniques and interpretations of stable isotopes in metallogenic studies evolve over time. Techniques being developed at present could strongly influence the future direction of stable isotope research in metallogenic studies. These include the development of new techniques (e.g. clumped isotopes), the ability to integrate data from different isotope systems (both stable and radiogenic) on the same rock, mineral or spot, and the increasing capability for inexpensive and rapid analyses.

Like the discovery of significant mass-independent sulfur isotope fractionation in the Archean, the development of analytical techniques to measure fractionation of clumped isotopes has the potential to provide a new tool to understand ore genesis. Rather than considering variations in the isotope abundance of an individual element, clumped isotopes measurements identify isotope variations in molecules. For example the CO_2 molecule can vary in atomic mass from 44 to 49, with mass number 44 ($^{12}C^{16}O_2$) being the most abundant (98.40%) and mass number 49 ($^{13}C^{18}O_2$) the least (44.5 ppb) (Eiler 2007). Just like isotopes, these "isotopologues' also fractionate with temperature (Ghosh et al. 2006). At present, clumped isotope geochemistry has been extensively used in paleoenvironmental, paleobiological and related studies (Huntington et al. 2011; Eagle et al. 2011), but the application to fluid flow and metamorphism (Swanson et al. 2012; Lloyd et al. 2017) suggest that they might also be useful in ore genesis studies where they can determine the temperature and isotope (i.e. $\delta^{18}O$) composition of ore fluids (Mering et al. 2018). Clumped isotopes may be particularly useful in low temperature systems, such as those hosted in basins, where estimating temperature and determining the origin of fluids are problematic. Interpretation of these data does not require knowledge of other information such as salinity and the isotopic signature is less susceptible to later alteration.

To date, the application of B-isotope studies of ore deposits has been almost exclusively based on tourmaline (Trumbull et al. 2020). White mica has been targeted before in studies of metamorphism (Konrad-Schmolke and Halama 2014), but it is virtually unexplored in ore deposit studies even though mica, particularly white mica, contains the highest concentrations of boron of all common rock-forming minerals (Harder 1974). White mica is even more common than tourmaline in alteration zones, and in cases where both minerals coexist, the difference in fluid-mineral B-isotope fractionation between them can be used for geothermometry. A case study of the Panasqueira Sn-W deposit by Codeço et al. (2019) showed that mineralization temperatures derived from $\Delta^{11}B$ in coexisting mica and tourmaline are consistent with other geothermometers and can be used to track cooling of the mineralizing system.

The development of microanalytical methods of isotope analysis has allowed for the first time determination of a large suite of isotope ratios, both stable and radiogenic, from essentially the same spot or the same mineral. Integration of such information can lead to conclusions that are more robust and far reaching than conclusions from each isotope system individually. A good example of this is the microanalytical collection of U–Pb, Lu–Hf and oxygen isotope data from zircon. The ability to collect such data individually has been around for 10–30 years, but only recently have these data been collected from the same spots, allowing synergy of interpretation. As an example, for magmatic zircons from an ore-related granite, the U–Pb system allows age determination (Chelle-Michou and Schaltegger 2023), the Lu–Hf system provides information about source of the magma (e.g. crustal versus mantle: Waltenberg 2023), and the oxygen system provides information about modifications of the source (e.g. metasomatism prior to magma generation: Valley 2003). Collectively, these data can provide information about the mineral system not available individually from the separate isotope information. For example, the data could indicate if a magma originated from the mantle and if the mantle had been metasomatized, potentially important information in determining magma fertility.

Another area with growth potential is the combination of isotope systems including the light stable isotopes described above (B, O, H, C, and S; also Li though not discussed herein), as well as heavier stable metal isotopes (e.g. Cu) and radiogenic isotopes. Many of these can be analysed in situ and with good precision by SIMS and/or LA-ICP-MS, which opens the door to detailed, petrographically-controlled analyses of zoning and overprinting relationships in ore and gangue minerals. Multiple-isotope analyses would remove much of the current ambiguity in identifying fluid and metals origin by isotope fingerprinting. However, a prerequisite for expanding the scope of in-situ analysis is the availability of homogeneous and matrix-matched reference materials for a wide range of mineral groups (oxides, sulfides, silicates, carbonates etc.). There needs to be heightened awareness of the importance for development and distribution of quality reference materials.

As discussed above and by Barker et al. (2013), the development of rapid, inexpensive methods of stable isotope data acquisition would enhance the utility of stable isotope data in mineral exploration. Although stable isotope data can provide vectors toward ore, the time required for analysis and the cost has generally restricted stable isotope (and other isotope) data to academic studies and not exploration. Continued development of rapid methods of analysis would increase the update by the exploration industry.

References

Adlakha EE, Hattori K, Davis WJ, Boucher B (2017) Characterizing fluids associated with the McArthur River U deposit, Canada, based on tourmaline trace element and stable (B, H) isotope compositions. Chem Geol 466:417–435

Aggerwal SK, You C-F (2016) A review on the determination of isotope ratios of boron with mass spectrometry. Mass Spec Rev 36:499–519

Akagi T, Franchi IA, Pillinger CT (1993) Oxygen isotope analysis of quartz by Nd/YAG-laser/ fluorination. Analyst (london) 118:1507–1510

Alirezaei S, Cameron EM (2001) Variations of sulfur isotopes in metamorphic rocks from Bamble Sector, southern Norway: a laser probe study. Chem Geol 181:23–45

Baksheev IA, Prokofiev VY, Trumbull RB, Wiedenbeck M, Yapaskurt VO (2015) Geochemical evolution of tourmaline in the Darasun gold district, Transbaikal region, Russia: evidence from chemical and boron isotopic compositions. Mineral Deposita 50:125–138

Barker SL, Dipple GM, Hickey KA, Lepore WA, Vaughan JR (2013) Applying stable isotopes to mineral exploration: teaching an old dog new tricks. Econ Geol 108:1–9

Barnes HL (ed) (1979) Geochemistry of hydrothermal ore deposits, 2nd edn. Wiley, New York, p 789

Barnes HL (ed) (1997) Geochemistry of hydrothermal ore deposits, 3rd edn. Wiley, New York, p 992

Barton LL, Fauque GD (2009) Biochemistry, physiology and biotechnology of sulfate-reducing bacteria. Adv Appl Microbiol 68:41–98

Bekker A, Barley ME, Fiorentini M, Rouxel OJ, Rumble D, Bersford SW (2009) Atmospheric sulfur in Archean komatiite-hosted nickel deposits. Science 326:1086–1089

Beaudoin G, Taylor BE (1994) high precision and spatial resolution sulfur isotope analysis using MILES laser microprobe. Geochim Cosmochim Acta 58:5055–5063

Beaudoin G, Therrien P (2009) The updated web stable isotope fractionation calculator. In: de Groot PA (ed) Handbook of stable isotope analytical techniques, vol II. Elsevier, Amsterdam, pp 1120–1122

Bendall C, Lahaye Y, Fiebig J, Weyer S, Brey GP (2006) In situ sulfur isotope analysis by laser ablation MC-ICPMS. Appl Geochem 21:782–787

Bigeleisen J, Perlman ML, Prosser HC (1952) Conversion of hydrogenic materials to hydrogen for isotope analysis. Anal Chem 24:1356–1357

Broecker W, Oversby V (1971) Chemical equilibria in the Earth. McGraw-Hill, New York, p 318

Burnham CW (1979) Magmas and hydrothermal fluids. In: Barnes HL (ed) Geochemistry of hydrothermal ore deposits, 2nd edn. Wiley, New York, pp 71–136

Caruso S, Fiorentini ML, Champion DC, Lu Y, Ueno Y, Smithies RH (2022) Sulfur isotope systematics of granitoids from the Yilgarn Craton sheds new light on the fluid reservoirs of Neoarchean orogenic gold deposits. Geochim Cosmochim Acta 326:199–213

Catanzaro EJ, Champion CE, Garner EL, Marinenko G, Sappenfield KM, Shields WR (1970) Boric acid; isotope, and assay standard reference materials. United States National Bureau of Standards Special Publication 260–17

Catling DC (2014) The great oxidation event transition. Treatise Geoch, 2nd edn 6:177–195. https://doi.org/10.1016/B978-0-08-095975-7.01307-3

Chacko T, Riciputi LR, Cole DR, Horita J (1999) A new technique for determining equilibrium hydrogen isotope fractionation factors using the ion microprobe: application to the epidote-water system. Geochim Cosmochim Acta 63:1–10

Chadwick J (1932) Possible existence of a neutron. Nature 129:312

Chelle-Michou C, Schaltegger U (2023) U-Pb dating of mineral deposits: from age constrains to ore-forming processes. In: Huston DL, Gutzmer J (eds) Isotopes in economic geology, metallogensis and exploration, Springer, Berlin, this volume

Clayton RN (2003) Oxygen isotopes in meteorites. Treatise Geoch 1:129–142

Clayton RN, Kieffer SW (1991) Oxygen isotope thermometer calibrations. Geochem Soc Spec Pub 3:3–10

Clayton RN, Mayeda TK (1963) The use of bromine pentafluoride in the extraction of oxygen from oxides and silicates for isotope analysis. Geochim Cosmochim Acta 27:43–52

Clayton RN, Grossman L, Mayeda TK (1973) A component of primitive nuclear composition in carbonaceous meteorites. Science 339:780–785

Clayton RN, Onuma N, Grossman L, Mayeda TK (1977) Distribution of the presolar component in allende and other carbonaceous chondrites. Earth Planet Sci Lett 34:209–224

Codeço MS, Weis P, Trumbull R, Glodny J, Wiedenbeck M, Romer RL (2019) Boron isotope muscovite-tourmaline geothermometry indicates fluid cooling during magmatic-hydrothermal W-Sn ore formation. Econ Geol 114:153–163

Cole DR, Ripley EM (1998) Oxygen isotope fractionation between chlorite and water from 170–350°C: a preliminary assessment based on partial exchange and fluid/rock experiments. Geochim Cosmochim Acta 63:449–457

Conrad R (2005) Quantification of methanogenic pathways using stable carbon isotope signatures: a review and a proposal. Org Geochem 36:739–752

Coplen TB (2011) Guideline and recommended terms for expression of stable-isotope-ratio and gas-ratio measurement results. Rapid Commun Mass Spectrom 25:2538–2560

Coplen TB, Kendall C, Hopple J (1983) Comparison of stable isotope reference samples. Nature 302:236–238

Craig H (1961) Isotope variations in meteoric waters. Science 133:1702–1703

Crowe DE, Valley JW, Baker KL (1990) Micro-analysis of sulfur-isotope ratios and zonation by laser microprobe. Geochim Cosmochim Acta 54:2075–2092

de Hoog JCM, Savov IP (2018) Boron isotopes as a tracer for subduction zone processes. Adv Isot Geochem 7:217–247

de Groot PA (ed) (2004) Handbook of stable isotope analytical techniques, vol 1. Elsevier, Amsterdam

de Groot PA (ed) (2009) Handbook of stable isotope analytical techniques, vol 2. Elsevier, Amsterdam

Detmers J, Brüchert V, Habicht KS, Juever J (2001) Divrsity of sulfur isotope fractionations by sulfate-reducing prokaryotes. Appl Environ Microbiol 2001:888–894

Ding T, Valkiers S, Kipphardt H, De Bièvre P, Taylor PDP, Gonfiantini R, Krouse R (2001) Calibrated sulfur isotope abundance ratios of three IAEA sulfur isotope reference materials and V-CDT with a reassessment of the atomic weight of sulfur. Geochim Cosmochim Acta 65:2433–2437

Ding X, Ripley EM, Shirey SB, Li C, Moore CH (2012) Os, Nd, O and S isotope constraints on the importance of country rock contamination in the generation of the conduit-related Eagele Ni-Cu-(PGE) deposit in the Midcontinent Rift System, Upper Michigan. Geochim Cosmochin Acta 89:10–30

Drivenes K, Larsen RB, Müller A, Sørensen BE, Wiedenbeck M, Raanes MP (2015) Late-magmatic immiscibility during batholith formation: assessment of B isotopes and trace elements in tourmaline from the land's end granite. SW England. Contrib Min Petrol 169:56. https://doi.org/10.1007/s00410-015-1151-6

Eagle RA, Tütken T, Martin TS, Tripati AK, Fricke HC, Connely M, Cifelli RL, Eiler JM (2011) Dinosaur body temperatures determined from isotope (^{13}C–^{18}O) ordering in fossil biominerals. Science 333:443–445

Eastoe CJ, Guilbert JM (1992) Stable chlorine isotopes in hydrothermal processes. Geochim Cosmochim Acta 56:4247–4255

Eiler JM (2007) "Clumped-isotope" geochemistry—the study of naturally-occurring, multiply-substituted isotopologues. Earth Planet Sci Lett 262:309–327

Eldridge CS, Compston W, Williams IS, Walshe JL, Both RA (1987) In situ microanalysis for $^{32}S^{34}S$ ratios using the ion microprobe SHRIMP. Intl J Mass Spectr Ion Proc 76:65–83

Eldridge DL, Guo W, Farquhar J (2016) Theoretical estimates of equilibrium sulfur isotope effects in aqueous sulfur systems: Highlighting the role of isomers in the sulfite and sulfoxylate systems. Geochim Cosmochim Acta 195:171–200

Elsenheimer D, Valley JW (1992) In situ oxygen isotope analysis of feldspar and quartz by Nd:YAG laser microprobe. Chem. Geol. (isotope Geoscience Section) 101:21–42

Engel AEJ, Clayton RN, Epstein S (1958) Variations in isotope composition of oxygen and carbon in Leadville Limestone (Mississippian, Colorado) and in its hydrothermal and metamorphic phases. J Geol 66:374–393

Engel AEJ, Engel CG (1958) Progressive metamorphism and granitization of the major paragneiss, NW Adirondack Mtns. N. y. Geol Soc Am Bull 69:1369–1413

Epstein S, Buchsbaum HA, Lowenstam H, Urey HC (1953) Revised carbonate-water isotope temperature scale. Geol Soc Am Bull 64:1315–1326

Farquhar J, Bao H, Thiemens M (2000) Atmospheric influence of Earth's earliest sulphur cycle. Science 289:756–758

Farquhar J, Savarino J, Airieau S, Thiemens MH (2001) Observation of wavelength sensitive mass-independent sulphur isotope effects during SO_2 photolysis: implications for the early atmosphere. J Geophys Res 106:1–11

Farquhar J, Peters M, Johnston DT, Strauss H, Masterson A, Wiechert U, Kaufman AJ (2007) Isotopic

evidence for Mesoarchaean anoxia and changing atmospheric sulphur chemistry. Nature 449:706
Ferry JM (1981) Petrology of graphitic sulfide-rich schists from south-central Maine: an example of desulfidation during prograde regional metamorphism. Am Mineral 66:908–931
Finch EG, Tomkins AG (2017) Pyrite-pyrrhotite stability in a metamorphic aureole: implications for orogenic gold genesis. Econ Geol 112:661–674
Fiorentini ML, Bekker A, Rouxel O, Wing BA, Maier W, Rumble D (2012) Multiple sulfur and iron isotope composition of magmatic Ni-Cu-(PGE) sulfide mineralization from eastern Botswana. Econ Geol 107:105–116
Foster GL, Hoenisch B, Paris G, Dwyer GS, Rae JWB, Elliott T, Gaillardet J, Hemming N, Louvat P, Vengosh A (2013) Interlaboratory comparison of boron isotope analyses of boric acid, seawater and marine $CaCO_3$ by MC-ICPMS and NTIMS. Chem Geol 358:1–14
Foster GL, Marschall HR, Palmer MR (2018) Boron isotope analysis of geological materials. Adv Isot Geochem 7:13–31
Friedman L, Lipman P, Obranovich JD, Gleason JD, Christiansen RL (1974) Meteoric waters in magmas. Science 184:1069–1072
Ghosh P, Adkins J, Affek H, Balta B, Guo W, Schauble EA, Schrag D, Eiler JM (2006) ^{13}C–^{18}O bonds in carbonate minerals: a new kind of paleothermometer. Geochim Cosmochim Acta 70:1439–1456
Giesemann A, Jäger HJ, Norman AL, Krouse HR, W.A. Brand WA, (1994) On-line sulfur-isotope determination using an elemental analyzer coupled to a mass spectrometer. Anal Chem 66:2816–2819
Gilg HA, Sheppard SMF (1996) Hydrogen isotope fractionation between kaolinite and water revisited. Geochim Cosmochim Acta 60:529–533
Goldhaber MB, Orr WL (1995) Kinetic controls on thermochemical sulfate reduction as a source of sedimentary H_2S. ACS Symp Ser 612:412–425
Graham CM, Sheppard SMF, Heaton THE (1980) Experimental hydrogen isotope studies: I. Systematics of hydrogen isotope fractionation in the systems epidote-H_2O, zoisite-H_2O and AlO(OH)-H_2O. Geochim Cosmochim Acta 44:353–364
Grinenko VA (1962) The preparation of sulphur dioxide for isotope analysis. Zh Neorgan Khim 7:2478–2483
Hagemann S, Hensler A-S, Figueiredo e Silva RC, Tsikos H (2023) Light stable isotope (O, H, C) signatures of BIF-hosted iron ore systems: implications for genetic models and exploration targeting. In: Huston DL, Gutzmer J (eds), Isotopes in economic geology, metallogensis and exploration, Springer, Berlin, this volume
Harder H (1974) 5D. Abundance in rock-forming minerals, boron minerals. In: Kedepohl KH (Ed) Handbook of geochemistry. Berlin, Springer Verlag, 5-D-1–5-D-6
Heidenreich JE III, Thiemens MH (1986) A non-mass dependent oxygen isotope effect in the production of ozone from moleular oxygen: the role of moleular symmetry in isotope chemistry. J Chem Phys 84:2129–2136
Health Physics Society (2011) Tritium. Health Physics Society Fact Sheet. 5 p. http://hps.org/documents/tritium_fact_sheet.pdf
Hildreth W, Christiansen RL, O'Neil JR (1984) Catastrophic isotope modificatin of rhyolite magmas at times of caldera subsidence, Yellowstone Plateau volcanic field. J Geophys Res 89:10153–10192
Hoefs J (1997) Stable isotope geochemistry, 4th edn. Springer, Berlin, p 201
Hoefs J (2021) Stable isotope geochemistry, 9th edn. Springer, Berlin, p 504
Honlet R, Gasparrini M, Muchez P, Swennen R, John CM (2018) A new approach to geobarometry by combining fluid inclusion and clumped isotope thermometry in hydrothermal carbonates. Terra Nova 2018:1–8. https://doi.org/10.1111/ter.12326
Horita J, Wesolowski DJ (1994) Liquid-vapor fractionation of oxygen and hydrogen isotopes of water from the freezing to the critical temperature. Geochim Cosmochim Acta 58:3425–3437
Hulston JR, Thode HG (1965) Variations in S^{33}, S^{34}, and S^{36} contents of meteorites and their relation to chemical nuclear effects. J Geophys Res 70:3475–3484
Huntington KW, Budd DA, Wernicke BP, Eiler JM, JM, (2011) Use of clumped-isotope thermometry to constrain the crystallization temperature of diagenetic calcite. J Sed Res 81:656–669
Huston DL, Champion DC (2023) Applications of lead isotopes to ore geology, metallogenesis and exploration. In: Huston DL, Gutzmer J (eds), Isotopes in economic geology, metallogensis and exploration, Springer, Berlin, this volume
Huston DL, Bolger C, Cozens G (1993) A comparison of mineral deposits at the Gecko and White Devil deposits: implications for ore genesis in the Tennant Creek district, Northern Territory, Australia. Econ Geol 88:1198–1225
Huston DL, Relvas JMRS, Gemmell JB, Drieberg S (2011) The role of granites in volcanic-hosted massive sulphide ore-forming systems: an assessment of magmatic–hydrothermal contributions. Mineral Deposita 46:473–507
Huston DL, LaFlamme C, Beaudoin G, Piercey S (2023) Light stable isotopes in volcanic-hosted massive sulfide ore systems. In: Huston DL, Gutzmer J (eds), Isotopes in economic geology, metallogensis and exploration, Springer, Berlin, this volume
Ireland TR (1995) Ion microprobe mass spectrometry: techniques and applications in cosmochemistry, geochemistry, and geochronology. Adv Anal Geochem 2:1–118
Ireland TR (2012) Oxygen isotope tracing of the solar system. Austr J Earth Sci 59:225–236
Ireland TR, Schram N, Holden P, Lanc P, Ávila J, Armstrong R, Amelin Y, Latimore A, Corrigan D, Clement S, Foster J, Compston W (2014) Charge-

mode electrometer measurements of S-isotopic compositions on SHRIMP-SI. Int J Mass Spectrometry 359:26–37

Jia Y, Kerrich R (1999) Nitrogen isotope systematics of mesothermal lode gold deposits: metamorphic, granitic, meteoric water, or mantle origin? Geology 27:1051–1054

Johnston DT (2011) Multiple sulfur isotopes and the evolution of Earth's surface sulfur cycle. Earth Sci Rev 106:161–183

Kelley SP, Fallick AE (1990) High precision spatially resolved analysis of $\delta^{34}S$ in sulphides using a laser extraction technique. Geochim Cosmochim Acta 54:883–888

Kennedy CB, Gault AG, Fortin D, Clark ID, Pedersen K, Scott SD, Ferris FG (2010) Carbon isotope fractionation by circumneutral iron-oxidizing bacteria. Geology 38:1087–1090

Kerrich R (1987) The stable istoope geochemistry of Au-Ag vein deposits in metamorphic rocks. Mineral Assoc Can Short Course Handb 13:287–336

Kharaka YK, Hanor JS (2003) Deep fluids in the continents: I. Sedimentary Basins. Treatise Geochem 5:1–48. https://doi.org/10.1016/B0-08-043751-6/05085-4

Kieffer SW (1982) Thermodynamics and lattice vibration of minerals: 5. Application to phase equilibria, isotope fractionation, and high pressure thermodynamic properties. Rev Geophys Space Phys 20:827–849

Kita NT, Ushikubo T, Fu B, Valley JW (2009) High precision SIMS oxygen isotope analysis and the effect of sample topography. Chem Geol 264:43–57

Kita NT, Huberty JM, Kozdon R, Beard BL, Valley JW (2010) High-precision SIMS, sulfur and iron stable isotope analyses of geological materials: accuracy, surface topography and crystal orientation. Surf Interface Anal 2010. https://doi.org/10.1002/sia.3424

Konrad-Schmolke M, Halama R (2014) Combined thermodynamic–geochemical modeling in metamorphic geology: Boron as tracer of fluid–rock interaction. Lithos 208–209:393–414

Korff SA, Danforth WE (1939) Neutron measurements with boron-trifluoride counters. Phys Rev 55:980

Kowalski P, Wunder B (2018) Boron-isotope fractionation among solids-fluids-melts: experiments and atomic modelling. Adv Isot Geochem 7:33–70

Kyser TK, ed (1987) Stable isotope geochemistry of low temperature fluids. Min Assoc Can Short Course 13

Kyser K (1986) Stable isotope variations in the mantle. Rev Mineral 16:141–164

LaFlamme C, Fiorentini ML, Lindsay MD, Bui TH (2018a) Atmospheric sulfur is recycled to the crystalline continental crust during supercontinent formation. Nature Commun 9:4380

LaFlamme C, Sugiono D, Thébaud N, Caruso S, Fiorentini M, Selvaraja V, Jeon H, Voute F, Martin L (2018b) Multiple sulfur isotopes monitor fluid evolution of an Archean orogenic gold deposit. Geochim Cosmochim Acta 222:436–446

Lambert-Smith JS, Rocholl A, Treloar PJ, Lawrence DM (2016) Discriminating fluid source regions in orogenic gold deposits using B-isotopes. Geochim Cosmochim Acta 194:57–76

Large RR (1975) Zonation of hydrothermal minerals at the Juno mine, Tennant Creek goldfield, central Australia. Econ Geol 70:1387–1413

Lesher CM, Campbell IH (1993) Geochemical and fluid dynamic modeling of compositional variations in Archean komatiite-hosted nickel sulfide ores in Western Australia. Econ Geol 88:804–816

Li YB, Liu JM (2006) Calculation of sulfur isotope fractionation in sulfides. Geochim Cosmochim Acta 70:1789–1795

Lloyd MK, Eiler JM, JM, Peter I. Nabelek PI, (2017) Clumped isotope thermometry of calcite and dolomite in a contact metamorphic environment. Geochim Cosmochim Acta 197:323–344

Lobato LM, Figueiredo e Silva RC, Angerer T, Mendes M, Hagemann S (2023) Fe isotopes applied to BIF-hosted iron ore deposits. In: Huston DL, Gutzmer J (eds), Isotopes in economic geology, metallogensis and exploration, Springer, Berlin, this volume

Lyons T (2007) Mass-independent fractionation of sulfur isotopes by isotope-selective photodissociation of SO_2. Geophys Res Lett 34:L22811

Machel HG (2001) Bacterial and thermochemical sulfate reduction in diagenetic environments – old and new insights. Sed Geol 140:143–175

Machel HG, Krause HR, Sassen R (1995) Products and distinguishing criteria of bacterial and thermochemical sulfate reduction. Appl Geochem 10:373–389

Marschall HR, Jiang S-Y (2011) Tourmaline isotopes: no element left behind. Elements 7:313–319

Marschall HR, Meyer C, Wunder B, Ludwig T, Heinrich W (2009) Experimental boron isotope fractionation between tourmaline and fluid: confirmation from in situ analyses by secondary ion mass spectrometry and from Rayleigh fractionation modelling. Contrib Mineral Petrol 158:675–681

Marschall HR, Foster GL (2018) Boron isotopes - the fifth element. Advances in Geochemistry 7:249–272

Mason PR, Košler J, de Hoog JC, Sylvester PJ, Meffan-Main S (2006) In situ determination of sulfur isotopes in sulfur-rich materials by laser ablation multiple-collector inductively coupled plasma mass spectrometry (LA-MC-ICP-MS). J Anal at Spectrom 21:177–186

Mathur R, Zhao Y (2023a) Copper isotopes used in mineral exploration. In: Huston DL, Gutzmer J (eds), Isotopes in economic geology, metallogensis and exploration, Springer, Berlin, this volume

Mauersberger K (1987) Measurement of heavy ozone in the stratosphere. Geophys Res Lett 8:80–83

McCrea JM (1950) On the isotope chemistry of carbonates and a paleotemperature scale. J Chem Phys 18:849–857

McKinney CR, McCrea JM, Epstein S, Allen HA, Urey HC (1950) Improvements in mass spectrometers for the measurement of small differences in isotope abundance ratios. Rev Sci Instrum 21:724–730

Mercadier J, Richard A, Cathelineau M (2012) Boron- and magnesium-rich marine brines at the origin of giant unconformity-related uranium deposits: $\delta^{11}B$ evidence from Mg-tourmalines. Geology 40:231–234

Mering JA, Barker SLL, Huntington KW, Simmons S, Dipple G, Andrew B, Schauer A (2018) Taking the temperature of hydrothermal ore deposits using clumped isotope thermometry. Econ Geol 113:1671–1678

Meshoulan A, Ellis GS, Ahmad WS, Deev A, Sessions AL, Yang Y, Adkins JF, Liu J, Gilhooly WP III, Aizenshtat Z, Amrani A (2016) Study of thermochemical sulfate reduction mechanism using compound specific sulfur isotope analysis. Geochim Cosmochim Acta 188:73–92

Mikova J, Kosler J, Wiedenbeck M (2014) Matrix effects during laser ablation MC ICP-MS analysis of boron isotopes in tourmaline. J Anal at Spectrom 29:903–914

Miller C, Halley S, Green G, Jones M (2001) Discovery of the West 45 volcanic-hosted massive sulfide deposit using oxygen isotopes and REE geochemistry. Econ Geol 96:1227–1238

Moran AE, Sisson VB, Leeman WP (1992) Boron depletion during progressive metamorphism: implications for subduction processes. Earth Planet Sci Lett 111:331–349

Muehlenbachs K, Anderson AT Jr, Sigvaldason GE (1974) Low O-18 basalts from Iceland. Geochim Gosmoshim Acta 38:577–588

Nabalek PI, Labotka C, O'Neil JR, Papike JJ (1984) Contrasting fluid/rock interaction between the Notch Peak granitic intrusion and argillites and limestones in western Utah: Evidence from stable isotopes and phase assemblages. Contrib Mineral Petrol 86:25–34

Naito K, Fukahori Y, He P, Sakurai W, Shimazaki H, Matsuhisa Y (1995) Oxygen and carbon isotope zonations of wall rocks around the Kamioka Pb-Zn skarn deposits, central Japan: application to prospecting. J Geochem Explor 54:199–211

Nier AO (1937) A mass-spectrographic study of the isotopes of Hg, Xe, Kr, Be, I, As, and Cs. Phys Rev 52:933

Ohmoto H (1972) Systematics of sulfur and carbon isotopes in hydrothermal ore deposits. Econ Geol 67:551–579

Ohmoto H (1986) Stable isotope geochemistry of ore deposits. Rev Mineral 16:491–559

Ohmoto H, Goldhaber RO (1997) Sulfur and carbon isotopes. In: Barnes HL (ed) Geochemistry of hydrothermal ore deposits, 2nd edn. Wiley, New York, pp 517–612

Ohmoto H, Lasaga AC (1982) Kinetics of reactions between aqueous sulfate and sulfides in hydrothermal systems. Geochim Cosmochim Acta 46:1727–1745

Ohmoto H, Rye RO (1979) Isotopes of sulfur and carbon. In: Barnes HL (ed) Geochemistry of hydrothermal ore deposits, 2nd edn. Wiley, New York, pp 509–567

Pack A, Herwartz D (2014) The triple oxygen isotope composition of the Earth mantle and understanding $\Delta^{17}O$ variations in terrestrial rocks and minerals. Earth Planet Sci Lett 390:138–145

Pal DC, Trumbull R, Wiedenbeck M (2010) Chemical and boron isotope compositions of tourmaline from the Jaduguda U (-Cu-Fe) deposit, Singhbhum shear zone, India: implications for the source and evolution of the mineralizing fluid. Chem Geol 277:245–260

Palmer MR, Slack JF (1989) Boron isotope composition of tourmaline from massive sulfide deposits and tourmalinites. Contrib Mineral Petrol 103:434–451

Paterson BA, Riciputi LR, McSween HY Jr (1997) A comparison of sulfur isotope ratio measurement using two ion microprobe techniques and application to analysis of troilite in ordinary chondrites. Geochim Cosmochim Acta 61:601–609

Peters ST, Alibabaie N, Pack A, McKibbin SJ, Raeisi D, Nayebi N, Torab F, Ireland T, Lehmann B (2020) Triple oxygen isotope variations in magnetite from iron-oxide deposits, central Iran, record magmatic fluid interaction with evaporite and carbonate host rocks. Geology 48:211–215

Phillips GN, Groves DI (1983) The nature of Archaean gold-bearing fluids as deduced from gold deposits in Western Australia. J Geol Soc Austr 30:25–39

Pichlmayer F, Blochberger K (1988) Isotopenhäufigkeitsanalyse von kohlenstoff, stickstoff und schwefel mittels gerätekopplung elementaranalysator-massenspektrometer. Fresenius' Zeitschrift Für Analytische Chemie 331:196–201

Pitcairn IK, Olivo GR, Teagle DAH, Craw D (2010) Sulfide evolution during prograde metamorphism of the Otago and Alpine schists. Can Mineral 48:1267–1295

Puchelt H, Sabels BR, Hoering TC (1971) Preparation of sulfur hexafluoride for isotope geochemical analysis. Geochim Cosmochim Acta 35:625–628

Quesnel B, Scheffer C, Beaudoin G (2023b) The light stable isotope (H, B, C, N, O, Si, S) composition of orogenic gold deposits. In: Huston DL, Gutzmer J (eds), Isotopes in economic geology, metallogensis and exploration, Springer, Berlin, this volume

Rafter TA (1957) Sulphur isotope variations in nature. Part I-the preparation of sulphur dioxide for mass spectrometer examination. NZ J Sci Technol 838:849–857

Robinson BS, Kusakabe M (1975) Quantitative preparation of sulfur dioxide, for $^{34}S/^{32}S$ analyses, from sulfides by combustion with cuprous oxide. Anal Chem 47:1179–1181

Rosman KJR, Taylor PDP (1998) Isotope composition of the elements 1997. Pure Appl Chem 70:217–235

Rotherham J, Blake KL, Cartwright I, Williams P (1998) Stable isotope evidence for the origin of the Mesoproterozoic Starra Au-Cu deposit, Cloncurry District, Northwest Queensland. Econ Geol 93:1435–1449

Rumble D, Hoering TC, Palin JM (1993) Preparation of SF_6 for sulfur isotope analysis by laser heating sulfide minerals in the presence of F_2 gas. Geochim Cosmochim Acta 57:4499–4512

Rye RO (2005) A review of the stable-isotope geochemistry of sulfate minerals in selected igneous environments and related hydrothermal systems. Chem Geol 215:5–36

Rye RO, Bethke PM, Wasserman MD (1992) The stable isotope geochemistry of acid sulfate alteration. Econ Geol 87:225–262

Sah RH, Brown PH (1997) Boron determination — a review of analytical methods. Microchem J 56:285–304

Sasaki A, Arilawa Y, Folinsbee RE (1979) Kiba reagent method of sulfur extraction applied to isotope work. Bull Geol Surv Japan 30:241–245

Schmitt AK, Simon JI (2004) Boron isotope variations in hydrous rhyolitic melts - a case study from Long Valley, California. Contrib Mineral Petrol 146:590–605

Seal RR II (2006) Sulfur Isotope Geochemistry of Sulfide Minerals: Rev Mineral 61:633–677

Shanks WC III (2014) Stable isotope geochemistry of mineral deposits. Treatise Geochem, 2nd edn 13:59–85. https://doi.org/10.1016/B978-0-08-095975-7.01103-7

Sharp ZD (1990) A laser-based microanalytical method for the in situ determination of oxygen isotope ratios of silicates and oxides. Geochim Cosmochim Acta 54:1353–1357

Sharp ZD, Gibbons JA, Maltsev O, Atudorei V, Pack A, Sengupta S, Shock EL, Knauth LP (2016) A calibration of the triple oxygen isotope fractionation in the SiO_2-H_2O system and applications to natural samples. Geochim Cosmochim Acta 186:105–119

Sheppard SMF, Nielsen RL, Taylor HP Jr (1969) Oxygen and hydrogen isotope ratios of clay minerals from porphyry copper deposits. Econ Geol 64:755–777

Sheppard SMF, Nielsen RL, Taylor HP Jr (1971) Hydrogen and oxygen isotope ratios in minerals from porphyry copper deposits. Econ Geol 66:515–542

Sheppard SMF (1986) Characterization and isotope variations in natural waters. Rev Mineral 16:165–184

Shimizu N, Hart SR (1982) Isotope fractionation in seeondary ion mass spectrometry. J Appl Phys 53:1303–1311

Siegel K, Wagner T, Trumbull RB, Jonsson E, Matalin G, Wälle M, Heinrich CA (2016) Stable isotope (B, H, O) and mineral-chemistry constraints on the magmatic to hydrothermal evolution of the Varutrask rare-element pegmatite (northern Sweden). Chem Geol 421:1–16

Simmons SF, Sawkins FJ, Schlutter DJ (1987) Mantle-derived helium in two Peruvian hydrothermal ore deposits. Nature 329:429

Simmons SF, White NC, John DA (2005) Geological characteristics of epitherm precious and base metal deposits. Econ Geol 100th Anniv Vol, p 485–522

Slack JF, Trumbull RB (2011) Tourmaline as a recorder of ore-forming processes. Elements 7:321–326

Soddy F (1913) Intra-atomic charge. Nature 92:399–400

Swanson EM, Wernicke BP, Eiler JM, Losh S (2012) Temperatures and fluids on faults based on carbonate clumpled isotope thermometry. Am J Sci 312:1–21

Taylor HP Jr (1979) Oxygen and hydrogen isotope relationships in hydrothermal mineral deposits. In: Barnes HL (ed) Geochemistry of hydrothermal ore deposits, 2nd edn. Wiley, New York, pp 236–277

Taylor BE (1986a) Magmatic volatiles: isotope variation of C, H and S. Rev Mineral 16:185–225

Taylor HP Jr (1986b) Igneous rocks: II. Isotope case studies of circumpacific magmatism. Rev Mineral 16:273–318

Taylor BE (1987) Stable isotope geochemistry of ore-forming fluids. Mineral Assoc Can Short Course Handb 13:337–445

Taylor HP Jr (1997) Oxygen and hydrogen isotope relationships in hydrothermal mineral deposits. In: Barnes HL (ed) Geochemistry of hydrothermal ore deposits, 3rd edn. Wiley, New York, pp 229–302

Taylor HP Jr, Sheppard SMF (1986) Igneous rocks: 1. Processes of isotope fractionation and isotope systematics. Rev Mineral 16:227–272

Thode HG, Monster J, Durford HB (1961) Sulphur Isotope Geochemistry. Geochim Cosmochi Acta 25:159–174

Thom J, Anderson GM (2008) The role of thermochemical sulfate reduction in the origin of Mississippi Valley-type deposits. I. Experimental Results. Geofluids 8:16–26

Thomson JJ (1913) Rays of positive electricity. Proc Royal Soc A 89:1–20

Trumbull RB, Slack JF (2018) Boron isotopes in the continental crust: granites, pegmatites, felsic volcanic rocks and related ore deposits. Advances in Geochemistry 7:249–272

Trumbull RB, Codeco MS, Jiang S-Y, Palmer MR, Slack JF (2020) Application of boron isotopes in tourmaline to understanding hydrothermal ore systems. Ore Geol Rev 125:103682

Ueno Y, Johnson MS, Danielache SO, Eskebjerg C, Pandey A, Yoshida N (2009) Geological sulfur isotopes indicate elevated OCS in the Archean atmosphere, solving faint young sun paradox. Proc Natl Acad Sci USA 106:14784–14789

Ueno Y, Aoyama S, Endo Y, Matsu'ura F, Foriel J, (2015) Rapid quadruple sulfur isotope analysis at the sub-micromole level by a flash heating with CoF_3. Chem Geol 419:29–35

Urey HC (1947) The thermodynamics of isotope substances. J Chem Soc 1947:562–581

Valley JW (1986) Stable isotope geochemistry of metamorphic rocks. Rev Mineral 16:445–490

Valley JW (2003) Oxygen isotope in zircon. Rev Mineral 53:343–385

Valley JW, Cole D, eds (2001) Stable isotope geochemistry. Rev Mineral 43:531 p

Valley JW, Graham CM (1991) Ion microprobe analysis of oxygen isotope ratios in metamorphic magnetite-diffusion reequilibration and implications for thermal history. Contr Mineral Petrol 109:38–52

Valley JW, Taylor Jr HP, O'Neill JR, eds (1986) Stable isotopes in high temperature geological processes. Rev Mineral 16:570 p

Vennemann TW, O'Neil JR (1993) A simple and inexpensive method of hydrogen isotope and water analyses of minerals and rocks based on zinc reagent. Chem Geol (isot Geosci Sect) 103:227–234

Waltenberg K (2023) Application of Lu-Hf isotopes to ore geology, metallogenesis and exploration. In: Huston DL, Gutzmer J (eds), Isotopes in economic geology, metallogensis and exploration, Springer, Berlin, this volume

Wedepohl KH (1969a) Compositon and abundance of common sedimentary rocks. In: Wedepohl KD (ed) Handbook of geochemistry. Springer-Verlag, New York, pp 250–271

Wedepohl KH (1969b) Compositon and abundance of common igneous rocks. In: Wedepohl KD (ed) Handbook of geochemistry. Springer-Verlag, New York, pp 227–249

Werner RA, Brand WA (2001) Referencing strategies and techniques in stable isotope ratio analysis. Rapid Commun Mass Spectrom 15:501–519

Whitehouse MJ (2013) Multiple sulfur isotope determination by SIMS: evaluation of reference sulfides for Δ^{33}S with observations and a case study on the determination of Δ^{36}S. Geostand Geoanal Res 37:19–33

Whiticar MJ (1999) Carbon and hydrogen isotope systematics of bacterial formation and oxidation of methane. Chem Geol 161:291–314

Williams N (2023) Light-element stable isotope studies of the clastic-dominated lead-zinc mineral systems of northern Australia & the North American Cordillera: implications for ore genesis and exploration. In: Huston DL, Gutzmer J (eds) Isotopes in economic geology, metallogensis and exploration, Springer, Berlin, this volume

Wilkinson JJ (2023) The potential of Zn isotopes in the science and exploration of ore deposits. In: Huston DL, Gutzmer J (eds) Isotopes in economic geology, metallogensis and exploration, Springer, Berlin, this volume

Xavier RP, Wiedenbeck M, Trumbull RB, Dreher AM, Monteiro LVS, Rhede D, de Araújo CEG, Torresi I (2008) Tourmaline B-isotopes fingerprint marine evaporites as the source of high salinity ore fluids in iron oxide-copper-gold deposits, Carajás Mineral Province (Brazil). Geology 36:743–746

Young ED, Galy A, Nagahara H (2002) Kinetic and equilibrium mass-dependent isotope fractionation laws in nature and their geochemical and cosmochemical significance. Geochim Cosmochim Acta 66:1095–1104

Zheng Y-F (1993a) Calculation of oxygen isotope fractionation in anhydrous silicate minerals. Geochim Cosmochim Acta 57:1079–1091

Zheng Y-F (1993b) Calculation of oxygen isotope fractionation in hydroxyl-bearing silicates. Earth Planet Sci Lett 120:247–263

Zheng Y-F (1999) Oxygen isotope fractionation in carbonate and sulfate minerals. Geochem J 33:109–126

Zheng YF, Simon K (1991) Oxygen isotope fractionation in hematite and magnetite: a theoretical calculation and application to geothermometry of metamorphic iron-formation. Eur J Mineral 3:877–886

Zhong R, Brugger J, Tomkins AG, Chen Y, Li W (2015) Fate of gold and base metals during metamorphic devolatilization of a pelite. Geochim Cosmochim Acta 171:338–352

Zhou T, Yuan F, Yue S, Liu X, Zhang X, Fan Y (2007) Geochemistry and evolution of ore-forming fluids of the Yueshan Cu–Au skarn-and vein-type deposits, Anhui Province, South China. Ore Geol Rev 31:279–303

Light Stable Isotopes in Volcanic-Hosted Massive Sulfide Ore Systems

David L. Huston, Crystal Laflamme, Georges Beaudoin, and Stephen Piercey

Abstract

Volcanic-hosted massive sulfide (VHMS) deposits, the ancient analogues of "black smoker" deposits that currently form on the seafloor, are the products of complex mineral systems involving the interaction of seawater with the underlying volcanic pile and associated magmatic intrusions. Light stable isotopes, particularly those of oxygen, hydrogen and sulfur, have had a strong influence in determining sources of ore fluids and sulfur as well as elucidating geological processes important in the VHMS mineral systems. Oxygen and hydrogen isotope data indicate that evolved seawater was the dominant ore-forming fluid in VHMS mineral systems through geological time, although a small proportion of deposits, including high sulfidation and tin-rich deposits, may have a significant, or dominant, magmatic-hydrothermal fluid component. Higher-temperature (> 200 °C) interaction of evolved seawater alters the rock pile below the seafloor, producing $\delta^{18}O$ depletion anomalies at the deposit and district scales that can be used as a vector to ore. In contrast, lower-temperature hydrothermal alteration results in $\delta^{18}O$-enriched zones that commonly cap mineralized positions. An apparent decrease in the degree of high temperature ^{18}O depletion with time may relate to the increasing importance of felsic-dominated host successions in younger deposits. $\delta^{18}O$ anomalies have potential as an exploration tool, and have contributed directly to discovery. The other important contribution of stable isotopes to understanding the VHMS mineral system is quantification of the contribution of sulfur sources. Conventional $\delta^{34}S$ data, when combined with $\Delta^{33}S$ data acquired using recently developed technologies, indicate that the dominant sulfur source is igneous sulfur, either leached from the volcanic pile or introduced as a magmatic volatile (these sources are not distinguishable). The thermochemical reduction of seawater sulfate is also an important, but subordinate, sulfur source.

Supplementary Information The online version contains supplementary material available at https://doi.org/10.1007/978-3-031-27897-6_9.

D. L. Huston (✉)
Geoscience Australia, GPO Box 378, Canberra, ACT 2601, Australia
e-mail: David.Huston@ga.gov.au

C. Laflamme · G. Beaudoin
Département de Géologie Et de Génie Géologique, Centre de Recherche Sur La Géologie Et L'ingénierie Des Ressources Minérales (E4m), Université Laval, 1065, Avenue de La Médecine, Québec, QC G1V 0A6, Canada

S. Piercey
Department of Earth Sciences, Memorial University of Newfoundland, Room 4063, Alexander Murray Building, 9 Arctic Avenue, St. John's, NL 1B 3X5, Canada

D. Huston and J. Gutzmer (eds.), *Isotopes in Economic Geology, Metallogenesis and Exploration*, Mineral Resource Reviews, https://doi.org/10.1007/978-3-031-27897-6_9

Estimation of the proportion of seawater sulfate with geological age indicate that, on average, it has increased from 5–10% in the Archean to 20–25% in the Phanerozoic. This most likely reflects the increase in seawater sulfate contents through geological time. Although untested as an exploration tool, variations in sulfur isotope data may have utility is discriminating fertile from barren sulfide accumulations or providing vectors to ores at the deposits scale. As exploration tools, light stable isotopes suffer from a relatively high cost and slow turn-around time. If these limitations can be overcome, and new analytical methods can be developed, light stable isotopes may emerge as another tool for exploration, particularly as discoveries are made at greater depth and under cover.

1 Introduction

As discussed elsewhere in this volume, light stable isotope data has proven key to understanding the genesis of a range of mineral systems (Hagemann et al. 2023; Huston et al. 2023; Quesnel et al. 2023; Williams 2023). Although these data have been particularly useful in understanding the sources of ore fluids and sulfur, the data also have proven useful in exploration (Miller et al. 2001). This chapter presents an overview of variations in stable isotope geochemistry in the volcanic-hosted massive sulfide (VHMS) mineral system and how this data has been used to understand this system and where it may be practical for exploration vectoring. It follows and builds on previous syntheses of stable isotopes in VHMS systems by Huston (1999), Shanks (2014) and Leybourne et al. (2022) and of active seafloor systems by Zierenberg and Shanks (1988) and Shanks (2001). Studies of active systems provide information such as direct isotopic measurements of venting fluids and the effect of changes during the timespan of the mineral system not available from studies of ancient systems.

2 The Volcanic-hosted Massive Sulfide Mineral System

The VHMS mineral system is arguably one of the best documented and understood mineral system. As VHMS deposits are preserved in many different parts of the world and formed through much of Earth's history, they provide, collectively, information on geological processes through time, including changes in hydrothermal and environmental processes. Despite these changes, VHMS deposits through time share many characteristics, both at the deposit- and district- to province-scale.

The understanding of the VHMS mineral system has come about not only from ancient deposits (Franklin et al. 1981, 2005; Huston et al. 2006) but also the discovery of "black smoker" deposits in modern oceanic basins (e.g., Hannington et al. 2005), which are thought to be modern analogues of VHMS deposits. Currently, hydrothermal fluids circulate extensively through the upper crust, particularly in submarine extensional zones associated with mid-oceanic ridges on divergent margins, or back-arc basins/rifted arcs along convergent margins. This circulation not only cools the oceanic upper crust, controlling the Earth's surficial heat budget (Stein 1995), but the fluids leach metals from the underlying rock and transfer these metals into the hydrosphere, having a major impact on the seawater metal budget. Although most metal is lost to the seawater column when these fluids vent (Converse et al. 1984), some of the metal is trapped in massive sulfide deposits that form at or just below the seafloor at venting sites (Herzig and Hannington 1995); the vents are also sites of abundant biological activity (Hannington et al. 2011). The growth of seafloor massive sulfide deposits involves dissolution, alteration and replacement of rock and massive sulfide within and below the mounds, forming complex mixtures of sulfide and altered host rock known as stockworks (Petersen et al. 2018).

Studies indicate that geologically ancient VHMS deposits formed mostly in back-arc basins and rifted arcs along convergent margins (e.g., Franklin et al. 2005). This differs from modern black smokers, which are known to form along both mid-oceanic ridges as well as back-arc basins and rifted arcs (Hannington et al. 2005). This difference arises due the low likelihood of preserving oceanic crust in the ancient record—most of this crust is lost to subduction. Figure 1 illustrates an idealized asymmetric, VHMS-bearing back-arc basin that has rifted pre-existing continental crust to the left and an arc that developed on this crust to the right. Geochemical analyses of volcanic successions that host VHMS deposits (Lesher et al., 1986; Kerrich and Wyman 1997; Hart et al. 2004; Piercey 2011) indicate that both back-arc and arc signatures are present in VHMS districts.

Volcanism and the associated magmatism are integral components of the VHMS mineral system, providing heat, fluids, metals and/or sulfur. The left-hand side of Fig. 1 illustrates a VHMS system associated with a semi-conformable, subvolcanic intrusion that has provided heat to drive fluid flow. The associated volcanic pile provided metals and sulfur to the mineralizing fluids. The right-hand side illustrates a situation in which a cross-cutting intrusion, possibly arc-related, has provided heat, metals, sulfur and fluids to the evolving VHMS system. These two systems should be considered end-member variants of the VHMS mineral system.

The other major source of system components in VHMS systems is the overlying seawater column. Based mostly on stable isotope evidence seawater is thought to be the dominant fluid source in most VHMS systems and an important source of sulfur in many, particularly younger, systems. The left-hand inset in Fig. 1 illustrates a mineral system in which emplacement of a semi-conformable, subvolcanic intrusion (c.f. Galley 2003) at a depth of 3–5 km drives convection of seawater in the volcanic pile. This produces semi-conformable alteration zones above the intrusion that become cross-cutting along syn-volcanic structures to become the proximal alteration zones associated with most VHMS deposits. This alteration pattern records the convection of evolving seawater that interacts with

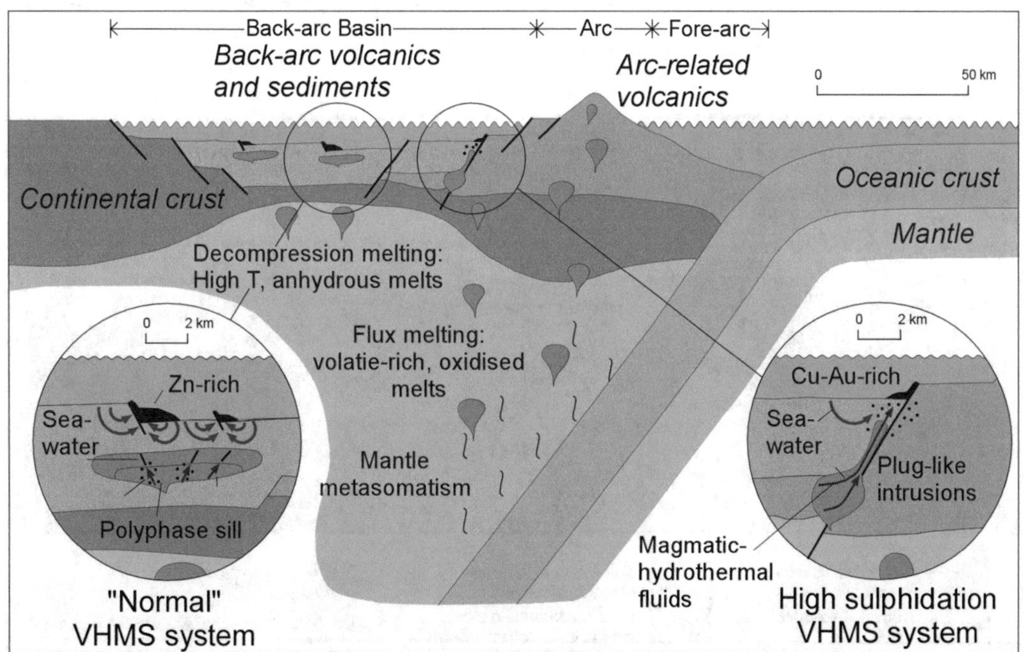

Fig. 1 Volcanic-hosted massive sulfide mineral system. Reproduced with permission from Huston et al. (2011); Copyright 2011 Geoscience Australia

the volcanic pile to extract metal and sulfur then moves upwards through the pile and deposits sulfur and metals to form a VHMS deposit. If the seawater was sulfate-bearing, reduction to sulfide by high temperature water–rock interaction was a second source of sulfur in the system.

Volcanic-hosted massive sulfide deposits form at and just below the seafloor and comprise three types of mineralized rock, stockwork, massive sulfide and exhalite (Fig. 2; Franklin et al. 1981, 2005; Huston et al. 2006). The stockwork zone consists of anastomosing veins, or "stockworks", of sulfide-bearing veins that also commonly contain quartz, carbonate, chlorite, sericite and, in some deposits, sulfate minerals. The stockworks typically vein strongly altered volcanic rocks. This "proximal" alteration typically is dominated by chlorite, sericite, quartz and iron sulfide minerals. In some cases, proximal alteration can contain advanced argillic assemblages (e.g., pyrophyllite, kaolinite) although such zones are much less common than the typical quartz-chlorite-sericite assemblage. Stockwork and proximal alteration zones generally grade outwards over a distance of 10–200 m to lower temperature "distal" alteration zones by either a decrease in alteration intensity or mineralogical changes.

Stockwork zones commonly grade upwards into massive sulfide zones, defined as rock that contains more than 60% sulfide minerals (Sangster and Scott 1976). Although early models interpreted that massive sulfide zones formed at the seafloor (e.g., Hutchinson 1973; Ohmoto 1986; Lydon 1988), more recent work has indicated that massive sulfide zones can also form by replacement of host rocks (Doyle and Allen 2003) or by increasing intensity of stockwork (Fig. 2). Massive sulfide zones are commonly zoned mineralogically and compositionally, typically from Cu-rich at the base to Zn-rich at the top; many Phanerozoic deposits have sulfate-rich zones at the very top. In some cases, massive sulfide zones are flanked by exhalite, chemical sediments formed from hydrothermal fluids that have interacted with seawater. Exhalites can be siliceous and/or iron-rich, can be enriched in many hydrothermal elements, including base metals, and can be deposited hundreds of meters to even kilometers away from massive sulfide deposits (Peter 2003 and references therein). In many, though not all, districts, exhalites mark

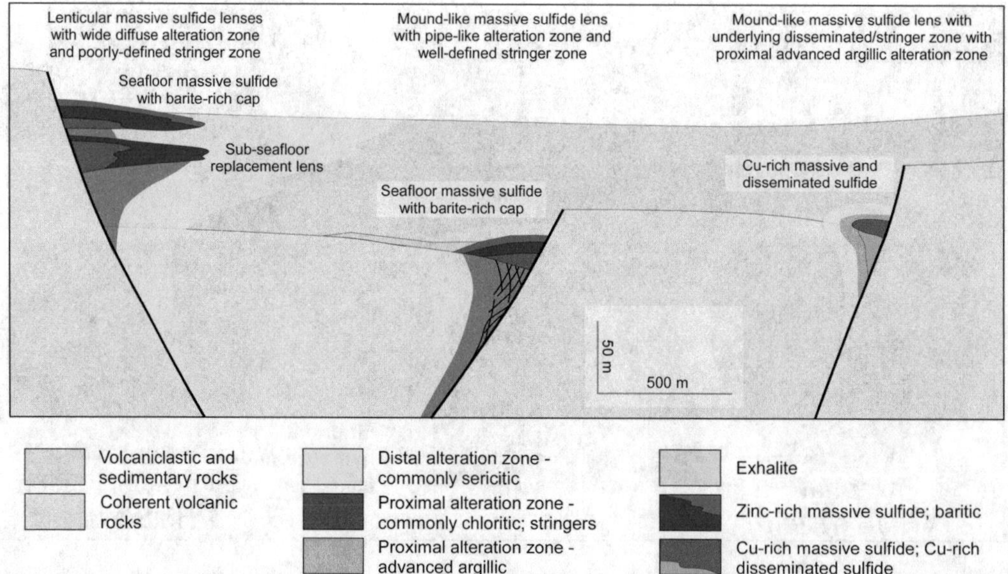

Fig. 2 Idealized cross-section of volcanic-hosted massive sulfide deposits showing different ore and alteration zones

stratigraphic positions prospective for VHMS mineralization.

The right hand inset in Fig. 1 shows a scenario in which the magmatic chamber that drove circulation in the VHMS system also concentrated magmatic-hydrothermal fluids and these were introduced into the circulating VHMS fluid system. In this case both seawater and magmatic volatiles would be fluid sources, and magmatic-hydrothermal sulfur and metals could form a significant component of ore fluid, with the rest derived either from convecting seawater or leaching of the volcanic pile. A mixed fluid with seawater and magmatic-hydrothermal components would then rise to the seafloor, where rapid cooling upon mixing with ambient seawater would cause metal deposition at or just below the seafloor.

3 Stable Isotope Terminology

In this chapter conventional isotopic results are reported relative to internationally recognized standards using standard δ notation: V-SMOW (Vienna-standard mean oceanic water: Werner and Brand 2001) for oxygen and hydrogen and V-CDT (Vienna- Canyon Diablo troillite: Ding et al 2001). Typical 1σ uncertainties in measurements are 0.2‰ for $\delta^{18}O$, 3‰ for δD and 0.2‰ for $\delta^{34}S$. Details of analytical techniques, standards and uncertainties are summarized by Huston et al. (2023).

4 Stable Isotopes and the Genesis of Volcanic-hosted Massive Sulfide Systems

Stable isotope data have played a key role in constraining and understanding fluid and sulfur sources in VHMS mineral systems. Because the oxygen and hydrogen isotope composition of seawater is thought to have been relatively constant over geological time ($\delta^{18}O \sim -3$ to 0‰ and $\delta D \sim -30$ to 0‰: Sheppard 1986), and this value is significantly different to magmatic-hydrothermal fluids ($\delta^{18}O \sim 5.5$ to 10‰ and $\delta D \sim -50$ to -35‰: Taylor 1987), these data can be used to assess the relative importance of evolved seawater and magmatic-hydrothermal fluids in individual VHMS ore systems at the deposit and district scales. Similarly, as the $\delta^{34}S$ composition of seawater sulfate ($\delta^{34}S$ + 10 to + 35‰; Crockford et al. 2019) differs to that of sulfur derived from volcanic or magmatic-hydrothermal sources ($\delta^{34}S$ −5 to + 8‰; Chaussidon et al. 1989), sulfur isotope data have been used to infer the importance of seawater-derived sulfur to the VHMS mineral system through geological time.

4.1 Oxygen and Hydrogen Isotopes

As seawater has oxygen and hydrogen isotope compositions that are significantly different to the rocks that underlie and host VHMS deposits, alteration of these rocks by (evolved) seawater can significantly affect the isotopic composition of the altered rocks. This interaction occurs not only in proximity to the ore zones, but also more regionally. In both cases, oxygen isotope variations have been used to map hydrothermal alteration zones.

4.1.1 Ore-proximal Patterns in Oxygen and Hydrogen Isotopes

Over the last five decades numerous studies (Table E1: updated from Huston 1999) have documented variations in $\delta^{18}O$ and δD in the immediate vicinity of VHMS deposits. These studies have indicated systematic patterns, such as a decrease in $\delta^{18}O$ toward ore (Fig. 3a), that can be used for exploration (e.g., Miller et al. 2001). There are fewer studies documenting variations in δD, and these studies do not show as consistent of a pattern and are likely to be affected by metamorphism. However, despite this, most of the deposits shown in Fig. 3b are characterized by an increase in δD proximal to ore.

Beaty and Taylor (1982) first illustrated these relationships in a diagram schematically illustrating variations in $\delta^{18}O$ away from proximal alteration zones for a range of deposits. Figure 3 is an update of this original diagram that includes many historic studies as well as more recent

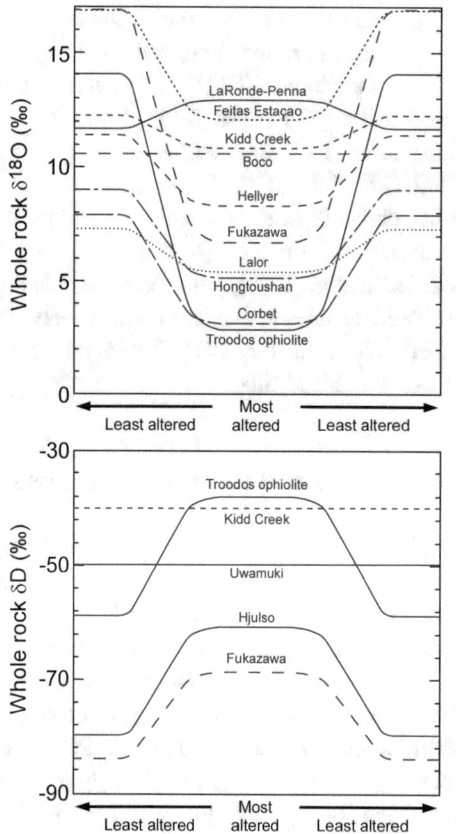

Fig. 3 Schematic diagrams showing variations in whole-rock $\delta^{18}O$ and δD values around VHMS deposits (updated from original diagrams in Beaty and Taylor 1982, using data from Table E1). Because of the large number of studies, this diagram mostly shows deposits with large number of analyses, but also includes some of the classic deposits (e.g., Beaty and Taylor 1982) where the relationship was first demonstrated. The width of the most intense isotopically-altered zone ranges from a few tens to a few hundreds of meters. The scale of the isotopic zonation is up to 1000 m, but more typically a few hundreds of meters

studies that are based on a large number of analyses (most deposits with a small dataset have been excluded, but are summarised in Table E1). At most deposits, there is a consistent pattern of decreasing whole rock $\delta^{18}O$ toward the core of the proximal alteration zone (Fig. 3a). In detail, however, the magnitude and variation between distal and proximal $\delta^{18}O$ varies between deposits and, possibly, through time. Deposits associated with advanced argillic alteration assemblages (red lines in Fig. 3a), however, are characterized by proximal $\delta^{18}O$ values that are similar to or even higher than distal $\delta^{18}O$ values.

To assess possible reasons for this variability, Fig. 4 plots the variation of (a) average distal whole-rock $\delta^{18}O$, (b) average proximal whole-rock $\delta^{18}O$ and (c) the isotopic contrast (i.e. $\delta^{18}O_{distal}$ – $\delta^{18}O_{proximal}$) with geological time. The figure also shows the variations in (d) $\delta^{18}O_{distal}$ versus the isotopic contrast, and (e) $\delta^{18}O_{distal}$ and (f) $\delta^{18}O_{proximal}$ versus 100Cu/(Cu + Zn) of the deposit. No correlation is evident between $\delta^{18}O_{proximal}$ and isotopic contrast, so this diagram is not shown. Although the certainty is limited due to a small sample size, there appears to be a broad decrease in $\delta^{18}O$ values from distal zones and, possibly, proximal zones, with age. Variations in $\delta^{18}O$ in altered rocks are controlled by the temperature of the isotopic exchange, the water/rock (W/R) ratio, $\delta^{18}O$ of the altering fluid and $\delta^{18}O$ and chemical and mineralogical composition of the altered rock.

Figure 4e and 4f indicate that although there is not a relationship between $\delta^{18}O_{distal}$ and 100Cu/(Cu + Zn), low $\delta^{18}O_{proximal}$ alteration zones tend to be associated with Cu-rich deposits. As 100Cu/(Cu + Zn) is a proxy for the overall temperature of the hydrothermal system (e.g., Large 1977; Lydon 1988), this suggests that $\delta^{18}O_{proximal}$ reflects, to a large degree, the overall temperature of the hydrothermal system, particularly in hydrothermal upflow zones with very high W/R. Overall, this temperature has decreased over geological time. The observation that $\delta^{18}O_{distal}$ is unrelated to 100Cu/(Cu + Zn) suggests, however a different control for the isotopic composition of the outer, lower temperature and lower W/R portions of the VHMS mineral system.

A possible reason for the disconnect between $\delta^{18}O_{distal}$ and the overall temperature of the hydrothermal system is that $\delta^{18}O_{distal}$ records a different type of fluid-rock interaction to the high temperature ore fluid. Under conditions of low W/R, the isotopic composition of fluids is largely controlled by the isotopic composition of the rocks with which they interact, in this case the bulk isotopic composition of the rock pile underlying the VHMS deposit. Figure 4a shows that rock pile in which felsic volcanic rocks

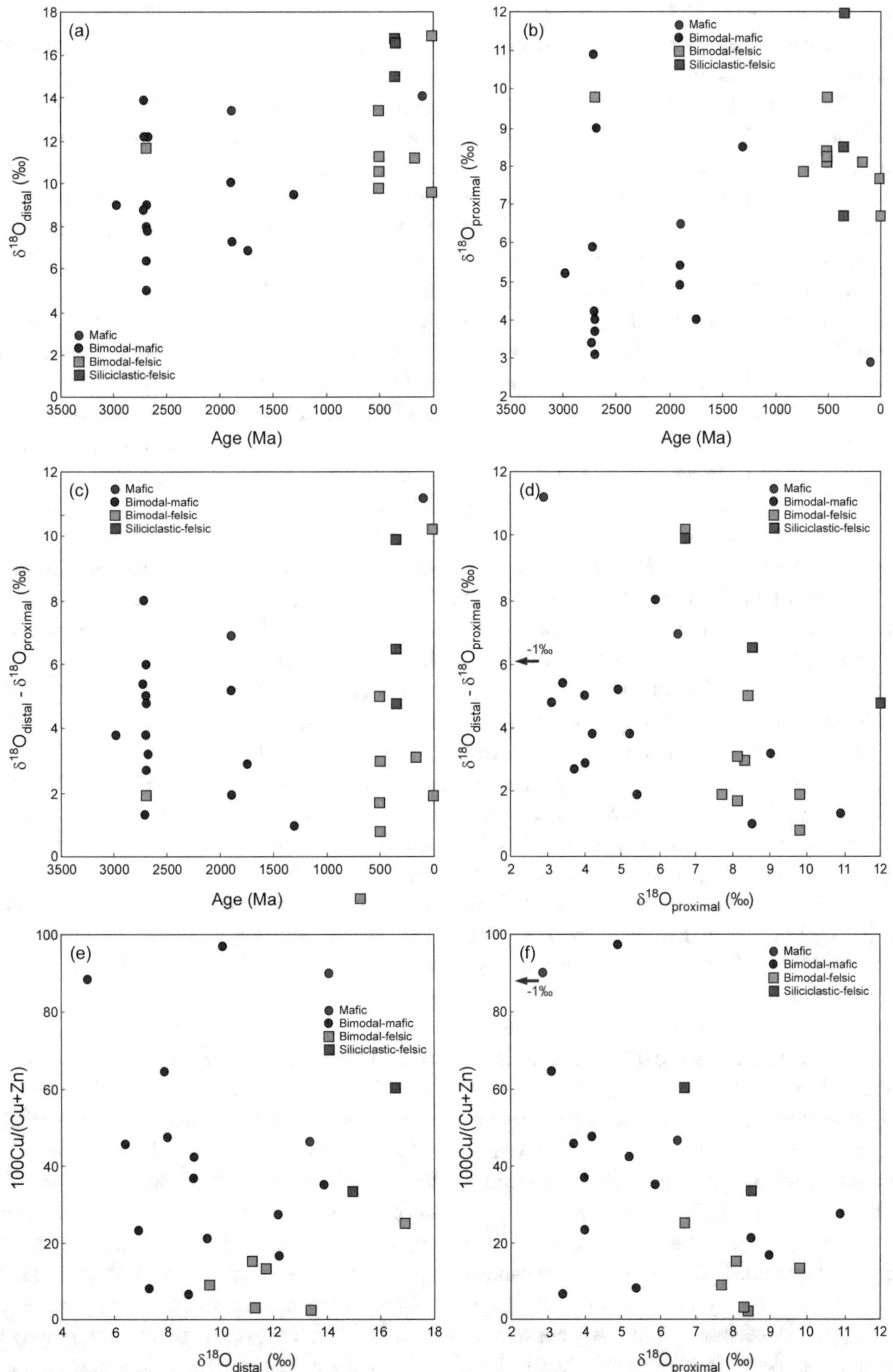

Fig. 4 Plots showing the variation of (a) average distal whole-rock $\delta^{18}O$, (b) average proximal whole-rock $\delta^{18}O$ and (c) the isotopic contrast (i.e. $\delta^{18}O_{distal} - \delta^{18}O_{proximal}$) with geological time, (d) distal $\delta^{18}O$ versus isotopic contrast, and (e) and (f) $\delta^{18}O_{distal}$ and $\delta^{18}O_{proximal}$ versus 100Cu/Cu + Zn) of the associated deposit. Data are from Table E1

dominate have higher $\delta^{18}O_{distal}$ those in which mafic volcanic rocks dominate. As unaltered felsic volcanic rocks (dacite and rhyolite) on average have higher $\delta^{18}O$ (6–9‰) than mafic rocks (basalt: 5–7‰) (Taylor and Sheppard 1986), low temperature, ambient fluids in equilibrium with felsic dominated rock piles under low W/R conditions would have higher $\delta^{18}O$ than fluids equilibrated with mafic-dominated piles. Interaction of such fluids in altered zones distal to the main hydrothermal channelway may account for the observed differences between $\delta^{18}O_{diistal}$ in felsic-dominated and mafic-dominated VHMS systems.

These mechanisms, however, do not directly account for the apparent correlations of $\delta^{18}O_{distal}$ and $\delta^{18}O_{proximal}$ with time (Figs. 4a and 4b); these correlations may be indirect. As shown by Barrie and Hannington (1999) and Huston et al. (2010), deposits in felsic-dominated succession are much more common in the Phanerozoic than in the Paleoproterozoic or Archean; conversely, deposits hosted by mafic-dominated successions are more common in the Paleoprotoerozoic and Archean. Hence, the correlation of $\delta^{18}O_{distal}$ and $\delta^{18}O_{proximal}$ with geological time may relate to the changing relative abundance of mafic- and felsic-dominated volcanic successions as VHMS hosts through time. A possible corollary to this inference is that, on average, older VHMS deposits formed at higher temperatures than younger deposits. Hence, it is important to consider the lithology of the altered host succession when interpreting whole rock $\delta^{18}O$ patterns in the context of genesis and exploration, an issue highlighted by many authors (e.g., Taylor et al. 2015). It also must be stressed that distal values are not those of pristine, unaltered volcanic rocks, but those of rocks that have experienced low temperature, pervasive alteration.

Metamorphic grade also does not seem to affect $\delta^{18}O$ patterns, since regional metamorphism most commonly did not involve significant water advection, and is considered isochemical (Riverin and Hodgson 1980); thus, variations in $\delta^{18}O$ see through regional metamorphism and reflect original fluid-rock interaction patterns and pathways.

As discussed below the difference in the isotopic patterns associated with deposits associated with advanced argillic alteration zones to those deposits associated with the typical chlorite and/or sericite dominated alteration assemblages could be caused by interaction of lower temperature hydrothermal fluids or by ore fluids with higher $\delta^{18}O$, such as magmatic-hydrothermal fluids (e.g., Beaudoin et al. 2014). In high sulfidation epithermal deposits, which are typically associated with advanced argillic alteration assemblages, the advanced argillic assemblages are typically interpreted to be the consequence of acidic fluids formed by the dissociation of SO_2 at high temperature (> 350 °C) in magmatic-hydrothermal fluids (Ohmoto and Rye 1979; Rye et al. 1992). Similar processes are envisioned to have occurred in VHMS deposits that have been influenced by magmatic-hydrothermal fluids, leading to hydrothermally altered rocks with elevated $\delta^{18}O$ (e.g., Sillitoe et al. 1996; Huston et al. 2011).

In general, oxygen and, to a lesser extent, hydrogen isotopes are zoned about VHMS deposits due to hydrothermal alteration associated with the ore-forming fluids. In detail, however, the patterns can be complicated and can be affected by the protolith lithological variability of the altered rocks. To illustrate this we summarize the results of some of the more detailed oxygen and hydrogen isotope studies to illustrate the similarities, and differences, in the isotopic patterns.

Hokuroku district, Japan. In the Miocene Hokuroku district of Japan, Hattori and Muehlenbachs (1980), Green et al. (1983) and Urabe et al. (1983) documented variations around bimodal felsic VHMS deposits of this district. Of the deposits described, the most information is available from the Fukazawa deposit (Green et al. 1983: Fig. 5a). Green et al. (1983) documented a zone with $\delta^{18}O$ below 8‰, which corresponds almost directly to the sericite-chlorite alteration zone that cuts stratigraphy and forms a pipe that extends below and above the orebody. This zone is surrounded by a 500–1500-m-wide, montmorillonite-altered halo that is characterized by $\delta^{18}O$ values between 8‰ and

14‰. Lateral to the montmorillonite zone, the rocks have been quartz- and zeolite-altered with $\delta^{18}O$ values above 14‰ (Date et al. 1983; Green et al. 1983).

Noranda district, Québec, Canada. Beaty and Taylor (1982) documented variations in whole rock $\delta^{18}O$ data along a traverse across the alteration pipe (in a similar position to stockwork zones in other VHMS deposits) at the Amulet deposit in the Noranda district, Québec (Fig. 5b), showing a pattern about the alteration pipe, with the central, most intensely, chlorite-altered part

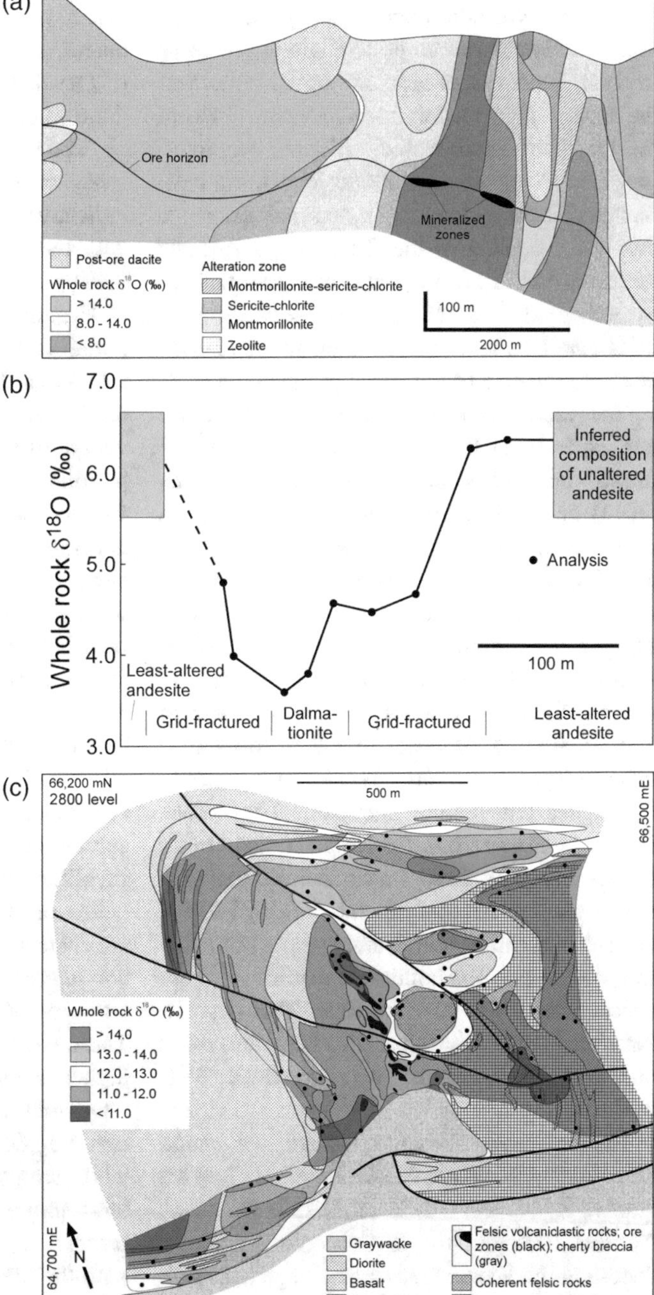

Fig. 5 Sections and plans showing proximal whole-rock $\delta^{18}O$ variations around VHMS deposits: (a) Fukazawa deposit, Hokuroku district, Japan (modified after Green et al. 1983), (b) Amulet deposit, Quebec, Canada (modified after Beaty and Taylor 1983) and (c) Kidd Creek deposit, Ontario, Canada (modified after Huston and Taylor 1999). "Dalmationite" in (b) refers to a metamorphosed, intensely altered zone consisting of coarse-grained cordierite porphyroblasts set in a fine-grained chlorite-quartz-anthophyllite matrix. "Grid-fractured" refers to a biotite-rich rock with a grid-like network of quartz-filled fractures

of the pipe having a $\delta^{18}O$ values of 3.6–4.0‰ that increased over a distance of 150 m outward to 6.0–6.7‰ in least altered volcanic rocks. Beaty and Taylor (1982) also demonstrated that, with the exception of the Kidd Creek deposit (see below), this pattern was consistent through most deposits that had been studied at that time.

In the New Vauze-Norbec area in the central part of the Noranda volcanic complex, Paradis et al. (1993) showed an up stratigraphy increase in $\delta^{18}O$, from values near 2‰ at the bottom of the stratigraphic section, to values near 14‰ at the top. This stratigraphic variation was interpreted to have formed under an increasing thermal gradient and increasing water/rock exchange during the cooling of the Flavrian syn-volcanic intrusion. Paradis et al. (1993) proposed that the hydrothermal fluid was Archean seawater with $\delta^{18}O$ near 0‰ that evolved through water–rock isotope exchange to a heavier composition.

The variations in $\delta^{18}O$ described above are one-dimensional transects or drill holes or two dimensional plans or sections. Taylor et al. (2014) determined three-dimensional variations in whole rock $\delta^{18}O$ at the Horne and Quemont deposits within the Noranda district, which combined have produced over 10 Moz of gold. The surface expression of the Horne deposit is associated with a zone of low $\delta^{18}O$ (< 6‰), as seen in other deposits discussed above. Zones of high (> 9‰) $\delta^{18}O$ distal to the deposits may indicate low temperature alteration peripheral to the higher temperature alteration. These data, data from elsewhere in the Noranda district (see below) and from Kidd Creek indicate that persistent zones of relatively high $\delta^{18}O$ may indicate the lack of significant high temperature fluid flow and downgrade exploration potential of these zones, although it must be stated that deposits associated with advanced argillic alteration zones may not be associated with proximal $\delta^{18}O$ depletion zones.

Kidd Creek deposit, Ontario, Canada. The ~ 2714 Ma Kidd Creek deposit (147.88 Mt grading 2.31% Cu, 6.18% Zn, 0.22% Pb, 87 g/t Ag and 0.01 g/t Au) is the only significant deposit in the Kidd-Munro assemblage in eastern Ontario, Canada. It consists of several steeply-plunging orebodies that are hosted by a mainly rhyolitic volcaniclastic unit within a succession dominated by mafic volcanic rocks (Huston and Taylor 1999: Fig. 5c). The orebodies occur along the southeastern limb of a tight, steeply-plunging anticline.

The first oxygen isotope study by Beaty et al. (1988) at Kidd Creek demonstrated that $\delta^{18}O$ values in proximal quartz-rich alteration assemblages, at 10–12‰, were much higher than the $\delta^{18}O$ values of other VHMS deposits, and that these values increased stratigraphically below the ore zone to values of 13–16‰ in felsic rocks. Beaty et al. (1988) interpreted these results as indicative of a two-stage hydrothermal system, with the main stage fluids having values of 6–9‰.

Huston and Taylor (1999) followed these initial studies with a more detailed study. They found that the $\delta^{18}O$ values are influenced by the original lithology of altered rocks; Fig. 4c shows variations in $\delta^{18}O$ only of samples with rhyolitic protoliths as determined by lithological observations (e.g., the presences of quartz phenocrysts) or the geochemistry of immobile elements. The lowest $\delta^{18}O$ values (<11‰) at Kidd Creek are associated with the "cherty breccia" alteration facies that is closely associated with the ore lenses, a result consistent with Beaty et al. (1988). Huston and Taylor (1999) also found that $\delta^{18}O$ values increase both laterally away from and stratigraphically above the ore zones to values that exceed 14‰. The host succession includes both fragmental and coherent rhyolitic rocks. Coherent units stratigraphically below or at the same position as the ore lenses are characterised by high δO values (> 13‰) that are interpreted to be the consequence of low temperature alteration that silicified these units. These rocks then became impermeable to later, higher temperature fluid flow, preserving the ^{18}O-enriched signature and focussing fluid flow into fragmental rocks. These fragmental rocks were strongly affected by these later high-temperature fluids, which produced the cherty breccia and the ^{18}O-depleted zone associated with the ore lenses (Fig. 5c: Huston and Taylor 1999). This may have produced the isotopic disequilibria that Beaty et al. (1988) interpreted to indicate two hydrothermal stages.

Stratigraphically above the ore position, $\delta^{18}O$ data define an ^{18}O-enriched ($\delta^{18}O$ > 13‰) zone that forms a carapace that has been folded by the later anticline. This zone is interpreted to have formed as the result of lower temperature alteration associated with the cooling of the Kidd Creek hydrothermal system. This carapace is only breached above the southwest ore lens (Huston and Taylor 1999), and may indicate limited high-temperature fluid flow after the Kidd Creek deposit was covered by later volcanic rocks. $\delta^{18}O$ patterns around the Kidd Creek deposit not only provide vectors toward ore, but they also provide constraints on ore fluid hydrology before, during and after ore formation.

LaRonde deposit, Québec, Canada. The LaRonde deposit (58.76 Mt grading 0.33% Cu, 2.17% Z, 45 g/t Ag and 4.31 g/t Au) is hosted by dacitic to rhyolitic rocks of the Bousquet Formation, Blake River Group (2699–2697 Ma, Mercier-Langevin et al. 2007). The deposit consists of four stacked semi- to massive sulfide lenses within complexly zoned alteration assemblages that has been metamorphosed at lower amphibolite facies to assemblages of mostly quartz, biotite, chlorite, garnet, muscovite, staurolite, rutile and pyrite. An unusual characteristic of the deposit is the presence of an aluminous alteration zone consisting of quartz, kyanite, andalusite, muscovite and pyrite (Dubé et al. 2007). $\delta^{18}O$ values of the altered host rocks range between 9.0 and 14.2 ‰ for all but 2 analyses, with average $\delta^{18}O$ values of different alteration facies between 9.8 and 12.9‰ and the aluminous facies having the highest $\delta^{18}O$ (Beaudoin et al. 2014).

Fluid-rock modelling indicates that the high $\delta^{18}O$ values at LaRonde can be explained by reaction of volcanic rocks with an initial $\delta^{18}O$ value of 7–9‰ with a fluid with initial $\delta^{18}O$ close to 5‰, at temperatures of 100–200°C, under W/R ratios up to 50:1 (Beaudoin et al. 2014). These W/R ratios are typical of proximal alteration zones, which are fluid dominated, as shown by Green et al. (1983), Shanks (2014) and many others. As these proximal alteration zones define fluid conduits, they have much higher W/R ratios than the VHMS system as a whole.

The initial fluid composition of 5‰ was interpreted to have been evolved seawater perhaps mixed with a magmatic water component exsolved from dacitic to rhyolitic magma, consistent with the aluminous alteration interpreted to be metamorphosed advanced argillic alteration analogous to subaerial high-sulfidation epithermal environments (Dubé et al. 2007).

4.1.2 District-scale Patterns in Oxygen Isotopes

Although most $\delta^{18}O$ patterns have been established in the immediate vicinity of VHMS deposits, there are several more regional studies, including studies of the ~ 400 Ma West Shasta district in California, USA (Taylor and South 1985), the ~ 2698 Ma Noranda district in Quebec, Canada (Cathles 1993), the ~ 2714 Ma Kidd-Munro district in Ontario, Canada (Taylor and Huston 1999), the ~ 3240 Ma Panorama district in Western Australia (Brauhart et al. 2000), the ~ 2745 Ma Sturgeon Lake district in Ontario, Canada (Holk et al. 2008) and the ~ 2681 Ma Izok Lake district in Nunavut, Canada (Taylor et al. 2015). Of these the two best documented districts, Noranda and Panorama, are described below.

Noranda, Québec, Canada. The Noranda district in the Superior Province, Québec, is a classical VHMS district that rests undeformed since its formation (~2700 Ma), with only a shallow tilt to the east. The district is hosted by a sequence of bimodal mafic-felsic volcanic rocks of the Blake River Group. The Noranda Volcanic Complex hosts 18 VHMS deposits that have yielded 48.5 Mt of ore, in addition to the Horne deposit (54.3 Mt) and its undeveloped Horne 5 zone (~150 Mt; Kerr and Gibson 1993). A seminal study by Cathles (1993) documented the plan view of the VHMS system from 588 oxygen isotope analyses (Fig. 6) and was augmented by 599 analyses by Taylor and Timbal (2002). These studies show that the syn-volcanic Flavrian pluton is surrounded by an annular zone with $\delta^{18}O$ values < 6‰ that extend up section as six fingers that point toward most of the VHMS

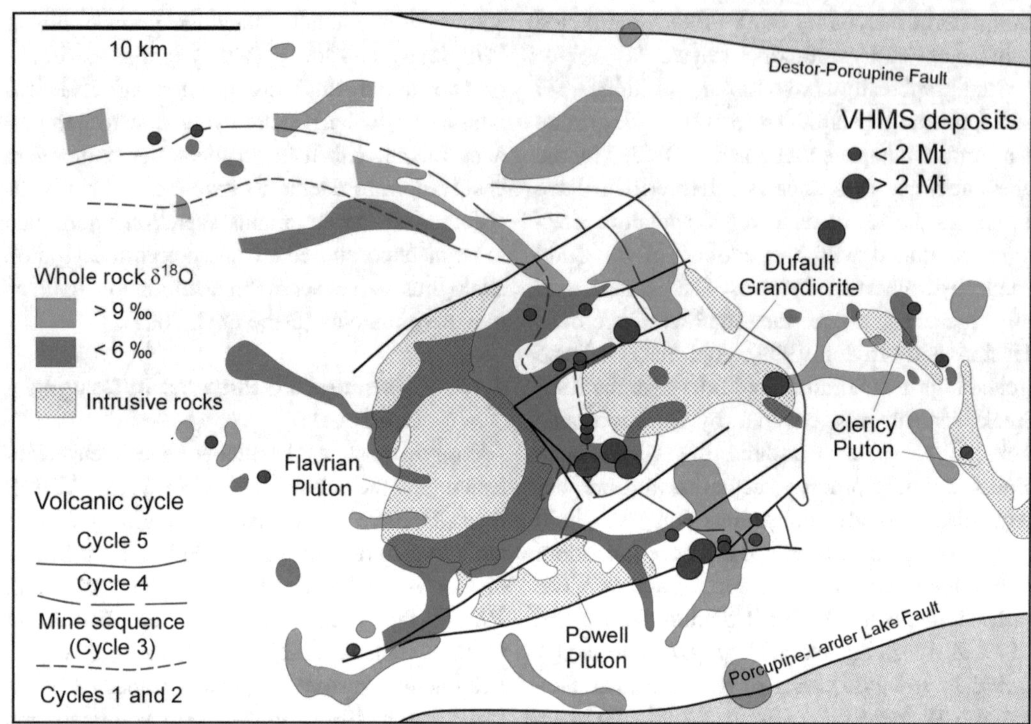

Fig. 6 Whole-rock $\delta^{18}O$ variations in the Noranda district, Quebec, Canada (modified after Cathles 1993, to incorporate geological data from Kerr and Gibson 1993)

deposits of the district (Fig. 6: Cathles 1993). Stratigraphically above the low $\delta^{18}O$ fingers and most VHMS deposits, the volcanic rocks have $\delta^{18}O$ values > 9‰, interpreted to represent the reaction of the volcanic rocks above the deposits with cooled hydrothermal fluids, thus forming a cap rock indicative of underlying higher temperature hydrothermal fluid activity (Cathles 1993; Taylor and Holk 2002).

Panorama, Western Australia. The ~ 3238 Ma (Buick et al. 2002) Panorama district in the East Pilbara Terrane of Western Australia is the oldest significant VHMS district known, and also one of the best preserved. The district contains two economically significant deposits including Sulphur Springs (13.8 Mt grading 3.8% Zn, 0.2% Pb, 1.5% Cu and 17.0 g/t Au) and Kangaroo Caves (3.8 Mt grading 6.0% Zn, 0.3% Pb, 0.8% Cu and 15.0 g/t Au) among several smaller deposits and prospects hosted at or near the top of the bimodal volcanic Kangaroo Caves Formation. Although the district has seen only sub-greenschist metamorphism, the host succession has been structurally tipped on its side, providing an oblique cross-section of the underlying sub-volcanic Strelley Granite, through the host Kangaroo Caves Formation, and into the overlying turbidites of the Soanesville Group (Fig. 7a). This volcanic succession was extensively altered, with systematic zonation in alteration assemblages at the district scale. Brauhart et al. (1998) found that the lower part of the volcanic succession is dominated by a semi-conformable chlorite-quartz alteration zone that becomes transgressive, forming crosscutting pipes below major deposits such as Sulphur Springs and Kangaroo Caves. This chlorite-rich zone grades upward from the semi-conformable zone and outward from the crosscutting pipes into semi-conformable sericite-quartz and feldspar-sericite-quartz zones.

As part of a district-scale alteration study, Brauhart et al. (2000) analyzed 188 volcanic and granite samples for both oxygen isotopes and whole rock geochemistry. Figure 7b shows the

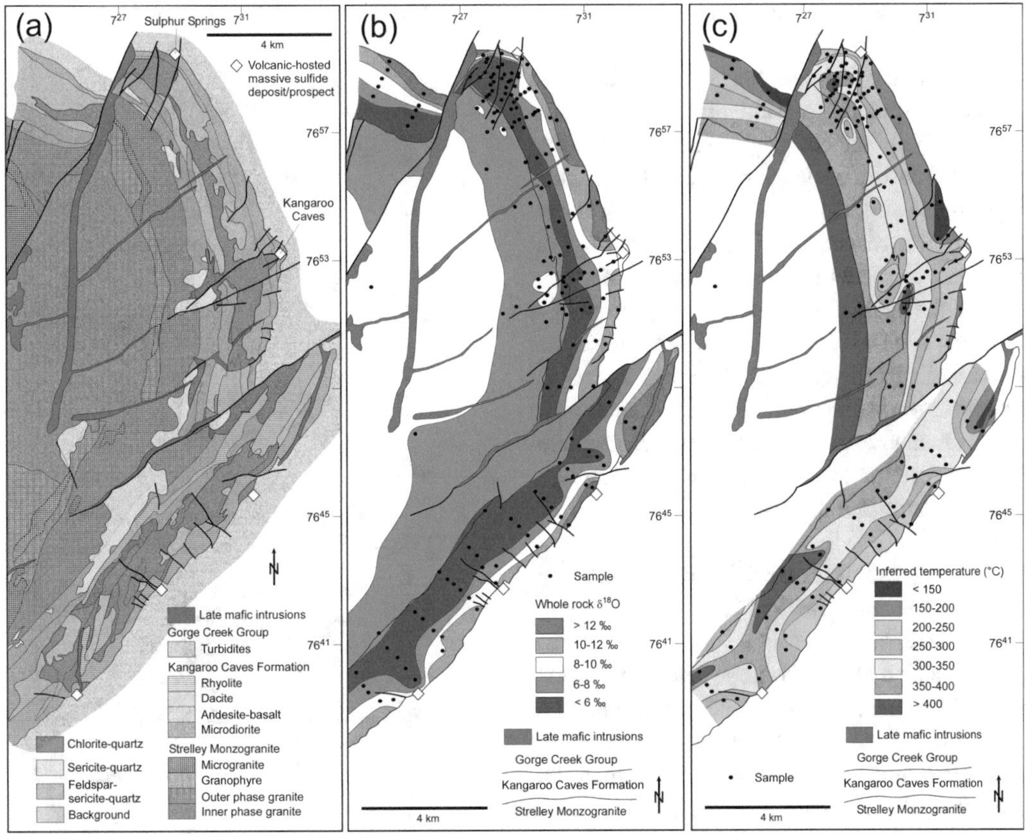

Fig. 7 Maps of the Panorama VHMS district showing (a) geology and alteration zonation, (b) whole-rock $\delta^{18}O$ variations, and (c) variations in estimated hydrothermal temperatures calculated from whole-rock $\delta^{18}O$ and geochemical data (modified after Brauhart et al. 2000). The "background" alteration assemblage refers to spilitic assemblages in basalt or keratophyric assemblges in felsic rocks and typified by the presence albite-, K-feldspar-chlorite- and/or ankerite-quartz-pyrite-dominant assemblages (Brauhart et al. 1998)

distribution of $\delta^{18}O_{\text{whole rock}}$ from these data. The zone with the lowest $\delta^{18}O$ (< 6‰) corresponds closely with the semi-conformable chlorite-quartz alteration zone in the lower part of the volcanic pile. $\delta^{18}O$ increases upwards through the volcanic pile to values of 12–14‰ at the top. The higher $\delta^{18}O$ values correspond to the sericite-quartz and feldspar-sericite-quartz alteration zones. Like the chlorite-quartz alteration zones, pipes of low $\delta^{18}O$ values extend from the basal zone below the locations of known VHMS deposits. The pipe that extends below Sulphur Springs has $\delta^{18}O$ values of < 6‰, whereas the pipe that extends below Kangaroo Caves has values of 6–8‰. This observation is consistent with deposit-scale patterns in which Cu-rich deposits are associated with stronger $\delta^{18}O$ anomalies than Zn-rich deposits.

Following the method of Miller et al. (2001) and using mineral abundances calculated from the whole rock geochemical data, Brauhart et al. (2000) estimated the fractionation function between the altered rock and altering fluid ($\Delta^{18}O_{\text{whole rock-fluid}}$). They then calculated the temperature of alteration using the $\Delta^{18}O_{\text{whole rock-fluid}}$ functions and assuming $\delta^{18}O_{\text{fluid}}$ (Fig. 7c). Using a $\delta^{18}O_{\text{fluid}}$ of 2‰ suggested that alteration temperatures locally exceeded 400 °C in the high temperature, semi-conformable chlorite-rich part of the alteration system, within and just above the Strelley Granite. The alteration temperature decreased, in general, upward to the contact with the overlying Soanesville Group, where the temperature was generally below 200 °C. Immediately below the Sulphur Springs and Kangaroo deposits, the chlorite-rich alteration zones become transgressive, connecting the semi-conformable zone at the base of the volcanic pile with individual deposits. The $\delta^{18}O_{whole\ rock}$ and calculated temperature also reflect this alteration pattern, with these data also defining transgressive zones connecting the high temperature semi-conformable alteration zone with the proximal alteration zones associated with the deposits. Although a $\delta^{18}O_{fluid}$ value of 2‰ was assumed in Fig. 6c, changing this assumed value does not change the relative temperature patterns even though the absolute temperatures estimated do change.

4.1.3 Implications of δ^{18}O-δD Data to Determining Fluid Sources

Oxygen and hydrogen isotope data were a key piece of evidence that was used to argue that ore fluids that formed VHMS deposits were evolved seawater (Ohmoto et al. 1983). Since then, the gathering of new data has indicated a more complicated picture, although evolved seawater still appears to be the most important ore fluid. Table 1 (updated from Huston 1999; this table excludes deposits metamorphosed at grades of amphibolite or higher and some deposits (e.g., Raul, Peru) for which a VHMS origin has been questioned) and Fig. 8 summarize the δ^{18}O and δD characteristics of VHMS ore fluids as deduced from deposits that have not undergone high-grade metamorphism (amphibolite or greater) that may disturb the original isotopic characteristics.

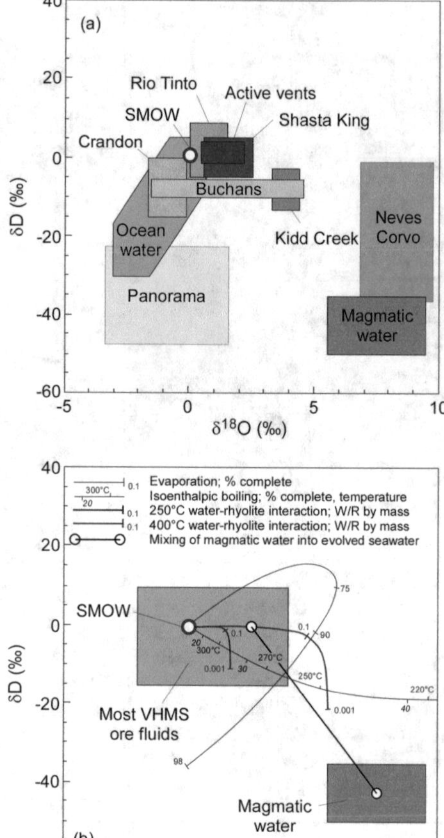

Fig. 8 δ^{18}O-δD diagrams showing (a) the fields of selected VHMS ore-forming fluids (from Table 1) and the fields of seawater and magmatic water (Sheppard, 1986; Taylor, 1992), and (b) paths showing the evolution of seawater undergoing evaporation (Knauth and Beeunus, 1986), open-system water–rock interaction (e.g., Taylor 1987), adiabatic boiling (using data of Friedman and O'Neil (1977) and Keenan et al. (1969)), and mixing with magmatic water (updated from Huston 1999)

Most VHMS deposits have fluid δ^{18}O-δD signatures similar to seawater. In general, δ^{18}O values are within 2‰ and δD values are within 20‰ of standard mean oceanic water (SMOW). These values are consistent with evolved seawater being the dominant ore fluid (Fig. 8b).

As discussed above, currently active seafloor systems provide information, such as direct measurements of the isotopic composition of mineralizing fluids and the isotopic response (and geochemical response in general) of these fluids to geological perturbations of the mineral system. Although much of the data come from active deposits not associated with convergent margins, the data provide general guidance for seafloor mineral systems. The vent fluids tend to have δ^{18}O 0.5–1.0‰ heavier than, and δD values very similar to ambient seawater (Shanks 2001), which supports the inference (e.g., Ohmoto et al. 1983) that evolved seawater was the dominant fluid source in most VHMS mineral systems.

There are, however, several important exceptions to the inference that seawater is the dominant fluid source, including several deposits in the ca 350 Ma Iberian Pyrite Belt, and the ca 2698 Ma Horne, ca 2714 Ma Kidd Creek and ca 2736 Ma South Bay deposits of the Superior Province, Canada. These data may indicate a significant contribution of other fluid sources, particularly magmatic-hydrothermal fluids, in a limited number of deposits (Fig. 8b).

The best case for a significant magmatic-hydrothermal component is the Neves Corvo deposit in the Iberian Pyrite Belt, which has the highest estimated $\delta^{18}O_{fluid}$ from a VHMS deposit at 8.3 ± 1.5‰ (Relvas et al. 2006). This deposit is a highly unusual for VHMS deposits in being very Sn-rich, with abundant pyrrhotite, not unlike some carbonate-replacement tin deposits that are typically thought to be magmatic-hydrothermal in origin (e.g., Blevin and Chappell 1995). Some of the other deposits with high $\delta^{18}O_{fluid}$ are associated with advanced argillic alteration assemblages, for example the Boco deposit in Tasmania (Herrmann et al. 2009), an alteration assemblage possibly indicative of a magmatic-hydrothermal contribution (Huston et al. 2011). The unusually ^{18}O-enriched signature of advanced argillic assemblages at the LaRonde-Penna district could also be the product of magmatic-hydrothermal contribution (see above and Beaudoin et al. 2014). Hence it would appear that there is a significant subclass of VHMS deposits in which a significant magmatic-hydrothermal contribution is likely, but for most of these deposits the most likely fluid source is evolved seawater. A small contribution of magmatic-hydrothermal fluids, however, would be diluted by the dominant evolved seawater contribution and not visible in stable isotopic data. Such a contribution may be visible in other data, for example metal budgets, and remains an important line of enquiry for future research on VHMS deposits.

In other districts, however, seawater-dominated hydrothermal systems, which form VHMS deposits, are physically separated from coeval hydrothermal systems developed in subvolcanic intrusions. Drieberg et al. (2013) showed this in the Panorama district. The ore fluids that formed the VHMS deposits had lower salinity, density and $\delta^{18}O_{fluid}$ (Table 1) than fluids that formed Sn-Cu–Zn and Mo mineral occurrences near the upper margin of the granite. Drieberg et al. (2013) concluded that density differences and low permeability barriers prevented the mixing of the seawater-dominated hydrothermal system in the volcanic pile with the magmatic-hydrothermal system in the granite.

4.1.4 Application of Oxygen and Hydrogen Isotope Geochemistry to Exploration and Ore Genesis

Data from well-studied deposits described above, combined with additional data presented in Table E1, indicate that oxygen isotope data define high-temperature fluid pathways associated with VHMS deposits at the deposit scale. For most deposits, high temperature alteration zones are associated with ^{18}O depletion anomalies (low $\delta^{18}O$) that can extend up to several hundreds of meters laterally from ore and greater distances with depth. In some cases the low $\delta^{18}O$ anomalies also extend above the deposit to define a hanging wall alteration zone.

The data also indicate changes in isotopic patterns with geological time and deposit type. $\delta^{18}O$ values of proximal alteration zones are higher for deposits in felsic-dominated succession than those hosted by mafic-dominated successions. This may simply reflect the temperature of upflow, which, based on 100Cu/(Cu + Zn) is likely to be higher in the mafic-dominated systems. The variation in time may relate to the greater abundance of felsic-dominated, lower temperature systems at younger times in Earth's history.

The main exceptions to the general patterns described above are deposits in which a significant magmatic-hydrothermal component in inferred (see also below). In these cases (Kidd Creek, LaRonde and Boco: Table 1), the ^{18}O depletion anomaly is suppressed or can present as a weak ^{18}O enrichment anomaly (Fig. 3). Hence, it is important to interpret oxygen isotope data in a geological context including the age, host succession and type of deposit being targeted.

Studies of the Noranda and Panorama districts demonstrate that district-scale variations in whole-rock $\delta^{18}O$ data correlate to variations in alteration assemblages and provide district-scale vectors to ore. The variations most likely relate to temperature gradients within the hydrothermal system. Other studies, including those at West Shasta (Taylor and South 1985), Izok Lake (Taylor et al. 2014), Kidd-Munro (Taylor and Huston 1999), and Sturgeon Lake (Holk et al. 2008) districts, demonstrate broadly similar relationships. The relative intensity of the anomaly at the district scale may also be indicative of the Cu:Zn ratio of associated deposits as it is at the deposit scale.

In many cases low-$\delta^{18}O$, high temperature alteration zones associated deposits are surrounded or capped by zones of higher $\delta^{18}O$. These zones, which are not readily identified by mineralogical or geochemical alteration mapping, represent lower temperature alteration zones, and regional gradients from these zones toward higher-temperature zones can be used as district-scale vectors.

Overall, whole rock oxygen isotopes allow detection of the infiltration of hydrothermal fluids, and mapping of the fluid pathways, which enable targeting for VHMS mineralization. Several characteristics of VHMS systems can be uniquely recognized using stable isotopes: (1) a syn-volcanic intrusion driving hydrothermal fluid flow during cooling will be overlain by a high-temperature alteration with low $\delta^{18}O$ values, providing a regional exploration target for a potentially fertile syn-volcanic intrusion; (2) discordant zones of low $\delta^{18}O$ values map the zones of hydrothermal fluids flow up-section on top of which VHMS deposits are most likely located; (3) discordant to semi-concordant zones of high $\delta^{18}O$ values record the waning upflow of hydrothermal fluids at lower temperatures, which sits above of the higher temperature up-flow zone in the underlying stratigraphic section of volcanic rocks, and thus enabling targeting of blind high temperature (low $\delta^{18}O$) up-flow zones.

4.2 Sulfur Isotopes

Sulfur is an important component of the VHMS mineral system for two reasons. First, as iron and base metal sulfides are very insoluble at low to moderate temperatures, the presence of H_2S precipitates iron and base metal sulfides when high temperature, H_2S-rich ore fluids quench as they reach the seafloor. Second, some metals, for instance gold, are transported by sulfide complexes. In reduced hydrothermal upper crustal fluids, sulfur (as H_2S) can be sourced from sulfide leached from the underlying volcanosedimentary succession, thermochemical sulfate reduction of seawater (TSR), either at depth or in the near-surface during fluid upflow, and/or disproportionation of magmatic SO_2 (c.f. Ohmoto 1996). The sulfur isotope composition of sulfides in all mineral systems reflects not only the sulfur source but geochemical interactions between magmas, fluids, host rocks, and seawater, that lead to precipitation of sulfide minerals (Ono et al. 2007; Peters et al. 2010; LaFlamme et al. 2018; Martin et al. 2021; Huston et al. 2023). Owing to the high temperature of the upflow zones of VHMS deposits (Eldridge et al. 1983; Ohmoto 1986), it is unlikely that bacterial sulfate reduction (BSR) played a role in an important role in the core of VHMS systems, although in some deposits BSR may have been an important processes in the lower temperature peripheries to VHMS systems (see below).

4.2.1 Secular Variations in Sulfur Isotope Ratios from Volcanic-hosted Massive Sulfide Deposits

Figure 9 illustrates changes in $\delta^{34}S$ of sphalerite and sulfate minerals from VHMS deposits with geological time based upon a compilation of data by Huston et al. (2022: Table E2). Sphalerite was used for this analysis as it forms almost entirely as the result of hydrothermal activity and therefore records the characteristics of hydrothermal sulfur. Although it would be ideal to also consider $\delta^{34}S$ variations of chalcopyrite with geological time, as it reflects higher temperature parts of the VHMS system, an analogous

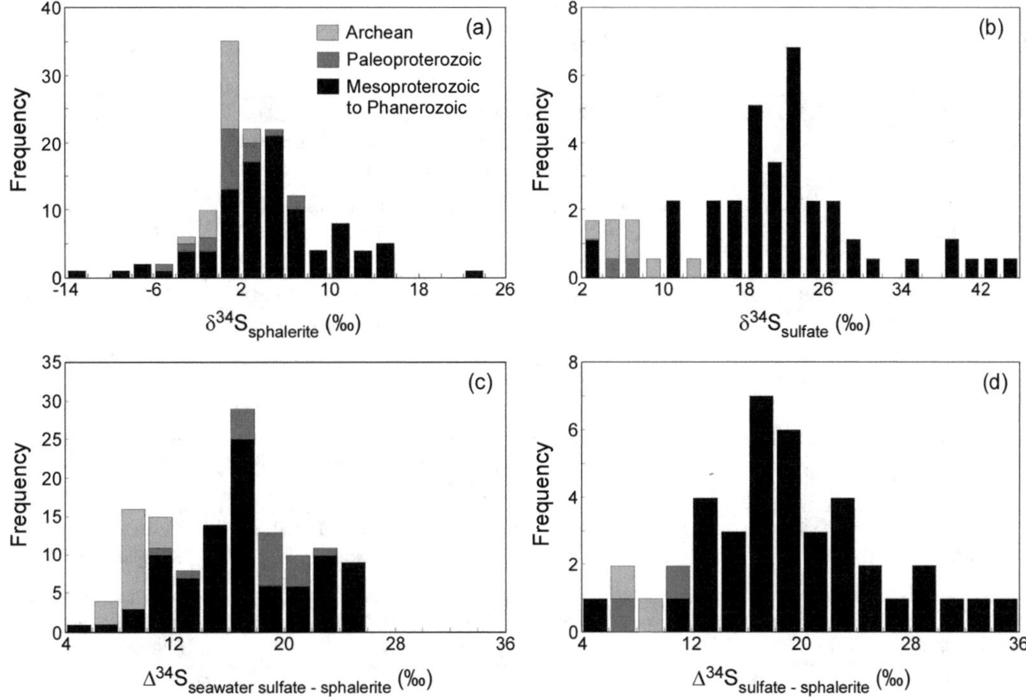

Fig. 9 Histograms showing (a) variations in median δ^{34}S values of sphalerite, (b) variations in median δ^{34}S values of sulfate minerals, (c) $\Delta^{34}S_{seawater\ sulfate-sphalerite}$ (estimated from secular seawater sulfate variations and median sphalerite values), and (d) $\Delta^{34}S_{sulfate-sphalerite}$ (using median sulfate and sphalerite values for individual deposits). Median mineral δ^{34}S data are from Huston et al. (2022), and seawater sulfate data are after Crockford et al. (2019) supplemented by Neo- and Paleoarchean barite data summarised in Huston et al. (2022)

compilation of chalcopyrite does not exist. As minerals such as pyrite and pyrrhotite do not form exclusively from hydrothermal processes they can reflect the input of non-hydrothermal sulfur produced by biological and other processes.

Although limited by the concentration of deposits into restricted time intervals, there are some relatively consistent patterns in the data. The most prominent pattern is the relatively consistent values of median $\delta^{34}S_{sphalerite}$ of Paleoproterozoic (-4.6‰ to 6.5‰, median = 1.7‰, n = 18: Fig. 9a) and Archean (-2.8‰ to 2.4‰, median = 0.8‰, n = 20) deposits at around 0‰. This contrasts with the much greater variability of deposits of Mesoproterozoic-Phanerozoic age (-13.7‰ to 23.9‰, median = 4.6, n = 96). $\delta^{34}S_{sulfate}$ shows a similar pattern, although shifted to higher values (Fig. 8b). For Mesoproterozoic-Phanerozoic deposits, the range is 3.5‰ to 44.4‰ with a median of 22.2‰ (n = 56), for Paleoproterozoic deposits the range is 4.4‰ to 11.0‰ with a median of 8.8‰ (n = 4), and for Archean deposits the range is 3.8‰ to 13.2‰ with a median of 6.2‰ (n = 7).

Sangster (1968) was the first to document the variability in δ^{34}S data in Phanerozoic VHMS and sediment-hosted deposits. He found that $\delta^{34}S_{sulfate}$ in these deposits was similar to the $\delta^{34}S_{sulfate}$ of coeval seawater. The right-hand side of Fig. 10 illustrates this relationship. Sangster (1968) also noted that $\delta^{34}S_{sulfide}$ values are 17.5 $\pm$ 2.5‰ (1σ) lower than coeval seawater, a relationship that has held up even as the sulfur isotope database has increased (Fig. 9c; see also Huston 1999). The similarity between VHMS $\delta^{34}S_{sulfate}$ and coeval seawater $\delta^{34}S_{sulfate}$ is most simply interpreted to indicate that the sulfur in VHMS sulfate minerals, for the most part (see discussion of exceptions below), was derived from coeval seawater sulfate.

Table 1 Estimated $\delta^{18}O$ and δD values for VHMS ore fluids

Deposit	Age (Ma)	$\delta^{18}O$ (‰)	δD (‰)	T (°C)	Method of estimation	Source
Active sea floor vents	0.0	1.3 ± 0.9	1 ± 3	220–400	Measurements of venting flids adjusted to remove influence of entrained seawater	DeRonde (1995)
Hokuroku district, Japan		-0.7 ± 1.6	-30 to -10	230–270	Fluid inclusion analysis forδD; quartz$\delta^{18}O$ and fluid inclusion T_h for$\delta^{18}O$	Hattori and Sakai (1979)
Kosaka		1.0 ± 2.0	-30 to 15	220–330	Fluid inclusion analysis for δD; quartz $\delta^{18}O$ and fluid inclusion T_h for $\delta^{18}O$	Pisutha-Arnond and Ohmoto (1983); Hattori and Muehlenbachs (1980)
Iwami, Japan		-2.5 ± 0.9	-55 to -35	230–270	Fluid inclusion analysis for δD; quartz $\delta^{18}O$ and fluid inclusion T_h for $\delta^{18}O$	Hattori and Sakai (1979)
Troodos Ophiolite, Cyprus		0.5 ± 1.0	0 ± 5	350	Quartz and whole-rock analyses, assuming temperature	Heaton and Sheppard (1977)
Buchans, Newfoundland		1.5 ± 3.0	-8 ± 2	240–370	Mineral separate analyses and geothermomtry	Kowalik et al. (1981)
Iberian Pyrite Belt						
Aljustrel		3.3 ± 1.8	-1 to 18	160–270	Mineral separate analysis and geothermometry; the calculated $\delta^{18}O$ values range from 0.0‰ to 5.7‰, with modes at 0.0–1.7‰ and 3.1–5.7‰. The $\delta^{18}O$ value range in the mean ± one standard deviation	Barriga and Kerrich (1984); Munha et al. (1986)
Rio Tinto		0.7 ± 0.7	-5 to 8	210–230	Mineral separate analysis and geothermometry; the calculated $\delta^{18}O$ values range from 0.0‰ to 1.3‰	Munha et al. (1986)
Chanca		0.9	-10 to 0	220	Mineral separate analyses and geothermometry	Munha et al. (1986)
Salgadinho		4.0	-10 to 0	230	Mineral separate analyses and geothermometry	Munha et al. (1986)
Feitas		4.2–5.2		270–315	Quartz $\delta^{18}O$ and fluid inclusion T_h	Inverno et al. (2008)
Neves Corvo		8.3 ± 1.5	-37 to -11	300–400	Mineral separate analyses and geothermometry	Relvas et al. (2006)
Boco, Tasmania, Australia		3.2–5.7		270–360	Pyrophllite $\delta^{18}O$ value and thermal stability of pyrophyllite	Herrmann et al. (2009)
Baiyinchang, China		-5.3 to 3.1		160–280	Quartz $\delta^{18}O$ and fluid inclusion T_h	Hou et al. (2008)
Bruce, Arizona, USA		1.5 ± 0.5		250–300	Extrapolated chlorite value, assuming temperature	Larson (1984)
Crandon, Wisconsin, USA		-0.9 ± 0.8	-15 to 0	220–290	Mineral separate analyses and geothermometry; the calculated $\delta^{18}O$ values range from -2.1‰ to 0.1‰	Munha et al. (1986)

(continued)

Table 1 (continued)

Deposit	Age (Ma)	$\delta^{18}O$ (‰)	δD (‰)	T (°C)	Method of estimation	Source
Mattagami Lake, Quebéc, Canada		1.5 ± 1.0	1 ± 3	240–350	Mineral separate $\delta^{18}O$ and fluid inclusion T_h	Costa et al. (1983)
Noranda district, Quebéc, Canada						
Horne		3.0 ± 1.5	-40 to -30	250–350	Mineral separate analyses and geothermometry	Maclean and Hoy (1991)
Mobrun		2.0 ± 2.0		150–250	Mineral separate analyses and geothermometry	Hoy (1993)
Norbec		1.0 ± 2.0		200–300	Mineral separate analyses and geothermometry	Hoy (1993)
Amulet		0.5 ± 1.0		250–350	Mineral separate analyses and geothermometry	Beaty and Taylor (1982)
Ansil		-0.5 ± 1.0		200–350	Mineral separate analyses and geothermometry	Hoy (1993)
Corbet		-2.0 ± 2.0		250–300	Mineral separate analyses and geothermometry	Hoy (1993)
Kidd Creek, Ontario, Canada		3.8 ± 0.5	-8 ± 5	300–350	Chlorite analyses, assuming temperature	Huston and Taylor (1999)
South Bay, Ontario, Canada		3.3 ± 1.2		300	Quartz analyses, assuming temperature	Urabe and Scott (1983)
Panorama, Western Australia						
VHMS	3240	-0.8 ± 2.6	-48 to -23	90–270	Mineral separate $\delta^{18}O$ and fluid inclusion T_h; fluid inclusion waters and mineral separates for δD	Drieberg et al. (2013)
Granite-hosted veins (greisen)	3240	4.1 to 9.9 (9.3 ± 0.6)	-48 to -18	240–590 (590)	Mineral separate $\delta^{18}O$ and fluid inclusion T_h; fluid inclusion waters and mineral separates for δD	Drieberg et al. (2013)

This is in accord with paragenetic observations and thermodynamic-based models suggesting the sulfate minerals in VHMS deposits form when upwelling hydrothermal fluids mix with sulfate-bearing seawater (cf. Eldridge et al. 1983; Ohmoto et al. 1983).

In contrast, the offset between $\delta^{34}S_{sphalerite}$ and seawater $\delta^{34}S_{seawater\ sulfate}$ in Phanerozoic deposits could be the results of several different processes. For Phanerozoic deposits $\Delta^{34}S_{seawater\ sulfate\text{-}sphalerite}$ ranges from 3‰ to 30‰, with a strong mode at 15–20‰ (median = 19.5‰: Fig. 9c). A similar pattern is seen for $\Delta^{34}S_{sulfate\text{-}sphalerite}$ using data from Phanerozoic deposits for which $\delta^{34}S$ data are available for both sulfate minerals and sphalerite (Fig. 9d). The range in $\Delta^{34}S_{seawater\ sulfate\text{-}sphalerite}$ for Phanerozoic deposits has been explained by several processes, including: (1) biogenic reduction of seawater sulfate (Sangster 1968), (2) derivation of sulfur from a deep-seated (magmatic) source (Ishihara and Sasaki 1978), or (3) partial or complete reduction of seawater sulfate during circulation of evolving seawater (Sasaki 1970; Solomon et al. 1988). A more detailed discussion of these alternatives is presented by Huston (1999), and it is likely that all three, and possibly other, processes have contributed to secular variability in sulfur isotope characteristics of Phanerozoic deposits. It is important to stress that the values used in constructing Figs. 9 and 10 are median values and are indications of the "average" composition of deposits. Some deposits have very large ranges individually, and it is likely that non-hydrothermal processes, such as biogenic processes, have contributed to the variability in $\delta^{34}S_{sulfide}$ in these deposits, particularly at the micro-scale (see below).

The restricted, near-0‰ range of $\delta^{34}S_{sphalerite}$ and near-10‰ range of $\delta^{34}S_{sulfate}$ of Archean deposits, combined with evidence of only minor contributions of seawater sulfur from multiple sulfur isotope studies (see below) suggest that sulfur sources of VHMS deposits early in Earth's history differ from those later in Earth's history. The near-0‰ and, particularly, the uniform character of $\delta^{34}S_{sphalerite}$ in Archean deposits

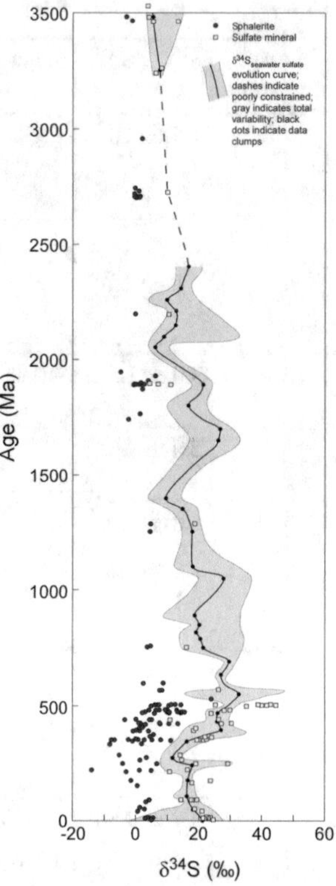

Fig. 10 Variations in $\delta^{34}S$ of sphalerite and sulfate minerals from VHMS deposits through time. Data shown are median values for individual deposits from Huston et al. (2022). Seawater sulfate curve (heavy line with light gray field showing full variability) is based on compilation of Crockford et al. (2019) supplemented by Neo- and Paleoarchean barite data of Huston et al. (2022)

suggests a dominant (leached) igneous origin of sulfide in these deposits.

To test this hypothesis the relative contribution of reduced seawater sulfate versus magmatic sulfur were determined using the median $\delta^{34}S_{sphalerite}$ data (Huston et al. 2022) and the estimated $\delta^{34}S$ of coeval seawater sulfate determined from the $\delta^{34}S_{seawater\ sulfate}$ curve of Claypool et al. (1980) and from a second $\delta^{34}S_{seawater\ sulfate}$ curve determined from evaporite data in Crockford et al. (2019). For periods of time not covered by the respective curves, estimates of $\delta^{34}S_{seawater\ sulfate}$ were made based on $\delta^{34}S_{sulfate}$ of Strauss (2004) and $\delta^{34}S_{barite}$ data from Huston et al. (2022). Numeric values for both curves are tabulated in Huston et al. (2022).

A small proportion of median $\delta^{34}S_{sphalerite}$ values were negative. If the value was between 0‰ and -5‰ (i.e. within the typical range of magmatic sulfur), a $\delta^{34}S_{sphalerite}$ value of 0‰ was assigned to the deposit (n = 17). Deposits with median $\delta^{34}S_{sphalerite}$ values below -5‰ were excluded from the analysis (n = 4).

The deposits were split into three groups by age: Archean (> 2500 Ma), Paleoproterozoic (2500–1600 Ma) and Mesoproterozoic-Phanerozoic (< 1600 Ma). Based on both the Claypool (not shown) and the Crockford (Fig. 11) $\delta^{34}S_{seawater\ sulfate}$ curves, the "typical" contribution (as measured by both median and mean estimates) of reduced seawater sulfate to VHMS sulfur budgets is significantly higher (20–25%) in Mesoproterozoic to Phanerozoic deposits than in Archean deposits (5–10%). The Mesoproterozoic-Phanerozoic distribution is characterized by a strong, tight mode at 0–5%, a much broader mode with a peak at 15–30% and a tail of values up to ~ 75% (Fig. 11a). In contrast, the Archean distribution is characterized by a strong mode with a peak at 0–15% and a tail with values to ~ 25% (Fig. 11d). The Archean distribution is consistent with estimates of seawater sulfate contributions based on multiple sulfur isotope data (0% - Panorama (3238 Ma): Golding et al. 2011; 3% - Kidd Creek (2714 Ma): Jamieson et al. 2013; < 5% - cauldron margin deposits and < 15% main cauldron and post cauldron deposits, Noranda district (2698 Ma): Sharman et al. 2015; 15% - Teutonic Bore and 18% Bentley (2694 Ma); Chen et al. 2015). The data are also consistent with interpretations from multiple sulfur isotope data that Neoarchean seawater contained much lower concentrations of sulfate than modern seawater (80 μmol/l at 2714 Ma versus 28,000 μmol/l presently: Jamieson et al. 2013).

Analysis of the data using the two different $\delta^{34}S_{seawater\ sulfate}$ curves yielded somewhat different results for Paleoproterozoic deposits: both

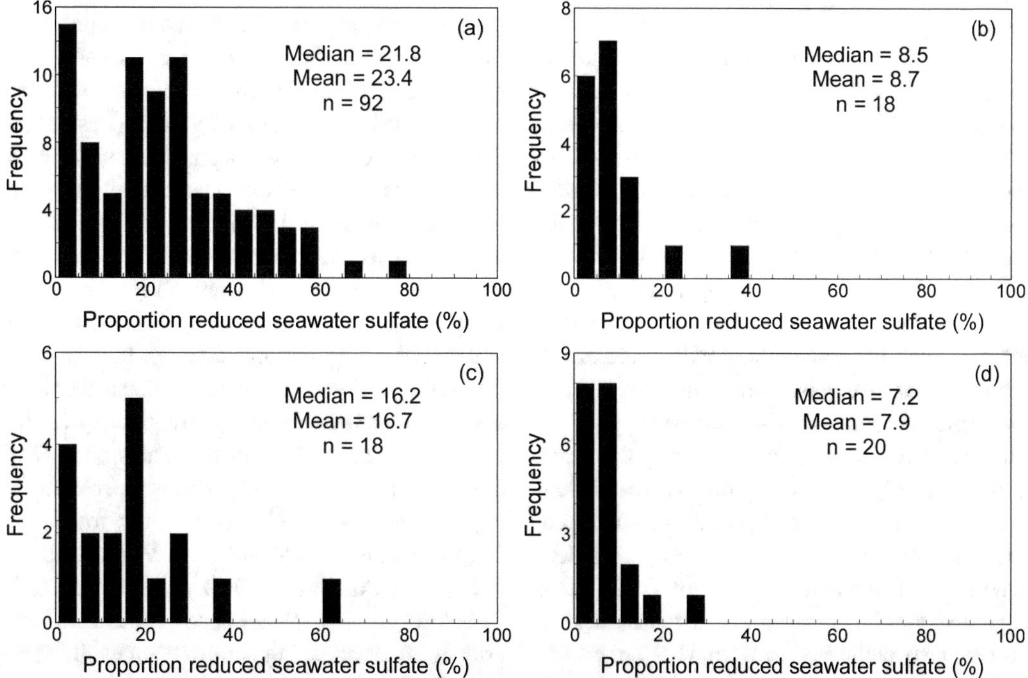

Fig. 11 Histograms showing the estimation contribution of reduced seawater sulfate to sulfur budgets for VHMS deposits that formed during the: (a) Mesoproterozoic to Phanerozoic, (b) and (c) Paleoproterozoic, and (d) Archean. Histogram (b) was calculated using the sulfate data from Crockford et al. (2019) whereas (c) was calculated using Paleoproterozoic barite data from Huston et al. (2022)

indicate a smaller contribution of seawater sulfate than in the Mesoproterozoic-Phanerozoic period, but differed relative to the Archean. Use of the Crockford et al. (2019)-based curve yielded a "typical" seawater sulfate contribution for Paleoproterozoic deposits nearly identical to that of Archean deposits (Fig. 11b), with a major peak at 0–15% and a tail with values to ~ 25%. In contrast use of $\delta^{34}S_{seawater\ sulfate}$ values from ore-related barite yields "typical" seawater sulfate contributions of ~ 16–17% (Fig. 11c). This discrepancy is due to the much higher $\delta^{34}S_{seawater\ sulfate}$ values based on the Crockford et al. (2019) data. A seawater sulfate contribution to the Paleoproterozoic VHMS budget of ~ 16% is consistent with results obtained from the ca 2027 Ma DeGrussa deposit in Western Australia (LaFlamme et al. 2021; see below). This analysis suggests more data are required to better calibrate the Paleoproterozoic (and Meso- to Paleoarchean) $\delta^{34}S_{seawater\ sulfate}$ curve. Analysis is ongoing as to the control of other factors, such as deposit type and alteration assemblage, on the seawater sulfate contributions to the VHMS sulfur budget.

The dominance of a magmatic sulfur source in all ages of deposits does not necessarily imply a magmatic-hydrothermal origin for the sulfur. Rather, given that the host succession of VHMS deposits is generally dominated by volcanic rocks, a more likely source for the sulfur in the deposits is leached sulfur from the underlying volcanic pile, as documented by Brauhart et al. (2001) in the Panorama district, and discussed below. The inference that Paleoproterozoic-Archean VHMS deposits are dominated by leached volcanic sulfur is also consistent with the likely composition of seawater during this time. Holland (1972) originally proposed that Paleoproterozoic and Archean seawater was reduced, iron-rich and sulfate-poor relative to modern seawater. Calculations by Huston (1999) also suggested that Paleoproterozoic-Archean seawater was also sulfide-poor, with H_2S concentrations in the parts per billion range. If Paleoproterozoic-Archean seawater was indeed sulfate- and sulfide-poor, the sulfur source of VHMS deposits would have been dominated by leached volcanic or magmatic sulfide, a conclusion supported by multiple sulfur isotope studies of the Neoarchean Kidd Creek (Jamieson et al. 2013) and Jaguar (Chen et al. 2015) deposits (see also below).

Following a compilation of the occurrence of sulfate minerals in volcanic-hosted massive sulfide deposits, Huston and Logan (2004) observed that Paleoarchean deposits commonly contain sulfate minerals within mineralized zones, whereas Mesoarchean to Paleoproterozoic deposits were rarely sulfate-bearing. A more comprehensive compilation (data from Huston et al. 2022) indicates that 64% (7 out of 11) of Paleoarchean deposits contain sulfate minerals, a higher rate than even Phanerozoic deposits (44%: 232 of 530), and that only 6% (16 of 280) of Neoarchean to Paleoproterozoic deposits contain sulfate minerals. Huston and Logan (2004) interpreted the rare occurrence of sulfate minerals in Neoarchean to Paleoproterozoic deposits as the result of the low concentration of sulfate in reduced seawater of this age, following Holland (1972). In contrast, the common occurrence of sulfate minerals in Phanerozoic deposits is interpreted to be the result of the common presence of sulfate in Phanerozoic oceans, with relatively restricted periods of anoxia (Eastoe and Gustin 1996). The most surprising result of these studies was the common presence of sulfate in Paleoarchean (>3200 Ma) deposits, which was interpreted as the consequence of either a thin sulfate-bearing surficial oceanic layer (c.f. Huston and Logan, 2004) or local concentrations of sulfate caused by volcanic activity. In both cases the sulfate is interpreted to be the consequence of rainout of sulfate produced by photolytic disproportionation of SO_2 in the atmosphere (Farquhar et al. 2000), a hypothesis developed to explain anomalous $\Delta^{33}S$ (see below for definition) signatures of Paleoarchean VHMS sulfates (Golding et al. 2011). The presence of $\Delta^{33}S$ signatures in Neoarchean deposits has also been critical in establishing sulfur sources in these deposits (see below). An unresolved issue is the lack of sulfates in Meso- to Neoarchean deposits as photolytic disproportionation of atmospheric SO_2 is thought to have continued until the Great

Oxidation Event at ~ 2400 Ma (Farquhar et al. 2000, 2013).

4.2.2 Deposit and District-scale Variations in $\delta^{34}S$ from Phanerozoic Deposits

Conventional $\delta^{34}S$ analyses from Phanerozoic VHMS deposits commonly have variabilities of 15‰ or more (e.g., Ducktown, Tennessee: LeHuray 1984; Bathurst, New Brunswick: Goodfellow and Peter 1999; Ming, Newfoundland: Brueckner et al. 2015; Hercules, Tasmania: Khin Zaw and Large 1992). These large ranges most likely reflect the complexities of sulfur sources and geochemical interactions, as discussed above, within the underlying rock pile and at or near the depositional site. Below we describe the $\delta^{34}S$ characteristics of three Phanerozoic provinces, Cambro-Ordovician deposits in the Dunage Zone, Newfoundland, Cambrian deposits in the West Tasmania Terrane, and the Iheya black smoker deposit south of Japan, as well as active seafloor systems in general. These provinces were chosen not only to illustrate $\delta^{34}S$ variability, but to discuss processes that cause the variability.

Deposits in the Dunnage Zone, Newfoundland. Cambro-Ordovician VHMS deposits of the Newfoundland Appalachians occur within volcanic and volcanosedimentary units in the Dunnage Zone (Fig. 12a). This zone hosts numerous VHMS deposits, including Cambrian mafic, bimodal-mafic, bimodal-felsic and felsic-siliciclastic deposits (Kean et al. 1995; Hinchey 2011; Pilote et al. 2014; Cloutier et al. 2015), and Ordovician bimodal-mafic and bimodal-felsic deposits (Dunning et al. 1987; MacLachlan and Dunning 1998). The $\delta^{34}S$ signatures of Dunnage Zone VHMS deposits are highly variable and include data from the Ming ($\delta^{34}S$ = 1–20‰; Brueckner et al. 2015), Whalesback ($\delta^{34}S$ = 1–6‰; Cloutier et al. 2015), Lemarchant ($\delta^{34}S$ = -38 to + 14‰; Lode et al. 2017), and mafic deposits from Notre Dame Bay ($\delta^{34}S$ = 2–20‰; Bachinski 1978; Toman 2013). The characteristics of the Ming and Lemarchant deposits are described in more detail below.

The ~ 487 Ma Ming deposit (21.9 Mt at 1.49% Cu, 0.19% Zn, 3.21 g/t Ag, and 0.61 g/t Au) is spatially associated with boninitic to arc tholeiitic mafic rocks and hosted by tholeiitic felsic rocks the Pacquet Complex in the Baie Verte Peninsula, Newfoundland (Pilote et al. 2017). The deposit contains variably Au-Cu-bearing massive sulfide that is locally capped by a strongly quartz-altered rhyolite, which are together underlain by a footwall zone with Cu-rich chalcopyrite-pyrite-pyrrhotite-chlorite-rich stringers (Brueckner et al. 2015). In situ sulfur isotopic data for pyrite, chalcopyrite, pyrrhotite and arsenopyrite in massive and semi-massive sulfide mineralization have $\delta^{34}S$ between 2.8‰ and 12.0‰, whereas stringer sulfides have $\delta^{34}S$ values of 6‰ to 16‰; sulfides in the quartz-altered rhyolite cap have heavier $\delta^{34}S$ values of 5.9‰ to 19.6‰ (Fig. 12b; Brueckner et al. 2015). Notably, there are decreases in $\delta^{34}S$ of mineral assemblages interpreted to have formed at higher temperatures (e.g., > 300 °C chalcopyrite-pyrrhotite-pyrite assemblages), whereas those forming most proximal to the massive sulfide-seawater interface (e.g., those in the quartz-altered rhyolite cap) have higher $\delta^{34}S$ values (Fig. 12b; Brueckner et al., 2015).

The ca 513–509 Ma Lemarchant deposit is a polymetallic bimodal felsic deposit (2.28 Mt at 0.64% Cu, 9.87% Zn, 1.51% Pb, 99.9 g/t Ag, and 1.26 g/t Au) hosted by felsic rocks of the Bindons Pond formation in the Tally Pond Group (Cloutier et al. 2017). The deposit contains barite, sphalerite, galena, pyrite, and chalcopyrite, as well as abundant sulfosalt minerals, with enrichment in epithermal suite elements (Gill et al. 2016, 2019). The ores are overlain by sulfide-rich hydrothermal mudstones genetically related to mineralization (Cloutier et al. 2017).

Ores and hydrothermal mudstones from Lemarchant have the widest known range in $\delta^{34}S$ for VHMS deposits, ranging from -38.8‰ to 15.1‰ (Lode et al. 2017; Gill et al. 2019). Within the mineralized system, the $\delta^{34}S$ values of pyrite and chalcopyrite (0.3‰ to 10.6‰) overlap with the values for galena (-6.4‰ to 15.1‰: Fig. 12c). In contrast, $\delta^{34}S$ of sulfide

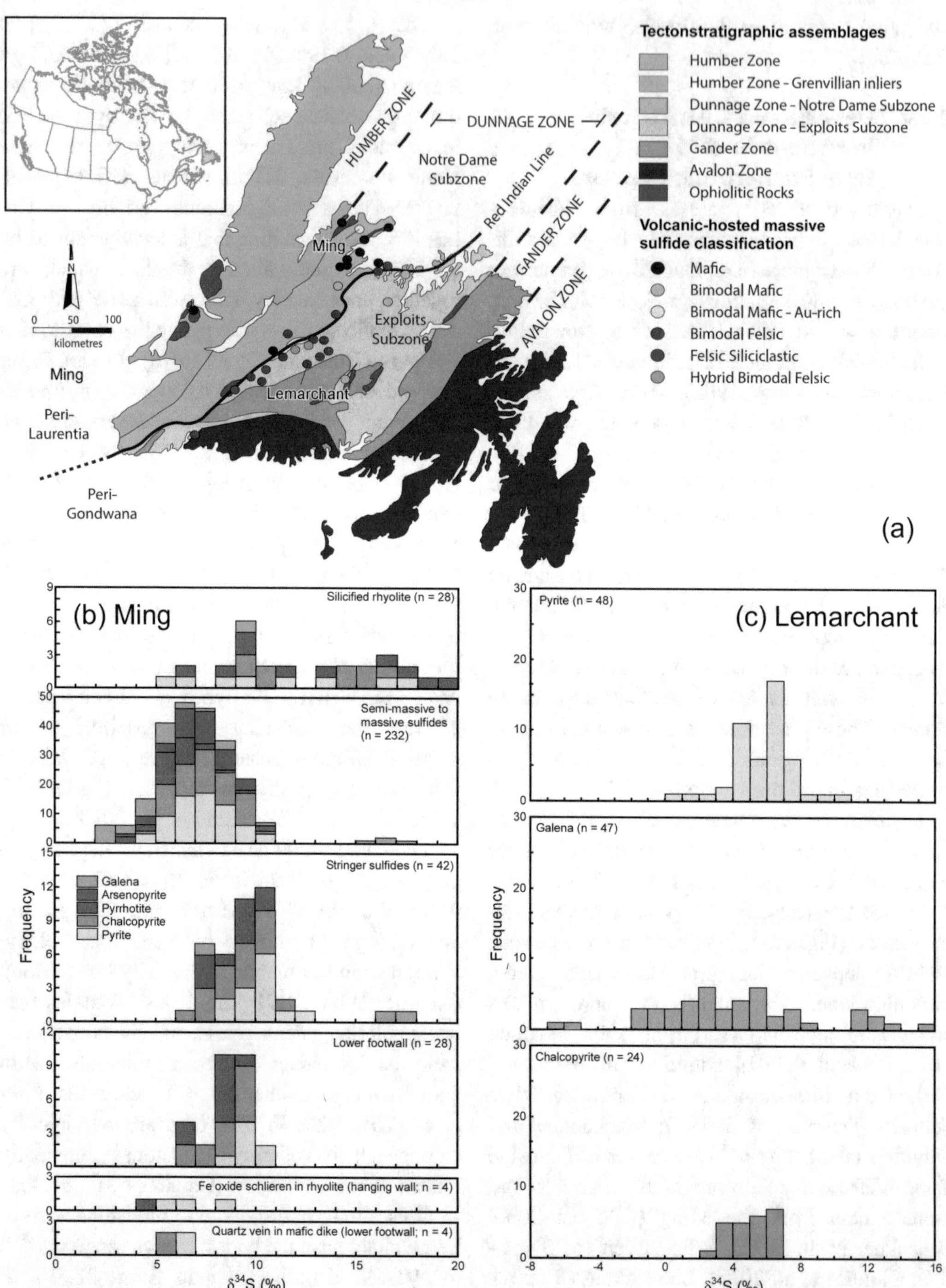

Fig. 12 Sulfur isotope characteristics of volcanic-hosted massive sulfide deposits in the Dunnage Zone, Newfoundland: (a) location of Dunnage Zone and major deposits (modified after Lode et al. 2017), (b) histogram showing the distribution of $\delta^{34}S$ in sulfide minerals from the Ming deposit (modified after Brueckner et al. 2015), (c) histogram showing the distribution of $\delta^{34}S$ in sulfide minerals from the ore zone of the Lemarchant deposit (modified after Gill et al. 2019)

minerals in the hydrothermal mudstones ranges from -38.8 to 14.4‰, with an average value of -12.6‰ (not shown). Proximal to mineralization, however, $\delta^{34}S$ tends to be higher and overlaps values from massive sulfide (Lode et al. 2017).

Overall, these results indicate that the relative proportion of TSR-derived H_2S and leached igneous H_2S ($\pm$ magmatic-hydrothermal sulfur) from the surrounding host rocks is variable within each deposit. For instance, Brueckner et al. (2015) and Cloutier et al. (2015) showed that the vast majority (>50%) of sulfur in the Ming and Whalesback deposits was from leaching of igneous basement (with a possible contribution of magmatic-hydrothermal sulfur for Ming) with much lesser contributions coming from TSR-derived H_2S, particularly for assemblages that were Cu-rich and/or those with enrichments in epithermal suite elements, including precious metals. Gill et al. (2019) showed that there was H_2S derived from both leaching and TSR in the Lemarchant deposit, but they also argued that some of the low $\delta^{34}S$ values (< 0‰) found were the result of magmatic-hydrothermal SO_2 disproportionation upon cooling and condensing, and the generation of light signatures in sulfides that crystallized incorporated the magmatic-hydrothermal derived H_2S. Mudstones from this same deposit have a much more complex history of biogenic sulfate reduction, microbial sulfide oxidation, and microbial disproportionation of intermediate sulfur compounds; however, with proximity to mineralization $\delta^{34}S$ increased to values near of above 0‰, indicating a greater hydrothermal input and possible magmatic-hydrothermal input. Higher grade copper zones are associated with $\delta^{34}S$ close to 0‰, Zn-Pb-rich zones are often associated $\delta^{34}S$ signatures consistent with mixed igneous and TSR-derived H_2S, whereas deposits enriched in Au, Ag and related elements can have igneous, TSR, or, possibly, magmatic-hydrothermal H_2S with low $\delta^{34}S$ values. These findings highlight the utility of $\delta^{34}S$ of an exploration vector to high grade ore in the VHMS setting, and are discussed further below.

High sulfidation deposits. Sillitoe et al. (1996) observed that a small proportion of VHMS and black smoker deposits contain significant amounts of minerals, such as bornite and tennantite, that are characteristic of hypogene high sulfidation hydrothermal conditions. Many of these deposits are also characterized by hypogene advanced argillic alteration assemblages, defined by the presence of minerals such as kaolinite, pyrophyllite and diaspore. Following work by Gamo et al. (1997), Herzig et al. (1998) and Gemmell et al. (2004), Huston et al. (2011) compared the $\delta^{34}S$ signature of high sulfidation deposits with that of more typical deposits (Fig. 13), for both modern black smokers and Cambrian VHMS deposits in western Tasmania. They found that high sulfidation deposits are characterized anomalously light $\delta^{34}S_{sulfide}$ and $\delta^{34}S_{sulfur}$ signatures, generally between -10‰ and 0‰, which is significantly lower than the $\delta^{34}S$ range seen in coeval deposits that lack high sulfidation mineralogy and advanced argillic alteration assemblages (mostly 0–10‰ for black smoker deposits and 8–19‰ for the western Tasmanian deposits: Fig. 13). Moreover, $\delta^{34}S$ of barite associated with the high sulfidation assemblage in the Tasmanian deposits (19–30‰: Walshe and Solomon 1981) and at Hine Hina (16–17‰: Herzig et al. 1988) are lower than coeval seawater and 20–30‰ heavier than coeval sulfide minerals (i.e. $\Delta^{34}S_{barite\text{-}sulfide} \sim$ 20–30‰).

The sulfur isotope systematics at Hine Hina were interpreted by Herzig et al. (1998) to indicate the incorporation of significant quantities of H_2S produced by disproportionation of magmatic SO_2 to form H_2S and H_2SO_4 and highly acidic hydrothermal fluids. The data from high sulfidation deposits in general can be interpreted similarly, as discussed by Huston et al. (2011), who concluded that "the presence of a major population of anomalously low $\delta^{34}S_{sulfide}$ can be suggestive of disproportionation of magmatic-hydrothermal SO_2, particularly if $\Delta^{34}S_{sulfate\text{-}sulfide} \sim$ 20–30‰, $\Delta^{34}S_{seawater\ sulfate\text{-}ore\ sulfate}$ is non-zero (i.e. $|\Delta^{34}S_{seawater\ sulfate\text{-}ore\ sulfate}| > 5$‰)

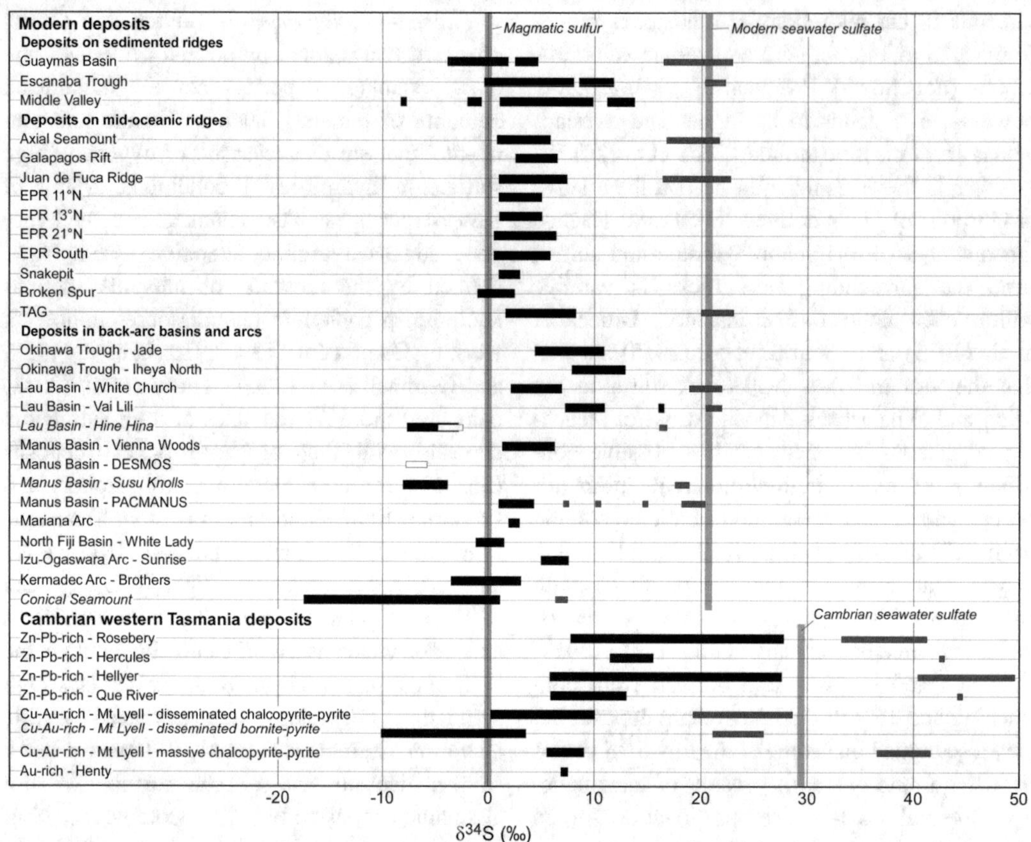

Fig. 13 Comparison of $\delta^{34}S$ values for hydrothermal sites on midocean ridges and arc environments on the modern seafloor and for deposits from the Cambrian Mount Read district, Tasmania. Modified from Herzig et al. (1998) and Gemmell et al. (2004) to include data from Kim et al. (2004), Solomon et al. (1969, 1988), Green et al. (1981) and Walshe and Solomon (1981). Intalicized names indicate high sulfidation deposits

and, most importantly, advanced argillic alteration assemblages are present."

Iheya North black smoker deposit. The Iheya North hydrothermal field (Fig. 14a) is located within the Okinawa Trough, interpreted to be an incipient intracontinental arc-back-arc basin (Shinjo and Kato 2000). Nine hydrothermal mounds make up the field, occurring within bimodal basaltic-rhyolite volcanic rocks, overlain by terrigenous sediments (Ishibashi et al. 2015). The Integrated Ocean Drilling Program Expedition 331 completed 5 drill holes into and surrounding the North Big Chimney (Fig. 14b; shows three of five holes). Sulfur isotope analyses of hydrothermal pyrite in the five drill holes at varying depths demonstrates that the massive sulfides at the North Big Chimney yield consistent $\delta^{34}S = 11.9 \pm 1.1‰$ (1σ: Figs. 14c-d), near identical to the $\delta^{34}S$ composition of the vent fluid (Aoyama et al. 2014). However, progressively outwards from the main vent site hydrothermal pyrite returns lower and more variable $\delta^{34}S$ values of $7.0 \pm 3.8‰$ (1σ) (LaFlamme et al. 2018: Fig. 14e).

The spatial variation within the hydrothermal system indicates that the lower temperature surroundings of the hydrothermal system precipitate hydrothermal pyrite with an increasingly depleted $\delta^{34}S$ value. LaFlamme et al. (2018) demonstrate that this pattern can only be accounted for by way of increased leaching of $\delta^{34}S$-depleted sedimentary pyrite originally deposited by biotic

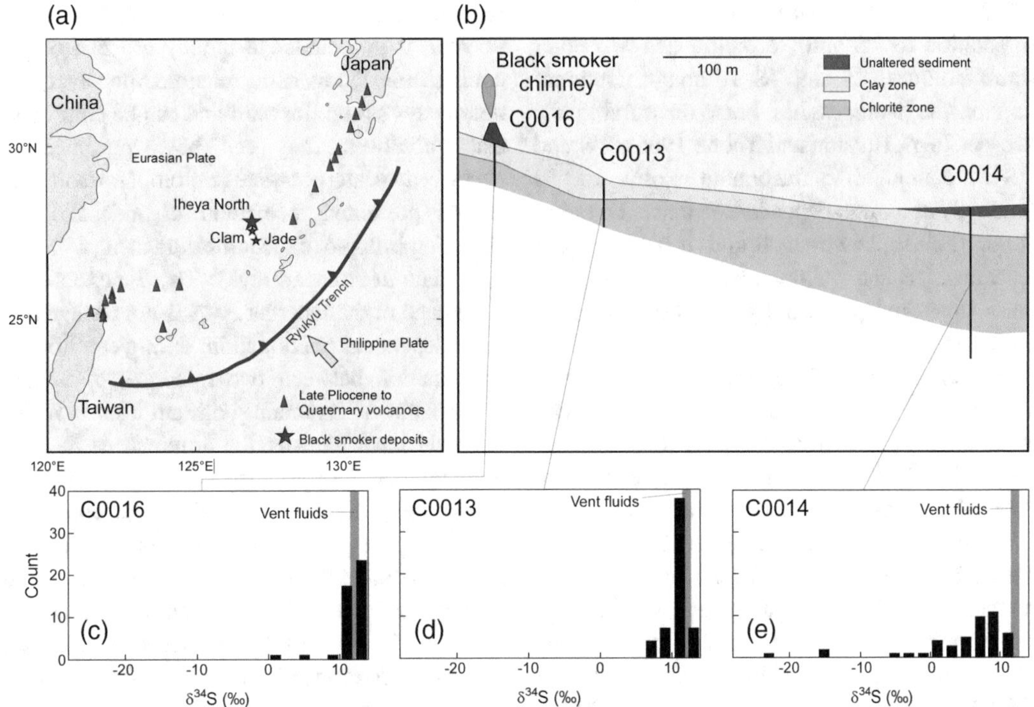

Fig. 14 Diagrams showing (a) the location of the Iheya North black smoker deposit, (b) a geological cross section of the deposit showing the locations of the drill drill holes, and (c) to (e) $\delta^{34}S$ distributions from the three drill holes showing proximal to distal variations in $\delta^{34}S$ (modified after LaFlamme et al 2018)

metabolisms. As the most significant metal enrichments (Fe, Zn, Cu, Bi, Tl, and Cd) are associated with sulfides at the North Big Chimney containing $\delta^{34}S$ values near-identical to the vent fluid, it is clear that sulfur isotopes can vector toward metals in sedimented seafloor hydrothermal systems (LaFlamme et al. 2018). This knowledge can then be applied in metal vectoring within VHMS deposits.

Modern seafloor systems on divergent margins. Prior to the discovery of seafloor hydrothermal systems along convergent margins in the 1990s, a large amount of data, including $\delta^{34}S$ data of venting fluids and ore minerals, had been collected from seafloor systems on divergent margins (Shanks 2001 and references therein). Although collected from a different tectonic setting, these data provide insights into sulfur sources and processes that cause $\delta^{34}S$ variability in seafloor systems. One of the more puzzling results was the observation that in the 9–10°N East Pacific Rise (9–10°N EPR) vent field, minerals on the inside walls of venting chimneys appear to be out of equilibrium with the venting fluids: $\Delta^{34}S_{\text{vent-inner wall sulfide}} \sim 1.8$–4.0‰, much greater the fractionation expected at the temperatures of venting (Shanks 2001).

Shanks (2001) also indicate that geological events can change isotopic signatures of the vent fluid. Immediately following a seismic (cracking) event at depth in the 9–10°N EPR vent field, $\delta^{34}S$ values of the vent fluid at the P vent decreased sharply by ~ 3‰, coinciding with significant changes in vent chemistry (although not $\delta^{18}O$). Hence it appears that geological events can cause significant changes in the isotopic (and chemical) characteristics of active systems, changes that cannot be documented in ancient systems due to the lack of precise temporal information available for modern systems.

4.2.3 Multiple Sulfur Isotopes

In addition to ^{32}S and ^{34}S, sulfur has two other stable isotopes, ^{33}S and ^{36}S. Although fractionation of the isotopes has been measurable for decades (e.g., Hulston and Thode 1965), ^{33}S and ^{36}S were thought to fractionate relative to ^{32}S (and ^{34}S) in a mass-dependent manner. Farquhar et al. (2000), however, found that, for some reactions, ^{33}S and ^{36}S fractionate independent of mass (mass-independent fractionation or MIF). Farquhar et al. (2000) and subsequent workers (Farquhar and Wing 2003; Johnston et al. 2007) showed that MIF occurs during photolytic breakdown of atmospheric SO_2 restricted to before the Great Oxidation Event (older than ca 2400 Ma). Mass independent fractionations of ^{33}S and ^{36}S are measured by $\Delta^{33}S$ and $\Delta^{36}S$, which indicate deviation from mass-dependent fractionation. These two parameters were defined by Farquhar et al. (2000) and processes that produce MIF are reviewed by Farquhar and Wing (2003), Johnston (2011) and Huston et al. (2023).

Multiple sulfur isotope studies have become important in genetic studies of VHMS deposits, enabling more precise estimates of the relative sulfur inputs from seawater sulfate, sulfur leached from sediments containing pyrite derived by bacterial sulfate reduction, and igneous sulfur in different environments (Ono et al. 2007; Peters et al. 2010; McDermott et al. 2015; Martin et al. 2021), and improving the understanding of the involved in this input. We illustrate uses of multiple sulfur isotopes in VHMS research using three examples, (1) assessment of isotopic and paragenetic equilibrium, (2) assessment of seawater-derived sulfur in Archean deposits, and (3) determination of sulfur sources and mineralizing processes at the DeGrussa deposit, Western Australia.

Assessing isotopic and paragenetic equilibrium. One of the more difficult aspects of interpreting isotopic data is determining if mineral assemblages analyzed are in isotopic equilibrium. If this can be established, isotopic fractionation between minerals can be used to estimate the temperature of mineral crystallization and the isotopic composition of the ore fluid. Typically, this is determined based on paragenetic relationships: if the minerals are in textural equilibrium, it is assumed that they are in isotopic equilibrium. Paragenetic relations, however, are not always straightforward and can be controversial; moreover, they can be overprinted or destroyed by later recrystallization. Jamieson et al. (2006) presented a method of independently assessing isotopic (dis)equilibrium using $\delta^{34}S$ and $\Delta^{33}S$ data in Archean rocks. These authors took advantage of the fact that $\Delta^{33}S$ is not changed by mass dependent fractionation; during equilibrium fractionation between two minerals, $\delta^{34}S$ fractionates but $\Delta^{33}S$ remains constant. Hence, if two minerals have statistically different $\Delta^{33}S$, it is unlikely that they are in isotopic equilibrium. Conversely, similar $\Delta^{33}S$ values are consistent with isotopic equilibrium, although crystallization from two different fluids with similar $\Delta^{33}S$ would also appear to be in isotopic equilibrium. Jamieson et al. (2006) used this method to assess eight mineral pairs from the Kidd Creek deposit. Two mineral pairs were not considered to be in isotopic equilibria, and a further three were considered to be in equilibria but small fractionation factors between the two minerals precluded the calculation of meaningful temperatures. One of the mineral pairs interpreted to be out of isotopic equilibrium from the $\Delta^{33}S$ data was clearly out of equilibrium from the $\delta^{34}S$ data. The three remaining mineral pairs in apparent isotopic equilibria yielded depositional temperatures within the expected ranges, consistent with being in true isotopic equilibrium. Although this technique is time restricted largely to the Archean, it provides independent criteria for assessing isotopic (dis)equilibria. The method is also applicable to assessing in situ data collected using secondary ion mass spectrometry (SIMS: Whitehouse 2013).

Sulfur sources in Archean deposits. In the Archean, prior to significant fractionation in $\delta^{34}S$ of VHMS deposits, the $\Delta^{33}S$ record of these deposits has proven to be useful (e.g., Jamieson et al. 2013; Caruso et al. 2019). At that time, the ferruginous ocean chemistry reflected the low redox state of the atmosphere and ocean basins incorporated atmosphere-derived elemental sulfur and sulfate formed by mass independent fractionation of sulfur (MIF-S) with large positive and negative $\Delta^{33}S$ anomalies (> ± 0.2‰; Farquhar et al. 2013). Therefore, the recycling of sulfur derived from the Archean ocean seawater sulfate may be traced using this signature. Neoarchean VHMS deposits, including the Kidd Creek (Jamieson et al. 2013) and Noranda (Sharman et al. 2015) deposits of the Superior Craton, as well as the Teutonic Bore complex (Chen et al. 2015) and Nimbus deposits (Caruso et al. 2019) of the Yilgarn Craton, generally preserve $\Delta^{33}S < 0$‰. This predominantly negative $\Delta^{33}S$ signature is interpreted to indicate hydrothermal fluids incorporated a small component of Archean seawater sulfate (3–18%), which recycled the oxidised form of MIF-S. Barré et al. (2022) suggested that in the Noranda and Matagami districts of the Abitibi Subprovince multiple sulfur isotope signatures can be directly correlated with metal endowment of deposits. Massive Zn-Cu-sulfide sub-seafloor replacement deposits incorporate less seawater sulfate and preserve a near igneous signature, whereas less-endowed Fe-exhalites yield a higher proportion of seawater sulfate and an increasingly enriched $\delta^{34}S$ and depleted $\Delta^{33}S$ signature, reflecting the isotopic composition of the Archean ocean seawater sulfate reservoir. Exhalites are intimately linked to massive sulfide occurrences, and so their sulfur isotope composition may be utilised for within-camp targeting.

DeGrussa deposit, Western Australia. The ~ 2.03 Ga DeGrussa Cu-Au–Ag deposit (and satellite Monty deposit) is located in the Bryah Basin of the Paleoproterozoic Capricorn Orogen of Western Australia (Hawke et al. 2015: Fig. 15). The deposits contain 9 Mt at 4.5% Cu and 1.5 g/t gold (Hilliard et al. 2017). Massive sulfide ore lenses are associated with gabbroic sills that intrude tubiditic volcaniclastic rocks, interpreted to have formed in a continental rift setting (Occhipinti et al. 2017). Massive sulfide ore lenses are commonly associated with magnetite and consist of pyrite with lesser pyrrhotite and chalcopyrite (LaFlamme et al. 2021).

An integrated dataset of in situ multiple sulfur isotope data and conventional fluorination-IRMS multiple sulfur isotope analyses identifies an overprinting non-VHMS sulfide mineralizing event with $\delta^{34}S > 8$‰. The original VHMS mineralizing event at Degrussa, however, yields $\delta^{34}S$ from 2.9‰ to 3.6‰, and $\Delta^{33}S$ from -0.08‰ to 0.00‰. A two component $\delta^{34}S$-$\Delta^{33}S$ mixing model indicates 11% of H_2S is derived from thermochemically reduced seawater sulfate mixed with magmatic H_2S, either derived from leaching of volcanic rock or from magmatic-hydrothermal fluids.

4.2.4 Application of Sulfur Isotope Geochemistry to Exploration and Ore Genesis

Volcanosedimentary basins commonly contain a range of sulfide accumulations (Fig. 16), some related to VHMS mineral systems (e.g., VHMS deposits and exhalites) and some not (e.g., pyritic black shale). Even among VHMS-related accumulations, sulfur sources are diverse, with the source potentially having implications to fertility. As presented above, sulfur isotope data can aid in distinguishing the origin and sulfur source of sulfide accumulations.

An important potential application of sulfur isotopes in VHMS exploration is to distinguish Fe-sulfide accumulations formed solely through sedimentary or diagenetic processes or through metamorphic processes from those that included exhalative hydrothermal sulfur (Brueckner et al. 2015; Lode et al. 2017; LaFlamme et al. 2021). Sulfur isotope characteristics, based on our review and shown in Fig. 16, appear to be sufficiently distinctive to distinguish exhalative components and may allow vectoring once exhalative horizons are determined.

Once a VHMS-related hydrothermal occurrence has been identified, sulfur isotope data may aid potential assessment and ranking during exploration programs. Different parts of a VHMS mineral system may have distinctive $\delta^{34}S$ signatures due to the varying incorporation of sulfur from different sources through the system (Fig. 16). The massive sulfide and stringer zones, produced by deposition of hydrothermal sulfide

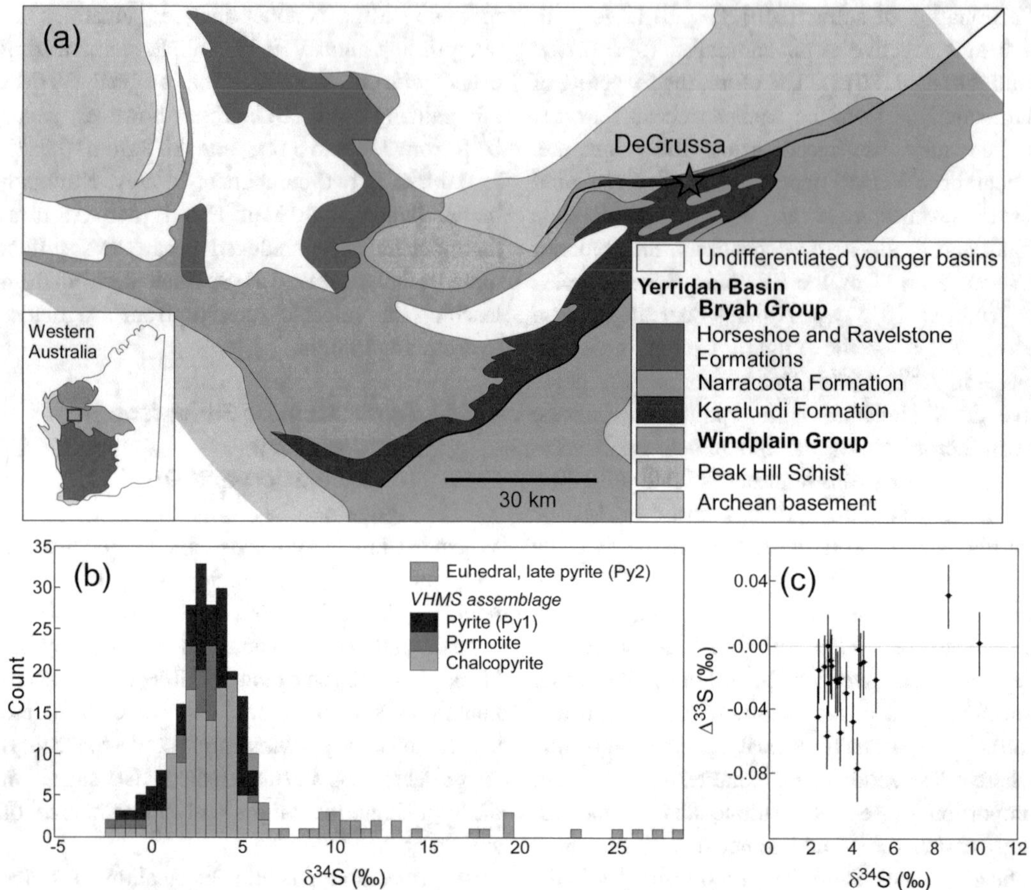

Fig. 15 Diagrams showing (a) the location and regional geology of the DeGrussa deposit, (b) a histogram showing variations in in situ $\delta^{34}S$ analyses of sulfide minerals, and (c) a scattergram showing the relationship between $\delta^{34}S$ and $\Delta^{33}S$ using fluorination analysis. Reproduced with permission and with small modification from LaFlamme et al. (2021); Copyright 2021 Elsevier

at or below the seafloor, have $\delta^{34}S$ signatures that reflect varying proportions of sulfide produced by thermochemical sulfate reduction of seawater sulfate ($\delta^{34}S \sim \delta^{34}S_{\text{coeval seawater sulfate}}$), leached from igneous sulfur deep in the VHMS system ($\delta^{34}S \sim 0‰$) or generated by disproportionation of magmatic SO_2 ($\delta^{34}S \sim$ -15‰ to -5‰). During early-stage prospect evaluation and near-mine exploration, these signatures may assist in discriminating the genesis and potential of prospects or new lens: for example, a significant population ore minerals with $\delta^{34}S$ below 0‰ may indicate the presence of Cu-Au-rich high sulfidation potential, particularly if accompanied by advanced argillic alteration assemblages ore an ore assemblage enriched in metals such as Bi, Te, Mo, Sn and Se..

A second tracer in sulfur isotope space, $\Delta^{33}S$, can further elucidate seawater contribution of sulfur to the system, especially in Precambrian deposits, in which the conventional seawater sulfur isotope signature is less certain (see LaFlamme et al. 2021). Mixing models in $\delta^{34}S$-$\Delta^{33}S$ space, can better predict the multiple sulfur isotope signature of massive sulfide ore. Preliminary work has highlighted that at the district scale, $\delta^{34}S$- $\Delta^{33}S$ signature of massive sulfide ore lenses may be directly correlated to Au grade (Sharman et al. 2015). Therefore, more studies concentrated on the potential of near-mine

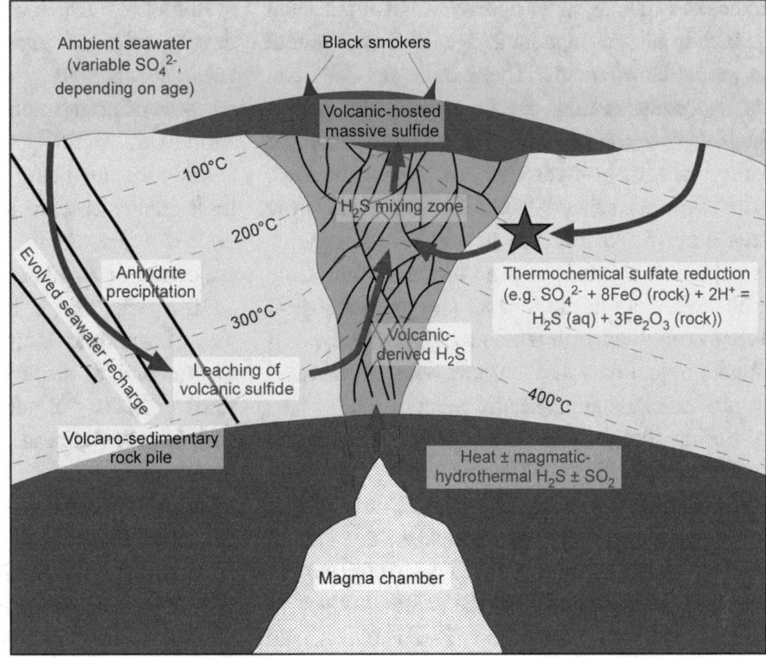

Fig. 16 Model of a seafloor hydrothermal system (not to scale) showing sulfur sources and sinks, and reactions involving that occur during fluid circulation below and on the seafloor (modified after LaFlamme et al. 2021)

multiple sulfur isotope vectoring within VHMS deposits of various ages and tectonic setting are warranted.

5 Application of and Impediments to the Use of Light Stable Isotopes in Exploration

Although stable isotope studies have been critical to characterizing and understanding ore-forming processes in VHMS systems, they have not been extensively utilized by the exploration industry, largely because of the slow turn-around time and high costs relative to other analytical tools. The review above has suggested that there are isotopic patterns that are potentially useful for exploration at the district to deposit scale. Further, the development of in-situ methods at relatively reasonable cost and with high volume analysis, compared to conventional methods, suggests the potential for more widespread usage with continued analytical development into the future.

The only example where isotopic data has been directly used to discover a VHMS deposit is the study by Miller et al. (2001), who combined whole rock oxygen isotope with other geochemical data to map paleothermal gradients and discover the West 45 lens at the ~ 480 Ma Thalanga deposit, Queensland. This illustrates the utility of whole rock oxygen isotope data at the deposit scale, and studies such as those by Cathles (1993) and Brauhart et al. (2000) illustrate the potential use of whole rock oxygen isotope data at the district scale.

6 Conclusions and Future Directions

Since recognition of VHMS deposits as a class, light stable isotopes, particularly those of oxygen, hydrogen and sulfur, have played a key role in understanding the hydrothermal system that formed these deposits. Estimates of fluid oxygen and hydrogen isotope data, determined from mineral isotopic data, suggest that in most cases evolved seawater was the dominant hydrothermal fluid. There is, however, sound evidence that magmatic-hydrothermal contributions may be important in some VHMS deposits, particularly when isotopic data are supported by other indicators of magmatic-hydrothermal input such as certain alteration assemblages and enrichment in particular elements. These latter deposits can be important as they can be enriched in metals such as tin, gold and other elements associated with magmatic-hydrothermal contributions. Alteration associated with hydrothermal flow leaves consistent patterns in whole rock oxygen isotope data that can be used to define fluid flow pathways, both at the deposit and district scales. When combined with whole rock geochemical or mineralogical data, the isotopic data can be used to produce paleotemperature maps.

Sulfur isotope data also provide important insights into the VHMS system, particularly addressing the sources of sulfur. Conventional $\delta^{34}S$ and multiple isotope data (e.g., $\Delta^{33}S$ data) indicate that although reduced seawater sulfate has been a component of the sulfur budget of VHMS systems since the Paleoarchean, its importance has increased in general from 5–10% in the Archean to 15–20% in the Paleoproterozoic and to 20–25% in the Phanerozoic; within these time intervals, however, the proportion can vary significantly between deposits. The other component is magmatic sulfur, which can be derived by leaching of volcanic rocks or through input of magmatic-hydrothermal H_2S or SO_2. The latter may be important to form high sulfidation VHMS deposits (c.f. Sillitoe et al. 1996), which are commonly enriched in gold.

Possibly because of their relatively high cost and/or slow turn-around time, isotopic data have not been employed extensively in exploring for VHMS deposits. Despite this, oxygen isotopes have directly contributed to the discovery of a new ore lens at the Thalanga deposit in Queensland (Miller et al. 2001), and they clearly have potential to produce paleotemperature and fluid flow maps at the deposit and district scales. Sulfur isotope data have utility in fingerprinting mineralized from barren sulfide bodies, and may also provide vectors to ore.

The last twenty years has seen many significant improvements in analytical technologies and in the understanding of stable isotope systematics. The most important has been the development of microanalytical tools such as LA-ICP-MS that allow rapid in situ isotopic analysis of petrographically constrained samples. Moreover, rapid and inexpensive analytical techniques are being developed for a range of light stable isotopic systems (see Huston et al. 2023 for discussion) that may allow the rapid collection of large datasets useful in exploration.

The other major leap has come through the use of multiple isotopic systems such as the oxygen and, as discussed here, the sulfur system. Prior to 2000, ^{33}S and ^{36}S were rarely analyzed and largely ignored. The discovery of mass-independent sulfur isotope fractionation has resulted in major changes in our understanding of the sulfur cycle and new insights into ancient environments on Earth. These discoveries have also impacted our understanding of VHMS systems, including evidence of the importance of reduced seawater sulfate even before the Great Oxidation Event.

New technologies and the understanding derived from them will continue to impact our understanding of the VHMS mineral system. These new isotopic data will lead to new discoveries, either directly as costs come down, or indirectly, through better understanding of mineralizing processes.

Acknowledgements The authors thank David Champion, Jens Gutzmer and an anonymous reviewer, for their reviews that have significantly improved this contribution. This contribution is published with permission of the Chief Executive Officer of Geoscience Australia.

References

Addy SK, Ypma PJM (1977) Origin of massive sulfide deposits at Ducktown, Tennessee: an oxygen, carbon and hydrogen isotope study. Econ Geol 72:1245–1268

Aggarwal PK, Longstaffe FJ (1987) Oxygen-isotope geochemistry of metamorphosed massive sulfide deposits of Flin Flon—Snow Lake belt, Manitoba. Contrib Mineral Petrol 96:314–325

Aoyama S, Nishizawa M, Katai K, Ueno Y (2014) Microbial sulfate reduction within the Iheya North subseafloor hydrothermal system constrained by quadruple sulfur isotopes. Earth Planet Sci Lett 398:113–126. https://doi.org/10.1016/j.epsl.2014.04.039

Aruajo SM, Scott SD, Longstaffe FJ (1996) Oxygen isotope composition of alteration zones of highly metamorphosed volcanogenic massive sulfide deposits; Geco, Canada, and Palmeiropolis, Brazil. Econ Geol 91:697–712

Bachinski DJ (1978) Sulfur isotopic composition of thermally metamorphosed cupriferous iron sulfide ores associated with cordierite-anthophyllite rocks, Gull Pond, Newfoundland. Econ Geol 73:64–72

Barré G, LaFlamme C, Beaudoin G, Goutier J, Cartigny P (2022) The application of multiple sulphur isotopes to determine the architecture of Archean VMS deposits. Geol Assoc Can-Mineral Assoc Can Abstr 45:50

Barrie CT, Hannington MD (1999) Classification of VMS deposits based on host rock composition. Rev Econ Geol 8:1–12

Barriga FJAS, Kerrich R (1984) Extreme ^{18}O-enriched volcanics and ^{18}O-evolved marine water, Aljustrel, Iberian Pyrite Belt: transition from high to low Rayleigh number convective regimes. Geochim Cosmochim Acta 48:1021–1031

Beaty DW, Taylor HP Jr (1982) Some petrologic and oxygen isotope relationships in the Amulet mine, Noranda, Quebec, and their bearing on the origin of Archean massive sulfide deposits. Econ Geol 77:95–108

Beaty DW, Taylor HP Jr, Coads PR (1988) An oxygen isotope study of the Kidd Creek, Ontario, volcanogenic massive sulfide deposit: evidence for a high ^{18}O ore fluid. Econ Geol 83:1–17

Beaudoin G, Mercier-Langevin P, Dube B, Taylor BE (2014) Low-temperature alteration at the world-class LaRonde Penna Archean Au-rich volcanogenic massive sulfide deposit, Abitibi Subprovince, Quebec, Canada: evidence from wholerock oxygen isotopes. Econ Geol:167–182

Blevin PL, Chappell BW (1995) Chemistry, origin, and evolution of mineralized granites in the Lachlan fold belt, Australia: the metallogeny of I- and S-type granites. Econ Geol 90:1604–1619

Brauhart CW, Groves DI, Morant P (1998) Regional alteration systems associated with volcanogenic massive sulfide mineralization at Panorama, Pilbara, Western Australia. Econ Geol 93:292–302

Brauhart CW, Huston DL, Andrew A (2000) Definition of regional alteration in the Panorama VMS district using oxygen isotope mapping: implications for the origin of the hydrothermal system and applications to exploration. Mineral Deposita 35:727–740

Brauhart CW, Huston DL, Groves DI, Mikucki EJ, Gardoll SJ (2001) Geochemical mass transfer patterns as indicators of the architecture of a complete volcanic-hosted massive sulfide hydrothermal alteration system in the Panorama district, Pilbara, Western Australia. Econ Geol 96:1263–1278

Brueckner S, Piercey S, Layne G, Piercey G, Sylvester P (2015) Variations of sulphur isotope signatures in sulphides from the metamorphosed Ming Cu(−Au) volcanogenic massive sulphide deposit, Newfoundland Appalachians, Canada. Mineral Deposita 50:619–640

Buick R, Brauhart CW, Morant P, Thornett JR, Maniw J, Archibald NJ, Doepel M (2002) Geochronology and stratigraphic relationships of the Sulphur Springs Group and Strelley Granite: a temporally distinct igneous province in the Archaean Pilbara craton, Australia. Precambrian Res 114:87–120

Caruso S, Fiorentini ML, Barnes SJ, LaFlamme C, Martin LA (2019) Microchemical and sulfur isotope constraints on the magmatic and hydrothermal evolution of the Black Swan Succession, Western Australia. Mineral Deposita 317:211–229

Cathles LM (1993) Oxygen isotope alteration in the Noranda mining district, Abitibi greenstone belt, Quebec. Econ Geol 88:1483–1511

Chaussidon M, Albarede F, Sheppard SMF (1989) Sulphur isotope variations in the mantle from ion microprobe analyses of micro-sulphide inclusions. Earth Planet Sci Lett 92:144–156

Chen M, Campbell IH, Xue Y, Tian W, Ireland TR, Holden P, Cas RAF, Hayman PC, Das R (2015) Multiple sulfur isotope analyses support a magmatic model for the volcanogenic massive sulfide deposits of the Teutonic Bore Volcanic Complex, Yilgarn Craton, Western Australia. Econ Geol 110:1411–1423

Claypool GE, Holser WT, Kaplan IR, Sakai H, Zak I (1980) The age curves of sulfur and oxygen isotopes in marine sulfate and their mutual interpretation. Chem Geol 28:199–260

Cloutier J, Piercey SJ, Layne G, Heslop J, Hussey A, Piercey G (2015) Styles, textural evolution, and sulfur isotope systematics of Cu-rich sulfides from the Cambrian Whalesback volcanogenic massive sulfide deposit, central Newfoundland, Canada. Econ Geol 110:1215–1234

Cloutier J, Piercey SJ, Lode S, Vande Guchte M, Copeland DA (2017) Lithostratigraphic and structural reconstruction of the Zn-Pb-Cu-Ag-Au Lemarchant volcanogenic massive sulphide (VMS) deposit, Tally Pond group, central Newfoundland, Canada. Ore Geol Rev 84:154–173

Converse DR, Holland HD, Edmond JM (1984) Flow rates in the axial hot springs of the East Pacific Rise (21°N): implications for the heat budgets and the formation of massive sulfide deposits: Earth Planet Sci Lett 69:159–175

Costa UR, Barnett RL, Kerrich R (1983) The Mattagami Lake mine Archean Zn-Cu sulfide deposit, Quebec: hydrothermal coprecipitation of talc and sulfides in a sea-floor brine pool—evidence from geochemistry, $^{18}O/^{16}O$, and mineral chemistry. Econ Geol 78:1144–1203

Crockford PW, Kunzmann M, Bekker A, Hayles J, Bao H, Halverson GP, Peng Y, Bui TH, Cox GM, Gibson TM, Wörndle S, Rainbird R, Lepland A, Swanson-Hysell NL, Master S, Sreenivas B, Kuznetsov A, Krupenik V, Wing BA (2019) Claypool continued: extending the isotopic record of sedimentary sulfate. Chem Geol 513:200–225

Date J, Watanabe Y, Saeki Y (1983) Zonal alteration around the Fukazawa kuroko deposits, Akita Prefecture, northern Japan. Econ Geol Mon 5:365–386

de Groot PA (1993) Stable isotope (C, O, H), major- and trace element studies on hydrothermal alteration and related ore mineralization in the volcano-sedimentary belt of Bergslagen, Sweden. Geologica Ultraiectina, 98, 181 pp

de Ronde CEJ (1995) Fluid chemistry and isotope characteristics of seafloor hydrothermal systems and associated VMS deposits: potential for magmatic contributions. Mineral Assoc Can Short Course Ser 23:479–509

Ding T, Valkiers S, Kipphardt H, De Bièvre P, Taylor PDP, Gonfiantini R, Krouse R (2001) Calibrated sulfur isotope abundance ratios of three IAEA sulfur isotope reference materials and V-CDT with a reassessment of the atomic weight of sulfur. Geochim Cosmochim Acta 65:2433–2437

Doyle MG, Allen RL (2003) Subsea-floor replacement in volcanic-hosted massive sulfide deposits. Ore Geol Rev 23:183–222

Drieberg SL, Hagemann SG, Huston DL, Landis G, Ryan CG, Van Achterbergh E, Vennemann T (2013) The interplay of evolved seawater and magmatic-hydrothermal fluids in the 3.24 Ga Panorama volcanic-hosted massive sulfide hydrothermal system, north Pilbara Craton. Western Australia. Econ Geol 108:79–110

Dubé B, Gosselin P, Mercier-Langevin P, Hannington M, Galley A (2007) Gold-rich volcanogenic massive sulphide deposits. Mineral Deposits Division, Geol Assoc Can Spec Publ 5:75–94

Dunning GR, Kean BF, Thurlow JG, Swinden HS (1987) Geochronology of the Buchans, Roberts Arm, and Victoria Lake groups and Mansfield Cove Complex, Newfoundland. Can J Earth Sci 24:1175–1184

Eastoe CJ, Gustin MM (1996) Volcanogenic massive sulfide deposits and anoxia in the Phanerozoic oceans. Ore Geol Rev 10:179–197

Eldridge CS, Barton PB, Ohmoto H (1983) Mineral textures and their bearing on formation of the Kuroko orebodies. Econ Geol Mon 5:241–281

Farquhar J, Bao H, Thiemens M (2000) Atmospheric influence of Earth's earliest sulfur cycle. Science 28:756–758

Farquhar J, Cliff J, Zerkle AL, Kamyshyn A, Poulton SW, Adams CM, Harms DB (2013) Pathways for Neoarchean pyrite formation constrained by mass-independent sulfur isotopes. Proc National Acad Sci 110:17638–17643

Farquhar J, Wing BA (2003) Multiple sulfur isotopes and the evolution of the atmosphere. Earth Planet Sci Let 213:1–13

Franklin JM, Lydon JW, Sangster DM (1981) Volcanic-associated massive sulfide deposits. Econ Geol 75th Anniv Vol, pp 485–627

Franklin JM, Gibson HL, Jonasson IR, Galley AG (2005) Volcanogenic massive sulfide deposits. Econ Geol 100th Anniv Vol, pp 523–560

Friedman I., O'Neil JR (1977) Compilation of stable isotope fractionation factors of geochemical interest. US Geol Surv Prof Pap 440-KK:1–12

Galley AG (2003) Composite synvolcanic intrusions associated with Precambrian VMS-related hydrothermal systems. Mineral Deposita 38:443–473

Gamo T, Okamura K, Charlou J, Urabe T, Auzende J, Ishibashi J, Shitashima K, Chiba H, Binns RA, Gena K, Henry K, Matsubayashi O, Moss R, Nagaya Y, Naka J, Ruellan E (1997) Acidic and sulfate-rich hydrothermal fluids from the Manus back-arc basin, Papua New Guinea. Geology 25:139–142

Gemmell JB, Sharpe R, Jonasson I, Herzig P (2004) Sulfur isotope evidence for magmatic contribution to subaqueous and subaerial epithermal mineralisation: Conical Seamount and Ladolam Au deposit, Papua New Guinea. Econ Geol 99:1711–1725

Gill SB, Piercey SJ, Layton-Matthews D (2016) Mineralogy and metal zoning of the Cambrian Zn-Pb-Cu-Ag-Au Lemarchant volcanogenic massive sulfide (VMS) deposit, Newfoundland. Can Mineral 54:1307–1344

Gill SB, Piercey SJ, Layne GD, Piercey G (2019) Sulphur and lead isotope geochemistry of sulphide minerals from the Zn-Pb-Cu-Ag-Au Lemarchant volcanogenic massive sulphide (VMS) deposit, Newfoundland, Canada. Ore Geol Rev 104:422–435

Golding SD, Duck LJ, Young E, Baublys KA, Glikson M, Kamber BS (2011) Earliest seafloor hydrothermal systems on Earth: comparisons with modern analogues. In: Golding SD, Glikson M (eds) Earliest life on Earth: habitats, environments and methods of detection. Springer, Dordrecht, pp 15–49

Goodfellow W, Peter J (1999) Sulphur isotope composition of the Brunswick No. 12 massive sulphide deposit, Bathurst Mining Camp, New Brunswick: implications for ambient environment, sulphur source, and ore genesis. Can J Earth Sci 36:127–134

Green GJ, Taheri J (1992) Stable isotopes and geochemistry as exploration indicators. Bull Geol Surv Tasmania 70:84–91

Green GR, Solomon M, Walshe JL (1981) The formation of the volcanic-hosted massive sulfide at Rosebery, Tasmania. Econ Geol 76:304–338

Green GR, Ohmoto H, Date J, Takahashi T (1983) Whole-rock oxygen isotope distribution in the Fukazawa-Kosaka area, Hokuroku district, Japan, and its potential application to mineral exploration. Econ Geol Mon 5:395–411

Hagemann S, Hensler A-S, Figueiredo e Silva RC, Tsiko H (2023) Light stable isotope (O, H, C) signatures of BIF-hosted iron ore systems: implications for genetic models and exploration targeting. In : Huston DL, Gutzmer J (eds) Isotopoes in economic geology, meallogenesis and exploration, Springer, Berlin, this volume

Hannington MD, de Ronde CEJ, Petersen S (2005) Seafloor tectonics and submarine hydrothermal systems. Econ Geol 100th Anniv Vol, pp 111–141

Hannington M, Jamieson J, Monecke T, Petersen S, Beaulieu S (2011) The abundance of seafloor massive sulfide deposits. Geology 39:1155–1158

Hart H, Gibson HL, Lesher CM (2004) Trace element geochemistry and petrogenesis of felsic volcanic rocks associated with volcanogenic Cu–Zn–Pb massive sulfide deposits. Econ Geol 99:1003–1013

Hattori K, Muehlenbachs K (1980) Marine hydrothermal alteration at a kuroko deposit, Kosaka, Japan. Contrib Mineral Petrol 74:285–292

Hattori K, Sakai H (1979) D/H ratios, origins, and evolution of the ore-forming fluids for the Neogene veins and kuroko deposits of Japan. Econ Geol 74:535–555

Hawke ML, Meffre S, Stein H, Hilliard P, Large R, Gemmell B (2015) Geochronology of the DeGrussa volcanic-hosted massive sulfide deposit and associated mineralisation of the Yerrida, Bryah and Padbury basins, Western Australia. Precambrian Res 267:250–284

Heaton THE, Sheppard SMF (1977) Hydrogen and oxygen isotope evidence for sea-water-hydrothermal alteration and ore deposition, Troodos complex, Cyprus. Volcanic processes in ore genesis. Institute of Mining and Metallurgy, London, pp 42–57

Herrmann W, Green GR, Barton MD, Davidson GJ (2009) Lithogeochemical and stable isotopic insights into submarine genesis of pyrophyllite-altered facies at the Boco prospect, western Tasmania. Econ Geol 104:775–792

Herzig PM, Hannington MD (1995) Polymetallic massive sulfides at the modern seafloor. a review. Ore Geol Rev 10:95–115

Herzig PM, Hannington MD, Arribas A Jr (1998) Sulfur isotope composition of hydrothermal precipitates from the Lau back-arc: implications for magmatic contributions to seafloor hydrothermal systems. Mineral Deposita 33:226–237

Hilliard P, Adamczyk KE, Hawke ML (2017) DeGrussa copper-gold deposit. Austr Inst Mining Metall Mon 32:393–400

Hinchey JG (2011) The Tulks volcanic belt, Victoria Lake Supergroup, central Newfoundland – geology, tectonic setting, and volcanogenic massive sulfide mineralization. Nfld Labrador Dep Nat Res Geol Surv St. John's NL Can Rep 2011–02, 167 p

Holk GJ, Taylor BE, Galley AG (2008) Oxygen isotope mapping of the Archean Sturgeon Lake caldera complex and VMS-related hydrothermal system, Northwestern Ontario, Canada. Mineral Deposita 43:623–640

Holland HD (1972) The geologic history of sea water—an attempt to solve the problem. Geochim Cosmochim Acta 36:637–651

Hou Z, Zaw K, Rona P (2008) Geology, fluid inclusions, and oxygen isotope geochemistry of the Baiyinchang pipe-style volcanic-hosted massive sulfide Cu deposit in Gansu Province, northwestern China. Econ Geol 103:269–292

Hoy LD (1993) Regional evolution of hydrothermal fluids in the Noranda district, Quebec: evidence for $\delta^{18}O$ values from volcanogenic massive sulfide deposits. Econ Geol 88:1526–1541

Hulston JR, Thode HG (1965) Variations in the S33, S34, and S36 contents of meteorites and their relation to chemical and nuclear effects. J Geophys Res 70:3475–3484

Huston DL (1999) Stable isotopes and their significance for understanding the genesis of volcanic-hosted massive sulfide deposits: a review. Rev Econ Geol 10:151–180

Huston DL, Logan GA (2004) Barite, BIFs and bugs: evidence for the evolution of the Earth's early hydrosphere. Earth Planet Sci Lett 220:41–55

Huston DL, Taylor BE (1999) Genetic significance of oxygen and hydrogen isotope variations at the Kidd Creek volcanic-hosted massive sulphide deposit, Ontario, Canada. Econ Geol Mon 10:335–350

Huston DL, Stevens B, Southgate PN, Muhling P, Wyborn L (2006) Australian Zn–Pb–Ag ore-forming systems: a review and analysis. Econ Geol 101:1117–1158

Huston DL, Pehrsson S, Eglington BM, Zaw K (2010) The geology and metallogeny of volcanic-hosted massive sulfide deposits: variations through geologic time and with tectonic setting. Econ Geol 105:571–591

Huston DL, Relvas JMRS, Gemmell JB, Drieberg S (2011) The role of granites in volcanic-hosted massive sulphide ore-forming systems: an assessment of magmatic–hydrothermal contributions. Mineral Deposita 46:473–507

Huston DL, Eglington B, Pehrsson S, Piercey SJ (2022) Global database of zinc-lead-bearing mineral deposits. Geosci Austr Rec 2022/10

Huston DL, Trumbull RB, Beaudoin G, Ireland T (2023) Light stable isotopes (H, B, C, O and S) in ore studies —methods, theory, applications and uncertainties. In : Huston DL, Gutzmer J (eds) Isotopoes in economic geology, meallogenesis and exploration, Springer, Berlin, this volume

Hutchinson RW (1973) Volcanogenic sulfide deposits and their metallogenic significance. Econ Geol 68:1223–1246

Inverno CMC, Solomon M, Barton MD, Foden J (2008) The Cu-stockwork and massive sulfide ore of the Feitais volcanic-hosted massive sulfifide deposit, Iberian Pyrite Belt, Portugal: a mineralogical, fluid inclusion, and isotopic investigation. Econ Geol 103:241–267

Ishibashi J-I, Ikegami F, Tsuji T, Urabe T (2015) Hydrothermal activity in the Okinawa Trough backarc basin: geological background and hydrothermal mineralisation. In: Ishibashi J-I (ed) Subseafloor biosphere linked to hydrothermal systems. Springer, Berlin, pp 337–359

Ishihara S, Sasaki A (1978) Sulfur of kuroko deposits—a deep seated origin. Mining Geol 28:361–367

Jamieson JW, Wing BA, Hannington MD, Farquhar J (2006) Evaluating isotopic equilibrium among sulfide mineral pairs in Archean ore deposits: case study from the Kidd Creek VMS deposit, Ontario, Canada. Econ Geol 101:1055–1061

Jamieson JW, Wing BA, Farquhar J, Hannington MD (2013) Neoarchaean seawater sulphate concentrations from sulphur isotopes in massive sulfide ore. Nature Geosci 6:61–64

Johnston DT (2011) Multiple sulfur isotopes and the evolution of Earth's surface sulfur cycle. Earth-Sci Rev 106:161–183

Johnston DT, Farquhar J, Canfield DE (2007) Sulfur isotope insights into microbial sulfate reduction: when microbes meet models. Geochim Cosmochim Acta 71:3929–3947

Kean BF, Evans DTW, Jenner GA (1995) Geology and mineralization of the Lushs Bight Group, Geol Surv Nfld Labrador Mineral Dev Div St John's NL Can Rep 95–2, 204 p

Keenan JH, Keyes FG, Hill PG, Moore JG (1969) Steam tables—thermodynamic properties of water including vapor, liquid, and solid phases (international edition—metric units). Wiley, New York, p 162

Kerr DJ, Gibson HL (1993) A comparison of the Horne volcanogenic massive sulfide deposit and intracauldron deposits of the Mine Sequence, Noranda, Quebec. Econ Geol 88:1419–1442

Kerrich R, Wyman D (1997) Review of developments in trace-element fingerprinting of geodynamic settings and their implications for mineral exploration. Austr J Earth Sci 44:465–488

Zaw K, Large RR (1992) The precious metal-rich South Hercules mineralization, western Tasmania: a possible subsea-floor replacement volcanic-hosted massive sulfide deposit. Econ Geol 87:931–952

Kim J, Lee I, Lee K-Y (2004) S, Sr, and Pb isotopic systematics of hydrothermal chimney precipitates from the eastern Manus Basin, western Pacific: evaluation of magmatic contribution of hydrothermal system. J Geophys Res 109:B12210

Knauth LP, Beeunus MA (1986) Isotope geochemistry of fluid inclusions in Permian halite with implications for the isotopic history of ocean water and the origin of saline formation waters. Geochim Cosmochim Acta 50:419–433

Kowalik J, Rye RO, Sawkins FJ (1981) Stable-isotope study of the Buchans, Newfoundland, polymetallic sulphide deposits. Geol Assoc Can Spec Pap 22:229–254

LaFlamme C, Hollis S, Jamieson J, Fiorentini M (2018) Three-dimensional spatially-constrained sulfur isotopes highlight processes controlling sulfur cycling in the near surface of the Iheya North hydrothermal system, Okinawa Trough. Geochem Geophys Geosys 19:2798–2812

LaFlamme C, Barré G, Fiorentini ML, Beaudoin G, Occhipinti S, Bell J (2021) A significant seawater sulfate reservoir at 2.0 Ga determined from multiple sulfur isotope analyses of the Paleoproterozoic Degrussa Cu-Au volcanogenic massive sulfide deposit. Geochim Cosmochim Acta 295:178–193

Large RR (1977) Chemical evolution and zonation of massive sulfide deposits in volcanic terrains. Econ Geol 72:549–572

Larson PB (1984) Geochemistry of the alteration pipe at the Bruce Cu-Zn volcanogenic massive sulfide deposit, Arizona. Econ Geol 79:1880–1896

Lerouge C, Deschamps Y, Joubert M, Bechu E, Fouillac AM, Castro JA (2001) Regional oxygen isotope systematics of felsic volcanics; a potential exploration tool for volcanogenic massive sulphide deposits in the Iberian Pyrite Belt. J Geochem Explor 72:193–210

LeHuray AP (1984) Lead and sulfur isotopes and a model for the origin of the Ducktown deposit, Tennessee. Econ Geol 79:1561–1573

Lesher CM, Goodwin AM, Campbell IH, Gorton MP (1986) Trace element geochemistry of ore-associated and barren, felsic metavolcanic rocks in the Superior Province, Canada. Can J Earth Sci 23:222–237

Leybourne MI, Peter JM, Kidder JA, Layton-Matthews D, Petrus JA, Rissmann CFW, Voinot A, Bowell R, Kyser TK (2022) Stable and radiogenic isotopes in the exploration for volcanogenic massive sulfide deposits. Can Mineral 60:433–468

Lode S, Piercey SJ, Layne GD, Piercey G, Cloutier J (2017) Multiple sulphur and lead sources recorded in hydrothermal exhalites associated with the Lemarchant volcanogenic massive sulphide deposit, central Newfoundland, Canada. Mineral Deposita 52:205–128

Lydon JW (1988) Ore deposit models #14. Volcanogenic massive sulphide deposits Part 2: Genetic models. Geosci Can 15:43–65

MacLachlan K, Dunning G (1998) U-Pb ages and tectono-magmatic evolution of Middle Ordovician volcanic rocks of the Wild Bight Group, Newfoundland Appalachians. Can J Earth Sci 35:998–1017

MacLean WH, Hoy LD (1991) Geochemistry of hydrothermally altered rocks at the Horne mine, Noranda, Quebec. Econ Geol 86:506–528

Martin AJ, McDonald I, Jenkin GRT, McFall KA, Boyce AJ, Jamieson JW (2021) A missing link between ancient and active mafic-hosted seafloor hydrothermal systems – magmatic volatile influx in the exceptionally preserved Mala VMS deposit, Troodos. Cyprus. Chem Geol 567:120127

McDermott JM, Ono S, Tivey MK, Seewald JS, Shanks WC III, Solow AR (2015) Identification of sulfur sources and isotopic equilibria in submarine hot-springs using multiple sulfur isotopes. Geochim Cosmochim Acta 160:169–187

Mercier-Langevin P, Dubé B, Hannington MD, Davis DW, Lafrance B, Gosselin G (2007) The LaRonde Penna Au-rich volcanogenic massive sulfide deposit, Abitibi Greenstone Belt, Quebec: Part I. Geology and Geochronology. Econ Geol 102:585–609

Mercier-Langevin P, Caté A, Ross P-S (2014) Whole-rock oxygen isotope mapping, Lalor auriferous VMS deposit footwall alteration zones, Snow Lake, west-central Manitoba (NTS 63K16). Report of Activities 2014, Manitoba Geological Survey, pp 94–103

Miller C, Halley S, Green G, Jones M (2001) Discovery of the West 45 volcanic-hosted massive sulfide deposit using oxygen isotopes and REE geochemistry. Econ Geol 96:1227–1237

Munhá J, Barriga FJAS, Kerrich R (1986) High ^{18}O ore-forming fluids in volcanic-hosted base metal massive sulfide deposits: geologic, $^{18}O/^{16}O$, and D/H evidence from the Iberian Pyrite Belt; Crandon, Wisconsin; and Blue Hill, Maine. Econ Geol 81:530–552

Occhipinti S, Hocking R, Lindsay M, Aitken A, Copp I, Jones J, Sheppard S, Pirajno F, Metelka V (2017) Paleoproterozoic basin development on the northern Yilgarn Craton, Western Australia. Precambrian Res 300:121–140

Ohmoto H (1986) Stable isotope geochemistry of ore deposits. Rev Mineral 16:491–559

Ohmoto H (1996) Formation of volcanogenic massive sulfide deposits: the kuroko perspective. Ore Geol Rev 10:135–177

Ohmoto H, Rye D (1979) Isotopes of sulfur and carbon. In: Barnes HL (ed) Geochemistry of hydrothermal ore deposits, 2nd edn. Wiley, New York, pp 509–567

Ohmoto H, Mizukami M, Drummond SE, Eldridge CS, Pisutha-Arnond V, Lenagh TC (1983) Chemical processes in Kuroko formation. Econ Geol Mon 5:570–604

Ono S, Shanks WC, Rouxel OJ, Rumble D (2007) S-33 constraints on the seawater sulfate contribution in modern seafloor hydrothermal vent sulfides. Geochim Cosmochim Acta 71:1170–1182

Paradis S, Taylor BE, Watkinson DH, Jonasson IR (1993) Oxygen isotope zonation and alteration in the northern Noranda district, Quebec: evidence for hydrothermal fluid flow. Econ Geol 88:1512–1525

Peter JM (2003) Ancient iron formations: their genesis and use in the exploration for stratiform base metal sulphide deposits, with examples from the Bathurst Mining Camp. Geotext 4:145–176

Peters M, Strauss H, Farquhar J, Ockert C, Eickmann B, Jost CL (2010) Sulfur cycling at the Mid-Atlantic Ridge: a multiple sulfur isotope approach. Chem Geol 268:180–196

Petersen S, Lehmann B, Murton BJ (2018) Modern seafloor hydrothermal systems: new perspective on ancient ore-forming processes. Elements 14:307–312

Piercey SJ (2011) The setting, style, and role of magmatism in the formation of volcanogenic massive sulfide deposits. Mineral Deposita 46:449–471

Pilote J-L, Piercey SJ, Mercier-Langevin P (2014) Stratigraphy and hydrothermal alteration of the Ming Cu-Au volcanogenic massive-sulphide deposit, Baie Verte Peninsula, Newfoundland. Geol Surv Can Curr Res 2014–7, 21 pp

Pilote J-L, Piercey SJ, Mercier-Langevin P (2017) Volcanic and structural reconstruction of the deformed and metamorphosed Ming volcanogenic massive sulfide deposit, Canada: implications for ore zone geometry and metal distribution. Econ Geol 112:1305–1332

Pisutha-Arnond V, Ohmoto H (1983) Thermal history, and chemical and isotopic compositions of the ore-forming fluids responsible for the kuroko massive sulfide deposits in the Hokuroku district of Japan. Econ Geol Mon 5:523–558

Quesnel B, Scheffer C, Beaudoin G (2023) The light stable isotope (H, B, C, N, O, Si, S) composition of orogenic gold deposits. In: Huston DL, Gutzmer J (eds) Isotopes in economic geology, metallogenesis and exploration, Springer, Berlin, this volume

Relvas JMRS, Barriga FJAS, Longstaffe F (2006) Hydrothermal alteration and mineralization in the Neves-Corvo volcanic-hosted massive sulfide deposit, Portugal: II. Oxygen, hydrogen and carbon isotopes. Econ Geol 101:791–804

Riverin G, Hodgson CJ (1980) Wall-rock alteration at the Millenbach Cu-Zn mine, Noranda, Quebec. Econ Geol 75:424–444

Rye RO, Bethke PM, Wasserman MD (1992) The stable isotope geochemistry of acid sulfate alteration. Econ Geol 87:225–262

Sangster DF (1968) Relative sulphur isotope abundances of ancient seas and strata-bound sulphide deposits. Proc Geol Assoc Can 19:79–91

Sangster DF, Scott SD (1976) Precambrian stratabound, massive Cu-Z-Pb sulfide ores of North America. In: Wolf KH (ed) Handbook of strata-bound and stratiform ore deposits. Elsevier, Amsterdam, pp 129–222

Sasaki A (1970) Seawater sulfate as a possible determinant for sulfur isotopic compositions of some strata-bound sulfide ores. Geochem J 4:41–51

Shanks WC III (2001) Stable isotopes in seafloor hydrothermal systems: vent fluids, hydrothermal deposits, hydrothermal alteration, and microbial processes. Rev Mineral Geochem 43:469–526

Shanks WC III (2014) Stable isotope geochemistry of mineral deposits. Treatise Geochem, 2nd edn 13:59–85. https://doi.org/10.1016/B978-0-08-095975-7.01103-7

Sharman ER, Taylor BE, Minarik WG, Dubé B, Wing BA (2015) Sulfur isotope and trace element data from ore sulfides in the Noranda district (Abitibi, Canada): implications for volcanogenic massive sulfide deposit genesis. Mineral Deposita 50:591–606

Sheppard SMF (1986) Characterization and isotopic variations in natural waters. Rev Mineral 16:165–183

Shinjo R, Kato Y (2000) Geochemical constraints on the origin of bimodal magmatism at the Okinawa Trough, an incipient backarc basin. Lithos 54:117–137. https://doi.org/10.1016/S0024-4937(00)00034-7

Sillitoe RH, Hannington MD, Thompson JFH (1996) High sulfidation deposits in the volcanogenic massive sulfide environment. Econ Geol 91:204–212

Solomon M, Rafter TA, Jensen ML (1969) Isotope studies on the Rosebery, Mount Farrell and Mount Lyell ores, Tasmania. Mineral Deposita 4:172–199

Solomon M, Eastoe CJ, Walshe JL, Green GR (1988) Mineral deposits and sulfur isotope abundances in the Mount Read Volcanics between Que River and Mount Darwin. Econ Geol 83:1307–1328

Stein CA (1995) Heat flow and hydrothermal circulation. Geophys Mon 91:425–445

Strauss H (2004) 4 Ga of seawater evolution: evidence from the sulfur isotopic composition of sulfate. Geol Soc Am Spec Pap 379:195–205

Taylor BE (1987) Stable isotope geochemistry of ore-forming fluids. Mineral Assoc Can Short Course Ser 13:337–445

Taylor BE (1992) Degassing of H_2O from rhyolitic magma during eruption and shallow intrusion, and the isotopic composition of magmatic water in hydrothermal systems. Geol Surv Jpn Rep 279:190–194

Taylor BE, Holk GJ (2002) Stable applications in the exploration for volcanic-associated massive sulphide deposits: a preliminary summary. Geol Surv Can Open File 4431:41–44

Taylor BE, Huston DL (1999) Regional ^{18}O zoning and hydrogen isotope studies in the Kidd Creek Volcanic Complex, Timmins, Ontario. Econ Geol Mon 10:351–378

Taylor HP Jr, Sheppard SMF (1986) Igneous rocks: I. Processes of isotopic fractionation and isotope systematics. Rev Mineral 16:227–272

Taylor BE, South BC (1985) Regional stable isotope systematics of hydrothermal alteration and massive sulfide deposition in the West Shasta district, California. Econ Geol 80:2149–2163

Taylor BE, Timbal A (2002) Regional stable isotope studies in the Noranda Volcanic. Geol Surv Can Open File 4431:243–252

Taylor BE, de Kemp E, Grunsky E, Martin L, Rigg D, Goutier J, Lauzière K, Dubé B (2014) 3-D visualization of the Archean Horne and Quemont Au-bearing VMS hydrothermal systems, Blake River Group, Québec. Econ Geol 109:183–203

Taylor BE, Peter JM, Laakso K, Rivard B (2015) Oxygen isotope zonation about the Izok Ag-VMS deposit, Slave Province, Nunavut: hanging-wall vector to mineralization. Geol Surv Can Open File 7853:27–44

Toman H (2013) Geology and metallogeny of north-central Newfoundland and the Little Deer VMS deposit. Unpublished M.Sc. thesis, Memorial University of Newfoundland, 184 p

Urabe T, Scott SD (1983) Geology and footwall alteration of the South Bay massive sulphide deposit, northwestern Ontario, Canada. Can J Earth Sci 20:1862–1879

Urabe T, Scott SD, Hattori K (1983) A comparison of footwall-rock alteration and geothermal systems beneath some Japanese and Canadian volcanogenic massive sulfide deposits. Econ Geol Mon 5:345–364

Walshe JL, Solomon M (1981) An investigation into the environment of formation and the volcanic-hosted Mt. Lyell copper deposits using geology, mineralogy, stable isotopes, and a six-component chlorite solid solution model. Econ Geol 76:246–284

Werner RA, Brand WA (2001) Referencing strategies and techniques in stable isotope ratio analysis. Rapid Commun Mass Spectrom 15:501–519

Whitehouse MJ (2013) Multiple sulfur isotope determination by SIMS: evaluation of reference sulfides for $\Delta^{33}S$ with observations and a case study on the determination of $\Delta^{36}S$. Geostand Geoanal Res 37:19–33

Williams N (2023) Stable isotope studies of the clastic-dominated lead-zinc (CD Pb-Zn) mineralization and their implications for ore genesis and exploration. In : Huston DL, Gutzmer J (eds) Isotopoes in economic geology, meallogenesis and exploration, Springer, Berlin, this volume

Zheng Y, Gu L, Tang XQ, Liu SH (2011) Oxygen isotope characteristics of the footwall alteration zones in the Hongtoushan volcanogenic massive sulfide deposit, Liaoning Province, China and restoration of their formation temperatures. Acta Geol Sin 85:683–693

Zierenberg RA, Shanks WC III (1988) Isotopic studies of epigenetic features in metalliferous sediment, Atlantis II Deep, Red Sea. Can Mineral 26:737–753

The Light Stable Isotope (Hydrogen, Boron, Carbon, Nitrogen, Oxygen, Silicon, Sulfur) Composition of Orogenic Gold Deposits

Benoît Quesnel, Christophe Scheffer, and Georges Beaudoin

Abstract

Orogenic gold deposits formed in various terranes of most ages since the Paleoarchean and generally consist of quartz veins hosted in shear zones formed at the ductile brittle transition under greenschist to lower amphibolite metamorphic conditions. Vein mineralogy is dominated by quartz with various amounts of silicates, carbonates, phyllosilicates, borates, tungstates, sulfides, and oxides. The isotopic composition of these minerals and fluid inclusions has been investigated since the 1960s to constrain the characteristics of orogenic fluid systems involved in the formation of gold deposits worldwide. This review is based on 8580 stable isotope analyses, including $\delta^{18}O$, δD, $\delta^{13}C$, $\delta^{34}S$ $\delta^{15}N$, $\delta^{11}B$, and $\delta^{30}Si$ values, from 5478 samples from 558 orogenic gold deposits reported in the literature from 1960 to 2010. This contribution describes the variability of the light stable isotopic systems as function of the minerals, the age of the deposits, their regional setting, and their country rocks. The temperature of isotopic equilibrium of orogenic gold veins is estimated from mineral pairs for oxygen and sulfur isotopes. Based on these temperatures, and on fractionation between mineral and fluid components (H_2O, CO_2 and H_2S), the isotopic composition of fluids is estimated to better constrain the main parameters shared by most of auriferous orogenic fluid systems. Orogenic gold deposits display similar isotopic features through time, suggesting that fluid conditions and sources leading to the formation of orogenic gold deposits did not change significantly from the Archean to the Cenozoic. No consistent secular variations of mineral isotope composition for oxygen ($-8.1‰ \leq \delta^{18}O$ 33‰, n = 4011), hydrogen ($-187‰ \leq \delta D$ −4‰, n = 246), carbon ($-26.7‰ \leq \delta^{13}C$ 12.3‰, n = 1179), boron ($-21.6‰ \leq \delta^{11}B$ 9‰, n = 119), and silicon ($-0.5‰ \leq \delta^{30}Si$ 0.8‰, n = 33) are documented. Only nitrogen ($1.6‰ \leq \delta^{15}N \leq 23.7‰$, n = 258) and sulfide sulfur from deposits hosted in sedimentary rocks ($-27.2‰ \leq \delta^{34}S \leq 25‰$, n = 717) display secular variations. For nitrogen, the change in composition is interpreted to record the variation of $\delta^{15}N$ values of sediments devolatilized during metamorphism. For sulfur, secular variations reflect

Supplementary Information The online version contains supplementary material available at https://doi.org/10.1007/978-3-031-27897-6_10.

B. Quesnel · C. Scheffer · G. Beaudoin (✉)
Département de Géologie et Génie Géologique, Centre de recherche sur la géologie et l'ingénierie des ressources minérales (E4m), Université Laval, Québec G1V 0A6, Canada
e-mail: georges.beaudoin@ggl.ulaval.ca

D. Huston and J. Gutzmer (eds.), *Isotopes in Economic Geology, Metallogenesis and Exploration*, Mineral Resource Reviews, https://doi.org/10.1007/978-3-031-27897-6_10

incorporation of local sedimentary sulfur of ultimate seawater origin. No significant variation of temperature of vein formation is documented for orogenic gold deposits of different ages. Quartz-silicate, quartz-carbonate and sulfide-sulfide mineral pairs display consistent temperatures of 360 ± 76 °C (1σ; n = 332), in agreement with the more common greenschist facies hostrocks and fluid inclusion microthermometry. Fluid sources for orogenic gold deposits are complex but the isotopic systems (hydrogen, boron, carbon, nitrogen, oxygen, sulfur) are most consistent with contributions from metamorphic fluids released by devolatilization of igneous, volcano-sedimentary and/or sedimentary rocks. The contribution of magmatic water exsolved from magma during crystallization is not a necessary component, even if permissible in specific cases. Isotopic data arrays can be interpreted as the result of fluid mixing between a high T (~550 °C)—high $\delta^{18}O$ (~10‰)—low δD (~−60‰) deep-seated (metamorphic) fluid reservoir and a low T (~200 °C)—low $\delta^{18}O$ (~2‰)—high δD (~0‰) upper crustal fluid reservoir in a number of orogenic gold deposits. The origin of the upper crustal fluid is most likely sea- or meteoric water filling the host rock porosity, with a long history of water–rock isotope exchange. Mixing of deep-seated and upper crustal fluids also explains the large variation of tourmaline $\delta^{11}B$ values from orogenic gold veins. Regional spatial variations of oxygen and hydrogen isotope compositions of deep-seated fluid reservoirs are documented between orogenic gold districts. This is the case for the Val-d'Or (Abitibi), Coolgardie and Kalgoorlie (Yilgarn) where the oxygen isotope composition of the deep-seated fluid end-member is 4‰ lower compared to that from the Timmins, Larder Lake, and Kirkland Lake districts (Abitibi). However, both mixing trends converge towards a common, low $\delta^{18}O$ upper crustal fluid end-member. Such variations cannot be related to fluid buffering at the site of deposition and suggest provinciality of the fluid source. The contribution of meteoric water is mainly recorded by fluid inclusions from Mesozoic and Cenozoic age deposits, but micas are not systematically in isotopic equilibrium with fluid inclusions trapped in quartz from the same vein. This suggests late involvement of meteoric water unrelated to deposit formation. Yet, a number of deposits with low δD mica may record infiltration of meteoric water in orogenic gold deposits. Isotope exchange between mineralizing fluid and country rocks is documented for oxygen, carbon, sulfur and silicon isotopes. Large variations (> 10‰) of sulfide $\delta^{34}S$ values at the deposit scale are likely related to evolving redox conditions of the mineralizing fluid during reaction with country rocks. Deposits hosted in sedimentary rocks show a shift to higher $\delta^{18}O$ values as a result of fluid/rock oxygen exchange with the regional sedimentary country rocks.

Keywords

Stable isotope · fractionation · temperature · fluid source · metamorphic crustal · hydrogen · oxygen · boron · carbon · nitrogen · sulfur · silicon

1 Introduction

Orogenic gold deposits comprise a diverse class of mineral deposits that contain the second largest resource of gold, after the Witwatersrand (Republic of South Africa) paleoplacers (Dubé and Gosselin 2007). In addition to their economic importance, orogenic gold deposits record important geological process active at mid-crustal levels, where the deformation behavior of rocks changes, with increasing depth, from brittle to ductile. The formation of orogenic gold deposits involves advection of significant volumes of crustal fluids, which carry large amounts of dissolved mass and energy at the scale of the crust.

The class of orogenic gold deposits was proposed by Groves et al. (1998) to combine gold deposits that shared numerous characteristics, yet had remained grouped under different deposit types in literature. The orogenic gold deposits were defined to include a range of epigenetic gold styles variably named "mesothermal", "quartz-carbonate", "lode gold", "greenstone gold", "Mother Lode", "turbidite-hosted", "BIF (Banded-Iron Formation)-hosted", etc. The class of orogenic gold deposits was defined to include deposits where gold is the major economic metal, and which formed in volcanic and/or sedimentary, accretionary, metamorphic terranes in convergent settings of all ages. The geological characteristics of orogenic gold deposits have been reviewed by Goldfarb et al. (2005) and Goldfarb and Groves (2015) amongst others, and are briefly summarized hereafter. Orogenic gold deposits consist of gold-bearing quartz-carbonate veins or disseminated sulfides, commonly representing less than 5 vol.% sulfides, most commonly pyrite and arsenopyrite. The veins display a range of structural features from foliated shear bands to brittle breccia and are characteristically associated to steeply-dipping reverse faults, related to major crustal shear zones typically marking terrane boundaries. The country rocks are varied and include almost all types of volcanic, plutonic, sedimentary rocks and their metamorphosed equivalents. Hydrothermal alteration is characteristically narrow and controlled by the vein structures, replacing peak-metamorphic minerals by white mica, Fe–Mg carbonate minerals and sodic feldspar, being the most common. The veins formed from low-salinity (3–12 wt.% NaCl eq.) carbonic-aqueous hydrothermal fluids with $X_{CO_2} < 0.3$, at temperatures typically about 350 °C. Orogenic gold deposits formed at depths of less than 20 km, typically near the base of the seismogenic continental crust, under high, to supra-lithostatic, fluctuating fluid pressure. Deposits formed deeper in the crust are characterized by ductile deformation, with gold mineralization mostly in disseminated sulfide minerals or less commonly in foliated veins, and formed at the higher temperature range. Those formed higher in the crust are characterized by more brittle structures, and a lower temperature of formation.

In spite of these well-defined geological characteristics, the origin of hydrothermal fluids that formed orogenic gold deposits remains controversial. Goldfarb and Groves (2015) reviewed sources for fluids in orogenic gold deposits, stating that the stable isotope composition of the minerals is inconsistent between deposits, and leads to equivocal interpretations about the source of fluids and dissolved components. This is particularly intriguing since the auriferous hydrothermal fluids are considered to have a well-defined composition that should enable determination of the source(s) of fluids. Goldfarb and Groves (2015) reiterated a long-held view that fluid-rock exchange, at the site of reaction, could mask part of the original isotopic signature (Ridley and Diamond 2000). If the association of the orogenic gold deposits with a major shear zone is a well-established fact, the role of the shear zone has been contested from being a deep-seated fluid channelling feature (e.g., McCuaig and Kerrich 1998) to a breach into a mid-crustal low permeability barrier (Beaudoin et al. 2006).

The first major compilation of oxygen, H, carbon, sulfur stable isotope data for Au–Ag deposits in metamorphic rocks, now defined largely as orogenic gold deposits, was undertaken by Kerrich (1987). This led to the concept of provinciality in the oxygen isotope composition of orogenic gold deposits, inferred to indicate large and uniform fluid source rock volumes, but different for each orogenic gold vein field (McCuaig and Kerrich 1998). High $\delta^{18}O$ values are characteristic of the auriferous fluids forming orogenic gold deposits, with a range of $\delta^{18}O_{H2O}$ values of 6 to 11‰ for Precambrian deposits (McCuaig and Kerrich 1998) and of 7 to 13‰ for Phanerozoic deposits (Bierlein and Crowe 2000), and as shown by the range of quartz $\delta^{18}O$ values of 6.9 to 25‰ for the 25 largest orogenic gold deposits (Goldfarb et al. 2005).

Goldfarb et al. (2005) compiled δD values ranging from −150 to −26‰ for mica and from −98 to −24‰ for fluid inclusions from the 25 largest orogenic gold deposits. Using the likely temperatures of isotopic equilibrium, the hydrothermal fluids δD values were suggested to range from −80 to −20‰, excluding low values from fluid inclusions of dubious or mixed origins

(McCuaig and Kerrich 1998; Goldfarb et al. 2005). Combined with the estimated fluid $\delta^{18}O$ values of 6 to 14‰, this has been taken as a strong indication for a metamorphic source for the auriferous fluids (McCuaig and Kerrich 1998).

The carbon isotope composition of hydrothermal carbonates associated with gold mineralization was shown to range mostly between −10 and 0‰, with outlier values as low as −32‰ (Goldfarb et al. 2005). McCuaig and Kerrich (1998) also remarked the narrow range in $\delta^{13}C$ values for a given deposit. The isotope composition of carbon has led to interpretation of a number of potential sources, from the mantle to magmas, to crustal devolatilization, mixing and redox reactions (McCuaig and Kerrich 1998). Likewise, the sulfur isotope composition of sulfides are summarized to mostly range from 0 to 9‰, but with a total range from −20 to 25‰, interpreted to reflect local sulfur sources (Goldfarb et al. 2005).

Despite the low variance in geological features, and in the chemical PVTX characteristics of auriferous fluids associated with orogenic gold deposits, the variation of the stable isotope composition of hydrothermal minerals and fluid inclusions has led to various, at times conflicting or equivocal, interpretations. Here, we attempt to address the variations in the stable isotope composition of hydrothermal minerals and fluid inclusions in orogenic gold deposits on the basis of a detailed analysis of an extensive compilation of data published between 1960 and 2010. The study of the database allows to outline major stable isotope compositional features for orogenic gold deposits to yield improved understanding of the sources of fluids, carbon, sulfur, nitrogen, boron and silicon for this deposit type.

2 Methodology

2.1 Database

The database integrates fifty years of stable isotope and orogenic gold deposit research published between 1960 and 2010, since which time data interpretation has been ongoing (Beaudoin 2011). Review of data published since 2010 shows no special features from more recent studies, but specific cases are discussed in relation to pre-2010 data. The first study of stable isotopes in orogenic gold systems was on the sulfur isotope composition of gold quartz veins of the Yellowknife district, Canada (Wanless et al. 1960). Because gold-bearing quartz ± carbonate veins have been described using a range of descriptive names from lode to quartz-carbonate to orogenic gold veins, the search in literature has used all known terms used to identify stable isotope research for this deposit type. The deposits compiled in the database comprise all the major orogenic gold deposits as defined by Goldfarb et al. (2005). Some of the deposits aggregated in the compilation have controversial metallogenic affinity. Nevertheless, the amount of data from deposits of controversial metallogenic affinity is small, such that it should not significantly impact on the overall stable isotope characteristics of orogenic gold deposits, unless otherwise noted. Data were compiled from 193 references, for which tabulated stable isotope values were reported (Supplementary Material Spreadsheet 1). Data reported in figures or as range of values, without individual mineral composition, were not considered in this study. The data were compiled dominantly from peer-reviewed journals, with the exception of a few major deposits for which data were only available in university theses.

The database contains 8580 stable isotope analyses for oxygen (n = 4014), hydrogen (n = 706), carbon (n = 1244), sulfur (n = 2215), nitrogen (n = 254), boron (n = 117) and silicon (n = 30) from 5478 samples (Supplementary Material Spreadsheet 1). Stable isotopic compositions are expressed in ‰, and are reported relative to VSMOW, VPDB, VCDT, AIR, NIST SRM 951, and NBS28 for $\delta^{18}O$ and δD, $\delta^{13}C$, $\delta^{34}S$, $\delta^{15}N$, $\delta^{11}B$ and $\delta^{30}Si$, respectively. Data reported in literature relative to SMOW, PDB and CDT are considered identical to the newer IAEA Vienna (V-) standards. Where samples have the same name, and several isotope or mineral analyses are reported, the results are deemed to be from the same sample even though textural equilibrium can seldom be verified.

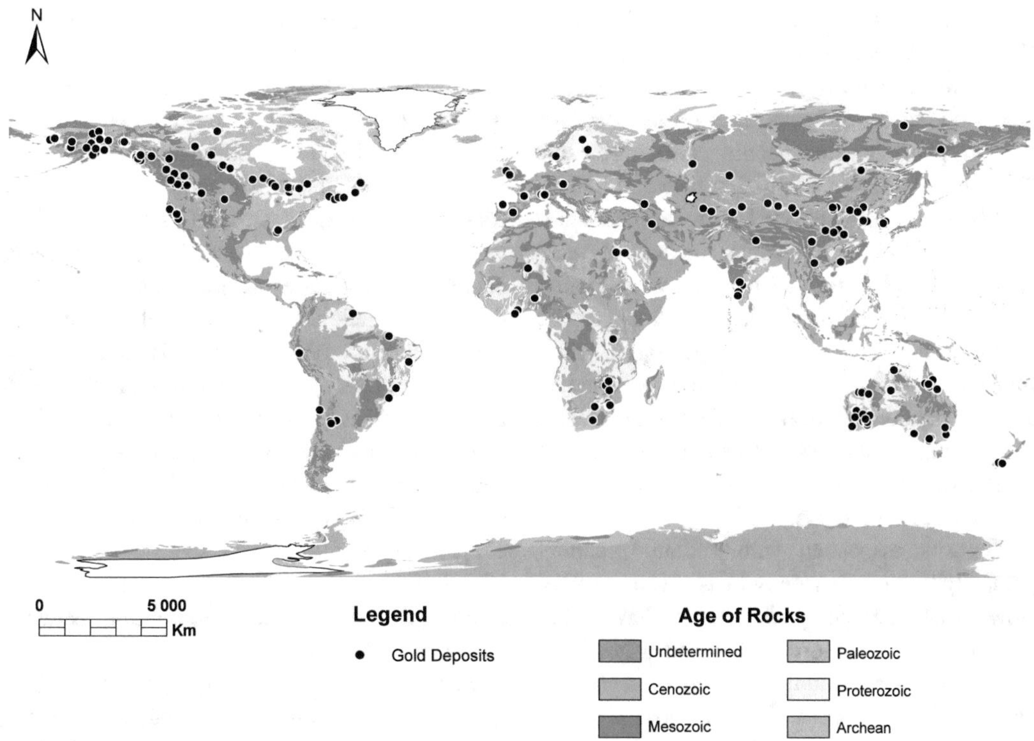

Fig. 1 Location of orogenic gold deposits with stable isotope data. The world geology map is prepared using digital data from Gosselin and Dubé (2005)

Oxygen isotope compositions have been measured in 22 minerals. Two studies report $\delta^{18}O$ values from fluid inclusions (FI) hosted by quartz or scheelite (Santosh et al. 1995; Billstrom et al. 2009), which are not considered further because the $\delta^{18}O$ values are most likely equilibrated with the host mineral oxygen. The hydrogen isotope composition has been measured in seven minerals and from FI in quartz. Mao et al. (2008) report seven δD values from "K-feldspar". Because the source of hydrogen in theses analyses is unspecified, but likely to be from FI, these data were not included in the database. Carbon isotopic compositions have been measured in seven minerals and in FI trapped in quartz and scheelite. Sulfur isotopic compositions have been measured in 13 minerals. Nitrogen and silicon isotopic composition have been measured in three and two minerals, respectively. Boron has been measured in tourmaline. The samples are from 564 deposits from all continents with the exception of Antarctica (Fig. 1), and range in age from the Archean to the Cenozoic, including 180 Archean, 68 Proterozoic, 104 Paleozoic, 129 Mesozoic, and 83 Cenozoic deposits. For Archean deposits, most of them are Neo-Archean (n = 164), with fewer Paleo- (n = 2) and Meso-Archean (n = 13). The age of the gold deposits is known with variable accuracy, such that geochronological data yield precise age for some deposits, whereas only age inferences can be made for others.

Gold deposits are hosted in a range of country rock types including sedimentary, volcano-sedimentary, or igneous rocks with variable chemical composition from felsic to ultramafic. For deposits hosted by several types of rocks, the dominant lithology/chemical composition is recorded in the database. Sedimentary rock-hosted deposits include unmetamorphosed clastic and/or chemical sedimentary rocks (greywackes, sandstones, shales, limestones, mudstones, argilites, BIF) or their metamorphosed equivalents (paragneiss, schist) which are all reported as "sedimentary" for simplicity. Volcano-sedimentary rock-hosted deposits are mainly represented by Archean and Proterozoic greenstone belts. Igneous rock-hosted deposits include plutonic or volcanic rocks. Volcanic rocks mainly consist of rhyolite to basaltic rocks in greenstone belts, as well as mafic and ultramafic dykes. Plutonic rocks include a wide variety of rocks such as tonalite-trondhjemite-granodiorite (TTG) batholiths as well as granitic—diorite intrusions in various geological environment. Based on the description provided in each publication, geochemical descriptors (felsic, intermediate, mafic and ultramafic) are used to summarize the composition of dominant intrusive and volcanic rocks.

Deposits associated with zeolite, prehnite-pumpellyite or sub-greenschists facies, and/or "low" metamorphic grade rocks, have been combined into a sub-greenschist category. Those hosted by lower, middle, or upper greenschist facies rocks have been classified into a greenschist category. Lower, middle, or upper amphibolite, and granulite metamorphic facies host rocks are categorized into a high-grade metamorphic category. The resulting categories must be used with caution since orogenic gold deposits commonly post-date regional metamorphism. Moreover, in numerous studies, the metamorphic grade is not accurately described or covers a range, such as "lower greenschist to amphibolite". In such cases, we arbitrarily attributed a deposit to the higher metamorphic grade. Finally, for most of deposits from the North China Craton, a metamorphic grade class is not attributed because the gold deposits formed during the Mesozoic, significantly younger than Precambrian regional granulite grade metamorphism.

2.2 Temperature of Equilibrium

There can be several isotopic fractionations for a pair of minerals (e.g. δ_{quartz}–$\delta_{mineral}$) (Beaudoin and Therrien 2009) such that the computed temperatures can vary depending on the selected fractionation. Oxygen isotope fractionations derived from previous studies by Vho et al. (2019) are used for temperature estimates from quartz-tourmaline, quartz-chlorite, quartz-biotite, quartz-muscovite, quartz-albite, quartz-plagioclase, quartz-feldspar, quartz-ankerite, quartz-calcite and quartz-dolomite mineral pairs and are reported with a 2σ confidence interval (Supplementary Material, Spreadsheet 2). For quartz-sericite,[1] no fractionation equation exists in literature and, consequently, we use the quartz-muscovite fractionation equation from Vho et al. (2019). For quartz-scheelite, only one fractionation equation is available (Zheng 1992), but it yields low temperatures (mainly between 110 and 200 °C), inconsistent with the greenschist facies conditions and other geothermometers. Consequently, a combination of two equations, quartz-H_2O from Clayton et al. (1972) and scheelite-H_2O from Wesolowski and Ohmoto (1986), has been used to estimate the temperature of equilibrium between quartz and scheelite.

For quartz-plagioclase, quartz-feldspar and quartz-carbonate pairs, no information about the composition of the mineral is provided. Consequently, data are not presented in δ_{quartz}–$\delta_{mineral}$ diagrams and temperatures have not been calculated.

For sulfide pairs, pyrite-galena, pyrite-chalcopyrite, pyrite-pyrrothite, pyrite-sphalerite, sphalerite-chalcopyrite and chalcopyrite-galena fractionation data are from Kajiwara and Krouse (1971). The fractionation from Liu et al. (2015) was used for sphalerite-galena. The data are plotted in $\delta_{mineral}$–$\delta_{mineral}$ diagrams overlain by isotherms ranging from 250 to 550 °C.

2.3 Oxygen and Hydrogen Isotope Composition of Water

Oxygen and hydrogen isotope compositions of H_2O are calculated from quartz and OH-bearing minerals pairs that yield equilibrium

[1] We use "sericite" herein as a general term for unspecified white mica, as this term is used in the cited reference, even though the term has been disapproved by the International Mineralogical Association. In most cases "sericite" refers to fined grained muscovite, phengite or related white micas.

temperatures in the range 250–550 °C. This temperature range is consistent with the accepted range in temperature of formation for orogenic gold deposits (Goldfarb and Groves 2015). Thus, it is presumed that those mineral pairs record O–H isotope equilibrium. The water isotopic composition has been calculated using various hydrogen mineral-H_2O fractionations and oxygen quartz-H_2O fractionation equation from Vho et al. (2019). The entire dataset is presented in Supplementary Material 2. For tourmaline and chlorite, the fractionations of Kotzer et al. (1993) and Graham et al. (1984) have been used, respectively. For biotite and muscovite (or sericite) the hydrogen fractionations from Suzuoki and Epstein (1976) were used.

2.4 Carbon Isotope Composition of CO_2

The carbon isotope composition of CO_2 was calculated using calcite-CO_2 and dolomite-CO_2 fractionations from Ohmoto and Rye (1979) assuming equilibrium temperatures in the range 250–550 °C as determined from quartz-calcite and quartz-dolomite oxygen fractionation data (Vho et al. 2019). Because there are no carbon ankerite-CO_2 fractionation in literature, we did not compute $\delta^{13}C$ of CO_2 from ankerite. The entire dataset is presented in Supplementary Material 2.

3 Stable Isotope Composition of Minerals and Fluid Inclusions

3.1 Oxygen Isotopes

3.1.1 Silicates, Tungstates, Borates and Oxides

The oxygen isotope composition of 2357 quartz samples from orogenic gold deposits ranges from 5.2 to 25.9‰ (Fig. 2). Quartz $\delta^{18}O$ values show a broad, bimodal, distribution with one mode near 11.5‰, and the second near 15‰ (Fig. 2).[2] Quartz from orogenic gold deposits formed during the Archean shows this bimodal distribution, whereas quartz from deposits formed during the Proterozoic are characterized by $\delta^{18}O$ values about the lower mode (11.6‰; Fig. 2a). Orogenic gold deposits formed during the Phanerozoic have quartz $\delta^{18}O$ values about the higher mode (15‰) and tailing towards the highest $\delta^{18}O$ values (Fig. 2a).

Figure 2b shows the distribution of quartz $\delta^{18}O$ values for orogenic terranes for which there is more than 50 analyses. Quartz from the Archean Yilgarn Craton (Australia) shows a normal distribution about the 11.5‰ mode, with few higher values. In contrast, quartz from the Archean Superior Craton (Canada) displays a bimodal distribution. Quartz from the Proterozoic Trans-Hudson Orogen (Canada) plots mostly about the 11.5‰ mode. Quartz from the Appalachian/Caledonian and Hercynian orogens (America and Europe) plots mostly near 13.5‰ whereas those from the Tasman Orogen (Australia) are characterized by higher values with a mode near 16.5‰. Mesozoic and Cenozoic orogenic gold deposits from the Cordillera (North America) are characterized by heavier quartz oxygen isotope compositions mainly between 12.5‰ and 19.5‰ (Fig. 2b). Mesozoic orogenic gold deposits from the Daebo Orogen and North China Craton (China, Korea) plot mostly near 13‰.

Figure 2c shows the distribution of all quartz $\delta^{18}O$ values as function of the principal country rock type. Quartz hosted in igneous rocks shows the lowest $\delta^{18}O$ values with a normal distribution centered at 11.5‰. Deposits hosted in sedimentary country rocks have higher $\delta^{18}O$ values with a normal distribution centered at 15‰. The $\delta^{18}O$ values for deposits hosted in volcano-sedimentary rocks display a bimodal distribution with modes centered at 12‰ and 15‰. Most of the high $\delta^{18}O$ values are from deposits hosted by sedimentary and volcano-sedimentary rocks. The distribution of $\delta^{18}O$ values for quartz hosted in igneous rocks, divided by geochemical affinity, is presented on

[2] The character (uni-, bi- or polymodal) of the distribution and the location of the modes were determined visually. Modes refer to the class with the most observations.

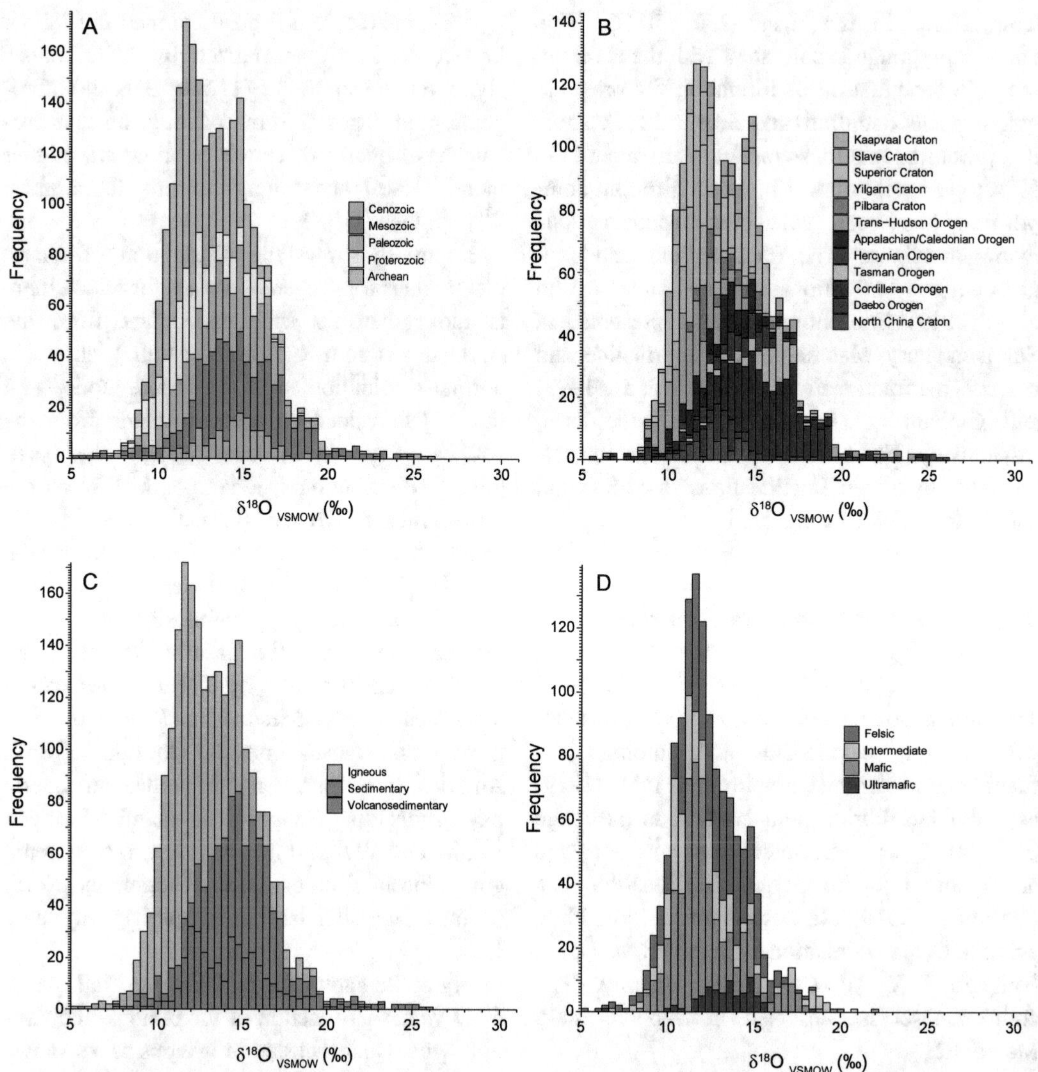

Fig. 2 Histograms showing $\delta^{18}O$ values of quartz forming veins from orogenic gold deposits based on: **a** the age of deposits; **b** the craton or orogen hosting the deposits (only craton or orogen with more than 50 $\delta^{18}O$ values are shown); **c** country rock of deposits; and **d** the geochemistry of igneous rocks hosting the deposits

Fig. 2d. No major difference is identified between mafic and felsic country rocks, both showing a unimodal distribution about the major 11.5‰ mode. Vein quartz hosted in ultramafic rocks ranges between 11.5‰ and 15.5‰, without a well-defined mode.

Quartz from orogenic gold districts from the Archean Superior Craton displays distinct oxygen isotopic compositions. The lower $\delta^{18}O$ values, between 9‰ and 14‰, with a mode centered on 11‰ are characteristic of the Val-d'Or district, whereas the higher $\delta^{18}O$ values between 13‰ and 16‰, with a mode at 15‰ are mostly from the Timmins district (Fig. 3a). There is no grouping of $\delta^{18}O$ values for other districts perhaps because of the smaller number of analyses. The difference in $\delta^{18}O$ values between the Val-d'Or and the Timmins districts correlates with the distribution of $\delta^{18}O$ values as function of the type of country rocks (Figs. 3a, b). Quartz from veins hosted in igneous rocks have $\delta^{18}O$ values ranging from 5 to 22‰ with a main mode

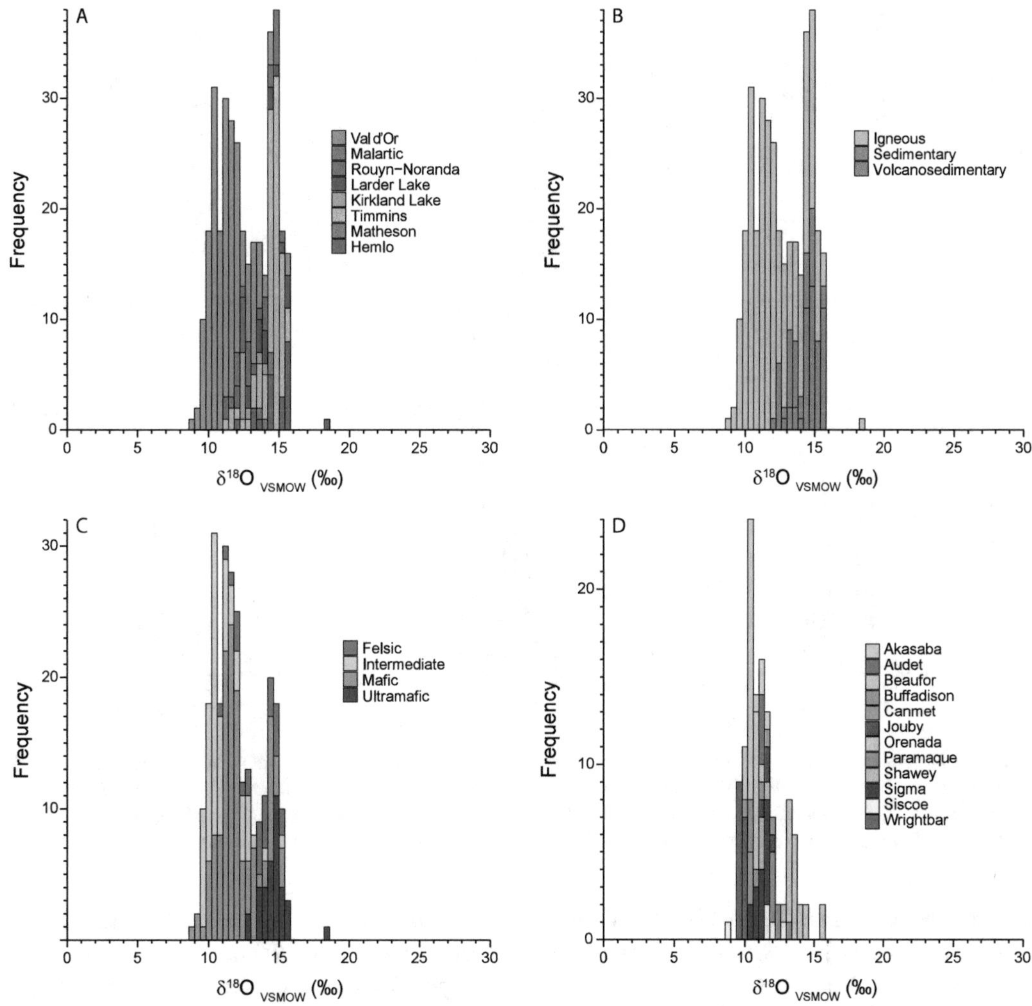

Fig. 3 Histograms showing $\delta^{18}O$ values of quartz forming veins from orogenic gold deposits from the Superior Craton based on: **a** district; **b** country rocks; **c** geochemistry of igneous rocks hosting deposits; and **d** deposits from the Val-d'Or district

at 11‰, whereas those hosted in sedimentary displays higher $\delta^{18}O$ values, between 5 and 27‰, with a mode at 14.5‰, and those hosted in volcano-sedimentary rocks have values between 7 and 24‰, with modes at 11.5 and 14.5‰. As shown in Fig. 3c, the igneous rocks hosting deposits with high quartz $\delta^{18}O$ values are mainly ultramafic.

For the Yilgarn Craton, quartz $\delta^{18}O$ values range mainly between ~10 and ~14‰, with no distinction between districts (Fig. 4a). Most of these veins are hosted in mafic igneous rocks and the few hosted in volcano-sedimentary rocks have similar $\delta^{18}O$ values (Figs. 4b, c). In districts from the Cordilleran Orogen for which there are more than 20 analyses, quartz displays $\delta^{18}O$ values mostly between ~12 and ~20‰ (Fig. 4d) higher than those documented from the Superior and Yilgarn cratons. In the Cordilleran Orogen, the Fairview, Klondike and Bridge River districts (12–14‰, 14–16‰ and 18–20‰, respectively) in Canada, the Berners Bay (12 to 16‰) and Mother Lode districts (16–19‰; United States of America—USA) display distinct ranges of $\delta^{18}O$ values. In contrast, the $\delta^{18}O$ values from the Nome district (USA) define a

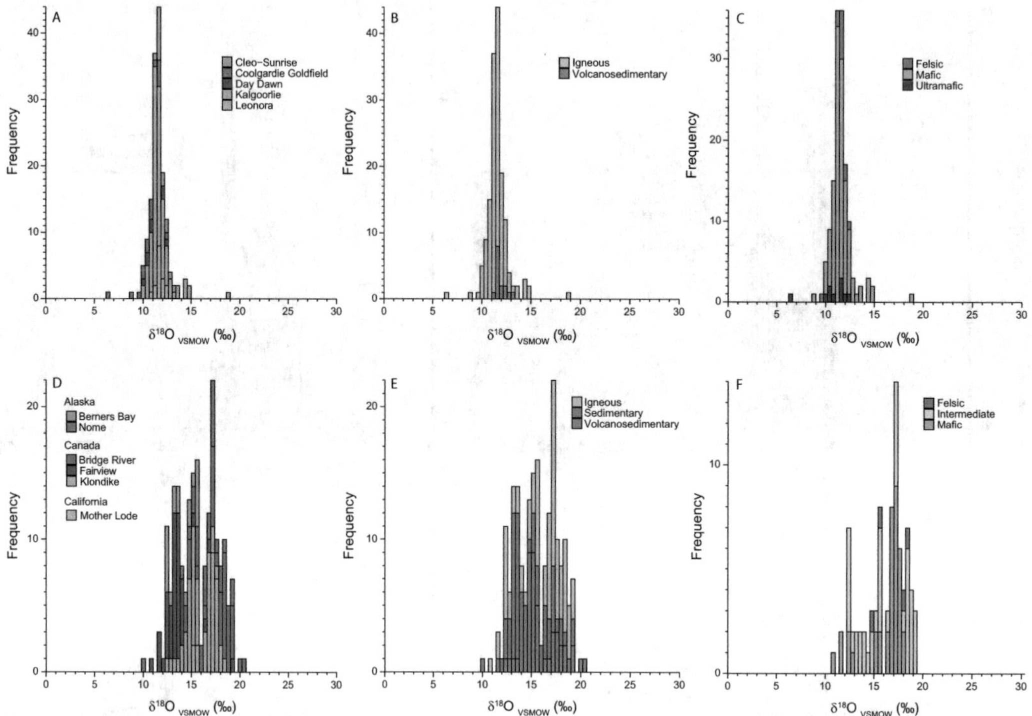

Fig. 4 Histograms showing δ^{18}O values of quartz forming veins from orogenic gold deposits from the Yilgarn Craton and Cordilleran Orogen based on: **a** district (Yilgarn Craton); **b** country rocks (Yilgarn Craton); **c** geochemistry of igneous rocks hosting deposits (Yilgarn Craton); **d** districts (Cordilleran Orogen); **e** country rocks (Cordilleran Orogen); and **f** geochemistry of igneous rocks hosting deposits (Cordilleran Orogen)

large range of values (10 to 20‰). All these districts are hosted in volcano-sedimentary and sedimentary rocks but in some of the districts, major deposits are hosted by granite (Fig. 4e). Data do not vary with the chemistry of the igneous rocks (Fig. 4f).

At the district scale, quartz from orogenic gold deposits is characterized by narrow ranges of δ^{18}O values (Fig. 3d). For example, in the Val-d'Or vein field, quartz from the Beaufor deposit is characterized by quartz oxygen isotope compositions in the range 9.9–11.1‰, with an average of 10.6 ± 0.2‰ (Beaudoin and Pitre 2005). Few studies allow comparison of several deposits in one district. In the Val-d'Or vein field, Beaudoin and Pitre (2005) and Beaudoin and Chiaradia (2016) showed no systematic variation in composition across vein thickness, along strike, or down-dip. In Val-d'Or, there is a gradual change in the average composition of quartz of a deposit at the scale of the district (Beaudoin and Pitre 2005).

Figure 5 presents the oxygen isotope composition of other silicates, tungstate, borates and oxides. Muscovite (n = 84) and sericite (n = 70) have δ^{18}O values ranging from 7.0 to 17.0‰ and 6.6 to 16.0‰, respectively (Fig. 5a), with one muscovite having a low δ^{18}O value of 0.7‰. Chlorite (n = 87) and biotite (n = 45) have similar δ^{18}O values ranging from 1.0 to 12.0‰ and from 2.0 to 12.6‰, respectively. Two talc samples yield δ^{18}O values of 7.0 and 7.9‰ (Fig. 5a). Scheelite (n = 54) has a small range of δ^{18}O values from 2.0 to 6.6‰, whereas tourmaline (n = 40) displays a wider range of δ^{18}O values from 4.9 to 11.0‰ (Fig. 5b). Actinolite (n = 19) has δ^{18}O values from 5.0 to 11.5‰, whereas two epidote samples yield δ^{18}O values of 7.7 and 9.6‰ (Fig. 5b). Undifferentiated feldspar (n = 20) and plagioclase (n = 20) have δ^{18}O

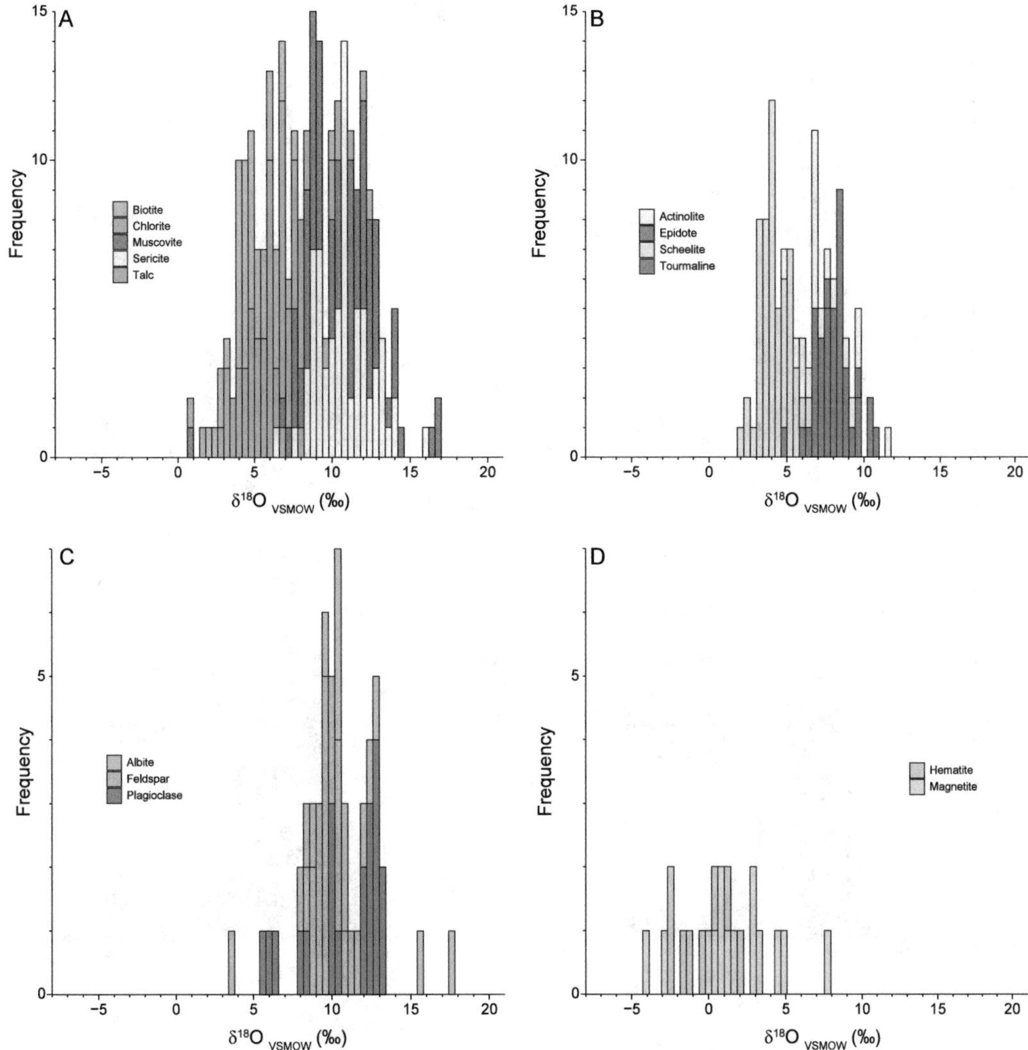

Fig. 5 Histograms showing $\delta^{18}O$ values of silicates and oxides minerals in deposits: **a** biotite, chlorite, muscovite, sericite and talc; **b** actinolite, epidote, scheelite, and tourmaline: **c** albite, feldspar and plagioclase; and **d** hematite and magnetite

values ranging from 3.6 to 11.3‰ and from 5.5 to 13.4‰, respectively, whereas albite (n = 17) displays $\delta^{18}O$ values from 8.3 to 17.8‰ (Fig. 5 c). Magnetite (n = 14) and hematite (n = 8) have low $\delta\ ^{18}O$ values from −4.1 to 7.6‰ and from −2.3 to 2.2‰, respectively (Fig. 5d).

3.1.2 Carbonates

Figure 6 shows the oxygen isotope composition of carbonate minerals, including calcite (n = 493), dolomite (n = 345), ankerite (n = 207), siderite (n = 53), magnesite (n = 8) and undifferentiated carbonate minerals (n = 42). Carbonate $\delta^{18}O$ values show three modes, with two principal modes centered on 10‰ and 12.5‰ (Fig. 6a), and a third mode centered on 24‰. Ankerite, calcite and dolomite have $\delta^{18}O$ values ranging from 7.7 to 25.0‰, from 6.3 to 27.5‰, and from 1.2 to 25.3‰, respectively, with the same bimodal distribution for ankerite and calcite. Dolomite displays a bimodal distribution with two modes centered on 12‰ and 16‰. Siderite has $\delta^{18}O$

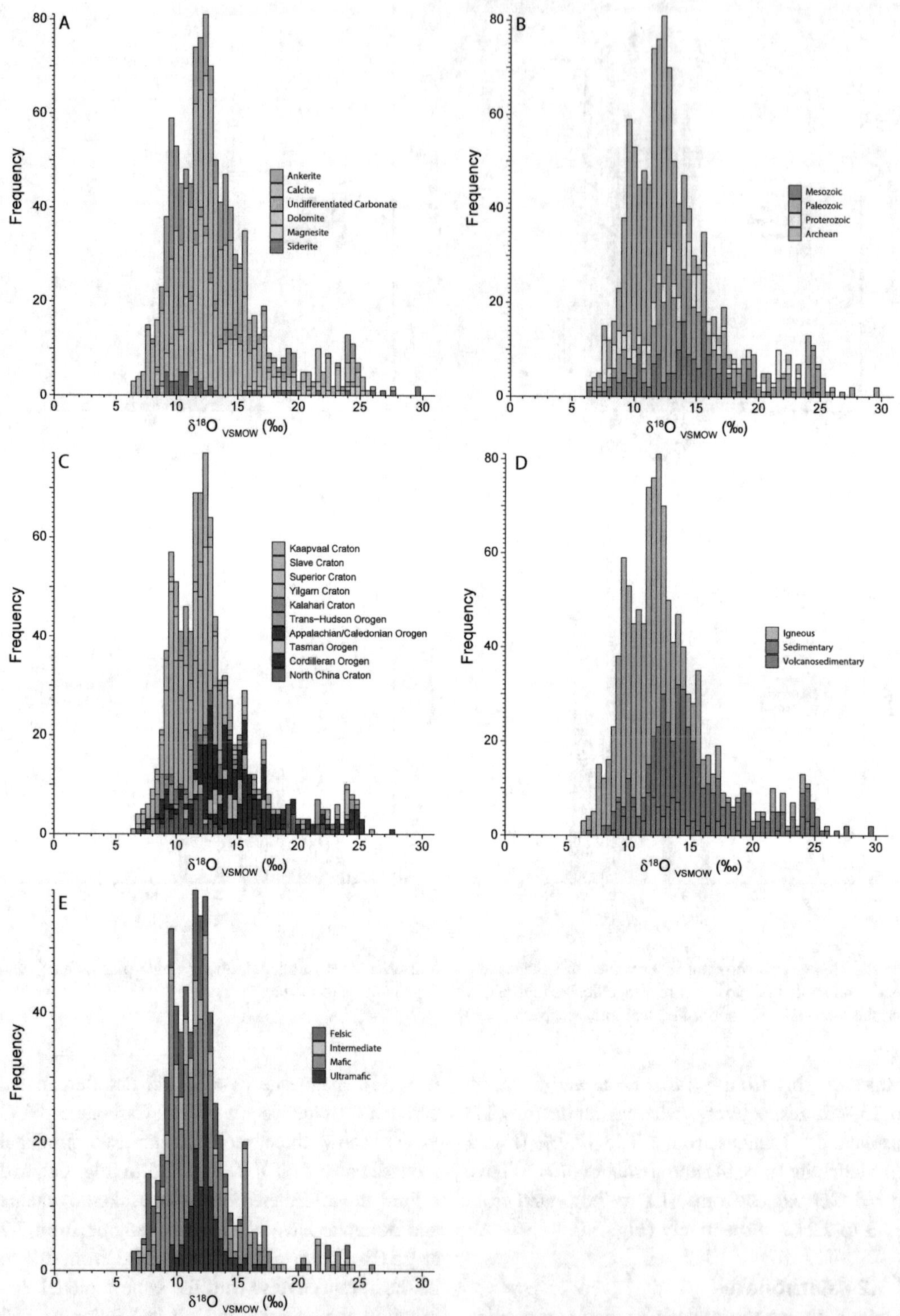

Fig. 6 Histograms showing $\delta^{18}O$ values of carbonate from veins in orogenic gold deposits according to: **a** carbonate mineral; **b** age of deposit; **c** craton or orogen hosting deposits (only craton or orogen with more than 30 $\delta^{18}O$ values are shown); **d** country rocks of deposits; and **e** geochemistry of igneous rocks hosting deposits

values ranging from 7.7 to 29.8‰, but most of data are around 11‰. The few $\delta^{18}O$ values of magnesite and undifferentiated carbonate minerals do not show consistent distribution patterns.

Archean deposits display a bimodal distribution of $\delta^{18}O$ values with two principal modes centered on 10‰ and 12.5‰ (Fig. 6b). Proterozoic deposits do not display a specific distribution of $\delta^{18}O$ values, whereas Paleozoic deposits display a bimodal distribution with two principal modes centered on 12.5‰ and 14.5‰. Mesozoic deposits have two main modes of $\delta^{18}O$ values at 14.5‰ and 16‰. Values higher than ~20‰, are mostly from Paleozoic deposits with only few Archean, Proterozoic and Mesozoic analyses. Figure 6c shows that $\delta^{18}O$ values for Archean cratons (Kaapvaal, Slave, Superior, Yilgarn) are mostly lower than ~15‰. The distribution of the $\delta^{18}O$ values from Yilgarn is centered on 10‰, whereas data from the Kaapvaal, Slave and Superior cratons are mostly centered on 12.5‰. For the Appalachian/Caledonian Orogen, $\delta^{18}O$ values are scattered about a large mode near 13‰. For others cratons/orogens, data are less numerous and scattered. Figure 6d displays carbonate $\delta^{18}O$ values as function of the country rock type. Deposits hosted in igneous rocks have a bimodal distribution with two modes about 10‰ and 12.5‰. Values for deposits hosted in sedimentary rocks display a normal distribution centered on ~14‰, whereas those hosted in volcano-sedimentary environment display three modes centered on ~10‰, ~12‰ and ~14‰. Figure 6e shows carbonate $\delta^{18}O$ values as function of the chemical composition of igneous country rocks. Deposits hosted in felsic to mafic rocks have two modes about 10‰ and 12.5‰. Deposits hosted in ultramafic rocks have a normal distribution of carbonate $\delta^{18}O$ values centered on 12.5‰.

3.1.3 Hydrogen Isotopes in Silicates

The hydrogen isotope composition (Fig. 7a) has been measured in muscovite (n = 86), sericite (n = 51), biotite (n = 32), chlorite (n = 42), actinolite (n = 9), epidote (n = 5), and tourmaline (n = 13). The range in δD values for vein minerals is remarkably wide, from −187 to −4‰ (Fig. 7a). The distribution of vein mineral δD values is bimodal, with a principal mode near -58‰, and a second mode near −85‰, with an asymmetric tail to low δD values. Epidote has the highest δD values (−30 to −4‰). Muscovite and sericite are characterized by δD values that spread mainly between −80 and −26‰, with a mode near −55‰, but with a wide spread to low δD values (−186‰; Fig. 7a). Chlorite displays a main mode at slightly lower values compared to muscovite and sericite (between −90‰ and −70‰), but also yields low δD values (−187‰). Biotite displays a relatively small range of δD values from −117 to −51‰ (Fig. 7a). Tourmaline and actinolite have δD values that span the range of the two main modes of hydrogen isotope compositions.

In Fig. 7b, δD values of minerals are reported relative to the age of the deposits. The bimodal distribution about the principal (−58‰) and the second (−85‰) modes encompasses δD values of all ages. However, Mesozoic and Cenozoic deposit δD values display a unimodal distribution centered around -58‰, with an asymmetric tail to the lowest δD values for deposits from the Cordillera Orogen. Deposits from Trans-Hudson Orogen display a unimodal distribution with values ranging between ~−120 and ~−30‰, and a mode at ~−90‰. For other cratons/orogens, data are less numerous and scattered.

Figures 7d and e show the distribution of δD values as function of country rock and of chemical composition of igneous rocks, respectively. No trends were identified, but the lowest values documented for Mesozoic and Cenozoic deposits from the Cordillera Orogen are from deposits hosted in sedimentary and volcano-sedimentary rocks.

3.1.4 Hydrogen Isotopes in Fluid Inclusions

Figure 8 presents the δD values of fluid inclusions extracted from quartz (n = 468). The range in δD values for fluid inclusions is from −179 to −7‰, with a wide and asymmetric distribution towards low δD values (Fig. 8a). The mode is centered on −70‰. Hydrogen isotope compositions of fluid inclusions from Archean,

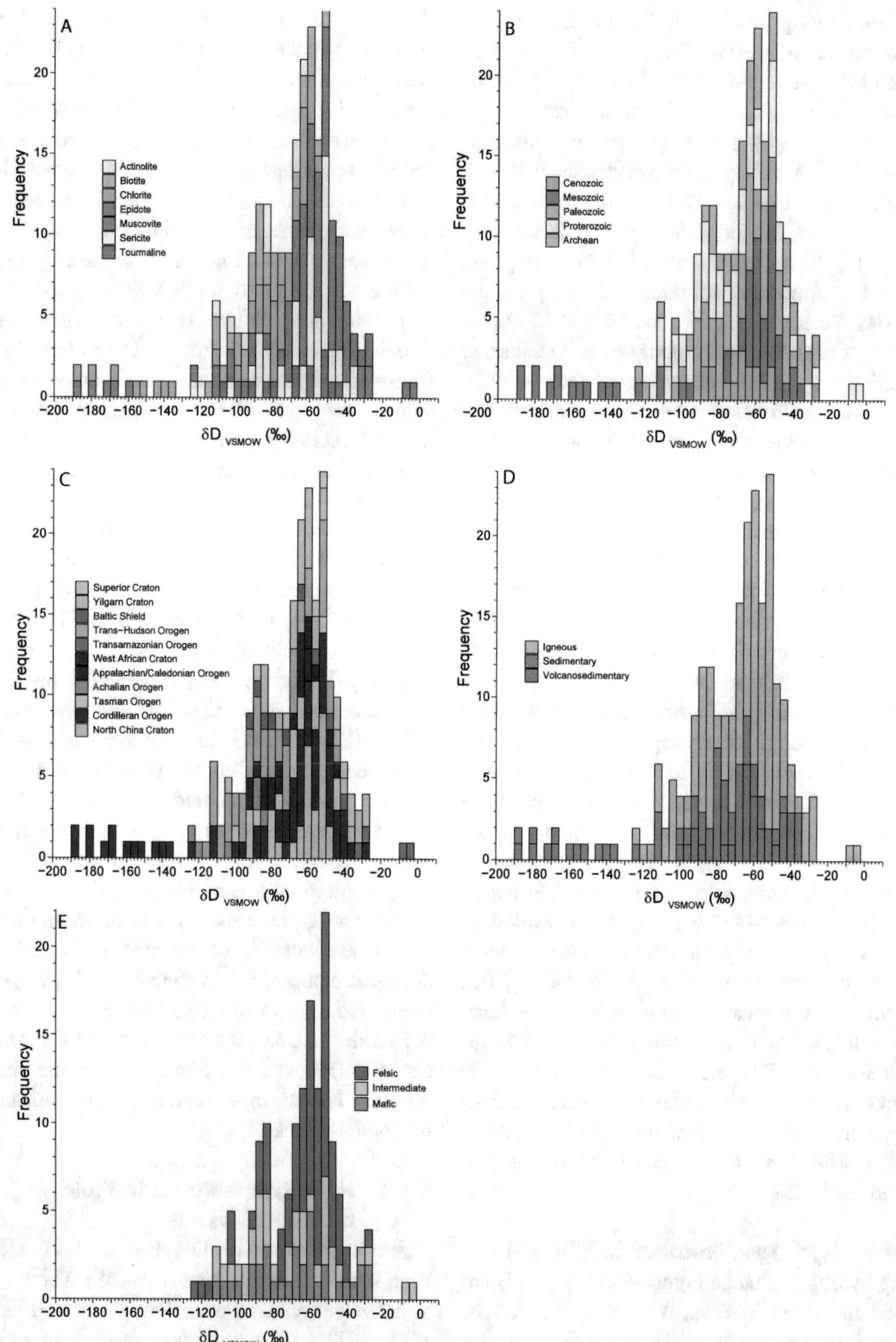

Fig. 7 Histograms showing δD values in OH-bearing minerals from veins from orogenic gold deposits based on: **a** nature of mineral; **b** age of deposit; **c** craton or orogen hosting deposits; **d** country rocks of deposits; and **e** geochemistry of igneous rocks hosting deposits

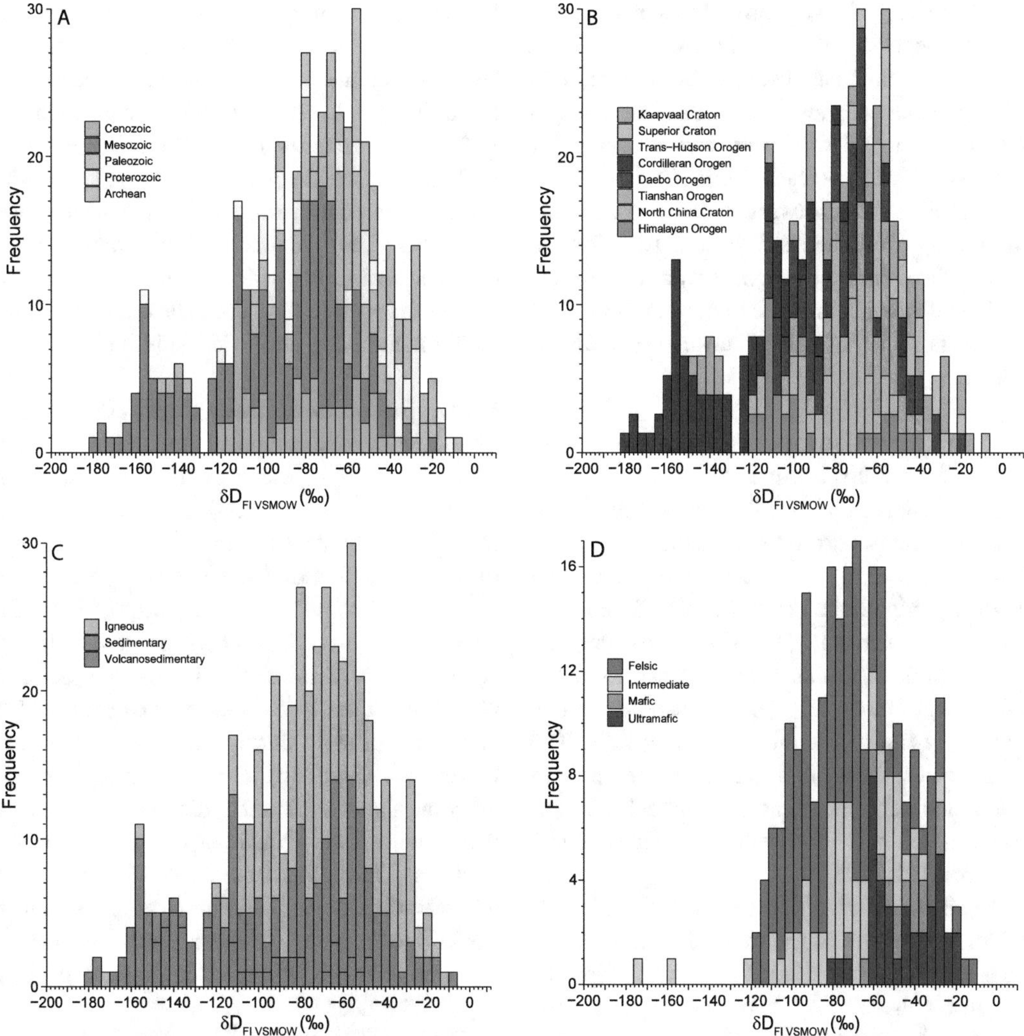

Fig. 8 Histograms showing δD values from fluid inclusion (FI) trapped in quartz from veins from orogenic gold deposits based on: **a** age of deposit; **b** craton or orogen hosting deposits (only craton or orogen with more than 20 δD values are shown); **c** country rocks of deposits; and **d** geochemistry of igneous rocks hosting deposits

Proterozoic, Paleozoic and Cenozoic orogenic gold deposits are typically heavier, from −7 to ∼−110‰, with few outlying values to ∼ −150‰ (Fig. 8a). Fluid inclusions δD values for Mesozoic orogenic gold deposits typically have values lower than −40‰, with a bimodal distribution with a mode near −70‰ and a second mode centered on −160‰ (Fig. 8a).

Figure 8b shows the distribution of δD values from fluid inclusions as function of craton/orogen. The main mode near −70‰ is dominated by deposits from the Daebo and Tianshan orogens, and from the North China Craton Craton. Deposits from the Himalayan Orogen display a wide range of δD values, with a secondary mode at ∼−110‰ from three deposits in Tibet (Supplementary material, Spreadsheet 1). The low δD values (< −120‰) are documented mainly for deposits hosted in the Cordillera Orogen, with a few measurements from deposits hosted in the Baltic Shield and the Trans-Hudson Orogen (Supplementary material, Spreadsheet 1).

Figure 8c shows δD values as function of the type of country rocks. The lowest δD values are characteristic of volcano-sedimentary rocks hosting deposits in the Mesozoic Cordillera Orogen. The dominant country rocks of deposits with δD values higher than −120‰ are igneous and sedimentary with only few analyses for deposits hosted in volcano-sedimentary rocks. In Fig. 8d, δD values between −120‰ and −60‰ are mainly from deposits hosted by felsic to intermediate igneous country rocks, whereas mafic and ultramafic hosted deposits have fluid inclusions with δD values between −60‰ and −20‰.

3.1.5 Carbon Isotopes in Minerals and Fluid Inclusions

Figure 9a shows the distribution of the carbon isotopic compositions for calcite (n = 513), dolomite (n = 344), ankerite (n = 214), siderite (n = 44), magnesite (n = 7), undifferentiated carbonate minerals (n = 47), and graphite (n = 2). Most δ^{13}C values range from −15 to 3‰ with a wide mode at −4‰, a secondary mode around −14‰ and a third one at −22‰. The δ^{13}C values from fluid inclusions trapped in quartz and scheelite show a unimodal distribution centered near −6‰ with rare values as low as −26‰ (Fig. 10a).

There is no significant difference of δ^{13}C values as function of the mineral or fluid inclusions (Figs. 9a and 10a). In Archean deposits, carbonates have δ^{13}C between −8 and 3‰, whereas Proterozoic, Paleozoic and Mesozoic deposits have similar δ^{13}C values distributed from −15 to 3‰ with a mode at −5‰ (Fig. 9b). However, low δ^{13}C values are documented for Paleozoic deposits of the Meguma district hosted in sedimentary rocks (Appalachian/Caledonian Orogen; Fig. 9b, c, d).

Carbonates from Archean Slave, Superior, Kaapvaal and Yilgarn cratons and from the Mesozoic North China Craton Craton display δ^{13}C values between −7 and 0‰, whereas the range of δ^{13}C values from the West African Craton is centered on −13‰ (Fig. 9c).

The carbonate δ^{13}C values from deposits hosted by igneous rocks show a unimodal distribution with a mode near −4‰ (Fig. 9d). Volcano-sedimentary rock-hosted deposits display a narrow range of δ^{13}C values, between −10 and 0‰. Sedimentary rock hosted deposits show a multimodal distribution with a main mode at −6‰, and three other modes around −3, −13 and −22‰. Felsic igneous rocks hosted deposits show a unimodal distribution with a main mode at −4‰. The carbonate δ^{13}C values from deposits hosted in mafic and ultramafic rocks, mainly from −9 to 0‰, each show a distribution with a main mode near −3‰ (Fig. 9e).

3.1.6 Sulfur in Sulfides and Sulfates

The sulfur isotope composition of sulfides ranges from −30 to 17‰, with a mode at ~ 3‰ and a tail to low δ^{34}S values. (Fig. 11a). Anhydrite (n = 3) and barite (n = 29) δ^{34}S values range from −2.9 to 18.2‰ with a mode at ~8‰. Most δ^{34}S values are from pyrite (n = 1305), arsenopyrite (n = 258) and galena (n = 199). No variation of δ^{34}S in relation to sulfide species is identified. The δ^{34}S values are similar for deposits formed during the Archean and Proterozoic (Fig. 11b). However, Paleozoic age deposits show a bimodal distribution of δ^{34}S values with a main mode at 0‰ and a second mode at 3‰. Mesozoic and Cenozoic age deposits display a distribution of δ^{34}S values similar to that of Archean and Proterozoic deposits, although with a wider range to low and high δ^{34}S values.

The distribution of δ^{34}S values is homogeneous for all cratons/orogens, and most of the data are spread along the entire range as illustrated for the Slave, Superior and Daebo cratons/orogen (Fig. 11c). The range of δ^{34}S values from the West African Craton is narrower, from −12 to −5‰. However, these data are from six deposits and have been measured mainly on arsenopyrite (Fig. 11a). Similarly, the Trans-Hudson Orogen and Dharwar Craton display a narrow distribution from 2 to 10‰. No significant difference has been identified for δ^{34}S values from deposits hosted by igneous, sedimentary or volcano-sedimentary rocks (Fig. 11d). However, sulfide minerals from deposits hosted in sedimentary rocks display

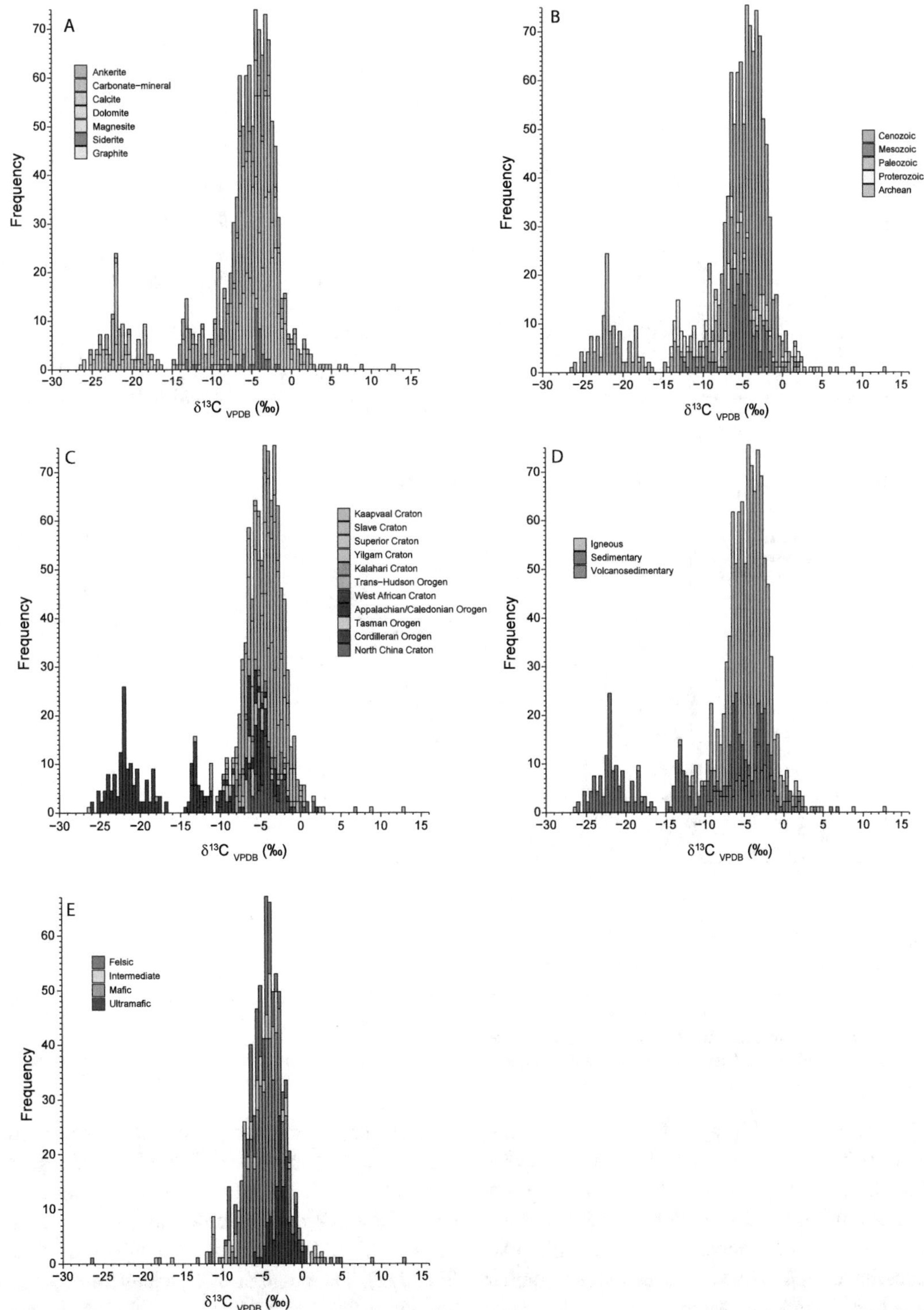

Fig. 9 Histograms showing $\delta^{13}C$ values of carbonate minerals and graphite from orogenic gold deposits based on: **a** nature of minerals; **b** age of deposit: **c** craton or orogen hosting deposits (only craton or orogen with more than 30 $\delta^{13}C$ values are shown); **d** country rocks of deposits; and **e** geochemistry of igneous rocks hosting deposits

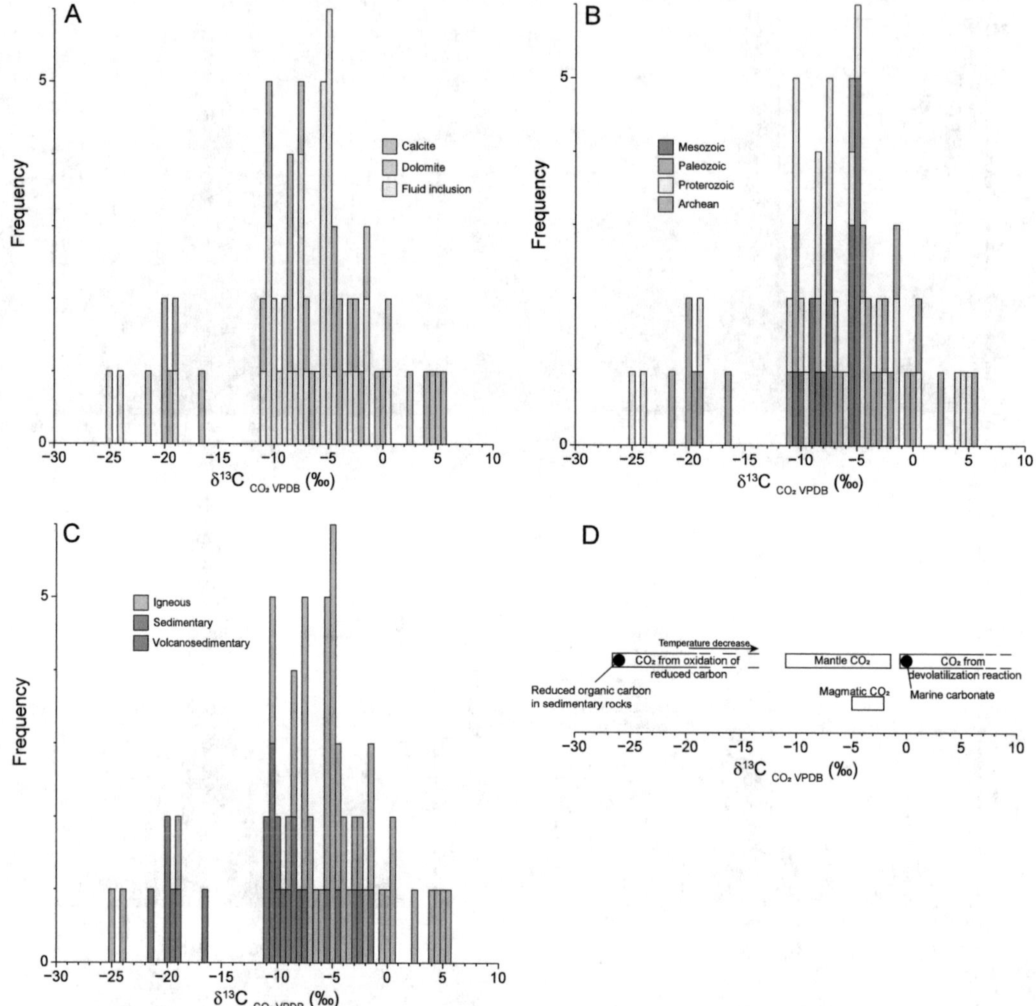

Fig. 10 $\delta^{13}C$ values of CO_2 from fluid inclusion (in quartz and scheelite) and CO_2 in equilibrium with calcite and dolomite using temperatures of equilibrium calculated using O isotope fractionation of Vho et al. (2019) on quartz-carbonate minerals pairs (see Fig. 13 and Supplementary material, spreadsheet 2). Histograms showing $\delta^{13}C_{CO_2}$ values are presented based on: **a** mineral; **b** age of deposit; and **c** country rocks of deposit. **d** shows $\delta^{13}C_{CO_2}$ values of various geological reservoirs (Taylor 1986; Kerrich 1989; Trull et al. 1993)

$\delta^{34}S$ variations with age (Fig. 11e). Archean deposits have $\delta^{34}S$ values with a unimodal distribution at 2.5‰, whereas Proterozoic deposits display a unimodal distribution at 4‰. Paleozoic deposits display a bimodal distribution with two main modes at 0‰ and 2.5‰ and with a number of values around 9‰. Mesozoic deposits display scattered $\delta^{34}S$ values with several data about 2.5‰ and tailing to low values. There are not enough data from Cenozoic age deposits to describe the distribution. The $\delta^{34}S$ values of minerals hosted by mafic to ultramafic igneous rocks show a bimodal distribution, with a main mode near 1.5‰ and a second mode at −5‰ (Fig. 11e). The range of $\delta^{34}S$ values for deposits hosted by felsic igneous rocks is slightly higher than for mafic rocks, between −5‰ and 15‰ with a mode at 3‰.

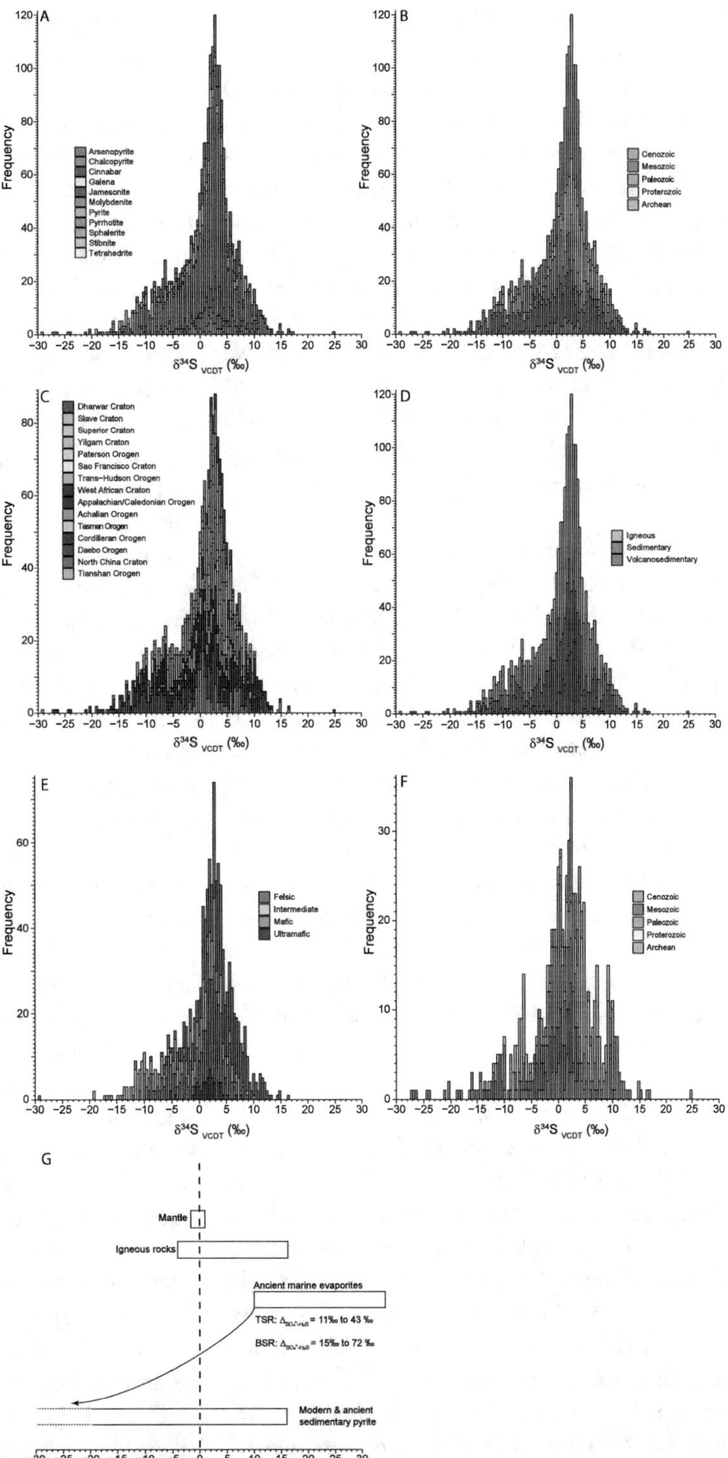

Fig. 11 Histograms showing $\delta^{34}S$ values of vein sulfide minerals from orogenic gold deposits based on: **a** mineral; **b** age of deposit; **c** craton or orogen hosting deposit (only craton or orogen with more than 50 $\delta^{34}S$ values are shown); **d** country rocks of deposits; **e** geochemistry of igneous rocks hosting deposits; and **f** age of deposits in sedimentary country rocks. **g** Variations in $\delta^{34}S$ values for various geological reservoirs (Machel et al. 1995; Wortmann et al. 2001; Seal 2006; Johnston et al. 2007; Labidi et al. 2013). $\Delta^{34}S^{2-}_{SO_4-H_2S}$ due to thermochemical sulfate reduction (TSR) calculateed at 500 °C and 100 °C (Eldridge et al. 2016)

3.1.7 Nitrogen in Silicates

The N isotope composition of biotite (n = 15), muscovite (n = 208) and K-feldspar (n = 31) range from 1 and 24‰ (Fig. 12a). Archean deposits have mostly $\delta^{15}N$ values higher than 10‰. Proterozoic deposits display a range between 7 to 12‰ whereas Phanerozoic deposits have mostly $\delta^{15}N$ values below 7‰ (Fig. 12b). Figure 12c shows that $\delta^{15}N$ values labelled by craton/orogen mimics the distribution as a function of the age of the deposits. Figure 12d shows that low $\delta^{15}N$ values are mostly documented from deposits hosted in volcano-sedimentary and sedimentary rocks whereas higher values are mainly from igneous rock hosted deposits.

3.1.8 Boron in Tourmaline

The B isotope composition of tourmaline (n = 117) ranges from −21.6 to 9‰ (Fig. 13a). The $\delta^{11}B$ values for Proterozoic deposits range from −16 to −9‰ (Figs. 13a, b). Archean deposits hosted in the Dharwar Craton have $\delta^{11}B$ values that spread mainly between −5 and 9‰, with a bimodal distribution with a main mode about −2.5‰ and a second mode about 4‰. The $\delta^{11}B$ values for the Archean Yilgarn Craton are the lowest and range between −24 and −18‰. These data are similar to $\delta^{11}B$ values of tourmaline compiled by Trumbull et al. (2020) from orogenic gold deposits worldwide, ranging between −24.8 and 19.8, with two modes at −15 and −2‰.

3.1.9 Silicon in Silicates

The silicon isotope composition of quartz (n = 19) and sericite (n = 11) range from −0.5 to 0.8‰ and from −0.3 to 0.4‰, respectively (Fig. 14a). The limited data is from two deposits, Hemlo (Superior Craton) of controversial origin (Ding et al. 1996), and Sawaya'erdun (Tianshan Orogen) (Liu et al. 2007), such that the representativeness of the range of $\delta^{30}Si$ values is limited. No difference in composition is identified for $\delta^{30}Si$ values in relation to host mineral and deposit age (Fig. 14b).

4 Isotope Equilibrium and Temperature of Formation

The isotope composition of a mineral depends on intrinsic and extrinsic factors. Intrinsic factors are related to the crystal structure, mass of the atoms forming molecular bounds in minerals, and isotope diffusion coefficients. These intrinsic factors determine partitioning of light and heavy isotopes of an element between two coexisting phases, the isotopic fractionation. The extrinsic factors comprise the temperature of the system, the pressure, particularly for gaseous phases, the bulk isotopic composition of the system and its components, and the isotope exchange reaction rate which, if fast, enables close system isotopic equilibrium to be reached or, if slow, will yield a kinetic open system in isotopic disequilibrium. The extrinsic factors determine the magnitude of the fractionation between isotopic species, temperature being the most important.

In order to make interpretations from the isotopic composition of minerals, it is thus essential to verify the assumptions about the state of equilibrium of the system in order to use theoretical or experimental fractionations to derive the temperature of equilibrium, or the bulk isotopic composition of the system and its components. Verification of equilibrium is best achieved comparing the composition of isotopic phases using δ–δ diagrams (Criss et al. 1987). In δ–δ diagrams, two coexisting phases in a closed system, minerals for example, will plot on the isotherm of the temperature of equilibrium, which is determined by the fractionation between the two species. If mineral pairs formed at the same temperature, but in isotopic systems with different bulk compositions, the data will plot along the same isotherm. If minerals formed at different temperatures in a system with a bulk constant isotopic composition, data will plot in an array orthogonal to isotherms.

δ–δ diagrams are drawn for quartz-silicate/borate/tungstate (Fig. 15), quartz-carbonate (Fig. 16), and sulfide-sulfide

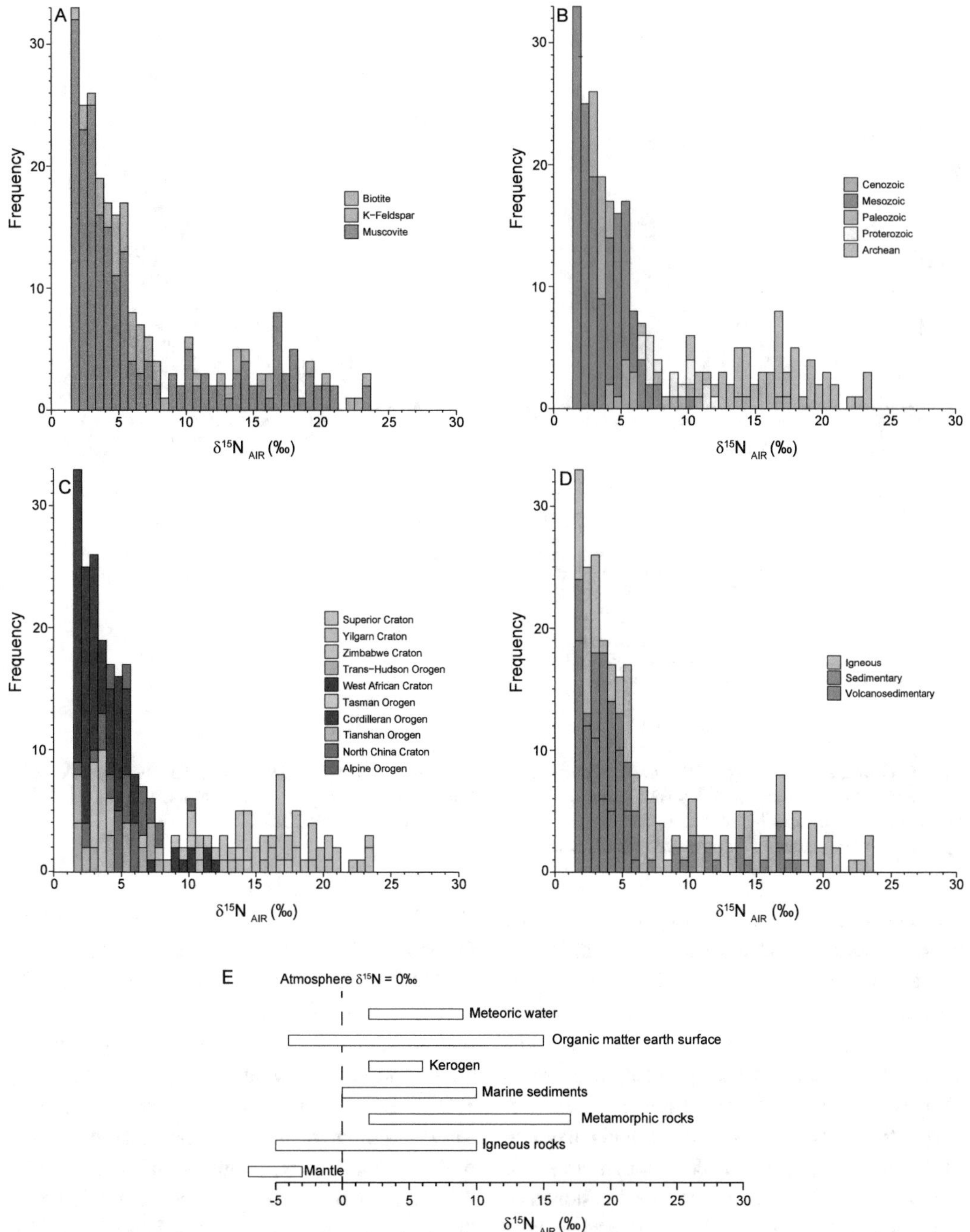

Fig. 12 Histograms showing $\delta^{15}N$ values of minerals from veins from orogenic gold deposits based on: **a** mineral; **b** age of deposit; **c** craton or orogen hosting deposit; and **d** country rock of deposits. **e** $\delta^{15}N$ values of various geological reservoirs are (Peters et al. 1978; Haendel et al. 1986; Minagawa and Wada 1986; Bebout and Fogel 1992; Compton et al. 1992; Williams et al. 1995; Wu et al. 1997; Ader et al. 1998; Jia and Kerrich 2000; Sephton et al. 2002; Gu 2009)

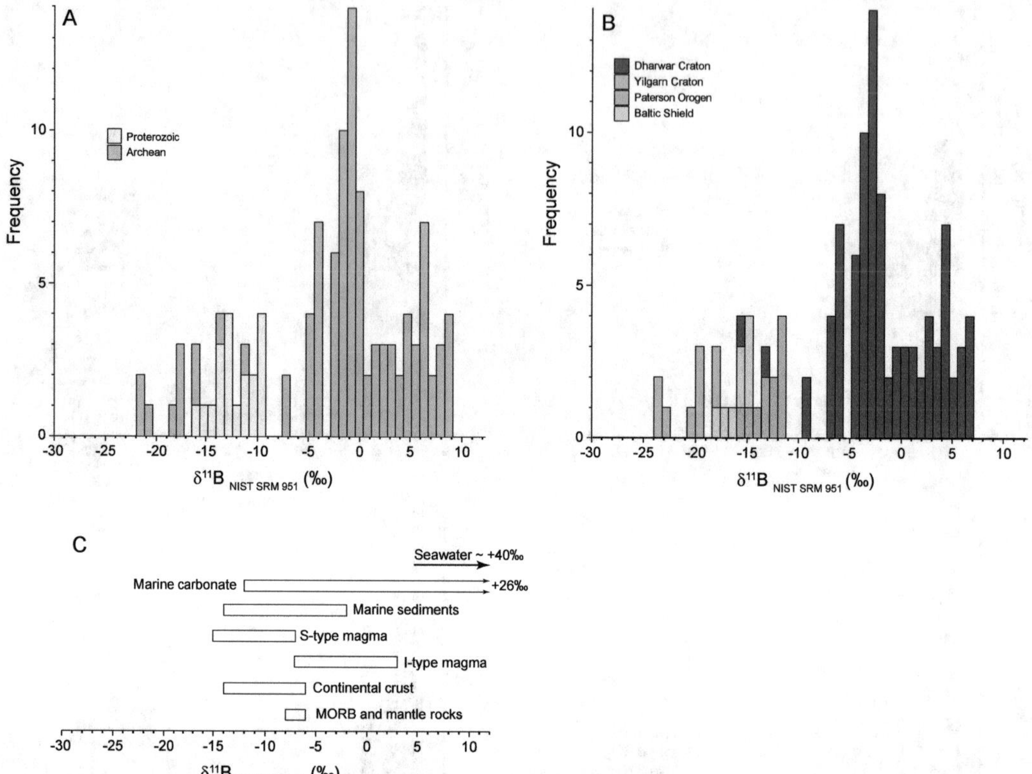

Fig. 13 Histograms showing $\delta^{11}B$ values of tourmaline from veins from orogenic gold deposits based on: **a** the age of the deposit; and **b** the craton or orogen hosting deposits. **c** $\delta^{11}B$ values of various geological reservoirs (Palmer and Swihart 1996; van Hinsberg et al. 2011; Lambert-Smith et al. 2016; Marschall et al. 2017; Trumbull and Slack 2018)

(Fig. 17). Quartz-muscovite mineral pairs from the same sample form a linear array that is parallel to isotherms (Fig. 15a). The data array is centered on the 350 °C isotherm calculated using Vho et al. (2019), which indicates that most quartz-muscovite pairs are approaching isotopic equilibrium near 350 °C. Few mineral pairs plot outside the array, which indicate either isotopic equilibrium at higher/lower temperatures, or isotopic disequilibrium. Using the fractionation of Vho et al. (2019), 60 of the 68 samples plot between the 250 and 550 °C isotherms. An apparent lower temperature of equilibrium is documented for Mesozoic age deposits (Fig. 15 a), but this trend is related to the fact that five of the seven quartz-muscovite with temperatures below 350 °C are from the Alleghany district, Cordilleran Orogen (Böhlke and Kistler 1986). Quartz-sericite pairs show a similar distribution (Fig. 15b), but data are mostly centered along the 450 °C quartz-muscovite isotherm. Quartz-biotite pairs (Fig. 15c) show more scatter than quartz-muscovite, but are mostly centered on the 350 °C isotherm. Quartz-chlorite pairs (Fig. 15d) show a similar distribution than that of quartz-biotite pairs, although a number of samples yield large $\Delta_{qz\text{-}chl}$, up to 12‰, which clearly indicates these mineral pairs were not in isotopic equilibrium. Quartz-albite (Fig. 15e) pairs plot mainly parallel to 250 °C isotherm. Quartz-tourmaline pairs (Fig. 15f) form an array centered on the 450 °C isotherm using Vho et al. (2019), but would be around the 350 °C isotherm using the fractionation of Kotzer et al. (1993), which has

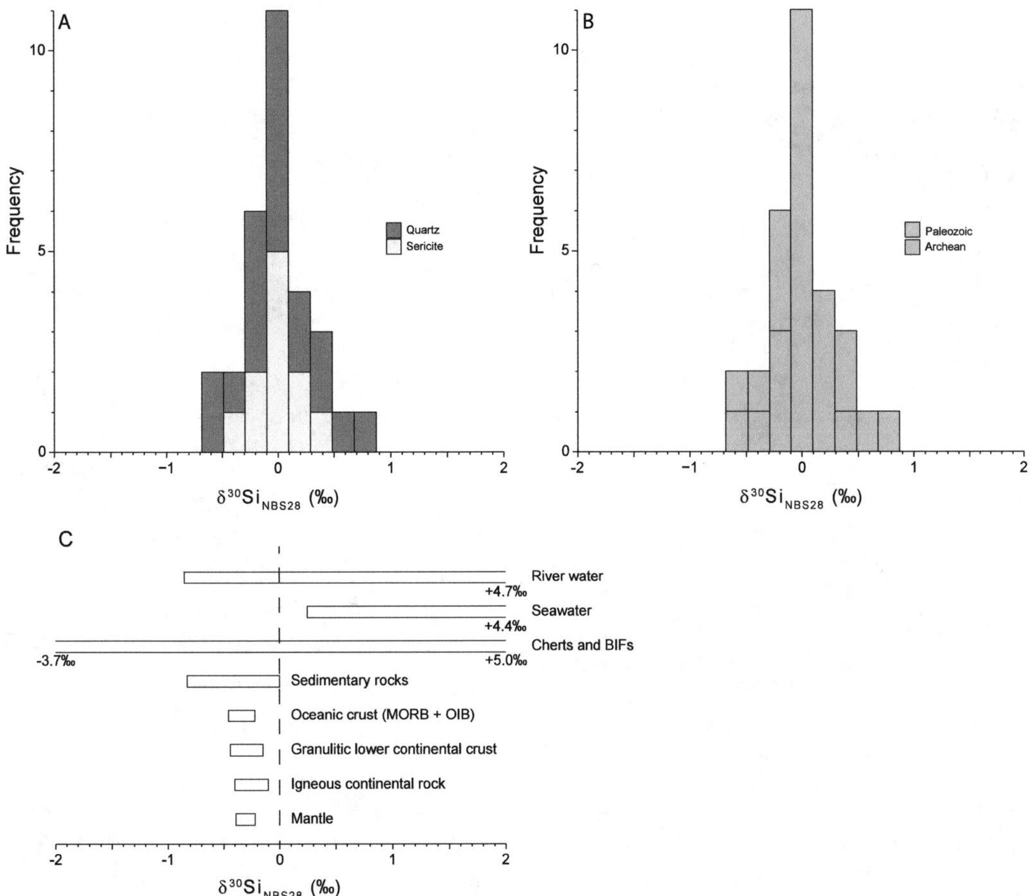

Fig. 14 Histograms showing δ^{30}Si values of minerals from vein from orogenic gold deposits based on: **a** mineral; and **b** age of deposit. **c** δ^{30}Si values of various geological reservoirs (Poitrasson 2017 and reference therein)

been defined using Proterozoic and Archean orogenic gold-bearing veins. Several samples have small ($\sim$1‰) or large (< 7‰) $\Delta_{qz\text{-}tur}$ values that indicate that these samples are not in isotope equilibrium. Finally, quartz-scheelite pairs scatter in the δ–δ diagram (Fig. 15 g), with several pairs with small ($\sim$3‰) or large (< 12‰) $\Delta_{qz\text{-}sch}$ values that indicate that these samples are not in isotope equilibrium. For all quartz-silicate/borate/tungstate pairs, there is no obvious trends with the composition of the country rocks, or age of the deposit.

Quartz-ankerite mineral pairs form a broad array parallel to isotherms (Fig. 16a), but with a large proportion of samples with fractionations too small or too large to record isotopic equilibrium at geologically reasonable temperatures (250–550 °C; (Goldfarb et al. 2005). Both quartz-calcite (Fig. 16b) and quartz-dolomite (Fig. 16c) also form broad arrays along isotherms, but data distribution also shows a trend of high δ^{18}O values for calcite and dolomite (< 25‰), that indicates isotope disequilibrium with coexisting quartz in these samples.

Because the magnitude of Δ^{34}S between sulfide minerals at equilibrium temperatures between 250 and 550 °C is small (e.g., 1.6 to 0.7‰ for pyrite-chalcopyrite), small Δ^{34}S variations yield large variations of the calculated temperature of equilibrium, such that temperature estimation using sulfur isotopes is less accurate. Pyrite-chalcopyrite (Fig. 17a) pyrite-sphalerite

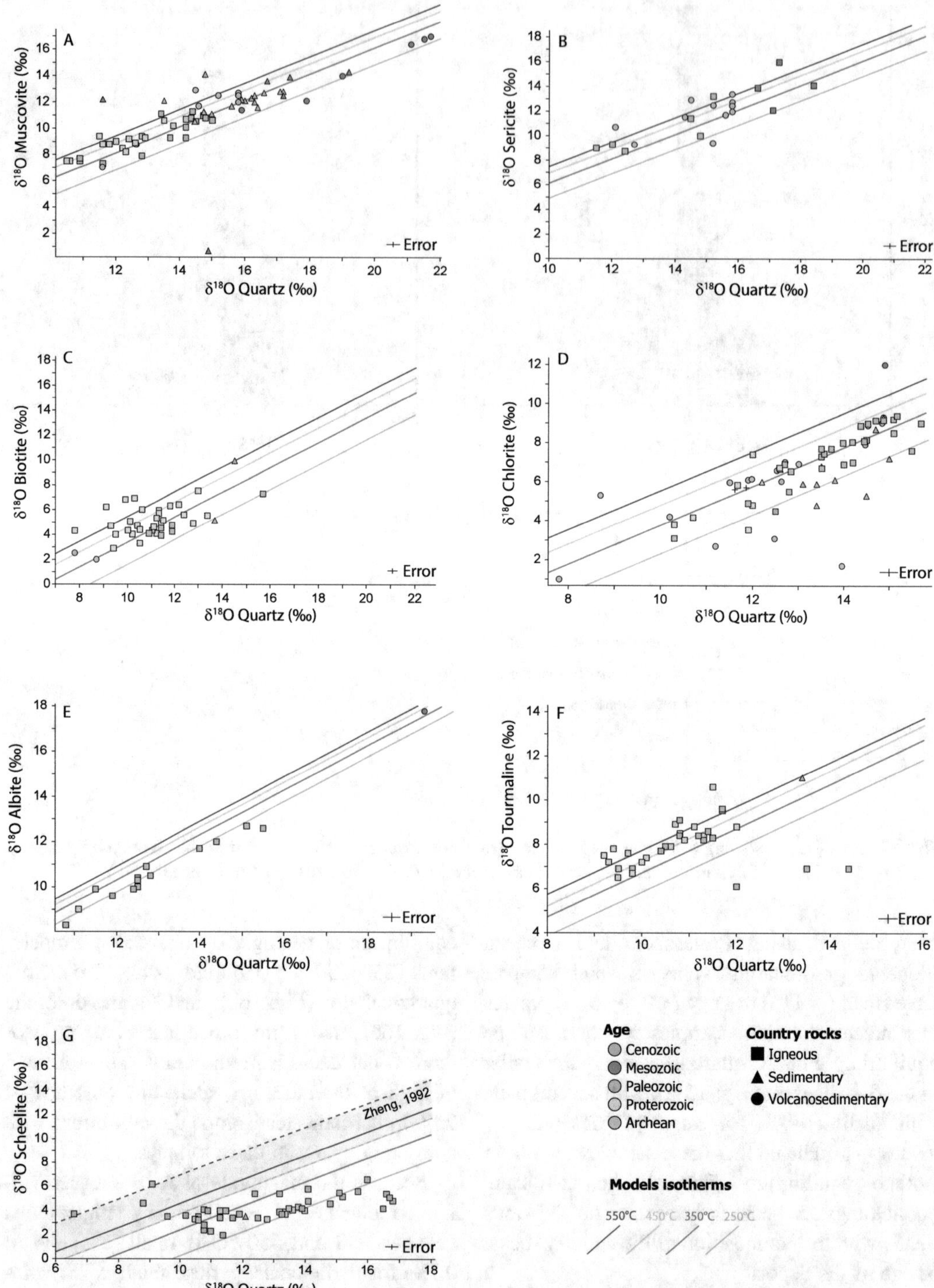

Fig. 15 $\delta^{18}O_{silicates}$ versus $\delta^{18}O_{quartz}$ of coexisting vein minerals with symbols indicating the age and the type of country rocks. Isotherms drawn using oxygen isotope fractionation equations from Vho et al. (2019) for **a**, **b**, **c**, **d**, **e**, **g** and **h** and from Clayton et al. (1972) and Wesolowski and Ohmoto (1986) for **f**

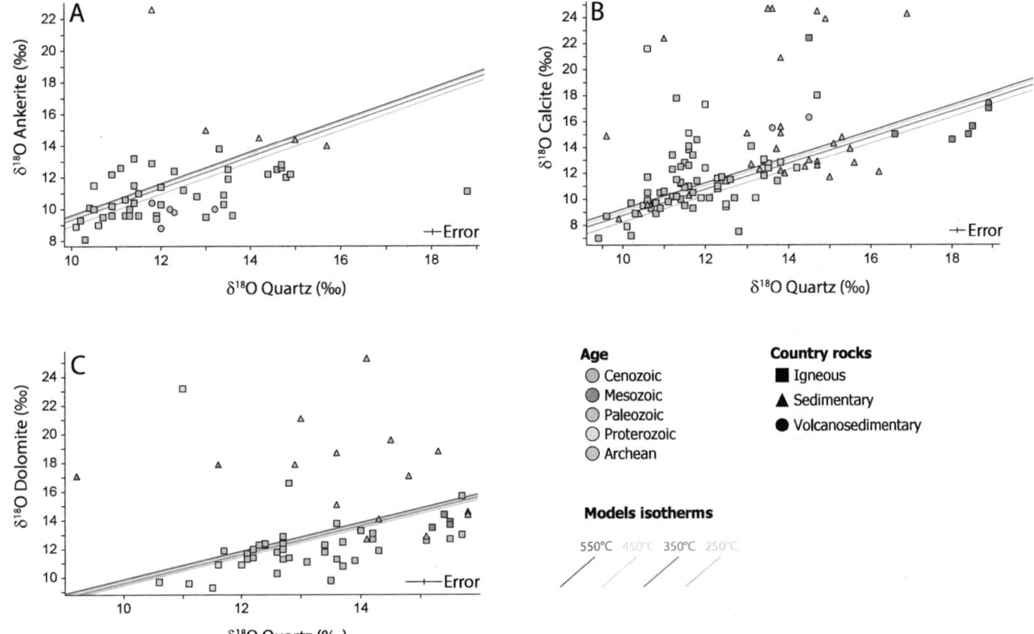

Fig. 16 $\delta^{18}O_{carbonate}$ versus $\delta^{18}O_{Oquartz}$ of coexisting vein minerals with symbols indicating the age and the type of country rocks. Isotherms drawn using O isotope fractionation equations from Vho et al. (2019) for **a**, and **c**, and from Clayton et al. (1972) and Wesolowski and Ohmoto (1986) for **b**

(Fig. 17b) and pyrite-galena (Fig. 17c) pairs similarly plot as broad arrays parallel to isotherms. For these mineral pairs, only a small proportion of samples show small or large fractionations that indicate sulfur isotope disequilibrium. Most data plot close to the 350 °C isotherm indicating isotope equilibrium. A few pyrite-pyrrhotite pairs scatter in the δ-δ diagram (Fig. 17d), indicating sulfur isotope disequilibrium. Sphalerite-chalcopyrite (Fig. 17e) and sphalerite-galena (Fig. 17f) pairs, similar to most pyrite-sulfide pairs, plot in broad arrays centered along the 350 °C isotherm. Finally, the galena-chalcopyrite δ–δ diagram (Fig. 17 g) shows a broad array parallel to isotherms, but at low calculated temperatures (circa 150 °C).

In summary, quartz-silicate, quartz-carbonate, and sulfide-sulfide mineral-pairs that display evidence for isotopic equilibrium between the 250–550 °C yield an average temperature of 360 ± 76 °C ($\sigma = 1$, $n = 332$) without evidence for secular change with the age of formation of then orogenic gold deposits.

5 Discussion

5.1 Secular Variations in Mineral and Fluid Inclusion Compositions

Figure 2a shows that $\delta^{18}O$ values of quartz from Archean (8–16‰) and Proterozoic (8–16‰) age deposits broadly overlap with Paleozoic (9–17.5‰), Mesozoic (10–22.5‰) and Cenozoic (10–17.5‰) age deposits, yet Phanerozoic age deposits yield slightly higher $\delta^{18}O$ values (9–22.5‰) than Precambrian age deposits (8–16‰). Figure 6b, likewise shows that carbonate minerals from Archean and Proterozoic age deposits typically have lower $\delta^{18}O$ values (7–15‰) than that of Paleozoic (9–25‰) and Mesozoic (8–20‰) age deposits. The change in quartz and carbonate oxygen isotope composition with age of deposit (Figs. 2a and 6b), is not continuous with a reversal to lower $\delta^{18}O$ values for Cenozoic age deposits, casting doubt on a secular evolution

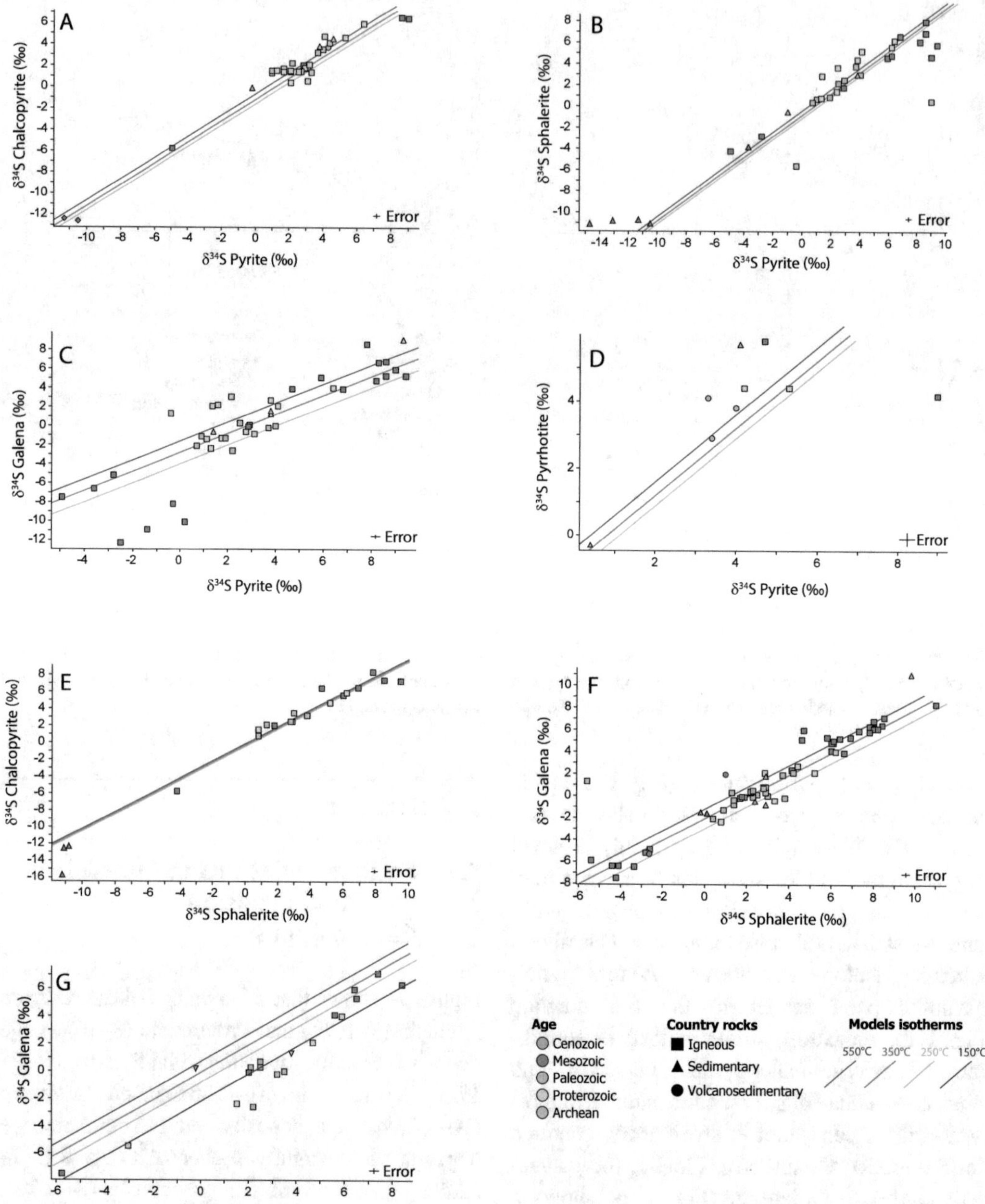

Fig. 17 $\delta^{34}S_{Sulfide}$ versus $\delta^{34}S_{Sulfide}$ of coexisting vein sulphide minerals with symbols indicating the age and the type of country rocks. Isotherms drawn using S isotope fractionation equations from Kajiwara and Krouse (1971) for a, b, d and f, from Li and Liu (2006) for c and e, and from Liu et al. (2015) for g

of the composition of hydrothermal fluids in orogenic gold deposits, as suggested by Goldfarb and Groves (2015). As shown in Fig. 2c, the lower Archean quartz $\delta^{18}O$ values are dominantly for orogenic gold deposits hosted by igneous rocks, whereas Phanerozoic age deposit higher $\delta^{18}O$ values are characteristic for deposits hosted by sedimentary and volcano-sedimentary country rocks. Figure 6d shows a similar pattern where carbonate with lower and higher $\delta^{18}O$ values correspond to Archean and Proterozoic age deposits dominantly hosted, respectively, by igneous or sedimentary and volcano-sedimentary country rocks.

Less abundant data for other silicate minerals do not allow analysis by age of deposit. In contrast, carbonate mineral $\delta^{13}C$ values do not display a distribution by age of deposit, with the exception of the low $\delta^{13}C$ values from deposits of the Appalachian Orogen, which indicates this a provincial feature (Fig. 9b). Similarly, silicate (Fig. 7b) and inclusion fluid[3] (Fig. 8b) δD values do not display different distributions based on the age of the deposits, with the exception of the low values from the Mesozoic Cordillera Orogen deposits. Sulfide sulfur in deposits hosted by igneous and volcano-sedimentary rocks does not display compositional variations related to the deposit age (Fig. 11). In contrast, deposits hosted in sedimentary rocks show different but overlapping ranges of $\delta^{34}S$ values with age. Chang et al. (2008) showed that Phanerozoic orogenic gold deposits hosted in sedimentary rocks have sulfide $\delta^{34}S$ values that closely track that of seawater sulfate at the age of the host rocks. This result was interpreted to record leaching of reduced seawater sulfate from the sedimentary country rocks. Likewise, Goldfarb et al. (1997) showed that the Juneau district Cenozoic orogenic gold deposits have $\delta^{34}S$ values that follow those of their Phanerozoic sedimentary host rocks. Thus, the sulfur isotope composition of orogenic gold deposits hosted in sedimentary rocks vary with the age of the host rocks following the secular variation of seawater sulfate, but with no evidence of secular variation related to the age of the deposits.

The less abundant data for N (Fig. 12b), B (Fig. 13a) and silicon (Fig. 14b) allows for less definitive assessments of secular variations in composition. The N isotope composition is characteristically higher for Archean (11–24‰) and Proterozoic (7–12.5‰) compared to Paleozoic- (2.5–14‰) and Mesozoic- to Cenozoic-aged deposits (5–7‰: Fig. 12b). The decrease in $\delta^{15}N$ values with decreasing deposit age has been ascribed to progressive mantle degassing ($\delta^{15}N = -5‰ + /-2‰$) into the atmosphere ($\delta^{15}N = 0‰$), and the gradual sequestration of heavier atmospheric N_2 into sedimentary rock micas (Jia and Kerrich 2004; Kerrich et al. 2006). However, other studies suggest that atmospheric $\delta^{15}N$ values were constant through time and that isotopic variations in sedimentary rocks reflect isotopic fractionation related to fluid-rock interaction and/or biogenic activity (Cartigny and Marty 2013).

The compilation shows that Archean age deposits have mostly higher $\delta^{11}B$ values than Proterozic age deposits (Fig. 13a). However, recent studies show that several Archean deposits also have low tourmaline $\delta^{11}B$ values, such as in Val-d'Or (Beaudoin et al. 2013; Daver et al. 2020) and the Hattu schist belt (Molnár et al. 2016). Finally, the limited amount of silicon isotope compositions does not display a trend with age (Fig. 14b).

5.2 Temperature Variations

5.2.1 Isotopic Disequilibrium

Most minerals pairs plot between the 250–550 °C isotherms, temperatures typical for the formation of orogenic gold deposits, thus suggesting widespread isotopic equilibrium for quartz-silicates/borate/tungstate and sulfide-sulfide pairs. In contrast, quartz-carbonate pair $\delta^{18}O$ values are scattered with only few pairs along 250–550 °C isotherms suggesting common isotopic disequilibrium.

Diachronous formation of quartz and carbonate could explain such disequilibrium, although quartz and carbonate are commonly paragenetically coeval in orogenic gold deposits. Isotopic exchange can also occur between mineral-pairs during cooling by volume diffusion (Giletti 1986). As shown by Sharp and Kirschner (1994), the degree of retrograde oxygen diffusion

[3] Factors controlling fluid inclusion isotopic data are complex and described in more detail below.

between quartz and calcite depends of the closure temperature of quartz which, in turn, depends of its grain size, oxygen diffusion coefficient and cooling rate. They showed that for quartz grains in a large reservoir of calcite, resetting below 400 °C will only occur for small quartz grain sizes. Consequently, retrograde diffusion exchange between quartz and calcite in veins is unlikely to explain scattered data in δ_{Qtz}–δ_{Carb} space (Fig. 16). Late infiltration of low temperature fluids can significantly alter the $\delta^{18}O$ value of calcite and disturb the initial isotopic equilibrium between co-genetic minerals (Sharp and Kirschner 1994). This low temperature resetting has been proposed to explain the common isotopic disequilibrium documented for carbonates with high $\delta^{18}O$ values compared to quartz (Kontak and Kerrich 1997; Beaudoin and Pitre 2005): Fig. 16). Because veins commonly experienced successive deformation events (as folded or sheared veins), such late fluid circulation can easily be channelized in newly formed fractures.

5.2.2 Secular Variations in Temperature of Formation of Orogenic Gold Deposits

Goldfarb et al. (2005) suggested that Phanerozoic and Paleoproterozoic orogenic gold deposits formed at lower temperatures (250–350 °C) that older Archean deposits (325–400 °C). However, they acknowledged that each age group of orogenic gold deposits displayed large, and overlapping, ranges of temperature of formation. For mineral pairs showing isotope equilibrium, there is no secular trend of changing equilibrium temperature with age of deposit (Figs. 15, 16 and 17). This observation is independent of the fractionation used to compute the temperature, as the different fractionations will only shift the nominal value of the temperature, not the data distribution along isotherms. Thus, the stable isotope composition of vein minerals does not support the interpretation of a lower temperature of formation for Proterozoic and Phanerozoic orogenic gold deposits, compared to higher temperature Archean age orogenic gold deposits.

5.3 Composition of Hydrothermal Fluids and Dissolved Elements

In the following sections, the $\delta^{18}O$ and δD values of H_2O and $\delta^{13}C$ values of CO_2 are calculated from minerals isotopic composition at temperatures of 250–550 °C for mineral pairs that show evidence of isotope equilibrium. $\delta^{34}S$ values of H_2S have not been calculated because isotope fractionation between sulfides and H_2S are small (< 1‰) between 250 and 550 °C (Li and Liu 2006). The source of H_2S is discussed based on $\delta^{34}S$ values of sulfides.

5.3.1 Water

Metamorphic versus magmatic water. The two most common sources of fluids proposed for orogenic gold deposits are magmatic and metamorphic water. As reviewed by Goldfarb and Groves (2015), the inconsistent timing between magmatism and orogenic gold deposit formation, and lack of a specific magmatic association, both argue against the ubiquity of magmatic fluids in the formation of orogenic gold deposits.

Most of the calculated water compositions plot in the field for metamorphic water, but a small proportion of analyses also plot in the field for magmatic water and a few plot outside both fields (Figs. 18a, b; Sheppard 1986). Magmatic water has been argued in some cases (Li et al. 2012; Zeng et al. 2014; Deng et al. 2015), but this interpretation is based on fluid $\delta^{18}O$ and δD values between −0.2 and 6‰ and −96 and −52‰, respectively, that plot largely outside of the magmatic water field, and are too low for the inferred granitic magma source. Thus, metamorphic water is the more common fluid source, and in most cases the only likely fluid source for the formation of orogenic gold deposits. The variation of oxygen and hydrogen isotopes composition of metamorphic water(s) documented on Fig. 18 probably reflect that fluids were sourced from metamorphism of variable proportions of sedimentary and igneous rocks in the crust.

Evidence of meteoric water in fluid inclusions. Fluid inclusions are trapped either during crystal

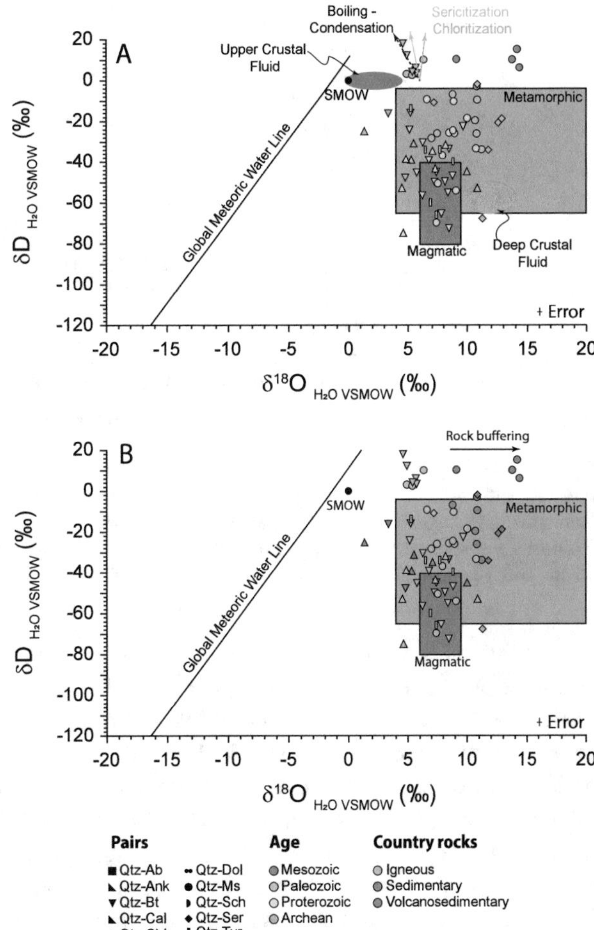

Fig. 18 Diagrams showing the variation of $\delta^{18}O$ and δD values of H_2O calculated from quartz and OH-bearing minerals. **a** δD_{H2O} versus $\delta^{18}O_{H2O}$ with symbols indicating the age of deposits, and **b** the type of country rocks. The high δD values are interpreted to reflect the effect of evaporation–condensation cycles, whereas higher $\delta^{18}O$ reflect O isotope buffering by fluid-rock interactions. The Global Meteoric Water Line is derived from Craig (1961), whereas the metamorphic and magmatic water boxes are from Sheppard (1986)

growth (primary) or in cracks during late fluid circulation (secondary). Caution must be exercised using isotope composition of fluid inclusions because the mechanical or thermal decrepitation leach methods extract all populations of fluid inclusions, some of which not associated with gold formation. It has been shown by Faure (2003) than thermal decrepitation yield inaccurate δD values, possibly because of uncontrolled reactions between released H_2O and Si–OH bounds in quartz, for example. Likewise, Gaboury (2013) showed that fluid inclusions have different hydrocarbon compositions as a function of the temperature of decrepitation. As an example, Foley et al. (1989) showed bulk extract contained two types of fluid inclusions with distinct δD values at Creede (Colorado, USA). This is particularly likely in orogenic gold deposits where veins are commonly deformed and thus may have trapped various fluids during and/or after mineralization. For example, in laminated veins, most fluid inclusions are secondary (Ridley and Diamond 2000). Pickthorn et al. (1987) reviewed evidence for discrepancy between fluid inclusion and mica δD values and concluded to mixing of various generations of fluid inclusions, including late meteoric water trapped in secondary inclusions in the host minerals. This was contested by Nesbitt et al. (1987) on the basis that various mixtures of fluid inclusions should yield a mixing line between two end members, contrary to the low variance of δD values.

As shown on Fig. 19, micas are commonly out of isotopic equilibrium with fluid inclusions trapped in quartz from the same vein. Isotopic

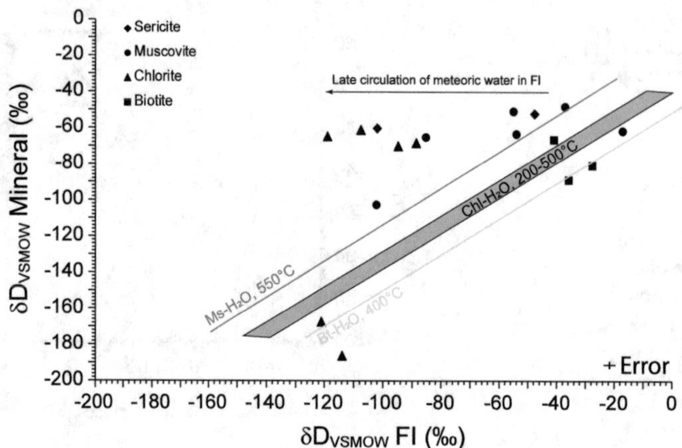

Fig. 19 $\delta D_{Mineral}$ versus δD_{FI} coexisting OH-bearing mineral and fluid inclusions from a same sample. Isotherms have been drawn using mineral-H_2O H isotope fractionations of Suzuoki and Epstein (1976) for muscovite and biotite and from Graham et al. (1987) for chlorite. The grey area represents the uncertainty associated to the chlorite-H_2O H isotope fractionation of Graham et al. (1987) between −40‰ and −30‰ between 200 and 500 °C

disequilibrium is well documented for chlorite and some sericite/muscovite displaying high δD values (~-60‰) compared to the δD values of coexisting fluid inclusion (<-60‰). Such disequilibrium has been interpreted to record infiltration of late, low δD meteoric water or a mixture of several types of fluid inclusions with different hydrogen isotope compositions (Goldfarb et al. 1991). Fluid inclusions (Figs. 8a) and OH-bearing minerals (Fig. 7b) low δD values are mainly documented from Mesozoic and Cenozoic deposits, and are commonly interpreted to result from late meteoric water infiltration disconnected with the formation of the deposits (Nesbitt et al. 1989; Zhang et al. 1989; Madu et al. 1990; Goldfarb et al. 1991; Shaw et al. 1991; Rushton et al. 1993; Apodaca 1994; Jia et al. 2003). However, similar low δD values are also documented for older deposits. In deposits from Trans-Hudson terrane (Figs. 8a, b), Liu (1992) considered that meteoric water circulated in the vein structure, but after the formation of the deposit. In contrast, Billstrom et al. (2009) suggest that low δD values in Baltic Shield Paleoproterozoic age deposits could reflect infiltration of surface water during gold deposition.

Fluid Mixing. Figures 20a, b, c and d show the variation of $\delta^{18}O_{H2O}$ and δD_{H2O} values as a function of the temperature of equilibrium between pairs of minerals. For both isotopes, data is spread along a linear trend between low $\delta^{18}O$ (~2‰), high δD (~10‰) values at low temperatures (~250 °C), and high $\delta^{18}O$ (~12‰), low δD (~-100‰) values at high temperatures (~550 °C) of equilibrium. This trend is similar to that documented by Beaudoin and Chiaradia (2016) for quartz and tourmaline from orogenic veins from the Val-d'Or vein field, which was interpreted to result from mixing between a low temperature upper crustal fluid and a high temperature deep-seated crustal fluid. Data for orogenic gold deposits (Figs. 20a,b,c,d) show that the same mixing trend between similar fluid end-members is recorded for deposits of all ages (Archean to Cenozoic), hosted in different country rock types, suggesting that fluid mixing between two common reservoirs may be an important process documented in orogenic gold deposits of all ages, worldwide, yet poorly documented in literature. The high $\delta^{18}O$—low δD—high T deep-seated crustal fluid is likely metamorphic in origin, whereas the low $\delta^{18}O$—high δD—low T upper crustal fluid is water of surficial origin with a long history of water–rock reactions, perhaps not unlike long residence water in cratons, as reviewed by Warr et al.

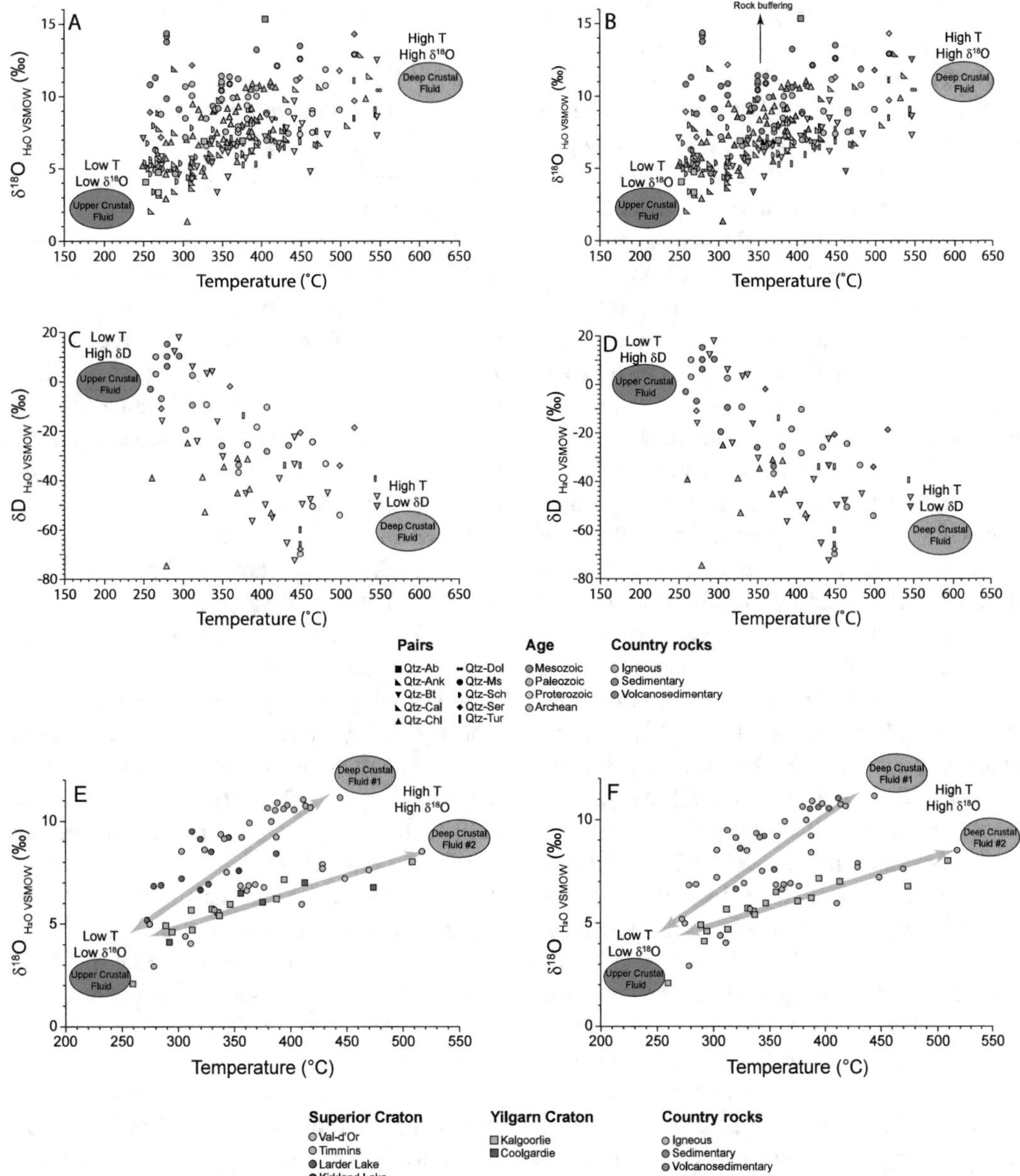

Fig. 20 Diagrams showing the variation of $\delta^{18}O$ of H_2O calculated from quartz and OH-bearing minerals as function of temperature calculated from mineral pairs based on: **a** the age of deposits; and **b** the type of country rocks. Diagrams showing the variation of δD of H_2O calculated from quartz and OH-bearing minerals as function of temperature calculated from mineral pairs based on: **c** the age of deposits and **d** the type of country rocks. The broad linear arrays in **a**, **b**, **c** and **d** suggest mixing between a low temperature, low $\delta^{18}O$, high δD (interpreted as an upper crustal) and a high temperature, high $\delta^{18}O$, low δD (interpreted as a deep crustal) fluids. Diagrams showing the variation of $\delta^{18}O$ of H_2O calculated from quartz and OH-bearing minerals as function of temperature calculated from mineral pairs based on: **e** the district; and **f** the type of country rocks. The shift in $\delta^{18}O$ values between different districts could suggest provinciality on the fluid, that is that the deep-seated fluid reservoir varies slightly in composition between orogenic gold vein districts

(2021). The ultimate origin of the surficial water, obscured by water–rock exchange, may be difficult to decipher in several deposits. High δD_{H2O} values, up to 0‰, in the Wiluna deposit (Yilgarn) have been interpreted to result from mixing of surface water, at shallow levels, with metamorphic-magmatic fluids (Hagemann et al. 1994). Boiron et al. (2003) showed that the deep-seated fluid was the dominant reservoir involved in the first step of the vein formation followed by progressive infiltration of surficial water during basement uplift. It is known that crustal rocks porosity contains water deep in the crust (Smithson et al. 2000). This water may have been trapped since rock formation or may have infiltrated from surface (seawater or meteoric) slowly reacting with country rocks along the fluid pathways. If fluid mixing has not been commonly identified in orogenic gold districts, the data compilation indicates that most deposits, in which we can interpret a temperature of equilibrium and a H–O fluid composition from mineral pairs, plot along a broad array that we interpret as a mixing line between two common reservoirs. Beaudoin et al. (2006) showed that the regional change in oxygen isotope composition of orogenic gold deposits of the Val-d'Or vein field can be reproduced by simulating fluid flow and oxygen isotope transport and reaction of the deep-seated fluid in rocks saturated by the upper crustal fluid end-members. Infiltration of the deep-seated fluids along the higher permeability shears and fractures of the orogenic gold deposits resulted in mixing with the upper crustal fluids filling the host rocks porosity, thus yielding the mixing lines shown in Fig. 20. Mixing of various proportions of fluids from different reservoirs also explains the gradual change in composition between deposits in one district, as shown in the Val-d'Or (Beaudoin and Pitre 2005) and Victoria (Gray et al. 1991) vein fields.

Fluid buffering. As shown in Figs. 2c, 6d, the highest quartz and carbonate $\delta^{18}O$ values are documented in deposits hosted dominantly in volcano-sedimentary and sedimentary rocks, consistent with the interpretation that isotopic oxygen exchange between fluids and sedimentary rocks leads to ^{18}O-enriched quartz (Böhlke and Kistler 1986; Goldfarb et al. 1991, 2004; Boer et al. 1995; Zhang et al. 2006) and carbonates (Böhlke and Kistler 1986; Steed and Morris 1997). For carbonates, some values enriched in ^{18}O also likely reflect a late, retrograde, and low-temperature re-equilibration with a surface-derived fluid.

Figures 18b and 20b shows that most $\delta^{18}O_{H2O}$ values higher than those along the mixing trend, are from deposits hosted in sedimentary or volcano-sedimentary rocks. This suggests oxygen isotope buffering of the fluids by sedimentary and volcano-sedimentary country rocks, which is detected in sedimentary rocks because of their heavier oxygen isotopic composition (Sheppard 1986).

Provinciality. Figures 20e and f present $\delta^{18}O_{H2O}$ values as function of temperature for the Archean Yilgarn (Kalgoorlie and Coolgardie districts) and Superior (Val-d'Or, Timmins, Larder Lake, and Kirkland Lake districts) cratons. Data displays two trends between a common low $\delta^{18}O_{H2O}$—high δD—low temperature upper crustal fluid endmember, and two slightly different deep-crustal fluid endmembers distinguished by their oxygen isotope composition. Deep crustal fluid #1 is characteristic of the Kirkland Lake, Larder Lake, and Timmins Superior Craton districts, and has a $\delta^{18}O_{H2O}$ about 4‰ higher than deep crustal fluid #2, which is defined by deposits of Superior Craton Val-d'Or, and Yilgarn Craton Kalgoorlie and Coolgardie, districts. This suggests that the deep-seated fluids were sourced from rocks at depth with a slightly different oxygen isotope composition, likely reflecting different proportions of igneous and sedimentary rocks in the crustal segment undergoing prograde metamorphism, such that the deep-seated fluid reservoir varies slightly in composition between orogenic gold vein districts. The difference in end-member composition cannot be an artifact of water–rock reactions because deposits from both trends are hosted dominantly by igneous rocks (Fig. 20f).

Origin of high δD fluids. The formation of the gold-bearing quartz vein is commonly associated with the precipitation of micas in the veins and host rock, or micas formed by hydration of host-rock minerals (Groves et al. 1998; Ridley and Diamond 2000; Eilu and Groves 2001; Craw et al. 2009). Muscovite and chlorite are the two main micaceous minerals in orogenic gold deposit. Here, we model the evolution of the H–O isotopic composition of water from which micas precipitate progressively in an open system following Rayleigh distillation to determine if mica precipitation can explain the deuterium enrichment, up to 20‰, of the hydrothermal fluids in orogenic system (Fig. 18e, f). Isotopic exchange between fluid and hostrock is not considered. Equations 1 and 2 are modified from Sharp (2017):

$$\delta_{Liq} = [\delta_{Liq} + 1000]F^{(\alpha_{Chl/Ms-Liq}-1)} - 1000, \quad (1)$$

and

$$\delta_{Chl/Ms} = \alpha_{Chl/Ms-Liq}\left(\delta_{Liq} + 1000\right) - 1000, \quad (2)$$

where F is the remaining water fraction, at 350 °C and using 1000ln $\alpha_{chlorite\text{-}H2O}$ = −35‰ (Graham et al. 1987) for H, 1000ln $\alpha_{chlorite\text{-}H2O}$ = −0.7‰ (Wenner and Taylor 1971) for O, 1000ln $\alpha_{muscovite\text{-}H2O}$ = −37.8‰ (Suzuoki and Epstein 1976) for H, and 1000ln $\alpha_{muscovite\text{-}H2O}$ = 1.3‰ (Vho et al. 2019) for oxygen.

The composition of water in equilibrium with chlorite and/or muscovite during progressive crystallization of micas follows a D-enrichment trend toward high δD values (Fig. 21c). A 20‰ enrichment in deuterium is reached after consumption of ~40–45% hydrogen to form chlorite or muscovite. During chlorite precipitation, there is a ~0.4‰ ^{18}O enrichment of the residual fluid, for a δD enrichment of ~20‰, whereas muscovite precipitation causes a decrease of fluid δ^{18}O values of about −0.7‰ for the same deuterium enrichment. In a δD–δ^{18}O space, muscovitization yields a trend with a steep negative slope whereas chlorite precipitation yields a steep positive slope. The change in δD of fluids during muscovite precipitation could contribute to explain high δD values of the hydrothermal fluid as shown in Fig. 18a.

Seismic rupture of fault veins occurs after a pre-seismic stage during which fluid pressure increases progressively to reach and overcome the lithostatic pressure (Sibson et al. 1988). Seismic rupture occurs once the increasing shear stress reaches a critical value, which then triggers progressive drainage of the overpressured fluid upward along the fault, and an abrupt fluid pressure decrease. This leads to increased effective stress on the fault plane that seals the fault, during which vein mineral precipitates, which reduces permeability of the fault and yield a new cycle of increasing fluid pressure. This cycle can be repeated numerous times, as attested by the typical crack and seal texture of orogenic veins, in the so-called fault-valve model (Sibson et al. 1988; Robert et al. 1995; Cox 2005). It has been shown that these pressure fluctuations can induce water vapor phase separation (Wilkinson and Johnston 1996; Ridley and Diamond 2000). It is likely that lower density vapor will migrate upward along a high permeability structure, separating from the residual liquid in a vein opening. Upon sealing of the vein, the vapor will condense under increasing pressure. The boiling-condensation cycles can be summarized as follow:

$$F_0 \rightarrow V_1 \rightarrow F_1 \rightarrow V_2 \rightarrow F_2 \rightarrow \ldots V_n \rightarrow F_n, \quad (3)$$

where F_0 is the initial fluid, V_n the composition of vapor separated from F_{n-1}, which then condenses in a new fluid F_n and n the number of boiling-condensation cycles (Fig. 21b).

We model the isotope effects during boiling-condensation open-system Rayleigh isotopic fractionation for vapor formation from a fraction of the liquid and assume complete condensation of vapor. This means that the new fluid F_n will have the same isotopic composition as the vapor V_n using Eqs. 4 and 5, modified after Sharp (2017):

$$\delta_{Liq} = \left[\delta_{Liq,i} + 1000\right]F^{(\alpha_{Vap-Liq}-1)} - 1000, \quad (4)$$

and

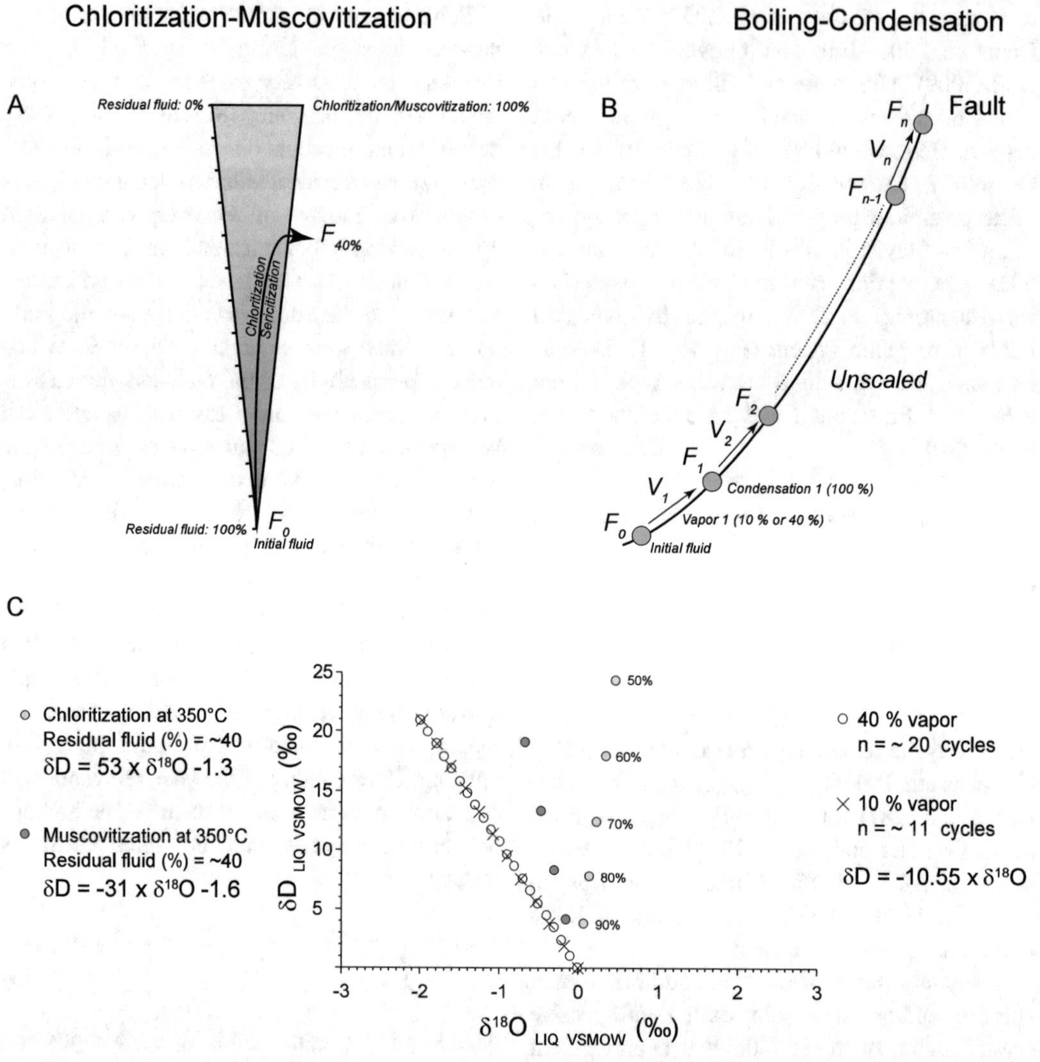

Fig. 21 **a** Conceptual model of the effect of chloritization/sericitization of country rock on the ore fluid. **b** Conceptual model of the successive boiling-condensation cycles related to the fault valve model (Robert et al. 1995). Vapor formation from an over-pressured fluid (F_0) can occur during seismic events. Vapor (V_1) is expected to migrate upward, faster than residual liquid and then to condense entirely (F_1) during re-sealing of fault. This process can be repeated numerous (n) times because of the successive re-opening of the fault during seismic cycles. **c** Quantification of the vector of enrichment in deuterium and depletion in ^{18}O of the evolved fluid (Fn) computed for the different models

$$\delta_{\mathrm{Vap}} = \alpha_{\mathrm{Vap-Liq}}\left(\delta_{\mathrm{Liq}} + 1000\right) - 1000. \quad (5)$$

Using isothermal conditions at 350 °C, considering 10% or 40% of vapor formation from liquid for each successive vapor separation episode, and oxygen and hydrogen isotope fractionation between vapor and liquid from Horita and Wesolowski (1994).

The two scenarios (10 and 40% vapor separation) yield similar linear trends with higher δD and lower δ^{18}O values for the condensed water, along a vector with a slope of −10.5 (Fig. 21c). The proportion of water evaporation for each cycle (10% or 40%) only changes the magnitude of the isotopic shift (Fig. 21c), and consequently the number of cycles required to reach 20‰

enrichment in deuterium of residual water, and a corresponding depletion of −2‰ in ^{18}O, similar to the maximum fluid δD values in orogenic gold deposits (Figs. 18a, b). Using 10% of evaporation, ~11 cycles are required to reach the δD value of 20‰, whereas ~20 cycles are needed with 40% evaporation steps (Fig. 21c). The modeled boiling-condensation trend is consistent with positive δD values (~20‰) combined with smaller ^{18}O depletion (~-4‰) of the fluid recorded by equilibration fractionation with hydrogen bearing minerals (Fig. 18a). The boiling-condensation trend is consistent with data with $\delta D_{H2O} > 0$‰ from the Val-d'Or vein field (Beaudoin and Chiaradia 2016) showing similar deuterium enrichment (δD from 0‰ to 29‰) and ^{18}O depletion ($\delta^{18}O$ from 5.9‰ to 2.9‰) of mineralizing fluid.

5.3.2 Source(s) of CO_2

CO_2 plays an important role in the formation of orogenic gold deposits as testified by (i) the common carbonation of country rocks, (ii) the association of gold with quartz-carbonate veins, and (iii) the occurrence of CO_2-rich fluid inclusions trapped in orogenic quartz veins (Smith et al. 1984; Ho et al. 1992). CO_2 has the capacity to buffer the pH of the fluid in a range where the gold complexation with reduced sulfur is increased (Phillips and Evans 2004). Nevertheless, the origin of CO_2 remains debated. Four main carbon reservoirs have been proposed to explain the wide variation of $\delta^{13}C$ values of CO_2; (i) reduced organic carbon in sedimentary rocks (such as graphite, organic matter) with low $\delta^{13}C$ values −26‰, Kerrich (1989); (ii) deep-seated mantle-derived carbon with $\delta^{13}C$ ranging from −10.8 to −1.6‰ (Trull et al. 1993), (iii) magmatic CO_2 ($\delta^{13}C$ from −5 to −2‰; (Taylor 1986), and (iv) seawater-derived carbonate with a $\delta^{13}C$ value near 0‰.

Figure 10 shows the distribution of $\delta^{13}C$ values CO_2 either from fluid inclusion trapped in quartz and scheelite or calculated from quartz-carbonate pairs with equilibrium temperatures in the range 250–550 °C. The $\delta^{13}C$ values of CO_2 display a similar range of values, between −25 and 6‰, as well as a similar distribution, mainly between −12 and 0‰ (Fig. 10a). No trend between $\delta^{13}C$ values and the age of deposit is documented, although all Archean age deposits have $\delta^{13}C_{CO2}$ values above −6‰.

The positive $\delta^{13}C$ values are commonly interpreted as metamorphic CO_2 released during decarbonation reaction of a source with $\delta^{13}C$ near 0‰, such as sedimentary carbonate rocks (Oberthuer et al. 1996; Chen et al. 2008; Scheffer et al. 2017). The low $\delta^{13}C_{CO2}$ values (< −8‰) are mainly documented from deposits hosted in sedimentary rocks (Fig. 10c). This is particularly well illustrated by the Meguma district from the Appalachian/Caledonian Orogen (Fig. 9c), where carbonate minerals have negative $\delta^{13}C$ values (−25.9‰ < $\delta^{13}C$ < −18.2‰), which was interpreted by Kontak and Kerrich (1997) to represent the formation of ^{13}C-depleted CO_2 by oxidation of graphite from the sedimentary country rocks (Fig. 9d). Most authors have argued that CO_2 with negative $\delta^{13}C_{CO2}$ values could formed by oxidation/hydrolysis of reduced carbon in sedimentary rocks during metamorphism (Oberthuer et al. 1996; Kontak and Kerrich 1997; Klein et al. 2008; Billstrom et al. 2009).

Nevertheless, some low $\delta^{13}C_{CO2}$ values are also documented from deposits hosted by igneous rocks. One value at −24.1‰ is from fluid inclusions in quartz from the Proterozoic Chega Tudo gold deposit (Brazil), where Klein et al. (2008) argued that the low $\delta^{13}C_{CO2}$ values reflect organic carbon from the carbonaceous schist hosting the porphyritic rocks. The two others values (−25.0‰ and −18.8‰) have been measured from fluid inclusions in scheelite from the Proterozoic intrusive-hosted Björkdal gold deposit, Sweden, (Billstrom et al. 2009), where they argued that these values indicate organic carbon derived from the greywacke-dominated sedimentary rocks that host the intrusion.

The most debated $\delta^{13}C_{CO2}$ values are those between −12 and −1‰ (Fig. 10). Several authors argued that these signatures indicate a deep-seated carbon source, such as sub-continental mantle or an igneous carbon source (Rye and Rye 1974; Shelton et al. 1988; Zhang et al. 1989; Santosh et al. 1995; Oberthuer et al. 1996;

Haeberlin 2002; So et al. 2002; Pandalai et al. 2003; Klein et al. 2007; Billstrom et al. 2009; Sun et al. 2009). Klein et al. (2007) concluded that the wide range of $\delta^{13}C_{CO2}$ values (−10.7 to −3.9‰) do not allow to discriminate a particular source and could reflect leaching of carbon of the country rocks along the fluid pathway. Alternatively, some authors argued that these signatures could also be derived from prograde metamorphism of reduced carbon-bearing sediments in the middle crust (Zhang et al. 1989; Madu et al. 1990; Oberthuer et al. 1996).

5.3.3 Source of H_2S

The sulfide minerals in orogenic gold deposits (Fig. 11) have $\delta^{34}S$ mostly from ~0‰ to 10‰ with a significant number of data tailing to values as low as −30‰. Mantle H_2S has a narrow range of compositions with $\delta^{34}S$ between ~−1.5‰ and ~1.0‰ (Labidi et al. 2013) whereas magmatic H_2S has $\delta^{34}S$ values typically ranging between ~−3.7‰ and ~16‰ (Seal 2006). Ancient marine evaporites record the marine sulfate reservoir with positive $\delta^{34}S$ values mainly between ~10 and ~35‰ (Claypool et al. 1980; Seal 2006). Finally, sedimentary pyrite defines a large range of $\delta^{34}S$ with low values, as low as −50‰, and positive values up to ~16‰. This large range of values reflects variable sulfur isotope fractionation during bacterial and thermochemical sulfate reduction (BSR and TSR). Sulfur isotope fractionation during TSR yields sulfides with 11–43‰ lower than sulfate $\delta^{34}S$ values at 500 °C and 100 °C respectively (Eldridge et al. 2016). During BSR, depletion of ^{34}S is up to 72‰ depending of environmental conditions (Machel et al. 1995; Wortmann et al. 2001; Johnston et al. 2007). In orogenic gold deposits, the range in H_2S sulfur isotope composition can result from mixing of different sulfur reservoirs or sulfur leaching during fluid/rock exchange along the fluid pathway, and changes in physico-chemical conditions (pressure, temperature, fO_2, pH) of the fluid (Schwarcz and Rees 1985; Fedorowich et al. 1991; Thode et al. 1991; Couture and Pilote 1993; Neumayr et al. 2008). Commonly, low $\delta^{34}S$ values (< 10‰) are attributed to BSR sources of reduced H_2S such as from sedimentary rocks (Nie and Bjorlykke 1994; Oberthuer et al. 1996; Nie 1998; Bierlein et al. 2004; Goldfarb et al. 2004; Chen et al. 2008), whereas higher $\delta^{34}S$ values (> 10‰) are attributed to TSR sources of H_2S (Kontak and Smith 1989; Tornos et al. 1997) or igneous sulfur (Hattori and Cameron 1986).

Investigation of mass independent fractionation of sulfur (MIF-S; $\Delta^{33}S$) in hydrothermal systems has been used as a complementary tracer of the source of H_2S. Whereas $\delta^{34}S$ values are sensitive to physico-chemical processes which can modify its isotopic composition along the fluid pathway or at the deposition site, MIF-S is a chemically conservative tracer. By combining $\delta^{34}S$ and $\Delta^{33}S$, several studies of Archean orogenic gold deposits from Yilgarn, Barberton and Abitibi have shown that H_2S is partly derived from Archean sedimentary rocks (Helt et al. 2014; Agangi et al. 2016; Selvaraja et al. 2017; LaFlamme et al. 2018; Petrella et al. 2020). The low variance of $\Delta^{33}S$ values in a deposit has been interpreted to indicate a homogeneous sedimentary H_2S reservoir, such that $\delta^{34}S$ variations in a deposit record variations of the oxidation state of the hydrothermal fluid (LaFlamme et al. 2018; Petrella et al. 2020) or redox reactions with country rocks, as suggested in Neumayr et al. (2008) for the St. Yves gold camp (Yilgarn). On the other hand, $\delta^{34}S$ variations at the deposit or district scales can record the incorporation of H_2S from the local country rocks as documented by Goldfarb et al. (1991); Goldfarb et al. (1997)

5.3.4 Nitrogen

Similar secular changes of the N isotope composition of mica and K-feldspar from orogenic gold veins and of sedimentary rocks suggest that hydrothermal fluids were dominantly derived from metamorphic dehydration of sedimentary rocks, consistent with the oxygen and hydrogen isotopic composition (Jia and Kerrich 1999, 2004; Jia et al. 2001, 2003), or that were equilibrated with sedimentary rocks along the fluid pathway (Kreuzer 2005). Nitrogen is incorporated into sedimentary rocks by biological activity and is thus mainly bound to organic matter. Organic matter displays a wide range of

$\delta^{15}N$ values (−4–15‰) (Minagawa and Wada 1986; Wu et al. 1997; Gu 2009). Kerogen in sedimentary rocks displays a more limited range of $\delta^{15}N$ values (2–6‰) (Williams et al. 1995; Ader et al. 1998; Sephton et al. 2002), whereas organic N in marine sediments ranges between 0‰ and 10‰ (Peters et al. 1978; Compton et al. 1992; Williams et al. 1995). During diagenesis and metamorphism, the maturation of organic matter releases ammonium (NH_4^+) that substitutes for K^+ in mica and K-feldspar. Metamorphic rocks display $\delta^{15}N$ values between 2 and 17‰ (Haendel et al. 1986; Bebout and Fogel 1992; Jia and Kerrich 2000). Some studies argued that metamorphism decreases the N concentration, and increase $\delta^{15}N$ values of the metamorphosed rocks (Haendel et al. 1986; Bebout and Fogel 1992), but the N isotope fractionation between source rock, hydrothermal fluid and newly formed minerals is small (Busigny et al. 2003; Jia et al. 2003; Jia and Kerrich 2004; Pitcairn et al. 2005; Kerrich et al. 2006).

5.3.5 Boron

The B isotope composition of tourmaline from orogenic gold deposits displays a wide range of values from −22‰ and 9‰ (Fig. 13). At temperatures between 250 and 550 °C, the tourmaline-fluid fractionation ranges from −4.5 and −1.6‰, (Meyer et al. 2008). Consequently, the large variation of 30‰ for $\delta^{11}B$ values cannot be the result of variation in the temperature of formation of orogenic gold deposits. The B isotope composition of the marine seawater reservoir has values near 40‰. In contrast, $\delta^{11}B$ values of others terrestrial reservoirs mostly overlap: continental crust (−6 to −14‰), I-type and S-type magmas (−7 to 3‰ and −7 to −15‰, respectively), MORB and mantle rocks (−8 to −6‰), marine sediments (−14 to −2‰), or marine carbonate (−12 to 26‰) (Fig. 13c) (Palmer and Swihart 1996; van Hinsberg et al. 2011; Lambert-Smith et al. 2016; Marschall et al. 2017; Trumbull and Slack 2018).

At the Paleoproterozoic Palokas gold deposit (Fennoscandian Shield), $\delta^{11}B$ values in the range of −4.5—1.0‰ are interpreted to result from a single magmatic-hydrothermal event (Ranta et al. 2017). In the Paleozoic Passagem de Mariana gold deposit (Brazil), Trumbull et al. (2019) reported $\delta^{11}B$ values between −11.5 and 7.1‰ for orogenic gold veins, and argued in favor to a crustal B source derived from the sedimentary rocks hosting the deposit. A similar hypothesis was discussed by Büttner et al. (2016) and Jiang et al. (2002) to explain the range of $\delta^{11}B$ values (−18.4 to −8.9‰ and −21.4 to −17.8‰, respectively) in Proterozoic gold deposits from the Twangiza-Namoya gold belt (Democratic Republic of Congo) and in the Archean Mount Gibson deposit (Yilgarn Craton), respectively.

The contribution of at least two different sources of B is invoked to explain the wide range of $\delta^{11}B$ recorded in some deposits. At Hira Buddini (India), Krienitz et al. (2008) documented tourmaline with $\delta^{11}B$ values between −4 and 9‰, interpreted to result from mixing between a metamorphic fluid ($\delta^{11}B \sim 0$‰), generated by devolatilization of volcanic rocks, and a magmatic fluid ($\delta^{11}B \sim 10$‰) exsolved from I-type granitic intrusions. A similar interpretation was advanced by Molnár et al. (2016) to explain the wide range of $\delta^{11}B$ values (−24.1 to −9.6‰) in the Archean Hattu schist belt (Finland), where the range in values was interpreted to reflect mixing between a heavy magmatic B with a lighter B reservoir from metamorphic devolatilization of sedimentary rocks.

Tourmaline from orogenic gold quartz-tourmaline veins from the Proterozoic Tapera Grande and Quartzito deposits (Brazil) shows $\delta^{11}B$ values ranging from −15.7 to −5.0‰ (Garda et al. 2009). The wide range of value is interpreted to result from mixing B from multiple sources, derived from volcano-sedimentary rocks (higher $\delta^{11}B$), and from granitic or sedimentary rocks with lower $\delta^{11}B$ values. Similarly, tourmaline from the Proterozoic Bhukia gold deposit (India) has $\delta^{11}B$ values in the range of −12.6 to −9.2‰, that have been interpreted to originate from two fluid sources, a granite-derived hydrothermal and a metapelite-derived metamorphic fluids (Hazarika et al. 2019). Lambert-Smith et al. (2016) also argued that the wide range of δ^{11}B values, from 3.5 to 19.8‰, in tourmaline from gold-rich veins in the Loulo mining district (Mali), resulted from mixing marine evaporite with a sedimentary-derived B. In the Val-d'Or district, Beaudoin et al. (2013) and Daver et al. (2020) concluded that tourmaline δ^{11}B values, between −15.6 and −7.7‰, are a result of mixing of a low δ^{11}B prograde metamorphic fluids released from sedimentary and/or volcanic rocks with a high δ^{11}B seawater-derived pore fluid. Trumbull et al. (2020) show a bimodal distribution of the δ^{11}B values for orogenic gold deposit tourmaline, consistent with two common B reservoirs, sourced from metamorphic/igneous rocks (low δ^{11}B) or from seawater-altered volcano-sedimentary rocks (high δ^{11}B).

In summary, B in orogenic gold deposits generally derived from multiple sources. It is generally accepted that B in orogenic deposits is derived from the devolatilization of volcano-sedimentary and/or sedimentary sequences during metamorphism (Jiang et al. 2002; Krienitz et al. 2008; Garda et al. 2009; Beaudoin et al. 2013; Büttner et al. 2016; Lambert-Smith et al. 2016; Molnár et al. 2016; Hazarika et al. 2019; Trumbull et al. 2019; Daver et al. 2020). The wide range of B isotopic values suggest mixing with other B source(s) that remains matter of debate. B-rich seawater-derived fluid (Beaudoin et al. 2013; Lambert-Smith et al. 2016; Daver et al. 2020) has been proposed as possible source, consistent with the upper crustal fluid identified using oxygen and hydrogen isotopes. B-rich magmatic-derived fluid are also suggested in rare cases (Krienitz et al. 2008; Molnár et al. 2016) but are unlikely to be a common boron reservoir.

5.3.6 Silicon

The silicon isotope composition of quartz is poorly documented in orogenic gold deposits. Only 30 δ^{30}Si values are compiled (Fig. 14a,b), from the Hemlo (Superior Craton) and the Sawaya'erdun (Tianshan Orogen, China) deposits (Ding et al. 1996; Liu et al. 2007). Natural variation of the δ^{30}Si values between various terrestrial reservoirs is small (Fig. 14c). Mantle peridotite, granulitic lower continental crust, igneous continental rocks, oceanic crust (MORB + OIB), and sedimentary rocks display overlapping ranges of δ^{30}Si values between −0.8 and 0‰ (Poitrasson 2017). Only chert-BIFs, and seawater stand out with values between −3.6–5.0‰ and 0.6–4.4‰, respectively (Poitrasson 2017). Silicon does not fractionate significantly during fluid-rock interaction at temperatures above 50 °C (Douthitt 1982; Ding et al. 1988; Geilert et al. 2014; Poitrasson 2017), and thus does not provide information on temperature and processes during quartz precipitation. For orogenic gold systems, δ^{30}Si is mainly used to discuss the source of silicon by comparing the δ^{30}Si values of the veins with those of the local country-rocks. Hence, (Liu et al. 2007) argued that the silica (δ^{30}Si = −0.5–0.5‰) associated with the formation of the Sawaya'erdun orogenic gold deposit was derived from sedimentary (δ^{30}Si = −0.5–1.9‰) and volcanic (δ^{30}Si = 0.1‰) country rocks.

6 Conclusions

The stable isotope geochemistry of veins minerals has been widely used to constrain various parameters related to fluid systems involved in the formation of orogenic gold deposits. In this review, we compare and integrate multi-isotopes data from numerous deposits worldwide of various ages and locations to identify global trends shared by most of orogenic gold deposits.

6.1 Temperature

Temperature estimates using quartz-silicate, quartz-carbonate and sulfide-sulfide mineral-pairs display consistent temperature ranges of 360 ± 76 °C (σ = 1, n = 332) with no secular changes, which are in agreement with temperatures expected for deposits formed at temperatures typical for the greenschist to lower amphibolite facies of the country rocks. Quartz-silicate pairs are mostly in isotopic equilibrium, suggesting that retrograde isotopic exchange between minerals and fluids did not disturb the bulk isotopic composition. In contrast, a significant proportion of quartz-carbonate pairs show isotopic disequilibrium. This could result from quartz and carbonate forming at different times, but later oxygen isotopic exchange of carbonate with a low-temperature fluid is more consistent with the paragenetic sequence of orogenic gold deposits. Sulfide-sulfide mineral-pairs are mostly in isotopic equilibrium.

6.2 Fluid Reservoirs

The oxygen and hydrogen isotope composition of fluids indicate that fluid mixing is a process documented in many orogenic gold deposits. The mixing endmembers are a low T—high δD—low $\delta^{18}O$ upper crustal and a high T—low δD—high $\delta^{18}O$ deep-seated fluid reservoirs.

The nature of the upper crustal fluid is surficial water of various origins having experienced long lived exchange with country rocks. Meteoric water is rarely documented and is commonly related to late infiltration disconnected from the formation of the gold deposits. This is likely the case for most low δD values (< −120‰) documented in OH-bearing mineral and fluid inclusion, mainly from Mesozoic deposits.

The deep-seated fluid endmember is likely metamorphic in origin as attested by hydrogen and oxygen isotope compositions of water plotting mostly in the field for metamorphic water, even if a few data plot in the overlapping field for magmatic and metamorphic waters. The $\delta^{11}B$ values also point a metamorphic source for the fluid that mixed with B-rich seawater-derived fluid, consistent with O–H isotope. The $\delta^{15}N$ values of orogenic gold vein minerals also favor a deep-seated fluid reservoir derived from metamorphic dehydration of sedimentary rocks. The contribution of magmatic water exsolved from magma is documented in some specific cases but is not an essential component for the formation of orogenic gold deposits.

Spatial variations of oxygen and hydrogen isotope compositions of deep-seated fluid between districts indicate variable proportions of igneous and sedimentary rocks in the crustal segment released water during prograde metamorphism. Such provinciality is documented for $\delta^{18}O$ values of quartz between cratons and districts.

Finally, a significant number of high δD values (up to 20‰) for water result of (i) muscovite precipitation during wallrock alteration and (ii) successive phase-separation of water associated to cyclic seismic rupture of fault-veins that displays trends of enrichment in deuterium that fit well the data.

6.2.1 Carbon Isotopes

No secular variations of mineral isotope composition are identified for carbon isotopes ($\delta^{13}C$). The $\delta^{13}C$ values in orogenic gold deposits suggest that CO_2 can be derived from the oxidation of reduced carbon and/or from decarbonation reaction of ~0‰ inorganic sedimentary rocks during metamorphism, or derived from a deep-seated carbon source such as the mantle. Wide ranges of $\delta^{13}C$ are commonly documented at the deposit scale that suggests a carbon contribution from multiple sources, or that fluid acquired its $\delta^{13}C$ isotopic signature along the fluid pathway through exchange with country rocks.

6.2.2 Sulfur Isotope

$\delta^{34}S$ values of sulfide minerals do not display secular variation or provinciality, but display wide ranges of values. Mixing of sulfur from different reservoirs and/or changes in physico-chemical conditions of the fluid at the site of deposition have been proposed to explain the $\delta^{34}S$ variation in a deposit. Commonly, low $\delta^{34}S$ values are attributed to BSR sulfur from sedimentary rocks, whereas positive $\delta^{34}S$ values are attributed to TSR sulfur or igneous sulfur. Orogenic gold deposits hosted by sedimentary rocks contain sulfur leached from the local country rocks. At the deposit scale, $\delta^{34}S$ values record changes in the oxidation state of the hydrothermal fluid related to redox reactions with country rocks along the fluid pathway or at the site of deposition.

Acknowledgements This research has been made possible by the scientific freedom permitted by Discovery Research Grants of the Natural Science and Engineering Council of Canada. We thank Richard Goldfarb for sharing his unique knowledge of orogenic gold deposits and unpublished data. Lydia Lobato and Kevin Andsell are thanked for providing data. The compilation of the database would not have been possible without the help of Émilie Bédard and Samuël Simard. We are grateful for the constructive comments by referee R.J. Goldfarb and a second anonymous reviewer, and Editor D. Huston, which led to significant improvements to the paper. This work is the Metal Earth contribution MERC-ME-2022-38.

References

Ader M, Boudou J-P, Javoy M, Goffe B, Daniels E (1998) Isotope study on organic nitrogen of Westphalian anthracites from the Western Middle field of Pennsylvania (U.S.A.) and from the Bramsche Massif (Germany). Org Geoch 29:315–323. https://doi.org/10.1016/S0146-6380(98)00072-2

Agangi A, Hofmann A, Eickmann B, Marin-Carbonne J, Reddy SM (2016) An atmospheric source of S in Mesoarchaean structurally-controlled gold mineralisation of the Barberton Greenstone Belt. Precambrian Res 285:10–20. https://doi.org/10.1016/j.precamres.2016.09.004

Apodaca LE (1994) Genesis of lode gold deposits of the Rock Creek area, Nome mining district, Seward Peninsula, Alaska. Unpubished PhD thesis, University of Colorado

Beaudoin G (2011) The stable isotope geochemistry of orogenic gold deposits. In: Barra F, Reich M, Campos E, Tornos F (eds) Proceedings of the eleventh Biennial SGA meeting, Antogfagasta, Chile, pp 556–558

Beaudoin G, Chiaradia M (2016) Fluid mixing in orogenic gold deposits: evidence from the H-O-Sr isotope composition of the Val-d'Or vein field (Abitibi, Canada). Chem Geol 437:7–18. https://doi.org/10.1016/j.chemgeo.2016.05.009

Beaudoin G, Pitre D (2005) Stable isotope geochemistry of the Archean Val-d'Or (Canada) orogenic gold vein field. Miner Deposita 40:59–75. https://doi.org/10.1007/s00126-005-0474-z

Beaudoin G, Therrien R, Savard C (2006) 3D numerical modelling of fluid flow in the Val-d'Or orogenic gold district: major crustal shear zones drain fluids from overpressured vein fields. Miner Deposita 41:82–98. https://doi.org/10.1007/s00126-005-0043-5

Beaudoin G, Rollion-Bard C, Giuliani G (2013) The boron isotope composition of tourmaline from the Val-d'Or orogenic gold deposits, Quebec, Canada. In: Mineral deposit research for a high-tech world 12th SGA Biennial meeting, Proceedings vol 3, pp 1090–1092

Beaudoin G, Therrien P (2009) The updated web stable isotope fractionation calculator. In: Handbook of stable isotope analytical techniques, Volume-II. De Groot, P.A. (ed.). Elsevier: 1120–1122

Bebout GE, Fogel ML (1992) Nitrogen-isotope compositions of metasedimentary rocks in the Catalina Schist, California: implications for metamorphic devolatilization history. Geochim Cosmochim Acta 56:2839–2849. https://doi.org/10.1016/0016-7037(92)90363-N

Bierlein FP, Crowe DE (2000) Phanerozoic orogenic lode gold deposits Rev. Econ Geol 13:103–139

Bierlein FP, Christie AB, Smith PK (2004) A comparison of orogenic gold mineralisation in central Victoria (AUS), western South Island (NZ) and Nova Scotia (CAN): implications for variations in the endowment of Palaeozoic metamorphic terrains. Ore Geol Rev 25:125–168

Billstrom K, Broman C, Jonsson E, Recio C, Boyce AJ, Torssander P (2009) Geochronological, stable isotopes and fluid inclusion constraints for a premetamorphic development of the intrusive-hosted Bjorkdal Au deposit, northern Sweden. Int J Earth Sci 98:1027–1052. https://doi.org/10.1007/s00531-008-0301-8

Böhlke JK, Kistler RW (1986) Rb-Sr, K-Ar, and stable isotope evidence for the ages and sources of fluid components of gold-bearing quartz veins in the northern Sierra Nevada foothills metamorphic belt, California. Econ Geol 81:296–322. https://doi.org/10.2113/gsecongeo.81.2.296

Boer RH, Meyer FM, Robb LJ, Graney JR, Vennemann TW, Kesler SE (1995) Mesothermal-type mineralization in the Sabie-Pilgrim's Rest gold field, South Africa. Econ Geol 90:860–876. https://doi.org/10.2113/gsecongeo.90.4.860

Boiron M-C, Cathelineau M, Banks DA, Fourcade S, Vallance J (2003) Mixing of metamorphic and surficial fluids during the uplift of the Hercynian upper crust: consequences for gold deposition. Chem Geol 194:119–141. https://doi.org/10.1016/S0009-2541(02)00274-7

Busigny V, Cartigny P, Philippot P, Ader M, Javoy M (2003) Massive recycling of nitrogen and other fluid-mobile elements (K, Rb, Cs, H) in a cold slab environment: evidence from HP to UHP oceanic metasediments of the Schistes Lustrés nappe (western Alps, Europe). Earth Planet Sci Lett 215:27–42. https://doi.org/10.1016/S0012-821X(03)00453-9

Büttner SH, Reid W, Glodny J, Wiedenbeck M, Chuwa G, Moloto T, Gucsik A (2016) Fluid sources in the Twangiza-Namoya Gold Belt (Democratic Republic of Congo): evidence from tourmaline and fluid compositions, and from boron and Rb–Sr isotope systematics. Precambrian Res 280:161–178. https://doi.org/10.1016/j.precamres.2016.05.006

Cartigny P, Marty B (2013) Nitrogen isotopes and mantle geodynamics: the emergence of life and the atmosphere–crust–mantle connection. Elements 9:359–366. https://doi.org/10.2113/gselements.9.5.359

Chang Z, Large RR, Maslennikov V (2008) Sulfur isotopes in sediment-hosted orogenic gold deposits: evidence for an early timing and a seawater sulfur source. Geology 36:971–974. https://doi.org/10.1130/G25001A.1

Chen Y-J, Pirajno F, Qi J-P (2008) The Shanggong gold deposit, Eastern Qinling Orogen, China: isotope geochemistry and implications for ore genesis. J Asian Earth Sci 33:252–266

Claypool GE, Holser WT, Kaplan IR, Sakai H, Zak I (1980) The age curves of sulfur and oxygen isotopes in marine sulfate and their mutual interpretation. Chem Geol 28:199–260. https://doi.org/10.1016/0009-2541(80)90047-9

Clayton RN, O'Neil JR, Mayeda TK (1972) Oxygen isotope exchange between quartz and water. J Geophys Res 1896–1977(77):3057–3067. https://doi.org/10.1029/JB077i017p03057

Compton JS, Williams LB, Ferrell RE (1992) Mineralization of organogenic ammonium in the Monterey Formation, Santa Maria and San Joaquin basins, California, USA. Geochim Cosmochim Acta 56:1979–1991. https://doi.org/10.1016/0016-7037(92)90324-C

Couture JF, Pilote P (1993) The geology and alteration patterns of a disseminated, shear zone-hosted mesothermal gold deposit; the Francoeur 3 Deposit, Rouyn-Noranda, Quebec. Econ Geol 88:1664–1684. https://doi.org/10.2113/gsecongeo.88.6.1664

Cox SF (2005) Coupling between deformation, fluid pressures, and fluid flow in ore-producing hydrothermal systems at depth in the crust. Econ Geol 100th Anniv 39–75

Craig H (1961) Isotopic variations in meteoric waters. Science 133:1702. https://doi.org/10.1126/science.133.3465.1702

Craw D, Upton P, Mackenzie DJ (2009) Hydrothermal alteration styles in ancient and modern orogenic gold deposits, New Zealand. NZ J Geol Geophys 52:11–26. https://doi.org/10.1080/00288300909509874

Criss RE, Gregory RT, Taylor HP (1987) Kinetic theory of oxygen isotopic exchange between minerals and water. Geochim Cosmochim Acta 51:1099–1108. https://doi.org/10.1016/0016-7037(87)90203-1

Daver L, Jébrak M, Beaudoin G, Trumbull RB (2020) Three-stage formation of greenstone-hosted orogenic gold deposits in the Val-d'Or mining district, Abitibi, Canada: evidence from pyrite and tourmaline. Ore Geol Rev 120:103449. https://doi.org/10.1016/j.oregeorev.2020.103449

Deng J, Liu X, Wang Q, Pan R (2015) Origin of the Jiaodong-type Xinli gold deposit, Jiaodong Peninsula, China: constraints from fluid inclusion and C-D-O-S-Sr isotope compositions. Ore Geol Rev 65:674–686. https://doi.org/10.1016/j.oregeorev.2014.04.018

Ding TP, Wan DF, Li JC, Jiang SY, Song HB, Li YH, Liu ZJ (1988) The analytical method of silicon isotopes and its geological application. Miner Deposits 7:90–95

Ding T, Jiang S, Wan D, Li Y, Song H, Liu Z, Yao X (1996) Silicon isotope geochemistry. Geological Publishing House, Beijing

Douthitt CB (1982) The geochemistry of the stable isotopes of silicon. Geochim Cosmochim Acta 46:1449–1458. https://doi.org/10.1016/0016-7037(82)90278-2

Dubé B, Gosselin P (2007) Greenstone-hosted quartz-carbonate vein deposits. Geol Assoc Can Mineral Dep Div Spec Publ 5:49–73

Eilu P, Groves DI (2001) Primary alteration and geochemical dispersion haloes of Archaean orogenic gold deposits in the Yilgarn Craton: the pre-weathering scenario. Geochem Explor Environ Anal 1:183. https://doi.org/10.1144/geochem.1.3.183

Eldridge DL, Guo W, Farquhar J (2016) Theoretical estimates of equilibrium sulfur isotope effects in aqueous sulfur systems: highlighting the role of isomers in the sulfite and sulfoxylate systems. Geochim Cosmochim Acta 195:171–200. https://doi.org/10.1016/j.gca.2016.09.021

Faure K (2003) δD values of fluid inclusion water in quartz and calcite ejecta from active geothermal systems: do values reflect those of original hydrothermal water? Econ Geol 98:657–660. https://doi.org/10.2113/gsecongeo.98.3.657

Fedorowich JS, Stauffer MR, Kerrich R (1991) Structural setting and fluid characteristics of the Proterozoic Tartan Lake gold deposit, Trans-Hudson Orogen, northern Manitoba. Econ Geol 86:1434–1467. https://doi.org/10.2113/gsecongeo.86.7.1434

Foley NK, Bethke PM, Rye RO (1989) A reinterpretation of the δD_{H2O} of inclusion fluids in contemporaneous quartz and sphalerite, Creede mining district, Colorado: a generic problem for shallow orebodies? Econ Geol 84:1966–1977

Gaboury D (2013) Does gold in orogenic deposits come from pyrite in deeply buried carbon-rich sediments?: insight from volatiles in fluid inclusions. Geology 41:1207–1210. https://doi.org/10.1130/g34788.1

Garda GM, Trumbull RB, Beljavskis P, Wiedenbeck M (2009) Boron isotope composition of tourmalinite and vein tourmalines associated with gold mineralization, Serra do Itaberaba Group, central Ribeira Belt, SE Brazil. Chem Geol 264:207–220. https://doi.org/10.1016/j.chemgeo.2009.03.013

Geilert S, Vroon PZ, Roerdink DL, Van Cappellen P, van Bergen MJ (2014) Silicon isotope fractionation during abiotic silica precipitation at low temperatures: inferences from flow-through experiments. Geochim Cosmochim Acta 142:95–114. https://doi.org/10.1016/j.gca.2014.07.003

Giletti BJ (1986) Diffusion effects on oxygen isotope temperatures of slowly cooled igneous and metamorphic rocks. Earth Planet Sci Lett 77:218–228. https://doi.org/10.1016/0012-821X(86)90162-7

Goldfarb RJ, Groves DI (2015) Orogenic gold: common or evolving fluid and metal sources through time. Lithos 233:2–26. https://doi.org/10.1016/j.lithos.2015.07.011

Goldfarb RJ, Newberry RJ, Pickthorn WJ, Gent CA (1991) Oxygen, hydrogen, and sulfur isotope studies in the Juneau gold belt, southeastern Alaska; constraints on the origin of hydrothermal fluids. Econ Geol 86:66–80. https://doi.org/10.2113/gsecongeo.86.1.66

Goldfarb RJ, Miller LD, Leach DL, Snee LW (1997) Gold deposits in metamorphic rocks of Alaska. Econ Geol Mon 9:151–190

Goldfarb RJ, Ayuso R, Miller ML, Ebert SW, Marsh EE, Petsel SA, Miller LD, Bradley D, Johnson C, McClelland W (2004) The Late Cretaceous Donlin Creek gold deposit, southwestern Alaska: controls on epizonal ore formation. Econ Geol 99:643–671. https://doi.org/10.2113/99.4.643

Goldfarb RJ, Baker T, Dubé B, Groves DI, Hart CJR, Gosselin P (2005) Distribution, character, and genesis of gold deposits in metamorphic terranes. Econ Geol 100th Anniv Vol, 407–450

Gosselin P, Dubé B (2005) Gold deposits and gold districts of the world. Geol Surv Can Open File 4895

Graham CM, Atkinson J, Harmon RS (1984) Hydrogen isotope fractionation in the system chlorite-water. Nat Environ Res Counc UK Publ Ser D 25:139–140

Graham CM, Viglino JA, Harmon RS (1987) Experimental study of hydrogen-isotope exchange between aluminous chlorite and water and of hydrogen diffusion in chlorite. Am Mineral 72:566–657

Gray DR, Gregory RT, Durney DW (1991) Rock-buffered fluid-rock interaction in deformed quartz-rich turbidite sequences, eastern Australia. J Geophys Res 96 (B12):19681–19704

Groves DI, Goldfarb RJ, Gebre-Mariam M, Hagemann SG, Robert F (1998) Orogenic gold deposits: a proposed classification in the context of their crustal distribution and relationship to other gold deposit types. Ore Geol Rev 13:7–27

Gu B (2009) Variations and controls of nitrogen stable isotopes in particulate organic matter of lakes. Oecologia 160:421–431. https://doi.org/10.1007/s00442-009-1323-z

Haeberlin Y (2002) Geological and structural setting, age, and geochemistry of the orogenic gold deposits at the Pataz Province, eastern Andean Cordillera, Peru. Unpublishe PhD thesis, Université de Genève

Haendel D, Mühle K, Nitzsche H-M, Stiehl G, Wand U (1986) Isotopic variations of the fixed nitrogen in metamorphic rocks. Geochim Cosmochim Acta 50:749–758. https://doi.org/10.1016/0016-7037(86)90351-0

Hagemann SG, Gebre-Mariam M, Groves DI (1994) Surface-water influx in shallow-level Archean lode-gold deposits in Western, Australia. Geology 22:1067–1070. https://doi.org/10.1130/0091-7613(1994)022%3c1067:swiisl%3e2.3.co;2

Hattori K, Cameron EM (1986) Archaean magmatic sulphate. Nature 319:45–47

Hazarika P, Bhuyan N, Upadhyay D, Abhinay K, Singh NN (2019) The nature and sources of ore-forming fluids in the Bhukia gold deposit, Western India: constraints from chemical and boron isotopic composition of tourmaline. Lithos 350–351:105227. https://doi.org/10.1016/j.lithos.2019.105227

Helt KM, Williams-Jones AE, Clark JR, Wing BA, Wares RP (2014) Constraints on the genesis of the Archean oxidized, intrusion-related Canadian Malartic gold deposit, Quebec, Canada. Econ Geol 109:713–735. https://doi.org/10.2113/econgeo.109.3.713

Ho SE, Groves DI, McNaughton NJ, Mikucki EJ (1992) The source of ore fluids and solutes in Archaean lode-gold deposits of Western Australia. J Volcanol Geotherm Res 50:173–196

Horita J, Wesolowski DJ (1994) Liquid-vapor fractionation of oxygen and hydrogen isotopes of water from the freezing to the critical temperature. Geochim Cosmochim Acta 58:3425–3437. https://doi.org/10.1016/0016-7037(94)90096-5

Jia Y, Kerrich R (1999) Nitrogen isotope systematics of mesothermal lode gold deposits: Metamorphic, granitic, meteoric water, or mantle origin? Geology 27:1051–1054. https://doi.org/10.1130/0091-7613(1999)027%3c1051:nisoml%3e2.3.co;2

Jia Y, Kerrich R (2000) Giant quartz vein systems in accretionary orogenic belts: the evidence for a metamorphic fluid origin from $\delta^{15}N$ and $\delta^{13}C$ studies. Earth Planet Sci Lett 184:211–224

Jia Y, Kerrich R (2004) Nitrogen 15–enriched Precambrian kerogen and hydrothermal systems. Geochem Geophys Geosys 5:Q07005. https://doi.org/10.1029/2004GC000716

Jia Y, Li XIA, Kerrich R (2001) Stable Isotope (O, H, S, C, and N) Systematics of quartz vein systems in the turbidite-hosted Central and North Deborah gold deposits of the Bendigo gold field, central Victoria, Australia: constraints on the origin of ore-forming fluids. Econ Geol 96:705–721. https://doi.org/10.2113/96.4.705

Jia Y, Kerrich R, Goldfarb R (2003) Metamorphic origin of ore-forming fluids for orogenic gold-bearing quartz vein systems in the North American Cordillera: constraints from a reconnaissance study of $\delta^{15}N$, δD, and $\delta^{18}O$. Econ Geol 98:109–123. https://doi.org/10.2113/98.1.109

Jiang S-Y, Palmer MR, Yeats CJ (2002) Chemical and boron isotopic compositions of tourmaline from the Archean Big Bell and Mount Gibson gold deposits, Murchison Province, Yilgarn Craton, Western Australia. Chem Geol 188:229–247. https://doi.org/10.1016/s0009-2541(02)00107-9

Johnston DT, Farquhar J, Canfield DE (2007) Sulfur isotope insights into microbial sulfate reduction: When microbes meet models. Geochim Cosmochim Acta 71:3929–3947. https://doi.org/10.1016/j.gca.2007.05.008

Kajiwara Y, Krouse HR (1971) Sulfur isotope partitioning in metallic sulfide systems. Can J Earth Sci 8:1397–1408. https://doi.org/10.1139/e71-129

Kerrich R (1987) The stable isotope geochemistry of Au-Ag vein deposits in metamorphic rocks. Miner Assoc Can Short Course Handb 13:287–336

Kerrich R (1989) Geochemical evidence on the sources of fluids and solutes for shear zone hosted mesothermal gold deposits. Geol Assoc Can Short Course Notes 6:129–198

Kerrich R, Jia Y, Manikyamba C, Naqvi SM (2006) Secular variations of N-isotopes in terrestrial reservoirs and ore deposits. Geol Soc Am Mem 198:81–104

Klein EL, Harris C, Giret A, Moura CAV (2007) The Cipoeiro gold deposit, Gurupi Belt, Brazil: geology, chlorite geochemistry, and stable isotope study. J S Am Earth Sci 23:242–255

Klein EL, Ribeiro JWA, Harris C, Moura CAV, Giret A (2008) Geology and fluid characteristics of the Mina Velha and Mandiocal orebodies and implications for the genesis of the orogenic Chega Tudo gold deposit, Gurupi Belt, Brazil. Econ Geol 103:957–980. https://doi.org/10.2113/gsecongeo.103.5.957

Kontak DJ, Smith PK (1989) Sulphur isotopic composition of sulphides from the Beaver Dam and other Meguma-Group-hosted gold deposits, Nova Scotia; implications for genetic models. Can J Earth Sci 26:1617–1629

Kontak DJ, Kerrich R (1997) An isotopic (C, O, Sr) study of vein gold deposits in the Meguma Terrane, Nova Scotia; implication for source reservoirs. Econ Geol 92:161–180. https://doi.org/10.2113/gsecongeo.92.2.161

Kotzer TG, Kyser TK, King RW, Kerrich R (1993) An empirical oxygen- and hydrogen-isotope geothermometer for quartz-tourmaline and tourmaline-water. Geochim Cosmochim Acta 57:3421–3426. https://doi.org/10.1016/0016-7037(93)90548-B

Kreuzer OP (2005) Intrusion-hosted mineralization in the Charters Towers goldfield, north Queensland: new isotopic and fluid inclusion constraints on the timing and rrigin of the auriferous veins. Econ Geol 100:1583–1603. https://doi.org/10.2113/100.8.1583

Krienitz M, Trumbull R, Hellmann A, Kolb J, Meyer F, Wiedenbeck M (2008) Hydrothermal gold mineralization at the Hira Buddini gold mine, India: constraints on fluid evolution and fluid sources from boron isotopic compositions of tourmaline. Miner Deposita 43:421–434. https://doi.org/10.1007/s00126-007-0172-0

Labidi J, Cartigny P, Moreira M (2013) Non-chondritic sulphur isotope composition of the terrestrial mantle. Nature 501:208–211. https://doi.org/10.1038/nature12490

LaFlamme C, Sugiono D, Thébaud N, Caruso S, Fiorentini M, Selvaraja V, Jeon H, Voute F, Martin L (2018) Multiple sulfur isotopes monitor fluid evolution of an Archean orogenic gold deposit. Geochim Cosmochim Acta 222:436–446. https://doi.org/10.1016/j.gca.2017.11.003

Lambert-Smith JS, Rocholl A, Treloar PJ, Lawrence DM (2016) Discriminating fluid source regions in orogenic gold deposits using B-isotopes. Geochim Cosmochim Acta 194:57–76. https://doi.org/10.1016/j.gca.2016.08.025

Li J-W, Bi S-J, Selby D, Chen L, Vasconcelos P, Thiede D, Zhou M-F, Zhao X-F, Li Z-K, Qiu H-N (2012) Giant Mesozoic gold provinces related to the destruction of the North China craton. Earth Planet Sci Lett 349–350:26–37. https://doi.org/10.1016/j.epsl.2012.06.058

Li Y, Liu J (2006) Calculation of sulfur isotope fractionation in sulfides. Geochim Cosmochim Acta 70:1789–1795. https://doi.org/10.1016/j.gca.2005.12.015

Liu P (1992) The origin of the Archean Jardine iron formation-hosted lode gold deposit, Montana. Unpublished MS thesis, Iowa State University

Liu J, Zheng M, Cook NJ, Long X, Deng J, Zhai Y (2007) Geological and geochemical characteristics of the Sawaya'erdun gold deposit, southwestern Chinese Tianshan. Ore Geol Rev 32:125–156

Liu S, Li Y, Liu J, Shi Y (2015) First-principles study of sulfur isotope fractionation in pyrite-type disulfides. Am Mineral 100:203–208. https://doi.org/10.2138/am-2015-5003

Machel HG, Krouse HR, Sassen R (1995) Products and distinguishing criteria of bacterial and thermochemical sulfate reduction. Appl Geochem 10:373–389. https://doi.org/10.1016/0883-2927(95)00008-8

Madu BE, Nesbitt BE, Muehlenbachs K (1990) A mesothermal gold-stibnite-quartz vein occurrence in the Canadian Cordillera. Econ Geol 85:1260–1268. https://doi.org/10.2113/gsecongeo.85.6.1260

Mao J, Wang Y, Li H, Pirajno F, Zhang C, Wang R (2008) The relationship of mantle-derived fluids to gold metallogenesis in the Jiaodong Peninsula: Evidence from D-O-C-S isotope systematics. Ore Geol Rev 33:361–381

Marschall HR, Wanless VD, Shimizu N, Pogge von Strandmann PAE, Elliott T, Monteleone BD (2017) The boron and lithium isotopic composition of mid-ocean ridge basalts and the mantle. Geochim Cosmochim Acta 207:102–138. https://doi.org/10.1016/j.gca.2017.03.028

McCuaig TC, Kerrich R (1998) P-T-t-deformation-fluid characteristics of lode gold deposits: evidence from alteration systematics. Ore Geol Rev 12:381–453

Meyer C, Wunder B, Meixner A, Romer RL, Heinrich W (2008) Boron-isotope fractionation between tourmaline and fluid: an experimental re-investigation. Contrib Miner Petrol 156:259–267. https://doi.org/10.1007/s00410-008-0285-1

Minagawa M, Wada E (1986) Nitrogen isotope ratios of red tide organisms in the East China Sea: a characterization of biological nitrogen fixation. Mar Chem 19:245–259

Molnár F, Mänttäri I, O'Brien H, Lahaye Y, Pakkanen L, Johanson B, Käpyaho A, Sorjonen-Ward P, Whitehouse M, Sakellaris G (2016) Boron, sulphur and copper isotope systematics in the orogenic gold deposits of the Archaean Hattu schist belt, eastern Finland. Ore Geol Rev 77:133–162. https://doi.org/10.1016/j.oregeorev.2016.02.012

Nesbitt BE, Muehlenbachs K, Murowchick JB (1987) Comment and reply on "Dual origins of lode gold deposits in the Canadian Cordillera." Geology 15:472–473

Nesbitt BE, Muehlenbachs K, Murowchick JB (1989) Genetic implications of stable isotope characteristics of mesothermal Au deposits and related Sb and Hg deposits in the Canadian Cordillera. Econ Geol 84:1489–1506. https://doi.org/10.2113/gsecongeo.84.6.1489

Neumayr P, Walshe J, Hagemann S, Petersen K, Roache A, Frikken P, Horn L, Halley S (2008) Oxidized and reduced mineral assemblages in greenstone belt rocks of the St. Ives gold camp, Western Australia: vectors to high-grade ore bodies in Archaean gold deposits? Mineral Deposita 43:363–371. https://doi.org/10.1007/s00126-007-0170-2

Nie F-J, Bjorlykke A (1994) Lead and sulfur isotope studies of the Wulashan quartz-K feldspar and quartz vein gold deposit, southwestern Inner Mongolia, People's Republic of China. Econ Geol 89:1289–1305. https://doi.org/10.2113/gsecongeo.89.6.1289

Nie F (1998) Geology and Origin of the Dongping alkalic-type gold deposit, northern Hebei Province, People's Republic of China. Resource Geol 48:139–158. https://doi.org/10.1111/j.1751-3928.1998.tb00013.x

Oberthuer T, Mumm AS, Vetter U, Simon K, Amanor JA (1996) Gold mineralization in the Ashanti Belt of Ghana; genetic constraints of the stable isotope geochemistry. Econ Geol 91:289–301. https://doi.org/10.2113/gsecongeo.91.2.289

Ohmoto H, Rye RO (1979) Isotopes of sulfur and carbon. In: Barnes HL (ed) Geochemistry of hydrothermal ore deposits, 2nd edn. Wiley, New York, pp 509–567

Palmer MR, Swihart GH (1996) Chapter 13. Boron isotope geochemistry: an oveview In: Palmer MR, Swihart (eds), Boron. De Gruyter, Berlin, Boston, pp 709–744

Pandalai HS, Jadhav GN, Mathew B, Panchapakesan V, Raju KK, Patil ML (2003) Dissolution channels in quartz and the role of pressure changes in gold and sulfide deposition in the Archean, greenstone-hosted, Hutti gold deposit, Karnataka, India. Miner Deposita 38:597–624. https://doi.org/10.1007/s00126-002-0345-9

Peters KE, Sweeney RE, Kaplan IR (1978) Correlation of carbon and nitrogen stable isotope ratios in sedimentary organic matter 1. Limnol Oceanogr 23:598–604. https://doi.org/10.4319/lo.1978.23.4.0598

Petrella L, Thébaud N, Laflamme C, Martin L, Occhipinti S, Bigelow J (2020) In-situ sulfur isotopes analysis as an exploration tool for orogenic gold mineralization in the Granites-Tanami Gold Province, Australia: learnings from the Callie deposit. J Geochem Explor 214:106542. https://doi.org/10.1016/j.gexplo.2020.106542

Phillips GN, Evans KA (2004) Role of CO_2 in the formation of gold deposits. Nature 429:860–863. https://doi.org/10.1038/nature02644

Pickthorn WJ, Goldfarb RJ, Leach DL (1987) Comment and reply on "Dual origins of lode gold deposits in the Canadian Cordillera." Geology 15:471–472

Pitcairn IK, Teagle DAH, Kerrich R, Craw D, Brewer TS (2005) The behavior of nitrogen and nitrogen isotopes during metamorphism and mineralization: evidence from the Otago and Alpine Schists, New Zealand. Earth Planet Sci Lett 233:229–246. https://doi.org/10.1016/j.epsl.2005.01.029

Poitrasson F (2017) Silicon isotope geochemistry. Rev Miner Geochem 82:289–344. https://doi.org/10.2138/rmg.2017.82.8

Ranta J-P, Hanski E, Cook N, Lahaye Y (2017) Source of boron in the Palokas gold deposit, northern Finland: evidence from boron isotopes and major element composition of tourmaline. Miner Deposita 52:733–746. https://doi.org/10.1007/s00126-016-0700-x

Ridley JR, Diamond LW (2000) Fluid chemistry of orogenic lode gold deposits and implications for genetic models. Rev Econ Geol 13:141–162

Robert F, Boullier A-M, Firdaous K (1995) Gold-quartz veins in metamorphic terranes and their bearing on the role of fluids in faulting. J Geophys Res Solid Earth 100:12861–12879. https://doi.org/10.1029/95JB00190

Rushton RW, Nesbitt BE, Muehlenbachs K, Mortensen JK (1993) A fluid inclusion and stable isotope study of Au quartz veins in the Klondike District, Yukon Territory, Canada; a section through a mesothermal vein system. Econ Geol 88:647–678. https://doi.org/10.2113/gsecongeo.88.3.647

Rye DM, Rye RO (1974) Homestake Gold Mine, South Dakota; I, Stable isotope studies. Econ Geol 69:293–317. https://doi.org/10.2113/gsecongeo.69.3.293

Santosh M, Nadeau S, Javoy M (1995) Stable isotopic evidence for the involvement of mantle-derived fluids in Wynad gold mineralization, south India. J Geol 103:718–727

Scheffer C, Tarantola A, Vanderhaeghe O, Rigaudier T, Photiades A (2017) CO_2 flow during orogenic gravitational collapse: syntectonic decarbonation and fluid mixing at the ductile-brittle transition (Lavrion, Greece). Chem Geol 450:248–263. https://doi.org/10.1016/j.chemgeo.2016.12.005

Schwarcz HP, Rees CE (1985) Grant 202 Sulphur isotope studies of Archean gold deposits. Ontario Geol Surv Misc Pap 127:151–156

Seal RR II (2006) Sulfur isotope geochemistry of sulfide minerals. Rev Mineral Geochem 61:633–677. https://doi.org/10.2138/rmg.2006.61.12

Selvaraja V, Caruso S, Fiorentini ML, LaFlamme CK, Bui T-H (2017) Atmospheric sulfur in the orogenic gold deposits of the Archean Yilgarn Craton,

Australia. Geology 45:691–694. https://doi.org/10.1130/G39018.1

Sephton MA, Amor K, Franchi IA, Wignall PB, Newton R, Zonneveld J-P (2002) Carbon and nitrogen isotope disturbances and an end-Norian (Late Triassic) extinction event. Geology 30:1119–1122. https://doi.org/10.1130/0091-7613(2002)030%3c1119:CANIDA%3e2.0.CO;2

Sharp ZD, Kirschner DL (1994) Quartz-calcite oxygen isotope thermometry: a calibration based on natural isotopic variations. Geochim Cosmochim Acta 58:4491–4501. https://doi.org/10.1016/0016-7037(94)90350-6

Sharp ZD (2017) Principles of stable isotope geochemistry, 2nd edn. Albuquerque, University of New Mexico. https://digitalrepository.unm.edu/unm_oer/1/

Shaw RP, Morton RD, Gray J, Krouse HR (1991) Origins of metamorphic lode gold deposits: implications of stable isotope data from the central Rocky Mountains, Canada. Mineral Petrol 43:193–209. https://doi.org/10.1007/BF01166891

Shelton KL, So C-S, Chang J-S (1988) Gold-rich mesothermal vein deposits of the Republic of Korea; geochemical studies of the Jungwon gold area. Econ Geol 83:1221–1237. https://doi.org/10.2113/gsecongeo.83.6.1221

Sheppard SMF (1986) Characterization and isotopic variations in natural waters. Rev Mineral 16:165–183

Sibson RH, Robert F, Poulsen KH (1988) High-angle reverse faults, fluid-pressure cycling, and mesothermal gold-quartz deposits. Geology 16:551–555. https://doi.org/10.1130/0091-7613(1988)016%3c0551:HARFFP%3e2.3.CO;2

Smith TJ, Cloke PL, Kesler SE (1984) Geochemistry of fluid inclusions from the McIntyre-Hollinger gold deposit, Timmins, Ontario, Canada. Econ Geol 79:1265–1285. https://doi.org/10.2113/gsecongeo.79.6.1265

Smithson SB, Wenzel F, Ganchin YV, Morozov LB (2000) Seismic results at Kola and KTB deep scientific borehole: velocities, reflections, fluids, and crustal composition. Tectonophysics 329:301–317

So C-S, Yung S-T, Shelton KL, Zhang D-q (2002) Geochemistry of the Youngbogari deposit, Republic of Korea: an unusual mesothermal gold–silver deposit of the Youngdong area. Geochem J 36:155–171

Steed GM, Morris JH (1997) Isotopic evidence for the origins of a Caledonian gold-arsenopyrite-pyrite deposit at Clontibret, Ireland. Trans Inst Mining Metall Sect B Appl Earth Sci B106:109–118

Sun X, Zhang Y, Xiong D, Sun W, Shi G, Zhai W, Wang S (2009) Crust and mantle contributions to gold-forming process at the Daping deposit, Ailaoshan gold belt, Yunnan, China. Ore Geol Rev 36:235–249

Suzuoki T, Epstein S (1976) Hydrogen isotope fractionation between OH-bearing minerals and water. Geochim Cosmochim Acta 40:1229–1240. https://doi.org/10.1016/0016-7037(76)90158-7

Taylor B (1986) Magmatic volatiles; isotopic variation of C, H, and S. Rev Mineral 16:185–225

Thode HG, Ding T, Crocket JH (1991) Sulphur-isotope and elemental geochemistry studies of the Hemlo gold mineralization, Ontario; sources of sulphur and implications for the mineralization process. Can J Earth Sci 28:13–25

Tornos F, Spiro B, Shepherd TJ, Ribera F (1997) Sandstone-hosted gold lodes of the southern West Asturian Leonese Zone (NW Spain): the role of depth in the genesis of the mineralization. Chronique De La Recherche Minière 528:71–86

Trull T, Nadeau S, Pineau F, Polvé M, Javoy M (1993) C-He systematics in hotspot xenoliths: Implications for mantle carbon contents and carbon recycling. Earth Planet Sci Lett 118:43–64. https://doi.org/10.1016/0012-821X(93)90158-6

Trumbull RB, Slack JF (2018) Boron isotopes in the continental crust: granites, pegmatites, Felsic Volcanic Rocks, and Related Ore Deposits. In: Marschall H, Foster G (eds) Boron isotopes: the fifth element. Springer International Publishing, Cham, pp 249–272

Trumbull RB, Garda GM, Xavier RP, Cavalcanti JAD, Codeço MS (2019) Tourmaline in the Passagem de Mariana gold deposit (Brazil) revisited: major-element, trace-element and B-isotope constraints on metallogenesis. Miner Deposita 54:395–414. https://doi.org/10.1007/s00126-018-0819-z

Trumbull RB, Codeço MS, Jiang S-Y, Palmer MR, Slack JF (2020) Boron isotope variations in tourmaline from hydrothermal ore deposits: a review of controlling factors and insights for mineralizing systems. Ore Geol Rev 125:103682

van Hinsberg VJ, Henry DJ, Marschall HR (2011) Tourmaline: an ideal indicator of its host environment. Can Miner 49:1–16. https://doi.org/10.3749/canmin.49.1.1

Vho A, Lanari P, Rubatto D (2019) An internally-consistent database for oxygen isotope fractionation between minerals. J Petrol 60:21012129. https://doi.org/10.1093/petrology/egaa001

Wanless RK, Rw B, Lowdon JA (1960) Sulfur isotope investigation of the gold-quartz deposits of the Yellowknife District [Northwest Territories]. Econ Geol 55:1591–1621. https://doi.org/10.2113/gsecongeo.55.8.1591

Warr O, Giunta T, Onstott TC, Kieft TL, Harris RL, Nisson DM, Lollar BS (2021) The role of low-temperature 18O exchange in the isotopic evolution of deep subsurface fluids. Chem Geol 561:120027. https://doi.org/10.1016/j.chemgeo.2020.120027

Wenner DB, Taylor HP (1971) Temperatures of serpentinization of ultramafic rocks based on O18/O16 fractionation between coexisting serpentine and magnetite. Contrib Mineral Petrol 32:165–185. https://doi.org/10.1007/BF00643332

Wesolowski D, Ohmoto H (1986) Calculated oxygen isotope fractionation factors between water and the

minerals scheelite and powellite. Econ Geol 81:471–477. https://doi.org/10.2113/gsecongeo.81.2.471
Wilkinson JJ, Johnston JD (1996) Pressure fluctuations, phase separation, and gold precipitation during seismic fracture propagation. Geology 24:395–398. https://doi.org/10.1130/0091-7613(1996)024%3c0395:PFPSAG%3e2.3.CO;2
Williams LB, Ferrell RE, Hutcheon I, Bakel AJ, Walsh MM, Krouse HR (1995) Nitrogen isotope geochemistry of organic matter and minerals during diagenesis and hydrocarbon migration. Geochim Cosmochim Acta 59:765–779. https://doi.org/10.1016/0016-7037(95)00005-K
Wortmann UG, Bernasconi SM, Böttcher ME (2001) Hypersulfidic deep biosphere indicates extreme sulfur isotope fractionation during single-step microbial sulfate reduction. Geology 29:647–650. https://doi.org/10.1130/0091-7613(2001)029%3c0647:HDBIES%3e2.0.CO;2
Wu J, Calvert SE, Wong CS (1997) Nitrogen isotope variations in the subarctic northeast Pacific: relationships to nitrate utilization and trophic structure. Deep Sea Res Part I Oceanogr Res Pap 44:287–314. https://doi.org/10.1016/S0967-0637(96)00099-4
Zeng Q, Wang Z, He H, Wang Y, Zhang S, Liu J (2014) Multiple isotope composition (S, Pb, H, O, He, and Ar) and genetic implications for gold deposits in the Jiapigou gold belt, Northeast China. Miner Deposita 49:145–164. https://doi.org/10.1007/s00126-013-0475-2
Zhang X, Nesbitt BE, Muehlenbachs K (1989) Gold mineralization in the Okanagan Valley, southern British Columbia; fluid inclusion and stable isotope studies. Econ Geol 84:410–424. https://doi.org/10.2113/gsecongeo.84.2.410
Zhang J, Chen Y, Zhang F, Li C (2006) Ore fluid geochemistry of the Jinlongshan Carlin-type gold ore belt in Shaanxi Province, China. Chin J Geochem 25:23–31
Zheng Y-F (1992) Oxygen isotope fractionation in wolframite. Eur J Miner 4:1331–1335

Light-Element Stable Isotope Studies of the Clastic-Dominated Lead–Zinc Mineral Systems of Northern Australia and the North American Cordillera: Implications for Ore Genesis and Exploration

Neil Williams

Abstract

Clastic-dominated lead–zinc (CD Pb–Zn) deposits are an important source of the world's Pb and Zn supply. Their genesis is contentious due to uncertainties regarding the time of ore formation relative to the deposition of the fine-grained carbonaceous strata that host CD Pb–Zn mineralization. Sulfur-isotopic studies are playing an important role in determining if ore minerals precipitated when hydrothermal fluids exhaled into the water column from which the host strata were being deposited, or when hydrothermal fluids entered the host strata during diagenesis or even later after lithification. Older conventional S-isotopic studies, based on analyses of bulk mineral-separate samples obtained by either physical or chemical separation methods, provided data that has been widely used to support a syngenetic-exhalative origin for CD Pb–Zn mineralization. However, with the advent in the late 1980's of in situ S-isotopic studies using micro-analytical methods, it soon became apparent that detailed S-isotopic variations of genetic importance are blurred in conventional analytical data sets because of averaging during sample preparation. Clastic-dominated Pb–Zn mineralization in the North Australian Proterozoic metallogenic province and the North American Paleozoic Cordilleran province has been the subject of many stable isotope studies based on both *bulk* and in situ analytical methods. Together with detailed mineral texture observations, the studies have revealed a similar sulfide mineral paragenesis in both provinces. The earliest sulfide phase in the paragenesis is fine-grained pyrite that sometimes has a framboidal texture. This pyrite typically has a wide range of $\delta^{34}S$ values that are more than 15‰ lower than the value of coeval seawater sulfate. These features are typical of, and very strong evidence for, pyrite formation by bacterial sulfate reduction (BSR) either syngenetically in an anoxic water column or during early diagenesis in anoxic muds. The formation of this early pyrite is followed by one or more later generations of pyrite that often occur as overgrowths around the early pyrite generation. The later pyrite generations have $\delta^{34}S$ values that are much higher than the early pyrite, often approaching the value of coeval seawater sulfate. Later pyrite formation has been variously attributed to BSR in a more restricted diagenetic environment, to sulfate driven-anaerobic oxidation of methane

N. Williams (✉)
Research School of Earth Sciences, Australian National University, Canberra, ACT, Australia
e-mail: williamsgeoscience@iinet.net.au

School of Earth, Atmosphere & Life Sciences, University of Wollongong, Wollongong, NSW, Australia

D. Huston and J. Gutzmer (eds.), *Isotopes in Economic Geology, Metallogenesis and Exploration*, Mineral Resource Reviews, https://doi.org/10.1007/978-3-031-27897-6_11

(SD-AOM) and to abiotic thermal sulfate reduction (TSR), with all three mechanisms again involving coeval seawater sulfate. The main sulfide ore minerals, galena and sphalerite, either overlap with or postdate later pyrite generations and are most often attributed to TSR of seawater sulfate. However, in comparison with pyrite, there is a dearth of in situ $\delta^{34}S$ data for galena and sphalerite that needs to be rectified to better understand ore forming processes. Importantly, the available data do not support a simple sedimentary-exhalative model for the formation of all but part of one of the Northern American and Australian deposits. The exception is the giant Red Dog deposit group in Alaska where various lines of evidence, including stable isotopic data, indicate that ore formation was protracted, ranging from early syn-sedimentary to early diagenetic sulfide formation through to late sulfide deposition in veins and breccias. The Red Dog deposits are the only example with early sphalerite with extremely low negative $\delta^{34}S$ values typical of a BSR-driven precipitation mechanism. By contrast, later stages of pyrite, sphalerite and galena have higher positive $\delta^{34}S$ values indicative of a TSR-driven precipitation mechanism. In CD Pb–Zn deposits in carbonate-bearing strata, carbon and oxygen isotope studies of the carbonates provide evidence that the dominant carbonate species in the ore-forming hydrothermal fluids was H_2CO_3, and that the fluids were initially warm ($\geq$ 150 °C) and neutral to acid. The $\delta^{18}O$ values of the hydrothermal fluids are $\geq$ 6‰, suggesting these fluids were basinal fluids that evolved through exchange with the basinal sedimentary rocks. Known CD Pb–Zn deposits all occur at or near current land surfaces and their discovery involved traditional prospecting, geophysical and geochemical exploration techniques. Light stable isotopes are unlikely to play a significant role in the future search for new CD Pb–Zn deposits deep beneath current land surfaces, but are likely to prove useful in identifying ore-forming hydrothermal fluid pathways in buried CD Pb–Zn systems and be a vector to new mineralization.

1 Introduction

Clastic-dominated lead–zinc (CD Pb–Zn) mineral deposits are the main source of the world's zinc and lead supply. They are therefore an attractive exploration target but their genesis, a critical element in exploration strategies, remains contentious (Huston et al. 2006; Magnall et al. 2021; Spinks et al. 2021). Reviews of the CD Pb–Zn deposit type include Gustafson and Williams (1981), Leach et al. (2005), Leach et al. (2010) and Wilkinson (2014). Common features of the deposit type are: (1) their occurrence in fine-grained carbonaceous and pyritic sedimentary rocks; (2) ores that are sedimentary in appearance with sulfide banding that parallels bedding in adjacent inter-ore strata; (3) a stratiform and stratabound relationship with their host sedimentary rock sequences; (4) stacked layers of mineralization; (5) high sulfide sulfur contents (typically $\geq$ 20 wt% S); (6) a simple sulfide mineralogy dominated by pyrite, sphalerite and galena and (7) low Cu ($\leq$ ~0.4 wt% Cu) and variable Ag ($\leq$ ~150 g/t Ag) contents. Deposits have an irregular age distribution. They first appeared ~1850 Ma but there are few in rocks aged between 1350 and 760 Ma. Barite is often associated with Paleozoic-age deposits but less so with Proterozoic examples.

This chapter discusses the genetic and exploration implications of the light-element (carbon, oxygen and sulfur) isotope geochemistry of CD Pb–Zn mineralization in the two most comprehensively studied CD Pb–Zn metallogenic provinces. These are the North Australian Proterozoic province (Huston et al. 2006) and the North American Paleozoic Cordilleran province (Leach et al. 2010). General information on light stable isotopes, including their chemistry, fractionation, analysis and applications can be found in Huston et al. (2023). Stable isotope applications of particular relevance to CD Pb–Zn deposits are highlighted in the following summary of CD Pb–Zn mineral systems.

2 Clastic-Dominated Pb–Zn Mineral Systems

The mineral system concept is growing in prominence in ore genesis studies and mineral exploration (Hagemann et al. 2016). The main components of a mineral system are the source of ore- and gangue-mineral constituents; the transport of these constituents to sites of ore formation; the trap where ore forms and the energy that drives the system.

The generic CD Pb–Zn mineral systems in Fig. 1 are based on the work of Goodfellow (1987), Hoggard et al. (2020), Huston et al. (2006, 2016), Manning and Embso (2018) and Sangster (2002, 2018). The sedimentary rock sequences in the system occur in sedimentary basins formed in extensional tectonic settings. Basin fill includes both shallow- and deeper-water facies material. During burial and compaction seawater-sulfate bearing basinal waters are thought to react with the basin fill and scavenge metals, generating the mineralizing hydrothermal fluids. Possible energy sources that drive systems include heat flow from the basement, gravity, compaction, seismic pumping, and compressional tectonism (Cooke et al. 1998), and the release of pressure in over-pressured fluid reservoirs triggered by extension (Vearncombe et al. 1996). The main channelways along which the hydrothermal fluids move from deep within host basins to trap sites are extensional growth faults.

The trap for CD Pb–Zn mineralization is the environment associated with the deposition and lithification of the fine-grained carbonaceous strata that host the mineralization. These host rocks are typically deposited in half-graben tectonic settings adjacent to the same extensional growth faults that may have been hydrothermal fluid conduits. Wall rocks adjacent to feeder faults are often hydrothermally altered and spatially associated with feeder-zone breccia- and vein-hosted mineralization.

Ore deposition occurs when the chemistry of the hydrothermal fluid changes upon arrival at the trap site. One possible trigger for change is the exhalation of the fluid into the host-sediment depositional environment (syngenetic ore formation), a scenario illustrated in Fig. 1a. If the hydrothermal fluid has a low salinity, ore-forming reactions would most likely be triggered by mixing between the hydrothermal fluid and seawater (Pathway 1, Fig. 1a). On the other hand, if the exhaling hydrothermal fluid has a high salinity, it would mostly like be denser than seawater and would pond in sea-floor depressions and form brine pools in which sulfide formation is triggered by changes associated with the development and evolution of the brine pool chemistry (Pathway 2, Fig. 1a). Figure 1b illustrates subsurface ore forming scenarios that are triggered by reactions between the hydrothermal fluid and components of the host strata during either diagenesis (syndiagenetic ore formation) or after lithification (epigenetic ore formation). The timing of CD Pb–Zn ore formation relative to the deposition of the host rocks has been debated for decades and remains a critical uncertainty in CD Pb–Zn mineral systems. In the case of syndiagenetic and epigenetic ore formation (Fig. 1b) an outflow zone for the hydrothermal fluids may occur in rocks downstream from the trap environment.

2.1 Applications of Stable Isotopes in Clastic-Dominated Pb–Zn Systems

Many of the important chemical reactions in CD Pb–Zn systems fractionate stable isotopes and carbon, oxygen and sulfur isotopic studies of systems have done much to improve our understanding of how ore forms within the systems.

2.1.1 Metal Source Identification

It is difficult to elucidate metal sources for CD Pb–Zn deposits using stable isotope evidence. However, some insights can be gained indirectly through stable isotope studies of possible source rocks. For example, Cooke et al. (1998) found that altered volcanic horizons stratigraphically below CD Pb–Zn mineralization in the North

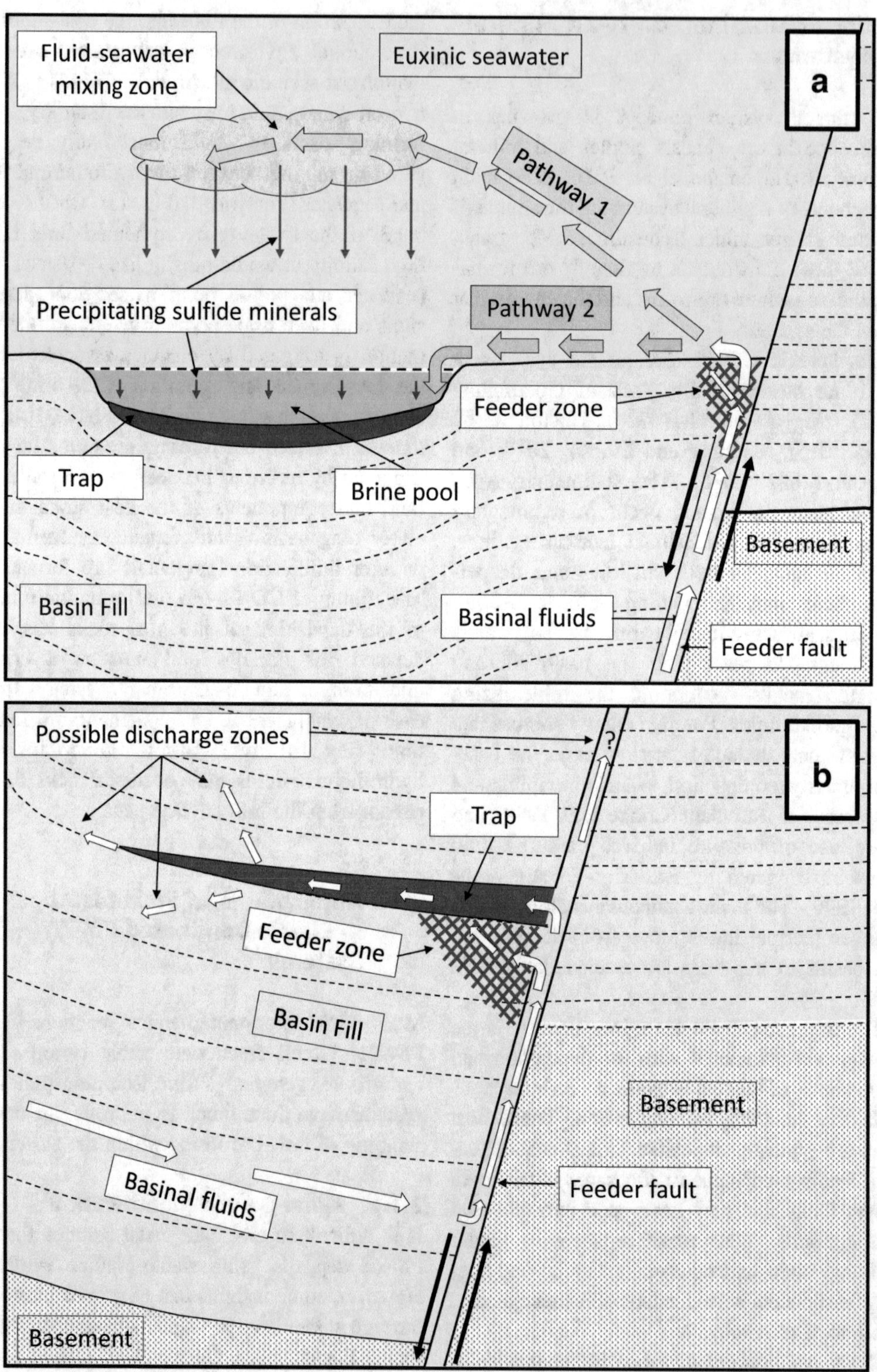

Fig. 1 Generic clastic-dominated Pb–Zn mineral systems showing **a** syngenetic ore formation, and **b** syndiagenetic to epigenetic ore formation. Pathways (1 & 2) in **a** are discussed in the text

Australian CD Pb–Zn province (Fig. 2) are almost 100% depleted in Pb, Zn and Cu and are therefore a possible metal source. Cooke et al. (1998) used fluid inclusion and bulk C- and O-isotope data from the altered volcanic rocks to conclude that the fluids which caused the alteration were warm (≈ 100 °C) and saline (> 20 wt % NaCl equivalent) and had a $\delta^{18}O_{SMOW}$ value of ≈ −1‰ and a $\delta^{13}C_{PDB}$ value of ≈ −7‰. These values are consistent with the idea that the metal-depleting fluids were originally meteoric waters ± seawater and that their salinity and ^{18}O content increased as they moved through overlying basinal carbonates and evaporites and deeper into underlying volcanic rocks. The question of whether or not the isotopic composition of these fluids prove that volcanic rocks were the source of metals in CD Pb–Zn mineralization higher in the stratigraphic succession is discussed later in this Chapter.

2.1.2 Sulfide-Sulfur Source Identification

In contrast to metal sources, sulfur isotope studies have proved very useful in identifying the source of sulfide-S in CD Pb–Zn mineralization. Sulfur isotope data, along with other geological evidence, suggest strongly that the sulfide-S source was seawater sulfate, and came either directly from seawater or indirectly via sulfate minerals such as gypsum, anhydrite or barite. The value of $\delta^{34}S_{seawater\ sulfate}$ has varied significantly through geological time, and the secular $\delta^{34}S_{seawater\ sulfate}$ curve was initially constructed using worldwide sulfur isotopic data from marine evaporites (Claypool et al. 1980). Evaporites are best preserved in Phanerozoic strata and the secular $\delta^{34}S_{seawater\ sulfate}$ curve is therefore best documented in the Phanerozoic, but less so in the Precambrian where evaporites are rare. During the Phanerozoic the curve varies between ~12 and 30‰ and the processes controlling $\delta^{34}S_{seawater\ sulfate}$ values are well understood (Bottrell and Newton 2006; Halevy et al. 2012; Canfield 2013). Although the Precambrian part of the curve is not so well documented, it continues to be improved using not only scant evaporite data, but also carbonate associated sulfate (CAS) data (Kah et al. 2004; Gellatly and Lyons 2005; Chu et al. 2007; Luo et al. 2015; Crockford et al. 2019; Turchyn and DePaolo 2019). In this Chapter the best estimates of the value of $\delta^{34}S_{seawater\ sulfate}$ at the time of deposition of the strata hosting CD Pb–Zn deposits are presented under the Source heading, whereas the applicability of this value to ore formation is

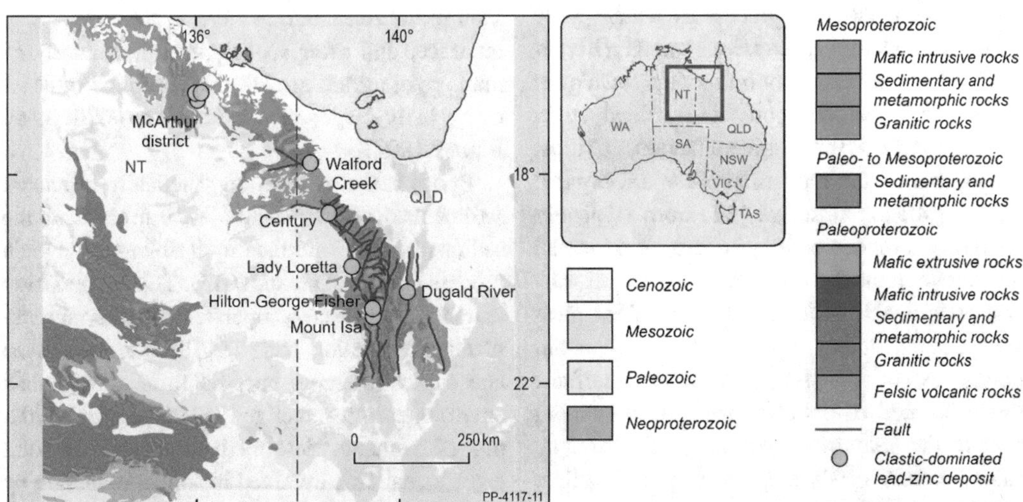

Fig. 2 Location of the significant clastic-dominated Pb–Zn deposits and prospects in the Carpentaria Zinc Belt of Northern Australia

discussed under the Trap heading because, as will become apparent later in this Chapter, one of the main issues relating to CD Pb–Zn ore genesis is how and where in the mineral system sulfate is reduced to sulfide.

2.1.3 Transport Mechanisms

In a seminal study of hydrothermal fluids capable of forming CD Pb–Zn mineralization, Cooke et al. (2000) used geochemical modelling to show that both cool ($\leq$ 150 °C) oxidized hydrothermal fluids (dominated by $SO_4^{2-} \pm HSO_4^-$) and warmer (~250 °C) reduced acidic fluids (dominated by H_2S or HS^-) can transport and deposit sufficient quantities of Pb and Zn to form CD Pb–Zn deposits. Stable isotopic studies can help determine which fluid type was involved in the formation of a particular deposit and examples are discussed in this Chapter.

2.1.4 Trap Mechanisms

The S-isotopic composition of galena, sphalerite and pyrite are important determinants of the processes by which the sulfides form in CD Pb–Zn deposits, the relative timing of sulfide mineral formation (paragenesis), and temperatures of ore formation.

A low-temperature (< 100 °C) process frequently mentioned in discussions of CD Pb–Zn ore formation is bacterial sulfate reduction (BSR). BSR is the main process by which pyrite forms in modern carbonaceous muds (Berner 1970), sediments that may be modern analogues of the pyritic carbonaceous fine-grained strata that host CD Pb–Zn mineralization. Critical components of the modern process are organic matter, sulfate, a source of iron (typically hydrous iron oxide coatings on detrital minerals) and sulfate reducing bacteria. In modern carbonaceous muds pyrite formation by BSR most often occurs during early diagenesis, within centimeters of the sediment–water interface. Where the rate of organic matter accumulation exceeds the rate of oxidation of the organic matter, anoxic conditions develop and sulfate reducing bacteria become active. The bacteria metabolize the organic matter and reduce sulfate that diffuses into the sediments from the overlying water column. The basic reaction in the process is:

$$2CH_2O_{\text{organic matter}} + SO_4^{2-} \rightarrow H_2S + 2\ HCO_3^- \quad (1)$$

The H_2S produced by BSR reacts with hydrous iron oxides in the muds to form either mackinawite (FeS) and/or greigite (Fe_3S_4). Some H_2S is bacterially oxidized to elemental sulfur which in turn reacts with the mackinawite and greigite to form pyrite. Pyrite formation by BSR is a low temperature process as most sulfate reducing bacteria live within a temperature range of 0° to ~48 °C with a few, the thermophiles, able to live in a range of ~30° to ~100 °C (Ohmoto and Goldhaber 1997). In detail, pyrite formation by BSR is a complex process and further information can be found in Berner (1970, 1984), Goldhaber and Kaplan (1974) and Rickard and Luther (2007). In modern environments pyrite sometimes forms by BSR in stratified water bodies where stagnant anoxic layers underlie oxic surface waters. In the anoxic layer, provided there are sufficient concentrations of SO_4^{2-}, suitable oxidants, reactive organic matter and iron, pyrite can form by the same BSR reaction pathway to that just described. The main locus of such pyrite formation is close to the oxic-anoxic interface, and after settling into the underlying muds pyrite grain growth may continue by BSR in the early diagenetic environment (Wilkin and Barnes 1997).

Pyrite can also form diagenetically in anoxic sediments deeper (meters to ~10 m) beneath the sediment–water interface by a different biogenic process (Borowski et al. 2013). This second low-temperature process operates in the sulfate-methane transition zone (SMT) where methane and sulfate-reducing bacteria form a diagenetic environment characterized by the anoxic oxidation of methane. Here methane rather than solid organic matter is involved in sulfate reduction by way of reactions of the type:

$$CH_4 + SO_4^{2-} \rightarrow HS^- + HCO_3^- + H_2O \quad (2)$$

Following Rieger et al. (2020) this sulfate driven-anaerobic oxidation of methane process is referred to in this Chapter as SD-AOM.

A third pathway for pyrite formation involves the inorganic reduction of sulfate, a reaction requiring temperatures higher than those for BSR and SD-AOM, and either organic carbon or another reductant. Such reduction is referred to as thermal sulfate reduction (TSR), a process that appears to become important at temperatures of 100–140 °C and greater (Huston et al. 2023).

The formation of pyrite, as well as galena and sphalerite, in CD Pb–Zn deposits has been variously attributed to BSR, SD-AOM and TSR, and sulfur isotope studies are one of the main ways distinguishing between these pathways. Details about the S-isotopic fractionations associated with BSR and TSR are discussed by Huston et al. (2023). Typically, $S_{sulfide} - S_{sulfate}$ fractionation by BSR ranges from ~15 to 60‰ while the kinetic isotope fractionation for TSR has a smaller range of ~10–20‰ (Kiyosu and Krouse 1990). As the $\delta^{34}S_{seawater\ sulfate}$ value through geological time is $\geq$ ~10‰ (Bottrell and Newton 2006), the $\delta^{34}S$ values of sulfides formed by the BSR pathway are typically low and negative, whereas those formed by the TSR pathway are high and positive. This difference is used often to distinguish between the two reduction pathways (Machel et al. 1995).

However, the SD-AOM study of Borowski et al. (2013) has complicated interpretations of $\delta^{34}S_{sulfide}$ ranges and values. Whereas in modern sediments pyrite formed by BSR is characterized by low negative $\delta^{34}S$ values, pyrite formed by SD-AOM is characterized by high $\delta^{34}S$ positive values, making it difficult to distinguish between TSR and SD-AOM using sulfur isotopic data alone. The interpretation of high positive $\delta^{34}S$ values is further complicated by two other processes that have been shown to produce high positive $\delta^{34}S$ values. One is when sulfides form by BSR in euxinic sulfate-limited conditions (Gomes and Hurtgen 2015) and the other is when sulfides form by BSR in a restricted or closed system in which Rayleigh fractionation becomes important (Ohmoto and Goldhaber 1997; Huston et al. 2023). Choosing between these possible causes of high positive $\delta^{34}S$ values requires other evidence besides sulfur isotope data and is a topic examined in many of the studies reviewed in this chapter.

Huston et al. (2023) used the exchange of ^{32}S and ^{34}S between galena and sphalerite (their Eq. 2) to illustrate their discussion of stable isotope fractionation. The magnitude of $\Delta^{34}S_{sphalerite\text{-}galena}$ is temperature dependent and can be used as a CD Pb–Zn ore-forming geothermometer provided the two sulfides coprecipitated in isotopic equilibrium or formed at different times under the same physicochemical conditions and assuming, in both instances, that the samples analysed are pure and are not contaminated with other sulfides. Throughout this chapter, for the sake of consistency, all temperatures derived using on the $\Delta^{34}S_{sphalerite\text{-}galena}$ geothermometer are based on the experimental calibration equation of Ding et al. (2003) which is almost identical to the preferred equation of Ohmoto and Rye (1979). The cited temperatures also assume the above-mentioned conditions have been met for accurate temperature determinations.

2.1.5 Isotopic Analytical Methods and Reporting Nomenclature

Stable isotopic analytical techniques have evolved through time (Huston et al. 2023). The older techniques required ~10 mg of sample for analysis and bulked mineral separates were typically obtained either by physical (e.g. hand picking) or selective chemical dissolution methods. By contrast, starting in the late 1980s, bulk techniques began giving way to in situ microanalytical techniques. To maintain consistency with Huston et al. (2023), isotopic values determined by the bulk analytical techniques are identified in this Chapter by the subscript "*bulk*" (e.g. $\delta^{34}S_{bulk\ pyrite}$), whereas values determined by in situ microanalytical techniques are identified by the subscript "in situ" (e.g. $\delta^{34}S_{in\ situ\ pyrite}$).

Both analytical techniques pose analytical challenges (Huston et al. 2023). Nevertheless, data acquired using both techniques have helped advance our understanding of CD Pb–Zn mineral systems. Because of the small grain sizes of minerals in many CD Pb–Zn deposits (diameters < 1 mm) *bulk* analytical techniques produce values that can be averages of many tens of millions of individual grains (Eldridge et al. 1993). As shown by Eldridge et al. (1993), the averaging aspect of *bulk* analyses has been found to have obscured genetically important isotopic complexities and Ohmoto and Goldhaber (1997) warned that interpretations developed using *bulk* data will not always be accurate. Therefore, where possible, this Chapter emphasizes conclusions reached using in situ data, with the implications of *bulk* data only discussed where they are important to the interpretation of in situ data or have had a major impact on our understanding of particular CD Pb–Zn mineral systems.

3 The Northern Australian Proterozoic Clastic-Dominated Pb–Zn Province

The sedimentary host-rock sequences of this province are Palaeo- to Mesoproterozoic in age (Southgate et al. 2000) and occur in the McArthur Basin in the north, and in the Mt Isa Inlier in the south (Fig. 2). The province hosts five of the world's ten largest CD Pb–Zn deposits based on Pb + Zn content (Leach et al. 2005). These, in order of decreasing Pb–Zn content, are HYC (3rd), Hilton (6th), Mt Isa (7th), George Fisher (9th) and Century (10th). All five deposits have supported significant mining operations. All but one of the five deposits occur in strata deposited between ~1665 and ~1635 Ma. The exception is Century which occurs in strata deposited ~1595 Ma (Page and Sweet 1998). The province also contains a number of smaller CD Pb–Zn type deposits, including Dugald River, Lady Loretta, Walford Creek, and Teena (Rohrlach et al. 1998; Williams 1998; Large et al. 2005; Hayward et al. 2021).

3.1 Clastic-Dominated Pb–Zn Deposits in the McArthur Basin

The largest CD Pb–Zn deposit in the McArthur Basin is the HYC deposit, sometimes referred to as the McArthur River deposit (Fig. 3a). The HYC deposit has been described by Croxford and Jephcott (1972), Murray (1975), Lambert (1976), Eldridge et al. (1993), Large et al. (1998), Ireland et al. (2004a,b) and Spinks et al. (2021). The HYC deposit is located in the Batten Fault Zone portion of the McArthur Basin, as are several smaller CD Pb–Zn deposits and occurrences (Fig. 3a), including Emu Plains (Ahmad et al. 2013; Walker et al. 1977), Mitchell Yard and Myrtle (Ahmad et al. 2013), the stratiform part of Ridge II (Williams 1978a), Teena (Hayward et al. 2021) and W-Fold (Murray 1975; Walker et al. 1977).

Almost all the CD Pb–Zn mineralization in the McArthur Basin occurs in carbonaceous and dolomitic mudstones of the 1640 ± 3 Ma HYC Pyritic Shale Member of the Barney Creek Formation in the McArthur Group (Fig. 3b). Away from CD Pb–Zn mineralization pyrite is common in the Barney Creek Formation, and sometimes in the overlying Reward Dolostone (Fig. 3b). The most pyritic and organic-rich portions of the two formations occur in McArthur Basin sub-basin depocenters.

A number of small discordant vein- and breccia-hosted Pb–Zn and Cu deposits also occur in the Batten Fault Zone in dolostone formations older than the Barney Creek Formation. Two of these, Cooley and Ridge (Fig. 3a), lie immediately east of, and appear to be a feeder-zone component of the HYC mineral system (Williams 1978a; Rye and Williams 1981).

3.2 Stable Isotope Studies in the McArthur Basin

3.2.1 Source

As noted earlier, Cooke et al. (1998) used fluid inclusion and bulk C–O (carbonate) and oxygen (silicate) isotope measurements of altered

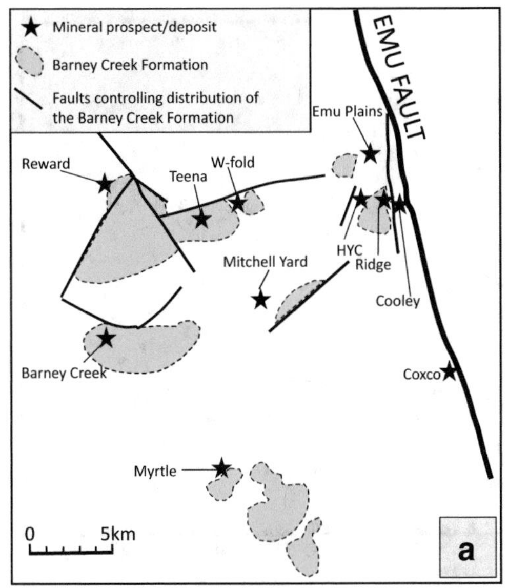

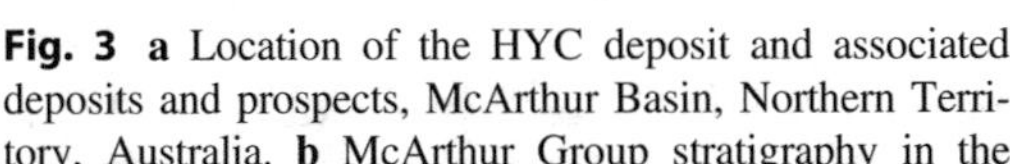
Fig. 3 **a** Location of the HYC deposit and associated deposits and prospects, McArthur Basin, Northern Territory, Australia. **b** McArthur Group stratigraphy in the HYC area. Stratigraphic units discussed in this Chapter are highlighted in color

volcanic rocks lying stratigraphically deep beneath the Barney Creek Formation to determine if they were the source of the metals in the McArthur CD Pb–Zn mineralization. The volcanic rocks studied by Cooke et al. (1998) are exposed in the Mallapunyah Dome region, ~80 km SSW of the HYC deposit. Bulk whole rock $\delta^{18}O_{SMOW}$ data from altered mafic volcanic rocks in the Dome region range from 15.7 to 19.7‰, a range used with other evidence to conclude that alteration occurred at ≈ 100 °C and involved saline (> 20 wt% NaCl equivalent) fluids with a $\delta^{18}O_{SMOW}$ value of ≈ −1‰ and a $\delta^{13}C_{PDB}$ value of ≈ −7‰. In contrast bulk whole rock $\delta^{18}O_{SMOW}$ data for similarly-altered mafic rocks between ~80 and 200 km east of the Mallapunyah Dome range from 6.0 to 10.4‰ (Champion et al. 2020), a difference attributed either to alteration at ≈ 250 °C by similar fluids (with a $\delta^{18}O_{SMOW}$ value of 1.7‰) to those involved in the Mallapunyah Dome alteration or to alteration by very different fluids. Although neither study proves conclusively that the metals in the McArthur Basin CD Pb–Zn mineralization came from the altered volcanic rocks, both postulate the stable isotopic composition of the hydrothermal fluids that might have formed the CD Pb–Zn mineralization.

The value of $\delta^{34}S_{seawater\ sulfate}$ at the time of deposition of the strata hosting McArthur CD Pb–Zn deposits was taken by Magnall et al. (2020b) to be, at a minimum, 25‰ based on the global comparisons of Li et al. (2015). This is in good agreement with the median $\delta^{34}S_{evaporative\ gypsum}$ value of 26.8‰ for the Myrtle Shale Member (Fig. 3b) of the McArthur Group (Crockford et al. 2019). However, the secular $\delta^{34}S_{seawater\ sulfate}$ curve of Chu et al. (2007), based on CAS data through the Jixian section, northern China, gives a $\delta^{34}S_{seawater\ sulfate}$ value of 20‰ for sedimentary rocks of the same age as the Barney Creek Formation. The sulfide-forming processes discussed in the McArthur *Trap* section below considers the genetic implications of both a 20‰ and a 26.8‰ value for McArthur $\delta^{34}S_{seawater\ sulfate}$.

3.2.2 Transport

Rye and Williams (1981) studied the *bulk* isotopic composition of ore-stage vein-fill dolomite in the Cooley and Ridge mineralization and the results are shown in Fig. 4. The $\delta^{13}C$ and $\delta^{18}O$

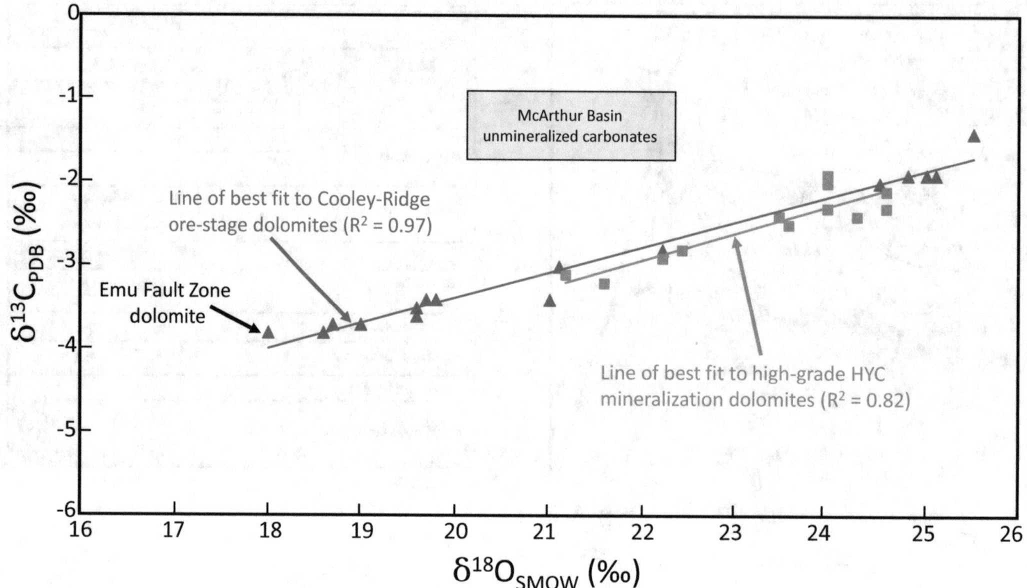

Fig. 4 $\delta^{18}O$ versus $\delta^{13}C$ values of dolomite in the discordant and concordant Ridge –Cooley mineralization and HYC deposit (after Rye and Williams 1981; Smith and Croxford 1975). The data field for the unmineralized McArthur Basin rocks represents, respectively, the 1st and 3rd quartiles of the $\delta^{18}O$ and $\delta^{13}C$ values determined by Kunzmann et al. (2019) for 485 drill-core samples remote from mineralization in the McArthur Basin

values correlate strongly, and dolomite from the Emu Fault zone (Smith and Croxford 1975) falls at the isotopically light end of the trend. Moreover, $\delta^{13}C$ and $\delta^{18}O$ values increase westward from the Emu Fault. Two explanations were proposed for the correlation using temperature-dependent fractionation modelling. The first is that the correlation reflects vein dolomite deposition from a hydrothermal fluid with $\delta^{13}C_{H2CO3}$ ~−2.75‰ and $\delta^{18}O_{SMOW}$ ~12.5‰ that cooled from ~325 to 100 °C as it flowed westwards from the Emu Fault. The isotopic composition of this fluid suggests it was a highly evolved basinal brine that had exchanged with a large marine carbonate reservoir (Taylor 1979). If the explanation is correct then the Emu Fault dolomite formed at ~325 °C, the Cooley dolomites formed between ~275 and ~185 °C, and the Ridge I dolomite formed between ~190 and ~175 °C. The second explanation is that the correlation again reflects a cooling trend, this time of a hydrothermal fluid that had a constant $\delta^{18}O_{SMOW}$ but a $\delta^{13}C$ that became progressively lower due to local oxidation of organic matter during hydrothermal alteration of the host rocks. This second explanation yields a cooling from ~220 to ~125 °C of a fluid that had initial $\delta^{13}C_{H2CO3}$ and $\delta^{18}O_{SMOW}$ values respectively of −4.5‰ and 8.1‰. By the time the fluid cooled to 125 °C the $\delta^{13}C_{H2CO3}$ value of the fluid would have decreased to −5.25‰. Rye and Williams (1981) noted that there were no unique fits of model curves to the dolomite isotope trend, and simply concluded that the data are consistent with the formation of ore-stage dolomite from a single hydrothermal fluid that cooled, and possibly changed in isotopic composition, as it moved westwards from the Emu Fault zone and deposited the Cooley and Ridge mineralization.

Importantly, Rye and Williams (1981) found that the C- and O-isotopic composition of dolomites in well-mineralized HYC samples (Smith and Croxford 1975) also correlate strongly and that the lines of best fit for the HYC and Cooley-Ridge data sets are almost indistinguishable (Fig. 4). The similarity is strong evidence that the Cooley and Ridge prospects and HYC deposit are genetically related and are parts of the one mineral system in which the discordant Cooley-Ridge mineralization is feeder-zone style mineralization. Dolomite vein-fill in the discordant Cooley-Ridge mineralization precipitated at warmer temperatures (~170–310 °C, 1st model or ~170–210 °C, 2nd model) from the same hydrothermal fluids that at lower temperatures (~170–240 °C, 1st model or ~125–175 °C, 2nd model) completely exchanged and isotopically equilibrated with pre-existing sedimentary dolomite in the HYC deposit and/or deposited hydrothermal dolomite in the HYC mineralization.

Large et al. (2001) studied the isotopic composition of carbonates in an interpreted hydrothermal alteration halo around the HYC deposit. The halo is characterized by anomalous whole-rock Zn, Pb and Tl values and, locally, Mn-rich ferroan dolomite-ankerite carbonates (Large et al. 2000). Carbonate C- and O-isotopic values in the halo define a field that is bordered by the HYC and Cooley-Ridge lines (Fig. 5) and extends away from the lines, and from the unaltered McArthur Basin dolomite field of Kunzmann et al. (2019), towards higher $\delta^{18}O_{dolomite}$ values and lower $\delta^{13}C_{dolomite}$ values. Using a temperature-dependent fluid-rock interaction modelling approach Large et al. (2001) found that the hydrothermal alteration was caused either by a HCO_3^-dominant neutral to alkaline fluid with a $\delta^{18}O_{SMOW}$ value of 5 ± 5‰ and a $\delta^{13}C$ value of −6‰ ± 1‰ over a temperature range of 50–120 °C, or a H_2CO_3-dominant neutral to acidic fluid with a $\delta^{18}O$ value of ~15‰ and a $\delta^{13}C$ value of −4 to −6‰ over a temperature of greater than 200 °C. Based on mineralogical features, interpreted to indicate alteration was a low-temperature event, Large et al (2001) concluded that a HCO_3^-dominant fluid caused the alteration. The conclusion is puzzling because the modelling of Cooke et al. (1998) showed that the isotopic composition of carbonates define a $\delta^{13}C - \delta^{18}O$ curve with a low positive slope like that shown in Fig. 4 when in equilibrium over a range of temperatures (25–200 °C) with H_2CO_3-dominant fluids, but not with HCO_3^--dominant fluids which define a U-shaped curve over the same temperature range

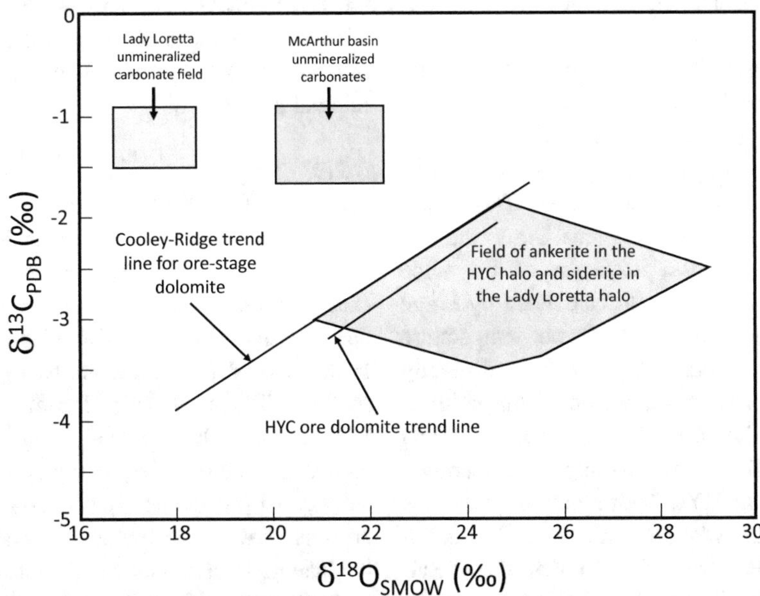

Fig. 5 $\delta^{18}O$ versus $\delta^{13}C$ values of carbonates in the alteration halo associated with the HYC deposit, together with carbonates from the HYC and Cooley-Ridge mineralization and unmineralized dolomitic rocks in the McArthur Basin and Lady Loretta areas (after Kunzmann et al. 2019; Large et al. 2001; McGoldrick et al. 1998). The data field for the unmineralized McArthur Basin rocks is from Fig. 4

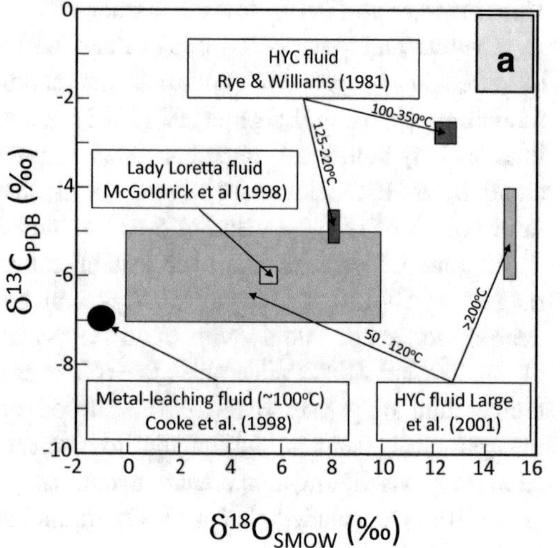

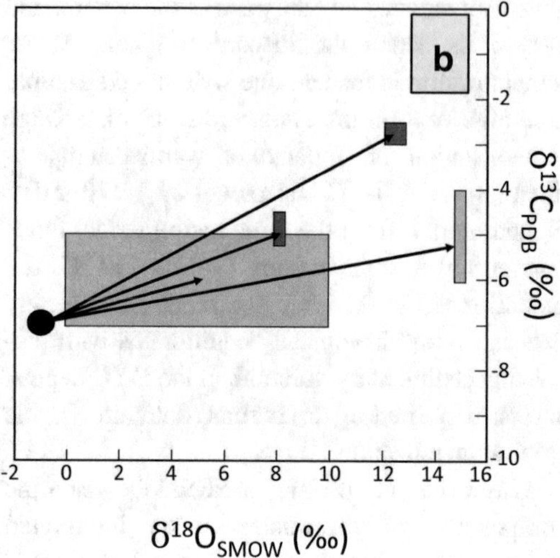

Fig. 6 δ^{18}O versus δ^{13}C values of the fluids postulated to have leached base metals volcanic rocks in the lower McArthur Basin (after Cooke et al. 1998) and to be possible hydrothermal fluids involved in the formation of the HYC and Lady Loretta clastic-dominated Pb–Zn deposits (after Large et al. 2001; McGoldrick et al. 1998; Rye and Williams 1981)

(Cooke et al. 1998: Fig. 19). Therefore the weight of evidence favors the formation of the HYC mineralization and associated hydrothermal alteration by a hot (initial temperature > 200 °C) neutral to acidic H_2CO_3-dominant hydrothermal fluid.

A comparison of the various C- and O-isotopic compositions postulated for the HYC ore-forming hydrothermal fluids with those postulated for the metal-leaching fluids by Cooke et al. (1998) is shown in Fig. 6a. The metal-leaching fluid is close in composition to the low $\delta^{13}C_{fluid}$ and $\delta^{18}O_{fluid}$ extremes of the wide isotopic compositional field calculated by Large et al. (2001) for their lower-temperature HCO_3^-dominant HYC fluid, but is distinctly lower isotopically to the isotopic compositions calculated by Large et al. (2001) and Rye and Williams (1981) for the three higher-temperature H_2CO_3dominant HYC hydrothermal fluids. As the HYC fluids were most likely H_2CO_3-dominant, it is clear from Fig. 6a that the metal-leaching fluids discussed earlier were not the immediate ore-forming fluids. However, if the metal-leaching fluids had continued to interact with McArthur Basin strata after leaching metal from volcanic material, they may have evolved to become the ore-forming fluids. As illustrated in Fig. 6b this evolution would require increasing temperatures and an increase in $\delta^{18}O_{fluid}$ values of between ~9 and ~16‰ and in $\delta^{13}C_{fluid}$ values of between ~2 and ~4‰. Further work is clearly needed to better understand the origin of the metals and hydrothermal fluids in the HYC mineral system.

3.2.3 Trap

As well as C- and O-isotopic studies, S-isotopic studies have also added to an understanding of CD Pb–Zn mineral systems in the McArthur Basin. Pyrite has been the studied isotopically in detail in the HYC and Teena deposits (Fig. 3a). In the HYC deposit there are two generations of pyrite (Williams 1978a; Eldridge et al. 1993; Large et al. 1998; Ireland et al. 2004b). The earlier generation, Py_1, comprises small (typically 5–10 μm diameter) euhedral to subhedral crystals that occur either in bedding parallel laminae or in spherical to ellipsoidal aggregates with diameters of up to ~100 μm. The second generation, Py_2, occurs either as overgrowths on individual Py_1 grains or as interstitial cement in Py_1 aggregates. Py_2 is optically duller and slightly browner and softer than Py_1. Chemically, the main chemical distinction between the pyrite generations appears to be an enrichment of Tl in Py_2 relative to Py_1 (Spinks et al. 2021).

In situ sulfur isotope analyses show that Py_1 and Py_2 are isotopically different (Eldridge et al. 1993). Py_2 has a broader spread of $\delta^{34}S_{\textit{in situ}\ \text{pyrite}}$ values than Py_1 and a median $\delta^{34}S_{\textit{in situ}\ \text{pyrite}}$ value ~16‰ higher (Fig. 7). If the value of $\delta^{34}S_{\text{seawater sulfate}}$ during the deposition of the Barney Creek Formation was 26.8‰, based on the Myrtle Shale Member median $\delta^{34}S_{\text{evaporative gypsum}}$ value discussed above, then $\Delta^{34}S_{\text{seawater sulfate-pyrite 1}}$ is ≤ 40‰ and that for Py2 is 30% (Fig. 7). Both fractionations are in the range of pyrite formation by BSR. Eldridge et al. (1993) concluded that Py_1 formed by BSR during early diagenesis when the system was open to sulfate, whereas Py_2 was formed during later diagenesis when the system was closed to sulfate, with $\delta^{34}S_{\text{sulfate}}$ increasing due to closed-system Rayleigh fractionation.

Magnall et al. (2020b) studied pyrite in and around the Teena deposit from three drill holes aligned in a roughly east–west direction over a distance of ~1.75 km across the Teena sub-basin (Hayward et al. 2021). Two holes intersect the mineralized Lower Pyritic Shale Member of the HYC Pyritic Shale and the third intersects the poorly-mineralized (< 6 wt% Zn) Lower Pyritic Shale Member ~115 m west of Teena. The study also included pyrite from the Middle HYC Pyritic Shale Member above the deposit. Two pyrite generations were identified at Teena, Py-1 and Py-2, each with two sub-types (a and b). Py-1a comprises < 5 µm grains and framboids, whereas Py-1b comprises > 5 µm idiomorphic euhedral pyrite that typically occurs on the margins of nodular carbonate. Py-2a comprises spherical and concentrically zoned crystals containing abundant host rock inclusions, and Py-2b comprises irregular, anhedral overgrowths on earlier Py-1 and contains interstitial sphalerite and galena. A minor, later pyrite type, Py-3, comprises coarse-grained euhedral crystals in late-stage sulfide-carbonate-quartz veins. The Teena Py-1 and Py-2 generations appear to be the same as the two HYC pyrite generations.

Py-1a is present throughout the stratigraphic interval studied at Teena and is abundant within correlative carbonaceous intervals in the three drill holes. Py-1a is interpreted to also predate the growth of diagenetic dolomite nodules. The restriction of Py-1b to the margins of the nodules is interpreted to indicate that Py-1b formed late during nodule growth. Py-2 is limited to regions of Pb–Zn mineralization. It is enriched in Pb and As and is interpreted to be hydrothermal in origin.

The Teena $\delta^{34}S_{\textit{in situ}\ \text{pyrite}}$ data of Magnall et al. (2020b) are summarized in Fig. 8. The data reveal that there are two classes of Py-1a at Teena, and these are shown separately in Fig. 8. The class with the lower median value has a large offset from coeval seawater sulfate (assumed here to be 26.8‰) of ~30‰, consistent with formation by BSR under relatively open system conditions. Py-1a is widespread both within, adjacent to, and stratigraphically above the Teena mineralization. In contrast the Py1a class with the higher median value of 33.3‰ is comparable in value to the assumed value of coeval seawater sulfate. This Py1a class is restricted to an interval of high pyrite abundance in the Middle HYC Pyritic Shale Member and it occurs in all three drill holes. Magnall et al. (2020b) concluded that the formation of this high-median-value Py-1a was not a localized process, but rather was a sub-basin wide process. Highly positive $\delta^{34}S_{\text{pyrite}}$ values are typical of sulfate reduction in a sulfate-limited environment that is widely interpreted to reflect water-mass restriction (e.g. Lyons et al. 2000). However, because the high positive Py-1a occurs in a part of the HYC Pyritic Shale that was laid down at a time of high relative sea level during a marine transgression (Kunzmann et al. 2019), Magnall et al. (2020b), concluded that sulfate limitation was caused by a high organic carbon flux that promoted BSR in an euxinic water column.

Py-1b generally has higher $\delta^{34}S$ values than those of the low-median value Py-1a class and is interpreted by Magnall et al. (2020b) as indicating that Py-1b formed in pore fluids during later diagenesis, either by BSR or SD-AOM. However, the range of Py-1b spans the entire range of

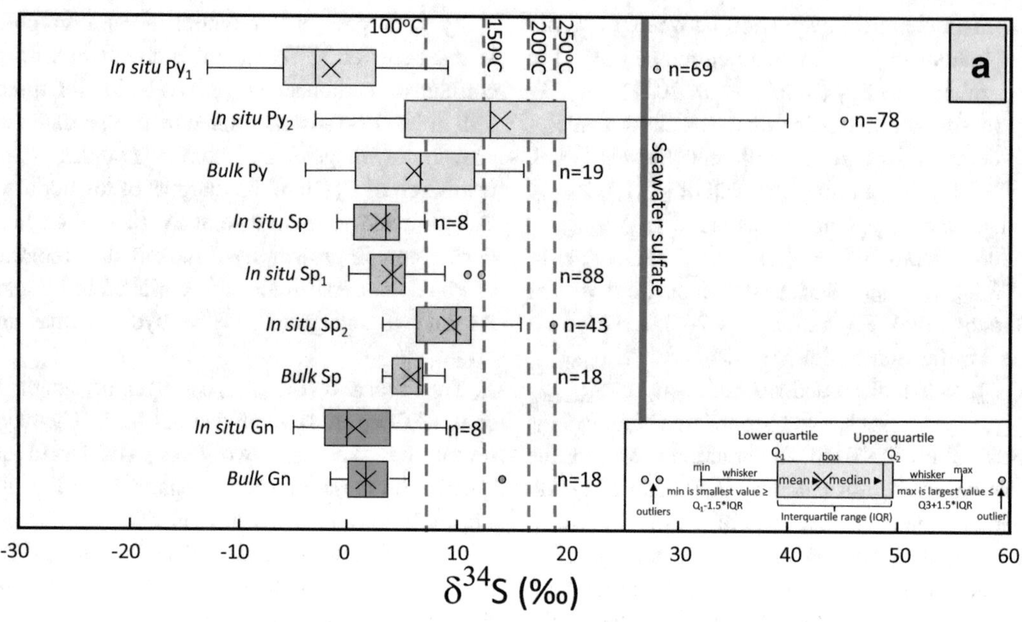

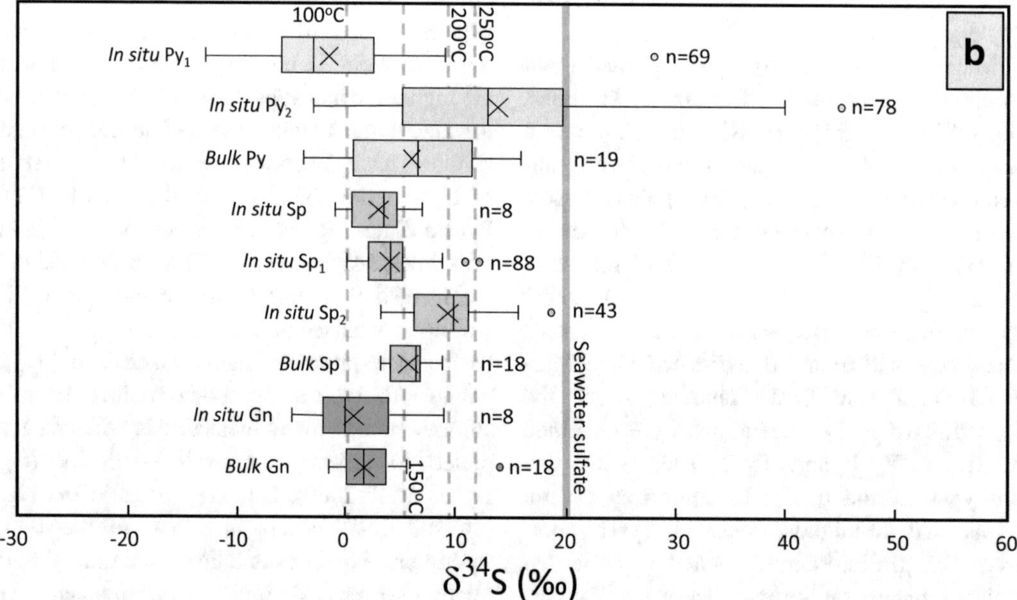

Fig. 7 Box and whisker diagrams of sulfur isotopic data for the HYC, Ridge and Cooley mineralization (after Smith and Croxford 1973; Rye and Williams 1981; Ireland et al. 2004b). Principal features of these and subsequent box and whisker plots are illustrated in the insert. The box shows the 25th (Q_1) to 75th (Q_3) percentile range of isotopic values in each data set. This range is the interquartile range (IQR). The vertical line within the box marks the median value of the data set and the cross marks the average value. The whiskers extend respectively each side of the box to the smallest data point $\geq Q_1 - 1.5$*IQR (the minimum) and to the largest data point $\leq Q_3 + 1.5$*IQR (the maximum). Outliers are those data point beyond the minimum and maximum limits and are shown at dots The dashed blue and orange vertical lines show the value of $\delta^{34}S_{sulfide}$ formed by thermochemical sulfate reduction at various temperatures. These lines have been determined using the fractionation equation of Kiyosu and Krouse (1990) assuming **a** seawater sulfate with a value of 26.8‰, and **b** seawater sulfate with a value of 20‰

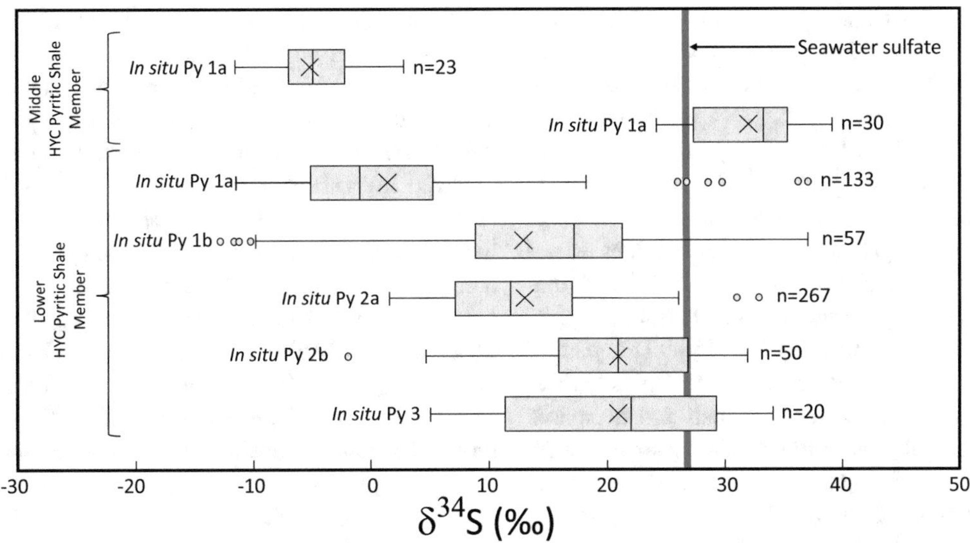

Fig. 8 Box and whisker diagram of sulfur isotopic data for the Teena mineralization (after Magnall et al. 2020b)

all the pyrite in the Lower HYC Pyritic Shale Member, suggesting the possibility that some of the pyrite classified as Py-1b on textural criteria may be Py-2, a possibility that requires further investigation. The $\delta^{34}S_{in\ situ\ Py2}$ values are intermediate between the py-1a end members, but showed no systematic trends within individual zoned aggregate samples. Because Py-2 only occurs within Pb–Zn mineralized rocks, and Py-2b sometimes has interstitial galena and sphalerite, Magnall et al. (2020b) concluded that Py-2 is part of the Teena hydrothermal event that postdates the early diagenetic Py-1 event. Unfortunately, Magnall et al. (2020b) did not measure the S-isotopic compositions of galena and sphalerite at Teena to determine the isotopic relationship of Py-2 to the two ore sulfides and further test their paragenetic sequence conclusion that Py-2, galena and sphalerite are cogenetic and formed during later diagenesis at Teena.

The S-isotope geochemistry of galena and sphalerite, however, has been studied in the HYC, Cooley and Ridge deposits. In a *bulk* isotope study Smith and Croxford (1973) reported $\delta^{34}S_{bulk}$ values for coexisting pyrite, galena and sphalerite in a vertical section through the HYC mineralization (Fig. 9). The $\Delta^{34}S_{bulk\ sphalerite\text{-}bulk\ galena}$ values are relatively

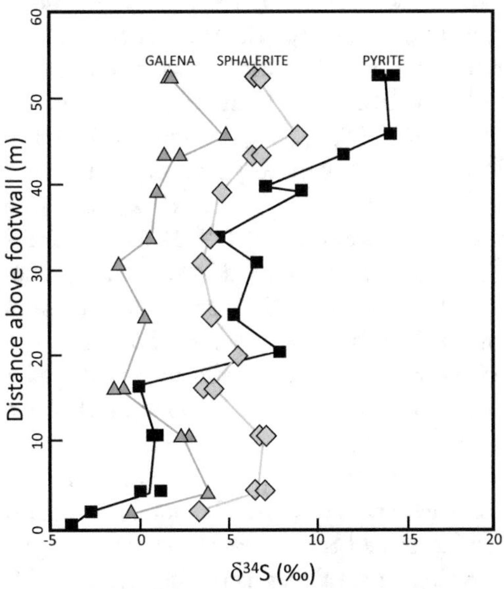

Fig. 9 Stratigraphic variations in *bulk* $\delta^{34}S$ values of pyrite, sphalerite and galena though the HYC deposit intersected in the V121 Shaft (after Smith and Croxford 1973)

uniform, suggesting the two ore minerals were in, or have approached, S-isotopic equilibrium. The fractionations give equilibration temperatures ranging from ~90 to ~240 °C. The $\delta^{34}S_{bulk\ pyrite}$ values of Smith and Croxford

(1973), as well as being shown in Fig. 9, are also included in Fig. 7. They reflect mixtures of Py_1 and Py_2 and it is therefore not surprising that the median $\delta^{34}S_{\mathit{bulk}\ \mathrm{pyrite}}$ value falls between those of Py_1 and Py_2. The $\delta^{34}S_{\mathit{bulk}\ \mathrm{pyrite}}$ values increase progressively upwards from −3.9‰ at the base of the deposit to 14.3‰ at the top, whereas galena and sphalerite values do not (Fig. 9). The pattern is strong evidence that galena and sphalerite are not in isotopic equilibrium with either Py_1 or Py_2. The pattern further suggests that the ratio Py_1/Py_2 is highest at the base of the HYC deposit and decreases systematically up through the deposit. Smith and Croxford's (1973) preferred genetic interpretation is that the HYC mineralization formed syngenetically by a dual S-source process in which Zn and Pb, either as sulfides or sulfide complexes, were introduced intermittently into a restricted local basin by exhaled basinal brines whereas Fe was contributed continuously as iron oxides or hydroxides that were converted to pyrite by BSR in a restricted environment in which the $\delta^{34}S_{\mathrm{seawater\ sulfate}}$ increased upward due to closed system Rayleigh fractionation.

In situ HYC $\delta^{34}S_{\mathrm{galena}}$ and $\delta^{34}S_{\mathrm{sphalerite}}$ data, reported by Eldridge et al. (1993), are included in Fig. 7. Unlike the *bulk* and in situ pyrite values, the *bulk* sphalerite and galena values are comparable to their in situ values. Eldridge et al. (1993) concluded that sphalerite and galena could not have formed from the same H_2S pool as the pyrite if all three sulfide minerals had had a common hydrothermal origin, and also that sphalerite and galena could not have formed from the same BSR generated H_2S pool as pyrite even if all three sulfides had a biogenic origin. To explain these conclusions Eldridge et al (1993) proposed a syndiagenetic model for the HYC mineralization in which the two pyrite generations formed at different stages of diagenesis by BSR whereas sphalerite and galena, as well as some gangue minerals, precipitated later from hydrothermal fluids that travelled parallel to bedding in the host sedimentary sequence before it was totally lithified. In this genetic model the two pyrite generations formed diagenetically by BSR from relatively cool fluids dominated by Fe, SO_4^{2-} and CO_2, whereas the base-metal sulfides formed later from a warmer Pb–Zn bearing hydrothermal fluid. The in situ data could not be used to differentiate between galena and sphalerite formation involving a sulfate-dominant oxidized hydrothermal fluid and inorganic sulfate reduction (i.e. TSR) as proposed by Williams (1978b), and precipitation from a H_2S-dominant reduced hydrothermal fluid. The genetic model is similar to the Teena model of Magnall et al. (2020b), the main difference being that Eldridge et al. (1993) regarded Py_2 to be an extension of the BSR diagenetic event that formed Py_1 rather than a part of the hydrothermal ore-forming event.

The possibility that galena and sphalerite formed by TSR is problematic if the sulfate being reduced is coeval seawater sulfate with the $\delta^{34}S$ value of ~26.8‰ discussed above. Figure 7a shows the $\delta^{34}S$ values of sulfides that would form by TSR from this sulfate at various temperatures, based on the kinetic fractionation equation of Kiyosu and Krouse (1990). The majority of the galena and sphalerite $\delta^{34}S$ values plot below 100 °C and below the temperature range for TSR, and below the temperature range suggested by the HYC $\Delta^{34}S_{\mathit{bulk}\ \mathrm{sphalerite}\text{-}\mathit{bulk}\ \mathrm{galena}}$ values. If the two ore minerals did form in the trap environment by TSR, then the sulfate involved was not coeval seawater sulfate with a $\delta^{34}S$ value of ~26.8‰, but rather was formed from sulfate with a lower $\delta^{34}S$ value. One possibility is that the sulfate had a $\delta^{34}S$ value of ~20‰, consistent with the secular $\delta^{34}S_{\mathrm{seawater\ sulfate}}$ curve of Chu et al. (2007). This possibility is illustrated in Fig. 7b. The resulting temperatures of galena and sphalerite formation are closer to those indicated by the galena-sphalerite sulfur isotope fractionation geothermometer. If the coeval seawater sulfate indeed had a $\delta^{34}S$ value of ~26.8‰ and was the source of reduced sulfur in the pyrite, then it is possible that galena and sphalerite formed by TSR from an oxidized hydrothermal fluid that transported sulfate with a $\delta^{34}S$ value of ~20‰, or from a reduced hydrothermal fluid in which the reduced sulfur had a value between ~0 and ~10‰. The possibility that the sulfide sulfur in the two ore minerals was not coeval seawater sulfate, and the oxidation state of the HYC hydrothermal fluid, are subjects that need further investigation.

The few *bulk* $\delta^{34}S$ values of coexisting sphalerite and galena in the Cooley and Ridge prospects (Rye and Williams 1981) are similar to those in the HYC, but $\Delta^{34}S_{bulk\ \mathrm{sphalerite}\text{-}bulk\ \mathrm{galena}}$ values are smaller and give a temperature range 130–300 °C. The temperatures fall westwards away from the Emu Fault, consistent with the temperatures indicated by the isotopic composition of the associated ore-stage dolomite.

Ireland et al. (2004b) measured the in situ sulfur isotopic composition of two grain-size variants of sphalerite in the HYC deposit, named Sp_1 and Sp_2 (Fig. 7). Sp_1 is the dominant type. It is very fine grained (1–10 μm) and occurs as irregular elongate aggregates (up to 200 μm long) that combine to form sphalerite laminae up to ~1 mm thick (Ireland et al. 2004b). The less abundant second type, Sp_2, occurs as partial to almost complete replacement of the nodular dolomite found predominately towards the fringes of the HYC deposit. Ireland et al. (2004b) only analysed Sp_1 and Sp_2 in the lower HYC ore lenses but sampling was widespread and included material from both the center and fringes of the deposit. Sp_1 $\delta^{34}S_{in\ situ}$ values are similar to the in situ sphalerite values reported by Eldridge et al. (1993), indicating that most of the sphalerite analysed in the previous study was Sp_1. The median values of $\delta^{34}S_{in\ situ\ \mathrm{Sp1}}$, $\delta^{34}S_{in\ situ\ \mathrm{Sp2}}$ and $\delta^{34}S_{bulk\ \mathrm{sphalerite}}$ (Fig. 7) suggest that Sp_1 represents ~60% of all the HYC sphalerite, rather than > 80% as estimated by Ireland et al. (2004b). The difference raises the possibility of there being a strong positively-skewed continuum of $\delta^{34}S_{\mathrm{sphalerite}}$ values across the whole of the HYC and that all the sphalerite formed during a single event, a possibility also requiring further investigation.

Ireland et al. (2004b) concluded that Sp_1 formed when pulsed distal exhalations of warm (100–150 °C) hypersaline sulfate-predominant metalliferous brines ponded in a local depression, producing a stratified brine pool with each stratum having different chemistries and temperatures. It was argued the brine pool was overlain by anoxic seawater and, above that layer, oxic seawater. To explain the lack of isotopic equilibrium between pyrite and sphalerite, Ireland et al. (2004b) postulated that Sp_1 formed in, and settled from, a layer near the base of the brine pool by reactions involving H_2S produced by BSR and minor amounts of H_2S generated by thermal sulfate reduction (TSR). In contrast, Py_1 formed at the same time as Sp_1 in a higher layer of the brine pool utilizing H_2S formed by BSR and more oxidized sulfur species (SO_4^{2-}, S^0 and $S_2O_3^{-1}$) produced by the cyclical sulfur oxidation and disproportionation in the oxic zone overlying the brine pool. By contrast Sp_2 and Py_2 are postulated to have formed diagenetically beneath the brine pool in a metalliferous brine environment that was partially closed and dominated by BSR-generated H_2S that produced heavier $\delta^{34}S$ values for Sp_2 and Py_2 than those for Sp_1 and Py_1 due to closed-system Rayleigh fractionation. No attempt is made to explain the disequilibrium isotopic relationship between Sp_2 and Py_2. The genetic model of Ireland et al (2004b) is complex and additional detail to that covered in the above summary can be found in the original reference.

Two more recent studies of the McArthur CD Pb–Zn mineralization, neither of which involve new stable isotope data, further supports the diagenetic/epigenetic ore-forming conclusions of Eldridge et al. (1993) and Magnall et al. (2020b) rather than the syngenetic-exhalative conclusions of Ireland et al. (2004b). Spinks et al. (2021) used Maia Mapper, a μXRF mapping system, and SEM and electron backscatter diffraction observations to show that galena and sphalerite in the HYC deposit formed by the hydrothermal replacement of dolomite. Similarly, Magnall et al. (2021), building on their earlier S-isotopic study at Teena, used other mineralogical and geochemical evidence to conclude that the Teena galena and sphalerite also formed by carbonate replacement.

3.3 Clastic-Dominated Pb–Zn Mineralization in the Southern Mt Isa Inlier

The most economically important CD Pb–Zn deposits in the Mt Isa Inlier are the Mt Isa, Hilton and George Fisher deposits (Fig. 2). All occur in the ca 1655 ± 7 Ma dolomitic Urquhart Shale (Page and Sweet 1998). The Urquhart Shale also hosts the large Mt Isa Cu orebodies that postdate the Pb–Zn orebodies (Kawasaki and Symons, 2011; Large et al. 2005; Perkins 1984). Unlike the McArthur Basin CD Pb–Zn mineralization, that in the Mt Isa district is complexly altered by faulting, folding, metamorphism and by hydrothermal alteration related to Cu mineralization. Much of the Mt Isa district CD Pb–Zn mineralization is banded like the McArthur ores but the banding and grain sizes are coarser. The main sulfide minerals are sphalerite, galena, pyrite and pyrrhotite. Deformed mineralization displays textures indicative of extensive sulfide remobilization, but some breccia and vein-hosted mineralization appears to have been introduced after the CD Pb–Zn mineralization (Chapman 2004).

3.3.1 Source

The value of $\delta^{34}S_{\text{seawater sulfate}}$ at the time of deposition of the strata hosting CD Pb–Zn deposits of the Mt Isa district is uncertain. The secular $\delta^{34}S_{\text{seawater sulfate}}$ curve of Chu et al. (2007) suggests the value was ~28‰, whereas in their study of the George Fisher deposit Rieger et al. (2020) assume a value between 18 and 25‰.

3.3.2 Transport

Carbon and oxygen isotope studies in the Mt Isa CD Pb–Zn district include Smith et al. (1978), Heinrich et al. (1989) and Waring et al. (1998). However, these studies are of little use in elucidating the nature of the fluids that formed the Mt Isa CD Pb–Zn deposits because, as was shown by Waring et al. (1998), the isotopic geochemistry of the Mt Isa carbonates reflects hydrothermal alteration associated with the later Mt Isa Cu mineralization, an event that overprinted C- and O-isotopic evidence of the earlier Pb–Zn mineralizing event.

3.3.3 Trap

Pyrite in the Urquhart Shale along a ~17 km interval of the formation stretching from ~4 km south of the Mt Isa deposit to a point ~14 km north of the deposit and ~5 km south of the George Fisher deposit has been studied by Painter et al (1999). The study included 104 $\delta^{34}S_{bulk}$ measurements of fine-grained (< 30 μm) pyrite, the earliest recognized pyrite generation in the Mt Isa region). Although the most pyritic portion of the studied interval is in the Mt Isa deposit itself, Painter et al. (1999) found no obvious difference in $\delta^{34}S_{bulk\ \text{pyrite}}$ values between mineralized and unmineralized Urquhart Shale.

The $\delta^{34}S_{bulk\ \text{pyrite}}$ values have a positively skewed distribution (Fig. 10), interpreted by Painter et al. (1999) as indicating the pyrite formed in a restricted environment during late diagenesis from sulfate-bearing fluids that flowed parallel to bedding. Petrographic and spatial arguments were used to conclude that the pyrite formed by TSR rather than BSR. Painter et al (1999) used geological and textural arguments and a statistical analysis of all previous *bulk* sulfur isotope results from the Mt Isa deposit (Fig. 10) to conclude that fine-grained pyrite predated the Pb–Zn mineralization and that the mineralizing hydrothermal fluids were dominated by reduced sulfur rather than by sulfate. This two stage syndiagenetic model is similar to that proposed for the HYC mineralization by Eldridge et al. (1993).

Further insights into the formation of pyrite in the Urquhart Shale and its CD Pb–Zn mineralization are provided by Rieger et al. (2020) in an in situ sulfur isotope study of pyrite in and near the George Fisher CD Pb–Zn deposit. The study includes data from unmineralized Urquhart Shale at Shovel Flats at the northern end of the study area of Painter et al. (1999). Four significant stages of sulfide deposition are recognized by Rieger et al. (2020). These are an early pre-ore pyrite stage, followed by three stages of hydrothermal sulfide deposition comprising

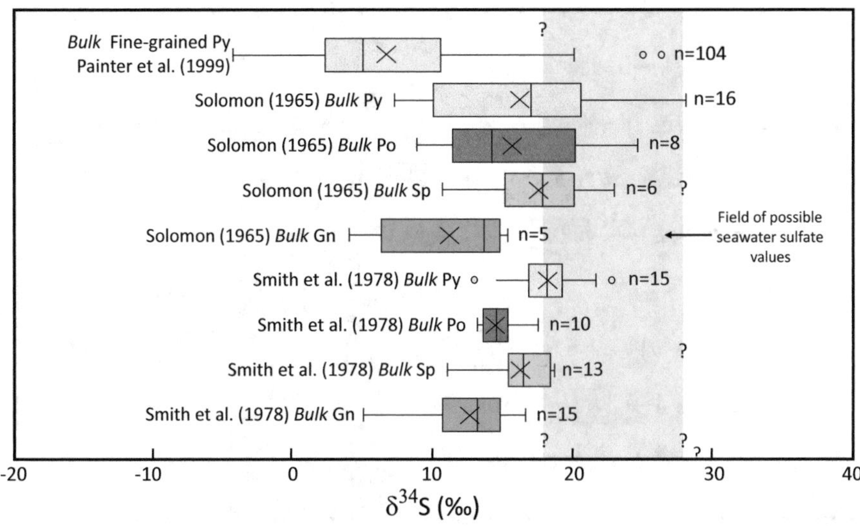

Fig. 10 Box and whisker diagram of sulfur isotopic data for sulfides in the Mt Isa CD Pb–Zn mineralization (after Painter et al. 1999; Smith et al. 1978; Solomon 1965)

pyrite, sphalerite, galena, pyrrhotite and chalcopyrite. The first hydrothermal stage is the dominant one and is the CD Pb–Zn stage. Five pyrite types are identified by Rieger et al. (2020). The earliest is Py-0, the fine-grained pyrite also studied by Painter et al. (1999). Rieger et al. (2020) divided Py-0 into two sub-types: Py-0a that occurs as the bright cores of zoned Py-0 crystals and Py-0b that occurs as duller overgrowths around Py-0a. The first hydrothermal pyrite phase, Py-1, is stratabound and coarser ($\leq$ several 100 μm) than Py-0. Py-1 is sometimes replaced by sphalerite or galena to give the pyrite atoll texture first described in the Mt Isa deposit by Grondijs and Schouten (1937). The second hydrothermal pyrite phase, Py-2, occurs in veins and breccias that cut and replace the host sediments and previous sulfide phases. The third hydrothermal pyrite stage, Py3, comprises stratabound, massive units or veins and breccias that cut all previous stages and host-rock bedding. Py-3 is often accompanied by pyrrhotite. Another pyrite type, euhedral pyrite (Py-euh), occurs as overgrowths on Py-0 and is the coarsest (up to several mm in grain size) pyrite type. It occurs at Shovel Flats as well as in the George Fisher deposit. Although Py-euh postdates Py-0, its relationship, if any, to either one or more of the hydrothermal pyrite phases is unclear. Further and very detailed descriptions of the textural relationship between the various sulfides in the George Fisher mineralization and enclosing sediments can be found in Rieger et al. (2020).

The in situ sulfur isotopic composition of the George Fisher pyrite types are shown in Fig. 11. Py-0 has the lowest isotopic values. In the George Fisher mineralization the median δ^{34}S value of Py-0a is −1.1‰ and that of Py-0b is 0.7‰, whereas at Shovel Flat the median Py-0a value is 4‰. The in situ Shovel Flats values are consistent with the *bulk* values of samples from the same area reported by Painter et al. (1999). The order of median δ^{34}S values is Py-0 < Py2 < Py-1 < Py-3. Py-euh at both George Fisher and Shovel Flats have similar wide distributions of δ^{34}S of values.

Because $\delta^{34}S_{in\ situ\ Py\text{-}0}$ values are offset from coeval seawater sulfate values (assumed to be between 18 and 25‰) by $\leq$ 25–32.5‰, Rieger et al. (2020) concluded Py-0 formed during early diagenesis by BSR under open-system conditions rather than during late diagenesis by TSR as argued by Painter et al. (1999). As Py-0b δ^{34}S values are slightly higher than those of Py-0a Rieger et al. (2020) also concluded that Py-0b formed under increasingly sulfate-limited conditions as diagenetic pore fluids became more restricted.

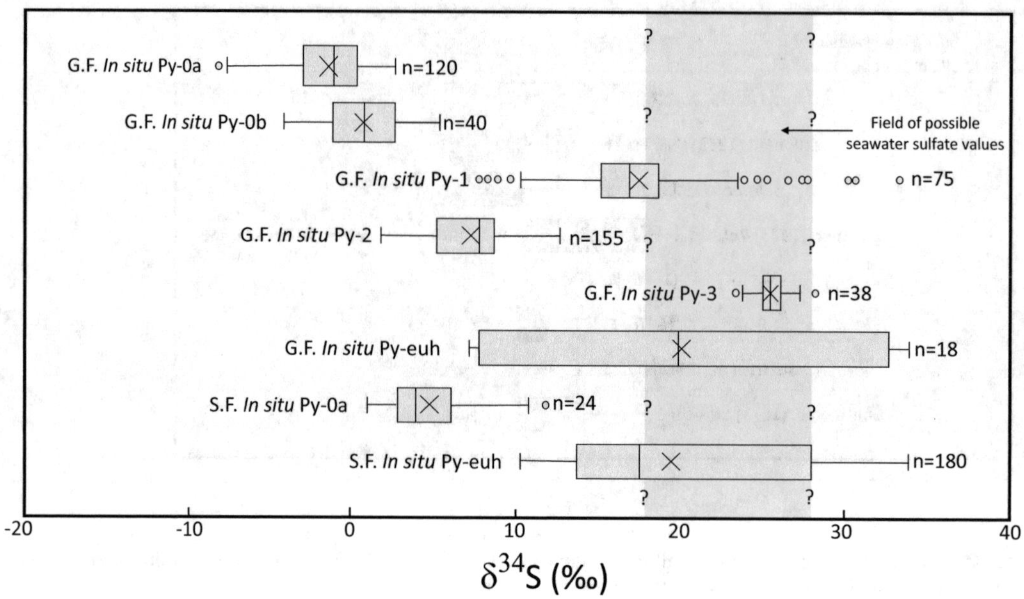

Fig. 11 Box and whisker diagram of sulfur isotope data for sulfides in the George Fisher (G.F.) deposit and unmineralized Urquhart Shale in the Shovel Flats (S.F.) area south the George Fisher (after Rieger et al. 2020)

Rieger et al. (2020) noted that the variability of Py-1 $\delta^{34}S$ values is at the hand-specimen scale, suggesting Py-1 formed a transport-limited environment that was not linked to the closing environment in which Py-0 formed because of the separation of the $\delta^{34}S$ values of the two pyrite types (Fig. 11). Given the high positive $\delta^{34}S$ values of Py-1 it was concluded it formed at a temperature of > 100 °C by TSR involving the oxidation of organic matter in the host Urquhart Shale. During deformation following the Stage 1 hydrothermal event it is argued that the earlier sulfides were recycled and replaced to form the Stage 2 mineralization, resulting in Py-2 being isotopically intermediate in values between Py-0 and Py-1. Py-3 is postulated to have formed by TSR during a later deformational event associated with the Cu mineralization event. Because Py-euh occurs in both mineralized and unmineralized samples Rieger et al. (2020) concluded that this pyrite type is unlikely to have had a hydrothermal origin. Py-euh could have formed by either TSR under conditions of variable rates of sulfate replenishment relative to reduction, or by sulfate reduction involving SD-AOM. Rieger et al. (2020) did not measure the isotopic composition of the sphalerite and galena in the George Fisher deposit to test their conclusion that the hydrothermal pyrite-forming mechanisms also extends to the formation of galena and sphalerite.

However, *bulk* S-isotopic values for galena and sphalerite, as well as for pyrite and pyrrhotite, have been reported from the Mt Isa CD Pb–Zn deposit by Solomon (1965) and Smith et al. (1978) and their data are shown in Fig. 10. The distributions of $\delta^{34}S$ values for the four minerals are similar in both studies, but those from Smith et al. (1978) are more useful from a genetic viewpoint as they include 14 measurements of coexisting galena, sphalerite, pyrrhotite and pyrite. Their $\delta^{34}S$ values, with one exception, decrease in the order pyrite > sphalerite > pyrrhotite > galena, suggesting sulfur isotope equilibrium was reached or approached between the four minerals. However, because of uncertainties regarding the purity of their mineral separates, and discrepancies in equilibration temperatures determined from $\Delta^{34}S$ values of the various sulfide pairs, Smith et al (1978) concluded that isotopic equilibrium was never fully established. The median values of pyrite measured by Solomon (1965) and Smith et al. (1978) are respectively 17.0 and 18.1‰ and are clearly different to the fine-grained pyrite studied by Painter et al. (1999). Rather the pyrite is similar to Py-1 at the George Fisher deposit which has a median value of 16.9‰. The similarity is further support for the conclusion of Rieger et al. (2020) that Py-1 formation is a part of the CD Pb–Zn ore-forming event.

The S-isotopic data of Smith et al. (1978) also show that the Py-1 type pyrite is more similar isotopically to coexisting galena and sphalerite than is the case with the pyrite coexisting with galena and sphalerite in the HYC ores. The difference suggests strongly that unlike the Mt Isa ores, the HYC ores do not contain significant quantities of hydrothermal pyrite, suggesting that the hydrothermal fluids that formed the Mt Isa mineralization had a higher abundance of Fe than the HYC hydrothermal fluids.

3.4 Clastic-Dominated Pb–Zn Mineralization in the Central Mt Isa Inlier

A small but interesting CD Pb–Zn deposit, Lady Loretta, occurs in the central Mt Isa Inlier (Fig. 2). The host is the 1653 ± 7 Ma Lady Loretta Formation which lies at the top McNamara Group (Page and Sweet 1998). The mineralization and its geological setting has been described by Carr (1984), Carr and Smith (1977) and Duffett (1998). Unlike the other North Australian CD Pb–Zn deposits, Lady Loretta has thick lenses of layered barite with sphalerite, pyrite and chert at the top of the mineralized sequence. The strata hosting Lady Loretta may have been formed in a lacustrine, rather than marine environment (Large and McGoldrick 1998). The strata are dolomitic, but the carbonate directly associated with the mineralization is siderite (Large and McGoldrick 1998). Lady Loretta is less metamorphosed than the ore deposits of the Mt Isa district, but more metamorphosed than those in the McArthur Basin (Carr and Smith 1977).

3.4.1 Source

A strong indication of the value of $\delta^{34}S_{water\ column\ sulfate}$ at the time of deposition of the strata hosting the Lady Loretta CD Pb–Zn deposit has been provided by Gellatly and Lyons (2005) who reported 23 CAS analyses from the Paradise Creek Formation that lies ~500 m beneath the Lady Loretta Formation in the McNamara Group (Southgate et al. 2000). The CAS $\delta^{34}S$ values range from 14.1 to 37.3‰ and have a median of 30.6‰ which is taken here to be the value of $\delta^{34}S_{water\ column\ sulfate}$ during Lady Loretta times.

3.4.2 Transport

Dolomite outside the Lady Loretta mineralization has *bulk* $\delta^{18}O_{SMOW}$ and $\delta^{13}C$ values respectively of 16.5–18.2‰ and −1.5 to −0.9‰ (McGoldrick et al. 1998). The $\delta^{13}C$ values fall in the range of unaltered McArthur Basin dolomites whereas the $\delta^{18}O_{SMOW}$ values are lower (Fig. 5), a difference attributed by Large et al. (2001) to a marine depositional environment in the McArthur Basin and a lacustrine one at Lady Loretta. Siderite within the Lady Loretta mineralization has *bulk* $\delta^{18}O$ and $\delta^{13}C$ values respectively of 23.2–28.3‰ and −3.2 to −1.9‰. McGoldrick et al. (1998) calculated that the siderite formed at temperatures near 100 °C from a fluid with a $\delta^{18}O_{SMOW}$ value of 6‰ and a $\delta^{13}C$ value of −6‰. Such a fluid falls in the low-temperature HYC hydrothermal fluid (Fig. 6) modelled by Large et al. (2001). The values are also consistent with the ore-forming hydrothermal fluid being an evolved basinal brine.

3.4.3 Trap

Bulk $\delta^{34}S$ values of pyrite, sphalerite, galena and barite at Lady Loretta are reported by Carr and Smith (1977). The median $\delta^{34}S$ values for the three sulfides are similar to those at Mt Isa, while barite has a median $\delta^{34}S$ value of 39.4‰ (Fig. 12). In the seven sets of $\delta^{34}S$ of coexisiting pyrite-sphalerite-galena measurements made by Carr and Smith (1977) only one set has the relationship $\delta^{34}S_{bulk\ pyrite} > \delta^{34}S_{bulk\ sphalerite} > \delta^{34}S_{bulk\ galena}$, suggesting the three sulfides are not in, or approached, isotopic equilibrium.

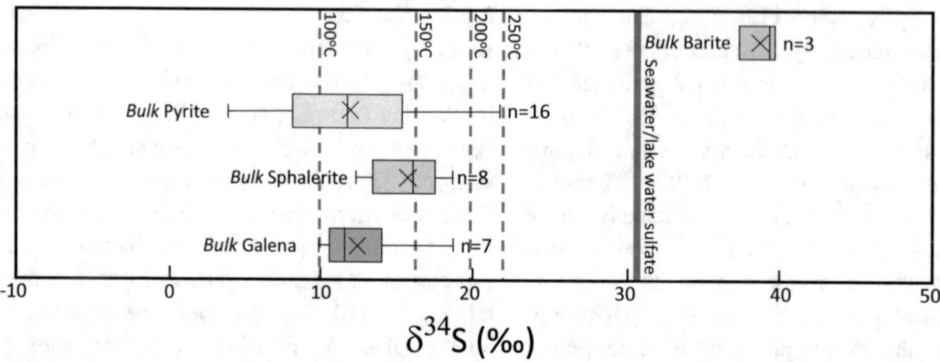

Fig. 12 Box and whisker diagram of sulfur isotopic data for sulfides and barite in the Lady Loretta deposit (after Carr and Smith 1977). The dashed blue, vertical lines show the value of $\delta^{34}S_{sulfide}$ formed by thermochemical sulfate reduction at various temperatures assuming a seawater sulfate with a value of 30.6‰

By contrast, in all seven sets $\delta^{34}S_{\textit{bulk}\ \mathrm{sphalerite}} > \delta^{34}S_{\textit{bulk}\ \mathrm{galena}}$, indicating the two ore minerals are in, or have approached, sulfur isotope equilibrium and that they are both in isotopic disequilibrium with pyrite. *Bulk* $\Delta^{34}S_{\mathrm{sphalerite\text{-}galena}}$ values range from +2.3 to +4.7‰, corresponding to a temperatures range of ~125–300 °C. However, this range should be treated with caution as Carr and Smith (1977) had difficulties obtaining pure mineral separates for analysis. Had galena and sphalerite formed by TSR of the ambient water column sulfate with a value of 30.6‰, then the range of ore-forming temperature, based on the fractionation equation of Kiyosu and Krouse (1990) would be ~110–160 °C (Fig. 12).

Carr and Smith (1977) noted a subtle increase in $\delta^{34}S_{\textit{bulk}\ \mathrm{pyrite}}$ values up section at Lady Loretta and suggested the trend reflected pyrite formation by BSR in a restricted environment with an increasingly limited supply of sulfate, as suggested by Smith and Croxford (1973) to explain a similar increase in $\delta^{34}S_{\textit{bulk}\ \mathrm{pyrite}}$ values up section in the HYC deposit. The suggestion is supported by the $\delta^{34}S_{\textit{bulk}\ \mathrm{barite}}$ values from the barite lenses at the top of the deposit which are ~9‰ higher than the assumed ambient water column sulfate value (Fig. 7). As sphalerite and galena appear isotopically unrelated to pyrite, Carr and Smith (1977) proposed a two S-source model for the formation of the Lady Loretta deposit in which Pb and Zn sulfides, or sulfide complexes of Pb and Zn, were exhaled into the restricted reservoir of seawater from which the pyrite formed independently by BSR, the same genetic model proposed for the HYC by Smith and Croxford (1973).

3.5 Clastic-Dominatd Pb–Zn Mineralization in the Northern Mt Isa Inlier

The large CD Pb–Zn Century deposit in the northern Mt Isa Inlier (Fig. 2) is hosted by the siliciclastic Lawn Hill Formation that was deposited at 1595 ± 6 Ma (Page and Sweet 1998). Although some pyrite occurs in a halo around the deposit, the mineralization mostly comprises sphalerite with only minor galena (Broadbent et al. 1998). Some 80–90% of the Century ores are finely banded and are interlayered with sideritic siltstone and black shale, with the remainder comprising later coarser-grained vein- and breccia-style mineralization.

There are two types of sphalerite in the banded mineralization, one with a porous texture and a high pyrobitumen content, and the other with a non-porous texture and a low pyrobitumen content (Broadbent et al. 1998). A number of minor vein and breccia-style occurrences of Pb–Zn mineralization occur in a 10 by 20 km area to the north, northwest and south of Century, and these, together with Century, comprise the Burketown mineral field (Polito et al. 2006).

3.5.1 Source

The value of $\delta^{34}S_{seawater\ sulfate}$ at the time of deposition of the strata hosting the Century deposit is uncertain. The secular $\delta^{34}S_{seawater\ sulfate}$ curve of Chu et al. (2007) indicates that the value changed suddenly from ~16 to ~22‰ around the Century host strata were being deposited.

3.5.2 Transport

Five textural types of siderite occur in the Century deposit. Broadbent et al. (1998) report $\delta^{18}O_{bulk}$ and $\delta^{13}C_{bulk}$ values for siderite mixtures that range respectively from 13.9 to 25.3‰ and −8.3 to 2.5‰ (Broadbent et al. 1998: Fig. 17). Because the analyses are of mixtures of siderite types, they could not be used to make quantitative estimates of the C- and O-isotopic composition of the hydrothermal fluids that formed the Century deposit. Broadbent et al. (1998) simply concluded that the siderites were deposited from an ^{18}O-enriched fluid, the source possibly being evaporative carbonate sequences deep beneath the Lawn Hill Formation. A subsequent fluid inclusion study (Polito et al. 2006) found that the porous sphalerite was deposited at temperatures between 120 and 160 °C from a saline basinal brine (~21% NaCl equivalent) with a δD_{fluid} range of −89 to −83‰. It was concluded that the fluids were highly evolved meteoric waters that came from siliciclastic strata some 8–10 km beneath the Century deposit site.

3.5.3 Trap

The *bulk* sulfur isotope values of sphalerite, galena and pyrite at Century measured by Broadbent et al. (1998) and Polito et al. (2006) are illustrated in Fig. 13. The median values of the pyrite, porous and non-porous sphalerite and galena in the stratiform mineralization reported by Broadbent et al. (1998) are similar (respectively 11.0‰, 10.2‰, 9.2‰ and 11.1‰) which, along with textural evidence, led Broadbent et al (1998) to conclude they are cogenetic. However, the sphalerite in the stratiform mineralization measured by Polito et al. (2006) has a median value of 13.2‰ indicating that the spread of $\delta^{34}S_{sulfide}$ values in the CD Pb–Zn mineralization at Century is greater than that found by Broadbent et al. (1998) and a good understanding of the formation of the Century mineralization

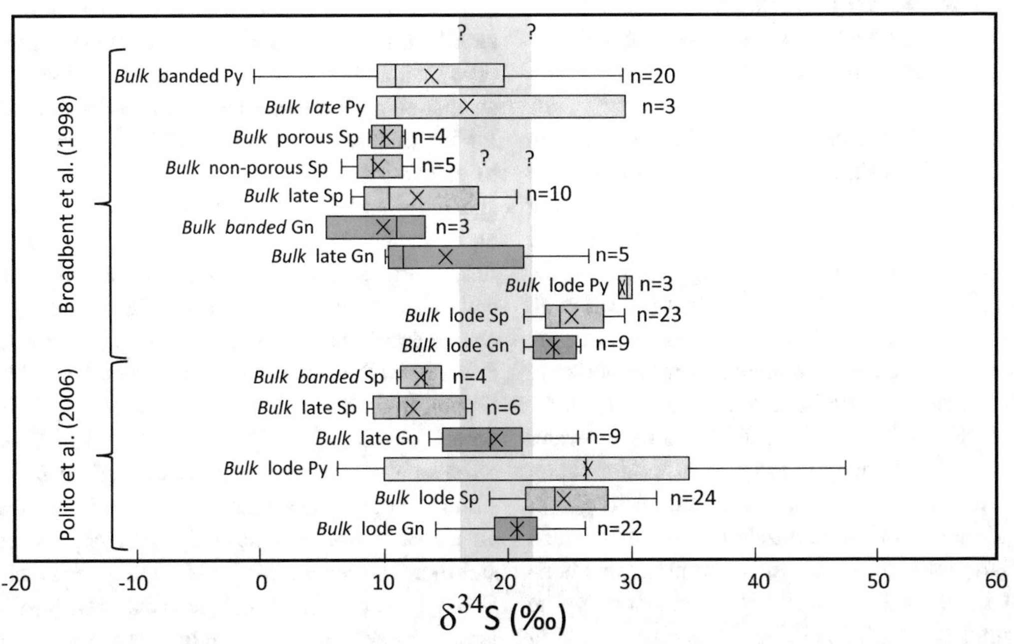

Fig. 13 Box and whisker diagram of sulfur isotopic data for sulfides in the Century deposit (after Broadbent et al. 1998; Polito et al. 2006)

requires a comprehensive *in-situ* S-isotope study of the deposit.

Sulfides in the Century vein and breccia mineralization have higher *bulk* values than those in the stratiform mineralization, while those in the nearby Burketown mineral field are even higher (Fig. 12). To explain this observation Broadbent et al (1998) argued that that the Century and Burketown deposits are part of a single mineralizing system involving a large closed-system sulfate-rich fluid reservoir that became progressively enriched in ^{34}S as sulfides formed by TSR over a long period. These authors also postulated that the Century CD Pb–Zn mineralization formed by the subsurface replacement of silica and infill of pore space in the host sediments during late diagenesis, with porous sphalerite forming by oil-mediated TSR reactions and the non-porous sphalerite by gas-mediated TSR reactions. However, Polito et al. (2006) concluded that the fluids that formed Century did not form the vein and breccia mineralization elsewhere in the Burketown mineral field.

4 The North American Paleozoic Cordillera Clastic-Dominated Pb–Zn Province—Canada

Several well-studied CD Pb–Zn deposits occur in the Selwyn Basin in the Canadian portion of the North American Cordillera (Fig. 14), including the Howards Pass deposit which is the second largest of the world's ten largest CD Pb–Zn deposits based on Pb + Zn content (Leach et al. 2005). The Selwyn Basin CD Pb–Zn deposits are hosted by strata ranging in age from Cambrian to Devonian. The Basin is well suited to sulfur isotope studies to elucidate the role seawater sulfate might have been played in the formation of these deposits because the secular $\delta^{34}S_{\text{seawater sulfate}}$ curve for the Paleozoic is well documented (Claypool et al. 1980); there are numerous horizons of sedimentary/early diagenetic pyrite within the basin, as well as horizons rich in barite, and there are three significant groups of CD Pb–Zn deposits of different ages in the basin. They are the Cambrian Anvil Range group, the Ordovician–Silurian Howards Pass group and the Late Devonian MacMillan Pass group. (Fig. 13). The Anvil Range deposits have not been studied using in situ techniques, unlike those MacMillan and Howards Pass deposits, and are therefore not considered further in this Chapter. However, information about their stable isotope characteristics can be found in Campbell and Ethier (1974) and Shanks et al. (1987).

In the first of two related papers that greatly influenced ideas on the formation of CD Pb–Zn mineralization worldwide, Goodfellow and Jonasson (1984) presented secular S-isotopic curves for pyrite and barite in the Selwyn Basin. The barite curve is similar in shape to the global seawater sulfate curve of Claypool et al. (1980), but is consistently 5–10‰ higher than the seawater sulfate curve. By contrast, the pyrite curve of Goodfellow and Jonasson (1984) is different and exceeds the $\delta^{34}S_{\text{seawater sulfate}}$ curve in three periods: the Late Ordovician, the Early Devonian and the Late Devonian. These three periods were interpreted to be times of almost complete reduction of seawater sulfate by BSR in restricted and stagnant deep euxinic layers of a stratified ocean. In the second paper Goodfellow (1987) noted that the interpreted periods of stagnation appear to coincide with periods of CD Pb–Zn mineralization and linked ore formation to stagnation. Because the isotopic curves for sphalerite and galena generally track the pyrite curve in the Selwyn Basin, Goodfellow (1987) further argued that the sulfide-S in the ore minerals also formed in the stagnant layer and concluded that CD Pb–Zn ores formed in the Selwyn Basin upon the exhalation of base-metal rich, but S-poor, hydrothermal fluids into stagnant euxinic ocean water (Fig. 1a, Pathway 1). This sedimentary-exhalative model is based on *bulk* sulfur isotopic measurements, but with the advent of in situ isotopic analytical technologies and their application to the MacMillan Pass and Howards Pass CD Pb–Zn deposits, a very different history of ore-forming processes has emerged.

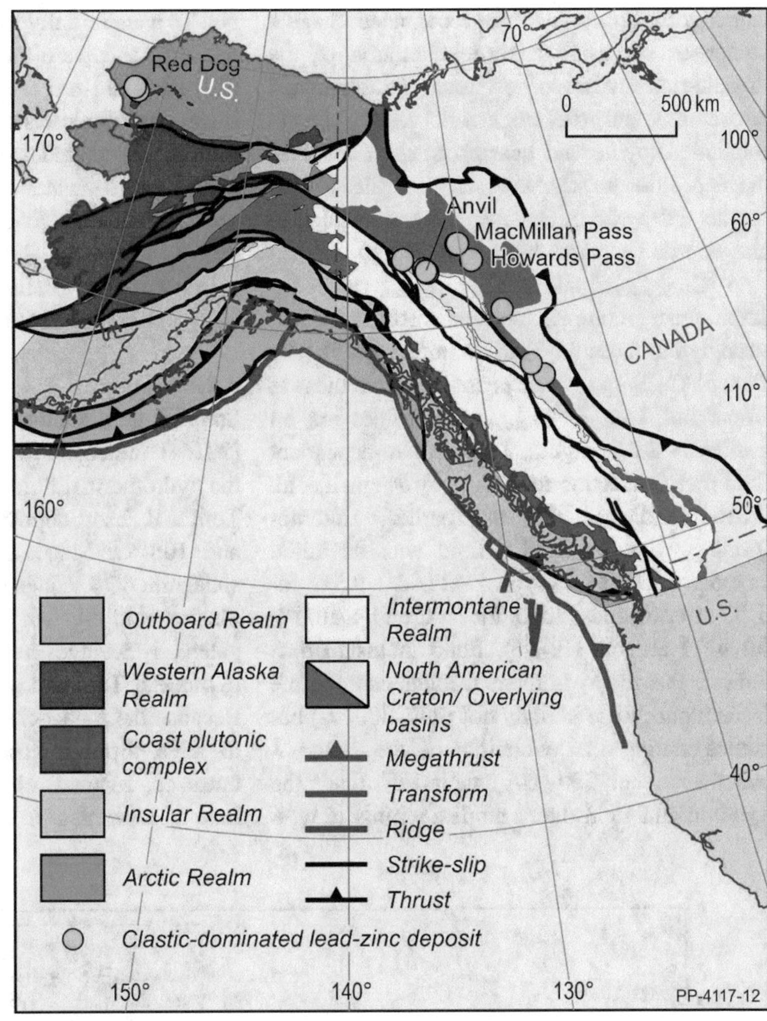

Fig. 14 Location of the most significant CD Pb–Zn deposit districts in the North American Cordillera CD Pb–Zn Province. The term Realm in the geological map legend refers to the various components of the pre-accretionary terranes that today comprise tectonic elements of the North American Cordillera

4.1 Clastic-Dominated Pb–Zn Mineralization in the MacMillan Pass District

The MacMillan Pass CD Pb–Zn mineralization occurs in the Late Devonian Earn Group. Although the host rocks have been slightly metamorphosed (to sub-greenschist facies) and deformed into tight folds, mineralization displaying well-preserved sedimentary and hydrothermal features occurs in the limbs of folds (Magnall et al. 2016b). There are two significant deposits at MacMillan Pass, Tom (Andsell et al. 1989) and Jason (Gardner and Hutcheon 1985). Tom is located ~6 km east of Jason. Both deposits have several zones of stratiform mineralization in a sequence of chert conglomerates and biosiliceous mudstones that accumulated in a submarine fan on the eastern edge of the Selwyn Basin. Both deposits have well-developed feeder zones.

4.1.1 Source

The value of $\delta^{34}S_{\text{seawater sulfate}}$ at the time of deposition of the Late Devonian strata hosting the Tom and Jason CD Pb–Zn deposits is ~24‰ (Magnall et al. 2016a).

4.1.2 Transport

A study of the Tom and Jason feeder zones by Magnall et al. (2016a) provides important insights into the nature of the fluids involved in the formation of the deposits. Two stages of feeder-zone development are recognized. Stage 1 comprises pervasive ankerite alteration of the organic-rich mudstone host rock and associated crosscutting ankerite stockwork veining (±py-robitumen, pyrite and quartz). Stage 2 involves the deposition of massive sulfide (galena, pyrrhotite and pyrite ± chalcopyrite and sphalerite) and siderite (±quartz and barytocalcite).

$\delta^{13}C_{bulk\ ankerite}$ and $\delta^{18}O_{bulk\ ankerite}$ values for Jason feeder complex correlate positively and strongly but those for Tom do not (Fig. 15). The Tom $\delta^{13}C_{bulk\ ankerite}$ values are similar to those at Jason, but Tom $\delta^{18}O_{bulk\ ankerite}$ values are on average ~2.5‰ higher. Temperature-dependent fluid-rock interaction modelling by Magnall et al. (2016a) indicates that the earliest ankerite-forming event involved a fluid with an initial isotopic value of −3‰ for $\delta^{13}C$ and + 9.5‰ for $\delta^{18}O$ at Jason and −2.5‰ for $\delta^{13}C$ and + 10.0‰ for $\delta^{18}O$ at Tom (Fig. 6). Fluid inclusion data indicate that the hydrothermal fluids entering the feeder zones were initially hot (270–300 °C) but cooled during the evolution of the Stage 1 ankerite event. It is concluded that the hydrothermal fluids had a modest salinity (6 wt% NaCl), were initially hot (270–300 °C) with a pH ($\leq$ 4.5) and were metal rich ($\gg$ 100 ppm Pb, Zn). As there are no evaporites in the Selwyn Basin, the modest salinity of the hydrothermal fluid is interpreted to have been derived originally from seawater trapped in Selwyn basinal rocks. Their high $\delta^{18}O$ values of ~10.0‰ suggest these basinal fluids evolved as they interacted with the basinal sedimentary lithologies (Magnall et al. 2016a).

4.1.3 Trap

Stage 2 fluid inclusion data of Magnall et al. (2016a) indicate that during sulfide deposition the hydrothermal fluids in the feeder zones to the Tom and Jason deposits cooled to between 215 and 108 °C. Stage 2 bulk pyrite, galena and sphalerite $\delta^{34}S$ values from feeder complexes are illustrated in Fig. 16. The Jason feeder zone bulk galena $\delta^{34}S$ values have limited range compared to those at Tom and are also significantly lower. Because the feeder-zone sulfide minerals appear to have formed close to the sediment–water interface, Magnall et al. (2016a) assumed that Late Devonian seawater sulfate with

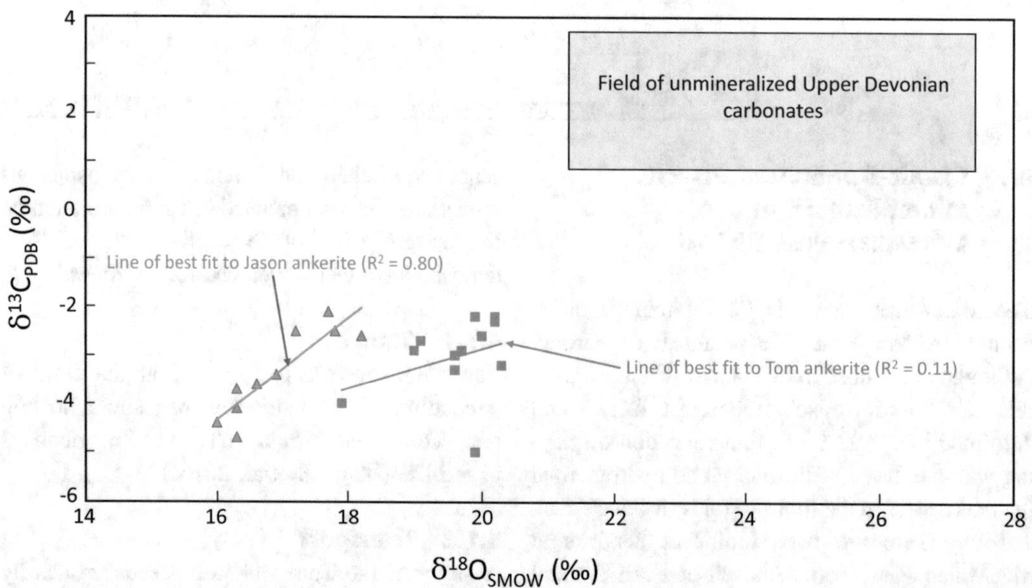

Fig. 15 The $\delta^{18}O$ versus $\delta^{13}C$ values of ankerite from the feeder zones to the Tom and Jason deposits, Macmillan Pass, Selwyn Basin (after Magnall et al. 2016a). The field of unmineralized Upper Devonian carbonates is after Carpenter et al. (1991)

$\delta^{34}S_{sulfate}$ ~24‰ was the source of the sulfide S. As the fluid inclusion data indicate that the temperature of sulfide formation was ≥ 108 °C, Magnall et al. (2016a) argued that the sulfide sulfur was generated by the TSR of seawater sulfate. Based on their sulfur isotope measurements and the kinetic isotope fractionation equation of Kiyosu and Krouse (1990) for TSR, Magnall et al. (2016a) found that the sulfides at Jason formed at temperatures broadly consistent with the temperatures derived from their fluid inclusion measurements. Given the contrasting S-isotopic composition of the Tom sulfides it was further suggested that the sulfide sulfur at Tom might have formed at cooler temperatures by BSR.

Turning to the stratiform CD Pb–Zn mineralization adjacent to the Tom and Jason feeder complexes, Magnall et al. (2015) observed that the ore-bearing sequences at Tom and Jason contain abundant preserved radiolarians, including beds up to 10-cm-thick with > 50% radiolarians that are partially preserved by an early generation of pyrite that is overprinted by base-metal sulfides. The radiolarian-rich lithologies are interpreted to be a favorable host for mineralization due to high porosities during early diagenesis. Because the lithology also contains abundant organic matter it has a high capacity for the generation of sulfide via in situ sulfate reduction.

The stable isotope geochemistry of the Tom and Jason CD Pb–Zn mineralization was studied by Magnall et al. (2016b). The study includes in situ $\delta^{34}S$ data for barite and pyrite from relatively undeformed drill core samples, as well as *bulk* $\delta^{34}S$ values for pyrite extracted from barren mudstones between the two deposits and above and below the Tom deposit. Magnall et al. (2016b) identified three phases of mineralization. Stage 1 includes > 7 μm framboidal pyrite (Py–I) and small (< 25 μm) anhedral crystals of barite (brt-I) that replace quartz in the host mudstones. Stage 2 is characterized by euhedral pyrite that occurs as stratiform accumulations (Py–IIa) or solitary crystals (Py–IIb), and by coarser-grained (> 25 μm) and equant barite (brt–II) that is often associated with Py-IIa. Stage 2 also has barite in veins (brt–III) as well as euhedral celsian. The main mineralization event is Stage 3. It includes pyrite, sphalerite, galena and witherite. Stage 3 pyrite (Py–III), occurs as large sub- to anhedral replacements of earlier barite, and witherite occurs as an accessory phase to sphalerite. Galena is the final phase in the Stage 3 event, occurring as anhedral, interstitial crystals that replace earlier barite and pyrite along stratiform horizons.

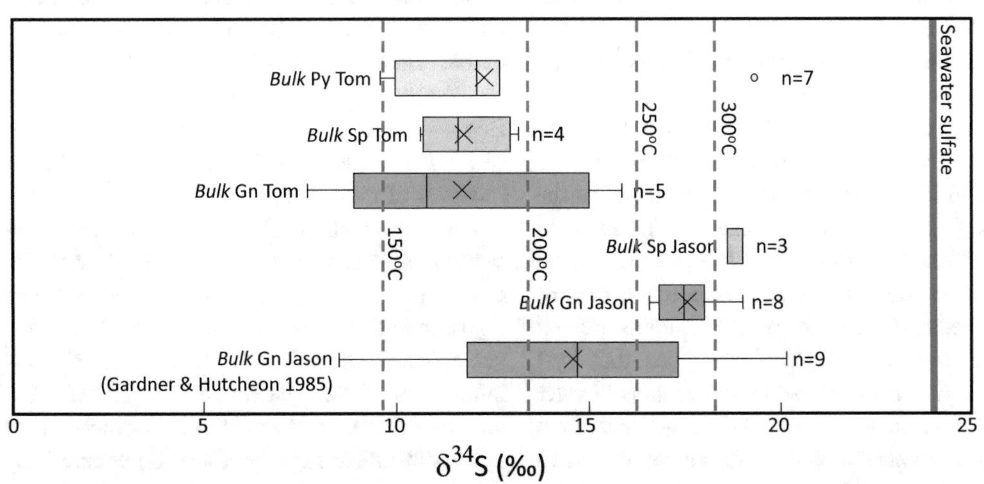

Fig. 16 Box and whisker diagram of sulfur isotopic data for sulfides in the feeder zones of the Tom and Jason deposits (after Gardner and Hutcheon 1985; Magnall et al. 2016a). The dashed blue, vertical lines show the value of $\delta^{34}S_{sulfide}$ formed by thermochemical sulfate reduction at various temperatures assuming a seawater sulfate with a value of 24‰

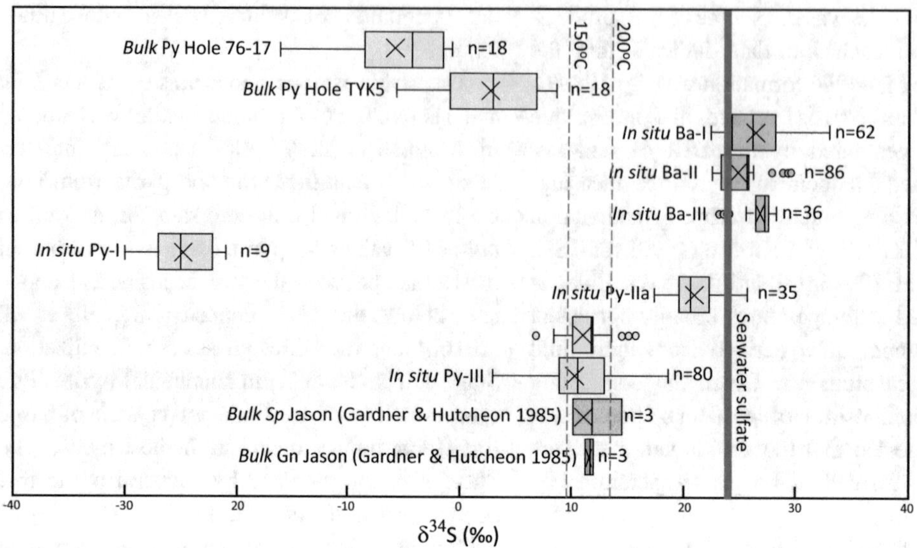

Fig. 17 Box and whisker diagram of sulfur isotopic data for sulfides and barite in the stratiform portions of the Tom and Jason deposits (after Gardner and Hutcheon 1985; Magnall et al. 2016b). The dashed blue, vertical lines show the value of $\delta^{34}S_{sulfide}$ formed by thermochemical sulfate reduction at various temperatures assuming a seawater sulfate with a value of 24‰

The in situ barite $\delta^{34}S$ values from the three stages are similar (Fig. 17) and Magnall et al. (2016b) concluded that the barite sulfate originated from modified Late Devonian seawater. They also observed that radiolarian tests in the host sediments have high Ba contents and concluded the enrichment reflected high levels of biological activity that predated diagenetic base-metal mineralization.

The various pyrite generations have distinctive sulfur isotopic compositions (Fig. 17). Py-I has the lowest $\delta^{34}S$ values, Py-IIa has the highest, and Py-IIb and Py-III have values between Py-I and Py-IIa. Py-I values are offset from the ambient seawater SO_4^{2-} value by between 44 and 54‰ and are interpreted to reflect formation by BSR, consistent with the framboidal texture of Py-I. Because the majority of framboids are > 7 μm in diameter it is further argued that Py-I formed diagenetically below the sediment–water interface rather than in a euxinic water column because framboids that form above the sediment–water interface typically have smaller diameters. Py–II is the heaviest isotopically and Magnall et al. (2016b) concluded from its isotopic composition and relationship with brt-II barite that it formed diagenetically below the sediment–water interface by SD-AOM. Py-III has a broad distribution of $\delta^{34}S$ values that overlap Py-II, but which extends to lower values. The formation of sphalerite and galena is not discussed in detail in Magnall et al. (2016b) other than a brief comment that barite replacement by these sulfides might be an important mineralizing pathway in those CD Pb–Zn deposits rich in barite. *Bulk* pyrite $\delta^{34}S$ values from barren mudstones range from −15.6 to +8.7‰ (n = 37). Because the range falls between the distributions of Py–I and II the *bulk* pyrite is interpreted to be mixtures of these two pyrite types.

Magnall et al. (2020c) presented a new subseafloor replacement model for the MacMillan Pass deposits using the stable isotope data and arguments discussed above, as well as new geological, geochemical and mineralogical information from Magnall et al. (2020a). The new model has ore formation occurring entirely in the subsurface from hot (300 °C) mineralizing fluids that thermally degraded organic matter, generating CO_2 that promoted barite dissolution and not only triggered sulfide precipitation by TSR but also increased porosity and permeability that further facilitated ore formation. Gardner and Hutcheon (1985) report *bulk* (physical separates) S-isotopic values of galena and sphalerite from both the feeder zone and the CD Pb–Zn mineralization at Jason (Figs. 16 and 17). The values are consistent with the derivation of their sulfide-S by TSR of Late Devonian seawater sulfate at temperatures $\geq \sim 150$ °C, with feeder zone having the higher temperatures (Fig. 16).

4.2 Clastic-Dominated Pb–Zn Mineralization in the Howards Pass District

CD Pb–Zn mineralization in the Howards Pass district occurs in the Ordovician–Silurian Duo Lake Formation. There are fourteen known deposits at Howards Pass, the largest being XY, Don, and Anniv. The main sulfides are sphalerite and galena, with pyrite a minor though ubiquitous component (Goodfellow and Jonasson 1986). The host sediments comprise carbonaceous and calcareous mudstones and some siliceous mudstones (Gadd et al. 2016). Descriptions of the geological and geochemical characteristics of the Howards Pass mineralization can be found in Gadd et al. (2016), Kelley et al. (2017), Martel (2017) and Slack et al. (2017). Following the *bulk sulfur* isotopic Howards Pass studies of Goodfellow and Jonasson (1984) and Goodfellow (1987) discussed above, Gadd et al. (2017) presented a more detailed study of the Howards Pass mineralization that included both in situ and *bulk* isotopic isotopic measurements. Johnson et al. (2018) subsequently published a new and very comprehensive *bulk* pyrite $\delta^{34}S$ stratigraphic profile for an unmineralized Duo Lake Formation section in the region of the XY, Central, Don and Don East mineralization.

4.2.1 Source

The value of $\delta^{34}S_{\text{seawater sulfate}}$ at the time of deposition of the Silurian strata hosting the Howards Pass CD Pb–Zn deposits is $\sim$24‰ (Gadd et al. 2017).

4.2.2 Trap

In a detailed textural study Gadd et al. (2016) showed that pyrite of the Howards Pass district had a protracted growth history. Py_1 is the earliest generation. It is framboidal and formed either in the water column or during very early diagenesis. During later diagenesis two younger pyrite generations (Py_{2a} and Py_{2b}) formed concurrently with sphalerite and galena. An even younger generation, Py_3, occurs as porphyroblasts in strain shadows in deformed rocks and is therefore related to later deformation rather than to the mineralizing event.

Gadd et al. (2017) measured the in situ sulfur isotopic composition of the four pyrite types and galena from across the Howards Pass district, as well as the *bulk* sulfur isotopic compositions of coexisting sphalerite and galena in the XY deposit. The data are summarized in Fig. 18. The $\delta^{34}S$ values of the framboidal Py_1 are all negative and $\Delta^{34}S_{\text{seawater sulfate-pyrite 1}}$ values range from $\sim$34 to 54‰. The range is strong evidence that Py_1 formed by BSR either syngenetically in the water column or during early diagenesis from coeval seawater sulfate with a $\delta^{34}S$ of $\sim$28‰. The $\delta^{34}S$ values of Py_{2a} and Py_{2b} are much higher than those of Py_1 and are interpreted to reflect formation during later diagenesis by TSR. Gadd et al. (2017) note that the Py_2 values mimic the secular Selwyn Basin pyrite curve that Goodfellow and Jonasson (1984) used to argue that the Howards Pass host rocks were deposited during a period of basin-wide stagnation when the oceanic environment was restricted and euxinic. However, as pointed out by Gadd et al. (2017), it is the isotopic composition of Py_1 that reflects conditions in the oceanic water column and the Py_1 data do not support the idea that the Howards Pass host rocks were deposited in a restricted euxinic environment.

The $\delta^{34}S_{\textit{in situ}\text{ galena}}$ values are significantly lighter than the $\delta^{34}S_{\textit{bulk}\text{ galena}}$ values (Fig. 18), but the reason for this difference remains unclear. Amongst the *bulk* sulfur isotope measurements there are 12 coexisting sphalerite-galena pairs, all with $\delta^{34}S_{\textit{bulk}\text{ sphalerite}} > \delta^{34}S_{\textit{bulk}\text{ galena}}$. Gadd et al.

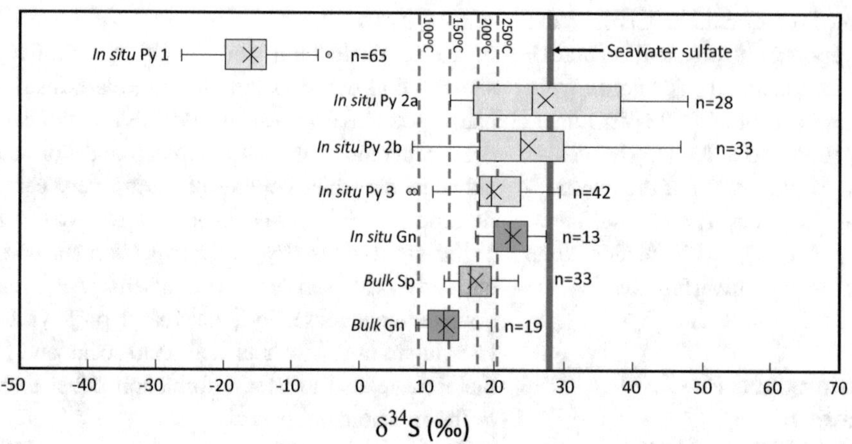

Fig. 18 Box and whisker diagram of sulfur isotopic data for sulfides in the Howards Pass deposits (after Gadd et al. 2017). The dashed blue, vertical lines show the value of $\delta^{34}S_{sulfide}$ formed by thermochemical sulfate reduction at various temperatures assuming a seawater sulfate with a value of 24‰

(2017) concluded that the two phases formed concurrently in isotopic equilibrium, with $\Delta^{34}S_{sphalerite\text{-}galena}$ yielding a median temperature of ~160 °C. It was further concluded galena and sphalerite formed by TSR involving coveal seawater sulfate with a $\delta^{34}S$ value of ~28‰, a conclusion further supported by the sulfide $\delta^{34}S$ values predicted by the kinetic fractionation equation of Kiyosu and Krouse (1990) for TSR involving sulfate with a $\delta^{34}S$ value of ~28‰ (Fig. 18).

Gadd et al. (2017) concluded that Howards Pass mineralization formed diagenetically beneath a brine pool in a depression floored (Fig. 1a Pathway 2) by water saturated, organic-rich hemipelagic sediments. Initial BSR of seawater sulfate triggered Py_1 precipitation at the top of the sediment pile or just above in the overlying water column. Subsequently exhaled dense warm (> 100 °C) hydrothermal fluids entered the system and ponded above the pyritic muds. These fluids then seeped down into the porous muds where TSR, initially catalyzed by BSR-produced H_2S, precipitated both Py_2 generations, sphalerite and galena deeper in the sediment pile. Py_3 is postulated to have formed later in response to metamorphism associated with Jurassic to Cretaceous orogenesis. In support of their argument that $Py_{2a \text{ and } 2b}$, sphalerite and galena formed diagenetically in the sediment pile, Gadd et al. (2017) noted that calcite associated with the mineralization has $\delta^{13}C$ values ranging from −1 to −6‰. Much of the calcite is nodular, suggesting it formed diagenetically, with its low $\delta^{13}C$ values coming from the oxidation of organic carbon (with $\delta^{13}C$ values of ~ −28‰) associated with the TSR event that formed Py_2, galena and sphalerite.

In an attempt to reconcile the older stagnant euxinic basin ore-forming model at Howards Pass with the new contradictory findings just discussed, Johnson et al. (2018) constructed a new *bulk* pyrite $\delta^{34}S$ stratigraphic profile through an unmineralized section of the Duo Lake Formation at Howards Pass. The new profile is shown in Fig. 19. The $\delta^{34}S$ values of pyrite range from ~−15‰ at the base of the section to 30‰ at the top, which is twice the range found by Goodfellow and Jonasson (1984). Johnson et al. (2018) argued that if the high positive values had been due to stagnation in a restricted anoxic oceanic basin, then the anoxic environment must have remained uninterrupted for some 20 Myr. The improbability of this situation, together with other geochemical evidence that the sequence was at least intermittently oxygenated (Slack et al. 2017), led Johnson et al. (2018) to propose that the base of the Duo Lake Formation was dominated by isotopically low Py_1, and that pyrite with highly positive $\delta^{34}S$ values became

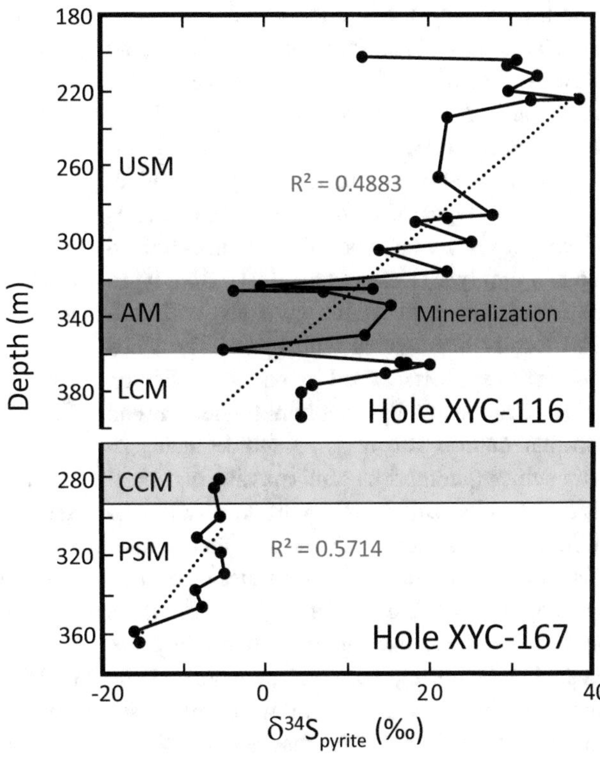

Fig. 19 Stratigraphic variations in $\delta^{34}S_{bulk\ pyrite}$ values in the five members of Duo Lake Formation, Selwyn Basin; the Pyritic Siliceous Mudstone (PSM), the Calcareous Mudstone (CCM), the Lower Cherty Mudstone (LCM), the Active Member (AM and host to the Howards Pass mineralization) and the Upper Siliceous Mudstone (USM) (adapted from Johnson et al. 2018)

increasingly abundant up section. It is unclear from the textural descriptions of the latter pyrite generation (predominantly subhedral to euhedral crystals) if it is Py_2 and/or Py_3 as defined by Gadd et al. (2016, 2017). Johnson et al. (2018) also found that $\delta^{13}C_{bulk\ carbonate}$ values decrease up section from ~−2.5 to ~−12.2‰, a trend that correlates with an increase in TOC. The trends are interpreted as indicating that beneath the Howards Pass mineralization pyrite formed by BSR, whereas starting immediately below the mineralized horizon, pyrite with high positive $\delta^{34}S$ values began forming by SD-AOM that also generated oxidized carbon complexes with low $\delta^{13}C$ values that concurrently formed low-$\delta^{13}C$ authigenic carbonate. Johnson et al. (2018) suggested the high positive isotopic composition of galena and sphalerite in the Howards Pass mineralization might also have formed by SD-AOM, rather than by TSR as proposed by Gadd et al. (2017). However, the various temperature estimates of galena and sphalerite at Howards Pass are inconsistent with SD-AOM being an important ore-forming process.

5 The North American Paleozoic Cordillera Clastic-Dominated Pb–Zn Province—United States

The Alaskan part of the Northern American Cordillera hosts the world's largest CD Pb–Zn deposit based on Pb + Zn content (Leach et al. 2005). It is the Red Dog deposit complex located in the Western Brooks Range in northern Alaska (Fig. 14). The complex comprises of four large CD Pb–Zn deposits (Qanaiyaq, Main, Aqqaluk and Paalaaq). They occur in organic-rich siliceous mudstones, shales, cherts and carbonates of the Carboniferous Kuna Formation (Leach et al. 2004). The deposits were originally one single deposit but were dismembered during the Mesozoic Brookian orogeny ~100 Ma after the deposition of the Kuna Formation.

The stratigraphic top of the deposits is marked by a sulfide-poor barite facies, beneath which, in descending order, lies a sulfide-bearing barite facies, a silica rock facies, a massive sulfide facies and, at the base, a sulfide vein facies

(Leach et al. 2004). About 20% of the Red Dog production has come from the sulfide-bearing barite facies, 60% from the massive sulfide facies, and 20% from the sulfide vein facies (Kelley et al. 2004).

The sulfide-bearing barite facies comprises two textural types of barite, a fine-grained (10–50 μm) equigranular type and a coarse-grained (up to 3 cm) type (Kelley et al. 2004). Both types are intergrown with sulfide minerals, with sulfides locally interlayered with barite. The silica rock facies is characterized by strongly silicified barite dominated by medium-grained zoned euhedral quartz. The massive sulfide facies (> 40% sulfides) comprises semi-massive and massive, unbedded sulfide mineralization, with the sulfides disseminated in a dense silica matrix. Within this facies there are rare zones of banded and fragmental sulfides. Although sulfide veins occur throughout the mineralization, they are most abundant at the base and periphery of the mineralization. Veins vary in width from ~1 mm to 1 m and in places the vein density is sufficient to constitute ore.

The main sulfides in the Red Dog deposits are sphalerite, galena, pyrite and marcasite. Kelley et al. (2004) divided the mineralization into 4 temporal stages, each characterized by different colors of sphalerite. The earliest stage, Stage 1, is characterized by early brown sphalerite, Stage 2 by yellow–brown sphalerite, Stage 3 by red-brown sphalerite and the final stage, Stage 4, by tan sphalerite.

5.1 Source

The value of $\delta^{34}S_{seawater\ sulfate}$ at the time of deposition of the Carboniferous strata hosting the Red Dog is 17.5 ± 1‰ (Johnson et al. 2004).

5.2 Transport

To date there have been no stable isotope studies directly relevant to the elucidation of the origin of the hydrothermal fluids that formed the Red Dog deposits. However, Leach et al. (2004) found that the Stage 4 red-brown sphalerite at Red Dog contains fluid inclusions suitable for study. They revealed that red-brown sphalerite formed between 100 and 180 °C from a fluid with a salinity of 14–19 wt% NaCl equivalent, consistent with the fluid being an evolved basin brine that obtained its salinity through the evaporation of seawater. No fluid inclusions data could be obtained from the earlier three sphalerite stages.

5.3 Trap

In situ sulfur isotopic values of sphalerite, pyrite and galena reported by Kelley et al. (2004) are shown in Fig. 20. As the value of $\delta^{34}S_{seawater\ sulfate}$ at the time of deposition of the Red Dog host rocks was ~17.5‰. $\Delta^{34}S_{sulfide\text{-}sulfate}$ values for Stage 1 sphalerite and pyrite range from ~40 to 55‰ providing good evidence that the two sulfides formed by BSR. However, given the possibility of barite-inclusion contamination in some samples, Kelley et al. (2004) noted that the fractionation could be considerably smaller—in the range 15–25‰. In either case, the Stage 1 sphalerite and pyrite data are consistent with BSR involvement in their formation. The rare Stage 1 galena has low positive $\delta^{34}S$ values, unlike the low negative values for sphalerite and pyrite, indicating that the ore-forming sulfides did not approach or achieve S-isotopic equilibrium. This isotopic relationship is interpreted as indicating rapid low temperature sulfide precipitation.

In contrast, Kelley et al. (2004) concluded that the formation of the Stage 2 and 3 mineralization was dominated by TSR based on $\Delta^{34}S_{sulfide\text{-}sulfate}$ values of ~15‰ for Stage 2 and 3 sulfides and Red Dog barite. Few available pyrite and galena sulfur isotope measurements from the Stage 2 and 3 mineralization lie within the ranges of the sphalerite measurements, suggesting that their formation also involved TSR. Sulfides in the Stage 4 mineralization are characterized by isotopic values similar to those in Stage 1 mineralization, suggesting BSR was also involved in their formation (Kelley et al. 2004).

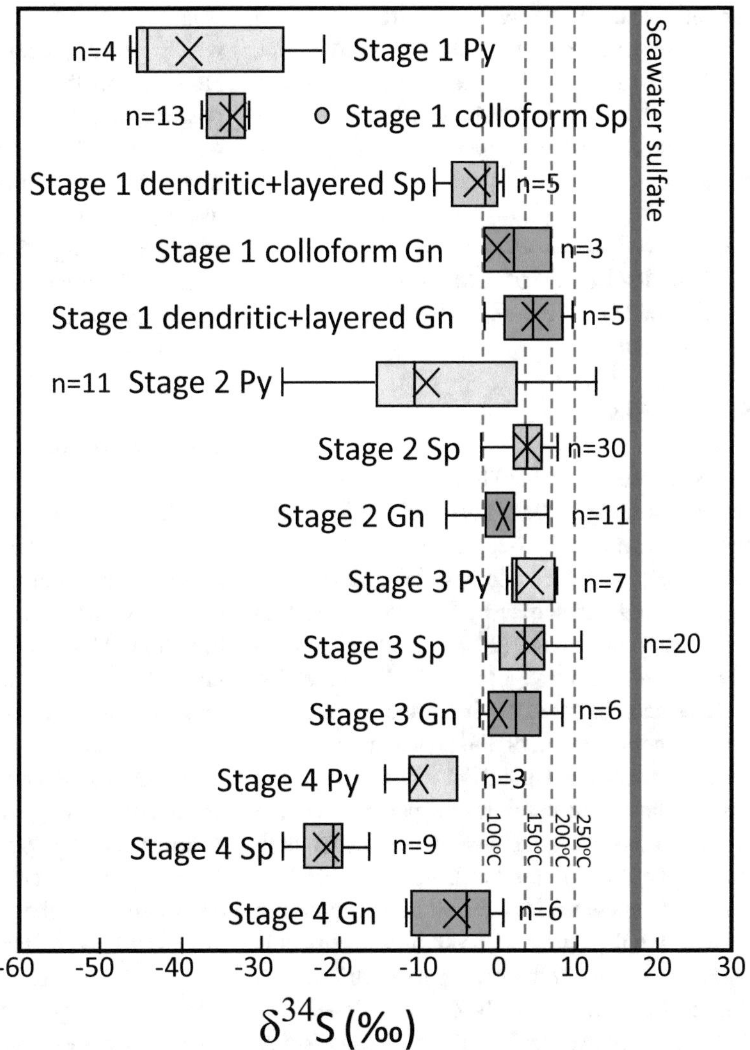

Fig. 20 Box and whisker diagram of in situ sulfur isotopic data for sulfides in the Red Dog deposits (after Kelley et al. 2004). The dashed blue, vertical lines show the value of $\delta^{34}S_{sulfide}$ formed by thermochemical sulfate reduction at various temperatures assuming a seawater sulfate with a value of 17.5‰

A comparison of the $\delta^{34}S$ values of sphalerite and galena with the sulfide $\delta^{34}S$ values predicted by the kinetic fractionation equation of Kiyosu and Krouse (1990) for TSR involving sulfate with a $\delta^{34}S$ value of ~17.5 (Fig. 20) are consistent with the Stage 1 colloform and dendritic galena, and Stages 2 and 3 galena and sphalerite all forming by TSR from coeval seawater sulfate.

In summary, Kelley et al (2004) concluded from geological, mineralogical, textural, geochemical and isotope data that the formation of the Red Dog mineralization was protracted and complex. During the early Stage 1 event barite, brown sphalerite, pyrite and galena formed by BSR in organic-rich muds at or below the seafloor at temperatures from cool (< 100 °C) hydrothermal fluids rich in Ba, Zn and Pb. Warmer (100–200 °C) hydrothermal fluids recrystallized barite and deposited Stage 2 yellow–brown sphalerite, pyrite and galena in partly-lithified organic-rich muds below the seafloor, with H_2S produced by TSR of sulfate sourced from either barite or seawater. During Stage 3 the continued alteration of mudstones produced methane, hydrofracturing and veins in which barite, red-brown sphalerite, pyrite or marcasite and galena were deposited from the evolving hydrothermal fluids. Red-brown sphalerite and pyrite also formed from the fluids by the replacement of earlier-deposited barite. Finally, Stage 4 event appears to be related to Brookian-age thrusting; it is thus not part of the main hydrothermal event that formed the original mineralization.

6 Implications for Clastic-Dominated Pb–Zn Mineral Systems

6.1 Source

The consensus view that Pb and Zn in CD Pb–Zn deposits is leached from host-basin sedimentary and/or volcanic rocks lying stratigraphy beneath the deposits is difficult to test using stable isotopes. At best, as shown by Cooke et al (1998) and Champion et al. (2020) in their source-related studies in the McArthur Basin, stable isotopes can be used to infer the C- and O-isotopic composition of fluids that might have leached metals from potential source rocks, but only in those rare cases where potential source rocks can be accessed and analysed. As both studies showed, the approach does not produce definitive outcomes. So far little attention has been paid to the source of the significant amounts of iron contained in the early pyrite attributed to BSR in pyrite-rich CD Pb–Zn deposits such as those in the McArthur Basin and Mt Isa district. Possibilities include an ordinary oceanic environment, a ferruginous ocean, and/or hydrothermal fluids. Which, if any, of these sources are important is a subject requiring further research.

The consensus view that contemporaneous seawater sulfate is the source of sulfide-S in CD Pb–Zn deposits has been clearly demonstrated in the case of the North American Paleozoic deposits. The involvement of seawater sulfate in the formation of the Proterozoic age Australian CD Pb–Zn deposits cannot be demonstrated so well, because the secular $\delta^{34}S_{\text{seawater sulfate}}$ curve for the Proterozoic is less well understood. Nevertheless the S-isotopic geochemistry of sulfides in the older Australian CD Pb–Zn deposits is consistent with their derivation from sulfate with large positive $\delta^{34}S$ values which in all probability was seawater sulfate. However, in the case of the HYC deposit, it is possible that the ore-forming reduced sulfur was generated by TSR at depth and transported to the trap in a reduced fluid. While this reduced sulfur was formed from seawater sulfate, there is some evidence to suggest it had a different $\delta^{34}S$ value to that of the sulfate in the water column from which the host strata of the HYC deposit were deposited.

6.2 Transport

Insights into the nature of the transporting hydrothermal fluids that form CD Pb–Zn deposits have come from carbon and oxygen isotopic composition of carbonates in and surrounding CD Pb–Zn mineralization. Fluid-mineral modelling of carbon and oxygen isotopic data typically yields results suggesting that the fluids had high $\delta^{18}O$ values ($\geq$ 6‰), typical of basinal-generated fluids that evolved geochemically through exchange with the basinal sedimentary rock pile (Fig. 6). In the HYC and Jason deposits where carbonate was either deposited from the hydrothermal fluids or fully exchanged with the fluids, $\delta^{13}C_{\text{carbonate}}$ and $\delta^{18}O_{\text{carbonate}}$ values correlate and have low positive slopes (Figs. 4 and 15) suggesting that the dominant carbonate species in the fluids was H_2CO_3 rather than HCO_3^- (Cooke et al. 1998: Fig. 19). The hydrothermal fluids that formed both the HYC and MacMillan Pass mineralization were therefore neutral to acidic.

6.3 Trap

The preferred trap-related ore-precipitating processes presented in the more comprehensive in situ studies discussed above are summarized in Table 1. The studies do not support the simple sedimentary-exhalative process favoured in the era preceding in situ sulfur isotope studies. Rather they indicate trap processes involving a protracted period of sulfide precipitation. In most

Table 1 Summary of genetic models arising from in situ light stable isotope results of comprehensively studied CD Pb–Zn deposits in the Northern Australian and North American Cordillera provinces

Deposit	Reference	Pre-mineralization events	Mineralization events
HYC	Eldridge et al. (1993)	Py_1 forms by BSR during early diagenesis in an environment open to seawater sulfate Py_2 forms by BSR during later diagenesis in an environment becoming increasingly closed to seawater sulfate	Sphalerite and galena form after Py_1 and Py_2 from a Pb–Zn rich hydrothermal fluid either by TSR from a sulfate-dominant Pb–Zn rich hydrothermal fluid or by direct precipitation from a H_2S dominant Pb–Zn rich hydrothermal fluid
	Ireland et al. (2004b)	Py_1 forms by BSR in the upper layers of a stratified brine pool Py_2 forms during early diagenesis in a sulfate-restricted environment beneath the brine pool by BSR and minor TSR	Sp_1 forms in the lower layers of a stratified brine pool by BSR Sp_2 forms during early diagenesis in the sulfate-restricted environment beneath the brine pool by BSR and minor TSR
Teena	Magnall et al. (2020b)	Py-1a forms both in and around the mineralized zone by BSR in an environment open to seawater sulfate Py-1b forms during diagenesis either by BSR or SD-AOM	Py-2 is part of the galena-sphalerite ore-forming event that post-dates the formation of Py-1a and Py1b. Magnall et al. (2020b). did not discuss in detail the processes involving the formation of galena and sphalerite
	Magnall et al. (2021)	As above	In a study unrelated to stable isotopes it is concluded that Py-2, galena and sphalerite form by carbonate replacement during the mid-stage of burial diagenesis, most likely by processes involving TSR
George Fisher	Rieger et al. (2020)	Py-0 forms during early diagenesis by BSR of seawater sulfate. The Py-0b variant forms under more sulfate-restricted conditions than Py-0a	Py-1 postdates the formation of Py-0 and forms by TSR during the CD Pb–Zn phase of galena and sphalerite formation. The formation of the later pyrite types, Py-2 and Py-3, are associated with the deformation and Cu-mineralizing events that postdate the CD–Pb-Zn ore formation. Py-3 is thought unlikely to have had a hydrothermal origin
MacMillan Pass	Magnall et al. (2016a)	In the Feeder Zone an early Stage 1event is characterized by pervasive ankerite alteration and ankerite-pyrite-quartz stockwork vein formation at ~220–270 °C	In the Feeder Zone a Later Stage 2 mineralizing event involves the deposition of galena, pyrrhotite and pyrite ± chalcopyrite and sphalerite at ~110–215. Sulfide S is generated from coeval seawater sulfate by TSR
	Magnall et al. (2016b)	In the CD Pb–Zn host sediments Py-I forms during early diagenesis by BSR. Py-II forms during later diagenesis by SD-AOM	Py-III, galena and sphalerite form by TSR
	Magnall et al. (2020c)	As above	Using a range of evidence including the stable isotope data presented in Magnall et al. (2016a, b) it was concluded that CD Pb–Zn ore formation occurred entirely in the subsurface from hot (300 °C) hydrothermal fluids that degraded organic matter, generating CO_2 that promoted barite dissolution and triggered sulfide precipitation by TSR, as well as increasing permeability that further facilitated ore formation

(continued)

Table 1 (continued)

Deposit	Reference	Pre-mineralization events	Mineralization events
Howards Pass	Gadd et al. (2017)	Py1 forms either syngenetically or during early diagenesis by BSR of coeval seawater	Py2, galena and sphalerite form during later diagenesis by TSR from hydrothermal fluids that seep downwards from a brine pool formed by the exhalation of dense hydrothermal fluids Johnson et al. (2018) concluded that at least some Py2, galena and sphalerite could have formed by SD-AOM
Red Dog	Kelley et al. (2004)	There is no pre-mineralization event as mineralization commences during sedimentation	Stage 1 mineralization: Brown sphalerite, pyrite, barite and rare galena form by BSR in unconsolidated organic matter-rich muds Stage 2 mineralization: Pyrite and yellow–brown sphalerite and formation involves TSR and the dissolution of pre-existing barite Stage 3 mineralization: Barite, red-brown sphalerite, pyrite and/or marcasite and chalcedonic quartz form in veins and replace pre-existing barite adjacent to the veins. Sulfide formation again involves TSR Stage 4 mineralization: This stage is related to Brookian-age thrusting that post-dates the Stage 1–3 events by ~100 my

of the recent studies it is argued that hydrothermal fluids did not exhale into the water column from which the host strata were deposited, but rather infiltrated the host strata after sedimentation. Only two of the recent studies (Ireland et al. 2004b; Gadd et al. 2017) retain an exhalative genetic component, and in both of these the critical trap element is a brine pool in which the exhalative fluids pond. It is argued by Ireland et al. (2004b) that sulfide precipitation is syngenetic and occurs mainly in the brine pool, whereas Gadd et al. (2017) argue that sulfides precipitate within the sedimentary pile beneath the pool.

In situ studies show that both the Northern Australian and Selwyn Basin CD Pb–Zn deposits have remarkably similar histories of sulfide deposition, commencing with pre-ore diagenetic pyrite formation involving BSR in a system open to sulfate and followed in most deposits by additional pyrite formation in more restricted environments as diagenesis and lithification progressed. This restriction leads to Rayleigh fractionation of sulfur isotopes and to higher $\delta^{34}S$ values in the later pyrite compared to the earlier pyrite. The process of later pyrite formation has been variously attributed to BSR (HYC: Eldridge et al. 1993), BSR or SD-AOM (Teena: Magnall et al. 2020b), SD-AOM (Tom and Jason: Magnall et al. 2020c) and to TSR (George Fisher: Rieger et al. 2020; Howards Pass: Gadd et al. 2017).

By comparison with pyrite, there is a lack of in situ sulfur isotope data available for galena and sphalerite. The main impact to date of in situ sulfur isotope studies on the formation of the ore minerals has been the detailed sulfide mineral textural observations required for good in situ analyses. These observations show that galena and sphalerite mostly form after early diagenetic pyrite and therefore do not have a syngenetic to earlier diagenetic origin. Processes by which the ore minerals form have been mostly attributed to TSR-generated reduced S, exceptions being BSR for Sp-1 at HYC as suggested by Ireland et al. (2004b) and the possibility of SD-AOM at

Howards Pass (Johnson et al. 2018). However, as noted earlier, ore-forming temperatures at Howards Pass appear too high for SD-AOM to have been a significant process in ore formation. Similarly, in the case of the HYC mineralization, $\Delta^{34}S_{sphalerite\text{-}galena}$ values and dolomite C-O isotopic modelling indicate that mineralization formed at temperatures $\geq$ 100 °C, consistent with the production of sulfide by TSR rather than BSR. This conclusion is further strengthened by a study, unrelated to stable isotopes, of the temperature of alteration of organic matter in and around the HYC mineralization by Vinnichenko et al. (2021).

The Red Dog district is the only example discussed in this Chapter that does not follow the same ore-forming mechanism of early pyrite formation involving BSR and the latter formation of galena and sphalerite involving TSR. For the Red Dog district there is good sulfur isotopic evidence that Stage 1 brown sphalerite formed by BSR in unconsolidated organic-rich muds. The subsequent main stages of base metal mineralization involved TSR, mostly likely associated with barite replacement, as is also postulated by Magnall et al. (2020c) for some of the Tom and Jason mineralization. In a much later stage of ore formation in the Red Dog district, some ~100 my younger than the main ore-forming event, sulfur isotopic data suggests sphalerite formation again involved BSR (Kelley et al. 2004).

6.3.1 Trap-Related Questions Still to be Answered

Cool-oxidized or warm-reduced hydrothermal fluids? It was noted at the beginning of this Chapter that Cooke et al. (2000) used geochemical modelling to show that both cool ($\leq$ 150 °C) oxidized hydrothermal fluids (dominated by $SO_4^{2-} \pm HSO_4^-$) and warmer (~250 °C) reduced acidic fluids (dominated by H_2S or HS^-) can transport and deposit sufficient quantities of Pb and Zn to form CD Pb–Zn deposits. The cool oxidized fluids were termed McArthur-type and Cooke et al. (2000) argued these fluids formed deposits such as HYC, Mt Isa and Hilton by in situ sulfate reduction and/or by low-temperature fluid mixing or interaction with earlier-formed pyrite. In contrast the warm reduced fluids were termed Selwyn-type and it was argued they formed the Selwyn Basin deposits at higher temperatures by processes like cooling, addition of H_2S and pH increases.

In the in situ isotopic studies reviewed above, the trap-related paragenetic sequences and their interpretation in the North Australian and Selwyn Basin CD Pb–Zn deposits are strikingly similar. The similarity suggests that the hydrothermal fluids in both metallogenic provinces were warm and very likely had similar chemistries, contrary to the conclusions of Cooke et al. (2000). However, in deposits such as Tom and Jason, where barite-replacement appears to be a significant ore-forming process, it is likely that TSR occurred within the trap itself, whereas in other deposits, such as HYC, the site of TSR is less obvious. Eldridge et al. (1993) suggested H_2S required for ore formation was introduced into the HYC trap by the hydrothermal fluid, implying that TSR occurred deeper in the mineral system whereas Williams (1978b), in a study not involving stable isotopes, favoured in situ TSR and a SO_4^{2}-dominant hydrothermal fluid. The site of TSR for CD Pb–Zn ore formation and the oxidation state of the hydrothermal fluids has not been discussed specifically in many of the studies reviewed in this Chapter, and is another subject worthy of further investigation to improve understanding of trap processes in the CD Pb–Zn mineral systems.

What is the temperature of clastic-dominated Pb–Zn ore-formation? A significant uncertainty in our understanding of CD Pb–Zn mineral systems, and one related to uncertainty about the oxidation state of the ore-forming hydrothermal fluids, is the temperature of ore formation. In the McArthur Basin, for example, Ireland et al. (2004b) argued that galena and sphalerite formed by BSR at temperatures < 100 °C whereas others (e.g. Eldridge et al.1993; Rye and Williams 1981; Magnall et al. 2020b) presented a range of evidence indicating temperatures were > 100 °C and therefore involved TSR. One line of evidence used to support temperatures of > 100 °C, $\Delta^{34}S_{sphalerite\text{-}galena}$ values, is permissive of warmer fluid temperatures but, as mentioned earlier, this geothermometer is based on assumptions that require further testing. There is a paucity of in situ sulfur isotopic data for galena and sphalerite in CD Pb–Zn deposits and more such data might help improve temperature estimates using the $\Delta^{34}S_{sphalerite\text{-}galena}$ fractionation geothermometer. However, irrespective of whether or not they do, new in situ galena and sphalerite sulfur isotope studies will almost certainly produce fresh and unexpected genetic insights as has already happened with in situ S-isotopic studies of pyrite in CD Pb–Zn deposits.

What is the correct explanation for the formation of sulfides with high $\delta^{34}S$ values? In many of the studies discussed in this Chapter sulfides with high $\delta^{34}S$ values have been variously attributed to sulfate-restricted conditions caused by BSR in a stagnant and restricted basin, BSR in an euxinic water column triggered by a high flux of organic carbon, BSR in a restricted diagenetic environment, SD-AOM during later diagenesis and TSR. All these processes have been shown elsewhere to produce sulfides with high $\delta^{34}S$ values, so high $\delta^{34}S$ values in and of themselves are not indicative of a particular process. The only S-isotopic signature that appears to be diagnostic of a particular process is a $\Delta^{34}S_{sulfide\,-\,sulfate}$ range of ~15–60‰, with much of the range having negative values. This S-isotope signature appears only to be displayed by sulfides formed by BSR in a system open to sulfate. Other processes are difficult to differentiate, although if the temperatures of sulfide reduction could be ascertained accurately, then choosing between biogenic and abiogenic processes would be possible. Clearly new and/or better diagnostic criteria would improve our understanding of the formation of sulfides with high $\delta^{34}S$ values in CD Pb–Zn deposits and help to remove uncertainties about the position of sulfides in the paragenetic sequence in those instances where mineral textures are inconclusive.

Where has all the oxidized organic carbon gone? Sedimentary organic matter is characterized by low negative $\delta^{13}C$ values (~−25‰), an isotopic signature that transfers to oxidized organic carbon species (Ohmoto and Goldhaber 1997). Low negative $\delta^{13}C$ values in carbonates are therefore a good indication that oxidized organic carbon is a significant component of those carbonates. Authigenic carbonates associated with both SD-AOM (Borowski et al. 2013) and TSR (Machel et al. 1995; Machel 2001) typically have low value $\delta^{13}C$ and is one line of evidence used to conclude that the respective sulfate-reduction mechanisms involved the oxidation of organic matter. Ireland et al. (2004b) used a lack of low $\delta^{13}C$ values in nodular dolomites in the HYC deposit to argue that their formation did not involve the oxidation of organic matter during ore formation by either BSR or TSR beneath the nodular dolomite-bearing layers as suggested by Logan et al. (2001). Similarly, Rye and Williams (1981) found no C-isotopic evidence for significant amounts of oxidized organic carbonate in the dolomites in and surrounding the HYC deposit. This isotopic evidence may be used to suggest that sulfide formation in the HYC deposit did not involve in situ sulfate reduction and concomitant oxidation of organic matter. This would be consistent with the model of Eldridge et al. (1993) in which reduced sulfur was transported into the HYC trap by the hydrothermal fluids and the model of Ireland et al. (2004b) in which the HYC ores formed syngenetically by BSR in a brine pool from which oxidized organic carbon was able to escape into the overlying water column. Alternatively, if ore formation did involve in situ sulfate reduction during diagenesis or later, then the large amounts of oxidized of organic carbonate that would have been generated during sulfate reduction (Williams 1978b) must have been removed through the outflow zone of the HYC CD Pb–Zn mineral system. In the other CD Pb–Zn deposits reviewed in this chapter, in which ore-formation is postulated to involve in situ TSR, little attention has been paid to the quantities and fate of the oxidized organic carbon that would have been generated as galena and sphalerite were precipitated. The fate of this carbon is another subject worthy of further investigation to improve our understanding of trap processes in the CD Pb–Zn mineral systems.

7 Implications for Future Exploration

The CD Pb–Zn deposits discussed in this chapter are located at, or close to, current land surfaces where traditional prospecting, geophysical and geochemical exploration methods have been effective. However, the search for new CD Pb–Zn deposits will become increasingly focused on buried deposits that have no near-surface expression. New exploration methods will be needed in the future and the stable isotope studies summarized in this chapter provide some clues as to how light stable isotopes might assist in the discovery of CD Pb–Zn mineralization at depth.

In the case of an undiscovered CD Pb–Zn deposit located deep beneath current land surfaces, the parts of the CD Pb–Zn mineral system most likely to have a surface or near-surface expression are extensions of the feeder faults above buried deposits and the discharge zones for the hydrothermal fluids, downstream from buried deposits (Fig. 1). If exploration is taking place in a metallogenic province containing previously discovered and well-studied CD Pb–Zn deposits, knowledge of the isotopic geochemistry of the hydrothermal fluids that formed the known deposits could help identify extensions of feeder faults above buried deposits. Such extensions might have hydrothermal alternation minerals and/or vein- and breccia-fill minerals with isotopic signatures consistent with alteration and formation from fluids matching those that formed the known deposits in the province. If an exploration strategy is based on a genetic ore-forming model in which galena and sphalerite formed in the trap environment by in situ TSR and the concomitant oxidation of organic matter, then authigenic or vein-fill carbonates with low negative $\delta^{13}C$ values in prospective terrains may help to identify the discharge zone of a CD Pb–Zn deposit and thus be a vector to mineralization at depth.

Both potential exploration applications of stable isotope geochemistry would require large numbers of analyses and, as discussed by Huston et al. (2023), the approach would require reductions in costs and the time required for analyses, and a significant increase in the number of analytical laboratories required to provide the analyses. These conditions would not be such a problem were the two applications used more selectively to analyse appropriate minerals in drill core from drill holes sited using other exploration tools.

8 Future Directions

Stable isotopes, particularly where integrated with other geochemical, mineralogical and geological techniques, are playing an increasing role in the study of CD Pb–Zn deposits and the trend will continue as stable isotope studies and technologies continue to evolve. As shown in this chapter, in situ stable isotope studies are proving to be particularly useful in genetic studies of CD Pb–Zn mineralization, but future in situ investigations would benefit from more integration with bulk isotopic and other whole-rock geochemical studies that would facilitate the quantification of ore-forming process. At present it is very difficult to estimate from existing data the relative importance of BSR, SD-AOM and TSR in ore formation. Also, integrated studies at a variety of sampling sizes would provide important new insights into the scales at which processes such as sulfate reduction operate during CD Pb–Zn ore formation.

Galena and sphalerite have been neglected relative to pyrite in in situ isotopic studies and the imbalance needs to be addressed given the uncertainties about galena and sphalerite formation in CD Pb–Zn deposits. Studies of the abundances of ^{33}S and ^{36}S, in addition to ^{32}S and ^{34}S, are providing new information about the formation of pyrite in sedimentary rocks unrelated to CD Pb–Zn mineralization as shown, for example, by Johnston et al. (2008) in their McArthur Basin pyrite study. Abundance data for all four sulfur isotopes in pyrite, galena and sphalerite in CD Pb–Zn mineralization should similarly refine our understanding of the ore-forming processes.

Huston et al. (2023) note that the development of analytical techniques to measure fractionations of "clumped isotopes" has the potential to become an important new ore genesis tool. Clumped isotope studies of carbonates in CD Pb–Zn deposits show promise as a much-needed and improved geothermometer that could help improve knowledge of the temperature of ore formation.

Stable isotopes will very likely play a greater role in CD Pb–Zn mineral exploration but, as is currently the case, that role is likely to remain subsidiary to the various geophysical and other geochemical exploration tools which are widely used today in the mineral exploration industry and which are being continually adapted to meet the challenges of locating ore at ever-increasing depths beneath the current land surface.

Acknowledgements I thank David Huston for his invitation to write this chapter and his support and advice as the chapter evolved, Jens Gutzmer for his wise counsel and guidance during the preparation of the chapter, and reviewers Joe Magnall and Raphael Baumgartner for their constructive criticism that helped greatly in the development of the final manuscript. Finally, I owe a great debt of gratitude to my erstwhile colleagues in the Mt Isa Mines group of companies, particularly Ross Logan and the late Bill Croxford, the late Bill Murray and late Tim Bennett who introduced me to the HYC, Teena, Lady Loretta and Mt Isa district Pb–Zn CD deposits and the many genetic, developmental and exploration challenges posed by these deposits.

References

Ahmad M, Dunster JN, Munson TJ (2013) Chapter 15: McArthur basin. Northern Territory Geol Surv Spec Publ 5:15:1–15:72

Andsell KA, Nesbitt BE, Longstaffe FJ (1989) A fluid inclusion and stable isotope study of the Tom Ba-Pb-Zn deposit, Yukon Territory, Canada. Econ Geol 84:841–856

Berner RA (1970) Sedimentary pyrite formation. Am J Sci 268:1–23

Berner RA (1984) Sedimentary pyrite formation: an update. Geochim Et Cosochim Acta 48:605–615

Borowski WS, Rodriguez NM, Paull CK, Ussler W III (2013) Are ^{34}S-enriched authigenic sulfide minerals a proxy for elevated methane flux and gas hydrates in the geologic record? Mar Pet Geol 43:381–395

Bottrell SH, Newton RJ (2006) Reconstruction of changes in global sulfur cycling from marine sulfate isotopes. Earth Sci Rev 75:59–83

Broadbent GC, Myers RF, Wright JV (1998) Geology and origin of shale-hosted Zn-Pb-Ag mineralization at the Century deposit, northwest Queensland, Australia. Econ Geol 93:1264–1294

Campbell FA, Ethier VG (1974) Sulfur isotopes, iron content of sphalerite, and ore textures in the Anvil ore body, Canada. Econ Geol 69:482–493

Canfield DE (2013) Sulfur isotopes in coal constrain the evolution of the Phanerozoic sulfur cycle. Proc Natl Acad Sci 110:8443–8446

Carpenter SJ, Lohmann KC, Holden P, Walter LM, Huston TJ, Halliday AN (1991) δ^{18}O, ^{87}Sr/^{86}Sr and Sr/Mg ratios of Late Devonian abiotic marine calcite: implications for the composition of ancient seawater. Geochim Et Cosochim Acta 55:1991–2010

Carr GR (1984) Primary geochemical and mineralogical dispersion in the vicinity of the Lady Loretta Zn–Pb–Ag deposit, North Queensland. J Geochem Expl 22:217–238

Carr GR, Smith JW (1977) A comparative isotopic study of the Lady Loretta zinc-lead-silver deposit. Mineral Deposita 12:105–110

Champion DC, Huston Dl, Bastrakov E, Siegel C, Thorne J, Gibson GM, Houser J (2020) Alteration of mafic igneous rocks of the southern McArthur Basin: comparison with the Mount Isa region and implications for basin-hosted base metal deposits. In: Exploring for the future: extended abstracts. Geoscience Australia, Canberra. https://doi.org/10.11636/134206

Chapman LH (2004) Geology and mineralization styles of the George Fisher Zn–Pb–Ag deposit, Mount Isa, Australia. Econ Geol 99:233–254

Chu X, Zhang T, Zhang Q, Lyons TW (2007) Sulfur and carbon isotope records from 1700 to 800 Ma carbonates of the Jixian section, northern China: implications for secular isotope variations in Proterozoic seawater and relationships to global supercontinental events. Geochim Cosochim Acta 71:4668–4692

Claypool GE, Holser WT, Kaplan IR, Sakai H, Zak I (1980) The age curves of sulfur and oxygen isotopes in marine sulfate and their mutual interpretation. Chem Geol 28:199–260

Cooke DR, Bull SW, Donovan S, Rogers JR (1998) K-metasomatism and base metal depletion in volcanic rocks from the McArthur Basin, Northern Territory—implications for base metal mineralization. Econ Geol 93:1237–1263

Cooke DR, Bull SW, Large RR, McGoldrick PJ (2000) The importance of oxidized brines for the formation of Australian Proterozoic stratiform sediment-hosted Pb–Zn (Sedex) deposits. Econ Geol 95:1–18

Crockford PW, Kunzmann M, Bekker A, Hayles J, Huiming B, Halverson GP, Peng Y, Bui TH, Cox GM, Gibson TM, Worndle S, Rainbird R, Lepland A, Swanson-Hysell NL, Master S, Sreenivas B,

Kuznetsov A, Krupenik V, Wing BA (2019) Claypool continued: extending the isotopic record of sedimentary sulfate. Chem Geol 513:200–225

Croxford NJW, Jephcott S (1972) The McArthur lead-zinc-silver deposit, Northern Territory, Australia. Proc Austr Inst Mining Metall. 243:1–26

Ding T, Zhang C, Wan D, Liu Z, Zhang G (2003) An experimental calibration on the sphalerite-galena sulfur isotope geothermometers. Acta Geol Sin 77:519–521

Duffett ML (1998) Gravity, magnetic and radiometric evidence for the geological setting of the Lady Loretta Pb–Zn–Ag deposit—a qualitative appraisal. Econ Geol 93:1295–1306

Eldridge CS, Williams N, Walshe JL (1993) Sulfur isotope variability in sediment-hosted massive sulfide deposits as determined using the ion microprobe SHRIMP: II a study of the H.Y.C. deposit at McArthur River, Northern Territory, Australia. Econ Geol 88:1–26

Gadd MG, Layton-Matthews D, Peter JM, Paradis SJ (2016) The world-class Howard's Pass SEDEX Zn-Pb district, Selwyn Basin, Yukon. Part I: trace element compositions of pyrite record input of hydrothermal, diagenetic, and metamorphic fluids to mineralization. Mineral Deposita 51:319–342

Gadd MG, Layton-Matthews D, Peter JM, Paradis SJ, Jonasson IR (2017) The world-class Howard's Pass SEDEX Zn-Pb district, Selwyn Basin, Yukon. Part II: the roles of thermochemical and bacterial sulfate reduction in metal fixation. Mineral Deposita 52:405–419

Gardner HD, Hutcheon I (1985) Geochemistry, mineralogy and geology of the Jason Pb–Zn deposits, MacMillan Pass, Yukon, Canada. Econ Geol 80:1257–1276

Gellatly AM, Lyons TW (2005) Trace sulfate in mid-Proterozoic carbonates and the sulfur isotope record of biospheric evolution. Geochim Et Cosochim Acta 69:3813–3829

Goldhaber MB, Kaplan IR (1974) The sulfur cycle. In: Goldberg ED (ed) The sea: 5 marine chemistry. Wiley, New York, pp 569–655

Gomes ML, Hurtgen MT (2015) Sulfur isotope fractionation in modern euxinic systems: implication for paleoenvironmental reconstructions of paired sulfate-sulfide isotope records. Geochim Cosmochim Acta 157:39–55

Goodfellow WD (1987) Anoxic stratified oceans as a source of sulphur in sediment-hosted stratiform Pb-Zn deposits (Selwyn Basin, Yukon, Canada). Chem Geol 65:359–382

Goodfellow WD, Jonasson IR (1984) Ocean stagnation and ventilation defined by δ^{34}S secular trends in pyrite and barite. Geology 12:583–586

Goodfellow WD, Jonasson IR (1986) Environment of formation of the Howards Pass (XY) Zn–Pb deposit, Selwyn Basin, Yukon. Can Inst Mining Metall Spec 37:19–50

Grondijs HF, Schouten C (1937) A study of the Mount Isa ores. Econ Geol 32:407–450

Gustafson LB, Williams N (1981) Sediment-hosted stratiform deposits of copper, lead and zinc. Econ Geol 75th Anniv Vol 139–178

Hagemann SG, Lisitsin VA, Huston DL (2016) Mineral system analysis: quo vadis. Ore Geol Rev 76:504–522

Halevy I, Peters SE, Fischer WW (2012) Sulfate burial constraints on the Phanerozoic sulfur cycle. Science 337:331–334

Hayward N, Magnall JM, Taylor M, King R, McMillan N, Gleeson SA, (2021) The Teena Zn–Pb deposit (McArthur Basin, Australia). Part 1: syndiagenetic base metal sulfide mineralization related to dynamic subbasin evolution. Econ Geol 116:1743–1768

Heinrich CA, Andrew AS, Wilkins RWT, Patterson DJ (1989) A fluid inclusion and stable isotope study of synmetamorphic copper formation at Mount Isa, Australia. Econ Geol 84:529–550

Hoggard MJ, Czarnota K, Richards FD, Huston DL, Jaques LA, Ghelichkhan S (2020) Global distribution of sediment-hosted metals controlled by craton edge stability. Nat Geosci 13:504–510

Huston DL, Stevens B, Southgate PN, Muhling P, Wyborn L (2006) Australian Zn–Pb–Ag ore-forming systems: a review and analysis. Econ Geol 101:1117–1157

Huston DL, Mernagh TP, Hagemann SG, Doublier MP, Fiorentini M, Champion DC, Jaques AL, Czarnota K, Cayley R, Skirrow R, Bastrakov E (2016) Tectono-metallogenic systems—the place of mineral systems within tectonic evolution, with an emphasis on Australian examples. Ore Geol Rev 76:168–210

Huston DL, Trumbull RB, Beaudoin G, Ireland T (2023) Light stable isotopes (H,B,C, O and S) in ore studies —methods, theory, applications and uncertainties. In: Huston DL, Gutzmer J (eds) Isotopes in economic geology, metallogensis and exploration. Springer, Berlin (this volume)

Ireland T, Bull SR, Large RR (2004a) Mass flow sedimentology within the HYC Zn–Pb–Ag deposit, Northern Territory, Australia: evidence for syn-sedimentary ore genesis. Mineral Deposita 39:143–158

Ireland T, Large RR, McGoldrick P, Blake M (2004b) Spatial distribution patterns of sulfur isotopes, nodular carbonate, and ore textures in the McArthur River (HYC) Zn–Pb–Ag deposit, Northern Territory, Australia. Econ Geol 99:1687–1709

Johnson CA, Kelley KD, Leach DL (2004) Sulfur and oxygen isotopes in barite deposits of the western Brooks Range, Alaska, and implications for the origin of the Red Dog massive sulfide deposits. Econ Geol 99:1435–1448

Johnson CA, Slack JF, Dumoulin JA, Kelley KD, Falck H (2018) Sulfur isotopes of host strata for Howards Pass (Yukon—Northwest Territories) Zn–Pb deposits implicate anaerobic oxidation of methane not basin stagnation. Geology 46:619–622

Johnston DT, Farquhar J, Summons RE, Shen Y, Kaufman AJ, Masterson AL, Canfield DE (2008) Sulfur isotope biogeochemistry of the Proterozoic McArthur Basin. Geochim Cosmochim Acta 72:4278–4290

Kah LC, Lyons TW, Frank TD (2004) Low marine sulphate and protracted oxygenation of the Proterozoic biosphere. Nature 431:834–838

Kawasaki K, Symons DTA (2011) Paleomagnetism of the Mt Isa Zn–Pb–Cu–Ag and George Fisher Zn–Pb–Ag deposits, Australia. Aust J Earth Sci 58:335–345

Kelley KD, Leach DL, Johnson CA, Clark JL, Fayek M, Slack JF, Anderson VM, Ayuso RA, Ridley WI (2004) Textural, compositional, and sulfur isotope variations of sulfide minerals in the Red Dog Zn–Pb–Ag Deposits, Brooks Range, Alaska: implications for ore formation. Econ Geol 99:1509–1532

Kelley KD, Selby D, Falck H, Slack JF (2017) Re-Os systematics and age of pyrite associated with stratiform Zn–Pb mineralization in the Howards Pass district, Yukon and Northwest Territories, Canada. Mineral Deposita 52:317–335

Kiyosu Y, Krouse RH (1990) The role of organic and acid in the sulfur abiogenic isotope reduction effect. Geochem J 24:21–27

Kunzmann M, Schmid S, Blaikie TN, Halverson GP (2019) Facies analysis sequence stratigraphy, and carbon isotope chemostratigraphy of a classic Zn-Pb host succession: the Proterozoic middle McArthur Group, McArthur Basin, Australia. Ore Geol Rev 106:150–175

Lambert IB (1976) The McArthur zinc-lead-silver deposit: features, metallogenesis and comparisons with some other stratiform ores. In: Wolf KH (ed) Handbook of stratabound and stratiform ore deposits, vol 6, pp 535–585

Large RR, McGoldrick PJ (1998) Lithogeochemical halos and geochemical vectors to stratiform sediment-hosted Zn-Pb-Ag deposits 1. Lady Loretta deposit, Queensland. J Geochem Expl 63:37–56

Large RR, Bull SW, Cooke DR, McGoldrick PJ (1998) A genetic model for the HYC deposit, Australia: based on regional sedimentology, geochemistry, and sulfide-sediment relationships. Econ Geol 93:1345–1368

Large RR, Bull SW, McGoldrick PJ (2000) Lithogeochemical halos and geochemical vectors to stratiform sediment hosted Zn–Pb–Ag deposits. Part 2: HYC deposit, Northern territory. J Geochem Exploration 64:105–126

Large RR, Bull SW, Winefield PR (2001) Carbon and oxygen isotope halo in carbonates related to the McArthur River (HYC) Zn–Pb–Ag deposit. Northern Australia: implications for sedimentation, ore genesis, and mineral exploration. Econ Geol 96:1567–1593

Large RR, Bull SW, McGoldrick PJ, Walters W, Derrick GM, Carr GR (2005) Stratiform and strata-bound Zn–Pb–Ag deposits in Proterozoic sedimentary basins, Northern Australia. Econ Geol 100th Anniv Vol 931–963

Leach DL, Marsh E, Emsbo P, Rombach CS, Kelley KD, Anthony M (2004) Nature of hydrothermal fluids at the shale-hosted Red Dog Zn–Pb–Ag deposits, Brooks Range, Alaska. Econ Geol 99:1449–1480

Leach DL, Bradley DC, Huston DL, Pisarevsky SA, Taylor RD, Gardoll SJ (2010) Sediment-hosted lead-zinc deposits in earth history. Econ Geol 105:593–625

Leach DL, Sangster DF, Kelley KD, Large RR, Garven G, Allen CR, Gutzmer J, Walters S (2005) Sediment-hosted lead-zinc deposits: a global perspective. Econ Geol 100th Anniv Vol 561–607

Li C, Planavsky NJ, Love GD, Reinhard CT, Hardisty D, Feng L, Bates SM, Huang J, Zhang Q, Chu X, Lyons, TW (2015) Marine redox conditions in the middle Proterozoic ocean and isotopic constraints on authigenic carbonate formation: Insights from the Chuanlinggou Formation, Yanshan Basin, North China. Geochim Cosmochim Acta 150:90–105

Logan GA, Hinman MC, Walter MR, Summons RE (2001) Biogeochemistry of the 1640 Ma McArthur River (HYC) lead-zinc ore and host-sediments, Northern Territory, Australia. Geochim Cosmochim Acta 65:2317–2336

Luo G, Ono S, Huang J, Algeo TJ, Li C, Zhou L, Robinson A, Lyons TW, Xie S (2015) Decline in oceanic sulfate levels during the early Mesoproterozoic. Precambrian Res 258:36–47

Lyons TW, Luepke JJ, Schreiber ME, Zieg GA (2000) Sulfur geochemical constraints on Mesoproterozoic restricted marine deposition: Lower Belt Supergroup, northwestern United States. Geochim Cosmochim Acta 64:427–437

Machel HG (2001) Bacterial and thermochemical sulfate reduction in diagenetic settings—old and new insights. Sediment Geol 140:143–175

Machel HG, Krouse HR, Sassen R (1995) Products and distinguishing criteria of bacterial and thermochemical sulfate reduction. Appl Geochem 10:373–389

Magnall JM, Gleeson SA, Paradis S (2015) The importance of siliceous radiolarian-bearing mudstones in the formation of sediment-hosted Zn-Pb±Ba mineralization in the Selwyn Basin, Yukon, Canada. Econ Geol 110:2139–2146

Magnall JM, Gleeson SA, Blamey NJF, Paradis S, Luo Y (2016a) The thermal and chemical evolution of hydrothermal vent fluids in shale hosted massive sulphide (SHMS) systems from the MacMillan Pass district (Yukon, Canada). Geochim Cosmochim Acta 193:251–273

Magnall JM, Gleeson SA, Stern RA, Newton RJ, Poulton SW, Paradis S (2016b) Open system sulphate reduction in a diagenetic environment – isotopic analysis of barite (δ^{34}S and δ^{18}O)and pyrite (δ^{34}S) from the Tom and Jason Late Devonian Zn–Pb–Ba deposits, Selwyn Basin, Canada. Geochim Cosmochim Acta 180:146–163

Magnall JM, Gleeson SA, Creaser RA, Paradis S, Glodny J, Kyle JR (2020a) The mineralogical evolution of the clastic dominant-type Zn–Pb±Ba deposits at Macmillan Pass (Yukon, Canada)—tracing subseafloor barite replacement in the layered mineralization. Econ Geol 115:961–979

Magnall JM, Gleeson SA, Hayward N, Rocholl A (2020b) Massive sulfide Zn deposits in the Proterozoic did not require euxinia. Geochem Persp Lett 13:19–24

Magnall JM, Gleeson SA, Paradis S (2020c) A new subseafloor replacement model for the Macmillan Pass clastic-dominant Zn-Pb-±Ba deposits (Yukon Canada). Econ Geol 115:935–959

Magnall JM, Hayward N, Gleeson SA, Schleicher A, Dalrymple I, King R, Mahlstadt N (2021) The Teena Zn-Pb deposit (McArthur Basin, Australia) Part II: carbonate replacement sulfide mineralization during burial diagenesis—implications for mineral exploration. Econ Geol 116:1769–1801

Manning AH, Embso P (2018) Testing the potential of brine reflux in the formation of sedimentary exhalative (sedex) ore deposits. Ore Geol Rev 102:862–874

Martel E (2017) The importance of structural mapping in ore deposits—a new perspective on the Howard's Pass Zn–Pb district, Northwest Territories, Canada. Econ Geol 112:1285–1304

McGoldrick PJ, Kitto PA, Large RR (1998) Variation of carbon and oxygen isotopes in the alteration halo to the Lady Loretta deposit—implications for exploration and ore genesis. In: Water-rock interaction: Proceeding of the 9th international symposium on water-rock interaction, 30 March–3 April 1998, Taupo, New Zealand, pp 561–564

Murray WJ (1975) McArthur River H.Y.C. lead-zinc and related deposits. Austr Inst Mining Metall Mon 5:329–339

Ohmoto H, Goldhaber MB (1997) Sulfur and carbon isotopes. In: Barnes HL (ed) Geochemistry of hydrothermal ore deposits, 3rd edn. Wiley, New York, pp 517–611

Ohmoto H, Rye RO (1979) Isotopes of sulfur and carbon. In: Barnes HL (ed) Geochemistry of hydrothermal ore deposits, 2nd edn. Wiley, New York, pp 509–567

Page RW, Sweet IP (1998) Geochronology of basin phases in the western Mt Isa Inlier and correlation with the McArthur Basin. Austr J Earth Sci 45:201–232

Painter MGM, Golding SD, Hannan KW, Neudert MK (1999) Sedimentologic, petrographic, and sulfur isotope constraints on fine-grained pyrite formation at Mt Isa Mine and environs, northwest Queensland, Australia. Econ Geol 94:883–912

Perkins WG (1984) Mount Isa silica-dolomite and copper orebodies. Econ Geol 79:601–637

Polito PA, Kyser TK, Golding SD, Southgate PN (2006) Zinc deposits and related mineralization of the Burketown mineral field, including the world class Century deposit, northern Australia: fluid inclusion and stable isotope evidence for basin fluid sources. Econ Geol 101:1251–1273

Rickard D, Luther GW III (2007) Chemistry of iron sulfides. Chem Rev 107:514–562

Rieger P, Magnall JM, Gleeson SA, Lilly R, Rocholl A, Kusebauch C, (2020) Sulfur isotope constraints on the conditions of pyrite formation in the Paleoproterozoic Urquhart Shale Formation and George Fisher Zn–Pb–Ag Deposit, northern Australia. Econ Geol 115:1003–1020

Rohrlach BD, Fu M, Clarke JDA (1998) Geological setting, paragenesis and fluid history of the Walford Creek Zn–Pb–Cu–Ag prospect, Mt Isa Basin, Australia. Austr J Earth Sci 45:68–81

Rye DM, Williams N (1981) Studies of the base metal sulfide deposits at McArthur River, Northern Territory, Australia: III. The stable isotope geochemistry of the H.Y.C., Ridge, and Cooley deposits. Econ Geol 76:1–26

Sangster DF (2002) The role of dense brines in the formation of vent-distal sedimentary-exhalative (SEDEX) lead-zinc deposits: field and laboratory evidence. Mineral Deposita 37:149–157

Sangster DF (2018) Toward an integrated genetic model for vent-distal SEDEX deposits. Mineral Deposita 53:509–527

Shanks WC III, Woodruff LG, Jilson GA, Jennings DS, Modene JS, Ryan BD (1987) Sulfur and lead isotope studies of stratiform Zn–Pb–Ag deposits, Anvil Range, Yukon: basinal brine exhalation and anoxic bottom-water mixing. Econ Geol 82:600–634

Slack JF, Falck H, Kelley KD, Xue GG (2017) Geochemistry of host rocks in the Howards Pass district, Yukon-Northwest Territories, Canada: implications for sedimentary environments of Zn–Pb and phosphate mineralization. Mineral Deposita 52:565–593

Smith JW, Croxford NJW (1973) Sulphur-isotope ratios in the McArthur lead-zinc-silver deposit. Nature 245:10–12

Smith JW, Croxford NJW (1975) An isotopic investigation of the environment of deposition of the McArthur mineralization. Mineral Deposita 10:269–276

Smith JW, Burns MS, Croxford NJW (1978) Stable isotope studies of the origins of mineralization at Mount Isa I. Mineral Deposita 13:369–381

Solomon PJ (1965) Investigations into sulfide mineralization at Mount Isa, Queensland Part 1: textural studies of the ores. Econ Geol 60:737–765

Southgate PN, Bradshaw BE, Domagala J, Jackson MJ, Idnurm M, Krasay AA, Page RW, Sami TT, Scott DF, Lindsay JF, McConachie BA, Tarlowski C (2000) Chronostratigraphic basin framework for Palaeoproterozoic rocks (1730–1575 Ma) in northern Australia and implications for base-metal mineralisation. Austr J Earth Sci 47:461–483

Spinks S, Pearce MA, Weihua L, Kunzmann M, Ryan CG, Moorhead GF, Kirkham R, Blaikie T, Sheldon HA, Schaubs PM, Rickard WDA (2021) Carbonate replacement as the principal ore formation process in the Proterozoic McArthur River (HYC) sediment-hosted Zn–Pb deposit, Australia. Econ Geol 116:693–718

Taylor HP (1979) Oxygen and hydrogen isotope relationships in hydrothermal mineral deposits. In: Barnes HL (ed) Geochemistry of hydrothermal ore deposits, 2nd edn. Wiley, New York, pp 236–277

Turchyn AV, DePaolo DJ (2019) Seawater chemistry through Phanerozoic time. Annu Rev Earth Planet Sci 47:197–224

Vearncombe JR, Chisnall AW, Dentith M, Dorling S, Rayner MN, Holyland PW (1996) Structural controls on Mississippi Valley-type mineralization, the southeast Lennard Shelf, Western Australia. Econ Geol Spec Publ 4:74–95

Vinnichenko G, Hope JM, Jarrett AJM, Williams N, Brocks JJ (2021) Reassessment of thermal preservation of organic matter in the Paleoproterozoic McArthur River (HYC) Zn-Pb ore deposit, Australia. Ore Geol Rev 133:104–129

Walker RN, Logan RG, Binnekamp JG (1977) Recent advances concerning the H.Y.C. and associated deposits, McArthur River, N.T. J Geol Soc Austr 24:365–380

Waring CL, Andrew AS, Ewers GR (1998) Use of O, C and S stable isotopes in regional mineral exploration. AGSO J Geol Geophys 17:301–313

Wilkin RT, Barnes HL (1997) Formation processes of framboidal pyrite. Geochim Cosmochim Acta 61:323–339

Wilkinson JJ (2014) Sediment-hosted zinc–lead mineralization. Treatise on geochemistry, 2nd edn, vol 3, pp 219–249

Williams N (1978a) Studies of the base metal sulfide deposits at McArthur River, Northern Territory, Australia: I. the Cooley and Ridge deposits. Econ Geol 73:1005–1035

Williams N (1978b) Studies of the base metal sulfide deposits at McArthur River, Northern Territory, Australia: II. The sulfide-S and organic-C relationships of the concordant deposits and their significance. Econ Geol 73:1036–1056

Williams PJ (1998) An introduction to the metallogeny of the McArthur River-Mount Isa-Cloncurry minerals province. Econ Geol 93:1120–1131

Light Stable Isotope (O, H, C) Signatures of BIF-Hosted Iron Ore Systems: Implications for Genetic Models and Exploration Targeting

Steffen Hagemann, Ana-Sophie Hensler, Rosaline Cristina Figueiredo e Silva, and Harilaos Tsikos

Abstract

Stable isotope data from hypogene (i.e., below the line of weathering) iron oxides and gangue minerals from BIF-hosted iron ore deposits in Australia, South Africa, and Brazil have significantly assisted in constraining different hydrothermal fluid sources and fluid flow models during the upgrade of BIF to iron ore. The $\delta^{18}O$ values on iron oxides from BIF and different paragenetic stages of enrichment display a consistent decrease from unenriched BIF (4–9‰) to as low as −10‰ for high-grade iron ore. This large shift in oxygen isotope values is interpreted as evidence for enormous incursion of 'ancient' meteoric water into fault and fracture zones at the time of iron enrichment during the Archean and Paleoproterozoic time. The $\delta^{18}O_{fluid}$ values of paragenetically early iron oxides of > 4‰ suggest the involvement of magmatic fluids in greenstone belt-hosted Carajás-type iron ore deposits, and basinal brines in basin-hosted Hamersley-type deposits. In contrast, the paragenetically late stage iron oxides in the metamorphosed, basin hosted iron ore deposits of the Quadrilátero Ferrífero display $\delta^{18}O_{fluid}$ values > 6‰. This reflects the renewed deep crustal, hypogene (metamorphic or magmatic) fluid influx. Carbon and oxygen isotope data on carbonates in BIF and hydrothermally altered iron ore indicate that carbon in the latter is not derived from BIF units, but represents either magmatic carbon in the case of the Carajás-type deposits or carbon within the underlying basin stratigraphy as in the case of the Hamersley-type iron deposits. The systematic decrease of $\delta^{18}O$ values in iron oxides from the early to late paragenetic stages and from the distal to proximal alteration zone, including the ore zone, may be used as a geochemical vector. In this case, oxygen isotope analyses on iron oxides provide a potential exploration tool, particularly for targeting the extension of iron ore bodies or entirely concealed high-grade iron ore deposits, in which hematite/magnetite are frequently the only mineral that can be readily analysed.

S. Hagemann (✉)
Centre for Exploration Targeting, School of Earth Sciences, University of Western Australia, Crawley, WA 6009, Australia
e-mail: Steffen.Hagemann@uwa.edu.au

A.-S. Hensler
Institute of Mineral Resources Engineering, RWTH Aachen University, 52062 Aachen, Germany

R. C. Figueiredo e Silva
Universidade Federal de Minas Gerais, Centro de Pesquisas Prof. Manoel Teixeira da Costa-Instituto de Geociências, Av. Antônio Carlos 6627, Campus Pampulha, Belo Horizonte, MG 31270.901, Brazil

H. Tsikos
Department of Geology, Rhodes University, Grahamstown 6140, South Africa

D. Huston and J. Gutzmer (eds.), *Isotopes in Economic Geology, Metallogenesis and Exploration*, Mineral Resource Reviews, https://doi.org/10.1007/978-3-031-27897-6_12

Keywords

BIF · Iron · Oxygen carbon isotopes · Genetic model · Targeting

1 Introduction

Stable isotope analyses are a relatively new approach in the Banded Iron Formation (BIF)-hosted iron ore system and have only been applied in the past three decades. Stable isotope analyses, however, can add fundamental information to our understanding of iron ore genesis, particularly because detailed deposit-scale investigations in Australia, Brazil and Africa have shown that hydrothermal fluids play a significant role in the upgrade of BIF to iron ore (e.g., Hagemann et al. 1999; Thorne et al. 2004; Cope et al. 2008, Rosière et al. 2008; Spier et al. 2008; Cabral and Rosiere 2013; Figueiredo e Silva et al. 2013; Hensler et al. 2015).

The purpose of this review is to summarize the impact of stable isotope geochemistry on the understanding of the BIF-hosted iron ore systems. In addition, potential applications of stable isotope analyses for the exploration of concealed iron deposits or extension of existing iron ore systems are discussed.

The first part of this review provides a summary of data for the major BIF-hosted iron systems in the world, namely the Hamersley (Australia), Transvaal and the Griqualand West (South Africa), Pic de Fon (West Africa), Quadrilátero Ferrífero (QF, Brazil), Carajás (Brazil), Urucum (Brazil), and Noamundi (India). The second part summarizes the genetic implications of the stable isotope data. The last part provides examples of how stable isotope data can be applied in exploration and discuss its limitations.

2 Review of Significant BIF-Hosted Iron Ore Districts and Deposits

The BIF-hosted iron ore system represents the world's largest and highest grade iron ore districts and deposits (Fig. 1). Banded iron-formation, the precursor to low- and high-grade BIF-hosted iron ore, consists of Archean and Paleoproterozoic Algoma-type BIF (e.g., Serra Norte iron ore district in the Carajás Mineral Province), Proterozoic Lake Superior-type BIF (e.g., deposits in the Hamersley province, and deposits in the Quadrilátero Ferrífero), and Neoproterozoic Rapitan-type BIF (e.g., the Urucum iron ore district). The following sections briefly review the major iron ore districts in the world; many of those are currently in production. Due to the different level of investigations, the information about different deposits and districts is inconsistent. All references to production or reserves below refer to Fe ore.

2.1 Precambrian BIF-Hosted Iron Ore Districts of Western Australia

The three major BIF-hosted iron ore provinces in Western Australia (Fig. 1; Hagemann et al. 2016, 2017) are the: (1) Hamersley province, which hosts the giant Mt Whaleback (1800 Mt), Mt Tom Price (900 Mt) and Paraburdoo (800 Mt) deposits, and the Mining Area C (950 Mt), and Chichester Range (2230 Mt) districts, (2) Yilgarn craton, which contains a large number of small to medium size (50–150 Mt) deposits such as Koolyanobbing (93 Mt; including Windarling), Wiluna West (78 Mt—not actively mined), Wiluna and Jack Hills (78 Mt—not actively mined), and (3) Pilbara craton with small to medium size iron ore districts in the central Pilbara, for example McPhee Creek (136 Mt), Mt Goldsworthy (122 Mt; including Nimingarra, Shay Gap, Sunrise Hill and Yarrie), Mt Webber (39 Mt) and Corunna Downs (37 Mt), and the western Pilbara that hosts lower-grade (when compared to the central part), primary magnetite-rich BIF, including Cape Lambert (485 Mt), Miaree (90 Mt), and Mt Oscar (43 Mt). Presently, only the Mt Webber deposit is in production.

The Hamersley province hosts two major BIF-hosted iron ore styles (Hagemann et al. 2017 and references therein): (1) high-grade (59–64 wt.-% Fe) hematite ore (commonly referred to as "blue" ore in the Hamersely Province of Western

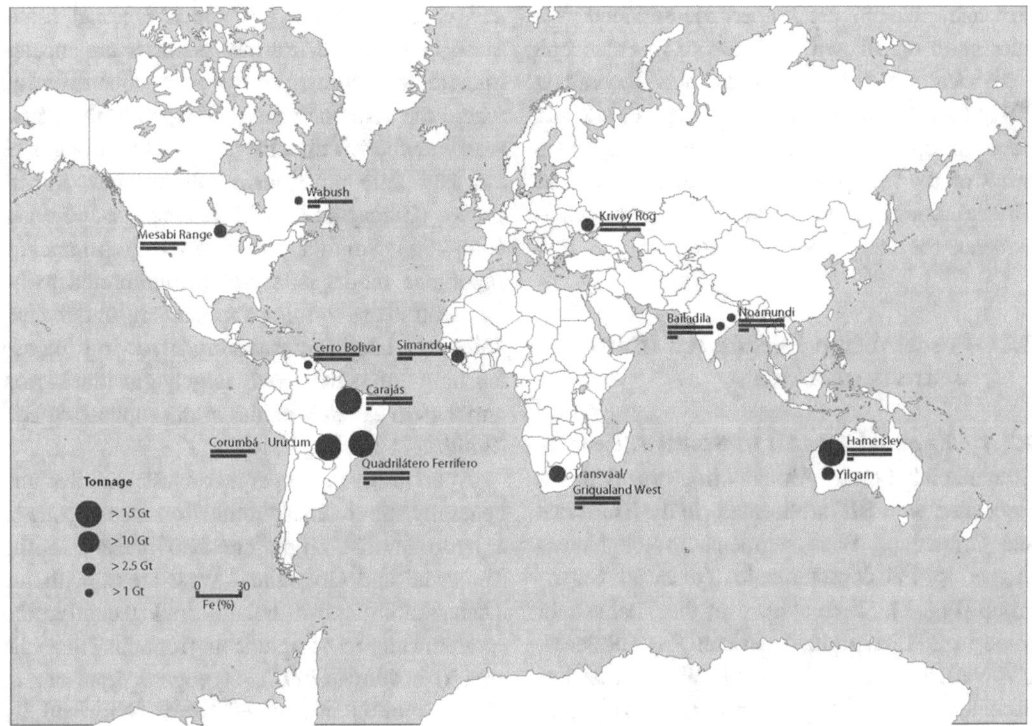

Fig. 1 Location of significant BIF-hosted iron ore districts and deposits. The cumulative length of the bar graphs beneath each locality indicates the Fe grade

Australia), which, in many localities, is accompanied by minor martite-microplaty hematite ore, but contains little or no goethite ore, is mostly confined to the Brockman Iron Formation; and (2) high-grade (58–62 wt.-% Fe) martite-goethite ores ("brown" ore) of the Marra Mamba and Brockman Iron Formation.

In the Yilgarn craton, martite-goethite ores are located at all major deposits; specular hematite ore is best developed at the Koolyanobbing and Weld Range deposits (Duuring et al. 2017a). Magnetite-martite ores are common at Koolyanobbing, Weld Range and Matthew's Ridge, whereas microplaty hematite-martite-specular hematite ore is more common at Windarling and Wiluna West (Duuring et al. 2017a).

In the Pilbara craton most deposits display broad, near-surface, supergene goethite ± martite ores that constitute the bulk of the high-grade iron ore, although narrower hypogene magnetite ± hematite ore zones are recognised in deeper parts of most iron ore deposits (Duuring et al. 2017b).

According to Hagemann et al. (2017) BIF-hosted iron ore deposits in the Hamersley province, Yilgarn and Pilbara cratons are classified as: (1) granite-greenstone belt hosted, fault zone controlled with mineralising fluids derived from magmatic (± metamorphic) sources, together with ancient, warm meteoric waters (most of the deposits in the Yilgarn and Pilbara cratons, such as Koolyanobbing or Mt Webber, belong to this type); (2) sedimentary basin-hosted that are fault controlled, (mostly normal faults), and mineralising fluids sourced from early basinal (evaporitic) brines and ancient, warm meteoric waters. These deposits contain martite-microplaty hematite as their main ore minerals and are exemplified by the giant deposits located in the Hamersley province, such as Mt Tom Price or Mt Whaleback; and (3) martite-goethite iron ore deposits, such as Mining Area C in the Hamersley province, which are supergene-enriched via interaction of BIF with Cretaceous to Paleocene cold meteoric waters (Morris and Kneeshaw 2011) that gained access to the deposit site via already emplaced ancient fault systems and as a result of the interplay between exhumation and surface modification. These deposits have no evidence for deep hypogene roots.

2.2 Precambrian BIF-Hosted Iron Ore Districts of Africa

2.2.1 Kapvaal Craton in South Africa

Commercial South African iron-ore deposits associated with BIF are located in the Transvaal and Griqualand West segments of the Neoarchaean to Palaeoproterozoic Transvaal Supergroup (Fig. 1). Since closure of the Thabazimbi Mine in 2015 (Anglo American Press Release, 16/07/2015), the Griqualand West area has become essentially the only productive region in terms of volume and spatial distribution of iron ore. Four deposits account for practically the entire annual production in South Africa, which at 2014 amounted to 78 Mt of high-grade ore (Smith and Beukes 2016). These deposits are Sishen and Kolomela (also previously known as Sishen South), operated by Kumba Iron Ore, and Khumani and Beeshoek, operated by Assmang Ltd. All deposits occur on a roughly N–S trend that connects the towns of Sishen and Postmasburg, the latter respectively situated at the northernmost and southernmost extremities of a doubly-plunging anticline locally known as the Maremane Dome. Several smaller deposits and occurrences of iron ore that have attracted attention in the broader Maremane area (Land 2013) include Demaneng, Kapstevel, Doornpan, Kareepan, MaCarthy, or Welgevonden.

The prevailing geological setting of the iron-ore deposits at the Maremane Dome involves karstification of Campbellrand dolostones at the stratigraphic base of the Transvaal Supergroup, followed by collapse of overlying Asbesheuwels BIF into karstic sinkholes and apparent replacement thereof by massive hematite. Texturally, the ores range from massive to laminated, breccia and conglomeratic types. On a regional scale, these karst-hosted iron ore deposits are located immediately below an angular unconformity, where the Asbesheuwels BIF (mainly the Kuruman Member) is directly overlain by red-beds of the late Paleoproterozoic Olifantshoek Supergroup (Gamagara-Mapedi formations: Beukes et al. 2003; Smith and Beukes 2016). An ancient supergene model is, therefore interpreted to be the central ore-forming mechanism, under conditions of lateritic weathering involving intense leaching of silica and largely residual iron enrichment as ferric oxides at the expense of BIF (Smith and Beukes 2016).

Apart from the supergene class, smaller and generally sub- to uneconomic iron ore deposits of a hydrothermal origin are also present in the Transvaal and Griqualand West areas, with the Thabazimbi deposit being, until recently, the economically most significant from the Transvaal area (Netshiowi 2002). Hypogene iron ore at Thabazimbi is interpreted to have formed by hydrothermal replacement of Penge BIF, temporally linked to folding and thrusting (Basson and Koegelenberg 2017). The Penge BIF is the stratigraphic equivalent to the Kuruman BIF from the Griqualand West Basin. Minor deposits of similar hydrothermal affinity are also known from Giqualand West in association with the Rooinekke BIF of the Koegas Subgroup (Smith and Beukes 2016).

2.2.2 West African Craton in Guinea

BIF-hosted iron ore deposits of West Africa are epitomised by the Pic de Fon deposit (Cope et al. 2008). The deposit has a strike length of 7.5 km and width of approximately 0.5 km, and is located at the southern end of the Simandou Range in SE Guinea. The ore-bearing stratigraphy consists of three BIF intervals of which the top two have undergone selective enrichment into massive iron oxide ore (up to 65 wt.-% Fe) over thicknesses as much as 250 m. Predominant iron oxide phases are, in paragenetic order, recrystallised martite (after magnetite), hematite overgrowths and bladed microplaty hematite replacing gangue. Chemical mass balance calculations suggest that iron ore formation at Pic de

Fon involved primarily conservative iron enrichment through silica leaching, although net addition of iron up to 36% is also locally observed. Fluid circulation at Pic de Fon was likely driven by post-Eburnean (2.1–2.0 Ga) orogenic collapse or later Proterozoic thermal events, with possible generation of late-stage microplaty hematite during the Pan-African orogeny at 750–550 Ma (Cope et al. 2008).

2.3 Precambrian BIF-Hosted Iron Ore Districts of Brazil

Brazil contains three major BIF-hosted iron ore districts: the Quadrilátero Ferrífero (QF) iron ore province, Carajás Mineral Province (CMP), and Urucum district. The following sections will provide a brief geological overview.

2.3.1 Quadrilátero Ferrífero Province

The QF iron ore province (Fig. 1) is the largest known accumulation of single itabirite-hosted iron ore bodies worldwide with ore reserves of about 9 Gt (52 wt-% Fe, Ministério de Minas e Energia 2009) and over 3.5 Gt of high-grade iron ore (> 65 wt-% Fe; Rosiere et al. 2008). Major ore deposits that are currently under exploitation are e.g. Casa de Pedra (>1.2 Gt high-grade ore), Capitão do Mato (>147 Mt high-grade ore) and Conceição (>470 Mt high-grade ore: Rosière et al. 2008). Most of the iron ore bodies of the QF consist of massive and schistose hypogene iron ore with subordinate supergene iron ore present in the weathering horizons. The hypogene-supergene iron ore is characterized by a complex iron (hydr-) oxide paragenetic sequence including: (itabirite-hosted) magnetite → (itabirite-hosted) martite → (massive ore-hosted) martite → (massive ore-hosted) granoblastic hematite → (massive and schistose-ore hosted) microplaty hematite → (schistose ore- and shear zone-hosted) specular hematite → (lateritic ore-hosted) goethite (e.g., Rosière et al. 2008; Spier et al. 2008; Cabral and Rosière 2013; Hensler et al. 2015). The different iron oxide generations are the result of either replacement (e.g. for martite, granoblastic hematite) or precipitation (e.g. specular hematite) processes (Rosière et al. 2008; Hensler et al. 2015; Oliveira et al. 2015).

Laser ablation-ICP-MS mineral chemical analysis of hypogene iron oxides from QF itabirite and related ore revealed trace element and REE chemical signatures that are distinct for the specific iron oxide generations (Hensler et al. 2015). Major mineralogical and chemical trends include: (i) martitisation, which is accompanied by an absolute depletion of Mg, Mn, Al, Ti and enrichment of Pb, As and light rare earth elements (LREE) in martite; (ii) recrystallisation of magnetite (and martite) to granoblastic and locally microplaty hematite is an "isochemical" process, in which major chemical changes are absent; and (iii) schistose specular hematite is trace element- and REE-depleted. The mineral chemical changes from one iron oxide generation to another one are interpreted to reflect distinct mineralisation events, changing fluid conditions and different fluid/rock ratios.

2.3.2 Carajás

The Carajás iron ore deposits are located in the eastern part of the state of Pará with reserves of 17 Gt (>64 wt.-% Fe). The deposits are hosted by an Archean metavolcano-sedimentary sequence, and protoliths to iron mineralization are jaspilites, under- and overlain by basalts, both of which are greenschist-facies metamorphosed. The major Serra Norte N1, N4E, N4W, N5E and N5S iron ore deposits in the CMP are distributed along, and structurally controlled by, the northern flank of the Carajás fold (Rosière et al. 2006). Varying degrees of hydrothermal alteration have affected jaspilites to form high-grade iron ores (Figueiredo e Silva et al. 2008, 2013; Lobato et al. 2008) from distal to proximal zones. The least-altered jaspilite contains portions of hematite-free, recrystallised chert in equilibrium with magnetite, interpreted as the early stage of hydrothermal alteration. Variably altered jaspilites may be brecciated, containing various amounts of hematite types (e.g. microplaty and anhedral), and vein-associated quartz, carbonate and sulphide minerals. A hydrothermal paragenetic sequence for the oxides is established from the earliest magnetite → martite → microplaty hematite → anhedral hematite (AnHem)- to the latest → euhedral-tabular hematite (EHem-Them).

Other mineralogical modifications are: (i) recrystallisation and cleansing of jasper with formation of chert and fine quartz; (ii) progressive leaching of chert and quartz, leaving oxides and a significant volume of empty space; (iii) silicification and dolomitisation with associated sulphides and oxides in veins, breccias and along jaspilite bands; (iv) advanced martitisation with the formation of AnHem, partial microcrystalline hematite recrystallisation to AnHem and partial open-space filling with microplaty/platy hematite; and (v) continued space-filling by comb-textured EHem and THem in veinlets and along bands.

2.3.3 Urucum

The Urucum manganese and iron ore mining district is located in the Mato Grosso state near the Brazilian-Bolivian border and consists of several open pit mines (e.g. Mineração Corumbaense Reunida, Urucum Mineração mine). In 2005, reserves were estimated at about 3.1 Gt of iron ore and 11 Mt of manganese ore (DNPM 2005; Walde and Hagemann 2007).

The iron ore is hosted in the Neoproterozoic metasedimentary rocks of the Santa Cruz Formation of the Jacadigo Group, which contains mainly hematite-rich BIFs and Mn ore horizons with subordinate interbedded arkosic sandstone (Urban et al. 1992; Walde and Hagemann 2007; Piacentini et al. 2013). The protore BIFs are interpreted to have been deposited in a glaciomarine environment (e.g. Urban et al. 1992; Klein 2005; Angerer et al. 2016). Recent petrographic-mineralogical studies of the Santa Cruz ore deposit (Angerer et al. 2016) distinguish between a dolomite-chert-hematite BIF (35–52 wt.-% Fe) in an upper and lower carbonaceous zone and chert-hematite BIF (45–56 wt.-% Fe) of an intermediate siliceous zone. Deci- to meter-thick diamictic hematite-rich chert (~16 wt.-% Fe) and mud (59–66 wt.-% Fe) layers are discontinuously intercalated in the BIFs (Angerer et al. 2016). Iron oxides of the BIFs are mainly fine-grained (grain sizes < 20 μm) anhedral to microplaty hematite. Locally microplaty hematite is oriented parallel to the bedding. Beside gangue minerals, diamictic hematite-rich chert layers contain microcrystalline (grain sizes of 10–20 μm) hematite "needles" and are interpreted to represent a replacement product of fibrous silicates. In the hematite-rich mud layers, different hematite variations are present ranging from relict hematite to granoblastic hematite.

The exact genesis of the ore deposits of the Urucum mining district is still ambiguous (e.g. Dorr 1964; Trompette et al. 1998; Walde and Hagemann 2007; Viehmann et al. 2016). This is mainly because most studies (e.g. Graf et al. 1994; Trompette et al. 1998; Freitas et al. 2011; Piacentini et al. 2013; Angerer et al. 2016; Viehmann et al. 2016) concentrated on the depositional conditions of the protore BIFs during the Neoproterozoic times rather than on the post-depositional modification and upgrade of the BIF to iron ore. Walde and Hagemann (2007), however, favoured a genetic ore model that includes the input of hypogene hydrothermal activity during the upgrade of the protore BIF and iron ore formation of Urucum, because of some mineralogical features of the ore, such as (1) the presence of braunite as indicator for elevated temperatures; (2) the presence of magnetite along fault zones; and (3) the emplacement of quartz-tourmaline veins crosscutting the Jacadigo Group. Gutzmer et al. (2008) argue based on detailed whole rock and trace element geochemistry analyses of high-grade ore samples that supergene processes played a major role in the formation of the Urucum iron ore district. However, most recent geochemical investigations of the BIFs from the Urucum district (Angerer et al. 2016; Viehmann et al. 2016) could not observe any geochemical evidence for a significant overprint and post-depositional alteration of the Urucum successions.

2.4 India

The high-grade (>60 wt.-%) iron ore deposits (e.g. Noamundi, Kiriburu, Megthaburu, Joda) of the Noamundi area are located in the north eastern part of India and hosted in the iron formations of the metasedimentary greenstone belt sequence of the Archean Western Iron Ore Group (Beukes et al. 2008; Roy and Venkatesh 2009; Bhattacharya and Ghosh 2012). The overall annual iron ore production of the entire mining area is about 12 Mt and resources are estimated to be about 3.3 Gt of high-grade iron ore (Bhattacharya and Ghosh 2012). The unaltered protore BIF is approximately 220 m thick and mainly consists of thin, alternating chert- and hematite-magnetite-rich mesobands. High-grade iron ore of the Noamundi mining area can be separated in upper goethite- and lower hematite-rich iron ore. Goethite-rich ore is mainly associated with reworked canga-type successions related to erosion of Cretaceous to early Cenozoic land surface. Hematite-rich ore accounts for the major part of the iron ore resources and can be subdivided into hard and soft-friable ore (Beukes et al. 2008; Bhattacharya and Ghosh 2012). The former is present particularly in the lower successions of the iron ore bodies, whereas the latter is located mainly in the saprolitic weathering zone of the ore body. The hematite-rich iron ore contains different generations of hematite (e.g. euhedral, anhedral, cryptoplaty and microplaty) with minor relict magnetite, martite and goethite.

Hard high-grade iron ore of the Noamundi mining area is interpreted to have formed by hydrothermal processes, involving the dissolution of quartz, likely cogenetic crystallisation of magnetite, subsequent martitisation and precipitation of microplaty hematite (Beukes et al. 2008; Roy and Venkatesh 2009; Bhattacharya and Ghosh 2012). Friable hematite-rich as well as goethite-rich ores are interpreted to have largely formed and/or been overprinted by supergene lateritic processes during Cretaceous-Cenozoic times (Beukes et al. 2008).

3 Oxygen, Hydrogen, Carbon Isotope Data of BIF and Related Iron Ore

Presently there are significant stable isotope data sets for BIF-hosted iron ore deposits and districts from Western Australia, South Africa, Guinea, Brazil and India (Fig. 1). The following sections review these data sets and their interpretation in the regional context. All $\delta^{18}O$ and δD values are presented in‰ relative to Vienna Standard Mean Ocean Water (VSMOW), whereas $\delta^{13}C$ and $\delta^{18}O$ on carbonates are shown in‰ relative to Vienna Pee Dee Belemnite (VPDB).

3.1 Western Australia

Stable isotope data sets, mainly oxygen on quartz and oxides, as well as carbon and oxygen on carbonates are available for the Hamersley province and the Yilgarn and Pilbara cratons (Fig. 2).

3.1.1 Oxygen Isotope Data on Iron Oxides and Silicates

Thorne et al. (2004) presented detailed oxygen isotope data sets on quartz and oxides from the Tom Price, Paraburdoo and Channar deposits. The $\delta^{18}O$ values of magnetite and hematite from hydrothermal alteration assemblages and high-grade iron ore range from −9.0 to −2.9‰, and thus are depleted up to 5–15‰ relative to the magnetite of the host BIF (Fig. 2).

Thorne et al. (2009) calculated the oxygen isotope composition of hydrothermal fluids ($\delta^{18}O_{fluid}$) in equilibrium with iron oxides using the isotope fractionation of Becker and Clayton (1976) for magnetite and of Yapp (1990) for hematite and temperatures based on the fluid inclusion trapping temperatures from BIF, hydrothermal alteration assemblages, and high-grade iron ores. Trapping temperatures were pressure-corrected using stratigraphic reconstructions of the Mt. Tom Price deposit (30 °C: Taylor et al. 2001; Thorne et al. 2004) and the Paraburdoo (40 °C) and Channar deposits (30 °C; Dalstra 2006) during the likely time of hydrothermal alteration in the Paleoproterozoic.

The range of $\delta^{18}O_{fluid}$ values for all of the Hamersley iron ore deposits is shown in Fig. 2. At Tom Price they range from 12.7 to 21.7‰ for magnetite of the Dales Gorge and Joffre BIF members, from −0.7 to −0.4‰ for magnetite in distal alteration zones, from −3.5 to 1.2‰ for

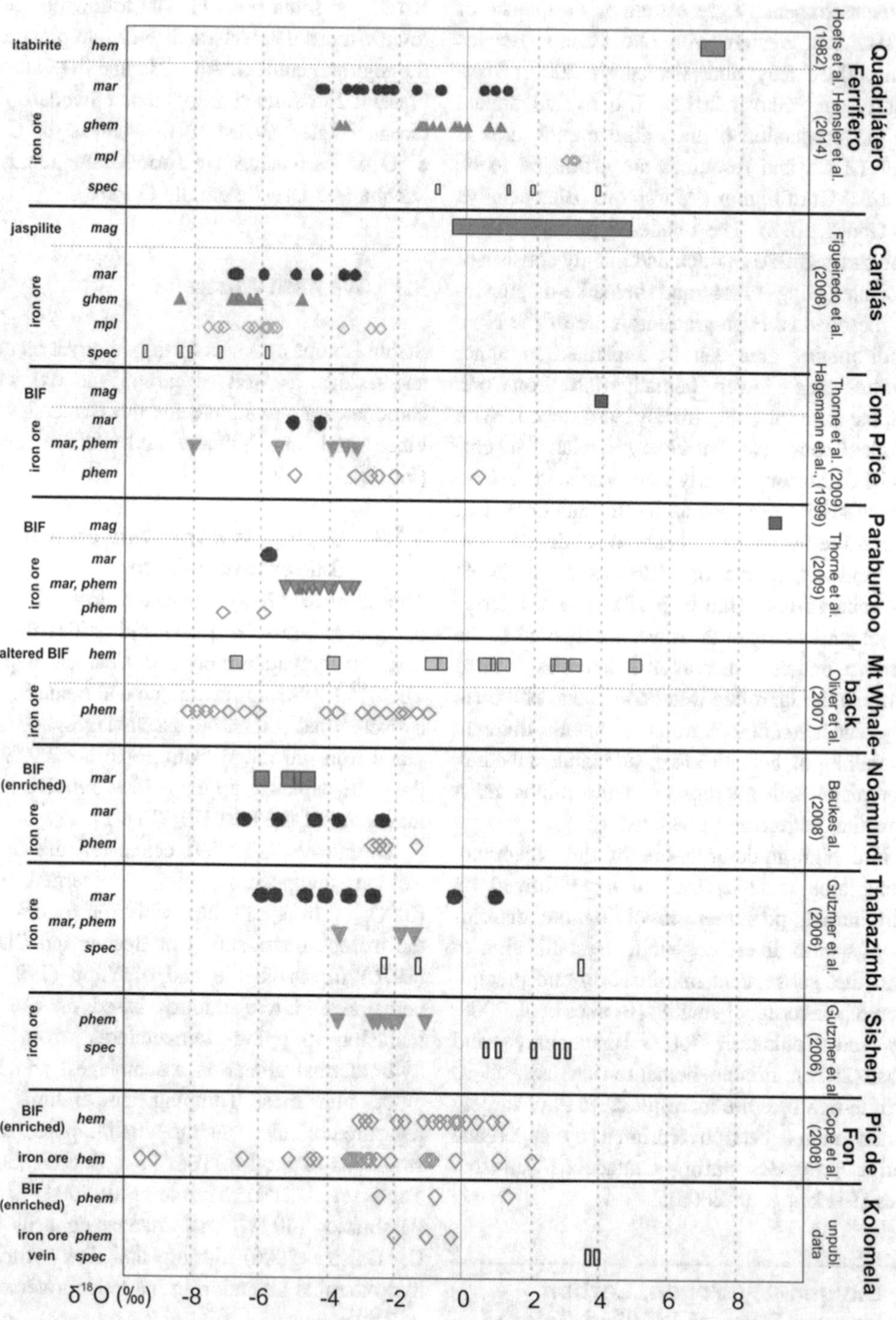

Fig. 2 Oxygen isotope values of different iron oxides (in ‰ relative to VSMOW) from various iron ore deposits worldwide (mag: magnetite, mar: martite, phem: platy hematite, mpl: microplaty hematite, spec: specular hematite) (modified after Hensler et al. 2014; Hagemann et al. 2016). Hematite oxygen isotope data from the Pic de Fon ore deposit were calculated on the basis of whole-rock and quartz oxygen isotope data (Cope et al. 2008). A depletion in ^{18}O from iron oxide of the least altered protore BIF/itabirite/jaspilite (grey box) to high-grade iron ore is documented in most ore deposits. Furthermore, across the paragenetic sequences of the Quadrilátero Ferrifero, Tom Price, Noamundi and Transvaal ores, earlier formed iron oxides (e.g. martite, granoblastic hematite) have mainly lower $\delta^{18}O$ values relative to later formed iron oxides (e.g. microplaty, platy and specular hematite)

hematite in the intermediate alteration zones and −8.3 to −5.8‰ and −6.7 to −4.3‰ for hematite from proximal alteration and high-grade martite-microplaty hematite ore, respectively (Thorne et al. 2004). Similar trends are observed at the Paraburdoo and Channar deposits (Fig. 2; Thorne et al. 2009).

The systematic decrease of $\delta^{18}O_{fluid}$ values from magnetite of the BIF to altered oxide-silicate-carbonate and to monomineralic oxide assemblages in the intermediate and proximal alteration zones, respectively, are interpreted by a variety of authors including Thorne et al. (2008), Angerer et al. (2014) and Hagemann et al. (2016) as evidence for the incursion of two hydrothermal fluids within the fault and shear zones. These fluids transformed BIF to high-grade iron ore via a series of complex fluid-rock reactions (cf. Hagemann et al. 2016). The ore-forming fluids are basinal brines ($\delta^{18}O_{fluid}$ values ranging between −2.0 and 6.0‰) and meteoric water ($\delta^{18}O_{fluid}$ values < −2‰).

Late-stage talc-bearing ore at the Mt. Tom Price deposit (Fig. 2) formed in the presence of a pulse of ^{18}O-enriched basinal brines, indicating that hydrothermal fluids may have repeatedly interacted with the BIFs during the Paleoproterozoic (Thorne et al. 2009).

Hydrothermal quartz-hematite veins from the Southern Batter fault zones within the Southern Ridge ore body at Tom Price show restricted $\delta^{18}O$ values for quartz from 16.1 to 18.3‰ (Hagemann et al. 1999). The $\delta^{18}O_{fluid}$ values were calculated using temperatures of 160 °C, estimated from microthermometric studies on the same quartz samples, and the quartz-H_2O fractionation curve (Matsuhisa et al. 1979); the $\delta^{18}O_{fluid}$ values range between −0.1 and 4.1‰. Hagemann et al. (1999) interpret these data as evidence for the influx of meteoric water or basinal brines that show minor fluid-rock reactions.

3.1.2 Spatial Distribution of $\delta^{18}O$ Values

At Mt. Tom Price there is a clear spatial relationship between the depletion in ^{18}O of hematite and magnetite relative to the BIF and the Southern Batter fault zone (Fig. 3). From a structural-fluid dynamic point of view the Southern Batter fault zone represents the main conduit for ascending and descending hydrothermal fluids in the Tom Price deposit (Hagemann et al. 1999; Taylor et al. 2001). The $\delta^{18}O$ values of samples proximal or within the Southern Batter fault zone are expected to be the most fluid dominated. For a given temperature their oxygen isotope compositions are likely to relate closely to those of the hydrothermal fluid, in this case meteoric water characterized by values from −8.6 to < 2‰ (Fig. 3). Farther from the fault zones, and at the periphery of the hydrothermal system, the $\delta^{18}O_{fluid}$ values of the iron oxides (−2 to 6.0‰) represent oxygen sourced from both the BIF and, to a lesser extent, the hydrothermal fluids.

Similar spatial zonation patterns were observed at Paraburdoo where $\delta^{18}O_{fluid}$ values of high-grade ore located within or close to the 4E Basal fault range between −9.6 and −6.0‰, whereas the proximal alteration zones bordering the fault zone display $\delta^{18}O_{fluid}$ between −0.9 and −0.6‰ (Thorne et al. 2009, 2014). The least altered Dales Gorge Member BIF display $\delta^{18}O_{fluid}$ values of 8.8 and 13.0‰ (Thorne et al. 2009, 2014).

3.1.3 Carbon and Oxygen Isotope Data on Carbonates

Detailed oxygen and carbon isotope analyses were performed on hydrothermally altered hematite-ankerite-magnetite in the North deposit, Tom Price (Thorne et al. 2004). Lower $\delta^{13}C$ values of ankerite ($\delta^{13}C$; −4.9 ± 2.2‰) from the hematite-ankerite-magnetite alteration zone, when compared to unmineralised BIF, indicate that the bulk of the carbon within the alteration zone is not derived from the BIF sequence (Thorne et al. 2004; Fig. 4). Similar oxygen isotope compositions, but increasingly heavier carbon isotopes from magnetite-siderite-iron silicate alteration (−8.8 ± 0.7‰) to hematite–ankerite–magnetite alteration zones (−4.9 ± 2.2‰), suggest the progressive exchange (mixing) with an external fluid with a heavy carbon isotope signature. It is likely that these saline

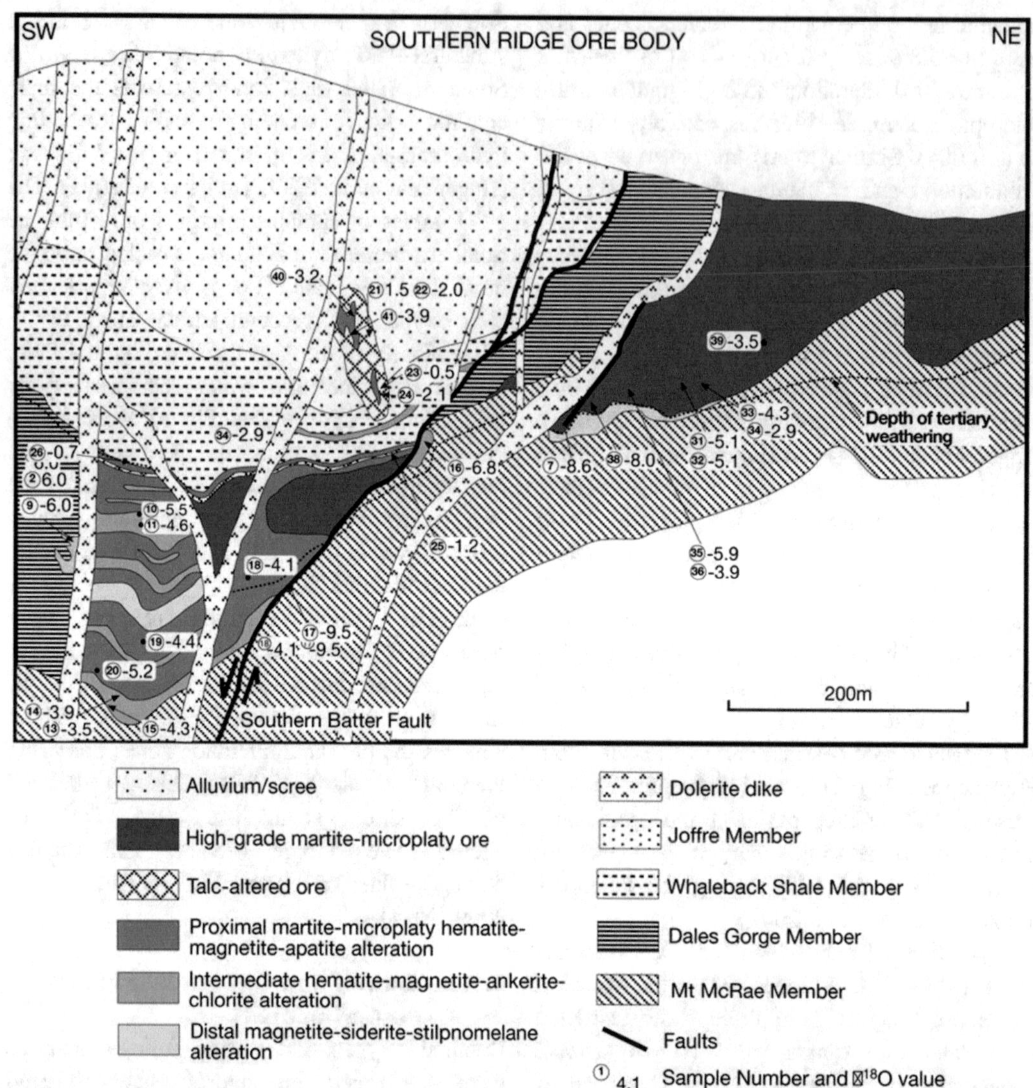

Fig. 3 Cross section of the Southern Ridge ore body at the Mt. Tom Price deposit displays the stratigraphic units and major normal faults, the spatial distribution of hydrothermal alteration assemblages and high-grade ore, and the $\delta^{18}O$ values (in ‰) of hematite and magnetite samples. Samples obtained from open-pit grab samples and diamond drill core. Note the low $\delta^{18}O$ values of samples proximal to fault zones. Section shown facing west. Reproduced with permission from Thorne et al. (2009); Copyright 2009 Society of Economic Geologists

fluids either reacted with the rocks from the Wittenoom Formation, or mixed with a fluid derived from the Wittenoom Formation (0.9 ± 0.7‰), and thus provided such a fluid composition (Fig. 4). Recent interpretation based on thermodynamic modelling of the de-silicification and carbonate addition processes at Tom Price questions the sole capacity of silica-dissolution of upward-flowing brines sourced from dolomite of the Wittenoom Formation, through shale-dominated Mt Sylvia and Mt McRae Formations (Evans et al. 2013). Instead, gravity-driven brines from an evaporitic source on an up-temperature pathway are a possible alternative solution to dissolve the large amount of chert required to form a giant hematite iron ore deposit such as at Mt. Tom Price (Evans et al. 2013). Likely, the de-silicification and carbonate

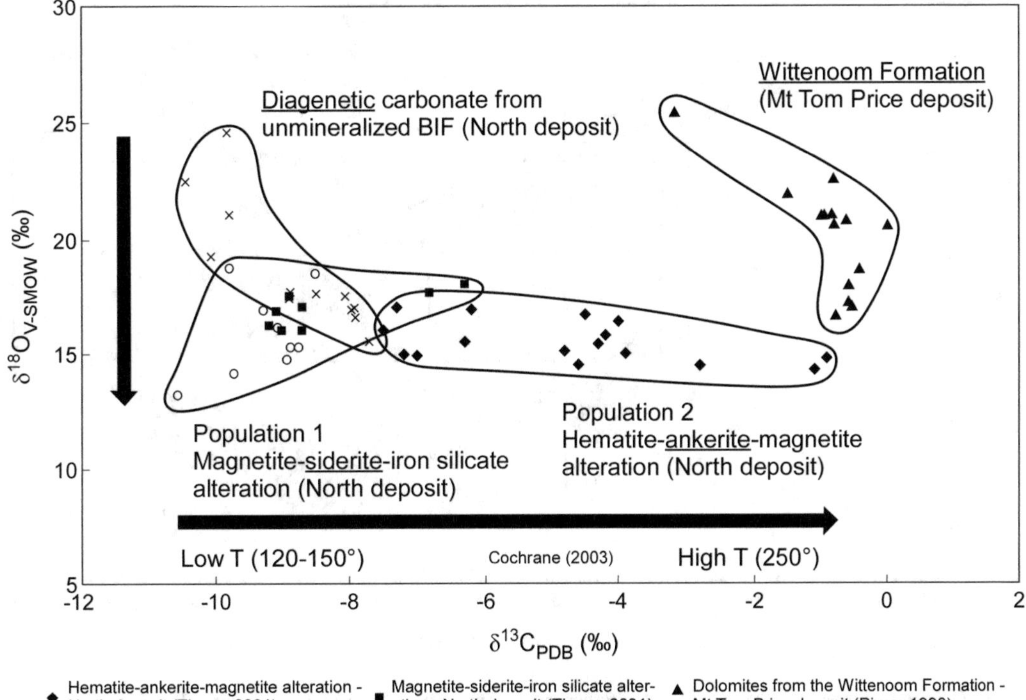

Fig. 4 Oxygen versus carbon isotope diagram showing the isotopic composition of diagenetic carbonates from unmineralised BIF, magnetite–siderite–iron silicate, hematite–ankerite–magnetite alteration and Wittenoom Formation in the Hamersley province, Western Australia. All samples are from the North deposit at the northern end of the Mt. Tom Price deposit, except from the Wittenoom Formation, which are from the Mt. Tom Price deposit. Temperatures of carbonate formation are obtained from Cochrane (2003). Modified after Thorne et al. (2014) and Hagemann et al. (2016).

addition processes in giant BIF-hosted deposits are complex and may contain both upwards and downwards flowing and/or possibly alternating, hydrothermal fluids.

Angerer et al. (2014) presents carbon and oxygen isotope data from early (co-magnetite) and late (co-specular hematite) carbonate generations documented at the Koolyanobbing, Beebyn and Windarling deposits in the Yilgarn craton (Fig. 5). The overall trend from heavy O and light C in early carbonate-magnetite stages to light O and heavy C in late carbonate-oxide stages is similar to the trends observed in the distinct alteration stages at the Mt. Tom Price deposit (Fig. 4; Taylor et al. 2001; Thorne et al. 2004; 2008; Angerer et al. 2014). The signature of the late-stage dolomite in Yilgarn deposits is compatible with the down flow of meteoric water and/or seawater into fault zones and associated ore bodies, assuming that $\delta^{13}C$ of Archean seawater is about 0‰, and that the O isotope signature of sea-/meteoric water remained constant during the hydrothermal process(es). The similarity of C and O isotope patterns between deposits in the Yilgarn (Fig. 5) and Hamersley province (Fig. 4) may be a coincidence, although at the least it suggests that contrasting processes and fluid sources can lead to similar isotopic values.

3.1.4 Hydrogen Isotope Data on Fluid Inclusions in Quartz

Hydrogen isotopes were obtained from fluid inclusions trapped in quartz from quartz-hematite veins of the Southern Batter fault zones in the Southern Ridge ore body at Tom Price. Fluid

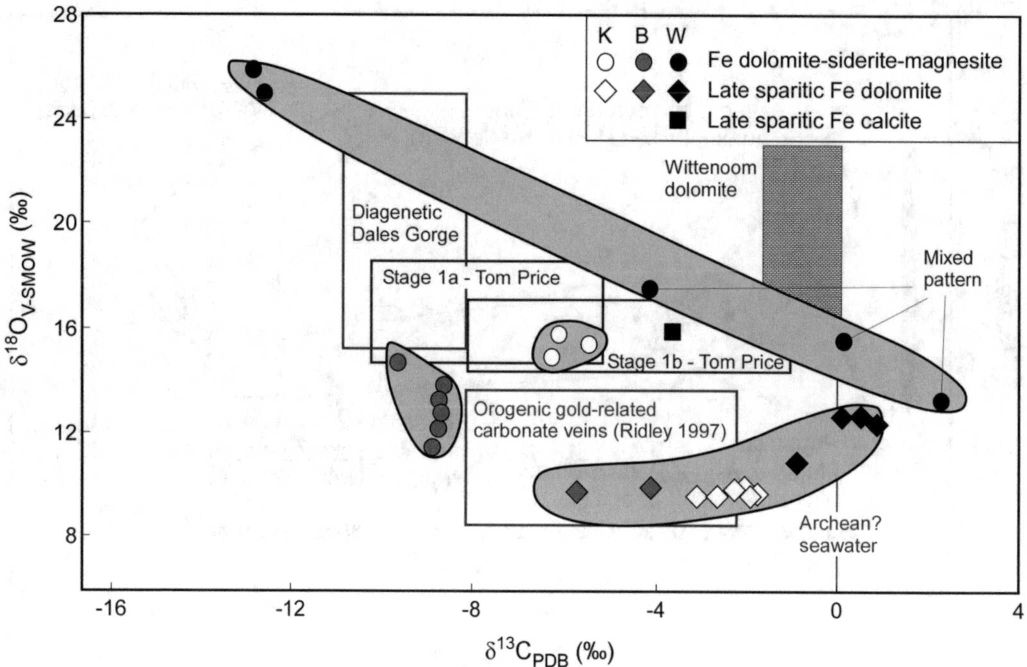

Fig. 5 Carbon and oxygen isotope data of hydrothermal carbonate in the Yilgarn craton deposits Koolyanobbing (K), Beebyn (B) and Windarling (W). Shown for comparison are data fields for diagenetic siderite/ankerite from the Dales Gorge BIF, hydrothermal ankerite from Tom Price stage 1a and 1b alteration zones, Paraburdoo Member of the Wittenoom dolomite (Taylor et al. 2001 and references therein), and carbonate veins in Yilgarn Craton orogenic gold deposits (Ridley 1997). Modified after Angerer et al. (2014)

inclusion waters yield a large range of δD values from −90‰ to −35‰ (Hagemann et al. 1999) with two populations at −53 to −66‰ and −34 to −43‰, and a single analysis of −90‰. These values were interpreted by Hagemann et al. (1999) as compatible with either seawater that was shifted to more negative values or, alternatively, with meteoric water or basinal brines that reacted with wallrocks.

3.2 Africa

3.2.1 Oxygen Isotope Data on Hematite

The majority of stable isotope data from South African BIF-hosted iron ores reported to date are essentially confined to bulk-rock measurements of massive hematite ore, irrespective of the textural variability of the ores. The general lack of carbonates in the ores and in adjacent rocks precludes them as targets for carbon and oxygen on carbonate isotope analyses. Limited data from early- and late-stage hematite from Thabazimbi have been reported by Beukes and Gutzmer (2008) and these are briefly referred to elsewhere in this paper when hydrothermal deposits are discussed in detail.

With respect to the predominant ancient-supergene type of ores at the Maremane Dome, the absence of systematic oxygen isotope studies is mainly due to the obvious lack of recognised alteration zones developed in BIF adjacent to the ores that would themselves provide potential first-order exploration vectors. Moreover, as indicated earlier, karst-hosted iron ore deposits at the Maremane Dome contain texturally diverse ore types across space, with a strong detrital component as encapsulated particularly in the breccia and conglomeratic ore types. In fact, pre-ore BIF may not be preserved at all in direct contact with some of these ore occurrences

(Moore et al. 2011). Such ores are invariably unsuitable for targeted sampling aimed at stable isotope analyses due to their inherently heterogeneous textural nature, and have thus led to the predominantly bulk sampling approach.

Figure 6 provides a comprehensive summary of oxygen isotope data for bulk hematite iron ore samples from three localities of ancient-supergene ore at the Maremane Dome of the Griqualand West area, namely Sishen, Beeshoek and Kolomela. The diagram also includes, for comparative purposes, data for the hydrothermal Thabazimbi ores from the Transvaal area. An obvious first observation on the diagram of Fig. 6 is the similarity in the range of $\delta^{18}O$ values for the supergene ores, between −4 and 3‰, with approximately two-thirds of the data falling in the range −4 to 0‰. Although the overall range in $\delta^{18}O$ values probably reflects the intrinsic sample heterogeneity behind the bulk ore measurements, the overwhelmingly light oxygen isotopic signature has been—and is being—used as strong indication for the meteoric character of the hydrothermal fluids implicated in ore formation and, by extension, for the paleo-supergene origin of the ores (Smith and Beukes 2016). The data range from the Thabazimbi ore is comparatively larger, with recorded $\delta^{18}O$ values as low as −6‰ and as high as 4‰. Interestingly, similar measured $\delta^{18}O$ values for hematite of the hydrothermal ores at Pic de Fon of Guinea also record a large range of low $\delta^{18}O$ between a maximum value of 1.3‰ and a minimum value of −8.9‰, suggestive of an ultimately paleo-meteoric source for the mineralising hydrothermal fluids.

Recent studies on altered BIF intersections adjacent to, and intercalated with, massive iron ore in the broader Kolomela area, have revealed several generations of microplaty to specular hematite veinlets and apparent replacement textures at the expense of the BIF (Papadopoulos 2016). Oxygen isotope data of micro-drilled hematite from such occurrences are included in Fig. 2. The data indicate that the vein and replacement type hematite records generally higher $\delta^{18}O$ values by comparison to adjacent massive hematite ore, and this is an observation that is in general agreement with similar signatures in hydrothermal ore types as discussed elsewhere in this paper. It is, therefore, possible that the iron ores at the Maremane Dome are not exclusively ancient-supergene in origin, and may contain a significant hydrothermal component to their formation, possibly in a fashion akin to supergene-modified genetic models.

Oxygen isotope data from the Simandou iron ore district of Guinea are solely obtained from the Pic de Fon deposit and can be categorised into two populations (Cope et al. 2008). Relatively higher values from least-altered BIF samples, when compared to BIF from the Hamersley or Kapvaal provinces, are interpreted to reflect fluid-rock interaction processes during retrograde, greenschist facies conditions. In contrast, samples of BIF that are significantly enriched in iron record a shift towards much lower $\delta^{18}O$ values (Fig. 2), comparable in their overall range to those from other hydrothermal iron ore deposits such as Thabazimbi (South Africa), Noamundi (India) and Mt. Tom Price (Australia). This is consistent with fluid-rock interaction involving large volumes of isotopically light, evolved meteoric water, at moderate temperatures (<380 °C). A slight decrease in the $\delta^{18}O$ values of coexisting quartz appears to conform to the same interpretation by virtue of its limited isotopic exchange with the same meteoric-sourced fluids.

3.3 Brazil

3.3.1 Oxygen Isotope Data on Iron Oxides

Oxygen isotope analyses across the complex iron oxide paragenetic sequence of the QF reveal distinct $\delta^{18}O$ signatures (Fig. 2: Hensler et al. 2014). With respect to the itabirite-hosted and ore forming iron oxide generations, differences in the oxygen isotope signature are particularly observed between the itabirite-hosted hematite, the early ore-forming iron oxides (martite and granoblastic hematite) and the late ore-forming iron oxides (microplaty and specular hematite). The $\delta^{18}O$ values of ore-hosted martite range

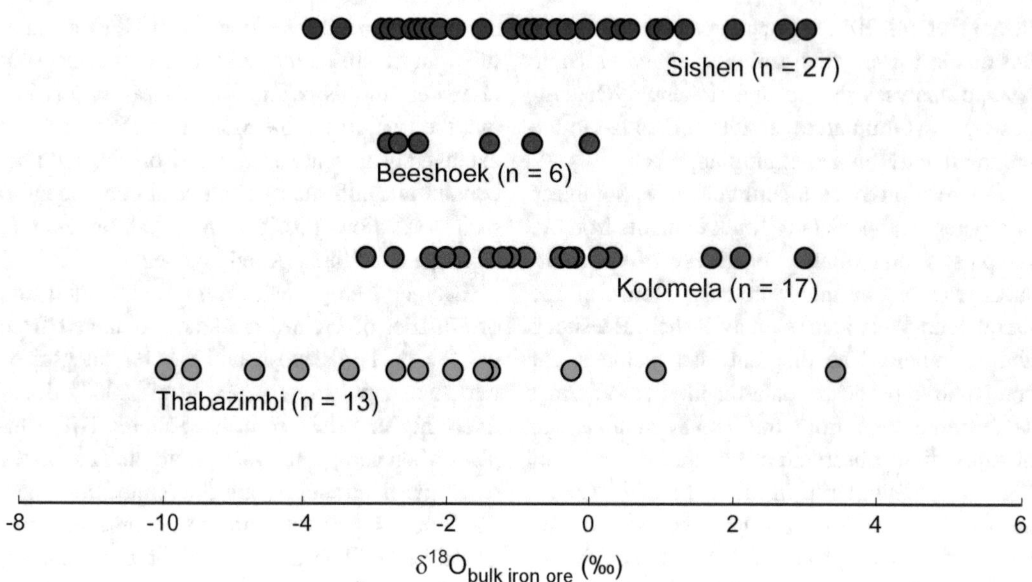

Fig. 6 Compilation of bulk hematite oxygen isotope data (versus VSMOW) for massive, laminated and breccia/conglomeratic-type iron ores of the Transvaal Supergroup, South Africa. The Sishen and Beeshoek deposits and the Kolomela prospect in the Griqualand West Basin represent examples of iron ore mineralisation of an ancient supergene origin. By contrast, the Thabazimbi deposit in the eastern Transvaal Basin is interpreted as hydrothermal in origin. Sishen data from Gutzmer et al. (2006); Beeshoek data from van Deventer (2009); Heuninkranz data from Land (2013) and Papadopoulos (2016); and Thabazimbi data from Netshiozwi (2002)

between −4.4 and 0.9‰ (Fig. 2) and are significantly depleted when compared to hematite from least-altered itabirite (~7‰: Hoefs et al. 1982). This strong depletion in ^{18}O is likely the result of the influx of large volumes of hydrothermal fluids with light ^{18}O signatures, and an extreme chemical fluid-rock exchange at low temperatures (<150 °C) and high fluid-rock ratios during the martitisation and cogenetic iron ore formation. The recrystallised ore-hosted granoblastic hematite has similar light $\delta^{18}O$ values compared to ore-hosted martite, ranging between −4.1 and 0.6‰ (Fig. 2). Similar $\delta^{18}O$ signatures of martite and granoblastic hematite assume that recrystallisation of granoblastic hematite took place under "isochemical" conditions with the martite, thus without major oxygen isotope exchange. This may be either due to the participation of only small fluid volumes and/or because the oxygen isotope composition of the fluid was in equilibrium with the earlier formed martite during recrystallisation. Relative to the early stage iron oxides (martite and granoblastic hematite), the later-stage microplaty and specular hematite that form the schistose ore have significantly heavier $\delta^{18}O$ values, ranging between 3.0–3.2‰ and −0.9–3.8‰, respectively (Fig. 2). This increase in the $\delta^{18}O$ values of the schistose, ore-forming hematite generations is likely to reflect the precipitation of microplaty and specular hematite from fluids with heavier oxygen isotope compositions during the later-stage of ore formation (Hensler et al. 2014). Interestingly, an increase in the $\delta^{18}O$ ratios from earlier stage hematite to later stage hematite is also recorded in other high-grade iron ore deposits worldwide, such as Noamundi, Thabazimbi, and Tom Price (Gutzmer et al. 2006; Beukes et al. 2008; Thorne et al. 2008, respectively).

Calculations of the $\delta^{18}O$ of fluids in equilibrium with the iron oxides of the QF were performed using the water-hematite curve after Yapp (1990) assuming temperatures of 145 °C (Rosière et al. 2008) for martite, granoblastic and microplaty hematite and at a temperature of 350 °C for specular hematite (Rosière et al.

2008). Fluids in equilibrium with martite and granoblastic hematite have relatively low average $\delta^{18}O_{fluid}$ values (approx. 0.8‰ and 1.6‰, respectively), whereas microplaty hematite and specular hematite have high average $\delta^{18}O_{fluid}$ values (6.0‰ and 9.6‰, respectively). With respect to the iron ore genesis and source of ore-forming fluids, the oxygen isotope compositions of the fluids in equilibrium with the distinct iron oxide paragenesis are interpreted to reflect: (i) the participation of isotopically light hydrothermal fluids (e.g. meteoric water or basinal brines) during the martitisation and likely cogenetic upgrade of itabirite to high-grade iron ore; and (ii) participation of modified meteoric water (with elevated $\delta^{18}O_{fluid}$ values) and/or infiltration of isotopically heavy fluids (e.g. metamorphic or magmatic water) into the ore system during the precipitation/crystallisation of microplaty and schistose specular hematite.

Figueiredo e Silva et al. (2008, 2013) selected samples for oxygen isotopes from the Carajás iron ores and least altered jaspilite from: (1) magnetite, (2) microplaty hematite, (3) anhedral hematite, (4) tabular hematite, (5) quartz from different vein breccias, and (6) least-altered (whole-rock basis) jaspilite containing the assemblage microcrystalline hematite ± martite; analyses of microcrystalline hematite were also performed. The $\delta^{18}O$ values display a large range (Fig. 2) from 15.2‰ for the least-altered jaspilites (whole-rock analyses; not shown in Fig. 2) to −9.5‰ for the paragenetically latest, euhedral-tabular hematite (Fig. 2) in the high-grade iron ore. The $\delta^{18}O_{fluid}$ values were calculated based on: (1) homogenization temperatures, after pressure correction, of fluid inclusions trapped in quartz and carbonate crystals that are in textural equilibrium with hematite and/or magnetite (Figueiredo et al. 2008, 2013; Lobato et al. 2008), and (2) the fractionation factors of Yapp (1990) for hematite, Zheng (1991, 1995) for magnetite, and Matsuhisa et al. (1979) for quartz. For magnetite, the average pressure-corrected trapping temperature of 245 °C, obtained from microthermometry analyses of fluid inclusions within vein type 1 quartz-carbonate crystals that are in textural equilibrium with magnetite, was used. For hematite, trapping temperatures (245–285 °C) of fluid inclusions trapped in rare quartz in equilibrium with hematite were used. The $\delta^{18}O_{fluid}$ values for the different hematite types increased about 6 to 8‰ with respect to the least altered jaspilite, whereas the average $\delta^{18}O_{fluid}$ value for magnetite increased by about 6‰ compared to the respective $\delta^{18}O$ values. There is a clear decrease in both the $\delta^{18}O_{fluid}$ and $\delta^{18}O_{fluid}$ values from the paragenetically earliest to the latest hematite types.

Oxygen isotope data on iron ore of the Urucum mining district have been performed on whole-rock banded iron ores, because mineral separation was not feasible due to diminutive hematite grain sizes (Hoefs et al. 1987). Calculations of the isotopic compositions of the two minerals comprising the BIF (namely hematite and quartz; Hoefs et al. 1987) reveal $\delta^{18}O$ values ranging between 21.3 and 26.4‰ for quartz and 0.8 to 6.5‰ for hematite. Temperature calculations using different hematite-quartz fractionation curves (Blattner et al. 1983; Matthews et al. 1983) propose that hematite (and quartz) crystallised at temperatures between 250 °C and 280 °C, and that the involved fluid had a $\delta^{18}O_{fluid}$ composition above 10‰ (Hoefs et al. 1987). With respect to the exceptional high oxygen isotopic fluid compositions, Hoefs et al. (1987) interpret that the isotopic composition of the iron ore reflect the diagenetic and burial metamorphic conditions, that have largely obliterated primary depositional oxygen isotope signatures of the BIF quartz and hematite. In a later study, Yapp (1990) used the same oxygen isotope data from the Urucum mining district that were originally published by Hoefs et al. (1987) in order to calculate the depositional temperatures of the Urucum BIF and related $\delta^{18}O_{fluid}$ values using the hematite–water and quartz-water fractionation curves of Yapp (1990) and Knauth and Epstein (1976), respectively. In contrast to the relatively high crystallisation temperatures of Hoefs et al. (1987), the calculated depositional temperatures (Yapp 1990) are low (ranging between 0 and 35 °C) and oxygen isotope signatures of involved fluids range between −8.8 and 0.0‰. Yapp (1990) interpreted this data as an oxygen isotope signature that reflects the primary glaciogene environment present during the BIF deposition rather than an overprint and modification of earlier isotopic signatures by post-depositional processes.

Using the isotopic composition of reworked rocks for the investigation of their depositional conditions assumes that isotopic exchange during diagenesis/metamorphism was negligible and that the primary isotopic composition was preserved during the diagenesis/metamorphism of the rocks.

3.3.2 Carbon and Oxygen Isotope Data on Carbonates

Except one recently published carbon isotope study by Figueiredo e Silva et al. (2013), that was carried out on hydrothermal BIF-hosted carbonates of the Carajás mining province, most of carbon and oxygen isotope data on carbonates from Brazilian BIF-hosted ore (e.g. from the QF and Urucum) have been performed on (least-altered) BIFs/itabirite in order to gain information about the depositional conditions and environment during which the BIFs were formed and not to track post-depositional alteration and ore genetic processes. Importantly, in order to allow a conclusion to what extent carbon and oxygen isotope signature at the QF and Urucum may reflect post-depositional alteration and ore genetic processes, detailed carbon and oxygen isotope studies need to be conducted also on highly altered (e.g., hydrothermal carbonate) and iron enriched BIF ore.

Detailed carbon and oxygen isotope data (Spier et al. 2007) on carbonates from the Cauê dolomitic itabirite sequence of the Águas Claras deposit reveal negative $\delta^{13}C$ values ranging between −2.5 and −0.8‰ and $\delta^{18}O$ values ranging between −8.5 and −2.4‰. Similar $\delta^{13}C$ values have been obtained in another carbon and oxygen isotope study on the carbonate-rich itabirites of the Aguas Claras deposit (Morgan et al. 2013), which reveal $\delta^{13}C$ values between −1.6 and 2.4‰. Respective $\delta^{18}O$ values differ from the study of Spier et al. (2007) ranging between −14.4 and −10.2‰. These differences in the $\delta^{18}O$ composition in the study by Spier et al. (2007) and Morgan et al. (2013) may be attributed to different sample locations within the Aguas Claras deposit and, accordingly, to a different mineralogical content of the analysed samples. In fact, Morgan et al. (2013) reports the $\delta^{18}O$ values becoming less negative with increasing depth, which may be related to different degree of weathering of the analysed samples. However, Spier et al. (2007), as well as Morgan et al. (2013), interpreted the $\delta^{13}C$ (and partially also the $\delta^{18}O$) signatures of the QF itabirite to represent typical primary isotopic signatures of marine carbonates, which precipitated in a shallow marine environment from a seawater-hydrothermal fluid.

With respect to the Urucum mining district, carbon isotopic analyses have been carried out on carbonates from BIF, Mn-rich ore zones and carbonate-rich veinlets in Mn ore (Klein and Ladeira 2004) and further on dolomite-chert-hematite BIF of the Santa Cruz deposit (Angerer et al. 2016). The results show relatively homogenous negative $\delta^{13}C$ values ranging between −7.0 and −3.4‰ and are interpreted to reflect the typical glaciomarine environment in which the BIFs were deposited during Neoproterozoic times (Klein and Ladeira 2004; Angerer et al. 2016). The primary depositional carbon isotopic signature of the Neoproterozoic BIFs of the Urucum mining district has, therefore not been obliterated by post-depositional alteration and/or ore-forming fluids.

For the case of Carajás, carbonate samples were selected from different hydrothermal vein breccia types located mainly in the N4E deposit (Figueiredo e Silva et al. 2013) and consist of: (1) calcite and kutnahorite in discordant V1 veins located in the distal alteration zone within jaspilites; (2) kutnahorite and dolomite in V4 breccias in high-grade ore, with kutnahorite in textural equilibrium with an assemblage of martite and microplaty hematite; (3) calcite from amygdaloidal basaltic wallrock; and (4) calcite veins in hydrothermally altered basaltic wallrocks. Calcite-kutnahorite $\delta^{13}C$ and $\delta^{18}O$ values from the distal alteration zones show a large $\delta^{13}C$ range of −5.5 to −2.4‰, and a relatively narrow $\delta^{18}O$ range of 9.3 to 11.7‰. However, dolomite

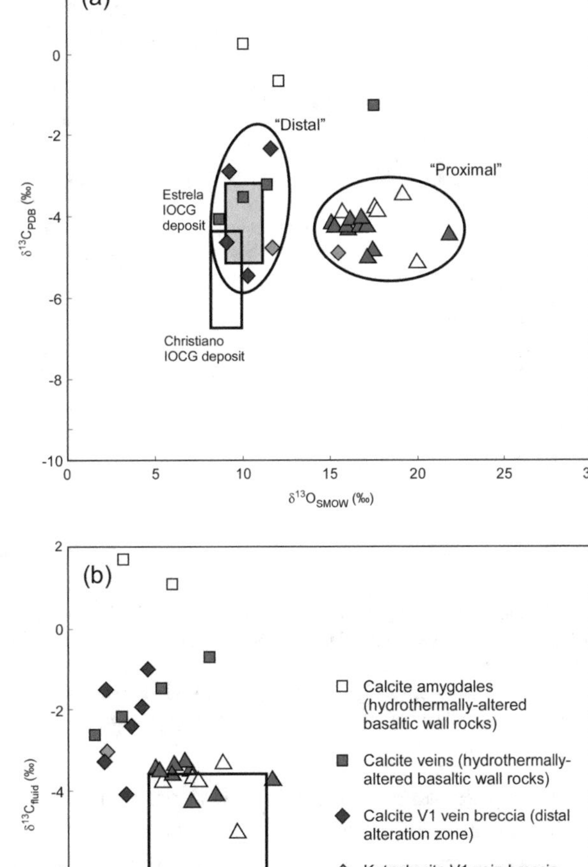

Fig. 7 Carbon and oxygen isotopes compositions of carbonates in calcite amygdale from the hydrothermally basaltic wall rock; calcite veins from distal alteration zone of basaltic wall rocks; calcite, kutnahorite from V1b vein-breccia in distal alteration zone in altered jaspilites, and dolomite and kutnahorite from V4 vein-breccia in proximal alteration zone. Symbols shown in (b) are also valid for (a). The two circles in (a) correspond to distal and proximal alteration zones, respectively; also shown are fields of oxygen and carbon isotopic composition of calcite from the Paleoproterozoic Cu (Mo–Au–Sn) Estrela deposit (Lindenmayer et al. 2005) (box in grey), and the Archean Cu(Au) Cristalino deposit (Ribeiro et al. 2007) (empty box). **b** Diagram showing calculated carbon and oxygen fluid values based on trapping temperatures based on fluid inclusions studies. Magmatic field according to Kyser and Kerrich (1990)

matrix breccias from the advanced hydrothermal zone, *i.e.*, ore, exhibit a wider $\delta^{18}O$ range from 15.1 to 21.8‰ and a more restricted $\delta^{13}C$ range from −5.0 to −3.9‰ (Fig. 7). This latter ranges point to a single carbon source of possible magmatic nature, whereas the larger $\delta^{18}O$ range suggests multiple oxygen sources.

3.3.3 Hydrogen Isotope Data on Fluid Inclusions

Hydrogen isotope data were obtained from fluid inclusions trapped in quartz from hydrothermal V1a, V2 and V3 vein types in the Serra Norte iron deposits, Carajás (Fig. 8). The hydrogen values from fluid inclusions trapped in V2 and V3 quartz veins are interpreted to reflect fluids synchronous with iron mineralization and are restricted to −20 to −43‰ with two outliers at −3 and −104‰ (Fig. 8).

3.3.4 Oxygen Isotope Data on Silicates and Oxides

Oxygen isotope analyses of iron oxides of the Noamundi mining area were performed on

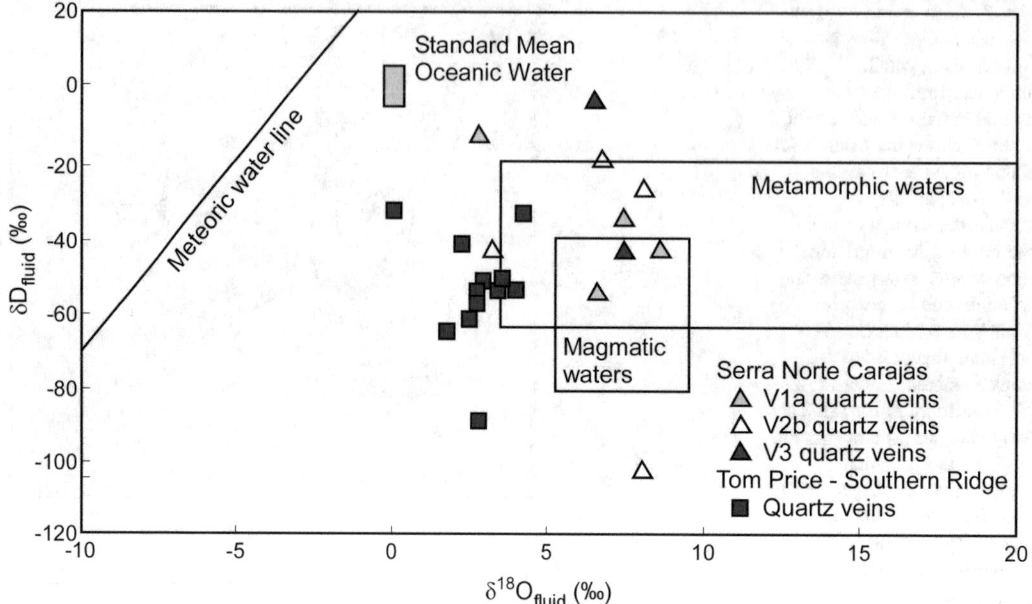

Fig. 8 Diagram displaying calculated $\delta^{18}O_{fluid}$ and δD isotope value for quartz and fluid inclusions, respectively from the Serra Norte iron deposits. Also presented are data from the Southern Ridge deposit at Tom Price in the Hamersley Province of Western Australia (Hagemann et al. 1999). The meteoric water line (Craig 1961), and fields of typical magmatic waters (Ohmoto 1986) and metamorphic waters as defined by Taylor (1974) are also shown.

cryptoplaty hematite from the hard iron ore and martite from partly mineralised BIFs and ore (Beukes et al. 2008). The oxygen isotope signature of cryptoplaty hematite ranges between −1.3 and 2.5‰. Relative to the ore-hosted cryptoplaty hematite, martite from the mineralised BIF and massive iron ore is depleted in ^{18}O and has $\delta^{18}O$ values ranging from −5.8 to −4.4‰ and from −6.7 to −2.6‰, respectively. The depleted $\delta^{18}O$ values of martite compared to cryptoplaty hematite is interpreted to reflect the formation of magnetite (as precursor mineral of martite) under higher temperatures than cryptoplaty hematite and a probable chemical inheritance of the oxygen isotope signature in martite from the magnetite precursor (Beukes and Gutzmer 2008). Additional interpretations about e.g. ore-forming fluid characteristics and sources have not been done so far by the means of the published oxygen isotope data (Beukes and Gutzmer 2008).

4 Interpretation of Oxygen, Carbon and Hydrogen Isotope Data with Respect to Hydrothermal Fluid Source(s), Fluid Flow and Processes

Oxygen and carbon isotope analyses on precursor BIF or, locally low-grade iron ore, have been completed since the early 1970’s with landmark papers by Becker (1971), Clayton et al. (1972) and Becker and Clayton (1976) on carbon and oxygen isotope ratios on BIF in the Hamersley of Western Australia. The first study on specifically iron ore was by Hoefs et al. (1982) who investigated oxygen isotope signatures on selected iron ore from the QF in Brazil. It was not until the end of the 1990’s that Hagemann et al. (1999) and Powell et al. (1999) reported oxygen isotopes on specific iron ore types from Hamersley iron deposits, and then Gutzmer et al. (2006) published a now seminal paper on oxygen isotopes of hematite and the genesis of high-grade BIF-hosted iron ores from a variety of deposits in South Africa, Brazil, India, and Australia.

These early studies were then followed up with detailed oxygen and carbon isotope investigations on iron oxides (both magnetite and hematite) formed during specific paragenetic ore stages in both the hypogene and supergene environment. Whereas the early oxygen isotope studies solely concentrated on the origin of the precursor BIF, the investigations at the end of the 1990's and thereafter also focused on the origin of the iron ore with respect to their crustal position, i.e., whether these ores are the product of hypogene hydrothermal alteration and/or weathering and supergene enrichment. In addition, other studies used the stable isotope (oxygen and carbon isotopes) signatures of iron ore and gangue carbonate to develop deposit-specific fluid flow paths and hydrothermal models for the origin of BIF-hosted iron ore.

With respect to the hydrogen isotope, only limited data is available, and only from fluid inclusions trapped in quartz veins. It is important to point out that hydrogen data may overlap both the magmatic and metamorphic fluid reservoirs, making these isotopic signature alone non-diagnostic.

The following sections will summarize the fundamental contributions of stable isotopes in the genesis of BIF-hosted iron ore but also provide a summary of the implications for the origin of iron ore in specific iron ore districts throughout the world.

4.1 Stable Isotope Analyses and Genetic Implications

Before about 1990's the most widely accepted BIF-hosted iron ore model was a combination of pure supergene enrichment via deep weathering processes during the Mesozoic to Paleogene (e.g. Dorr 1964) and diagenetic and metamorphic dehydration of Paleoproterozoic goethite via downwards penetration of supergene fluids up to the base of weathering, i.e., the classic CSIRO-AMIRA supergene-metamorphic model (Morris 1985). At the end of the 1990's these models were then challenged by the results of stable isotope and fluid inclusion studies on iron oxides and gangue minerals such as carbonates and quartz, which allowed the detection and quantification of a variety of hydrothermal fluids including basinal brines (Hagemann et al. 1999; Thorne et al. 2004), magmatic fluids (Figueiredo e Silva et al. 2008) and ancient meteoric waters (Hagemann et al. 1999; Thorne et al. 2004; Figueiredo e Silva et al. 2008; Cope et al. 2008). Much of the renewed interest in the genesis of BIF-hosted iron ore was sparked by the availability of deep diamond cores, intercepting fresh, hypogene iron ore, carbonates and quartz well beneath the weathering horizon, which required a re-think of the accepted geological models of the time. It was at this time that the application of stable isotopes, in combination with fluid inclusions, revealed fundamental new data and allowed new interpretations on the origin of BIF-hosted iron ore. Some of the fundamental new interpretations included:

1. The large spread of the oxygen isotope composition in iron oxides, in all of the deposits that have been analysed to date, and a general depletion of ^{18}O from BIF-hosted to iron ore-hosted iron oxides suggests the involvement of meteoric water ($\delta^{18}O_{fluid}$ between −10 and 2‰) at the time of mineralisation, i.e. during the upgrade of BIF to iron ore.
2. The presence of $\delta^{18}O_{fluid}$ data > 4‰ suggest the involvement of an additional hydrothermal fluid, besides meteoric water, which was determined to be of magmatic nature in the case of the greenstone belt-hosted Carajás-type iron ore deposits and basinal brines in the case of the sedimentary basin-hosted Hamersley-type deposits.
3. The spread in carbon isotope data indicates that carbon is not necessarily derived from carbonate-rich BIF units but also contains a component of "external" carbon; in the case of the Carajás deposit magmatic carbon, in the case of the Hamersley deposits sedimentary carbon from the underlying Wittenoom

dolomite, and in the case of Yilgarn deposits surface water (either seawater or meteoric water).

4.2 Regional Interpretations

Where stable isotope analyses on oxides or quartz associated with iron ore were conducted, the results played a significant role in the identification of hydrothermal fluid types and in the interpretation of fluid flow models for the respective iron ore deposits.

In the itabirite-hosted iron ore of the QF, two major trends are observed: (1) iron oxides that form the high-grade iron ore have lower $\delta^{18}O$ values than iron oxides from the protore itabirite; and (2) across the ore-forming paragenetic iron oxide sequence, a minor increase in the $\delta^{18}O$ values is recorded. Hensler et al. (2014) suggested that the strong decrease in $\delta^{18}O$ values is likely related to the influx of large volumes of meteoric water, with light oxygen isotopic signature, during the upgrade of itabirite to iron ore. The same interpretation is favoured for the Pic de Fon deposit in West Africa, where iron-enriched BIF records a substantial shift towards low $\delta^{18}O$ values compared to least-altered BIF (Cope et al. 2008). The development of large "feeder" systems, such as shear and fault zones, is likely fundamental for the circulation of those large fluid volumes and, therefore, for the significant geochemical processes (e.g. de-silicification and oxidation of magnetite to martite) during the formation of high-grade iron ore bodies. With respect to the iron ore, the minor increase of $\delta^{18}O$ values from paragenetically earlier to later iron oxides, thus from martite and granoblastic hematite to microplaty and specular hematite, may reflect the participation of an oxygen isotopic heavier water (e.g. modified heavier meteoric, metamorphic, or magmatic water) during the iron ore modification.

The large Tom Price and Paraburdoo iron ore deposits display a similar oxygen isotope trend (Fig. 2) as recorded in the QF and the Pic de Fon deposit, thus a strong $\delta^{18}O$ decrease from BIF-hosted magnetite to ore-hosted hematite. Meteoric water likely played also a significant role in shifting the $\delta^{18}O_{fluid}$ values of up to 8.5‰ for magnetite in unaltered BIF to as low as −8.9‰ for martite and platy hematite in high-grade iron ore of Tom Price, Paraburdoo and Pic de Fon (Thorne et al. 2014; Cope et al. 2008). Oxygen and hydrogen isotopes on quartz, hematite and fluid inclusions, respectively from the Southern Batter fault zone in the Southern Ridge deposit also show significant negative $\delta^{18}O_{fluid}$ values for quartz and hematite (up to −4‰) and, when paired with the hydrogen values, suggest significant surface water influx (likely meteoric water) into the fault zone and ore body (Hagemann et al. 1999). In the giant Carajás Serra Norte iron ore deposits there is a systematic decrease in $\delta^{18}O$ values of oxides from early to late paragenetic stages (Fig. 2), indicating a progressive increase in the influx of isotopically light, meteoric waters with time and proximity to the high-grade, hard-iron orebodies (Figueiredo e Silva et al. 2013). Figure 8 displays $\delta^{18}O_{fluid}$ versus δD isotope values for quartz and fluid inclusions, respectively. Some data points from the Serra Norte iron ore deposits fall within the field of primary magmatic fluids (Taylor 1974), with two samples that shifted the isotopic values towards lighter $\delta^{18}O$ and heavier δD compared to those that characterise average magmatic fluid. This shift may indicate an uncertain amount of meteoric water influx into the mineral system and subsequent fluid-rock reactions. Values of quartz samples from the Southern Ridge iron ore deposit in the Hamersley Province in Western Australia (Hagemann et al. 1999) are also shown in Fig. 8, and clearly display different populations in δD and $\delta^{18}O_{fluid}$ relative to Carajás, suggesting different fluid sources and/or fluid-rock reactions for these two datasets. Most of the Carajás data points fall within the magmatic/metamorphic water boxes, whereas the majority of the Southern Ridge data plot outside these boxes. The lack of lower ^{18}O values towards the meteoric water line in the Carajás samples could be explained by the absence of data from late-stage quartz veins (proximal alteration zone), which better reflect the influx of meteoric water.

Also, interestingly, the Carajás hydrogen isotope data set is more restricted than that for the Hamersley iron deposits (Hagemann et al. 1999; Fig. 8), suggesting that basinal brines were not involved (Figueiredo e Silva et al. 2008).

5 Implications for Exploration of BIF-Hosted Iron Ore Systems

The decrease of $\delta^{18}O$ values in iron oxides from the distal to proximal alteration zone including the orebody may be used as a geochemical vector and provides a potential exploration tool particularly for finding high-grade iron ore deposits (e.g. in the QF), in which hematite/magnetite are frequently the only mineral that can be readily analysed. In combination with magnetic and gravity imaging this geochemical vector may assist in the identification of extensions of existing iron orebodies and in finding new, deep-seated concealed iron ore mineralisation. The mapping and identification of major structural pathways, which facilitate the circulation of large volumes of meteoric water commonly reflected by light oxygen isotope values, will be crucial for identifying itabirite that is upgraded to high-grade iron ore.

The systematic decrease in $\delta^{18}O$ values of oxides from BIF-to iron ore-hosted iron oxides indicates a progressive increase in the influx of isotopically light, meteoric waters with time and proximity to the high-grade, hard iron orebodies.

6 Limitations and Problems of Stable Isotope Analyses on BIF-Related Iron Ores

One of the overarching features of BIF-hosted iron ores globally, irrespective of their exact genetic origin (i.e. entirely supergene versus mixed hypogene-supergene), is the relatively large range in $\delta^{18}O$ data recorded in the iron oxide fraction. This is particularly seen in bulk-ore hematite isotopic data, suggesting that large variations in $\delta^{18}O$ are probably the result of analyses of highly heterogeneous mixtures of multiple iron oxide generations. Such data are likely to obscure rather than illuminate the origin of the ores and their spatial and temporal development. Nevertheless, in general, lowest $\delta^{18}O$ values in iron oxides characterise the most enriched portions of the ores, and this appears to apply both in some mixed hypogene-supergene deposits (e.g. Pic de Fon) and in those interpreted to have formed by ancient supergene processes only (e.g. Sishen). By contrast, hydrothermally altered BIF adjacent to the massive ores contains often texturally discrete generations of hematite (e.g. specular, microplaty) that also display heavy oxygen isotopes. This evident isotopic trend may be associated with either multiple fluids (e.g. evolved meteoric fluids versus basinal brines) involved in iron remobilisation and enrichment during the complex history of the deposits, and/or isotopic exchange reactions in structurally-controlled, largely single-fluid systems under progressively different fluid/rock ratios and temperatures.

Irrespective of the causes for the isotopic variations in space within a given deposit or prospect, it is crucial to be able to constrain and ensure as much as possible the integrity of the stable isotope data obtained. In classic mixed hypogene-supergene deposits such as those of the Hamersley basin, a multi-isotopic approach (O, C, H) on different mineral species (oxides, carbonates and silicates) combined with fluid inclusion investigations has been demonstrated to provide robust enough constraints on the nature of the hydrothermal system and its development in space (e.g. Thorne et al. 2008). This approach, however, is limited with respect to carbon isotope analyses in supergene-only deposits, which are also typically carbonate-destructive.

The great majority of BIF-hosted deposits are of hybrid origin (i.e. hypogene ore overprinted by modern supergene enrichment or modified ancient supergene-only ores). These likely record a large variety of textural forms of iron oxides on an equally large variety of scales, both in the massive ores and in associated altered BIFs. Isotopic analyses of minerals from these ores, in the absence of rigorous petrographic and textural analysis and documentation of detailed paragenetic stages are likely to produce biased and misleading results. One must, therefore never underestimate the importance of detailed field work, petrography and mineral chemistry as the essential "backbone" of sound sample selection and robust geochemical and isotopic application in iron-ore exploration campaigns.

7 Conclusions

Stable oxygen and carbon isotope analyses in oxides, quartz and carbonate have significantly influenced the models for BIF-hosted iron ore worldwide.

Carbon and oxygen isotopes in carbonate associated with carbonate-rich, unaltered BIF and hydrothermal alteration carbonate suggest that $\delta^{13}C$ and $\delta^{18}O$ values of carbonates shift significantly from heavy to light $\delta^{13}C$ and light to heavy $\delta^{18}O$ values during hydrothermal alteration and mineralisation processes, thus suggesting an external (to the BIF) carbon source. Oxygen isotope studies on iron oxides from different ore deposits worldwide display a major decrease in the $\delta^{18}O$ values from BIF- to iron ore-hosted oxides and display the circulation and significance of large volumes of light oxygen isotope meteoric water during the upgrade of BIF to high-grade iron ore. Exploration of hidden BIF-hosted iron ore or extension of known iron ore bodies can be significantly assisted with stable isotope analyses of oxides, quartz and carbonate. Significant shifts in $\delta^{18}O$ and $\delta^{13}C$ values likely indicate massive hydrothermal fluid flow in the direction of high permeability fault zones, which in many deposits coincide spatially with zones of high-grade iron ore.

Acknowledgements The authors acknowledge David Huston who invited us to prepare this paper. This scientific work would have not been possible without the generous support of various mining and exploration companies including Rio Tinto, BHP, Atlas Iron, Vale, Vallourec, Kumba and Vectra. HT also likes to acknowledge "ASSMANG Ltd' for their past and ongoing support of the iron ore research group at Rhodes University. SGH, AH and RFS also thank Carlos Alberto Rosière, Phil Brown, Thomas Angerer and Paul Duuring for many stimulating discussions. SGH, AH and RFS also thank Lydia Lobato for many insightful discussions at the Lobato headquarters.

References

Angerer T, Duuring P, Hagemann SG, Thorne W, McCuaig TC (2014) A mineral system approach to iron ore in Archaean and Palaeoproterozoic BIF of Western Australia. Geol Soc London Spec Publ 393:81–115

Angerer T, Hagemann SG, Walde DHG, Halverson GP, Boyce AJ (2016) Multiple metal sources in the glaciomarine facies of the Neoproterozoic Jacadigo iron formation in the "Santa Cruz deposit", Corumbá, Brazil. Precambrian Res 275:369–393

Basson IJ, Koegelenberg C (2017) Structural controls on Fe mineralization at Thabazimbi Mine, South Africa. Ore Geol Rev 80:1056–1071

Becker RH (1971) Carbon and oxygen isotope ratios in iron-formation and associated rocks from the Hamersley Range of Western Australia and their implications. Unpublished PhD thesis, University of Chicago

Becker RH, Clayton RN (1976) Oxygen isotope study of a Precambrian banded iron formation, Hamersley Range, Western Australia. Geochim Cosmochim Acta 40:1153–1165

Beukes NJ, Gutzmer J (2008) Origin and paleoenvironmental significance of major iron formations at the Archean-Paleoproterozoic boundary. Rev Econ Geol 15:5–47

Beukes NJ, Gutzmer J, Mukhopadhyay J (2003) The geology and genesis of high-grade iron ore deposits. Trans Inst Mining Metall Sect B 112:B1–B25

Beukes NJ, Mukhopadyay J, Gutzmer J (2008) Genesis of high-grade iron ores of the Archean Iron Ore Group around Noamundi, India. Econ Geol 103:365–386

Bhattacharya HN, Ghosh KK (2012) Field and petrographic aspects of the iron ore mineralizations of Gandhamardan Hill, Keonjhor, Orissa and their genetic significance. J Geol Soc India 79:497–504

Blattner P, Braithwaite WR, Glover RB (1983) New evidence on magnetite oxygen isotope geothermometers at 175° and 112 °C in Wairakei steam pipelines (New Zealand). Chem Geol 41:195−204

Cabral AR, Rosière CA (2013) The chemical composition of specular hematite from Tilkerode, Harz, Germany: implications for the genesis of hydrothermal hematite and comparison with the Quadrilátero Ferrífero of Minas Gerais, Brazil. Miner Deposita 48:907–924

Clayton RN, O'Neil JR, Mayeda TK (1972) Oxygen isotope exchange between quartz and water. J Geophys Res 77:3057–3067

Cochrane N (2003) Phosphorus behavior during banded iron-formation enrichment. Unpub BSc thesis, University of Queensland, 86 p

Cope IL, Wilkinson JJ, Boyce AJ, Chapman JB, Herrington RJ, Harris CJ (2008) Genesis of the Pic de Fon iron oxide deposit, Simandou Range, Republic of Guinea, West Africa. Rev Econ Geol 15:197–222

Craig H (1961) Isotope variations in meteoric waters. Science 133:1702–1703

Dalstra HJ (2006) Structural controls of bedded iron ore in the Hamersley Province, Western Australia—an example from the Paraburdoo Ranges. Trans Inst Mining Metall Sect B 115:B139–B140

DNPM (2005) Brazilian Mineral Yearbook 2005:512, Brasilia, (DNPM)

Dorr JVN (1964) Supergene iron ores of Minas Gerais, Brazil. Econ Geol 59:1203–1240

Duuring P, Angerer T, Hagemann G (2017a) Iron ore deposits of the Yilgarn craton. Austr Inst Mining Metall Mon 32:181–184

Duuring P, Teitler Y, Hagemann SG (2017b) Iron ore deposits in the Pilbara craton. Austr Inst Mining Metall Mon 32:345–350

Evans K, McCuaig T, Leach D, Angerer T, Hagemann S (2013) Banded iron formation to iron ore: a record of the evolution of Earth environments? Geology 41:99–102

Figueiredo e Silva RC, Lobato LM, Rosière CA, Hagemann S, Zucchetti M, Baars FJ, Morais R, Andrade I (2008) Hydrothermal origin for the jaspilite-hosted, giant Serra Norte iron ore deposits in the Carajás mineral province, Para State, Brazil. Rev Econ Geol 15:255–290

Figueiredo e Silva RC, Hagemann SG, Lobato LM, Rosière CA, Banks DA, Davidson GJ, Vennemann TW, Hergt JM (2013) Hydrothermal fluid processes and evolution of the giant Serra Norte jaspilite-hosted iron ore deposits, Carajás Mineral Province, Brazil. Econ Geol 108:739–779

Freitas BT, Warren LV, Boggiani PC, De Almeida RP, Piacentini T (2011) Tectono-sedimentary evolution of the Neoproterozoic BIF-bearing Jacadigo Group, SW-Brazil. Sed Geol 238:48–70

Graf JL Jr, O'Connor EA, van Leeuwen P (1994) Rare earth element evidence of origin and depositional environment of Late Proterozoic ironstone beds and manganese-oxide deposits, SW Brazil and SE Bolivia. J S Am Earth Sci 7:115–133

Gutzmer J, Mukhopadhyay J, Beukes NJ, Pack A, Hayashi K, Sharp ZD (2006) Oxygen isotope composition of hematite and genesis of high-grade BIF-hosted iron ores. Geol Soc Am Mem 198:257–268

Hagemann SG, Barley ME, Folkert SL (1999) A hydrothermal origin for the giant BIF-hosted Tom Price iron ore deposit. In: Stanley CJ (ed) Mineral deposits, processes to processing. Balkema, Rotterdam, pp 41–44

Hagemann SG, Angerer T, Duuring P, Rosière CA, Figueiredo e Silva RC, Lobato L, Hensler AS, Walde D (2016) BIF-hosted iron mineral system: a review. Ore Geol Rev 76:317–359

Hagemann SG, Angerer T, Duuring P (2017) Iron ore systems in Western Australia. Austr Inst Mining Metall Mon 32:59–62

Hensler AS, Hagemann SG, Brown PE, Rosiere CA (2014) Using oxygen isotope chemistry to track hydrothermal processes and fluid sources in BIF-hosted iron ore deposits in the Quadrilátero Ferrífero, Minas Gerais, Brazil. Miner Deposita 49:293–311

Hensler AS, Hagemann SG, Rosière CA, Angerer T, Gilbert S (2015) Hydrothermal and metamorphic fluid-rock interaction associated with hypogene "hard" iron ore mineralisation in the Quadrilátero Ferrífero, Brazil: implications from in-situ laser ablation ICP-MS iron oxide chemistry. Ore Geol Rev 69:325–351

Hoefs J, Müller G, Schuster AK (1982) Polymetamorphic relations in iron ores from the Iron Quadrangle, Brazil: the correlation of oxygen isotope variations with deformation history. Contrib Miner Petrol 79:241–251

Hoefs J, Müller G, Schuster KA, Walde D (1987) The Fe-Mn ore deposits of Urucum, Brazil: an oxygen isotope study. Chem Geol 65:311–319

Klein C (2005) Some Precambrian banded iron-formations (BIFs) from around the world: their age, geologic setting, mineralogy, metamorphism, geochemistry, and origin. Am Miner 90:1473–1499

Klein C, Ladeira EA (2004) Geochemistry and mineralogy of Neoproterozoic banded iron-formations and some selected, siliceous manganese formations from the Urucum district, Mato Grosso do Sul, Brazil. Econ Geol 99:1233–1244

Knauth LP, Epstein S (1976) Hydrogen and oxygen isotope ratios in nodular and bedded cherts. Geochim Cosmochim Acta 40:1095–1108

Kyser TK, Kerrich R (1990) Geochemistry of fluids in tectonically active crustal regions. Mineral Assoc Can Short Course Ser 18:133–230

Land JS (2013) Genesis of BIF-hosted hematite iron ore deposits in the central part of the Maremane Anticline, Northern Cape Province, South Africa. Unpublished MSc thesis, Rhodes University, 112 p

Lindenmayer ZG, Fleck A, Gomes CH, Santos ABZ, Caron R, Paula FC, Laux JH, Pimentel MM, Sardinha AS (2005) Caracterização geológica do Alvo Estrela (Cu-Au), Serra dos Carajás, Pará. In: Marini OJ, Queiroz ET, and Ramos BW (eds) Caracterização de depósitos minerais em distritos mineiros da Amazônia. DNPM/CT-Mineral/FINEP/ADIMB, Brasília, pp 157–226

Lobato LM, Figueiredo e Silva RC, Hagemann S, Thorne W, Zucchetti M (2008) Hypogene alteration associated with high-grade BIF-related iron ore. Rev Econ Geol 15:107–128

Matsuhisa Y, Goldsmith JR, Clayton RN (1979) Oxygen isotopic fractionation in the system quartz-albite-anorthite-water. Geochim Cosmochim Acta 43:1131–1140

Matthews A, Goldsmith J, Clayton RN (1983) On the mechanisms and kinetics of oxygen isotope exchange in quartz and feldspars at elevated temperatures and pressures. Geol Soc Am Bull 94:396–412

Ministerio de Minas e Energia (2009) Quaresma LF, Produto 09—Minério de Ferro, Relatório Técnico 18, Perfil da Mineração de Ferro

Moore JM, Kuhn BK, Mark DF, Tsikos H (2011) A sugilite-bearing assemblage from the Wolhaarkop breccia, Bruce iron-ore mine, South Africa: evidence for alkali metasomatism and ^{40}Ar–^{39}Ar dating. Eur J Miner 23:661–673

Morgan R, Orberger B, Rosière CA, Wirth R, Carvalho CdM, Bellver-Baca MT (2013) The origin of coexisting carbonates in banded iron formations: a micromineralogical study of the 2.4 Ga Itabira Group, Brazil. Precambrian Res 224:491–511

Morris RC (1985) Genesis of iron ore in banded ironformation by supergene and supergene-metamorphic processes—a conceptual model. In: Wolf KH (ed) Handbook of strata-bound and stratiform ore deposits. Elsevier, Amsterdam, pp 73–235

Morris RC, Kneeshaw M (2011) Genesis modelling for the Hamersley BIF-hosted iron ores of Western Australia: a critical review. Austr J Earth Sci 58:417–451

Netshiozwi ST (2002) Origin of high-grade hematite ores at Thabazimbi Mine, Limpopo Province, South Africa. Unpublished MSc thesis, Rand Afrikaans University, 135 p

Oliveira LARd, Rosière CA, Rios FJ, Andrade S, Moraes Rd (2015) Chemical fingerprint of iron oxides related to iron enrichment of banded iron formation from the Cauê Formation—Esperança Deposit, Quadrilátero Ferrífero, Brazil: a laser ablation ICP-MS study. Braz J Geol 45:193–216

Ohmoto H (1986) Stable isotope geochemistry of ore deposits. Rev Mineral 16:491–560

Papadopoulos V (2016) Mineralogical and geochemical constraints on the origin, alteration history and metallogenic significance of the Manganore ironformation, Northern Cape Province, South Africa. Unpublished MSc thesis, Rhodes University, 201 p

Piacentini T, Vasconcelos PM, Farley KA (2013) $^{40}Ar/^{39}Ar$ constraints on the age and thermal history of the Urucum Neoproterozoic banded iron-formation, Brazil. Precambrian Res 228:48–62

Powell CM, Oliver NHS, Li Z-X, Martin DM, Ronaszeki J (1999) Synorogenic hydrothermal origin for giant Hamersley iron oxide ore bodies. Geology 27:175–178

Ribeiro A, Suita MTF, Sial AN, Fallick T, Ely F (2007) Geoquímica de isótopos estáveis (C, S, O) das rochas encaixantes e do minério de Cu(Au) do depósito Cristalino, Província Mineral de Carajás, Pará [abs.]: Congresso Brasileiro de Geoquímica, XI, Atibaia, Brazil vol 1, pp 1–4

Ridley JL (1997) Syn-metamorphic gold deposits in amphibolite and granulite facies rocks. Mitt Österr Miner Ges 142:101–110

Rosière CA, Baars FJ, Seoane JCS, Lobato LM, da Silva LL, de Souza SRC, Mendes GE (2006) Structure and iron mineralisation of the Carajás Province. Trans Inst Mining Metall Sect B 115:B126–B136

Rosière CA, Spier CA, Rios FJ, Suckau V (2008) The itabirites of the Quadrilátero Ferrífero and related high-grade iron ore deposits: an overview. Rev Econ Geol 15:223–254

Roy S, Venkatesh AS (2009) Mineralogy and geochemistry of banded iron formation and iron ores from eastern India with implications on their genesis. J Earth Sys Sci 118:619–641

Smith AJB, Beukes NJ (2016) Palaeoproterozoic banded iron formation-hosted high-grade hematite iron ore deposits of the Transvaal Supergroup, South Africa. Episodes 6:269–284

Spier CA, de Oliveira SMB, Sial AN, Rios FJ (2007) Geochemistry and genesis of the banded iron formations of the Caue Formation, Quadrilátero Ferrífero, Minas Gerais, Brazil. Precambrian Res 152:170–206

Spier CA, de Oliveira SMB, Rosiere CA, Ardisson JD (2008) Mineralogy and trace-element geochemistry of the high-grade iron ores of the Aguas Claras Mine and comparison with the Capao Xavier and Tamandua iron ore deposits, Quadrilátero Ferrifero, Brazil. Miner Deposita 43:229–254

Taylor HP Jr (1974) The application of hydrogen isotope problems of hydrothermal alteration and ore deposition. Econ Geol 69:843–883

Taylor D, Dalstra HJ, Harding AE, Broadbent GC, Barley ME (2001) Genesis of high-grade hematite orebodies of the Hamersley province, Western Australia. Econ Geol 96:837–873

Thorne WS, Hagemann SG, Barley M (2004) Petrographic and geochemical evidence for hydrothermal evolution of the North Deposit, Mt Tom Price, Western Australia. Miner Deposita 39:766–783

Thorne WS, Hagemann SG, Webb A, Clout J (2008) Banded iron formation-related iron ore deposits of the Hamersley Province, Western Australia. Rev Econ Geol 15:197–222

Thorne W, Hagemann S, Vennemann T, Oliver N (2009) Oxygen isotope compositions of iron oxides from high-grade BIF-hosted iron ore deposits of the central Hamersley Province, Western Australia: constraints on the evolution of hydrothermal fluids. Econ Geol 104:1019–1035

Thorne WS, Hagemann SG, Sepe D, Dalstra HJ, Banks DA (2014) Structural control, hydrothermal alteration zonation, and fluid chemistry of the concealed, high-grade 4EE iron orebody at the Paraburdoo 4E deposit, Hamersley Province, Western Australia. Econ Geol 109:1529–1562

Trompette R, de Alvarenga CJS, Walde D (1998) Geological evolution of the Neoproterozoic Corumbá graben system (Brazil): depositional context of the stratified Fe and Mn ores of the Jacadigo Group. J S Am Earth Sci 11:587–597

Urban H, Stribrny B, Lippolt HJ (1992) Iron and manganese deposits of the Urucum District, Mato Grosso do Sul, Brazil. Econ Geol 87:1375–1392

Van Deventer WF (2009) Textural and geochemical evidence for a supergene origin of the Paleoproterozoic high-grade BIF-hosted iron ores of the Maremane Dome, Northern Cape Province, South Africa. Unpublished MSc thesis, University of Johannesburg, 107 p

Viehmann S, Bau M, Bühn B, Dantas EL, Andrade FRD, Walde DHG (2016) Geochemical characterisation of Neoproterozoic marine habitats: evidence from trace elements and Nd isotopes in the Urucum iron and manganese formations, Brazil. Precambrian Res 282: 74–96

Walde DH, Hagemann SG (2007) The Neoproterozoic Urucum/mutun Fe and Mn deposits in W-Brazil/SE-

Bolivia: assessment of ore deposit models. Z Dtsch Ges Geowiss 158:45–55
Yapp CJ (1990) Oxygen isotopes in iron (III) oxides 1. Mineral-Water Fractionation Factors. Chem Geol 85:329–335
Zheng Y-F (1991) Calculation of oxygen isotope fractionation in metal oxides. Geochim Cosmochim Acta 55:2299–2307
Zheng Y-F (1995) Oxygen isotope fractionation in magnetites: structural effect and oxygen inheritance. Geochim Cosmochim Acta 121:309–316

Iron Isotopes Applied to BIF-Hosted Iron Deposits

Lydia Maria Lobato, Rosaline Cristina Figueiredo e Silva, Thomas Angerer, Mônica de Cássia Oliveira Mendes, and Steffen G. Hagemann

Abstract

Published and unpublished iron isotope data from banded iron formations (BIF) and their BIF-hosted hypogene (hydrothermal) iron ores from the Quadrilátero Ferrífero (itabirites), Corumbá, and Carajás iron districts in Brazil, as well as from the Hamersley province in Australia are presented and discussed. BIF constitutes a typically thinly bedded or laminated chemical sedimentary rock, with $\geq$ 15% Fe and layers of chert, chalcedony, jasper, or quartz, whereas itabirite is considered a laminated, metamorphosed iron formation rich in iron oxides, which may contain carbonate minerals, amphiboles, and abundant quartz. For the Paleoproterozoic Quadrilátero Ferrífero district, the range in δ^{56}Fe values of hypogene iron ores is similar to that of the metamorphosed BIFs, and iron isotope variations are better distinguished in different regional deformational domains. Light isotopic compositions dominate in the low deformation domain (δ^{56}Fe = −0.42 ± 0.12 to 0.29 ± 0.04‰), whereas the eastern, high-strain domain is characterized by heavy values (δ^{56}Fe = −0.09 ± 0.08 to 0.37 ± 0.06‰; Mendes et al., Mineral Deposita 52:159–180, 2017). Iron isotope composition for the Neoproterozoic iron formations of the Corumbá region (hematitic, dolomite-rich: −1.83 and −0.83‰; cherty-hematite: δ^{56}Fe −0.49‰) are controlled by: (1) primary seawater signature, (2) microbial activity, and (3) supergene goethite alteration. Hydrothermal alteration is reflected in the oxygen isotope data, but apparently not in the iron isotope fractionation. Iron and oxygen isotope pairing shows that δ^{56}Fe values increase, while δ^{18}O values decrease. In the Archean jaspilites of Carajás, hypogene ores tend to display lighter δ^{56}Fe values than their host BIF counterparts. Also, there is a correlation between coupled iron and oxygen isotope values that is clearer towards lighter isotopic values, especially for δ^{18}O. In the Paleoproterozoic Hamersley deposits, correlation between δ^{18}O and δ^{56}Fe values suggests a direct correlation of both isotope systems during low-grade, greenschist-facies metamorphism. On the other hand, despite the

L. M. Lobato (✉) · R. C. Figueiredo e Silva
Universidade Federal de Minas Gerais, Centro de Pesquisas Prof. Manoel Teixeira da Costa-Instituto de Geociências, Av. Antônio Carlos 6627, Campus Pampulha, Belo Horizonte, MG 31270-901, Brazil
e-mail: llobato.ufmg@gmail.com

T. Angerer
Institute of Mineralogy and Petrography, University of Innsbruck, Innrain 52, 6020 Innsbruck, Austria

M. de Cássia Oliveira Mendes
CDM—Centro de Desenvolvimento Mineral—Vale S.A, Santa Luzia, MG 33.040-900, Brazil

S. G. Hagemann
Centre for Exploration Targeting, School of Earth Sciences, University of Western Australia, Crawley, WA 6009, Australia

D. Huston and J. Gutzmer (eds.), *Isotopes in Economic Geology, Metallogenesis and Exploration*, Mineral Resource Reviews, https://doi.org/10.1007/978-3-031-27897-6_13

evident shift to negative $\delta^{18}O$ values and apparent preservation of the metamorphic $\delta^{56}Fe$ signature, iron ore and hydrothermally modified BIF show a correlation between $\delta^{18}O$ and $\delta^{56}Fe$ values. In contrast, in supergene-modified samples a negative correlation is apparent. The Carajás (+1.24 to + 0.44; one sample − 0.30‰) and Hamersley (+ 1.02 to − 0.29‰) hypogene ores display $\delta^{56}Fe$ in a similar interval, reaching positive values, whereas ores from the Quadrilátero Ferrífero show a tendency towards lower values (to − 0.80‰). This review indicates that the application of iron isotopes in exploration is presently limited mainly due to the restricted dataset available for ore samples. Nevertheless, and despite all local differences, there is a general tendency for hypogene ores to display moderately lighter $\delta^{56}Fe$ values for all deposits compared to precursor BIF. In contrast, a strong supergene imprint in ore leads to moderately heavier $\delta^{56}Fe$ values. As more data become available, and if these trends are confirmed, the use of this tool may be valuable in the future, for instance to decipher the hypogene or supergene origin of specific ore zones, and as a consequence the probable depth extension or interpretation of concealed, deep orebodies.

Keyword

Iron isotopes · Iron deposits · Quadrilátero Ferrífero · Corumbá · Carajás · Hamersley

1 Introduction

Stable isotope geochemistry has become extensively applied to geosciences in the last decades, with applications in a wide range of geological processes, including mineral deposits (e.g., Horn et al. 2006; Shanks 2014). Since the development of the high-resolution mass spectrometers with multiple collectors (MC-ICP-MS), analysis of non-traditional stable isotopes, such as iron isotopes, has become possible. In particular, iron isotope geochemistry is quite conspicuous because of its relatively high concentrations on Earth, making it a good tool for tracing the geochemical cycle of Fe (e.g., Johnson and Beard 2006; Dauphas et al. 2017).

Significant iron isotopic variability of natural materials is demonstrated by studies conducted in different geological environments (e.g., Levasseur et al. 2004; Williams et al. 2005b; Horn et al. 2006; Johnson and Beard 2006; Poitrasson 2006; Teng et al. 2008; Weyer 2008; Schoenberg et al. 2009; Xiaohu Li et al. 2017). Banded iron formations (BIF), for example, show a relatively large range in iron isotope compositions, varying from − 2.5 to + 2.2‰ in $\delta^{56}Fe$. In the study of BIF, these variations help discern iron sources and construct models for BIF genesis (e.g., Johnson et al. 2003, 2008a, b; Frost et al. 2007; Whitehouse and Fedo 2007; Fabre et al. 2011; Halverson et al. 2011; Planavsky et al. 2012; Czaja et al. 2013; Cox et al. 2016).

Despite the increase of iron isotope measurements for different geological settings, and for BIF themselves, there is little systematic application to hydrothermal BIF-hosted iron ores (e.g., Mendes et al. 2017). Examples include studies by Markl et al. (2006) of Fe-bearing minerals in the Variscan Schwarzwald hydrothermal vein deposits (Germany); magnetite in Kiruna-type (IOA Weis 2013; Weis et al. 2013; Bilenker 2015; Bilenker et al. 2016; Childress et al. 2016; Knipping et al. 2019; Troll et al. 2019); magnetite and sulfides in Archean Carajás IOCG deposits, Brazil (Santiago 2016); magnetite in the Chinese Xinqiao skarn deposit (Wang et al. 2011); and BIF-hosted magnetite ores of the Anshan-Benxi area in the North China craton (Dai et al. 2017). Limited data from hydrothermal iron deposits have shown that iron isotope ratios are partly controlled by hydrothermal fluid circulation (Horn et al. 2006; Markl et al. 2006), and by the hydrothermal mineralization conditions (Mendes et al. 2017).

The interaction of primary hematite with hydrothermal fluids in most cases enriches the lighter isotopes in the secondary hematite. For siderite, isotopes fractionate in the opposite direction (Markl et al. 2006). Hydrothermal fluids therefore preferentially scavenge the light iron from iron-bearing minerals from surrounding rocks (e.g., Rouxel et al. 2003; Markl et al. 2006). Consequently, if a hydrothermal fluid percolates through protore BIF, the isotope composition of the hydrothermal fluid and its eventual ores are expected to be relatively lower in iron isotope values, whereas the residual BIF tends towards higher δ^{56}Fe ratios. Precipitation of Fe-bearing minerals results in the iron isotope composition of residual fluids evolving with time. Precipitation of Fe(III) minerals leaves the remaining fluid enriched in light isotopes, resulting in progressively lighter precipitates from the same fluid.

The complexity of isotopic fractionation during hydrothermal mineralization was emphasized by Horn et al. (2006). The isotopic composition suggests that fractionation can take place during the precipitation of iron phases, with variation of the δ^{56}Fe values even on the scale of a single hematite crystal (δ^{56}Fe = − 0.5‰, in the core and − 1.8‰, in the rim), reflecting the relative increase in content of the lighter isotope in the residual fluid (see also Steinhoefel et al. 2009).

Iron isotopes potentially identify precursors to hydrothermal alteration products and their associated ores. During hydrothermal processes, isotopic fractionation and isotopic composition of iron ore minerals may indicate iron sources (i.e., sedimentary, magmatic or metamorphic, or fluid evolution), allowing the reconstruction of precipitation and redox processes (Markl et al. 2006) that take place during ore formation and/or alteration.

This paper reviews and presents published iron isotope data from the Brazilian Quadrilátero Ferrífero (QF), Urucum (in Corumbá) and Carajás (jaspilite) iron districts (Fig. 1a–c). New data from Corumbá and Carajás, as well as from the Hamersley (Fig. 1d) province in Australia (BIF and iron ores), are presented. New and previously reported oxygen isotope results are also introduced for some of the samples in order to investigate the behavior and potential coupling with iron isotope. The main goal of this contribution is to show the variation of iron isotopes from the protore BIF to their associated high-grade iron ores, as well as compare and contrast iron isotopes in the various iron ore districts. A brief review on iron isotope fractionation is included. Some future applications of iron isotope studies to mineral systems are suggested, with ideas about applying iron isotopes in the context of exploration.

2 Brief Summary on Iron Isotopes

In this section, we briefly review the most relevant information concerning iron fractionation in natural environments. Since the goal of this contribution is to evaluate how iron isotope variations take place during the hydrothermal upgrade of iron formations, the focus is directed mainly towards this latter subject.

Several reviews have been published over the past 15 years covering different aspects of iron isotope geochemistry (e.g., Beard and Johnson 2004; Johnson et al. 2004, 2008a; Dauphas and Rouxel 2006; Anbar and Rouxel 2007; Dauphas et al. 2017). In the present review, emphasis is given to the development of iron isotope systematics over the past decade, although still highlighting the important discoveries made before that time. The most important aspects in the past 10 years include the recognition that igneous rocks and minerals can display iron isotopic variations; a better understanding of the ancient iron marine cycle; and the first extensive use of iron isotope measurements in modern seawater to better understand the modern marine iron cycle. The past decade has also seen a large increase in the number of laboratory experiments aimed at determining equilibrium and kinetic fractionation factors needed to interpret iron isotope variations in natural samples.

Before examining iron isotope composition of iron ores, it is important to consider some aspects of the isotopic compositions of their protolith BIF. Iron is delivered to the depositional basins mainly through submarine hydrothermal vent fluids, with δ^{56}Fe composition being roughly –0.5 to 0.0‰ (Beard et al. 2003; Severmann

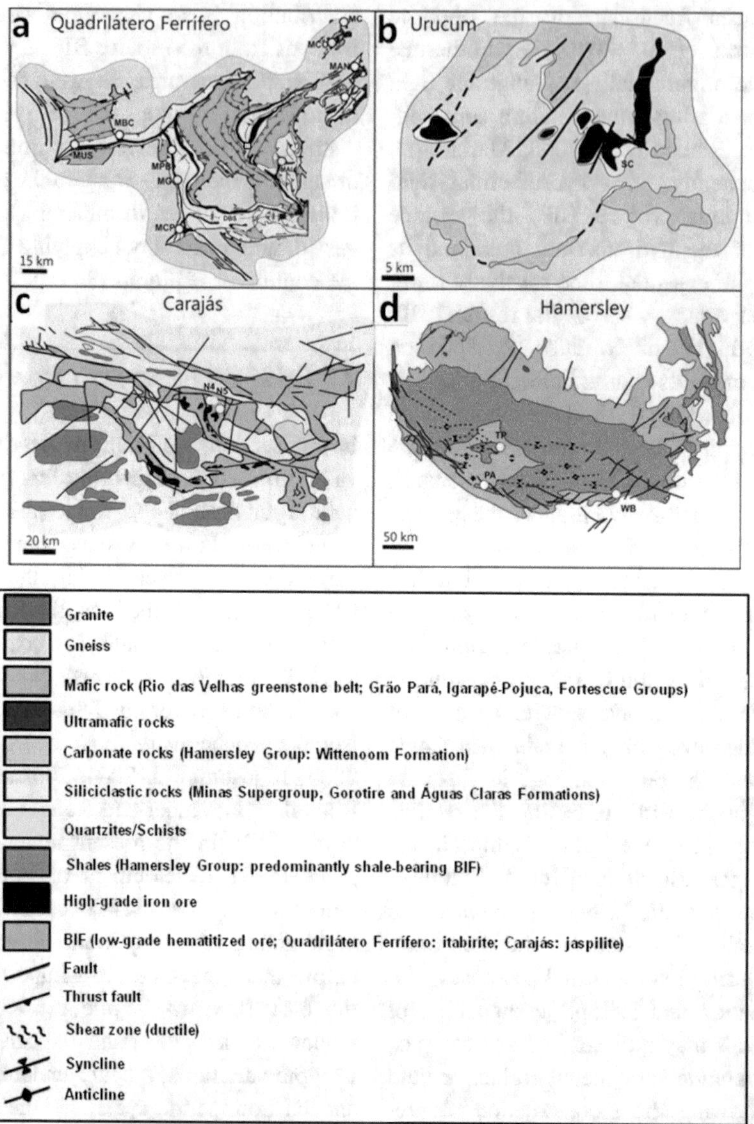

Fig. 1 Geological maps with sampling location for BIF-hosted iron ore districts and deposits (modified after Hagemann et al. 2016). **a** Quadrilátero Ferrífero (QF), Brazil (after Dorr 1969; Rosière et al. 2008). The open circles indicate locations of the Usiminas (MUS), Bocaina, (MBC), Pau Branco (MPB), Várzea do Lopes (MG), Casa de Pedra (MCP), Alegria (MAL), Morro Agudo (MAG), Andrade (MAN), Conceição (MCO), Cauê (MC) Mina de Fábrica (MF), and Dom Bosco Syncline (DBS) deposits. The yellow square indicates the southern Ouro Fino Syncline (OFS) portion of Gandarela Sycline (GS) (see Sampaio et al. 2018). **b** Urucum (Corumbá region), Brazil (after Walde and Hagemann 2007). Open circle with SC indicates the Santa Cruz deposit. **c** Serra Norte, Carajás mineral province, Brazil (after Figueiredo e Silva et al. 2008). Open circles indicate the N4 and N5 iron deposits. **d** Hamersley, Australia (after Angerer et al. 2012). Open cirlces indicate the TP Tom Price (TP), Paraburdoo (PA), and Whaleback (WB) deposits

et al. 2004; Johnson et al. 2008a). The deposition of iron oxides/hydroxides is influenced by the availability of oxygen in the ocean, which is reflected in the δ^{56}Fe signatures of iron formations (Johnson et al. 2008a; Planavsky et al. 2012). Quantitative oxidation of Fe(II) would result in zero fractionation between the hydrothermal iron source and iron oxyhydroxide minerals. In contrast, partial Fe(II) oxidation produces ferric oxides with positive δ^{56}Fe values (Johnson et al. 2008b; Planavsky et al. 2012), with consequent depletion of the ferrous iron in the iron pool (Dauphas et al. 2004; Rouxel et al. 2005; Planavsky et al. 2012). Archean BIFs, for example, tend to have positive δ^{56}Fe signatures (Whitehouse and Fedo 2007; Fabre et al. 2011; Czaja et al. 2013; Busigny et al. 2014) due to non-quantitative fractionation in response to limited availability of free oxygen. Similar signatures are also observed for the Neoproterozoic BIFs deposited during glaciation episodes (Busigny et al. 2018). With the rise of atmospheric oxygen, during the Great Oxidation Event (GOE) at *ca.* 2.4 Ga (eg. Holland 2006 and references therein), the δ^{56}Fe signatures of Paleoproterozoic BIF become similar to values of the hydrothermal source due to quantitative fractionation of iron (see recent discussion in e.g., Lantink et al. 2018).

When considering the iron isotope fractionation, both biological and abiological mechanisms must be taken into account. The largest iron isotopic fractionations are associated with biologically mediated or abiologically induced, redox transformations between ferrous and ferric iron (Johnson et al. 2002, 2008b; Beard and Johnson 2004; Anbar et al. 2005). Mass-dependent Rayleigh-type fractionation may arise during progressive removal of relatively ^{56}Fe-enriched iron oxides from the iron reservoir, resulting in the relative enrichment of ^{54}Fe in the residual pool (Rouxel et al. 2005; Planavsky et al. 2012).

Biologically induced dissimilatory iron reduction (DIR) produces some of the largest natural fractionations of stable iron isotopes (Crosby et al. 2007). This mechanism is driven by a bacterial metabolic process, with transference of electrons to Fe-oxyhydroxides (Lovley et al. 2004). The DIR produces aqueous ferrous iron ($Fe(II)_{aq}$) that is isotopically up to ~3‰ lighter than the ferric iron substrate (Beard et al. 1999, 2003; Johnson et al. 2005; Crosby et al. 2005, 2007; Kunzmann et al. 2017).

At higher temperatures, fractionation typical of hydrothermal fluids has been evaluated by experimental procedures through the investigation of the fractionation of iron between saline solutions and hematite (Saunier et al. 2011). These experiments demonstrate the absence of fractionation between fluid and precipitated hematite at 200 °C, and negative fractionation at even higher temperatures ($\Delta^{57}Fe_{fluid\text{-}hematite} \approx -0.5$‰, at 300 °C), suggesting preferential transport of isotopically light iron at temperatures > 200ºC. At 800 °C, Sossi and O'Neil (2017) show that experimentally determined fractionation factors between minerals and aqueous $FeCl_2$ fluid have a minimal fractionation between $^{VI}Fe^{2+}$ and the fluid. If this represents speciation of iron in fluids exsolving from magmas, the fractionation between them should be small, unless the iron is hosted in magnetite, with both ferrous and ferric iron.

3 Geological Background of Selected Iron-Ore Districts

3.1 Quadrilátero Ferrífero

The Quadrilátero Ferrífero (QF) province (Fig. 1a), located in Minas Gerais state, plays a significant role in the Brazilian mining industry, hosting important gold and iron ore deposits, besides manganese and bauxite resources. The Minas Gerais iron ore corresponded to *ca.* 52% of the total Brazil´s iron ore exports in 2014 (177.675 t, IBRAM 2015 report, http://www.ibram.org.br/sites/1300/1382/00005836.pd).

Iron orebodies are hosted in Paleoproterozoic banded iron formations (BIF) of the regionally extensive Cauê Formation (Itabira Group, Minas Supergroup: Dorr 1969; Fig. 1 of Mendes et al. 2017). The Minas Supergroup consists of a platformal sedimentary succession, including clastic (Caraça, Piracicaba and Sabará groups) and chemical (Itabira Group) units metamorphosed to low greenschist facies (Dorr 1969). The metamorphosed and oxidized Cauê BIFs are commonly referred to as itabirites, and are composed predominantly of iron oxides

Fig. 2 Photomicrographs, illustrating examples of different types of iron ores from districts discussed in the text. **a** Massive iron ore, formed predominantly by anhedral(A) (± subhedral(S)) hematite, Pau Branco deposit, Quadrilátero Ferrífero (QF). **b** Specularite-bearing, schistose iron ore from a high-strain ore zone, Várzea do Lopes deposit, QF. **c** Sample of hematite mud with compact, laminated texture and local chert lenses. **d** Brecciated martite-hematite ore from the N4W deposit, Carajás, where microcrystalline hematite is preserved. **e** Brecciated euhedral-tabular (ET) hematite ore from the N5E deposit, Carajás. **f** and **g** From Mt. Tom Price deposit, Hamersley province, the former showing aggregates of microplaty hematite (MpH) in high-grade ore, and the latter representing martite-microplaty hematite ore. Hem = hematite; Mt = martite; MHem = microcrystalline hematite; MpHem = microplaty hematite. Photomicrographs d, f, g taken under transmitted light; all others reflected light

(hematite, magnetite and martite, after magnetite) and quartz. Quartz itabirite is the most common, followed by calcareous (dolomite and ankerite, with minor calcite and siderite) and amphibolitic itabirites, the latter resulting from the metamorphism of calcareous itabirites (Rosière et al. 2008).

The QF region experienced important deformational events (e.g., Alkmim and Marshak 1998; Baltazar and Lobato 2020). The first was the Neoarchean Rio das Velhas event, followed by the 2.3–2.1 Ga Rhyacian-aged orogeny with the generation of NE-SW-trending fold and thrust belts verging to the NW. Then came the orogenic collapse at ca. 2.1 Ga (Alkmim and Marshak 1998). Finally, there was the Neoproterozoic Brasiliano orogeny, responsible for the formation of fold and thrust belts verging to the west (Alkmim and Marshak 1998), which imposed regional temperature and pressure gradients of 300–600 °C and 3–5 kbars, respectively, from west to east (Pires 1995). These tectono-metamorphic events resulted in a deformational and metamorphic gradient in the region that characterizes western, low-strain and eastern, high-strain domains, the latter most affected by Neoproterozoic tectonics (Rosière et al. 2001).

The evolution of the high-grade, hypogene ores is interpreted to have occurred synchronically to the Rhyacian-aged orogeny (2.03 Ga, Rosière et al. 2012), involving hydrothermal fluids of different origins and compositions, with formation of different generations of iron oxides (Rosière and Rios 2004; Rosière et al. 2008). During the first, contractional stage, reduced metamorphic fluids and oxidized connate water leached silica and carbonate from BIFs, and mobilized iron, leading to the formation of massive magnetite bodies, iron oxide veins and iron-enriched itabirites. The second stage involved low-temperature and low- to medium-salinity fluids, resulting in oxidation of magnetite to martite, and formation of anhedral and subhedral hematite (Fig. 2a). Granular hematite resulted from the recrystallization of these iron oxide types during metamorphism. The third mineralization stage involved percolation of saline hydrothermal fluids along shear zones, with the formation of tabular hematite, precipitated from low-temperature hydrothermal fluids. Specularite (Fig. 2b) subsequently formed along shear zones under high-strain and high-temperature conditions, overgrowing previous granular hematite. The Neoproterozoic, Brasiliano tectonic event was characterized by intense deformation in the eastern domain of the QF, where iron ores display a schistose structure, with well developed specularite crystals formed at high temperatures. These ores may have been formed under lower crustal conditions during the peak of the Rhyacian-aged orogeny, and then were thrust upwards to shallower levels during the Brasiliano Orogeny (Rosière et al. 2008).

3.2 Corumbá

Located near the Brazilian-Bolivian border (south of the town of Corumbá), the Urucum deposit lies between the southeastern edge of the Amazon craton and the northeastern corner of the Rio Apa block (Fig. 1 of Angerer et al. 2016).

The Banda Alta Formation of the Neoproterozoic Jacadigo Group in the Urucum massif, southwestern Brazil, represents one of the largest known Neoproterozoic hematite iron formations

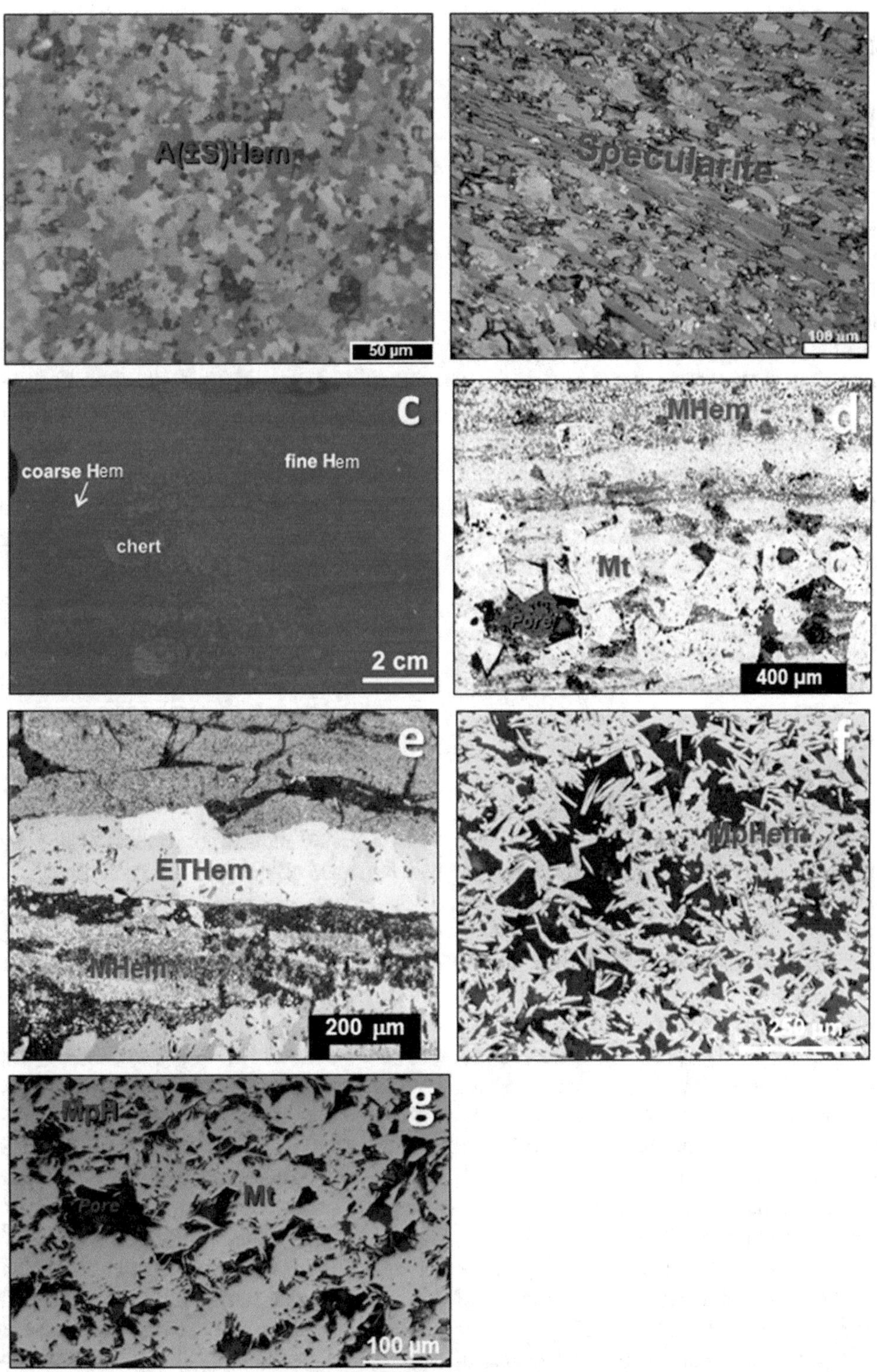
50 μm
Specularite
c
coarse Hem
fine Hem
chert
2 cm
MHem
d
Mt
Pore
400 μm
e
ETHem
MHem
200 μm
f
g
Mt
Pore
100 μm

(Fig. 1b). They are nodular and laminated, as well as banded, and are hereafter referred to as BIF following Walde and Hagemann (2007) and Angerer et al. (2016, and references therein). Several manganese oxide layers in the lower stratigraphic zone, as well as colluvial hematite ore (~45 to 62 wt% Fe) near the land surface, represent significant metal resources (Walde et al. 1981; Urban et al. 1992; Walde and Hagemann 2007). In the Urucum massif, the lowermost siliciclastic units of the Urucum Formation, the intermediate Córrego das Pedras Formation, and/or upper BIF and diamictite layers of the Banda Alta Formation, rest unconformably above a gneissic basement high of the Rio Apa Block (Dorr 1945). The depositional age is uncertain, and bracketed by a K–Ar cooling age of basement granites at 889 ± 44 Ma (Hasui and Almeida 1970) and an Ar–Ar age of diagenetic cryptomelane in the Mn-formation at 587 ± 7 Ma (Piacentini et al. 2013).

The banded and chemical-sedimentary nature of most of the hematite-rich rocks in the Banda Alta Formation has been established by several stratigraphic, petrographic and geochemical studies (Urban et al. 1992; Klein and Ladeira 2004; Freitas et al. 2011; Angerer et al. 2016; Viehmann et al. 2016). Stratigraphic studies have established a variable BIF stratigraphy across the Urucum region (Urban et al. 1992), with dolomite-rich BIF facies in the lower stratigraphic zone of the southeastern part of the massif (Angerer et al. 2016). A glaciomarine depositional environment has been invoked as a suitable setting for Neoproterozoic iron formations, and several examples are been correlated with Neoproterozoic ice ages (see reviews by Hoffman et al. 2011; Gaucher et al. 2015; Cox et al. 2013, 2016). According to Angerer et al. (2016), the Banda Alta Formation BIF was deposited in a redox-stratified marine sub-basin, which was strongly influenced by glacial advance/ retreat cycles with temporary influx of continental freshwater and upwelling seawater from deeper anoxic parts. Rare earth elements, base metal ratios, and ε_{Nd} suggest that dissolved metals in the Urucum seawater were most likely derived from a mix of terrigenous material of the nearby Neoproterozoic Brasília Belt (Viehmann et al. 2016) and low-temperature, hydrothermal fluids (Angerer et al. 2016). No evidence exists for high-temperature fluid source (Viehmann et al. 2016).

The supergene upgrade of the Corumbá hematite-rich BIF includes various significant processes that have led to the sought-after, colluvial high-grade iron ore (Walde and Hagemann 2007). The congruent dissolution of chert and carbonate from BIF forms the prominent, cavernous high-grade ore in the colluvium. The dissolution of dolomite from the upper carbonate-rich facies was probably the most efficient upgrade process, which is indicated by the absence of carbonate, but abundance of chert in BIF near the surface. Incongruent chert and dolomite dissolution, under the influence of Fe-rich weathering solutions, leaving fine, ochreous goethite residuals between partially dissolved hematite layers, was the second-most important process.

3.3 Carajás, Serra Norte Jaspilite-Hosted High-Grade Iron Ore Deposits

The Carajás mineral province, located in the eastern part of the state of Pará, Brazil, is the best-known, preserved Archean region of the Amazon Craton (Fig. 1 of Figueiredo e Silva et al. 2013). It is considered one of most important mineral provinces of the world, with regards to production and growing potential for Fe, Mn, Cu, Au, Ni, U, Ag, Pd, Pt, Os, Zn and W. The Carajás iron ore deposits (Fig. 1c) were discovered in 1967, with reserves estimated at 17 Gt of iron ore (> 64% Fe). The so-called Serra Norte iron deposits are hosted by the Archean metavolcanic-sedimentary sequence of the Grão Pará Group, Itacaiúnas Supergroup (DOCEGEO 1988). The protoliths to iron mineralization are jaspilites, under- and overlain by greenschist-facies metabasalts. The basal contact of high-grade iron ore is defined by a hydrothermally altered basaltic rock mainly composed of chlorite and microplaty hematite. The major Serra Norte N1, N4E, N4W, N5E and N5S iron ore deposits are distributed along, and structurally controlled by, the northern flank of the Carajás fold (Rosière et al. 2006).

Practically unmetamorphosed, Archean BIF (jaspilite) hosts high-grade iron ore (64% Fe) that is mainly composed of martite, and different types of hematite (Figueiredo e Silva et al. 2008, 2013). Three hydrothermal alteration zones are associated with the Serra Norte deposits: (1) A distal zone, which formed early (e.g., N4 deposit), is mainly characterized by recrystallization of jasper and removal of the original, fine-grained iron oxides, and by formation of euhedral to anhedral magnetite. This magnetite formed: (i) overgrowths on microcrystalline hematite (Fig. 2d; MHem) in jasper layers, (ii) grains in the nuclei of recrystallized chert, uniformly associated with crosscutting quartz veins, or (iii) grains in equilibrium with vein quartz and calcite. (2) An intermediate zone is characterized by widespread martitization (Fig. 2d), with common quartz-hematite veins that contain microplaty hematite and subordinate sulfides (e.g., N5S deposit). Precipitation of microplaty hematite and extensive martite alteration followed the formation of magnetite in the early alteration stage. (3) A proximal zone that was produced by later alteration events and was synchronous with the main ore-forming event. This zone represents the most advanced stage of hydrothermal alteration. It contains the high-grade iron ore, with varying types of hard and hard-porous ores. A complete oxide sequence defined by martite $\rightarrow$ microplaty hematite $\rightarrow$ anhedral hematite $\rightarrow$ euhedral-tabular hematite (Fig. 2e) is best documented in the N5E deposit, which contains the largest volume of hard, high-grade iron ore in Carajás (see also Lobato et al. 2008).

A dual magmatic-meteoric hydrothermal fluid-flow model is proposed for the high-grade iron ore (Figueiredo e Silva et. al. 2013; Hagemann et al. 2016), in which an early, saline, ascending modified magmatic fluid caused widespread oxidation of magnetite to hematite. Progressive influx of light ^{18}O meteoric water, mixing with the ascending magmatic fluids, is interpreted to have been initiated during the intermediate stage of alteration. The advanced and final hydrothermal stage was dominated by massive influx of low salinity, meteoric water, which maintained intermediate temperatures of 240–310 °C, and concomitant formation of the paragenetically latest, tabular hematite (Fig. 2e).

The giant Carajás iron deposits are unique in their setting within an Archean granite-greenstone belt and their modified magmatic-meteoric hydrothermal system, when compared to the other two end-member BIF iron deposit types, namely the basin-related Hamersley type and the metamorphosed metasedimentary, basin-related Quadrilátero-Ferrífero type (e.g., Hagemann et al. 2016).

The Carajás BIF iron ore system also constitutes a somewhat special case in that although the protolith jaspilite is Archean in age (Santos 2003), the upgrade of BIF to high-grade iron ore took place during Paleoproterozoic times (Figueiredo e Silva et al. 2008; Santos et al. 2010).

3.4 Hamersley High-Grade Iron Ore Deposits

The Hamersley province covers an area of about 80,000 km^2, and is situated between Archean basement complexes of the Yilgarn and Pilbara cratons (Fig. 1d). The stratigraphy in the province constitutes mostly Neo-Archean–Paleoproterozoic (2800–2300 Ma) sedimentary rocks of the Mount Bruce Supergroup resting on the Pilbara craton. The BIFs of the Hamersley Group are part of the Mount Bruce Supergroup, and host numerous bedded, high-grade orebodies within the ~2.6 Ga Marra Mamba Iron Formation (IF) and the ~2.5 Ga Brockman IF (for reviews see Thorne et al. 2008; Angerer et al. 2014, and references therein). The main types of BIF-hosted high-grade iron ores, presently produced from the Hamersley province, are martite-microplaty (Fig. 2f) hematite ore, hosted in the Brockman IF, and martite-goethite ore in the Brockman and Marra Mamba IFs.

The regionally abundant least-altered, yet low-grade metamorphosed BIF, show variably layered quartz-magnetite ± "early" hematite ±

Fe-silicate ± Fe-carbonate assemblage (Krapež et al. 2003; Klein 2005). Martite-microplaty (Fig. 2f and g) hematite ore consists almost entirely of martite bands or aggregates set in a porous (up to 30 vol. %), randomly orientated network of fine-grained (10–500 μm) hematite. In high-grade ore zones, BIF and locally associated shales are completely hematitized or leached leaving a martite residue. In peripheral zones, these rocks are hydrothermally altered, displaying distal carbonate–silicate-magnetite, intermediate carbonate–silicate-hematite, and proximal medium-grade martite-microplaty hematite associations (Thorne et al. 2008). The major martite-microplaty hematite orebodies include Mount Whaleback, Mount Tom Price, Paraburdoo, and Channar, with satellite deposits such as OB25 at Eastern Ridge and Hashimoto, both located in the South Ophthalmia Range (Thorne et al. 2008). The hematite orebodies are dominantly controlled by normal fault systems, i.e., the Southern Batter Fault in Mount Tom Price, the 4East Fault system in Paraburdoo, and the Central Fault and East Footwall Fault Zones in Mount Whaleback (Dalstra and Rosière 2008). Orebodies commonly extend below the weathering front to great depths, in places down to 500 m (e.g., Mount Whaleback). Goethite is present above the weathering front.

Martite-goethite ore consists of martite and various amounts of vitreous and ochreous goethite, with only very local microplaty hematite (Clout 2005). Important martite-goethite ore deposits include the numerous Marra Mamba iron formation (IF) orebodies in the Mining Area C, Hope Downs, and Chichester Range areas, as well as the Brockman BIF deposits in the South Ophthalmia Range, Mining Area C (Packsaddle), and Marillana area. The South Ophthalmia Range, Mining Area C, and Hope Downs orebodies are located in the structurally complex flanks of regional folds. On the other hand, the Chichester Range and Marillana deposits are extensive areas of almost flat-lying mineralized BIF lacking any structural complexity (Clout 2005). Generally, martite-goethite ore is limited to (paleo-) weathering zones and rarely reach deeper than 150 m.

The origin of the high-grade iron ore of the Hamersley province is subject to debate. Early work established the importance of supergene quartz leaching and goethite replacement in the upgrading of BIF to martite-goethite ore (Morris et al. 1980; Morris 1983; Morris and Horwitz 1983). Lascelles (2006) advocates a supergene-modified syngenetic model, whereas a supergene-metamorphic model was invoked by others (Morris and Fletcher 1987; Morris 1985, 2002) to explain the development of microplaty hematite-rich orebodies. Supergene-modified, hypogene fluid models, involving hot fluids and complex alteration, are suggested by various other researchers (Barley et al. 1999; Hagemann et al. 1999; Powell et al. 1999; Taylor et al. 2001; Webb et al. 2004).

4 Methods

Carajás and Hamersley samples were analyzed for Fe isotopes at the Geotop Isotope Laboratory, Université du Québec à Montréal, according to the methods used in Halverson et al. (2011). Approximately 30 mg of pulverized samples were weighed into Savillex™ teflon beaker and dissolved for 24 h at 80 °C in a mixture of double-distilled 6 M HCl, 7 M HNO_3, and 50% HF. The samples were then evaporated to dryness with excess HNO_3 to prevent the formation of apatite, then re-dissolved in 2.0 mL of concentrated aqua regia and dried down again. The samples were taken up again in 2.0 mL 2 M HCl and again dried down. The resulting salt was finally re-dissolved in 0.5 mL of 6 M HCl for ion exchange chromatography. Iron was separated using Bio Rad AG1 X4 (200–400 mesh) resin loaded into custom Teflon columns and separated from the matrix using 6 M HCl. Purified iron was eluted from the columns in 2 M HCl, which was then dried down and subsequently taken up in 0.5 M HNO_3 and diluted for isotopic measurement. Solutions were analyzed at the Geotop Isotope Laboratory at the Université du Québec à Montréal on a Nu Plasma II multi-collector inductively coupled plasma mass spectrometer (MC-ICP-MS) in high-resolution mode via wet sample introduction. Instrumental mass bias was corrected by using standard-sample-standard bracketing. Each sample was analyzed three times yielding typical 1-sigma errors of < 0.1 per mil for δ^{57}Fe and < 0.05 per mil for δ^{56}Fe. The data are reported in standard delta notation relative to the IRMM-14 reference standard.

Oxygen isotopes of samples from the Hamersley province and from Corumbá were analyzed at the Scottish Universities Environmental Research Centre (SUERC), University of Glasgow. Oxide separates were analyzed using a laser fluorination procedure, involving total sample reaction with excess ClF_3 using a CO_2 laser as a heat source (in excess of 1500 °C; following Sharp 1990). All combustions resulted in 100% release of O_2 from the silica lattice. This O_2 was then converted to CO_2 by reaction with hot graphite, then analyzed on-line by a VG Optima spectrometer. Reproducibility was within ± 0.2‰ (1σ), based on reproducibility of one international and two internal laboratory standards: UWG 2 (garnet = 5.8‰), GP147 (feldspar = 7.2‰) and SES 1 (quartz = 10.2‰). All results are reported in standard notation (δ^{18}O) as per mil (‰) deviations from the Standard Mean Ocean Water (V-SMOW) standard.

5 Published Iron and Oxygen Isotope Datasets

5.1 Quadrilátero Ferrífero (QF)

A series of deposits (Fig. 1a) located in different deformational domains of the QF have available iron isotope data (Table 1). These were obtained for both iron ores and their host itabirites and show significant variations (Mendes et al. 2017).

The QF itabirites display two groups (Fig. 3): quartz itabirites, which plot in two clusters, and carbonate itabirites that coincide with the lowest iron isotope cluster of the quartz itabirites. The δ^{56}Fe values of itabirites have very similar iron isotope ratios, varying from –0.95 to 0.27‰ (mean = –0.25‰; n = 14). The ranges for the different studied mines are: Usiminas (δ^{56}Fe from 0.02 ± 0.02‰ to 0.27 ± 0.07‰); Pau Branco (δ^{56}Fe from –0.09 ± 0.21 and 0.11 ± 0.09‰); Várzea do Lopes (δ^{56}Fe = –0.65 ± 0.04‰); Casa de Pedra (δ^{56}Fe = –0.95 ± 0.11 to –0.84 ± 0.08‰). Itabirites from Várzea do Lopes and Casa de Pedra have significantly lighter δ^{56}Fe values. Quartz itabirites from the eastern domain resulted in δ^{56}Fe varying from –0.12 ± 0.01 to 0.11 ± 0.03‰ (Conceição Mine and Morro do Agudo, respectively).

The deposition of the precursor sedimentary rocks to the QF itabirites was probably favored by the presence of an oasis of free oxygen in the Minas basin. The negative to low, near-zero positive δ^{56}Fe values for most of the quartz itabirites lie close or within the range reported for hydrothermal fluids (δ^{56}Fe ≈ –0.50 to 0‰), suggesting complete or near-complete oxidation of the dissolved Fe(II), which seems to be an emerging scenario for Paleoproterozoic BIFs (Planavsky et al. 2012). The most depleted δ^{56}Fe values (–0.95 to –0.65‰) of quartz itabirites may be explained by Rayleigh-type fractionation and BIF deposition further away from the hydrothermal source. Similarly, δ^{56}Fe values for carbonate and amphibole itabirites, deposited in shallower settings, were also a result of precipitation from a depleted source away from hydrothermal vents with negligible riverine Fe (Mendes et al. 2017).

Even though some inferences of positioning to the iron source can be made by the isotope signature of Cauê BIFs, it is not straightforward to define a stratigraphic sequencing, or the relative position on the depositional basin for the BIFs based on isotope data for the studied iron deposits, as a result of successive deformational events and stratigraphic inversions. Also, there is no record of a possible hydrothermal source for the iron deposited in the Minas Basin, making it impossible to determine the relative distance from the submarine volcanic centers. Nevertheless, the iron isotope values suggest that the Usiminas and Pau Branco deposits, and deposits from the eastern domains were probably located relatively closer to the iron sources than the Várzea do Lopes and Casa de Pedra deposits

Table 1 Sample/mineral list and isotope (iron and oxygen) composition of rock and mineral specimens from the Brazilian Quadrilátero Ferrífero (QF), Corumbá, and Carajás iron districts, as well as from the Hamersley province in Australia

Region	Classification	Sample type	Deposit	δ^{56}Fe (‰)	σ	δ^{57}Fe (‰)	σ	δ^{18}O (‰)	Total Fe_2O_3 (%)	Reference
QF	Quartz itabirite	Martite-mag-hem and qtz layers	Usiminas	0.08	0.03	0.11	0.04	–	50.9	Mendes et al. (2017)
QF	Quartz itabirite	Martite-mag-hem and qtz layers	Usiminas	0.27	0.07	0.35	0.21	–	57.2	Mendes et al. (2017)
QF	Quartz itabirite	Martite-mag-hem and qtz layers	Usiminas	0.02	0.02	− 0.07	0.14	–	66.3	Mendes et al. (2017)
QF	Massive iron ore	Anhedral hem and martite relicts	Usiminas	− 0.34	0.08	− 0.56	0.12	–	99.2	Mendes et al. (2017)
QF	Quartz itabirite	Martite-goethite-mag and qtz layers	Usiminas	0.16	0.04	0.19	0.07	–	47.2	Mendes et al. (2017)
QF	Massive iron ore	Anhedral granular mag	Bocaina	− 0.45	0.08	− 0.83	0.27	–	–	Mendes et al. (2017)
QF	Dolomitic itabirite	Dolomite and mag layers	Bocaina	− 0.61	0.13	− 0.93	0.11	–	–	Mendes et al. (2017)
QF	Specularite	Specular hem	Pau Branco	− 0.75	0.07	− 1.25	0.11	–	–	Mendes et al. (2017)
QF	Banded iron ore	Lamellar and granular hem	Pau Branco	− 0.22	0.04	− 0.33	0.09	–	–	Mendes et al. (2017)
QF	Specularite	Specular hem	Pau Branco	− 0.08	0.04	− 0.26	0.10	–	–	Mendes et al. (2017)
QF	Quartz itabirite	Martite-goethite-mag and qtz layers	Pau Branco	− 0.09	0.21	− 0.15	0.31	–	83.2	Mendes et al. (2017)
QF	Quartz itabirite	Martite-goethite and qtz layers	Pau Branco	0.11	0.09	0.13	0.15	–	40.3	Mendes et al. (2017)
QF	Massive iron ore	Anhedral-granular-lamellar hem	Pau Branco	− 0.26	0.00	− 0.45	0.10	–	99.1	Mendes et al. (2017)
QF	Banded iron ore	Lamellar-anhedral-granular hem and martite layers	Pau Branco	0.29	0.04	0.38	0.03	–	98.4	Mendes et al. (2017)
QF	Massive iron ore	Martite-granular-lamellar hem	Várzea do Lopes	− 0.69	0.03	− 0.99	0.06	–	–	Mendes et al. (2017)
QF	Schistose iron ore	Lamellar-granular-anhedral hem	Várzea do Lopes	− 0.42	0.12	− 0.66	0.18	–	–	Mendes et al. (2017)

(continued)

Table 1 (continued)

Region	Classification	Sample type	Deposit	$\delta^{56}Fe$ (‰)	σ	$\delta^{57}Fe$ (‰)	σ	$\delta^{18}O$ (‰)	Total Fe_2O_3 (%)	Reference
QF	Specularite	Specular hem	Várzea do Lopes	− 0.31	0.02	− 0.45	0.29	–	–	Mendes et al. (2017)
QF	Quartz itabirite	Martite-granular hem-goethite and qtz layers	Várzea do Lopes	− 0.65	0.04	− 0.99	0.08	–	75.2	Mendes et al. (2017)
QF	Specularite	Specular hem	Várzea do Lopes	0.83	0.38	1.29	0.58	–	–	Mendes et al. (2017)
QF	Quartz itabirite	Martite-hem-mag and qtz layers	Casa de Pedra	− 0.84	0.08	− 1.27	0.12	–	52.8	Mendes et al. (2017)
QF	Amphibolitic itabirite	Magnetitic and amphibole-carbonate layers	Casa de Pedra	− 0.95	0.09	− 1.45	0.14	–	–	Mendes et al. (2017)
QF	Brecciated/banded iron ore	Banded/brecciated martite-hem-mag	Casa de Pedra	− 0.51	0.04	− 0.79	0.06	–	98.5	Mendes et al. (2017)
QF	Specularite	Specular hem	Casa de Pedra	− 0.62	0.09	− 0.91	0.18	–	96.2	Mendes et al. (2017)
QF	Brecciated iron ore	Brecciated martite-mag-hem in goethite matrix	Casa de Pedra	− 0.31	0.02	− 0.45	0.29	–	–	Mendes et al. (2017)
QF	Brecciated iron ore	Martite-mag-hem in goethite matrix	Casa de Pedra	− 0.13	0.06	− 0.13	0.08	–	–	Mendes et al. (2017)
QF	Massive iron ore	Martite	Casa de Pedra	− 0.71	0.05	− 1.10	0.13	–	–	Mendes et al. (2017)
QF	Massive iron ore	Martite-mag-granular hem	Casa de Pedra	− 0.80	0.01	− 1.15	0.03	–	96.6	Mendes et al. (2017)
QF	Quartz itabirite	Martite-mag and qtz layers	Casa de Pedra	− 0.95	0.11	− 1.49	0.12	–	–	Mendes et al. (2017)
QF	Schistose iron ore	Lamellar-specular hem	Morro do Agudo	0.37	0.06	0.57	0.20	–	–	Mendes et al. (2017)
QF	Quartz itabirite	Lamellar-specular hem and qtz layers	Morro do Agudo	0.11	0.03	0.25	0.11	–	–	Mendes et al. (2017)
QF	Schistose iron ore	Lamellar-specular hematite	Alegria	− 0.09	0.08	− 0.13	0.14	–	–	Mendes et al. (2017)
QF	Schistose iron ore	Lamellar-specular hematite	Andrade	0.27	0.01	0.43	0.02	–	–	Mendes et al. (2017)

(continued)

Table 1 (continued)

Region	Classification	Sample type	Deposit	$\delta^{56}Fe$ (‰)	σ	$\delta^{57}Fe$ (‰)	σ	$\delta^{18}O$ (‰)	Total Fe_2O_3 (%)	Reference
QF	Quartz itabirite	Lamellar-specular hematite and qtz layers	Conceição	− 0.12	0.01	− 0.15	0.06	–	–	Mendes et al. (2017)
QF	Schistose iron ore	Lamellar-specular hematite	Conceição	0.04	0.08	0.04	0.23	–	–	Mendes et al. (2017)
QF	Quartz itabirite	Lamellar-specular hematite and qtz layers	Cauê	− 0.02	0.14	− 0.06	0.21	–	–	Mendes et al. (2017)
Corumbá	Reworked hem mud	Hematite	Santa Cruz	0.01	0.08	− 0.04	0.07	− 2.4	94.7	Angerer et al. (2016)
Corumbá	Reworked hem mud	Hematite	Santa Cruz	− 0.18	0.01	− 0.25	0.01	− 2.4	94.7	Angerer et al. (2016)
Corumbá	Chert-hem BIF	Hematite	Santa Cruz	− 0.49	0.06	− 0.71	0.07	4.6	65.3	Angerer et al. (2016)
Corumbá	Chert-dolomite-hem BIF	Hematite	Santa Cruz	− 0.83	0.02	− 1.22	0.04	3.4	62.7	Angerer et al. (2016)
Corumbá	Dolomite-chert-hem BIF	Hematite	Santa Cruz	− 1.83	0.02	− 2.62	0.03	1.3	62.1	Angerer et al. (2016)
Carajás	Jaspilite	Hem (iron-rich microband)	N4	1.13	0.17	1.67	0.23	–	94.7	Fabre et al. (2011)
Carajás	Jaspilite	Hem (iron-rich microband)	N4	1.07	0.04	1.61	0.06	–	89.6	Fabre et al. (2011)
Carajás	Jaspilite	Hem (iron-rich microband)	N4	1.16	0.12	1.74	0.12	–	94.2	Fabre et al. (2011)
Carajás	Jaspilite	Hem-goe (silica-rich microband)	N4	1.46	0.04	2.18	0.08	–	47.2	Fabre et al. (2011)
Carajás	Jaspilite	Hem-goe (iron-rich microband[2])	N4	1.41	0.08	2.07	0.01	–	19.0	Fabre et al. (2011)
Carajás	Jaspilite[1]	Hem-Goe (iron-rich microband)	N4	1.5	0.06	2.26	0.02	–	62.6	Fabre et al. (2011)
Carajás	Jaspilite	Hematite (silica-rich microband)	N4	1.75	0.04	2.58	0.21	–	32.0	Fabre et al. (2011)
Carajás	Jaspilite	Hematite (iron-rich microband)	N4	1.41	0.11	2.09	0.12	–	50.6	Fabre et al. (2011)
Carajás	Jaspilite	Hem-Mag (iron-rich microband)	N4	1.56	0.09	2.30	0.14	–	64.9	Fabre et al. (2011)
Carajás	High-grade ore	Anhedral hem	N5E	1.08	0.02	1.60	0.03	− 6.9	–	This study
Carajás	High-grade ore	Anhedral hem	N5E	0.74	0.04	0.69	0.01	–8.5	–	This study
Carajás	High-grade ore	Anhedral hem	N5E	0.51	0.01	0.74	0.11	–6.2	67.84	This study
Carajás	Jaspilite	Jaspilite	N4E	1.76	0.02	2.64	0.04	–	–	This study

(continued)

Table 1 (continued)

Region	Classification	Sample type	Deposit	δ^{56}Fe (‰)	σ	δ^{57}Fe (‰)	σ	δ^{18}O (‰)	Total Fe_2O_3 (%)	Reference
Carajás	Jaspilite	Jaspilite	N5S	0.64	0.01	0.92	0.01	10.7	–	This study
Carajás	Jaspilite	Magnetite	N5S	0.74	0.04	1.13	0.07	–	–	This study
Carajás	Jaspilite	Magnetite	N4W	0.95	0.01	1.37	0.03	3.8	–	This study
Carajás	Hydrothermally altered jaspilite	Magnetite (veins)	N4E	1.10	0.06	1.65	0.09	–0.4	–	This study
Carajás	Hydrothermally altered jaspilite	Magnetite (veins)	N4E	1.24	0.01	1.83	0.03	–0.1	–	This study
Carajás	High-grade ore	Martite	N4E	1.16	0.00	1.69	0.02	–4.3	61.95	This study
Carajás	High-grade ore	Martite	N5E	0.69	0.06	0.99	0.10	–5.9	96.51	This study
Carajás	High-grade ore	Martite (±MpHem-THem)	N5E	0.53	0.02	0.75	0.01	–	–	This study
Carajás	High-grade ore	Martite (±MpHem-THem)	N5E	0.62	0.01	0.92	0.01	–6.6	–	This study
Carajás	High-grade ore	Microplaty hem	N4E	0.67	0.02	0.97	0.02	–5.5	82.09	This study
Carajás	High-grade ore	Microplaty hem	N5E	0.62	0.01	0.93	0.02	–5.8	–	This study
Carajás	High-grade ore	Microplaty hem	N5S	0.46	0.01	0.69	0.01	–2.7	–	This study
Carajás	High-grade ore	Microplaty hem (±AHem)	N5S	1.11	0.02	1.63	0.00	–2.4	–	This study
Carajás	High-grade ore	Tabular hem	N5E	–0.30	–	–	–	–7.2	65.55	This study
Hamersley	Mt Newman BIF	Magnetite prox	Mesa Gap	0.19	0.02	0.28	0.03	3.29	53.87	This study
Hamersley	Dales Gorge BIF	Magnetite prox	Eastern Ridge	− 0.30	0.00	− 0.39	0.00	3.10	48.36	This study
Hamersley	Siderite breccia	Siderite	Marillana	− 1.24	0.01	− 1.84	0.02	26.34	57.40	This study
Hamersley	Siderite breccia	Siderite	Marillana	− 0.57	0.05	− 0.81	0.10	26.54	57.40	This study
Hamersley	Martite-Hem ore	MpHem-martite	Hashimoto	0.27	0.01	0.46	0.03	− 7.47	96.71	This study
Hamersley	Martite-Hem ore	MpHem-martite	Whaleback	− 0.29	0.01	− 0.40	0.01	− 7.57	98.30	This study
Hamersley	Martite-Hem ore	MpHem-martite	Whaleback	0.37	0.03	0.58	0.06	− 0.63	97.92	This study
Hamersley	Martite-Hem ore-Goe-ore	MpHem-martite	Paraburdoo	− 0.10	0.03	− 0.14	0.05	− 5.90	95.12	This study
Hamersley	Martite-Hem ore	MpHem-martite	Paraburdoo	0.35	0.02	0.53	0.04	− 5.80	95.12	This study
Hamersley	Tc-carb BIF	MpHem-martite	Tom Price	0.05	0.03	0.10	0.05	− 9.00	97.87	This study
Hamersley	Martite-Hem ore	MpHem-martite	Tom Price	− 0.16	0.01	− 0.25	0.03	− 5.50	97.87	This study

(continued)

Table 1 (continued)

Region	Classification	Sample type	Deposit	$\delta^{56}Fe$ (‰)	σ	$\delta^{57}Fe$ (‰)	σ	$\delta^{18}O$ (‰)	Total Fe_2O_3 (%)	Reference
Hamersley	Martite-Hem ore	MpHem-martite	Tom Price	1.02	0.10	1.60	0.10	− 5.10	97.87	This study
Hamersley	Martite-Hem ore	MpHem-martite	Tom Price	0.58	0.02	0.92	0.05	− 2.90	97.87	This study
Hamersley	Martite-Hem ore	MpHem-martite	Channar	0.22	0.01	0.32	0.04	n.a	–	This study
Hamersley	Dales Gorge BIF	Martite	Tom Price	1.09	0.01	1.61	0.02	n.a	–	This study
Hamersley	Dales Gorge BIF	Martite	Tom Price	1.00	0.02	1.45	0.08	8.80	–	This study
Hamersley	Martite ore	Martite	Packsaddle	0.81	0.02	1.18	0.04	n.a	44.13	This study
Hamersley	Martite ore	Martite	Marillana	0.78	0.02	1.16	0.03	− 3.12	86.57	This study
Hamersley	Dales Gorge BIF	Hematite	Paraburdoo	− 0.82 (− 0.9 to − 0.76)	0.05	–	–	–	–	Steinhoefel et al. (2010)
Hamersley	Dales Gorge BIF	Magnetite	Paraburdoo	− 0.87 (-0.94 to − 0.82)	0.06	–	–	–	–	Steinhoefel et al. (2010)
Hamersley	Dales Gorge BIF	Siderite	Paraburdoo	− 2.08 (− 2.18 to − 1.97)	0.15	–	–	–	–	Steinhoefel et al. (2010)
Hamersley	Dales Gorge BIF	Magnetite	Paraburdoo	− 0.01 (− 1.21 to 1.19)	0.49	–	–	–	–	Johnson et al. (2008a)
Hamersley	Dales Gorge BIF	Siderite	Paraburdoo	− 0.76 (− 2.06 to 1.00)	0.59	–	–	–	–	Johnson et al. (2008a)
Hamersley	Dales Gorge BIF	Magnetite	Wittenoom Gorge	0.37 (− 0.29 to 1.19)	0.38	–	–	–	–	Craddock & Dauphas (2011)
Hamersley	Dales Gorge BIF	Magnetite	Wittenoom Gorge	− 0.07	0.49	–	–	–	–	Craddock & Dauphas (2011)
Hamersley	Dales Gorge BIF	Siderite-min	Wittenoom Gorge	− 1.08	–	–	–	–	–	Craddock & Dauphas (2011)
Hamersley	Dales Gorge BIF	Siderite-max	Wittenoom Gorge	0.79	–	–	–	–	–	Craddock & Dauphas (2011)
Hamersley	Dales Gorge BIF	Magnetite	Yampire Gorge	0.51 (− 0.16 to 1.04)	0.35	–	–	–	–	Craddock & Dauphas (2011)
Hamersley	Dales Gorge BIF	Siderite	Yampire Gorge	0.22 (− 0.40 to 1.21)	0.45	–	–	6.8	–	Craddock & Dauphas (2011)

Abbreviations: AHem = anhedral hematite; amph = amphibole; avg = average; Goe = goethite; Hem = hematite; Mag = magnetite; max = maximum; min = minimum; MpHem = microplaty hematite; n.a. = not analyzed; Quartz = qtz; Tc = talc; THem = tabular hematite.

[1]: classification according to this study; [2]: despite considered "iron-rich" by Fabre et al. (2011), it is relatively poor in iron.

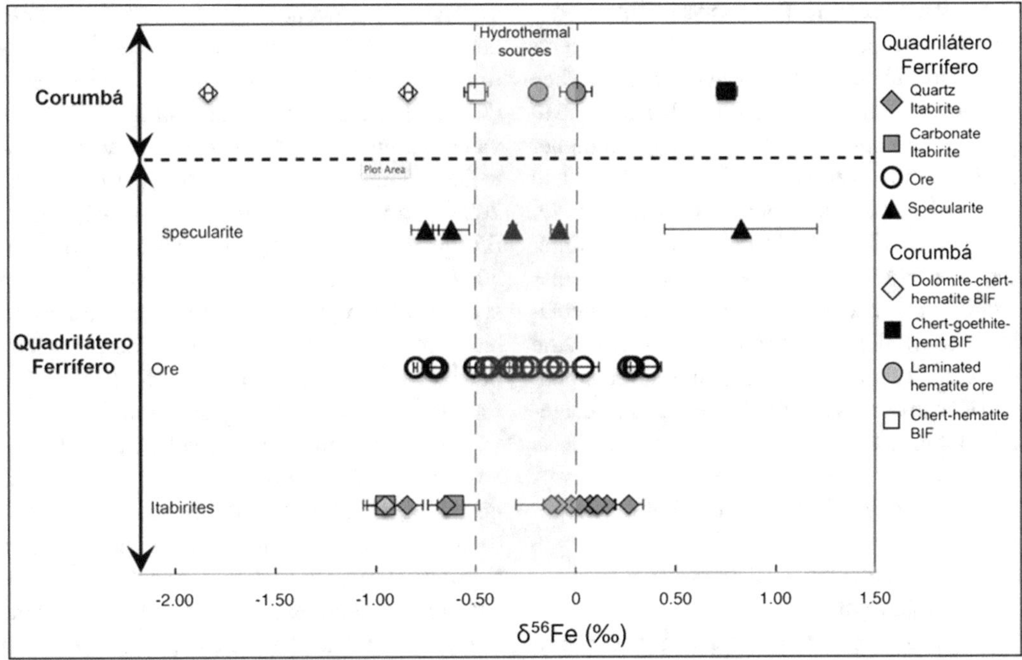

Fig. 3 Iron isotope data available for the Quadrilátero Ferrífero (Mendes et al. 2017) and Urucum (Corumbá) iron districts (Angerer et al. 2016), Brazil (Table 1). Field of hydrothermal source is δ^{56}Fe ≈ − 0.50 to 0‰ (Beard et al. 2003; Severmann et al. 2004; Johnson et al. 2008b; Planavsky et al. 2012) is shown

(location in Fig. 1a). Casa de Pedra must have been deposited closer to the shallower platform, as suggested by the presence of carbonate.

The similarity between the iron isotope composition of iron ores and their host itabirites suggests that hydrothermal mineralization did not significantly alter the protore composition. Nevertheless, the isotopic compositions of iron ores from the low- and high-strain domains are quite different, and may reflect distinct conditions of hydrothermal mineralization as a result of variations in the thermodynamic conditions of the fluid, such as acidity, salinity and temperature (Rosière and Rios 2004; Rosière et al. 2008). Their conclusions are mainly based on the comparison of fluid inclusion data for each domain. Although no additional, new data are available, it is feasible to envisage redox variations to explain the indicated isotopic differences. Besides, mechanisms of iron isotope fractionation may vary within the same domain.

Some other data for the QF are available from Sampaio et al. (2018) and Teixeira et al. (2017). The former evaluated δ^{56}Fe signatures of magnetite-amphibole-carbonate-bearing itabirites, as well as their hydrothermal and weathering products. They investigated two drill holes in the Ouro Fino (OFS) and Gandarela (GS) synclines (Fig. 1a). The δ^{56}Fe ratios average − 0.87‰ in the least altered samples. An analogous average of − 0.84‰ was calculated for samples with goethite (after amphibole), and of − 0.60‰ for those with martite, as well as in others with abundant goethite plus hematite (after martite). An average of + 0.41‰ characterizes samples with martite plus abundant microplaty hematite and goethite, and minor clay minerals; a similar signature is calculated for samples just beneath the surface ground. The work of Teixeira et al. (2017) focused on itabirite samples from the Alegria Mine (Fig. 1a); they show δ^{56}Fe between + 0.51 and + 1.33‰.

5.1.1 Iron Isotope Data from Corumbá

The iron isotope data (Table 1) from Angerer et al. (2016) show negative values for BIF. They are low in the dolomite-rich hematite BIF (δ^{56}Fe − 1.83 and − 0.83‰; Fig. 3), and higher both in chert-hematite BIF (δ^{56}Fe − 0.49‰) and associated (high-grade) hematite mud (Fig. 2c; δ^{56}Fe = − 0.18 and 0.01‰). The goethite-altered BIF has positive values (δ^{56}Fe = 0.76‰). Variable iron isotope values in hematite are compatible with complex processes in the Fe source (Tsikos et al. 2010), during precipitation (Kunzmann et al. 2017), and during diagenesis (Johnson et al. 2008b). This dataset reflects several iron isotopic fractionation processes (Angerer et al. 2016) (see Discussion).

5.1.2 Iron Isotope Background at Carajás

The only available iron isotope data from the Serra Norte iron deposits in Carajás (Fabre et al. 2011; Table 1) resulted from the investigation of the redox changes of Earth's ocean and atmosphere between 2.7 and 2.1 Ga ago. Data were obtained from the protore BIF of the N4 (Fig. 1c) deposit and show consistent positive δ^{56}Fe values from + 1.75 to + 1.07‰, with a mean value of + 1.38 ± 0.23‰ (Table 1; Fig. 4). There is no significant difference in isotope signatures between iron oxides in silica (jasper)- and iron (microcrystalline hematite)-rich microbands (Table 1). According to Fabre et al. (2011), the values are among the most positive ever measured in BIF (Johnson et al. 2003, 2008a, b; Rouxel et al. 2005), except for older BIF like Akilia/Isua (Dauphas et al. 2004; Whitehouse and Fedo 2007; Thomazo et al. 2009). However, Planavsky et al. (2012) have shown that Archean BIF can also have negative δ^{56}Fe values. These authors obtained a δ^{56}Fe range from − 1.53 to + 1.61‰ for Archean to early Paleoproterozoic BIFs.

Partial oxidation of aqueous Fe(II) into Fe(III) near the ocean surface, subsequent to the upwelling of deep anoxic seawater, may have been responsible for the positive iron isotopic signatures (Fabre et al. 2011). According to these authors, the "extremely high" δ^{56}Fe values in the Carajás jaspilite resulted from oxidation of a high degree of the upwelling Fe(II) mass (i.e., 44%, based on assumed initial hydrothermal Fe with δ^{56}Fe of 0.0‰). Planavsky et al. (2012) indicate that positive δ^{56}Fe values for Archean BIFs are considered to result from non-quantitative iron oxidation in an oxygen-poor ocean, taking to the preferential deposition of the isotopically heavier iron.

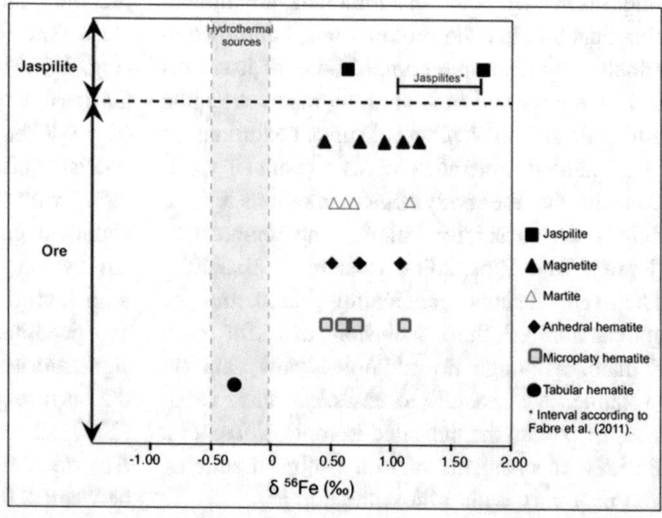

Fig. 4 Results of Fe isotope composition of jaspilites and iron ore oxides (this study) from the Serra Norte iron district, Carajás, Brazil (Table 1). Published results by Fabre et al. (2011) are also shown

6 New Iron Isotope Results and Iron-Oxygen Isotope Pairing

6.1 Corumbá: Iron and Oxygen Isotope Pairing

Both iron and oxygen isotopes are sensitive indicators for low- to high-temperature fluid-rock interaction. The BIF and iron ore at Corumbá have a simple modal mineral composition, consisting of ~50–99 vol % hematite, with the remainder being chert and/or ferroan dolomite (Angerer et al. 2016). The relationship of oxygen and iron isotopes in hematite during BIF formation and subsequent alteration (hypogene and supergene hydrothermal) are investigated for this review.

Unpublished oxygen and published iron (Angerer et al. 2016) isotope data have been paired either using the same or lithologically and stratigraphically closely associated samples (Table 1). The oxygen isotope signature for Neoproterozoic seawater is unknown, but most probably between − 8 and 0‰ (Jacobsen and Kaufman 1999; Veizer et al. 1999). However, it is noted that the carbonate mean values reveal a long-term increase of about + 8‰, in the Neoproterozoic and Cambrian to Present at 0‰ (Jacobsen and Kaufman 1999; Veizer et al. 1999). The $\delta^{18}O$ data reveal remarkable relationships with hematite $\delta^{56}Fe$ (Fig. 5). The chert-dolomite-hematite BIF samples (C-6, C-11, C-14; Fig. 5) show $\delta^{18}O > 0‰$, and a negative correlation between $\delta^{18}O$ and $\delta^{56}Fe$. The range of data implies non-equilibrium conditions. The recrystallized (high-grade) hematite muds (H-14, H-15; Fig. 5) are significantly depleted in $\delta^{18}O$ (< 0‰), while showing non-fractionated $\delta^{56}Fe$ (±0‰). Calculated $\delta^{18}O_{water}$ are up to 5‰ higher than that of $\delta^{18}O_{hematite}$. The large range of $\delta^{18}O_{water}$ results from uncertainties in the prevailing temperatures (see caption of Fig. 5 for temperature estimates and references of equations). Although there are no oxygen isotope data available for the goethite-altered BIF (H-17), a $\delta^{18}O$ value of < 0‰ may be assumed based on published data of world-wide supergene goethite-altered BIF (Gutzmer et al. 2006, and references

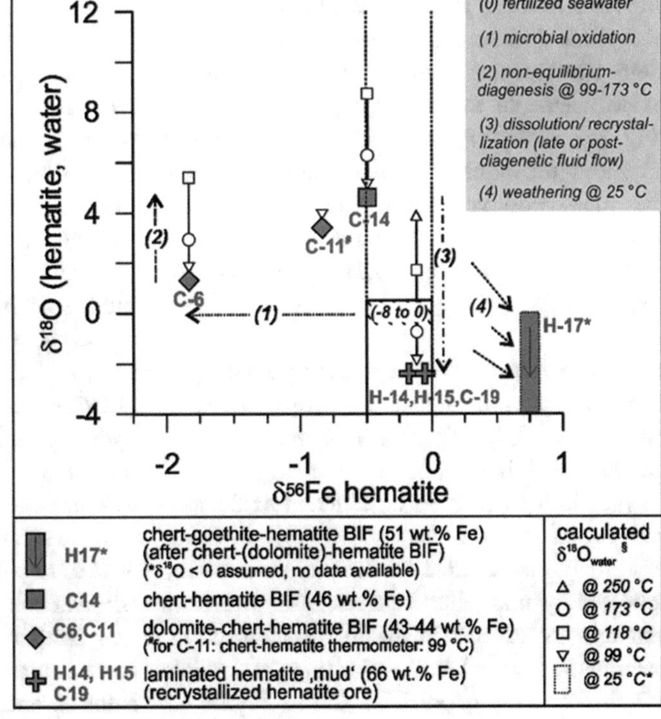

Fig. 5 Iron (Angerer et al. 2016) and oxygen (unpublished) isotope data for the Brazilian Urucum (Corumbá) iron district are shown in Table 1). Equilibrium temperatures 99 °C, 118 °C, 173 °C and $\delta^{18}O_{water}$ calculated with quartz-H_2O (Zheng 1993) and hematite-H_2O (Yapp 1990a,b) based on data from this study and from data of Hoefs et al. (1987)

therein). The data pairing reveals that during certain alteration events, the isotope systems are coupled, while in others they are decoupled (see Discussion).

6.2 Carajás

Twenty samples of massive, high-grade iron ore samples from the Serra Norte N4 and N5 deposits were analyzed including separates of martite, microplaty hematite, anhedral hematite and tabular hematite (Figueiredo e Silva et al. 2013; Table 1). Signatures are positive and vary from + 0.44 to + 1.24‰, in δ^{56}Fe (Fig. 4). One sample has a negative δ^{56}Fe signature of –0.30‰, which corresponds to the late-stage tabular hematite (Fig. 2e).

Two jaspilite samples show δ^{56}Fe of + 1.76 and + 0.64‰ (Table 1), which are similar to results by Fabre et al. (2011). Other two samples correspond to hydrothermally altered jaspilites (Table 1), where magnetite occurs along quartz-carbonate veins; δ^{56}Fe values are + 1.10 and + 1.24‰.

Figure 4 shows a decrease in isotopic signature from Archean jaspilite (Fabre et al. 2011) to high-grade iron ore samples. There is also a slight trend towards negative values from earlier oxide magnetite to the latest tabular hematite (Fig. 2e), indicating some fractionation during advanced ore mineralization. Magnetite shows the most positive values (up to + 1.24‰); anhedral and microplaty hematite have the lowest values of δ^{56}Fe at + 0.51 and + 0.46‰, respectively.

6.2.1 Iron and Oxygen Isotope Pairing

Published oxygen isotope data (Figueiredo e Silva et al. 2013) have been paired with the present unpublished iron isotope data (Fig. 6a). Oxygen isotope analyses on different oxide species reveal that the heaviest $\delta^{18}O_{SMOW}$ value of + 10.7‰ is recorded for the protore jaspilite, followed by magnetite, between –0.4 to 3.8‰, and then by different hematite species such as microplaty, anhedral and tabular (Fig. 2e), which fall in the range of –8.5 to –2.4‰. Figure 6a shows a progressive depletion in δ^{18}O values and δ^{56}Fe from the earliest hydrothermal oxide magnetite towards the latest tabular hematite.

6.3 Hamersley

Present iron isotope data from the Hamersley province derive from published data of the least-altered Dales Gorge Member BIF (Fig. 7), and from unpublished data of seventeen samples of altered BIF and hematite iron ore. The least-altered BIF samples come from the greater Paraburdoo region in the south (Johnson et al. 2008a; Steinhoefel et al. 2010), as well as from the Wittenoom Gorge and Yampire Gorge in the north of the province (Craddock and Dauphas 2011).

For the Dales Gorge Member, by far the most important host rock in the Brockman IF, samples include nine of hydrothermal martite-microplaty hematite iron ore; two of supergene martite ore; two of oxidized BIF; two of hydrothermally altered BIF; and two of early-siderite breccias. All rock samples were taken from diamond drill cores in various iron ore deposits (Table 1).

In BIF from the Paraburdoo-region, magnetite δ^{56}Fe values range from − 1.21 to + 1.19‰, (average − 0.01‰). Wittenoom Gorge magnetite BIF has δ^{56}Fe from − 0.29 to + 1.19‰ (average 0.37‰), and the Yampire Gorge from − 0.16 to + 1.04‰ (average 0.51‰). Taking only the interquartile ranges of each data subset into account, the entire BIF data range from − 0.3 to + 0.9‰ (Fig. 7). Associated anhedral hematite has slightly higher values (Table 1; Fig. 7).

Hematite in martite-microplaty hematite ore shows δ^{56}Fe values ranging from − 0.29 to + 1.02‰ (average 0.25‰). Magnetite in hydrothermally altered BIF, associated with the ore, shows δ^{56}Fe values ranging from − 0.30 to + 0.05‰. Martite from both oxidized BIF and iron ore shows distinctively heavier ^{56}Fe isotopes (+ 0.80 to + 1.10‰) compared with least-altered BIF and microplaty hematite ore (Fig. 7). A limiting factor prohibiting the exact measurement of isotope fractionation related to iron ore formation is the absence of a true precursor BIF sample for

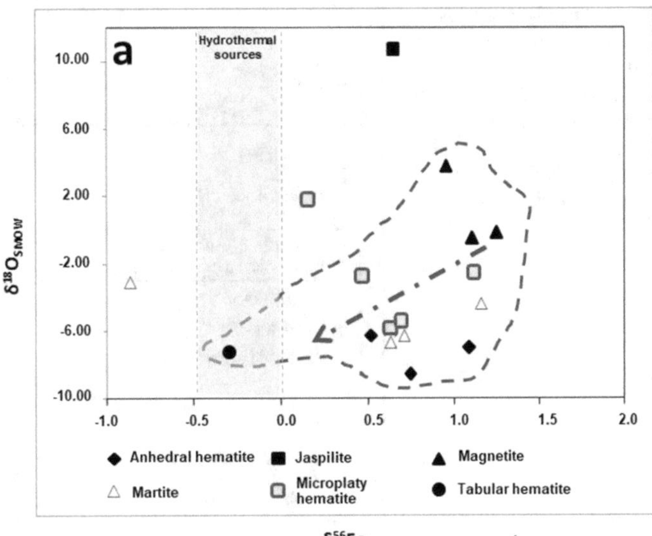

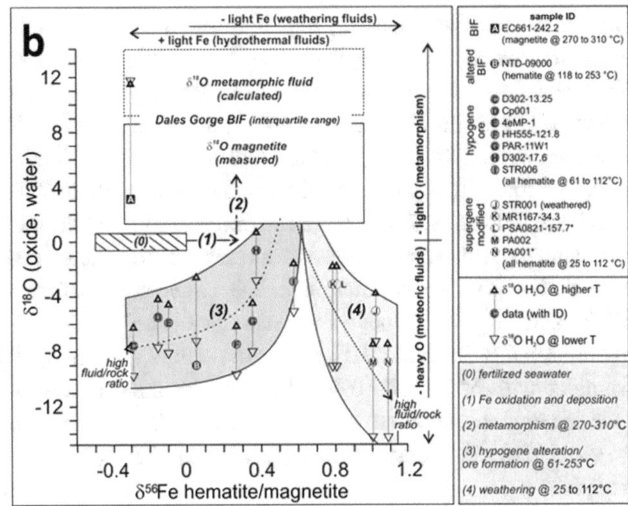

Fig. 6 Calculated δ^{56}Fe versus δ^{18}O for: **a** Carajás, and **b** Hamersley. Based on data available in Table 1

any of the ore samples. Nevertheless, the inclusion of oxygen isotope data allows deducing scenarios of iron isotopic fractionation.

6.3.1 Iron and Oxygen Isotope Pairing

Oxygen isotopes were obtained for a number of samples of the iron isotope data set to investigate the behavior and potential coupling of the Fe–O isotope system during hydrothermal and supergene alteration of BIF and iron ore development.

In order to establish a representative range of Fe–O isotopes for least-altered Dales Gorge BIF, separate published data sets on iron isotopes (Johnson et al. 2008a; Craddock and Dauphas 2011) and oxygen isotopes (Powell et al. 1999) were combined. These data are shown in Table 1 and Fig. 6b, together with published estimates of the pristine Paleoproterozoic seawater isotopic signature.

In unaltered BIF, anchizonal to low-greenschist facies metamorphism (270–310 °C, Becker and Clayton 1976) lead to significant higher δ^{18}O and δ^{56}Fe values, compared to primary seawater signature. In altered BIF and iron ore, the overall δ^{18}O values range from − 9 to 0‰ (Table 1, Fig. 6b). This marked difference (δ^{18}O up to 10‰) between the heavy oxygen isotope in metamorphic magnetite in the least-altered BIF

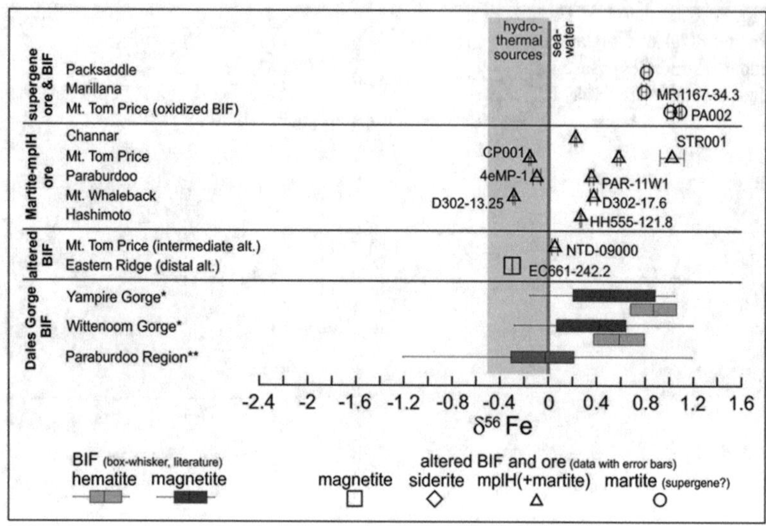

Fig. 7 Distribution of Fe isotope composition of oxides and siderite in iron ores and BIF of different ore deposits from the Hamersley iron district, Australia, including both published and unpublished data (data in Table 1)

(square in Fig. 6b; Table 1), and hydrothermal hematite (Powell et al. 1999) is in accordance with published data and models of influx of meteoric water depleted in $\delta^{18}O$ during ore formation (Gutzmer et al. 2006; Thorne et al. 2009).

No major shift is shown in iron isotope data, which are indistinguishable from the range of the least-altered BIF. It is, however, interesting to note that a positive correlation exists between $\delta^{18}O$ and $\delta^{56}Fe$ values for most of the hydrothermally altered BIF and ore samples (samples with $\delta^{56}Fe < 0.60‰$; Fig. 6b). For the four supergene modified samples (with $\delta^{56}Fe$ 0.79–1.19‰) the correlation appears to be negative.

Additional Fe–O isotope data also on Hamersley (Dales Gorge member) were combined by Li et al. (2013), who analyzed both magnetite and hematite in BIF. The iron and oxygen isotope signatures highlight a contrasting behavior of these isotopes in the two oxides.

7 Discussion

7.1 Quadrilátero Ferrífero

Although metamorphism did not modify the iron isotope composition of itabirites of the Cauê Formation, hydrothermal fluids related to iron mineralization generated ores that have somewhat lighter (lower) $\delta^{56}Fe$ ratios with respect to the original protore BIF.

The iron isotope composition of the iron ores is fairly similar to itabirites, suggesting that hydrothermal mineralization did not significantly alter the isotopic composition of the protores. One must also take into account that the chemical attributes of itabirites, the QF protores, no longer reflect those of the original banded iron formation (Harder 1914; Rosière et al. 2008), since modifications were attained via post-depositional alteration and multiple events of metamorphism (Rosière et al. 2008). On the other hand, iron ore iron isotope signatures vary across the QF (Figs. 1a and 3; Mendes et al. 2017), with differences between the western-low and eastern-high strain domains as defined by Rosière et al. (2001).

Mineralization in the western, low-strain domain is mainly characterized by percolation of low-temperature, saline fluids carrying isotopically light Fe(II). Such fluids precipitated ores with new iron oxides, depleted predominantly in ^{56}Fe ($\delta^{56}Fe = -0.80 \pm 0.01$ to $-0.13 \pm 0.06‰$), in open spaces evolved from the leaching of gangue minerals. Ores from this domain have an isotope signature mostly within the interval suggested for 'mineralizing' hydrothermal fluids ($\delta^{56}Fe \approx -0.50$ to 0‰; Rouxel et al.

2005; Markl et al. 2006; Fig. 3). According to Mendes et al. (2017), hydrothermal fluids must have become enriched in light iron isotopes by leaching of Fe(II) from the basement and country rocks (including host BIF), as suggested by Markl et al. (2006), during different stages of the Rhyacian-aged orogeny. The mechanisms of isotope fractionation may have varied in the western-low strain domain. Ores with δ^{56}Fe ratios equivalent to the hydrothermal fluids may have precipitated in equilibrium with these fluids. On the other hand, iron ore with positive δ^{56}Fe ratios is less common. Isotopic heavy iron ore may be a result of kinetic fractionation as a consequence of fast hematite precipitation as indicated by Skulan et al. (2002).

In the eastern, high-strain domain, high-temperature saline fluids, also carrying isotopically light Fe(II), precipitated oxides enriched in heavy iron isotopes. Ores in this domain are enriched in the heavier isotope (δ^{56}Fe = − 0.09 ± 0.08 to 0.37 ± 0.06‰; Mendes et al. 2017; Fig. 3), and this may have occurred due to iron isotope fractionation during redox transformation. This resulted in ores less depleted than those of the low-strain domain, and implies that both the isotopic composition of the mineralizing fluids and the mineralizing conditions differed from those in place in the western domain. Precipitation of hematite under P–T conditions in the higher strain zone may have occurred out of equilibrium with the hydrothermal fluid, resulting in preferential incorporation of the heavy iron isotope into hematite. An alternative suggestion by Mendes et al. (2017) is that positive δ^{56}Fe values of ores could reflect positive kinetic fractionation, resulting from rapid precipitation of hematite. Figure 8 shows that the QF data are compatible with those of the environmentally similar Hamersley BIF and iron ores.

The variation in δ^{56}Fe composition with depth in two drill holes (Fig. 1a) is used by Sampaio et al. (2018) to interpret results in more or less weathered samples, the latter considered to better preserve interaction with a hydrothermal fluid. In one of them (OFS), there is a shift from − 0.91‰ at depth to less negative values of − 0.60‰ near the surface; the most weathered samples have positive values of 0.30‰ and 1.33‰. In the second hole (GS), δ^{56}Fe increases linearly from − 0.11 to 1.23‰ towards the surface, although the most negative values − 0.70‰ and − 1.16‰ are near the top. The most negative δ^{56}Fe from the OFS can be explained by DIR

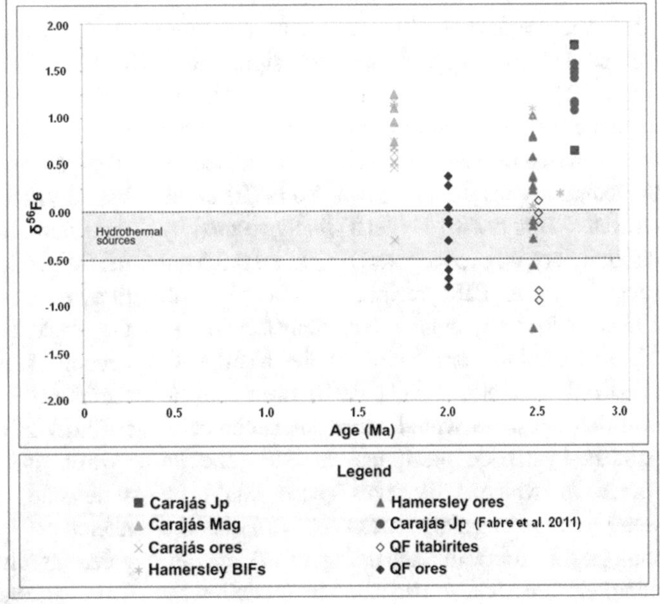

Fig. 8 Age versus δ^{56}Fe for all ore data (in Table 1)

and/or Rayleigh fractionation, while the more positive from the GS may reflect oxidation driven by interaction with a hydrothermal fluid enriched in iron.

7.2 Corumbá

Iron isotope fractionation discussed for the Corumbá BIF are controlled by: (1) the primary seawater isotopic signature, (2) microbial-aided Fe(III) to Fe(II) reduction in the source Fe pool, and (3) supergene Fe(III) to Fe(II) reduction and dissolution. Particular processes responsible for the oxygen isotope fractionation recorded in BIF are: (1) precipitation from seawater; (2) rock dehydration and recrystallization during protracted diagenesis; (3) interaction with hydrothermal alteration fluids during diagenetic gangue dissolution; and (4) interaction with supergene weathering solutions.

Low δ^{56}Fe values between − 0.7 and − 0.1‰ (Beard et al. 2003; Severmann et al. 2004; Johnson et al. 2008b; Klar et al. 2017) in chert-hematite BIF and hematite mud indicate that a deep ocean seawater was dominated by hydrothermal Fe(II), hence without significant iron isotope fractionation. Precipitation directly from seawater is postulated for the rocks themselves (Klein and Ladeira 2004; Angerer et al. 2016; Viehmann et al. 2016), and thus an oxygen isotope equilibration with seawater signature is likely. The much lower δ^{56}Fe recorded in hematite from carbonate-rich BIF (down to − 1.8‰ in sample C-6; Fig. 5) is explained by the temporary increase in low-δ^{56}Fe Fe(II) in the seawater. This low-δ^{56}Fe Fe(II) likely formed by dissimilatory iron reduction (DIR) of an oxidized source prior to BIF precipitation (Beard et al. 1999, 2003), or by abiotic reduction (Balci et al. 2006). Resulting fertilization of the basin with dissolved, isotopically light Fe(II) from a ferric iron-rich substrate would cause an isotopically light Fe-hydroxide precipitate in BIF. The Fe source in which DIR takes place could be weathered sedimentary rocks or BIF in the stratigraphic footwall. According to this model, microbial activity and associated fractionation by about − 0.3 to − 1.3‰ was highest shortly before and during precipitation of carbonate-rich BIF (Angerer et al. 2016). Carbonate spheroids in carbonate BIF layers are discussed as biogenic relics, and thus support microbiological activity in the basin (Angerer et al. 2016).

Late diagenesis occurred between ~ 100 and 180 °C, according to the hematite-quartz pair thermometer (Hoefs et al. 1987). Non-equilibrium oxygen isotope fractionation is reflected in a range of oxygen isotope data (samples C-6, C-11, C-14; Fig. 5). In contrast to oxygen isotopes, diagenesis does not seem to have any significant effect on the primary (i.e., hydrothermal and DIR influenced seawater) Fe-isotope fractionation, although a general possibility has been discussed by several authors (Johnson et al. 2008b; Kunzmann et al. 2017, and references therein). The observed negative Fe–O isotope correlation (Fig. 5) must be interpreted as a coincidental result of two distinct (decoupled) fractionation processes.

During a late diagenetic hydrothermal overprint, silica and carbonate dissolution caused small-scale upgrade to high-grade laminated hematite ore (Angerer et al. 2016). Hydrothermal alteration took place under fluid temperatures of 100–250 °C (Angerer et al. 2016). Based on the estimated temperature range, calculated $\delta^{18}O_{water}$ of fluids ranges between − 2 and + 4‰, respectively. Iron isotopes remain unfractionated (δ^{56}Fe ± 0‰) during this syn-diagenetic hydrothermal alteration (see laminated hematite ore in Fig. 3). Diagenetic re-equilibration of oxygen isotopes of hematite were accommodated by recrystallization, hence Fe_2O_3 redistribution at the microscale (Angerer et al. 2016). The contrasting lack of iron isotope fractionation shows that Fe_2O_3 behaved as an immobile component at the rock scale during these processes. The decoupled nature of isotopic fractionation is, therefore, a result of the different scales of mobility of oxygen and iron.

Iron isotope fractionation occurred during supergene alteration. Hematite from goethite-altered BIF shows heavier ^{56}Fe (0.76‰), compared to unweathered BIF and laminated ore. This indicates that hematite was not chemically inert during weathering. Abundant microporosity in the hematite matrix of saprolitic BIF indicates not only dissolution of silica and carbonate, but also partial dissolution of hematite. It is inferred that ^{56}Fe hematite represents a residue after preferential dissolution of ^{54}Fe. This is in accordance with the established understanding that weathering solutions have commonly light iron isotopes as a result of Fe(II) dissolution (Fantle and DePaolo 2004; Bergquist and Boyle 2006; Ingri et al. 2006). Weathering causes significant fractionation also of oxygen isotopes, where δ^{18}O values decrease by equilibration with the meteoric fluid (Fig. 5). The coupling of ^{56}Fe and ^{18}O isotopes supports the involvement of characteristically light oxygen-enriched and heavy iron-depleted supergene fluids.

7.3 Carajás

There is a general tendency for δ^{56}Fe of the Archean jaspilites from Carajás to show higher values (Figs. 4 and 6a) compared to hematite ores, particularly those of late paragenetic stage (Fig. 2e; Figueiredo e Silva et al. 2013). The iron isotope data for protore BIF (jaspilites) coincide with the heaviest δ^{56}Fe ore values (Fig. 4), suggesting that fluid evolution and mineralization did not alter significantly the iron isotope composition of the original oxides. However, a slight trend towards negative values exists from the early-stage magnetite to the latest tabular (Fig. 2e) hematite, indicating some fractionation during advanced ore mineralization, probably due to high influx of hydrothermal fluid in the waning mineralization stages.

Considering the dual magmatic-meteoric hydrothermal fluid flow model for Carajás (Figueiredo e Silva et. al. 2013; Hagemann et al. 2016), it is worth mentioning the study by Wawryk and Foden (2015). Wawryk and Foden (2015) show that oxidized magmas crystallize magmatic magnetite, which sequesters heavy Fe thus producing an isotopically light magmatic-hydrothermal fluid. This may explain the lower δ^{56}Fe of the Carajás ore samples, compared to protore BIF, as observed in Figs. 4 and 8.

The advanced alteration stage in high-grade ore displays the most depleted δ^{18}O and δ^{56}Fe values (Fig. 6a) and may represent the highest fluid-rock ratio during hydrothermal alteration as suggested by Figueiredo e Silva et al. (2013). This depletion is interpreted to result from the progressive mixture of descending, heated meteoric water with ascending modified magmatic fluids.

7.4 Hamersley

At first sight, martite-microplaty hematite ore δ^{56}Fe isotopes (2nd row Fig. 7; − 0.29 to + 1.02‰) are indistinguishable from the range of the least-altered BIF (3rd and 4th rows; Fig. 7) suggesting no significant fractionation throughout microplaty hematite ore formation. However, pairing of iron with oxygen isotope data reveals cryptic iron isotope fractionation trends.

The interquartile range of magnetite δ^{56}Fe, in the least-altered Dales Gorge Member BIF (from different localities) (Fig. 7), is from − 0.3 to + 0.9‰ (Fig. 7). The data set of Steinhoefel et al. (2010) from adjacent microbands in one sample (− 0.94 to − 0.82‰, average − 0.87‰) indicates that isotopic variability is low at the sample scale. However, in comparison with the other available data (Johnson et al. 2008a; Craddock and Dauphas 2011), it is evident that the variability of δ^{56}Fe values is high at the stratigraphic or regional scale. As yet, the reasons for this spread of iron isotopes in BIF are not fully understood (see "Brief summary on iron isotopes" section for references to literature focusing on isotope data related to BIF deposition).

While the prominent variability of iron isotope values in BIF, with a general shift towards heavier signature compared to seawater, most likely resulted from Fe source-related and depositional processes, it is unlikely that low-grade metamorphism changed significantly the iron isotopic budget of BIF. This is because Fe(III) reduction associated with hematite to magnetite replacement forming metamorphic assemblages cannot produce heavier isotope values (Fe reduction always decreases δ^{56}Fe; e.g., Beard et al. 1999, 2003). In contrast to iron isotopes, the oxygen isotopic signatures in magnetite are strongly dependent on metamorphic equilibration. In consequence, during deposition of BIF (see stage 1 arrow in Fig. 6b) and metamorphism (stage 2 arrow) iron and oxygen isotopes are not coupled.

However, this contrasts with subsequent carbonate-hematite alteration and hematite ore formation in BIF (Fig. 6b). Although any systematic modification of iron isotopes in ore is obscured by the large variety of δ^{56}Fe in least-altered BIF, a positive correlation between δ^{18}O and δ^{56}Fe (trend 3 arrow Fig. 6b) is evidence of significant fractionation towards lower isotopic values. Light δ^{18}O signatures of iron oxides result from intense meteoric water influx (Gutzmer et al. 2006), and isotopic equilibration with the ambient water, i.e., lowest values, is most complete in zones of highest fluid-rock interaction, for instances near ore-controlling structures (Thorne et al. 2009). The observed covariance of δ^{56}Fe with decreasing δ^{18}O thus means that iron isotope equilibration takes place during ore formation and is (as δ^{18}O) dependent on the intensity of fluid-rock interaction. According to Saunier et al. (2011), there is no iron isotope fractionation between hydrothermal fluid and precipitated hematite below 200 °C. The iron for microplaty hematite, locally sourced from dissolved iron oxides, Fe-rich carbonates and silicates, is thus characterized by ferrous Fe with low δ^{56}Fe values. Based on a carbonate-altered BIF (sample no. B, Fig. 6b), it can be speculated that hot basinal brines, which caused such carbonate alteration (Barley et al. 1999; Hagemann et al. 1999; Taylor et al. 2001; Thorne et al. 2004), were low in δ^{56}Fe and δ^{18}O, and thus developed light isotopic signatures in carbonates and oxides prior to the meteoric fluid flow. Alternatively, or additionally, incongruent dissolution/removal of heavy iron from Fe-rich phases during oxidation and hematite mineralization also caused low δ^{56}Fe values in iron ore samples. The latter fractionation process is invoked for magnetite oxidation, i.e., martitisation in ore. Although the involved meteoric water, at temperature of ~ 80–100 °C and low salinities (Thorne et al. 2008), cannot be very rich in iron, the fluid seems to be able to fractionate iron isotopes. In order to achieve this, mineral-fluid exchange of iron during protracted metasomatism at very high-fluid rock ratios is crucial. A similarly directed Fe–O trend in the Carajás hydrothermal hematite set (Fig. 6a) is striking and implies a Fe–O isotope coupling. However, this trend is formed by temporarily distinct (early to late) hematite stages and is thus discussed separately (see above). The Corumbá BIF samples, on the other hand, are characterized by disequilibrium during early microbial and late diagenetic hydrothermal processes and thus largely recorded uncoupled iron and oxygen isotope signatures.

Comparing the two Paleoproterozoic examples, QF and Hamersley, the range of δ^{56}Fe values are similar, and most host BIFs and hypogene iron ores fall within related ranges (Figs. 3, 7 and 8). In the Hamersley Province a two-stage model is invoked (Barley et al. 1999; Hagemann et al. 1999; Taylor et al. 2001; Thorne et al. 2004), and in the QF region Hensler et al. (2015) proposed a dual basinal-meteoric fluid mixing model. Mixing of meteoric fluids with deeper crustal brines is also implied for hematite precipitation in the Triassic Schwarzwald hydrothermal vein deposits (Germany), where δ^{56}Fe values range from − 0.49 to + 0.53 (Markl et al. 2006). This range partly overlaps those of the QF (− 0.80 to 0.37‰, excluding specularite) and Hamersley (− 0.30 to + 1.10‰) iron ores.

In contrast to the hydrothermal hematite stage, supergene-modified ore samples reveal a negative correlation of δ^{18}O and δ^{56}Fe (trend 4 arrow Fig. 6b). Hematite experienced isotopic fractionation to heavy Fe during incongruent dissolution of preferable light ^{56}Fe. The resulting weathering solutions are enriched in light iron isotopes (Fantle and DePaolo 2004; Bergquist and Boyle 2006; Ingri et al. 2006). This supergene fractionation process is the same one that has been inferred for goethite-altered Corumbá ore.

In summary, while oxygen fractionation follows a trend of decrease for both hydrothermal and supergene alteration, iron isotope fractionation shows marked differences: δ^{56}Fe decreases during hydrothermal alteration and increases during supergene alteration. The observed correlations suggest that during hypogene and supergene alteration processes, oxygen and iron isotope fractionations are coupled, and that fluid/rock ratios play significant roles for the magnitude of fractionation.

8 Conclusions

Based on iron isotope data, and coupled iron and oxygen isotope signatures of BIF-hosted iron ores from the iron districts considered in this contribution (Table 1), distributed in space and time, the following conclusions can be inferred:

1. Archean, jaspilite-hosted Carajás hypogene ores tend to display lower δ^{56}Fe values than their host BIF counterparts (Fig. 4). There is a correlation between coupled iron and oxygen isotope through the paragenetic sequence, progressively towards lower isotopic values, especially for δ^{18}O (Fig. 6a).
2. The Carajás (+ 1.16 to − 0.30‰) and the Paleoproterozoic Hamersley (+ 1.02 to − 0.29‰) hypogene ores display similar δ^{56}Fe intervals (Figs. 4 and 7). Ores from the Quadrilátero Ferrífero (QF) Paleoproterozoic Cauê Formation show a tendency towards lower values from + 0.37 to − 0.80‰ (Fig. 3).
3. In the Hamersley deposits, iron ore and hydrothermally modified BIF show a positive correlation between δ^{18}O and δ^{56}Fe (Fig. 6b), indicating that iron and oxygen isotope fractionation took place during oxidized meteoric water influx with variable fluid/rock ratios. In contrast, supergene-modified samples are characterized by a negative correlation of δ^{18}O and δ^{56}Fe, implying isotope coupling under distinct fluid-rock processes.
4. Quartz and carbonate itabirites from the QF have iron isotope signatures as a result of their distinct positioning in the depositional basin. The latter, with lower δ^{56}Fe ratios, is inferred to be more distal from the iron source than the former (Fig. 3).
5. The range in δ^{56}Fe values of hypogene iron ores is similar to that of the QF itabirites (Fig. 3). The iron isotope variations for the QF iron ores are better depicted when data are compared between the western, low-strain and the eastern, high-strain deformation domains, probably reflecting the different physical–chemical characteristics of the involved saline fluids that carried isotopically light Fe(II). Percolation of low-temperature fluids dominated the western domain, whereas the eastern domain was typified by high-temperature fluids. Precipitated oxides of this latter domain became enriched in the heavy iron isotopes due to iron isotope fractionation during redox transformation, resulting in ores less depleted than those of the low-strain domain.
6. Two processes controlled the iron isotopes in the Corumbá BIF (Fig. 3): a deep ocean seawater signature, characterized by a hydrothermal Fe fertilization, and a microbial or abiotic reduction in the iron source. This resulted in the formation of BIF with both non-fractionated δ^{56}Fe (± 0‰) and low δ^{56}Fe (− 0.5 to − 1.7‰; also Fig. 5). Local, diagenetic hydrothermal silica dissolution is reflected in the oxygen isotope data, but not in the iron isotope fractionation. Iron and oxygen isotope are coupled in the supergene stage: δ^{56}Fe values increase, while δ^{18}O values decrease (Fig. 5). This supports the involvement of light oxygen isotope-enriched and heavy iron isotope-depleted weathering solutions.
7. Overall, and despite all local differences, there is a general tendency for BIF δ^{56}Fe data to display moderately heavier values for all deposits compared to hypogene ores, which tend to shift towards slightly lower values. Direct coupling with oxygen isotopic values observed in Carajás, Hamersley, and Corumbá, indicate that δ^{56}Fe is a sensitive recorder for the intensity and type of fluid-rock interaction: lowest values are recorded in rocks of highest fluid-rock ratios during hydrothermal alteration and highest values are recorded in rocks of highest fluid-rock ratios during supergene/weathering alteration.

9 Implications for Exploration

The application of iron isotope in exploration is presently limited, although some promising applications are here indicated. For example, the correlation of increasing light oxygen and iron isotopes towards zones of high fluid-rock ratios may assist in targeting concealed shear zones, which could have acted as a significant fluid plumbing system, thereby facilitating the upgrade of BIF to high grade iron ore. Hand samples or drill core material from early, reconnaissance mapping or drilling could be analyzed for oxygen and iron isotopes. Any depleted values, when compared to least-altered BIF or itabirites, could be plotted into a geological map and contours of equal isotope values or ranges that may identify zones of oxygen and iron isotope depletion, hence increased hydrothermal fluid flow.

10 Future Work

Iron isotope studies in ore deposits have only been applied recently. Therefore, only limited case studies of iron isotopes in ore deposits or mineral systems have been conducted (e.g., Markl et al. 2006; Wang et al. 2011; Bilenker 2015; Santiago 2016). Most iron isotope studies have been applied to BIF systems where the redox environment is constrained and the potential for bacterial Fe(III) reduction has been established (Beard and Johnson 2004). Recently, Mendes et al. (2017) showed that the iron isotope composition of iron ore minerals varies systematically between deformational domains in the QF of Brazil. These authors link fluid temperatures and composition to different δ^{56}Fe values. By analogy, the δ^{56}Fe isotopic composition of iron oxide minerals in other iron ore systems hosted in itabirite, such as at Simandou (Guinea; Cope et al. 2008), could be analyzed in order to test whether metamorphism, or the lack thereof, and deformation in these deposits is linked to specific hydrothermal fluids, and whether δ^{56}Fe may be used to monitor and constrain such fluids. Iron isotope studies could be easily expanded to iron-oxide-copper–gold (IOCG) and iron-oxide-apatite (IOA) systems, where magnetite is a common alteration and, in rare cases, also an ore mineral. Some data are available in Santiago (2016) for IOCG deposits, and in Weis (2013), Weis et al. (2013), Bilenker (2015), Bilenker et al. (2016), Childress et al. (2016) Knipping et al. (2019), Troll et al. (2019) for IOA deposits.

Although many studies of IOCG systems have linked the origin of Cu to magmatic hydrothermal fluids (Williams et al. 2005a), the origin of Fe is largely unknown. Orogenic gold systems, in certain host rocks, contain magnetite as a characteristic hydrothermal alteration mineral, specifically in distal alteration zones (e.g., at the giant Golden Mile in Western Australia). Here the redox environment could be (better) constrained during hydrothermal fluid flow in distal portion of the ore system. Furthermore, in distal alteration zones of orogenic gold systems, the δ^{56}Fe composition of magnetite could be differentiated from that of the disseminated orthomagmatic magnetite, in host rock basalt or metamorphic equivalent, or metamorphosed syngenetic magnetite in gold-associated BIF deposits. Hydrothermal magnetite through fluid-rock interaction would produce fluids that have lower δ^{56}Fe values when compared to the orthomagmatic magnetite. In ancient, now land-based, volcanic-hosted massive sulfide systems, iron isotopes could be applied to vent fluids in order to better constrain the origin of the fluids, particularly the origin of iron in the various sulfides that characterize these systems.

Acknowledgements The present chapter was submitted in September 2019. Acceptance followed that same year. The final version was prepared for the publication of the book in September 2022. Subsequently, articles on iron isotopes applied to skans, VMS, polymetallic Fe-Cu-Pb-Zn, Kiruna IOA, and iron deposits were published, but could not be referenced.The authors wish to thank David Huston for inviting us to write this review contribution and also for reviewing and editing it. Thanks are due to Prof. G. Halverson for undertaking the iron isotope analyses. The SUERC, University of Glasgow, is acknowledged for the oxygen isotope analyses. Mining companies Rio Tinto Iron Ore, BHP-Billiton, Vale and Vetria Mineração are all acknowledged for allowing access to their mines and sampling. Two reviewers contributed with constructive comments and criticisms. LML, and RCFS are recipients of CNPq grants.

References

Alkmim FF, Marshak S (1998) Transamazonian orogeny in the southern São Francisco craton region, Minas Gerais, Brazil: evidence for Paleoproterozoic collision and collapse in the Quadrilátero Ferrífero. Precambrian Res 90:29–58

Anbar AD, Rouxel O (2007) Metal stable isotopes in paleoceanography. Annu Rev Earth Planet Sci 35:717–746

Anbar AD, Jarzecki AA, Spiro TG (2005) Theoretical investigation of iron isotope fractionation between Fe $(H_2O)_6{}^{3+}$ and $Fe(H_2O)_6{}^{2+}$: implications for iron stable isotope geochemistry. Geochim Cosmochim Acta 69:825–837

Angerer T, Hagemann SG, Danyushevsky LV (2012) Geochemical evolution of the banded iron formation-hosted high-grade iron ore system in the Koolyanobbing greenstone belt, Western Australia. Econ Geol 107:599–644

Angerer T, Duuring P, Hagemann SG, Thorne W, McCuaig TC (2014) A mineral system approach to iron ore in Archaean and Palaeoproterozoic BIF of Western Australia. Geol Soc London Spec Publ 393:81–115

Angerer T, Hagemann SG, Walde DN, Halverson GP, Boyce AJ (2016) Multiple metal sources in the glaciomarine facies of the Neoproterozoic Jacadigo iron formation in the "Santa Cruz deposit", Corumbá, Brazil. Precambrian Res 275:369–393

Balci N, Bullen TD, Witte-Lien K, Shanks WC, Motelica M, Mandernack KW (2006) Iron isotope fractionation during Fe(II) oxidation and Fe(III) precipitation microbially stimulated. Geochim Cosmochim Acta 70:622–639

Baltazar OF, Lobato LM (2020) Structural evolution of the Rio das Velhas greenstone belt, Quadrilátero Ferrífero Brazil: influence of Proterozoic orogenies on its western Archean gold deposits. Minerals 10:1–38

Barley ME, Pickard AL, Hagemann SG, Folkert SL (1999) Hydrothermal origin for the 2 billion year old Mount Tom Price giant iron ore deposit, Hamersley Province, Western Autralia. Miner Deposita 34:784–789

Beard BL, Johnson CM (2004) Inter-mineral Fe isotope variations in mantle-derived rocks and implications for the Fe geochemical cycle. Geochim Cosmochim Acta 68:4727–4743

Beard BL, Johnson CM, Cox L, Sun H, Nealson KH, Aguilar C (1999) Iron isotope biosignatures. Science 285:1889–1892

Beard BL, Johnson CM, Von Damm KL (2003) Iron isotope constraints on Fe cycling and mass balance in oxygenated Earth oceans. Geology 31:629–632

Becker RN, Clayton RN (1976) Oxygen isotope study of a Precambrian banded iron-formation, Hamersley Range, Western Australia. Geochim Cosmochim Acta 40:1153–1165

Bergquist BA, Boyle EA (2006) Iron isotopes in the Amazon River system: weathering and transport signatures. Earth Planet Sci Lett 248:54–68

Bilenker LD (2015) Elucidating igneous and ore-forming processes with iron isotopes by using experimental and field-based methods. Unpublished PhD thesis, The University of Michigan

Bilenker LD, Simon AC, Reich M, Lundstrom CC, Gajos N, Bindeman IN, Barra F, Munizaga F (2016) Fe-O stable isotope pairs elucidate a high-temperature origin of Chilean iron oxide-apatite deposits. Geochim Cosmochim Acta 177:94–104

Busigny V, Planavsky NJ, Jézéquel D, Crowe S, Louvat P, Moureau J, Viollier E, Lyons TW (2014) Iron isotopes in an Archean ocean analogue. Geochim Cosmochim Acta 133:443–462

Busigny V, Planavsky NJ, Goldbaum E, Lechte MA, Feng L, Lyons TW (2018) Origin of the Neoproterozoic Fulu iron formation, South China: insights from iron isotopes and rare earth element patterns. Geochim Cosmochim Acta 242:123–142

Childress TM, Simon AC, Day WC, Lundstrom CC, Bindeman IN (2016) Iron and oxygen isotope signatures of the Pea Ridge and Pilot Knob magnetite-apatite deposits, Southeast Missouri, USA. Econ Geol 111:2033–2044

Clout JMF (2005) Iron formation-hosted iron ores in the Hamersley Province of Western Australia. In: Iron Ore Conference 2005, Fremantle, WA, pp 19–21

Cope IL, Wilkinson JJ, Boyce AJ, Chapman JB, Herrington RJ, Harris CJ (2008) Genesis of the Pic de Fon iron oxide deposit, Simandou Range, Republic of Guinea, West Africa. Rev Econ Geol 15:339–360

Cox GM, Halverson GP, Minarik WG, Le Heron DP, Macdonald FA, Bellefroid EJ, Strauss JV (2013) Neoproterozoic iron formation: An evaluation of its temporal, environmental and tectonic significance. Chem Geol 362:232–249

Cox GM, Halverson GP, Poirier A, Le Heron D, Strauss JV, Stevenson R (2016) A model for Cryogenian iron formation. Earth Planet Sci Lett 43:280–292

Craddock PR, Dauphas N (2011) Iron and carbon isotope evidence for microbial iron respiration throughout the Archean. Earth Planet Sci Let 303:121–132

Crosby H, Johnson C, Roden E, Beard B (2005) Coupled Fe(II)–Fe(III) electron and atom exchange as a mechanism for Fe isotope fractionation during dissimilatory iron oxide reduction. Envir Sci Tech 39:6698–6704

Crosby HA, Roden EE, Johnson CM, Beard BL (2007) The mechanisms of iron isotope fractionation produced during dissimilatory Fe(III) reduction by *Shewanella putrefaciens* and *Geobacter sulfurreducens*. Geobiology 5:169–189

Czaja AD, Johnson CM, Beard BL, Roden EE, Li W, Moorbath S (2013) Biological Fe oxidation controlled deposition of banded iron formation in the ca. 3770 Ma Isua Supracrustal Belt (West Greenland). Earth Planet Sci Let 363:192–203

Dai Y, Zhu Y, Lianchang Z, Mingtian Z (2017) Meso- and Neoarchean banded iron formations and genesis of high-grade magnetite ores in the Anshan-Benxi Area, North China Craton. Econ Geol 112:1629–1651

Dalstra HJ, Rosière CA (2008) Structural controls on high-grade iron ores hosted by banded iron formation: a global perspective. Rev Econ Geol 15:73–106

Dauphas N, Rouxel O (2006) Mass spectrometry and natural variations of iron isotopes. Mass Spectr Rev 25:515–550

Dauphas N, van Zuilen M, Wahwa M, Davis AM, Marty B, Janney PE (2004) Clues from Fe isotope variations on the origin of early Archean BIFs from Greenland. Science 306:2077–2080

Dauphas N, Seth GJ, Rouxel O (2017) Iron isotope systematics. Rev Mineral Geochem 82:415–510

DOCEGEO (Rio Doce Geologia e Mineração S.A.) (1988) Revisão litoestratigráfica da Província Mineral de Carajás. In: Anais do XXXVI Congresso Brasileiro de Geologia, Belém, Brazil, Anexo, p 11–54

Dorr JVN (1945) Manganese and iron deposits of Morro do Urucum, Mato Grosso, Brazil. US Geol Surv Bull 946A:1–47

Dorr JVN (1969) Physiographic, stratigraphic and structural development of the Quadrilátero Ferrífero, Minas Gerais. US Geol Surv Prof Pap 641-A:110

Fabre S, Nédélec A, Poitrasson F, Strauss H, Thomazo C, Nogueira A (2011) Iron and sulphur isotopes from the Carajás mining province (Pará, Brazil): implications for the oxidation of the ocean and the atmosphere across the Archaean-Proterozoic transition. Chem Geol 289:124–139

Fantle MS, DePaolo DJ (2004) Iron isotopic fractionation during continental weathering. Earth Planet Sci Lett 228:547–562

Figueiredo e Silva RC, Lobato LM, Rosière CA (2008). A hydrothermal origin for the jaspilite-hosted giant Sierra Norte deposits in the Carajás Mineral Province, Pará State, Brazil. Rev Econ Geol 15:255–290

Figueiredo e Silva RC, Hagemann SG, Lobato LM, Rosière CA, Banks DA, Davidson GJ, Vennemann T, Hergt J (2013) Hydrothermal fluid processes and evolution of the giant Serra Norte jaspilite-hosted iron ore deposits, Carajás Mineral Province, Brazil. Econ Geol 108:739–779

Freitas BT, Warren LV, Boggiani PC, de Almeida RP, Piacentini T (2011) Tectono-sedimentary evolution of the Neoproterozoic BIF-bearing Jacadigo Group, SW-Brazil. Sedim Geol 238:48–70

Frost CD, von Blanckenburg F, Schoenberg R, Frost BR, Swapp SM (2007) Preservation of Fe isotope heterogeneities during diagenesis and metamorphism of banded iron formation. Contrib Mineral Petrol 153:211–235

Gaucher C, Sial AN, Frei R (2015) Chemostratigraphy of Neoproterozoic banded iron formation (BIF): types, age and origin. In: Ramkumar M (ed) Chemostratigraphy, concepts, techniques and applications. Elsevier, Amsterdam, pp 433–499

Gutzmer J, Mukhopadhyay J, Beukes NJ, Pack A, Hayashi K, Sharp ZD (2006) Oxygen isotope composition of hematite and genesis of high-grade BIF-hosted iron ores. Geol Soc Am Mem 198:257–268

Hagemann SG, Barley ME, Folkert SL, Yardley BBW, Banks DA (1999) A hydrothermal origin for the giant BIF-hosted Tom Price iron ore deposit. In: Stanley CJ (ed) Mineral Deposits—Process to Processing. Balkema, Rotterdam, pp 41–44

Hagemann SH, Angerer T, Duuring P, Rosière CA, Figueiredo e Silva RC, Lobato LM, Hensler AS, Walde DHG. BIF-hosted iron mineral system: a review (2016). Ore Geol Rev 76:317–359

Halverson GP, Poitrasson F, Hoffman PF, Nédélec A, Montel JM, Kirby J (2011) Fe isotope and trace element geochemistry of the Neoproterozoic syn-glacial Rapitan iron formation. Earth Planet Sci Lett 309:100–112

Harder EC (1914) The "Itabirite" iron ores of Brazil. Econ Geol 9:101–111

Hasui Y, Almeida F (1970) Geocronologia do centro-oeste brasileiro. Boletim Da Sociedade Brasileira De Geologia 19:5–26

Hensler AS, Hagemann SG, Rosière CA, Angerer T, Gilbert S (2015) Hydrothermal and metamorphic fluid-rock interaction associated with hypogene hard iron ore mineralization in the Quadrilátero Ferrífero, Brazil: Implications from in-situ laser ablation ICP-MS iron oxide chemistry. Ore Geol Rev 69:325–351

Hoefs J, Müller G, Schuster KA, Walde D (1987) The Fe-Mn ore deposits of Urucum, Brazil: an oxygen isotope study. Chem Geol Isotope Geoscience Section 65:311–319

Hoffman PF, Macdonald FA, Halverson GP (2011) Chemical sediments associated with Neoproterozoic glaciation: iron formation, cap carbonate, barite and phosphorite. Geol Soc London Mem 36:67–80

Holland HD (2006) The oxygenation of the atmosphere and oceans. Philos Trans R Soc Lond B Biol Sci 361 (1470):903–915

Horn I, von Blanckenburg F, Schoenberg R, Steinhoefel G, Markl G (2006) In situ iron isotope ratio determination using UV-femtosecond laser ablation with application to hydrothermal ore formation processes. Geochim Cosmochim Acta 70:3677–3688

Ingri J, Malinovsky D, Rodushkin I, Baxter DC, Widerlund A, Andersson P, Gustafsson O, Forsling W, Öhlander B (2006) Iron isotope fractionation in river colloidal matter. Earth Planet Sci Lett 245:792–798

Jacobsen SB, Kaufman AJ (1999) The Sr, C and O isotopic evolution of Neoproterozoic seawater. Chem Geol 161:37–57

Johnson CM, Beard B (2006) Fe isotopes: an emerging technique in understanding modern and ancient biogeochemical cycles. GSA Today 16:4–10

Johnson CM, Skulan JL, Beard BL, Sun H, Nealson KH, Braterman PS (2002) Isotopic fractionation between Fe(III) and Fe(II) in aqueous solutions. Earth Planet Sci Lett 195:141–153

Johnson CM, Beard BL, Beukes NJ, Klein C, O'Leary JM (2003) Ancient geochemical cycling in the Earth as inferred from Fe isotope studies of banded iron formations from the Transvaal Craton. Contrib Mineral Petrol 144:523–547

Johnson CM, Beard BL, Albarède F (eds) (2004) Geochemistry of non-traditional stable isotopes. Rev Mineral 55, 454 p

Johnson CM, Roden EE, Welch SA, Beard BL (2005) Experimental constraints on Fe isotope fractionation during magnetite and Fe carbonate formation coupled to dissimilatory hydrous ferric oxide reduction. Geochim Cosmochim Acta 69:963–993

Johnson CM, Beard BL, Klein C, Beukes NJ, Roden EE (2008a) Iron isotopes constrain biologic and abiologic processes in banded iron formation genesis. Geochim Cosmochim Acta 72:151–169

Johnson CM, Beard BL, Roden EE (2008b) The iron isotope fingerprints of redox and biogeochemical cycling in modern and ancient. Annu Rev Earth Planet Sci 36:457–493

Klar JK, Homoky WB, Statham PJ, Birchill AJ, Harris EL, Woodward EMS, Silburn B, Cooper MJ, James RH, Connelly DP, Chever F, Lichtschlag A, Graves C (2017) Stability of dissolved and soluble Fe(II) in shelf sediment pore waters and release to an oxic water column. Biogeochemistry 135:49–67

Klein C (2005) Some Precambrian banded iron-formations (BIFs) from around the world: their age, geologic setting, mineralogy, metamorphism, geochemistry, and origin. Am Mineral 90:1473–1499

Klein C, Ladeira EA (2004) Geochemistry and mineralogy of Neoproterozoic banded iron formations and some selected, siliceous manganese formations from the Urucum District, Mato Grosso do Sul, Brazil. Econ Geol 99:1233–1244

Knipping JL, Fiege A, Simon AC, Oeser M, Reich M, Bilenker LD (2019) In-situ iron isotope analyses reveal igneous and magmatic-hydrothermal growth of magnetite at the Los Colorados Kiruna-type iron oxide-apatite deposit, Chile. Am Min 104:471–484

Krapež B, Barley ME, Pickard AL (2003) Hydrothermal and resedimented origins of the precursor sediments to banded iron formation: sedimentological evidence from the Early Palaeoproterozoic Brockman Supersequence of Western Australia. Sedimentology 50:979–1011

Kunzmann M, Gibson TM, Halverson GP, Malcolm SWH, Bui TH, Carozza DA, Sperling EA, Poirier A, Cox GM, Wing BA (2017) Iron isotope biogeochemistry of Neoproterozoic marine shales. Geochim Cosmochim Acta 209:85–105

Lantinka ML, Oonka PBH, Floor GH, Tsikos H, Mason PRD (2018) Fe isotopes of a 2.4 Ga hematite-rich IF constrain marine redox conditions around the GOE. Precambrian Res 305:218–235

Lascelles DF (2006) The genesis of the Hope Downs iron ore deposit, Hamersley Province, western Australia. Econ Geol 101:1359–1376

Levasseur S, Frank M, Hein JR, Halliday AN (2004) The global variation in the iron isotope composition of marine hydrogenetic ferromanganese deposits: implications for seawater chemistry? Earth Planet Sci Lett 224:91–105

Li W, Huberty JM, Beard BL, Kita NT, Valley JW, Johnson CM (2013) Contrasting behavior of oxygen and iron isotopes in banded iron formations revealed by *in situ* isotopic analysis. Earth Planet Sci Lett 384:132–143

Li X, Wang J, Wang H (2017) Fe Isotopic compositions of modern seafloor hydrothermal systems and their influence factors. J Chem 2017:1417302. https://doi.org/10.1155/2017/1417302

Lobato LM, Figueiredo e Silva RC, Hagemann SG, Thorne WS, Zucchetti M, (2008) Hypogene alteration associated with high-grade banded iron formation-related iron ore. Rev Econ Geol 15:107–128

Lovley DR, Holmes DE, Nevin K (2004) Dissimilatory Fe(III) and Mn(IV) reduction. Adv Microb Physiol 49:219–286

Markl G, von Blanckenburg F, Wagner T (2006) Iron isotope fractionation during hydrothermal ore deposition and alteration. Geochim Cosmochim Acta 70:3011–3030

Mendes MCO, Lobato LM, Halverson GP, Kunsmann M, Rosière CA (2017) Iron isotope and REE+Y composition of the Paleoproterozoic banded iron formations and their related iron ores from the Quadrilátero Ferrífero, Brazil: Implications for their genesis. Mineral Deposita 52:159–180

Morris RC (1983) Supergene alteration of banded iron-formation. In: Trendall AF, Morris RC (eds) Iron-formation: facts and problems. Elsevier, Amsterdam, pp 513–534

Morris RC (1985) Genesis of iron ore in banded iron-formation by supergene and supergene-metamorphic processes; a conceptual model. In: Handbook of strata-bound and stratiform ore deposits vol 13, pp 73–235

Morris RC (2002) Genesis of high grade hematite orebodies of the Hamersley Province, Western Australia—a discussion. Econ Geol 97:177–181

Morris RC, Horwitz RC (1983) The origin of the iron-formation-rich Hamersley Group of Western-Australia—Deposition on a platform. Precambrian Res 21:273–297

Morris RC, Fletcher AB (1987) Increased solubility of quartz following ferrous-ferric iron reactions. Nature 330:250–252

Morris RC, Thornber MR, Ewers WE (1980) Deep-seated iron ores from banded iron-formation. Nature 288:250–252

Piacentini T, Vasconcelos PM, Farley KA (2013) $^{40}Ar/^{39}Ar$ constraints on the age and thermal history of the Urucum Neoproterozoic banded iron-formation, Brazil. Precambrian Res 228:48–62

Pires FRM (1995) Textural and mineralogical variations during the metamorphism of the Proterozoic Itabira

iron formation in the Quadrilátero Ferrífero, Minas Gerais, Brazil. An Acad Bras Ciênc 67:77–105
Planavsky NJ, Rouxel OJ, Bekker A, Hofmann A, Little CTS, Lyons TW (2012) Iron isotope composition of some Archean and Proterozoic iron formations. Geochim Cosmochim Acta 80:158–169
Poitrasson F (2006) On the iron isotope homogeneity level of the continental crust. Chem Geol 235:195–200
Powell CM, Oliver NHS, Li ZX, Martin DM, Ronaszeki J (1999) Synorogenic hydrothermal origin for giant Hamersley iron oxide ore bodies. Geology 27:175–178
Rosière CA, Rios FJ (2004) The origin of hematite in high-grade iron ores based on infrared microscopy and fluid inclusion studies: the example of the Conceição Mine, Quadrilátero Ferrífero, Brazil. Econ Geol 99:611–624
Rosière CA, Siemes H, Quade H, Brokmeier HG, Jansen EM (2001) Microstructures, textures and deformation mechanisms in hematite. J Struct Geol 23:1429–1440
Rosière CA, Baars FJ, Seoane JCS, Lobato LM, Silva LL, Souza SRC, Mendes GE (2006) Structure and iron mineralisation of the Carajás province. Trans Inst Mining Metall Sect B 115:B126–B136
Rosière CA, Spier CA, Rios FJ, Suckau VE (2008) The itabirites of the Quadrilátero Ferrífero and related high-grade iron ore deposits: an overview. Rev Econ Geol 15:223–254
Rosière CA, Sanglard JCD, Santos JOS, Mcnaughton N, Fletcher IR, Suckau VE, Spier CA (2012) Structural control and age of the high-grade iron ore of the Quadrilátero Ferrífero, Brazil. In: Peruvian Geological Congress and SEG 2012 Conference. Lima, Peru
Rouxel OJ, Dobbek N, Ludden J, Fouquet Y (2003) Iron isotope fractionation during oceanic crust alteration. Chem Geol 202:155–182
Rouxel OJ, Bekker A, Edwards KJ (2005) Iron isotope constraints on Archean and Paleoproterozoic ocean redox state. Science 307:1088–1091
Sampaio GMS, Pufahlb PK, Raye U, Kyser KT, Abreu AT, Alkmim AR, Nalini HA Jr (2018) Influence of weathering and hydrothermal alteration on the REE and δ^{56}Fe composition of iron formation, Cauê Formation, Iron Quadrangle, Brazil. Chem Geol 497:27–40
Santiago ESB (2016) Trace elements (in situ LA-ICP-MS) and stable isotopes (Δ^{33}S, Δ^{34}S, δ^{56}Fe, and δ^{18}O) in magnetite and sulphides: origin and evolution of the Neoarchean and Paleoproterozoic Cu-Au systems from the Carajás mineral province, Brazil. Unpublished PhD thesis, Universidade de Campinas
Santos JOS (2003) Geotectônica dos Escudos das Guianas e Brasil-Central. In: Bizzi LA, Schobbenhaus C, Vidotti RM, Goncalves JH (eds) Geologia, Tectônica e Recursos Minerais do Brasil. Companhia de Pesquisa e Recursos Minerais - CPRM. Brasília, Brazil, pp 169–226
Santos JOS, Lobato LM, Figueiredo e Silva RC, Zucchetti M, Fletcher IR, McNaughton NJ, Hagemann SG (2010) Two Statherian hydrothermal events in the Carajás province: evidence from Pb-Pb SHRIMP and Pb-Th SHRIMP dating of hydrothermal anatase and monazite. In: 7th South American Symp Isotope Geology (SSAGI), Brasília, Brazil
Saunier G, Pokrovski GS, Poitrasson F (2011) First experimental determination of iron isotope fractionation between haematite and aqueous solution at hydrothermal conditions. Geochim Cosmochim Acta 75:6629–6654
Schoenberg R, Marks MAW, Schuessler JA, von Blanckenburg F, Markl G (2009) Fe isotope systematics of coexisting amphibole and pyroxene in the alkaline igneous rock suite of the Ilímaussaq Complex, South Greenland. Chem Geol 258:65–77
Severmann S, Johnson CM, Beard BL, German CR, Edmonds HN, Chiba H, Green DRH (2004) The effect of plume processes on the Fe isotope composition of hydrothermally derived Fe in the deep ocean as inferred from the Rainbow vent site, Mid-Atlantic Ridge, 36°14′N. Earth Planet Sci Lett 225:63–76
Shanks WCP III (2014) Stable isotope geochemistry of mineral deposits. Treatise on Geochemistry 13:59–85
Sharp Z (1990) A laser-based microanalytical method for the in situ determination of oxygen isotope ratios of silicates and oxides. Geochim Cosmochim Acta 54:1353–1357
Skulan JL, Beard BL, Johnson CM (2002) Kinetic and equilibrium Fe isotope fractionation between aqueous Fe(III) and hematite. Geochim Cosmochim Acta 66:2995–3015
Sossi PA, O'Neill HS (2017) The effect of bonding environment on iron isotope fractionation between minerals at high temperature. Geochim Cosmochim Acta 196:121–143
Steinhoefel G, Horn I, von Blanckenburg F (2009) Micro-scale tracing of Fe and Si isotope signatures in banded iron formation using femtosecond laser ablation. Geochim Cosmochim Acta 73:5343–5360
Steinhoefel G, von Blanckenburg F, Horn I, Konhauser KO, Beukes NJ, Gutzmer J (2010) Deciphering formation processes of banded iron formations from the Transvaal and the Hamersley successions by combined Si and Fe isotope analysis using UV femtosecond laser ablation. Geochim Cosmochim Acta 74:2677–2696
Taylor D, Dalstra HJ, Harding AE, Broadbent GC, Barley ME (2001) Genesis of high-grade hematite orebodies of the Hamersley province, Western Australia. Econ Geol 96:837–873
Teixeira NL, Caxito FA, Rosière CA, Pecoits E, Vieira L, Frei R, Sial AN, Poitrasson F (2017) Trace elements and isotope geochemistry (C, O, Fe, Cr) of the Cauê iron formation, Quadrilátero Ferrifero, Brazil: Evidence for widespread microbial dissimilatory iron reduction at the Archean/Paleoproterozoic transition. Precambrian Res 298:39–55

Teng FZ, Dauphas N, Helz RT (2008) Iron isotope fractionation during magmatic differentiation in Kilauea Iki Lava Lake. Science 320:1620–1622

Thomazo C, Ader M, Farquhar J, Philippot P (2009) Methanotrophs regulated atmospheric sulfur isotope anomalies during the Late Archean (Tumbiana Formation, Western Australia). Earth Planet Sci Lett 279:65–75

Thorne WS, Hagemann SG, Barley M (2004) Petrographic and geochemical evidence for the hydrothermal evolution of the North deposit, Mt. Tom Price, Western Australia. Miner Deposita 39:766–783

Thorne WS, Hagemann SG, Webb A, Clout J (2008) Banded iron formation-related iron ore deposits of the Hamersley province, Western Australia. Rev Econ Geol 15:197–221

Thorne WS, Hagemann SG, Vennemann T, Oliver N (2009) Oxygen isotope compositions of iron oxides from high-grade BIF-hosted iron ore deposits of the central Hamersley Province, Western Australia. Constraints on the evolution of hydrothermal fluids. Econ Geol 104:1019–1035

Troll VR, Weis FA, Jonsson E, Andersson UB, Majidi SA, Högdahl K, Harris C, Millet M-A, Chinnasamy SS, Kooijman E, Nilsson KP (2019) Global Fe–O isotope correlation reveals magmatic origin of Kiruna-type apatite-iron-oxide ores. Nature Comm 10:1712. https://doi.org/10.1038/s41467-019-09244-4|www.nature.com/naturecommunications11234567890

Tsikos H, Matthews A, Yigal E, Moore JM (2010) Iron isotopes constrain biogeochemical redox cycling of iron and manganese in a Palaeoproterozoic stratified basin. Earth Planet Sci Lett 298:125–134

Urban N, Stribrny B, Lippolt HJ (1992) Iron and manganese deposits of the Urucum District, Mato Grosso do Sul, Brazil. Econ Geol 87:1375–1392

Veizer J, Ala D, Azmy K, Bruckschen P, Buhl D, Bruhn F, Carden GA, Diener A, Ebneth S, Godderis Y (1999) $^{87}Sr/^{86}Sr$, $\delta^{13}C$ and $\delta^{18}O$ evolution of Phanerozoic seawater. Chem Geol 161:59–88

Viehmann S, Bau M, Bühn B, Dantas EL, Andrade FR, Walde DN (2016) Geochemical characterisation of Neoproterozoic marine habitats: evidence from trace elements and Nd isotopes in the Urucum iron and manganese formations, Brazil. Precambrian Res 282:74–96

Walde DHG, Hagemann SG (2007) The Neoproterozoic Urucum/Mutún Fe and Mn deposits in W-Brazil/ SE-Bolivia: assessment of ore deposit models. Zdt Ges Geowiss 158:45–55

Walde DHG, Gierth E, Leonardos OH (1981) Stratigraphy and mineralogy of the manganese ores of Urucum, Mato Grosso, Brazil. Geol Rundsch 3:1077–1085

Wang Y, Zhu X-K, Mao J-W, Li Z-H, Chen Y-B (2011) Iron isotope fractionation during skarn-type alteration: A case study of Xinqiao Cu-S-Fe-Au deposit in the Middle-Lower Yangtze valley. Ore Geol Rev 43:194–202

Wawryk CM, Foden JD (2015) Fe-isotope fractionation in magmatic-hydrothermal mineral deposits: a case study from the Renison Sn-W deposit, Tasmania. Geochim Cosmochim Acta 150:285–298

Webb AD, Dickens GR, Oliver NHS (2004) Carbonate alteration of the Upper Mount McRae Shale beneath the martite-microplaty hematite ore deposit at Mount Whaleback, Western Australia. Mineral Deposita 39:632–645

Weis FA (2013) Oxygen and iron isotope systematics of the Grängesberg mining district (GMD), central Sweden. Unpublished PhD thesis, Uppsala Universitet

Weis FA, Troll VR, Jonsson E, Högdahl K, Barker A, Harris C, Millet MA, Nilsson KP (2013). Iron and oxygen isotope systematics of apatite-iron-oxide ores in central Sweden. In: Proceedings of the 12th Biennial SGA Meeting, vol 4, Mineral deposit research for a high-tech world. Uppsala, Sweden, pp 1675–1678

Weyer S (2008) What drives iron isotope fractionation in magma? Science 320:1600–1601

Whitehouse MJ, Fedo CM (2007) Microscale heterogeneity of Fe isotopes in >3.71 Ga banded iron formation from the Isua greenstone belt, southwest Greenland. Geology 35:719–722

Williams PJ, Barton MD, Johnson DA, Fontboté L, Haller A, Mark G, Oliver NHS, Marschik R (2005a) Iron oxide copper-gold deposits: Geology, space-time distribution, and possible modes of origin. Soc Econ Geol 100th Anniv Vol, pp 371–405

Williams HM, Peslier AH, McCammon C, Halliday AN, Levasseur S, Teutsch N, Burg JP (2005b) Systematic iron isotope variations in mantle rocks and minerals: the effects of partial melting and oxygen fugacity. Earth Planet Sci Lett 235:435–452

Yapp CJ (1990a) Oxygen isotopes in iron (III) oxides: 1 Mineral-Water Fractionation Factors. Chem Geol 85:329–335

Yapp CJ (1990b) Oxygen isotopes in iron (III) oxides: 2. Possible constraints on the depositional environment of a Precambrian quartz-hematite banded iron formation. Chem Geol 85:337–344

Zheng YF (1993) Calculation of oxygen isotope fractionation in anhydrous silicate minerals. Geochim Cosmochim Acta 57:1079–1091

Copper Isotopes Used in Mineral Exploration

Ryan Mathur and Yun Zhao

Abstract

The use of copper isotopes related to ore deposit location and genesis has greatly expanded over the past twenty years. The isotope values in ores, rocks, soils, and water range greater than 10‰ and provide ample isotopic variation to identify and interpret complex geological process. From an exploration standpoint, the copper isotope values in waters, sulfides and weathered rocks vector to mineralization at depth. Ground and surface waters display the greatest potential for both green and brownfields exploration, whereas Fe-oxides and other related ore minerals isotope compositions for exploration are nascent. From an ore genesis perspective, the copper isotopes serve as a redox proxy to aid in unraveling magmatic and hydrothermal processes related to metal sulfide precipitation. In summary, the use of copper isotopic approaches by the mining industry are ideal as they point to processes directly related to the metal of economic interest and should be employed in all stages of the mine life from exploration to extraction, and to environmental monitoring post-mining activities.

R. Mathur (✉)
Department of Geology, Juniata College, Huntingdon, PA 16652, USA
e-mail: mathurr@juniata.edu

Y. Zhao
State Key Laboratory of Geological Processes and Mineral Resources, China University of Geosciences, Beijing 100083, China

1 Introduction

Copper isotope analysis of different types of earth materials is an expanding field in the transition metal isotope geochemistry. Copper isotopes are highly sensitive to redox reactions and the majority of the studies completed to this point have used the copper isotope ratios to trace and understand geochemical reactions involving electron transfer. Other mechanisms have been proposed as causes for isotopic shifts such as biogeochemical processes, liquid–vapour transitions, mixing of different geochemical reservoirs and other physiochemical characteristics related to fluid-rock interactions. From a mineral exploration perspective, the copper isotope signatures in ores, rocks, fluids, and soils have been used as a means to vector to ores and understand the fundamental aspects of ore genesis.

The first reported copper isotope values of earth materials was by Shields et al. (1965). The errors were too large to identify the isotope differences among these materials. It was not until the introduction of advanced mass spectrometry instrumentation in the mid 1990s that analytical techniques allowed for more precise and accurate isotope measurements. The Multi-Collector Inductively-Coupled-Plasma Mass-Spectrometer

D. Huston and J. Gutzmer (eds.), *Isotopes in Economic Geology, Metallogenesis and Exploration*, Mineral Resource Reviews, https://doi.org/10.1007/978-3-031-27897-6_14

(MC-ICP-MS) permitted the simultaneous measurement of multiple metal isotopes with enhanced ionization of the metals to produce errors in the 4rd and 5th decimal place of isotope ratios (Halliday et al. 1995). This significant advancement led to the measurement of metal isotope variations that were initially predicted to be insignificant when isotopic fractionation principles were first developed in the 1940s (Urey 1947).

Copper ores received the majority of attention in the earliest studies of copper isotope because these materials contain large amounts of copper and were relatively simple to purify for isotopic analysis. It became apparent quickly within the literature that the degree of Cu isotope fractionation is significant and greater than the other isotope systems of similar mass (Fe, Ni and Zn). Thus, the mass spectrometry and purification of copper from different analytes received enhanced scrutiny to verify fractionation of greater than 10‰ (Fig. 1). The large range of isotopic compositions permits the ability to interpret and identify a variety of different geological processes.

Several different types of theoretical and experimental papers ranging from low to high temperature water–rock interactions have highlighted various mechanisms that lead to the fractionation of Cu isotopes (Zhu et al. 2002; Ehrlich et al. 2004; Weiss et al. 2004; Mathur et al. 2005; Markl et al. 2006; Seo et al. 2007; Pokrovsky et al. 2008; Balistrieri et al. 2008; Wall et al. 2011; Maher et al. 2011; Guo et al. 2020). Perhaps the most unique aspect of the studies of copper isotopes is the fact that redox reactions at low temperatures induce some of the largest fractionations measured in transition metals reported to date.

In this chapter we will provide an overview of the techniques used to measure copper isotopes, mechanisms that fractionate copper and most importantly, how these measurements can be used by exploration geologists in rocks, soils and solutions. Specifically, the organization of the chapter will focus on two themes: (1) the use of copper isotope signals in rocks, minerals and waters as a vector, and (2) the use of copper isotope signals in rocks and ores as a means to define source of metals and trace metallogenic processes. The techniques proposed here can be applied to detect and define systems containing copper rich minerals or minerals with higher concentrations of copper. This chapter illuminates the power of using copper isotope analysis for a diversity of deposit types and materials that formed in high to low temperature geological settings through a comparative summary of the salient points from published literature.

2 Analytical Methods

To obtain meaningful copper isotope measurements on MC-ICP-MS, copper must be purified from all other analytes in solution. Several papers provide liquid ion chromatographic protocols to separate copper from complex chemical matrices (Marechal et al. 1999; Zhu et al. 2000; Marechal and Albarede 2002; Mason et al. 2005; Mathur et al. 2005, 2012a; Chapman et al. 2006; Borrok et al. 2007; Petit et al. 2008; Lobo et al. 2014; Sossi et al. 2015; Liu et al. 2014a, 2015). Each of the papers have used anion exchange resins in the chloride form and have eluted copper with different normalities of hydrochloric acids. The process removes unwanted ions that could cause a mass bias or interference during the measurement.

The most common resin used to separate copper is the BioRad MP-1 (maro porous HCl form of the resin beads) in 10 ml BioRad columns. Most of the columns used for the protocol have been 10 ml BioRad spin chromatography columns. The aspect ratio of the tapered end of the column can impact the separation of copper. Depending on the matrix of the solution, the column procedure was repeated if the copper was not adequately purified after one pass through the column. The process uses relatively larger volumes of ultrapure acids and has been scaled based on the amount of ions placed on the resin. More recently, Kidder et al. (2020) demonstrated that using a clean-up cation exchange column in an automated ion exchange chromatography is ideal for high Na rich matrices.

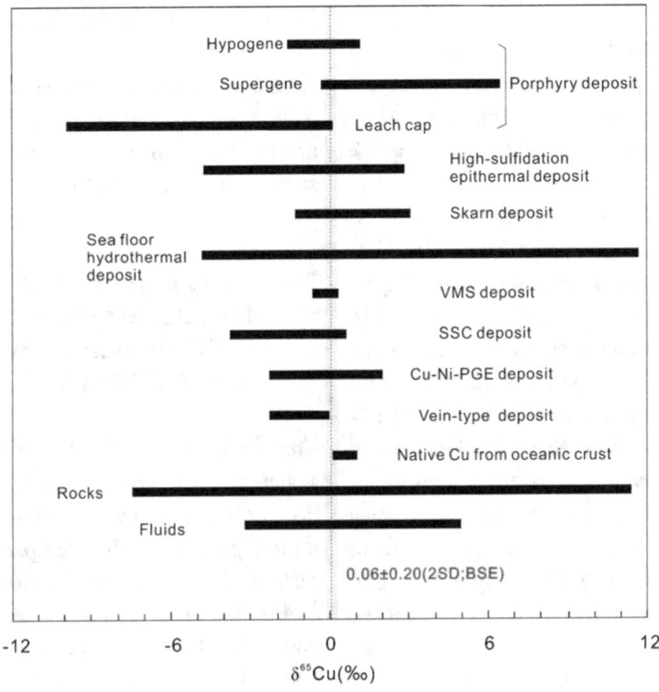

Fig. 1 Diagram showing variation in δ^{65}Cu values of different types of deposits, rocks and waters and soils, including porphyry deposits (Mathur et al. 2009; Wu et al. 2017), high-sulfidation epithermal deposits (Duan et al. 2015; Wu et al. 2017), skarn deposits (Maher and Larson 2007; Wang et al. 2017), sea floor hydrothermal deposits (Zhu et al. 2000; Rouxel et al. 2004), volcanic-associated massive sulfide (VMS) deposits (Mason et al. 2005; Housh and Çiftçi 2008; Ikehata et al. 2011), Cu–Ni deposits (Malitch et al. 2014; Ripley et al. 2015; Zhao et al. 2017, 2019, 2022; Tang et al. 2020), sediment-hosted stratiform copper (SSC) deposits (Asael et al. 2007, 2009, 2012), vein-type deposits (Haest et al. 2009), native Cu from oceanic crust (Dekov et al. 2013) as well as rocks and fluids (Mathur and Fantle 2015). The δ^{65}Cu value of bulk silicate Earth (BSE) is 0.06 ± 0.20‰ (2SD) (Liu et al. 2015)

Recovery of 90–100% of the copper during the chemical processing is critical to obtain meaningful results. Marechal et al. (1999) first pointed out that the resin can fractionate copper during the chromatographic separation on the order of 2–3‰ on synthetic solutions. They noted that ^{63}Cu is preferentially removed in the first 10 ml compared to the following 15 ml of copper eluted from the column. Different bonding strengths of the copper chlorides attached to the resin cause fractionation of copper during elution. Zhu et al. (2000) and Mathur et al. (2005) found the same relationships in natural rock and mineral matrices.

Copper minerals that possess 10's of % concentrations of copper do not require chemical purification by the ion exchange chromatography discussed above. Minerals such as chalcopyrite ($CuFeS_2$), chalcocite (Cu_2S), Cu-oxides (like copper carbonates such malachite and azurite), and native copper (Cu^o) have been dissolved and diluted for isotopic measurement both with and without chromatography and have generated the same copper isotope value. Therefore, measurement of these minerals does not require purification (Mathur et al. 2005; Zhang et al. 2020; Zhu et al. 2000).

The instrumentation setup and measurement of copper isotope on different MC-ICP-MS systems is relatively similar. The measurements were made in both wet and dry plasma in low-resolution mode. Solution concentrations range from 40 ppb in dry plasma mode up to 300 ppb in a wet plasma. The copper isotope voltages range from 2 to 10 V in this setup. On peak background blank subtraction and matching of standards and samples within 30% produces reliable and repeatable measurements. Most studies do not use gases other than Ar during the measuring session.

There are two general approaches for mass bias correction during measurement. The simplest is through using the standard-sample-standard bracketing technique (SSB). The NIST 976 is the most common standard used in the literature, and copper isotope values are presented in the traditional per mil (‰) format relative to NIST 976 isotope standard. The second method involves doping the samples with either Ni or Zn standards and correcting mass bias through the exponential law as presented in Marechal et al. (1999). Since copper has only two isotopes, same element spikes cannot be added to solve for mass bias and there is no means to identify mass dependence in the natural samples nor during the measurements of copper isotopes. In order for this method to work, the added element standard must ionize in the plasma in nearly the identical means as copper. Smaller errors on the measurements have been reported using these techniques (± 0.12‰ for SSB in comparison to ± 0.06‰ for doping, 2σ). Liu et al. (2014a) established a method using a combination of SSB and Zn-doping, yielding external reproducibility better than ± 0.05‰ (2σ). The best accuracy and precision reported for doped samples was by the use of allium reported in Sullivan et al. (2020; Yang et al. (2019) with errors on the order of 0.03 ‰.

Error estimations for the analyses have been assessed in various ways. The error associated with background/blank copper is minimal in comparison to the samples analysed and is not considered a large source of error. Nor is the instrumentation measurement error, the counting statistics associated with measurement of copper isotope ratios are normally in the 4th to 5th decimal place of the measured value. The repeatability of the measurement has been assessed by comparing whole procedural replicates, measuring the solution multiple times in one session, and monitoring the variations of the NIST standard throughout the session (in this method the standard is bracketed to the other standard nearest in the session). Of the three techniques, the largest errors on samples reported are the variation of the standard throughout the session (Liu et al. 2014a).

2.1 Cu Isotope Fractionation Factors, Evidence from Experimental Work and Field Studies

The Cu isotope fractionation factors in different systems are of great importance for the application of Cu isotopes tracking complicated geological processes. High temperature experimental results show that the Cu-isotopic fractionation factors between aqueous fluids and silicates are controlled by Cu speciation in the fluids (e.g. $CuCl(H_2O)$, $CuCl^{2-}$ and $CuCl_3{}^{2-}$) and silicate melts ($CuO_{1/2}$), with $\Delta^{65}Cu_{fluid\text{-}melt}$ ranging from 0.08 ± 0.01‰ to 0.69 ± 0.02‰ (Guo et al. 2020). The fractionation factor between liquid and vapour ($\Delta^{65}Cu_{liquid\text{–}vapour}$) is estimated to be 0.10 ± 0.07‰ at 400–450 °C (Rempel et al. 2012). Experimentally determined Cu isotope fractionation factors between sulfide melt and silicate magma are controlled by Ni contents in sulfide, with high Ni (~25 wt%, $\Delta^{65}Cu_{sulfide\ melt\text{-}magma}$ = ~ −0.1‰;) showing smaller fractionation than sulfides with low Ni [0.1 to 1.2 wt%, $\Delta^{65}Cu_{sulfide\ melt\text{-}magma} = 0.77 \times 10^6/T^2 - 4.46 \times 10^{12}/T^4$, where T is in °K (Xia et al. 2019)]. In the experiment work of Savage et al. (2015), sulfide phases also preferentially incorporate light Cu isotopes relative to bulk Cu ($\Delta^{65}Cu_{sulfide\ melt\text{-}bulk\ Cu}$ = ~ −0.1‰). Numerical Rayleigh fractionation modeling based in massive sulphide ores in the Tulaergen magmatic Ni–Cu deposit indicates that $\Delta^{65}Cu_{residual\ sulfide\ melt\text{–}MSS}$ is approximately 1.0011 (Zhao et al. 2019). Finally, Ni et al. (2021) demonstrated through evaporation by the use of a laser, large copper fractionation factors result in 10's of ‰ fractionation during planetary processes because of volatile nature of copper in ultrahigh temperature processes.

For lower temperature applications copper isotope fractionation factors have been defined in great detail for redox, phase change, adsorption and other mechanisms. The oxidative dissolution of copper sulphide minerals results in the heavier isotope concentrating in the aqueous phase as pointed out in multiple contributions (Zhu et al. 2000, 2002; Marechal and Albarede 2002; Ehrlich et al. 2004; Mathur et al. 2005; Markl et al. 2006; Asael et al. 2007; Fernandez and Borrok 2009). Sherman (2013) provides a detailed theoretical argument that demonstrates how kinetics modelled with Rayleigh distillation models explain values measured in supergene sulfides. The lack electron transfer does not induce nearly the same magnitude of copper isotope fractionation as seen in the Cu oxide weathering experiments in Plumhoff et al. (2021). Both biological and adsorption mechanisms have been studied in great detail in association with soil and environmental geochemical where copper can be fractionated at relatively smaller degree in comparison to the sulphide redox reactions (Pokrovsky et al. 2008; Bigalke et al. 2009, 2010, 2011; Weinstein et al. 2011; Liu et al. 2014b; Fekiacova et al. 2015; Babcsányi et al. 2016; Li et al. 2016; Song et al. 2016; Guinoiseau et al. 2017; Kusonwiriyawong et al. 2017; Dótor-Almazán et al. 2017; Blotevogel et al. 2018; Mihaljevič et al. 2018). These studies have clearly shown how various mechanisms for copper isotope fractionation operate in conjunction in the low temperature interactions characteristic of the critical zone.

2.2 Copper Isotopes Used as a Vector and for Source Information

2.2.1 Copper Isotope Signatures in Minerals

Multiple studies have used the copper isotope compositions to assess the lateral and vertical variations as a means to use the copper isotope values as a vector to track concealed orebodies. The vector is developed through plotting isotope values on maps or within cross sections to deduce patterns through contouring of the data. Most of these studies were performed on mineralization systems that operated at high temperatures. Here, we will summarize the findings and implications of the how and why copper isotope values change in these economically significant deposits.

The highest temperature systems concerning sulfide mineralization are mafic–ultramafic intrusions and associated deposits. In a large set of peridotites, basalts and subduction related andesites and dacites, Cu isotope fractionation is insignificant during mantle partial melting and magmatic differentiation, and the bulk silicate Earth (BSE) has an average δ^{65}Cu value of around 0.06‰ (Liu et al. 2015; Savage et al. 2015; Huang et al. 2016). However, mantle metasomatism with involvement of recycling crustal materials strongly fractionates Cu isotopes, spanning a wide range of δ^{65}Cu values from − 0.64‰ to + 1.82‰ (Liu et al. 2015). It seems that the addition of recycling crustal materials into the mantle source would result in obvious Cu isotope heterogeneity in the mantle source. The magmas derived from mantle partial melting are expected to inherit the Cu isotopic compositions of their sources.

Recent studies show that significant Cu isotope fractionation ($\sim$4‰ for δ^{65}Cu value) have been involved during the formation of the magmatic Ni–Cu deposits (Fig. 2; Malitch et al. 2014; Ripley et al. 2015; Zhao et al. 2017, 2019, 2022; Brzozowski et al. 2020; Tang et al. 2020). Many factors have been employed to explain the large variation of Cu isotopes ($\sim$4‰ for δ^{65}Cu value) in these deposits, including multiple mantle sources, variable degree of mantle melting, different magmatic processes, crustal contamination, R factor, and/or redox reactions (Malitch et al. 2014; Ripley et al. 2015; Zhao et al. 2017, 2019; Brzozowski et al. 2020; Tang et al. 2020). The generation of magmatic Ni–Cu deposits involves a series of stages: (1) mantle melting; (2) magma ascent; (3) segregation of sulfide melt from silicate melt and sulfide enrichment; and (4) silicate mineral crystallization and possible internal fractionation within segregated sulfide melt (Barnes and Lightfoot 2005; Naldrett 2010). Malitch et al. (2014) assert that the Cu isotopic variation observed in the Noril'sk deposits region may result from multiple magmatic processes, different magma pulses, or assimilation of Cu from external sources. The distinct difference in δ^{65}Cu values between the sheet-style and conduit-style mineralization in the Lake Superior area is likely caused by mantle sources, variable degrees of melting, sulfide retention, and sulfide liquid fractionation (Ripley et al. 2015). It is difficult to precisely evaluate the contribution of the factors causing significant variations of Cu isotope compositions in the mafic–ultramafic intrusions of the Noril'sk region and Lake Superior area, since these intrusions are in different locations with diverse country rocks. Zhao et al. (2017 2019) reported a systematical Cu and Fe isotopes study of the Tulaergen Ni–Cu deposit, China. In this deposit, > 2‰ variation of δ^{65}Cu values has been observed, and the potential influence of crustal contamination, hydrothermal overprinting, diffusion, mantle sources or magmatic processes are excluded (Zhao et al. 2017, 2019). On the basis of negative correlation between δ^{65}Cu and δ^{56}Fe values of chalcopyrite and positive correlation between δ^{65}Cu values of chalcopyrite and δ^{56}Fe values of whole-rocks, redox reactions are expected to be the main factor governing the Cu isotope fractionation in magmatic Ni–Cu mineralization system (Zhao et al. 2017, 2019). However, Tang et al. (2020) proposed that Cu isotopic ratios show no correlation with intrusion/orebody shape, location of sample in the orebody, lithofacies or type of mineralization in the Kalatongke and Baishiquan deposits. The large variation of Cu isotopic compositions in these deposits is attributed to the degree of partial melting in the mantle source and associated magmas (Tang et al. 2020).

Even though debates addressing processes responsible for Cu isotope fractionation in mafic–ultramafic intrusions need to be addressed in further studies, Cu isotopes shed a new light on mineral explorations. Malitch et al. (2014) show a progression from lower copper isotope values in distal mineralization (Kharaelakh deposit) and the Talnakh occurrences to the higher values proximal to the Noril'sk-1 intrusion. The negative correlation of S and Cu isotope compositions for an individual intrusion in the Noril'sk region may be a useful indicator of the potential for hosting Ni-Cu-PGE sulfide deposits (Malitch et al. 2014). The distinct difference of Cu isotopic compositions between sulfides in disseminated sheet style and conduit style mineralization may be indicative of different types of mineralization, even though assimilation of Cu from external sources should be carefully evaluated (Ripley et al. 2015). Zhao et al. (2017) reported the Cu isotopic compositions in the Tulaergen deposit and point out that the proximal parts of the system tend to enrich light Cu isotopes than the distal samples. Zhao et al. (2019) further modelled the Cu isotope variation with the combination of PGE tenors, and conclude that sulfide segregation enriches light Cu isotopes, which is consistent high-temperature experimental work (Xia et al. 2019). Numerical calculations also indicate that significant Cu isotope fractionation occurs between monosulfide solid-solution (MSS) and evolved sulfide melt, with heavy Cu isotopes preferring evolved liquid to MSS (Fig. 3; Zhao et al. 2019). Considering that Ni and IPGE are preferentially incorporated in MSS, and Cu, Ni and PPGE tend to enriched in evolved sulfide melt (e.g., Li et al. 1996; Barnes and Lightfoot, 2005; Naldrett, 2010, 2011), Zhao et al. (2019) proposed that the mafic–ultramafic intrusions with light Cu isotopes may be favourable hosts for Ni and PGE mineralization, whereas elevated δ^{65}Cu values are potentially indicative of enrichment in Cu, Ni, Pt, Pd, and Au (Fig. 3). We notice that the Cu isotopes of the Baishiquan deposit are slightly heavier than the Kalatongke deposit (Tang et al. 2020), which is potentially related to different mineralization processes. Further work needs to be done to clarify how to use Cu isotopes as a vector for mineral exploration in different mafic–ultramafic intrusions.

Perhaps the most studied mineralized systems by Cu isotopes are porphyry copper and hydrothermally related mineralization such as skarns and epithermal gold deposits. Figure 4 shows the general patterns recorded in chalcopyrites from porphyry copper deposits.

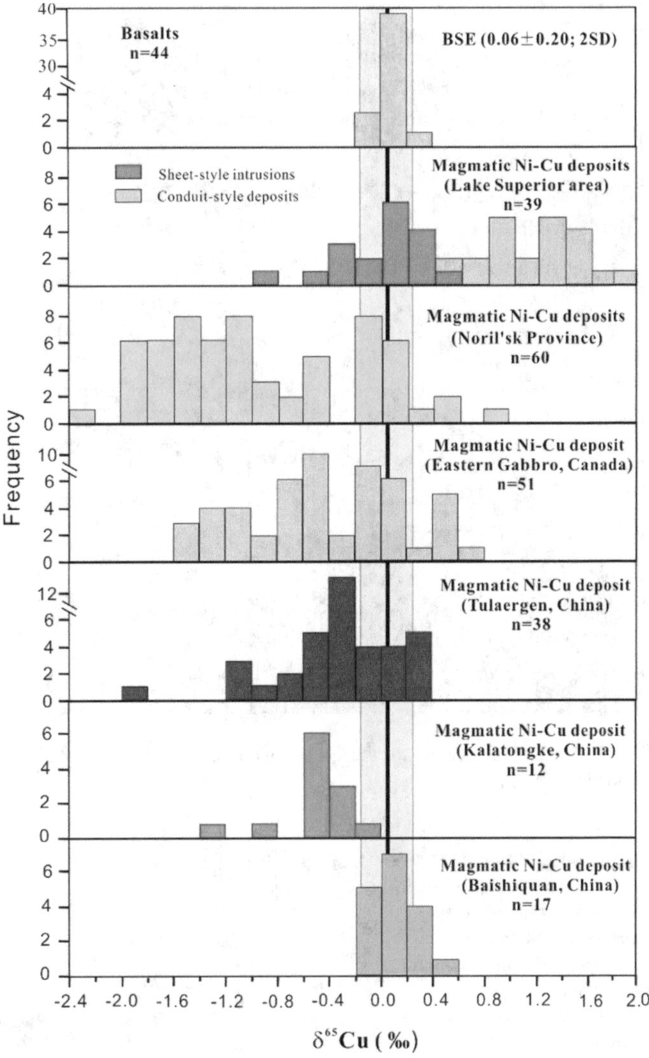

Fig. 2 Histograms of copper isotopic compositions of various types of basalts (Liu et al. 2015), and magmatic Ni–Cu deposits in the Lake Superior area (Ripley et al. 2015), in the Noril'sk Province (Malitch et al. 2014), in the Eastern gabbro (Brzozowski et al. 2020), at the Tulaergen deposit (Zhao et al. 2017, 2019), and at the Kalatongke and Baishiquan deposits (Tang et al. 2020). The δ^{65}Cu value of bulk silicate Earth (BSE) is 0.06 ± 0.20 ‰ (2SD) (Liu et al. 2015)

Chalcopyrite in veins from cores of the systems associated with higher temperature alteration silicate minerals like biotite and potassium feldspar have lower copper isotope values that cluster around 0‰ (Mathur et al. 2009, 2010, 2018; Mirnejad et al. 2010; Wall et al. 2010; Braxton and Mathur 2011; Asadi et al. 2015; Graham et al. 2004; Li et al. 2010; Duan et al. 2016; Gregory and Mathur 2017). In contrast, chalcopyrite in veins from distal part of the systems associated with lower temperature alteration silicate minerals like phyllite and kaolinite have higher copper isotope values. Other copper rich minerals like bornite also show similar patterns to chalcopyrite (Dendas 2011).

Skarns display an inverse pattern where the higher copper isotope values are found in the cores of the systems and the lower values are found in the outer parts of the system (Maher et al. 2003; Larson et al. 2003; Yao et al. 2016). This inverse relationship is most likely related to the mechanisms that are fractionating copper during the formation of copper minerals from the hydrothermal solution. Experimental studies indicated that pH, fO_2 and salinity related to fluid-vapour separation controlled copper species

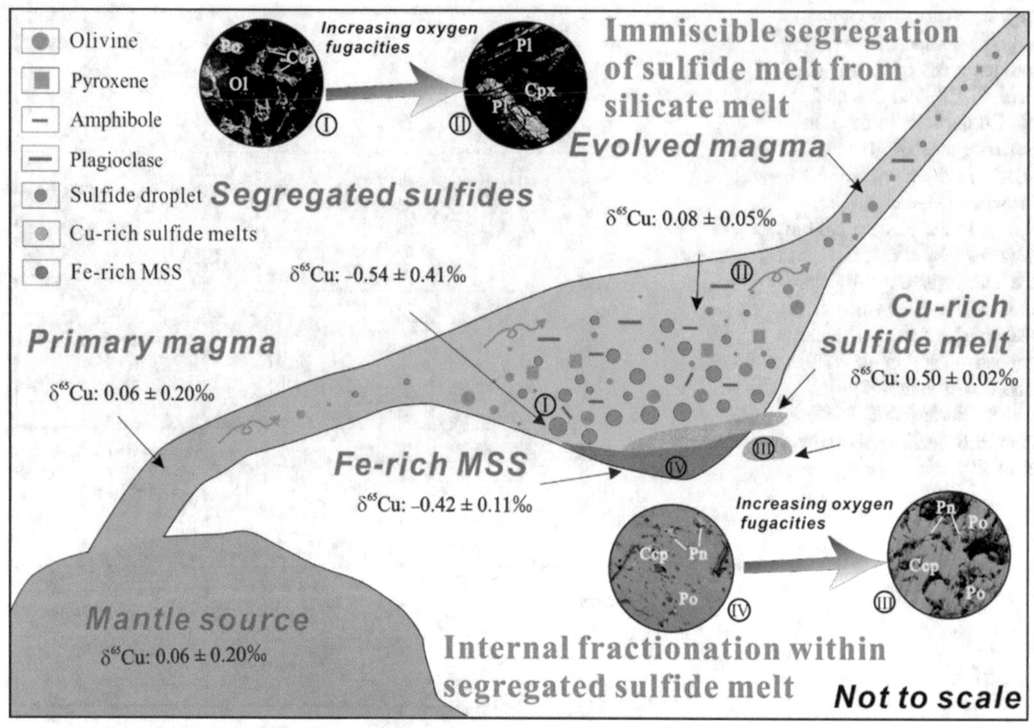

Fig. 3 A schematic model for the Cu isotope fractionation in the Tulaergen magmatic Ni–Cu deposit (modified from Zhao et al. 2019)

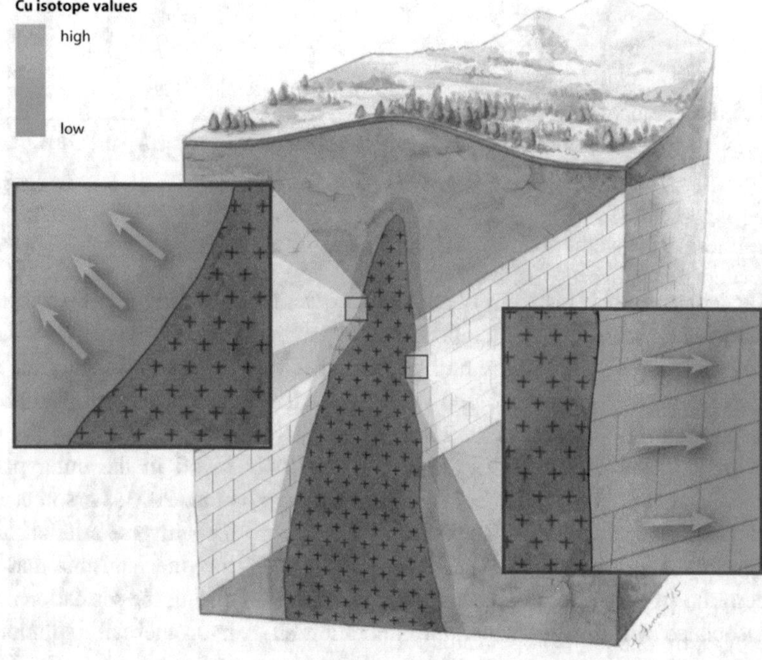

Fig. 4 Copper isotope values show pattern from proximal and distal locations in the stockwork and skarns associated with the intrusion of porphyritic rocks. Reproduced with permission from Mathur and Wang (2019); Copyright 2019 American Geophysical Union

into the vapour phase (Maher et al. (2011). Thus, perhaps the difference in fractionation patterns between these two systems is related to metal transport with different ligands. For instance, CuCl species form in the acidic fluids associated with porphyry copper deposits in silicate rocks in contrast to CuCO species which form in buffered reactions associated with skarn mineralization. The different bonds strength associated with the metal–ligand could lead to the different fractionation factors during mineral precipitation. Seo et al. (2007) provided theoretical calculations relating different bond strength calculations that predicted the isotopic shifts that would be potentially happening in this system.

Epithermal Au deposits show a similar pattern to porphyry copper deposits (Duan et al. 2016) and the gold rich ores in these systems also have values that cluster around 0‰ (Saunders et al. 2015; Mathur et al. 2012b). Interestingly, the pattern of lower values in the core of the deposits and higher values on the borders is identical to porphyry copper deposits. If porphyry copper deposits are related to epithermal systems at depth, the isotopic values at the cores of the systems should be higher than that found in porphyries, instead they are identical. To date, the copper isotopes for a porphyry copper deposit beneath an epithermal system was only carried out at Yanacocha (Condon et al. 2012). At Yanacocha, the values of copper isotopes in chalcopyrite are lowest in the deepest parts of the porphyry and become higher to the base of the epithermal system. The two chalcopyrite and chalcocite values at the base of the epithermal system return 0‰ and then values become higher to the outer portions of the epithermal system. These preliminary data suggest that the two hydrothermal systems are unique and not linked to the same hydrothermal event.

Volcanogenic massive sulfide display the largest range of copper isotope variations within the higher temperature hydrothermal systems. Rouxel et al. (2004) presented values from an active VMS site in the Mid-Atlantic and reported ranges of copper isotope values of up to 3‰ which was attributed to reworking and later stage mobilization of copper. Berkenbosch et al. (2015) found smaller variations in a modern VMS in the Pacific and noted that the liquid vapour partitioning could explain the fractionation as vapours emitted from the chimneys are metal rich. Ancient VMS systems have not displayed as much copper isotope variations in the massive ores of the systems with values ranging around 1‰. These studies did not find simple, systematic spatial variations with copper isotopes in massive sulphides (Ikehata et al. 2011; Mason et al. 2005).

The lower temperature systems like sedimentary copper and supergene copper deposits display much larger copper isotopic variations than PCDs and LMIs. The interpretations for the copper isotope data involve understanding redox reactions associated with copper migration at lower temperatures. For sedimentary copper deposits, classic deposits of the Kuperschifer (Asael et al. 2007, 2009, 2012; Pękala et al. 2011) and the sedimentary copper deposits in the Democratic Republic of Congo (Haest et al. 2009) have copper isotope values in sulfide and oxide minerals that trace redox reactions and reflect fluid pathways in these systems (Fig. 5). Similar conclusions can be deduced from copper isotope values of native copper in large occurrence of native copper deposits in Michigan, USA (Bornhorst and Mathur 2017) and the native Cu deposits in Brazil (Baggio et al. 2018). Luczaj and Huang (2018) indicate little copper isotope fractionation in MVT ores of the central USA and point to the lack of redox reactions as the cause for this homogeneity.

Copper isotope fractionation is greatest at the lowest temperature deposits associated with supergene processes. Copper isotope values for sulfide minerals and residual Fe-oxides in leached zones range from − 10‰ to + 15‰. A general relationship of lower copper isotope values in leach cap minerals in comparison to the higher isotope values in the enrichment zones point to oxidation associated with the migration of the water table as the cause for the pattern. Multiple studies have demonstrated the use of copper isotope values in ores and leach cap minerals can be used to assess the degree of weathering (Mathur et al. 2005, 2009, 2010; Mirnejad et al. 2010; Asadi et al. 2015).

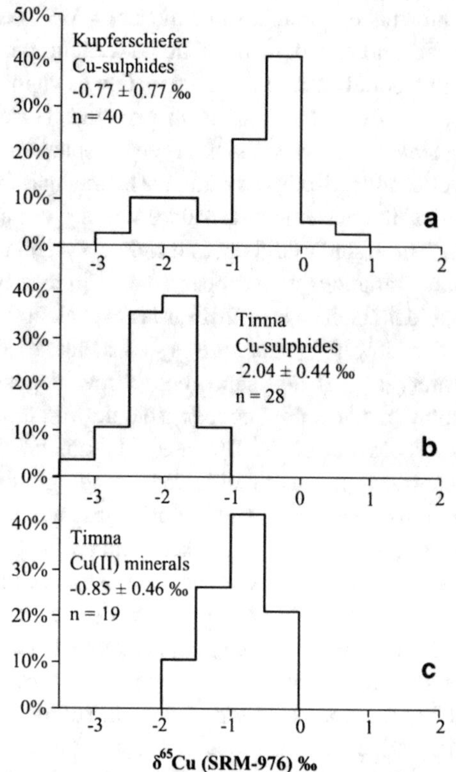

Fig. 5 Copper isotope variations in sedimentary copper deposits. Reproduced with permission from Asael et al. (2009); Copyright 2009 Elsevier

Specific examples of using the copper isotopes in exploration of supergene deposits are present in the following two contributions. Braxton and Mathur (2011) provide an excellent example of how the copper isotope values in the leach cap minerals can be used to identify the presence of multiple weathering steps of previous enrichment events. They demonstrated higher copper isotope values in the Fe-oxides of the leach cap correlate with larger concentrations of copper at depth. Equally important, they pointed to the fact that the copper isotope values of chalcocite change down hydrogeologic gradients and thus point to a flow path for the mineralization fluid. Braxton and Mathur (2014) show that in an area covered with Fe-oxide but no mineralization at surface, copper isotopes of Fe-oxide minerals vector to areas in the subsurface where supergene chalcocite has been weathered and concentrated. Both studies point to the value in using copper isotope values in Fe-oxides on the surface, which are abundant and useful in identification of buried exploration targets.

Given that lower temperature process can cause greater degrees of fractionation coupled with the redox sensitivity of the Cu isotopes, Mathur et al. (2018) recognized that the copper isotope signature of chalcocite could be used to trace its origin. The influence or overprint of supergene processes in many mineral deposits makes identification of the process of chalcocite origins enigmatic. These authors suggested (as shown in Fig. 6) that lower copper isotope values in chalcocite originated from Cu-chloride complexes which formed sedimentary copper deposit chalcocite, whereas supergene chalcocite had higher copper isotope values because it formed in oxidative environments. Magmatic chalcocite has values hovering around 0‰ as smaller degrees of fractionation occur at higher temperatures (Fig. 6).

2.2.2 Copper Isotope Signatures in Rocks

The use of rocks in their relationship to ores has not been explored in any great detail. The main purpose of these studies would be a means to link the copper mineralization to the source materials in some fashion. There is not a large variation of copper isotopes compositions found in igneous, metamorphic and sedimentary rocks, and the range of copper isotope values is similar to that of chalcopyrite formed at high temperatures at 0 ± 1‰ (Chapman et al. 2004; Sossi et al. 2015; Fru et al. 2016; Wang et al. 2019; Liu et al. 2019, 2014a; Huang et al. 2016; Busigny et al. 2018). Zheng et al. (2019) offered the first argument that relates higher copper isotope values found in rocks associated with the PCDs in Tibet. Their argument focuses on a few values of intermediate igneous rocks that slightly heavier (by approximately + 0.3‰) in which they argue the higher signatures are originated from weathered copper that is isotopically heavier copper transported by subducted materials. In another contribution, Wang et al. (2019) points to a similarity of the

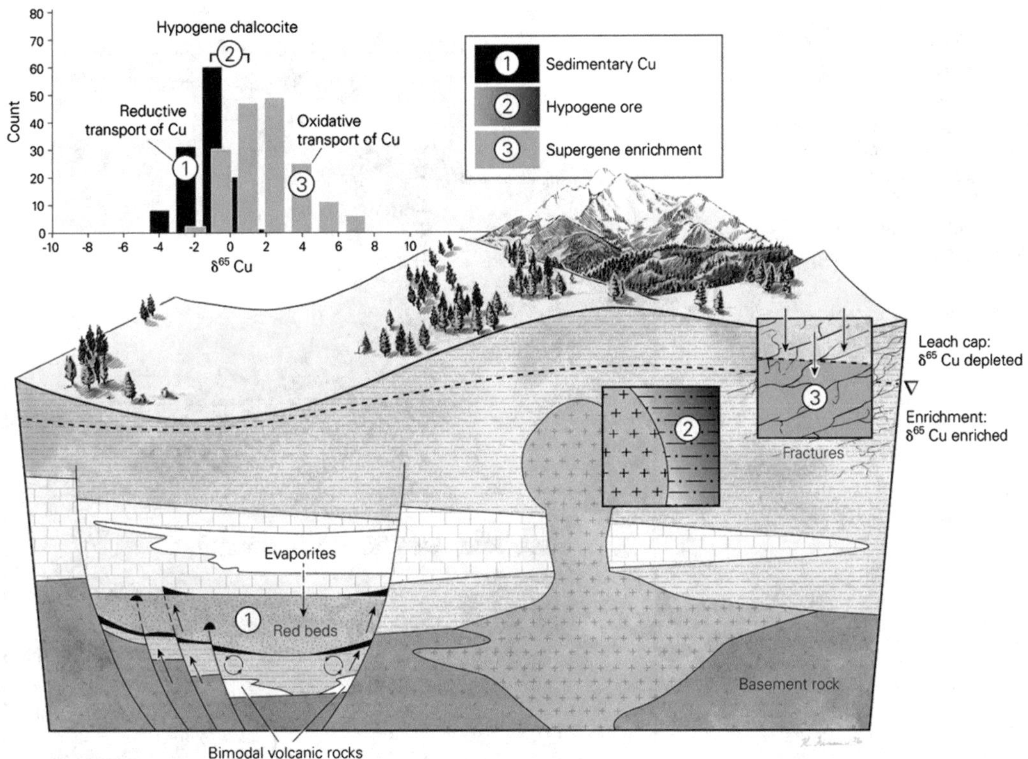

Fig. 6 The origins of chalcocite formation inferred through copper isotope analysis of chalcocite. The copper isotope compositions are linked to the different genetic processes associated with the formation of chalcocite. Reproduced from Mathur et al. (2018) (CC BY 4.0)

arc volcanic rock and sulfides in a non-economic type sulfide deposit. This suggests ores and rocks could have similar signatures and be used to trace metals.

There are no systematic studies that relate the mantle and metasomatized mantle to the genesis of magmas that are metal rich and are transported from the mantle through the lower crust into the upper crust. Several studies have shown that the metasomatized mantle (Liu et al. 2015; Zweifelhofer et al. 2018) could have highly variable copper isotope values, thus the starting point for the magmas is not homogeneous. Complicating the processes that could change the copper isotope values during petrogenesis is the fact that sulfide segregation at any point during ascent could greatly impact the copper isotope ratios found in the igneous rocks. At this point, what exactly the copper isotope values indicate in igneous rocks is not well constrained and more study is needed to further interpret and understand the values measured in the rocks associated with ores at surface. One study by Höhn et al. (2017) provides copper isotope variations in pyrites from metamorphic deposits, yielding a narrow δ^{65}Cu range of − 0.26 to 0.36‰. Thus, how copper moves and what happens in the mantle and metamorphic rocks still deserves greater attention to resolve how the copper isotope values in these rocks can be interpreted.

Slightly larger isotope variations have been documented in sedimentary rocks. Again the variations are on the order of 1‰ with no studies to date that link the sedimentary rock copper isotope signatures to ores directly. Although Chi Fru et al. (2016) use the copper isotope signature as a redox signal that marks the start of the Great Oxidation Event in the Archean, to date not many other applications to sedimentary systems are available in the literature. Nonetheless, the

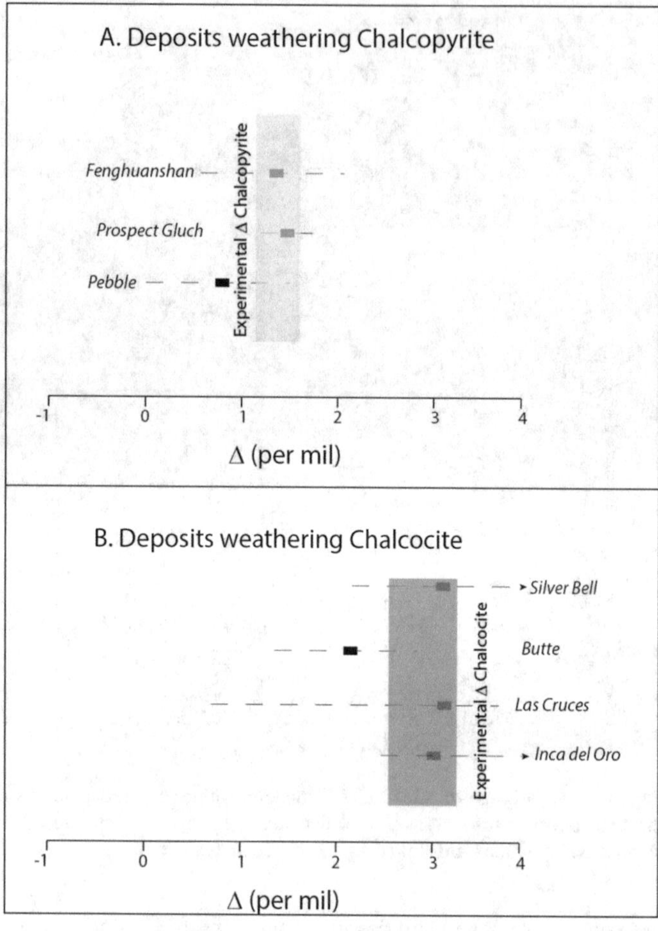

Fig. 7 Diagrams showing that Cu isotope compositions of waters can be used to infer the types of copper minerals weathering if the starting copper isotope composition of the minerals are constrained. Reproduced with permission from Mathur et al. (2014); Copyright 2014 Elsevier

application to finding paleofliud paths in sedimentary rocks with residual copper isotope ratios has potential in finding and understanding sedimentary copper and petroleum systems.

2.2.3 Cu Isotope Signatures in Waters and Soils

Copper isotope values in water has the greatest exploration potential because waters can sample the subsurface due to water rock interaction. Isotopic information from groundwater is far cheaper to obtain than data from drill core samples. Importantly, waters that have dissolved copper originating from sulfide-water interaction show unique copper isotope signatures. This is due to the fact that ^{65}Cu is favoured in the oxidation of copper due to bonding energies (Mathur et al. 2005; Pokrovsky et al. 2008; Ehrlich et al. 2004; Wall et al. 2011). Thus, the oxidation product, in this case a solution, is relatively enriched in ^{65}Cu. For instance modern rivers and oceans have copper isotope values of 1.1 ± 0.3‰ (2σ) (Bermin et al. 2006; Vance et al. 2008, 2016; Little et al. 2014), and copper derived from leaching of copper sulfides has values of 1.2 ± 0.3‰ (2σ) (Mathur et al. 2005, 2012a, 2014; Kimball et al. 2009; Mathur and Fantle 2015). Kinetic processes clearly play an important role in the extent of reactions and the rates of sulfide dissolution, which controlled the Cu isotope fractionation in the critical zone.

An example of this in nature is the case study of the Pebble deposit in Alaska. This study demonstrated that waters collected from seeps within the 0.5% Cu grade contour (defined by extensive drilling) has distinctly higher copper isotope values (2.6 ± 1.0‰, 2σ) than waters derived from seeps outside of this zone (Mathur et al. 2013). The range of the 0.5% copper is within 1.5 km distance from the mineralization center in this deposit, and the copper isotopic compositions are similar in these two mineralization zones. Kidder et al. (2021) show that the same patterns exists in groundwaters near PCDs and exotic Cu mineralization.

Importantly, the waters are approximately 1.5‰ heavier than the surrounding waters. It has been shown that the leaching of chalcopyrite results in waters that are 1.5‰ heavier than the leached chalcopyrite (Mathur et al. 2005; Kimball et al. 2009; Borrok et al. 2008). Leaching of other copper sulfide minerals also results in relatively different copper isotope compositions in waters. For instance, weathering of chalcocite which results in waters that 3‰ heavier than the leached chalcocite. Thus, using an estimation of the starting copper isotope composition of an ore, the copper isotope composition of the fluids not only indicates the weathering of copper sulfide, but also allows for the interpretation of the type of mineral providing copper to solution (Mathur et al. 2014).

The copper isotope composition of surface fluids can also be complicated by multiple types of other geochemical processes. For instance, waters interacting with soils and biological materials can impact the copper isotope composition of the fluids. Extensive studies on the biological and mineralogical impacts of clays on the copper isotope composition of fluids should also be considered during interpretation of the copper isotope values of waters (Pokrovsky et al. 2008; Bigalke et al. 2009, 2011; Weinstein et al. 2011; Liu et al. 2014b; Fekiacova et al. 2015; Babcsányi et al. 2016; Guinoiseau et al. 2017). Real time changes in copper isotope compositions of fluids and soils on a flood plain have demonstrated daily 0.5‰ shift in the copper isotope values as reported by Kusonwiriyawong et al. (2017). These studies illustrate that other processes may also impact the copper isotope signal in natural waters, however none of these processes fractionate copper in same direction and magnitude as oxidation of copper sulfides. For instance, at low temperature, no biological, adsorption or temperature related process produces + 3‰ isotopic shifts.

Use of soils to target mineralization at depth has not received much attention. The variety of the studies of copper isotope and soils focus on how the copper isotope signals can be used to trace metal contamination (Sivry et al. 2008; Bigalke et al. 2010; Li et al. 2016; Song et al. 2016; Blotevogel et al. 2018; Mihaljevič et al. 2018). Some hope does exist for using copper isotope signatures in soils, as Mathur et al (2012a, b) pointed out that the weathering of soils with sulfides produces similar vertical profiles of copper isotope signatures as seen in leach cap environments. However, more study is needed to clearly identify the potential of copper isotope values in soils in mineral exploration.

3 Conclusions

Copper isotopic fractionation in ores and waters have potential as vectors to ore deposits. The greatest limitations of using the techniques is that the vector can point to the presence of copper sulfide mineralizaton, but cannot clearly be used to indicate the amount of copper present in the system. Used in tandem with other geophysical and geochemical evidence the copper isotopic data obtained clearly add valuable information for the exploration geologist. Further study of copper isotopic compositions of soils and rocks will lead to further our understanding of how these materials could be used in the search of ore systems.

Aside from exploration vectors, the copper isotopic evidence summarized above demonstrate that the analysis provides a means to quantify the degrees of weathering, different types of metal transport in high and low temperature systems, and electron transfer in all type of geochemical reactions. Data used in this light has clearly added value to ore genesis models and the ability to capture reaction dynamics and kinetics which was previously not quantified and/or identified.

Acknowledgements Yun Zhao acknowledges funding under the National Natural Science Foundation of China (41803013), the Open Research Project from the State Key Laboratory of Geological Processes and Mineral Resources (GPMR202107, GPMR202116), and the State Key Laboratory for Mineral Deposits Research (2021-LAMD-K10). We also acknowledge NSF proposal 1924177 for support for this contribution.

References

Asadi S, Mathur R, Moore F, Zarasvandi A (2015) Copper isotope fractionation in the Meiduk porphyry copper deposit, Northwest of Kerman Cenozoic magmatic arc, Iran. Terra Nova 27:36–41

Asael D, Matthews A, Bar-Matthews M, Halicz L (2007) Copper isotope fractionation in sedimentary copper mineralization (Timna Valley, Israel). Chem Geol 243:238–254

Asael D, Matthews A, Oszczepalski S, Bar-Matthews M, Halicz L (2009) Fluid speciation controls of low temperature copper isotope fractionation applied to the Kupferschiefer and Timna ore deposits. Chem Geol 262:147–158

Asael D, Matthews A, Bar-Matthews M, Harlavan Y, Segal I (2012) Tracking redox controls and sources of sedimentary mineralization using copper and lead isotopes. Chem Geol 310–311:23–35

Babcsányi I, Chabaux F, Granet M, Meite F, Payraudeau S, Duplay J, Imfeld G (2016) Copper in soil fractions and runoff in a vineyard catchment: Insights from copper stable isotopes. Sci Total Environ 557–558:154–162

Baggio SB, Hartmann LA, Lazarov M, Massonne H-J, Opitz J, Theye T, Viefhaus T (2018) Origin of native copper in the Paraná volcanic province, Brazil, integrating Cu stable isotopes in a multi-analytical approach. Mineral Deposita 53:417–434

Balistrieri LS, Borrok DM, Wanty RB, Ridley WI (2008) Fractionation of Cu and Zn isotopes during adsorption onto amorphous Fe(III) oxyhydroxide: experimental mixing of acid rock drainage and ambient river water. Geochim Cosmochim Acta 72:311–328

Barnes S-J, Lightfoot PC (2005) Formation of magmatic nickel-sulfide ore deposits and processses affecting their copper and platinum-group element contents. Econ Geol 100th Anniv Vol 100:179–213

Berkenbosch HA, de Ronde CEJ, Paul BT, Gemmell JB (2015) Characteristics of Cu isotopes from chalcopyrite-rich black smoker chimneys at Brothers volcano, Kermadec arc, and Niuatahi volcano, Lau basin. Mineral Deposita 50:811–824. https://doi.org/10.1007/s00126-014-0571-y

Bermin J, Vance D, Archer C, Statham PJ (2006) The determination of the isotopic composition of Cu and Zn in sea water. Chem Geol 226:280–297

Bigalke M, Weyer S, Wilcke W (2009) Stable copper isotopes: a novel tool to trace copper behavior in hydromorphic soils. Soil Sci Soc Am J 74:60–73. https://doi.org/10.2136/sssaj2008.0377

Bigalke M, Weyer S, Kobza J, Wilcke W (2010) Stable Cu and Zn isotope ratios as tracers of sources and transport of Cu and Zn in contaminated soil. Geochim Cosmochim Acta 74:6801–6813

Bigalke M, Weyer S, Wilcke W (2011) Stable Cu isotope fractionation in soils during oxic weathering and podzolization. Geochim Cosmochim Acta 75:3119–3134

Blotevogel S, Oliva P, Sobanska S, Viers J, Vezin H, Audry S, Prunier J, Darrozes J, Orgogozo L, Courjault-Radé P, Schreck E (2018) The fate of Cu pesticides in vineyard soils: a case study using δ65Cu isotope ratios and EPR analysis. Chem Geol 477:35–46. https://doi.org/10.1016/j.chemgeo.2017.11.032

Bornhorst T, Mathur R (2017) Copper isotope constraints on the genesis of the Keweenaw Peninsula native copper district, Michigan, USA. Minerals 7:185

Borrok DM, Wanty RB, Ridley WI, Wolf R, Lamothe PJ, Adams M (2007) Separation of copper, iron, and zinc from complex aqueous solutions for isotopic measurement. Chem Geol 242:400–414

Borrok DM, Nimick DA, Wanty RB, Ridley WI (2008) Isotopic variations of dissolved copper and zinc in stream waters affected by historical mining. Geochim Cosmochim Acta 72:329–344

Braxton D, Mathur R (2011) Exploration applications of copper isotopes in the supergene environment: a case study of the Bayugo Porphyry Copper-Gold Deposit, Southern Philippines. Econ Geol 106:1447–1463. https://doi.org/10.2113/econgeo.106.8.1447

Braxton DP, Mathur R (2014) Copper isotopic vectors to supergene enrichment: leaches cap iostopic footprint of the Quellaveco porphyry copper deposit, southern Peru. In: SEG conference proceedings: SEG 2014: building exploration capability for the 21st century

Brzozowski MJ, Good DJ, Wu C, Li W (2020) Cu isotope systematics of conduit-type Cu–PGE mineralization in the Eastern Gabbro, Coldwell Complex, Canada. Mineral Deposita 56:707–724. https://doi.org/10.1007/s00126-020-00992-8

Busigny V, Chen J, Philippot P, Borensztajn S, Moynier F (2018) Insight into hydrothermal and subduction processes from copper and nitrogen isotopes in oceanic metagabbros. Earth Planet Sci Lett 498:54–64

Chapman JB, Mason TFD, Weiss DJ, Coles BJ, Wilkinson JJ, Anonymous (2004) An adapted column chemistry procedure for separation of Fe, Cu and Zn from geological matrices,and natural Zn isotopic variations in geological standard reference materials BCR-027, BCR-030 and NOD-P-1. Geol Soc Am Abstr Programs 36:448

Chapman JB, Mason TFD, Weiss DJ, Coles BJ, Wilkinson JJ (2006) Chemical separation and isotopic variations of Cu and Zn from five geological reference materials. Geostand Geoanal Res 30:5–16

Condon D, Mathur R, Simpson T, Mendoza N (2012) Cu isotope fractionation used to determine supergene processes at Yancocha mine in Northern Peru. Geol Soc Am Abstr Programs 46

Dekov VM, Rouxel O, Asael D, Hålenius U, Munnik F (2013) Native Cu from the oceanic crust: isotopic insights into native metal origin. Chem Geol 359:136–149. https://doi.org/10.1016/j.chemgeo.2013.10.001

Dendas M (2011) Copper isotopes tracers of fluid flow at Bingham Canyon, Utah. Unpub MSc thesis, University of Arizona

Dótor-Almazán A, Armienta-Hernández MA, Talavera-Mendoza O, Ruiz J (2017) Geochemical behavior of Cu and sulfur isotopes in the tropical mining region of Taxco, Guerrero (southern Mexico). Chem Geol 471:1–12

Duan J, Tang J, Li Y, Liu S-A, Wang Q, Yang C, Wang Y (2015) Copper isotopic signature of the Tiegelongnan high-sulfidation copper deposit, Tibet: implications for its origin and mineral exploration. Mineral Deposita 51:591–602. https://doi.org/10.1007/s00126-015-0624-x

Duan J, Tang J, Li Y, Liu S-A, Wang Q, Yang C, Wang Y (2016) Copper isotopic signature of the Tiegelongnan high-sulfidation copper deposit, Tibet: implications for its origin and mineral exploration. Mineral Deposita 51:591–602. https://doi.org/10.1007/s00126-015-0624-x

Ehrlich S, Butler I, Halicz L, Rickard D, Oldroyd A, Matthews A (2004) Experimental study of the copper isotope fractionation between aqueous Cu (II) and covellite, CuS. Chem Geol 209:259–269

Fekiacova Z, Cornu S, Pichat S (2015) Tracing contamination sources in soils with Cu and Zn isotopic ratios. Sci Total Environ 517:96–105. https://doi.org/10.1016/j.scitotenv.2015.02.046

Fernandez A, Borrok DM (2009) Fractionation of Cu, Fe, and Zn isotopes during the oxidative weathering of sulfide-rich rocks. Chem Geol 264:1–12. https://doi.org/10.1016/j.chemgeo.2009.01.024

Fru EC, Rodríguez NP, Partin CA, Lalonde SV, Andersson P, Weiss DJ, El Albani A, Rodushkin I, Konhauser KO (2016) Cu isotopes in marine black shales record the great oxidation event. Proc Natl Acad Sci 113:4941–4946. https://doi.org/10.1073/pnas.1523544113

Graham S, Pearson N, Jackson S, Griffin W, O'Reilly SY (2004) Tracing Cu and Fe from source to porphyry; in situ determination of Cu and Fe isotope ratios in sulfides from the Grasberg Cu-Au deposit. Chem Geol 207:147–169

Gregory MJ, Mathur R (2017) Understanding copper isotope behavior in the high temperature magmatic-hydrothermal porphyry environment. Geochem Geophys Geosys 18:4000–4015. https://doi.org/10.1002/2017gc007026

Guinoiseau D, Gélabert A, Allard T, Louvat P, Moreira-Turcq P, Benedetti MF (2017) Zinc and copper behaviour at the soil-river interface: new insights by Zn and Cu isotopes in the organic-rich Rio Negro basin. Geochim Cosmochim Acta 213:178–197. https://doi.org/10.1016/j.gca.2017.06.030

Guo H, Xia Y, Bai R, Zhang X, Huang F (2020) Experiments on Cu-isotope fractionation between chlorine-bearing fluid and silicate magma: implications for fluid exsolution and porphyry Cu deposits. Natl Sci Rev 7:1319–1330

Haest M, Muchez P, Petit JCJ, Vanhaecke F (2009) Cu isotope ratio variations in the Dikulushi Cu-Ag deposit, DRC: of primary origin or induced by supergene reworking. Econ Geol 104:1055–1064

Halliday AN, Lee D-C, Christensen JN, Walder AJ, Freedman PA, Jones CE, Hall CM, Yi W, Teagle D (1995) Recent developments in inductively coupled plasma magnetic sector multiple collector mass spectrometry. Int J Mass Spectrom Ion Process 146:21–33

Höhn S, Frimmel HE, Debaille V, Pašava J, Kuulmann L, Debouge W (2017) The case for metamorphic base metal mineralization: pyrite chemical, Cu and S isotope data from the Cu–Zn deposit at Kupferberg in Bavaria, Germany. Mineral Deposita. https://doi.org/10.1007/s00126-017-0714-z

Housh TB, Çiftçi E (2008) Cu isotope geochemistry of volcanogenic massive sulphide deposits of the eastern Pontides, Turkey. In: IOP conference series: earth and environmental science. IOP Publishing, 012025

Huang J, Liu S-A, Wörner G, Yu H, Xiao Y (2016) Copper isotope behavior during extreme magma differentiation and degassing: a case study on Laacher See phonolite tephra (East Eifel, Germany). Contrib Mineral Petrol 171:76

Ikehata K, Notsu K, Hirata T, Navarrete JU, Borrok DM, Viveros M, Ellzey JT (2011) Copper isotope characteristics of copper-rich minerals from Besshi-type volcanogenic massive sulfide deposits, Japan, determined using a femtosecond LA-MC-ICM-MS. Econ Geol 106:307–316

Kidder J, Voinot A, Leybourne M, Layton-Matthews D, Bowell R (2021) Using stable isotopes of Cu, Mo, S, and 87Sr/86Sr in hydrogeochemical mineral exploration as tracers of porphyry and exotic copper deposits. Appl Geochem 128:104935

Kidder JA, Voinot A, Sullivan KV, Chipley D, Valentino M, Layton-Matthews D, Leybourne M (2020) Improved ion-exchange column chromatography for Cu purification from high-Na matrices and isotopic analysis by MC-ICPMS. J Anal Atomic Spectrom 35:776–783

Kimball BE, Mathur R, Dohnalkova AC, Wall AJ, Runkel RL, Brantley SL (2009) Copper isotope fractionation in acid mine drainage. Geochim Cosmochim Acta 73:1247–1263

Kusonwiriyawong C, Bigalke M, Abgottspon F, Lazarov M, Schuth S, Weyer S, Wilcke W (2017) Isotopic variation of dissolved and colloidal iron and copper in a carbonatic floodplain soil after experimental flooding. Chem Geol 459:13–23. https://doi.org/10.1016/j.chemgeo.2017.03.033

Larson PB, Maher K, Ramos FC, Chang Z, Gaspar M, Meinert LD (2003) Copper isotope ratios in magmatic and hydrothermal ore-forming environments. Chem Geol 201:337–350

Li C, Barnes S-J, Makovicky E, Rose-Hansen J, Makovicky M (1996) Partitioning of nickel, copper, iridium, platinum, and palladium between monosulfide solid solution and sulfide liquid: effects of composition and temperature. Geochim Cosmochim Acta 60:1231–1238

Li S-Z, Zhu X-K, Wu L-H, Luo Y-M (2016) Cu isotopic compositions in Elsholtzia splendens: influence of soil condition and growth period on Cu isotopic fractionation in plant tissue. Chem Geol 444:49–58. https://doi.org/10.1016/j.chemgeo.2016.09.036

Li W, Jackson SE, Pearson NJ, Graham S (2010) Copper isotopic zonation in the Northparkes porphyry Cu–Au deposit, SE Australia. Geochim Cosmochim Acta 74:4078–4096

Little SH, Vance D, Walker-Brown C, Landing WM (2014) The oceanic mass balance of copper and zinc isotopes, investigated by analysis of their inputs, and outputs to ferromanganese oxide sediments. Geochim Cosmochim Acta 125:673–693. https://doi.org/10.1016/j.gca.2013.07.046

Liu S-A, Li D, Li S, Teng F-Z, Ke S, He Y, Lu Y (2014a) High-precision copper and iron isotope analysis of igneous rock standards by MC-ICP-MS. J Anal Atomic Spectrom 29:122–133. https://doi.org/10.1039/C3JA50232E

Liu S-A, Teng F-Z, Li S, Wei G-J, Ma J-L, Li D (2014b) Copper and iron isotope fractionation during weathering and pedogenesis: Insights from saprolite profiles. Geochim Cosmochim Acta 146:59–75

Liu S-A, Huang J, Liu J, Wörner G, Yang W, Tang Y-J, Chen Y, Tang L, Zheng J, Li S (2015) Copper isotopic composition of the silicate Earth. Earth Planet Sci Lett 427:95–103

Liu S-A, Liu P-P, Lv Y, Wang Z-Z, Dai J-G (2019) Cu and Zn isotope fractionation during oceanic alteration: implications for oceanic Cu and Zn cycles. Geochim Cosmochim Acta 257:191–205

Lobo L, Degryse P, Shortland A, Eremin K, Vanhaecke F (2014) Copper and antimony isotopic analysis via multi-collector ICP-mass spectrometry for provenancing ancient glass. J Anal Atomic Spectrom 29:58–64

Luczaj J, Huang H (2018) Copper and sulfur isotope ratios in Paleozoic-hosted Mississippi Valley-type mineralization in Wisconsin, USA. Appl Geochem 89:173–179

Mah er KC, Ramos FC, Larson PB, Anonymous (2003) Copper isotope characteristics of the Cu (+Au, Ag) skarn at Coroccohuayco, Peru. Geol Soc Am AbstrPrograms 35:518

Maher KC, Larson PB (2007) Variation in copper isotope ratios and controls on fractionation in hypogene skarn mineralization at Coroccohuayco and Tintaya, Perú. Econ Geol 102:225–237

Maher KC, Jackson S, Mountain B (2011) Experimental evaluation of the fluid–mineral fractionation of Cu isotopes at 250 °C and 300 °C. Chem Geol 286:229–239

Malitch KN, Latypov RM, Badanina IY, Sluzhenikin SF (2014) Insights into ore genesis of Ni–Cu–PGE sulfide deposits of the Noril'sk Province (Russia): evidence from copper and sulfur isotopes. Lithos 204:172–187

Marechal C, Albarede F (2002) Ion-exchange fractionation of copper and zinc isotopes. Geochim Cosmochim Acta 66:1499–1509

Marechal CN, Telouk P, Albarede F (1999) Precise analysis of copper and zinc isotopic compositions by plasma-source mass spectrometry. Chem Geol 156:251–273

Markl G, Lahaye Y, Schwinn G (2006) Copper isotopes as monitors of redox processes in hydrothermal mineralization. Geochim Cosmochim Acta 70:4215–4228

Mason TFD, Weiss DJ, Chapman JB, Wilkinson JJ, Tessalina SG, Spiro B, Horstwood MSA, Spratt J, Coles BJ (2005) Zn and Cu isotopic variability in the Alexandrinka volcanic-hosted massive sulphide (VHMS) ore deposit, Urals, Russia. Chem Geol 221:170–187

Mathur R, Wang D (2019) Transition metal isotopes applied to exploration geochemistry: insights from Fe, Cu, and Zn. In: Decrée S, Robb LR (eds) Ore deposits: origin, exploration, and exploitation. American Geophysical Union, pp 163–184

Mathur R, Ruiz J, Titley S, Liermann L, Buss H, Brantley SL (2005) Cu isotopic fractionation in the supergene environment with and without bacteria. Geochim Cosmochim Acta 69:5233–5246

Mathur R, Titley S, Barra F, Brantley S, Wilson M, Phillips A, Munizaga F, Maksaev V, Vervoort J, Hart G (2009) Exploration potential of Cu isotope fractionation in porphyry copper deposits. J Geochem Explor 102:1–6

Mathur R, Dendas M, Titley S, Phillips A (2010) Patterns in the copper isotope composition of minerals in Porphyry Copper deposits in Southwestern United States. Econ Geol 105:1457–1467

Mathur R, Jin L, Prush V, Paul J, Ebersole C, Fornadel A, Williams JZ, Brantley S (2012a) Cu isotopes and concentrations during weathering of black shale of the Marcellus formation, Huntingdon County, Pennsylvania (USA). Chem Geol 304–305:175–184. https://doi.org/10.1016/j.chemgeo.2012.02.015

Mathur R, Ruiz J, Casselman MJ, Megaw P, van Egmond R (2012b) Use of Cu isotopes to distinguish primary and secondary Cu mineralization in the Cañariaco Norte porphyry copper deposit, Northern Peru. Mineral Deposita 47:755–762

Mathur R, Munk L, Nguyen M, Gregory M, Annell H, Lang J (2013) Modern and paleofluid pathways revealed by cu isotope compositions in surface waters and ores of the Pebble Porphyry Cu–Au–Mo Deposit, Alaska. Econ Geol 108:529–541. https://doi.org/10.2113/econgeo.108.3.529

Mathur R, Munk LA, Townley B, Gou KY, Gómez Miguélez N, Titley S, Chen GG, Song S, Reich M, Tornos F, Ruiz J (2014) Tracing low-temperature aqueous metal migration in mineralized watersheds

with Cu isotope fractionation. Appl Geochem 51:109–115
Mathur R, Fantle MS (2015) Copper isotopic perspectives on supergene processes: implications for the global Cu cycle. Elements 11:323–329
Mathur R, Falck H, Belogub E, Milton J, Wilson M, Rose A, Powell W (2018) Origins of chalcocite defined by copper isotope values. Geofluids 2018:5854829. https://doi.org/10.1155/2018/5854829
Mihaljevič M, Jarošíková A, Ettler V, Vaněk A, Penížek V, Křibek B, Chrastný V, Sracek O, Trubač J, Svoboda M, Nyambe I (2018) Copper isotopic record in soils and tree rings near a copper smelter, Copperbelt, Zambia. Sci Total Environ 621:9–17
Mirnejad H, Mathur R, Einali M, Dendas M, Alirezaei S (2010) A comparative copper isotope study of porphyry copper deposits in Iran. Geochem Explor Environ Anal 10:413–418
Naldrett AJ (2010) From the mantle to the bank: the life of a Ni-Cu-(PGE) sulfide deposit. S Afr J Geol 113: 1–32
Naldrett AJ (2011) Fundamentals of magmatic sulfide deposits. Rev Econ Geol 17:1–50
Ni P, Macris CA, Darling EA, Shahar A (2021) Evaporation-induced copper isotope fractionation: insights from laser levitation experiments. Geochim Cosmochim Acta 298:131–148
Pękala M, Asael D, Butler IB, Matthews A, Rickard D (2011) Experimental study of Cu isotope fractionation during the reaction of aqueous Cu(II) with Fe(II) sulphides at temperatures between 40 and 200 °C. Chem Geol 289:31–38
Petit JC, De Jong J, Chou L, Mattielli N (2008) Development of Cu and Zn isotope MC-ICP-MS measurements: application to suspended particulate matter and sediments from the Scheldt Estuary. Geostand Geoanal Res 32:149–166
Plumhoff AM, Mathur R, Milovský R, Majzlan J (2021) Fractionation of the copper, oxygen and hydrogen isotopes between malachite and aqueous phase. Geochim Cosmochim Acta 300:246–257
Pokrovsky OS, Viers J, Emnova EE, Kompantseva EI, Freydier R (2008) Copper isotope fractionation during its interaction with soil and aquatic microorganisms and metal oxy(hydr)oxides; possible structural control. Geochim Cosmochim Acta 72:1742–1757
Rempel KU, Liebscher A, Meixner A, Romer RL, Heinrich W (2012) An experimental study of the elemental and isotopic fractionation of copper between aqueous vapour and liquid to 450 °C and 400 bar in the CuCl–NaCl–H_2O and CuCl–NaHS–NaCl–H_2O systems. Geochim Cosmochim Acta 94:199–216
Ripley EM, Dong S, Li C, Wasylenki LE (2015) Cu isotope variations between conduit and sheet-style Ni–Cu–PGE sulfide mineralization in the Midcontinent Rift System, North America. Chem Geol 414:59–68
Rouxel O, Fouquet Y, Ludden JN (2004) Copper isotope systematics of the Lucky Strike, Rainbow, and Logatchev sea-floor hydrothermal fields on the Mid-Atlantic Ridge. Econ Geol 99:585–600
Saunders JA, Mathur R, Kamenov GD, Shimizu T, Brueseke ME (2015) New isotopic evidence bearing on bonanza (Au–Ag) epithermal ore-forming processes. Mineral Deposita 51:1–11
Savage PS, Moynier F, Chen H, Shofner G, Siebert J, Badro J, Puchtel I (2015) Copper isotope evidence for large-scale sulphide fractionation during Earth's differentiation. Geochem Perspect Lett 1:53–64
Seo JH, Lee SK, Lee I (2007) Quantum chemical calculations of equilibrium copper (I) isotope fractionations in ore-forming fluids. Chem Geol 243:225–237
Sherman DM (2013) Equilibrium isotopic fractionation of copper during oxidation/reduction, aqueous complexation and ore-forming processes: predictions from hybrid density functional theory. Geochim Cosmochim Acta 118:85–97
Shields W, Goldich S, Garner E, Murphy T (1965) Natural variations in the abundance ratio and the atomic weight of copper. J Geophys Res 70:479–491
Sivry Y, Riotte J, Sonke JE, Audry S, Schäfer J, Viers J, Blanc G, Freydier R, Dupré B (2008) Zn isotopes as tracers of anthropogenic pollution from Zn-ore smelters The Riou Mort-Lot River system. Chem Geol 255:295–304
Song S, Mathur R, Ruiz J, Chen D, Allin N, Guo K, Kang W (2016) Fingerprinting two metal contaminants in streams with Cu isotopes near the Dexing Mine, China. Sci Total Environ 544:677–685
Sossi PA, Halverson GP, Nebel O, Eggins SM (2015) Combined separation of Cu, Fe and Zn from rock matrices and improved analytical protocols for stable isotope determination. Geostand Geoanal Res 39:129–149
Sullivan K, Layton-Matthews D, Leybourne M, Kidder J, Mester Z, Yang L (2020) Copper isotopic analysis in geological and biological reference materials by MC-ICP-MS. Geostand Geoanal Res 44:349–362
Tang D, Qin K, Su B, Mao Y, Evans NJ, Niu Y, Kang Z (2020) Sulfur and copper isotopic signatures of chalcopyrite at Kalatongke and Baishiquan: insights into the origin of magmatic Ni–Cu sulfide deposits. Geochim Cosmochim Acta 275:209–228
Urey HC (1947) The thermodynamics of isotope substances. J Chem Soc 1947:562–581
Vance D, Archer C, Bermin J, Perkins J, Statham PJ, Lohan MC, Ellwood MJ, Mills RA (2008) The copper isotope geochemistry of rivers and the oceans. Earth Planet Sci Lett 274:204–213
Vance D, Matthews A, Keech A, Archer C, Hudson G, Pett-Ridge J, Chadwick OA (2016) The behaviour of Cu and Zn isotopes during soil development: controls on the dissolved load of rivers. Chem Geol 445:36–53
Wall A, Heaney P, Mathur R, Gammons C, Brantley S (2010) Cu isotope systematics of the Butte mining district. Montana. Geochim Cosmochim Acta 74: A1093
Wall AJ, Heaney PJ, Mathur R, Post JE, Hanson JC, Eng PJ (2011) A flow-through reaction cell that couples time-resolved X-ray diffraction with stable isotope analysis. J Appl Crystallogr 44:429–432

Wang D, Sun X, Zheng Y, Wu S, Xia S, Chang H, Yu M (2017) Two pulses of mineralization and genesis of the Zhaxikang Sb–Pb–Zn–Ag deposit in southern Tibet: constraints from Fe–Zn isotopes. Ore Geol Rev 84:347–363

Wang Z, Park J-W, Wang X, Zou Z, Kim J, Zhang P, Li M (2019) Evolution of copper isotopes in arc systems: insights from lavas and molten sulfur in Niuatahi volcano, Tonga rear arc. Geochim Cosmochim Acta 250:18–33

Weinstein C, Moynier F, Wang K, Paniello R, Foriel J, Catalano J, Pichat S (2011) Isotopic fractionation of Cu in plants. Chem Geol 286:266–271

Weiss DJ, Mason TFD, Tessalina SG, Horstwood MSA, Wilkinson JJ, Chapman JB, Parrish RR, Anonymous (2004) Controls of Cu and Zn isotope variability within a VHMS deposit. Geol Soc Am Abstr Programs 36:515

Wu S, Zheng Y, Wang D, Chang H, Tan M (2017) Variation of copper isotopes in chalcopyrite from Dabu porphyry Cu–Mo deposit in Tibet and implications for mineral exploration. Ore Geol Rev 90:14–24

Xia Y, Kiseeva ES, Wade J, Huang F (2019) The effect of core segregation on the Cu and Zn isotope composition of the silicate Moon. Geochem Perspect Lett 12:12–17

Yang S-C, Welter L, Kolatkar A, Nieva J, Waitman KR, Huang K-F, Liao W-H, Takano S, Berelson WM, West AJ, Kuhn P, John SG (2019) A new anion exchange purification method for Cu stable isotopes in blood samples. Anal Bioanal Chem 411:765–776

Yao J, Mathur R, Sun W, Song W, Chen H, Mutti L, Xiang X, Luo X (2016) Fractionation of Cu and Mo isotopes caused by vapor-liquid partitioning, evidence from the Dahutang W-Cu–Mo ore field. Geochem Geophys Geosyst 17:1725–1739

Zhang Y, Bao Z, Lv N, Chen K, Zong C, Yuan H (2020) Copper isotope ratio measurements of Cu-dominated minerals without column chromatography using MC-ICP-MS. Front Chem 8:609

Zhao Y, Xue C, Liu S-A, Symons DTA, Zhao X, Yang Y, Ke J (2017) Copper isotope fractionation during sulfide-magma differentiation in the Tulaergen magmatic Ni–Cu deposit, NW China. Lithos 286–287:206–215

Zhao Y, Xue C, Liu S-A, Mathur R, Zhao X, Yang Y, Dai J, Man R, Liu X (2019) Redox reactions control Cu and Fe isotope fractionation in a magmatic Ni–Cu mineralization system. Geochim Cosmochim Acta 249:42–58

Zhao Y, Liu SA, Xue C, Li ML (2022) Copper isotope evidence for a Cu-rich mantle source of the world-class Jinchuan magmatic Ni–Cu deposit. Am Mineral 107:673–683

Zheng Y-C, Liu S-A, Wu C-D, Griffin WL, Li Z-Q, Xu B, Yang Z-M, Hou Z-Q, O'Reilly SY (2019) Cu isotopes reveal initial Cu enrichment in sources of giant porphyry deposits in a collisional setting. Geology 47:135–138

Zhu XK, O'Nions RK, Guo Y, Belshaw NS, Rickard D (2000) Determination of natural Cu-isotope variation by plasma-source mass spectrometry; implications for use as geochemical tracers. Chem Geol 163:139–149

Zhu XK, Guo Y, Williams RJP, O'Nions RK, Matthews A, Belshaw NS, Canters GW, de Waal EC, Weser U, Burgess BK, Salvato B (2002) Mass fractionation processes of transition metal isotopes. Earth Planet Sci Lett 200:47–62

Zweifelhofer G, Kempton PD, Mathur R, Brueseke ME (2018) Cu-isotope heterogeneity in the lithospheric mantle: evidence from type I and type II peridotite xenoliths from the geronimo volcanic field, SE Arizona. Geol Soc Am Abstr Programs 50(4). https://doi.org/10.1130/abs/2018NC-312631

The Potential of Zn Isotopes in the Science and Exploration of Ore Deposits

Jamie J. Wilkinson

Abstract

Since the turn of the Century, the growth in development and application of zinc isotopes to multiple fields in terrestrial and planetary sciences has been exponential. The potential for the application of zinc isotope systematics to ore deposit formation processes was obvious from the outset, given that they represent the most significant concentrations of zinc on Earth and because this approach allowed, for the first time, direct assessment of zinc metal origins and transport. This contribution presents a brief summary of the notation and analytical procedures for analysis of zinc isotopes and summarizes the terrestrial data reported to date. These results show that the variation in zinc isotope composition in rocks and ore systems is in fact rather small (< 2 ‰), linked, at least in part, to the single oxidation state in which zinc occurs in nature. Based on an assessment of the literature, the principal mechanisms for causing isotopic fractionation are all relatively low temperature processes: (i) biogenic; (ii) supergene dissolution-reprecipitation; (iii) adsorption–desorption reactions; and (iv) hydrothermal precipitation. High temperature igneous processes do not appear to produce significant isotopic variations. In ore deposit studies, it currently appears unlikely that zinc isotopes can be used to constrain potential metal sources, apart from zinc derived from carbonate host rocks which tends to be isotopically heavy. However, there are a number of systems in which systematic variation in δ^{66}Zn of sulfides suggests that Rayleigh-type fractionation during ore mineral precipitation occurs, opening up the possibility of using zinc isotopes to trace flow paths and vector in towards mineralized centers. Modeling of such hydrothermal processes is currently hindered by a paucity of experimentally-determined fractionation factors, but as such work is done, our ability to better understand and utilize zinc isotopic zonation patterns for the purposes of mineral exploration will be progressively enhanced.

1 Introduction

Over the past decade and a half, new insights into metal transport pathways and mechanisms in Earth and Planetary systems have been made possible by the development and application of non-traditional stable isotopes (Albarède 2004; Johnson et al. 2004). This burgeoning field has been driven by technological developments in

J. J. Wilkinson (✉)
London Centre for Ore Deposits and Exploration (LODE), Natural History Museum, Cromwell Road, London SW7 5BD, UK
e-mail: j.wilkinson@nhm.ac.uk

Department of Earth Science and Engineering, Imperial College London, Exhibition Road, London SW7 2AZ, UK

D. Huston and J. Gutzmer (eds.), *Isotopes in Economic Geology, Metallogenesis and Exploration*, Mineral Resource Reviews, https://doi.org/10.1007/978-3-031-27897-6_15

multicollector inductively-coupled-plasma mass spectrometry (MC-ICP-MS). Utilizing plasma ionization and multiple collection of ions, the technique has enabled small isotopic variations in poorly ionized elements to be resolved, a goal that was not achievable with traditional thermal ionization mass spectrometry (TIMS). Using MC-ICP-MS, analytical precision is typically around or below 0.05 ‰ (δ value), several tens or hundreds of times smaller than the isotopic range measured to date in terrestrial rocks, minerals and sediments for a range of elements. Nonetheless, analytical problems remain and there are many studies that have focused on their resolution (e.g. Moynier et al. 2017).

The ability to directly determine the isotopic composition of ore-forming metals has opened up a number of potential applications in studies of hydrothermal ore systems that could shed light on several key parts of the source-transport-trap cycle (Fig. 1). There is the possibility that isotope compositions may be able to pinpoint metal *sources* in contrast to conventional stable isotope systems such as O, H and S which can help to identify sources of water or sulfur but may only indirectly suggest where metals originated. Pb isotope tracing of metal sources is hindered by the complexities of correcting for radiogenic decay. Ore metal isotope studies may also help to understand the processes by which metals are *mobilized* from source rocks or magmas and what characteristics (e.g. mineralogy, pH, fO_2) of rocks and fluids contribute to efficient extraction; metal stable isotopes may be especially powerful for tracking redox changes because these are a major control of isotopic fractionation. Fertile hydrothermal fluids may leave a recognizable metal isotope imprint of their passage that could enable *flow pathways* to be identified. Finally, in the ore deposit environment itself, isotopic variations may be useful for understanding the controls of *ore mineral deposition*. Thus, the application of these stable isotopes could impact on every aspect of the source-transport-trap paradigm. In the longer term, improved understanding of these processes, especially where constrained by experimental data on fractionation behavior, will lead to refinement of deposit models and ultimately influence exploration methods. More immediately, however, the potential exists for metal isotopes to fractionate in systematic ways in hydrothermal environments, leaving either zoning patterns that can be directly utilized for vectoring towards hydrothermal centres, or a diagnostic fingerprint that may reflect efficient ore deposition—i.e. a signature of system *fertility*.

In this paper, I summarize work done to date on hydrothermal ore deposits, focusing on the transition metal Zn. This is not intended to be an exhaustive review; for this, the reader is referred to Johnson et al. (2004), Cloquet et al. (2008), Moynier et al. (2017) and Mathur and Wang (2019). Rather, I focus on findings reported in the literature and results from our own laboratory that are beginning to advance our understanding of processes of zinc transport and deposition in ore systems, particularly in relation to possible applications in mineral exploration.

2 Zinc Stable Isotopes

Zinc has five stable isotopes ^{64}Zn, ^{66}Zn, ^{67}Zn, ^{68}Zn and ^{70}Zn (Table 1) with average natural abundances of 48.63, 27.90, 4.10, 18.75 and 0.62% respectively (Rosman and Taylor 1998). $\delta^{66}Zn$ is normally reported in the literature, relating to the $^{66}Zn/^{64}Zn$ ratio, where the subscript "X" refers to an unknown and "STD" to a reference material:

$$\delta^{66}Zn_X = \left(\frac{Zn_X^{66/64} - Zn_{STD}^{66/64}}{Zn_{STD}^{66/64}}\right) \cdot 10^3 \quad (1)$$

Thus, $\delta^{66}Zn$ refers to the $^{66}Zn/^{64}Zn$ isotope ratio relative to the standard. There is currently no universal standard for zinc isotope measurements, with some data reported with respect to IRMM 3702, some to the Johnson Matthey Corporation zinc standard used by the Lyon group (Lyon JMC 3-0749-L Zn) and, more recently, to the ETH standard AA-ETH (Archer et al. 2017). Data reported relative to IRMM

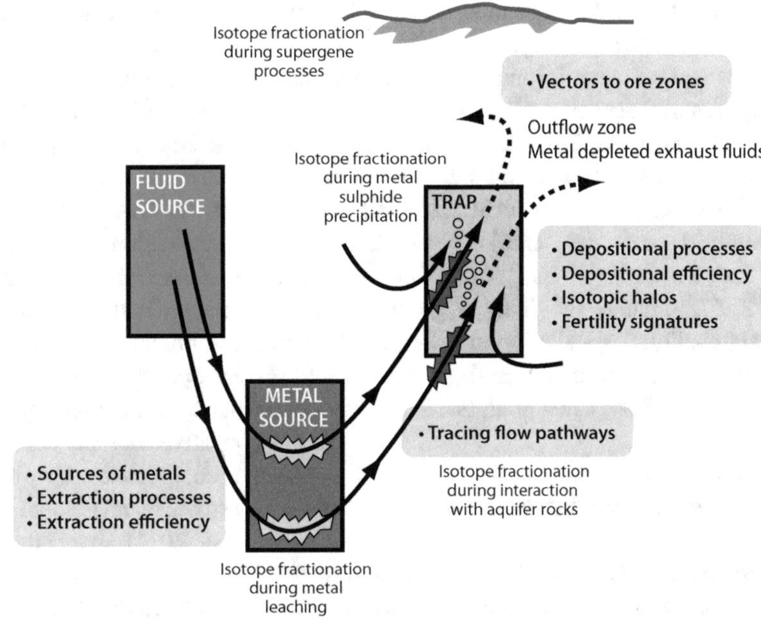

Fig. 1 Hydrothermal system schematic highlighting environments where fractionation of zinc isotopes may occur

Table 1 List of isotopes of zinc found in nature and statistics on the range of $\delta^{66}Zn$ values reported in the literature from a compilation of 483 analyses

Element	Isotopes	Delta notation	N	Mean (‰)	Min (‰)	Max (‰)	Range (‰)	Average precision (‰)
Zn	^{64}Zn (48.63%) ^{66}Zn (27.90%) ^{67}Zn (4.10%) ^{68}Zn (18.75%) ^{70}Zn (0.62%)	$\delta^{66}Zn$ (66/64)	483	0.33	−0.43	1.34	1.77	0.07

3702 (=AA-ETH) can be recalculated to Lyon JMC-Zn using the published offset between the two standards of 0.28 ‰ ($\delta^{66}Zn_{IRMM\ 3702} = \delta^{66}Zn_{AA\text{-}ETH} = \delta^{66}Zn_{JMC\text{-}Zn} + \quad + 0.28$ ‰; Ponzevera et al. 2006; Cloquet et al. 2008; Moeller et al. 2012; Archer et al. 2017). All analyses referred to in this paper are relative to the JMC-Zn standard.

3 Measurement of Zinc Isotopes

Zinc isotope analyses are normally carried out by introduction of a solution into the MC-ICP-MS and require a sample purification step to remove potentially interfering contaminants from the sample matrix (Albarède and Beard 2004). This is done via column chemistry separation methods using an ion exchange resin such as BioRad MP-1 or AG1X8 (Mathur and Wang 2019). Such methods can be quite involved and may need to be developed for specific sample types to avoid incomplete yields of the metal of interest as this can lead to fractionation (Sossi et al. 2015). Until recently, isotope measurements were typically made using a copper spike in the zinc solution to correct for instrumental mass bias and samples were interspersed with standards (standard-sample bracketing) in order to correct for drift. More recently, double-spike Zn isotopic measurements have been introduced and have yielded consistent results with those obtained by standard bracketing techniques (Moynier et al. 2017). Advantages of the double spike technique are that it provides high precision absolute elemental abundances together with the isotope ratios and accounts for mass discrimination during chemical separation. This has been important for the analysis of Zn in difficult matrixes such as seawater (e.g. Conway and John 2015). Expected, natural, mass-dependent fractionation relationships can be checked by comparing $^{66}Zn/^{64}Zn$ with $^{67}Zn/^{64}Zn$, $^{68}Zn/^{64}Zn$ and $^{70}Zn/^{64}Zn$ ratios, providing a useful check on data quality. Typical long-term 2σ precision of measurements is around ± 0.04 ‰ (e.g. Chapman et al. 2006; Chen et al. 2013).

4 Natural Variation in Zinc Isotopes

Overall, relatively small variations in zinc isotope compositions have been reported in terrestrial rock and mineral samples. In a compilation of over 400 analyses there is a total range of < 2 ‰ and the majority of data (25th–75th percentiles) fall in the $\delta^{66}Zn$ range 0.08–0.36 ‰ (Fig. 2). This is due, at least in part, to the single oxidation state in which zinc occurs, unlike other transition metals like molybdenum, iron and copper, for which redox changes can cause significant isotopic fractionation.

There is limited evidence for possible fractionation of Zn by igneous processes: single measurements of granite, granodiorite (Viers et al. 2007) and andesite (Bentahila et al. 2008) gave $\delta^{66}Zn$ values of 0.47, 0.41 and 0.55 ‰ respectively, mostly slightly higher than basalts that range from 0.2 to 0.5 ‰ (Cloquet et al. 2008) and which have an average of 0.32 ‰ (Fig. 2). However, the few more felsic *volcanic* rocks analyzed appear to be depleted in the heavier zinc isotopes (Fig. 2). The basalt standard BCR-1 is the most analyzed rock, with a reported mean $\delta^{66}Zn$ value of 0.25 ± 0.09 ‰ (2σ). The average igneous rock composition is 0.30 ± 0.07 ‰ (n = 77, 1σ; Moynier et al. 2017) and Bulk Earth is thought to have a $\delta^{66}Zn$ value of 0.28 ± 0.05 ‰ (Chen et al. 2013).

Clastic sedimentary rock data (as a proxy for average continental crust) fall in a similar range (average values of 0.12–0.53 ‰; Fig. 2) but carbonates tend to be shifted towards higher $\delta^{66}Zn$, e.g. 0.91 ± 0.48 ‰ (2σ) for recent deep-sea carbonates from the Pacific (Pichat et al. 2003) and 0.04–0.87 ‰ (mean = 0.34 ± 0.35 ‰, 2σ) for cap dolostones formed just after the Marinoan snowball Earth (Kunzmann et al. 2013). This is most likely due to a biogenic fractionation by marine microorganisms that preferentially take up light zinc isotopes—thereby leading to their depletion in seawater—potentially coupled with removal of light zinc isotopes by burial in organic matter (e.g. Vance et al. 2006; John et al. 2007, 2018). However, positive isotope fractionations of ~1 ‰ have been predicted by ab initio modeling for zinc carbonate relative to the parent solution (Fujii et al. 2011), so an inorganic control of part of this shift is perhaps possible.

We do not know much about the behavior of zinc isotopes during supergene weathering but measurable fractionations do occur. In their study of a range of secondary zinc minerals from several different base metal ore deposit types, Mondillo et al. (2018) showed that willemite has the greatest compositional variability ($\delta^{66}Zn$ −0.42 to 1.39 ‰), spanning the entire range of terrestrial variation in Zn isotopes recorded to date. Both positive and negative fractionation directions were recorded relative to the precursor phase for all the minerals studied. The observed data were explained with a model of isotopic fractionation in which partial dissolution of primary sphalerite was followed by precipitation of an initial secondary phase that preferentially incorporated heavy Zn isotopes due to the development of strong Zn–O bonds. Progressive depletion of the residual fluid by this process led to distillation effects ultimately resulting in the formation of small amounts of late-stage, isotopically-light, supergene zinc minerals.

Ferromanganese nodules also form in a surficial environment where dissolution-reprecipitation reactions, adsorption and biological activity may play a role in isotopic fractionation. Maréchal et al. (2000) concluded that the high $\delta^{66}Zn$ values observed in nodules (0.53–1.16 ‰; see Fig. 2) were related to marine biological productivity in a similar way to marine

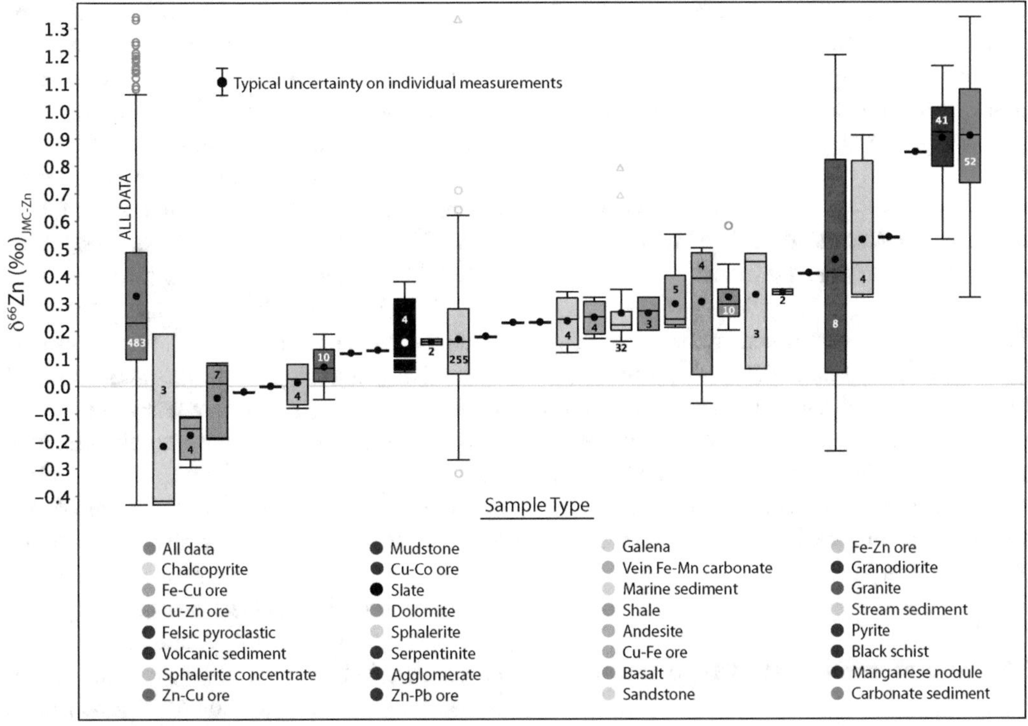

Fig. 2 Summary of terrestrial δ^{66}Zn data from rocks, ores and minerals, ordered in terms of increasing mean value. The grey box on the left represents the entire dataset. Sample types shown in black in the legend are where there is only one analysis available; for other sample types, the numbers of analyses are given on the boxes. The legend is listed in the same order as the boxes on the figure. Box boundaries = 25th and 75th percentiles; line = median; dot = mean; whiskers = the most extreme values that are not outliers; circles = outliers greater than 1.5 × the interquartile range away from the 25th or 75th percentile; triangles = outliers greater than 3 × the interquartile range away from the 25th or 75th percentile

carbonates. The observed global distribution of higher δ^{66}Zn in nodules from higher latitudes could be explained by seasonal productivity fluctuations. However, Little et al. (2014) showed that preferential uptake of heavy zinc isotopes is likely during adsorption from seawater onto manganese oxides (specifically δ-MnO_2) so that the coupling of different effects, both biological and inorganic, is likely.

More recently, Spinks and Uvarova (2019) showed that terrestrial Fe–Mn crusts were also enriched in isotopically heavy zinc, with those overlying unmineralized sources more enriched (by up to 0.54 ‰) than those overlying isotopically light sulfide mineralization (by up to 0.37 ‰). They suggested the difference was due to different zinc complexing behavior in the two environments and noted that the isotopic contrast could potentially be useful in mineral exploration.

Finally, zinc isotope variability has been recorded in modern submarine hydrothermal and volcanic hot spring systems indicating the likelihood of fractionation during hydrothermal ore deposit formation. Fluid δ^{66}Zn values measured at a range of submarine vent sites ranged from 0.00 to 1.04 ‰, with the higher and more variable values being linked to lower temperatures (John et al. 2008). It was suggested that subsurface precipitation of isotopically light sphalerite was the primary control of the variation in fluid composition, supported by the observation of vent chimney sulfide with lower or similar δ^{66}Zn to the fluids. A similar mechanism of sphalerite precipitation at depth was proposed to explain the high δ^{66}Zn ($\sim$0.70 ‰) of thermal spring waters at La Soufrière volcano, Guadeloupe (Chen et al. 2014), in agreement with isotopic fractionations calculated for aqueous complexes of Zn using molecular dynamics (Black et al. 2011; Ducher et al. 2018).

5 Zinc Isotopes in Ore-Forming Hydrothermal Systems

Hydrothermal processes can produce variations in sulfide δ^{66}Zn by up to 0.9 ‰ (Wilkinson et al. 2005a; Mason et al. 2005; John et al. 2008). Yet, there is still limited understanding of the controls of isotopic fractionation. Below, I consider the results reported to date from a range of hydrothermal ore deposits and summarize the current state of knowledge.

5.1 Sediment-Hosted Zn–Pb Deposits

Several studies have investigated hypogene Zn isotope variation in sediment-hosted base metal deposits. The first such study investigated possible controls on the zinc isotope composition of sphalerite in the Irish Zn-Pb orefield (Wilkinson et al. 2005a). Four possible controls on zinc isotope variation were evaluated: (1) compositional variability of the metal source terrain(s); (2) multiple zinc sources; (3) precipitation temperature; and (4) precipitation mechanism. The δ^{66}Zn values obtained ranged from −0.17 to 0.64 ‰, with one outlier at 1.33 ‰, encompassing the terrestrial variation (Fig. 2). Although largely overlapping with other sulfide data, the average composition of the Irish sphalerites (0.15 ± 0.19 ‰, 1σ; omitting outlier) is isotopically light compared to most rock samples.

This study showed that there was no clear relationship between zinc isotopic composition of sphalerite and fluid temperature, salinity or likely source rocks. Wilkinson et al. (2005a) concluded, based on the spatial variability in δ^{66}Zn, that fractionation was induced via sphalerite precipitation with an inferred negative kinetic isotopic effect during rapid supersaturation. Thus, fluids evolved to heavier isotopic compositions via a Rayleigh fractionation process as they migrated upward and outward through the ore deposits, generating late and/or distal sphalerite with high δ^{66}Zn (Fig. 3).

Gagnevin et al. (2012) followed up this work with a specific investigation of the Navan deposit in Ireland and found that δ^{66}Zn was correlated with δ^{56}Fe and δ^{34}S in sphalerite. They interpreted this to be a result of a negative kinetic Zn (and Fe) isotope fractionation during sphalerite precipitation, coupled with the linked process of fluid mixing involving fluids of contrasting δ^{34}S. Their sphalerite bulk concentrate samples gave a mean δ^{66}Zn of 0.01 ‰ confirming the relatively light average isotopic composition of Irish sphalerites and suggesting that average ore has lower δ^{66}Zn than that generally found in the coarser crystalline sphalerite grains that are more amenable to drilling out for isotopic analysis.

In a study of the Red Dog sediment-hosted Zn–Pb district, Kelley et al. (2009) reached similar conclusions regarding the control of zinc isotope fractionation in sphalerite. Samples from shale-hosted massive sulfide and stratigraphically underlying vein-breccia deposits showed a range of δ^{66}Zn values from 0.00 to 0.60 ‰. The lowest values were observed in the vein-breccias and the stratigraphically overlying (but structurally displaced) ore deposits showed a systematic trend of increasing δ^{66}Zn from south to north (Main-Aqqaluk-Paalaaq-Anarraaq). The δ^{66}Zn values were inversely correlated with sphalerite Fe/Mn and tended to be higher in low-Cu sphalerite, consistent with precipitation of sphalerite with lower δ^{66}Zn closer to the principal hydrothermal fluid conduits. It was inferred that Rayleigh fractionation during sulfide precipitation led to the preferential incorporation of light zinc isotopes in the earliest sphalerite forming at deeper levels (vein-breccias) and close to the principal fluid conduits in the orebodies. This was followed by precipitation of sulfides with higher δ^{66}Zn in shallower and/or more distal parts of the flow path.

More recently, Gao et al. (2018) reported similar trends from the giant Dongshengmiao sediment-hosted Zn–Pb deposit, China. δ^{66}Zn varied from 0.17 to 0.40 ‰ and showed a lateral trend of increasing δ^{66}Zn from southwest to

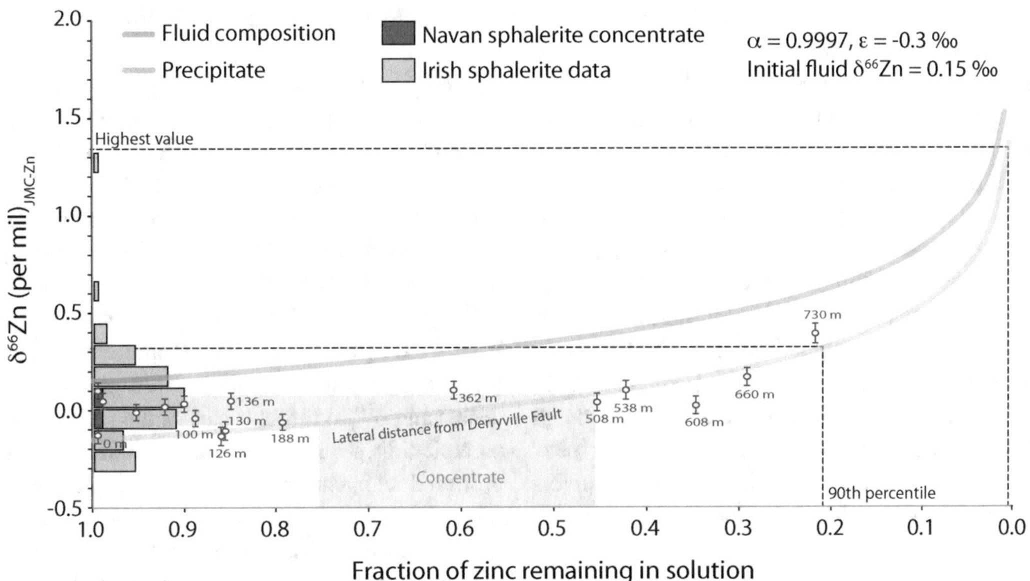

Fig. 3 Open system Rayleigh fractionation model to explain range and spatial distribution of δ^{66}Zn values of sphalerite from the Irish Zn–Pb orefield. The model is constrained by an imposed requirement to encompass the total range of measured δ^{66}Zn, assumes a starting fluid composition that is the same as that determined for Lower Paleozoic basement greywacke samples (Crowther 2007), assuming 100% extraction from these inferred source rocks (Wilkinson et al. 2005b) and an instantaneous equilibrium sphalerite-fluid fractionation factor ε ($\delta^{66}Zn_{mineral} - \delta^{66}Zn_{fluid}$) of −0.3 after the experimental data of Veeramani et al. (2015). Sample data from Wilkinson et al. (2005a; unpublished), Gagnevin et al. (2012) and Crowther (2007). The histogram shows the range of measured δ^{66}Zn for sphalerite with the 90th percentile highlighted; the majority of the ore falls within this range and, based on the model, could have formed after precipitation of between 0 and ~80% of the zinc in solution. The "ore average" compositions given by the bulk sphalerite concentrate data of Gagnevin et al. (2012) indicate that the average ore composition is consistent with having precipitated after ~25–55% of the zinc in solution had been removed. Spatially-constrained samples from a traverse away from the Derryville Zone feeder fault at the Lisheen deposit (Wilkinson, unpublished), arbitrarily plotted on a linear distance scale on the x-axis in red, can be interpreted in terms of progressive precipitation of sphalerite from batches of fluid during lateral flow, with the most distal sphalerite (at 730 m, located at the edge of the orebody) having a composition consistent with formation after about 80% of the zinc had already been precipitated from batches of fluid arriving at that point, i.e. the orebody as a whole represents a depositional efficiency of 80%. The most extreme δ^{66}Zn value measured, 1.33 ‰ from the Galmoy deposit, can be explained by relatively proximal but very late-stage precipitation of very small amounts of sphalerite in vugs from fluids that had already deposited most (~99%) of the original zinc in solution. Data plotting above the model line in proximal samples can be accounted for by rapid, disequilibrium precipitation of high proportions of the zinc in solution in the core of the ore body, with sphalerite compositions therefore approaching the initial solution δ^{66}Zn value. Although the model is subject to many uncertainties, it demonstrates how spatial and temporal trends of increasing δ^{66}Zn can be explained by Rayleigh fractionation processes during flow. The observed pattern of increasing δ^{66}Zn from the centre of hydrothermal deposits outwards thus has a firm theoretical basis and can be utilized in exploration. The discovery of minor, isotopically heavy sphalerite (or whole rocks with high δ^{66}Zn controlled by trace amounts of sphalerite) could therefore be indicative of the extensive precipitation of sphalerite elsewhere in the vicinity. A gradient of decreasing δ^{66}Zn towards ore would be predicted

northeast within the main ore body. The lead isotopic homogeneity of ore sulfides suggested that there was only one significant source for metal, thus precluding mixing of multiple metal sources as a control of this pattern. It was concluded that the most likely control on spatial variations was Rayleigh fractionation during hydrothermal fluid flow, with lighter Zn isotopes preferentially incorporated into the earliest sulfides to precipitate.

Zhou et al. (2014a, b) studied the Tianqiao, Banbanqiao and Shanshulin carbonate-hosted Pb–Zn deposits from the Sichuan–Yunnan–Guizhou Pb–Zn metallogenic province, southwest China. These deposits are associated with dolomitization related to reverse faults and are probably of the Mississippi Valley-type. Sphalerite from successive stages of deposition in all three deposits exhibited increasing δ^{66}Zn values in the range −0.26 to 0.71‰. Furthermore, δ^{66}Zn values in the core of the Tianqiao No. 1 orebody were lower than those on the periphery. Both trends are consistent with a kinetic Rayleigh fractionation effect during progressive precipitation, with earlier/deeper sphalerite being preferentially enriched in lighter zinc isotopes.

5.2 Volcanic-Hosted Massive Sulfide Deposits

In their study of the undeformed Devonian Alexandrinka volcanic-hosted massive sulfide (VHMS) deposit in the Urals, Mason et al. (2005) found that chalcopyrite-rich samples from vein stockwork had lower δ^{66}Zn by ~0.4 ‰ (−0.43 to −0.10 ‰) relative to sphalerite-bearing samples (−0.03 to 0.08 ‰) and attributed this to equilibrium partitioning of isotopically light Zn into chalcopyrite during its precipitation. Alternatively, it is possible that the zinc measured in the chalcopyrite-rich samples ("Cu–Zn ore"; Fig. 2) was principally contained within minor sphalerite; in this case, the lower δ^{66}Zn could be explained by preferential precipitation of light zinc isotopes in more proximal and higher temperature stockwork sphalerite. In a vent chimney sample, δ^{66}Zn increased toward the chimney rim by 0.26 ‰ (−0.03 to 0.23 ‰), which may have been caused by changing temperature (hence fractionation factor), or Rayleigh distillation. Clastic sulfide samples, mostly containing < 15% sphalerite, were isotopically light (−0.30 to −0.05 ‰) possibly due to chalcopyrite predominance as a Zn host, or post-depositional seafloor oxidative dissolution and re-precipitation which led to systematic negative shifts in δ^{66}Zn relative to the primary sulfides.

5.3 Vein-Hosted Zn–Pb Deposits

The only vein-type deposit to have been investigated is the Zhaxikang Sb–Pb–Zn–Ag deposit in Southern Tibet (Wang et al. 2017, 2018). The origin of this fault/vein-controlled deposit is controversial with both magmatic-hydrothermal (Duan et al. 2016) and magmatic-hydrothermally overprinted SEDEX (Wang et al. 2017) origins being proposed. δ^{66}Zn values of sulfides range from −0.03 to 0.38 ‰ (Fig. 2) and cannot distinguish between these alternatives. A trend of decreasing δ^{66}Zn through part of the paragenetic precipitation sequence was noted, although the proposed explanation of boiling coupled with mineral precipitation as a control (Wang et al. 2018) is inconsistent with the Rayleigh distillation models presented and no independent evidence of phase separation was recorded.

5.4 Porphyry-Type and Other Ore Deposits

To date, only a single determination of δ^{66}Zn has been made in a porphyry system, from a chalcopyrite separate from Chuquicamata, Chile. This returned a typical crustal value of 0.19 ‰ (Maréchal et al. 1999). In porphyry deposits, the focus has been on the primary ore metals, Cu and Mo (see Cooke et al. 2014; Mathur and Zhao 2023). At the time of writing, no other deposit types containing zinc as a major or minor component have been investigated.

6 Discussion

6.1 Identifying Metal Sources

Despite suggestions to the contrary, it is not considered here very likely that—at least at the present time—zinc isotope measurements can be used to identify the source of metals for hydrothermal ore deposits. The principal reason is because of the narrow compositional variability of terrestrial reservoirs and the apparently greater variability that is introduced during depositional processes. Added to this, there is almost no information on how zinc isotopes may be fractionated during extraction from a source by hydrothermal fluids or potentially modified during transport. Thus, it is difficult to identify the primary source signal from samples where various fractionation steps may have occurred along the system flow path (Fig. 1). The one exception might be where zinc is sourced from carbonate rocks because these tend to have a fairly distinctive, isotopically heavy composition (Fig. 2). Consequently, questions such as the possible importance of host rock sourcing of metals for skarn deposits and the potential derivation of metals from different aquifer compositions as an influence on the metal tenor of MVT deposits—in particular the apparent relationship between carbonate aquifers and more Zn-rich deposits (Sverjensky 1984)—may be tractable.

6.2 Precipitation Efficiency

If source fluid compositions are known, or can be inferred, the isotopic composition of precipitated ore minerals could provide a clue to depositional efficiency—a key parameter in ore studies, yet one that has been almost impossible to constrain to date. Any fractionation between fluid and mineral will generate a progressive shift in isotopic composition of the mineral being deposited. Because of the reservoir effect, locally high degrees of precipitation in an ore body from a batch of fluid will result in the average isotopic composition of the ore approaching the input fluid composition (Fig. 3). By contrast, lower degrees of precipitation could produce a broad, highly-skewed distribution, which, with a negative $\Delta_{\text{fluid-mineral}}$ fractionation would produce proximal ore with a lower δ^{66}Zn than the input fluid and more distal ore, precipitated later, with higher δ^{66}Zn than the initial fluid. For example, in the Irish sediment-hosted Zn–Pb deposits described above, the dispersion in zinc isotope compositions and slight bias towards isotopically heavy compositions within an ore lens could reflect intermediate precipitation efficiency in an open system.

6.3 Hypogene Isotopic Zonation

From the global compilation presented in Fig. 2, it is possible to make the general observation that δ^{66}Zn from copper-bearing ores (mixed mineralogy samples), dominated by Fe–Cu or Cu–Zn sulfides is consistently low. Given that chalcopyrite separates from the Alexandrinka VHMS deposit returned some of the lowest δ^{66}Zn values measured to date (Mason et al. 2005), one could conclude that Zn substituted into the chalcopyrite structure may be in a bonding environment that favours the light isotopes. This would mean that measurements of Cu–Zn ores may show a systematic shift to lower δ^{66}Zn towards the core of systems due to an increase in chalcopyrite abundance at higher temperatures. As a tool, this has limited use because assay and sulfide zonation patterns would be simpler and cheaper ways to observe the same trends. However, if it is trace sphalerite in such samples that is controlling the signature then it implies that sphalerite can record a systematic zoning from the high temperature cores of hydrothermal systems out into the typically lower temperature domains where sphalerite becomes dominant.

Based on existing studies of low temperature, sediment-hosted Zn–Pb systems, it appears that Zn isotopes are fractionated systematically in this way during ore formation. In all case studies completed to date, it has been inferred that light zinc isotopes are preferentially incorporated into sphalerite ($\alpha_{\text{sphalerite-fluid}} < 1$) so that fluids—and later sphalerite precipitated from them—evolve towards heavier compositions in time and space (Figs. 3, 4). This behavior is consistent with what has been observed in modern hydrothermal environments (e.g. John et al. 2008). Such systematic fractionation provides a theoretical basis for the development of exploration tools that can utilize zonation patterns of δ^{66}Zn in sphalerite (and possibly altered rocks containing trace sphalerite or other zinc-bearing minerals) around ore bodies (Fig. 4). In the case of the Irish ore system, there is evidence that δ^{66}Zn varies both vertically up through the systems and laterally across individual orebodies, a trend that can be modeled using a Rayleigh-type fractionation approach (Fig. 3).

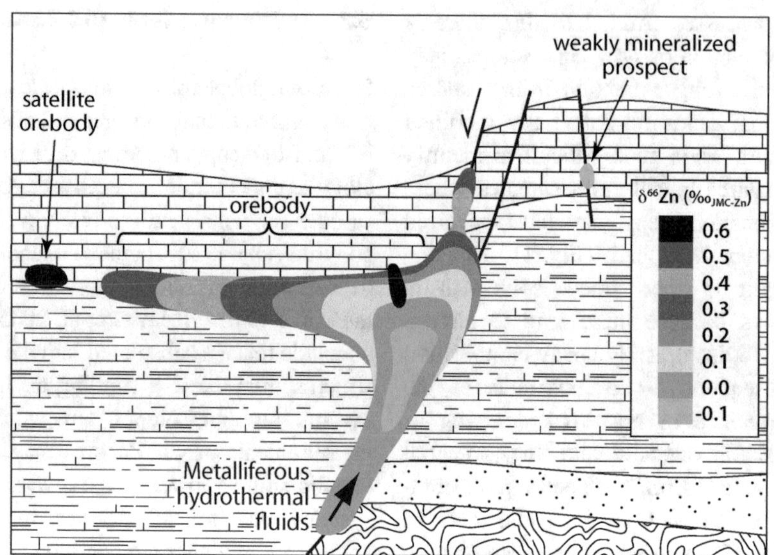

Fig. 4 Cartoon cross-section illustrating possible zinc isotope zonation that might develop in a hydrothermal zinc deposit. Geology based loosely on a SEDEX-type system, but trends would be applicable to vein-hosted hydrothermal deposits. Color scale illustrates pattern of δ^{66}Zn that may be observed in sphalerite mineral separates, or in whole rock where major to trace sphalerite dominates the Zn budget of the sample. Distal mineralization, at the margin of the orebody or developed along escape structures, is fractionated towards heavier isotopic compositions. Late-stage, overprinting mineralization may also be isotopically heavy. Weakly-mineralized satellite prospects may have compositions similar to proximal sphalerite from an orebody if they are formed by limited precipitation from similar source fluids

6.4 Supergene Isotopic Zonation

Similarly, in the supergene modification of pre-existing sulfide deposits, the prediction of an increase in δ^{66}Zn in most supergene minerals where Zn is tetrahedrally-coordinated in relatively strong Zn–O bonds ($\alpha_{secondary\ mineral\text{-}sphalerite} > 1$) can produce a supergene ore that is isotopically heavy but has peripheral or paragenetically-late supergene Zn minerals that have low δ^{66}Zn values because of the preferential removal of heavy zinc isotopes from the fluid by earlier precipitation (Mondillo et al. 2018).

7 Summary and Conclusions

The study of non-traditional stable isotopes in general and as applied to ore deposits in particular remains at a fledgling stage. It has become apparent that wider variations in isotopic compositions are present in hydrothermal ore deposits than in any other terrestrial environments; this makes them particularly interesting for further investigation. Several studies have inferred the probability of Rayleigh distillation processes in the systematic fractionation of zinc during precipitation of ore minerals. If this general process is confirmed it will open up many possible applications, with the isotope systematics potentially tracking flow pathways through deposits and out into the "exhaust" region (Fig. 1), the domain of spent ore fluids.

At the present time, models to explain isotope fractionation patterns of ore metals are underconstrained, particularly because of the limited experimental data on fluid-mineral fractionations and lack of knowledge on whether equilibrium or kinetic fractionations are likely to prevail. There is the additional complication of biologically-mediated fractionation of Zn in low temperature systems (e.g. Wanty et al. 2013). Consequently, current interpretations are somewhat speculative and well-designed experimental studies to measure isotopic fractionations under relevant hydrothermal conditions of metal extraction, transport and metal sulfide deposition (Fig. 1) are needed. Once these are available, quantitative modeling will be better able to test alternative hypotheses for natural datasets.

Despite the current limitations, the data summarized here for zinc provide an indication of the types of new insights that studies of ore metal stable isotopes may provide. In time, these isotope systems are likely to provide powerful new tools for testing models of metal transport and deposition and thereby improve the ore deposit models that essentially underpin all modern mineral exploration. More directly, the operation of Rayleigh distillation processes may produce patterns that help to unravel complex flow paths and produce isotopic gradients that can be utilized for vectoring towards the centre of hydrothermal systems. It may be possible to identify mixing between magmatic and country rock-derived metals on the fringes of magmatic-hydrothermal systems if there is an isotopic contrast, such as for magmas intruding carbonate sequences.

Notwithstanding these opportunities, the fact remains that for a geochemical tool to achieve widespread take up in the mineral exploration industry it needs to be cheap, rapid and not require a high level of technical knowledge to apply. Zinc isotopes measurements and the subsequent interpretation of the data do not presently meet these requirements. This means that it is unlikely that vectoring using zinc isotope zonation will become a widely utilized method for the foreseeable future. Even effective tools based on mineral trace element chemistry, that are far easier to apply, are still only being taken up gradually by major explorers (e.g. Cooke et al. 2020a,b; Wilkinson et al. 2020) because they are perceived to be relatively high cost and slow to implement (~6 week turnaround time). Probably the most significant contribution that zinc isotopes can make to studies of hydrothermal systems in the immediate future is to help resolve outstanding questions about hydrothermal processes that conventional datasets have been unable to solve, such as understanding extraction of metals from source rocks, tracing flow paths and identifying the controls of depositional efficiency.

Acknowledgements The author would like to thank the numerous researchers involved in the development and application of zinc isotopes to ore systems in the MAGIC Laboratories, Imperial College London, without whom this review would not have been written, namely: Prof. Dominik Weiss, Prof. Mark Rĕhkamper, Dr. Thomas Mason, Barry Coles and Dr. Helen Crowther. Their contributions are acknowledged; however, any mistakes or shortcomings in this paper are my own. Finally, I would like to thank Marcus Kunzmann and Ryan Mathur for their helpful reviews that improved the manuscript.

References

Albarède F (2004) The stable isotope geochemistry of copper and zinc. Rev Mineral Geochem 55:409–427

Albarède F, Beard BL (2004) Analytical methods for nontraditional isotopes. Rev Mineral Geochem 55:113–152

Archer C, Andersen MB, Cloquet C, Conway TM, Dong S, Ellwood M, Moore R, Nelson J, Rehkämper M, Rouxel O, Samanta M, Shin K-C, Sohrin Y, Takano S, Wasylenki L (2017) Inter-calibration of a proposed new primary reference standard AA-ETH Zn for zinc isotopic analysis. J Anal at Spectrom 32:415–419

Bentahila Y, Ben Othman D, Luck J-M (2008) Strontium, lead and zinc isotopes in marine cores as tracers of sedimentary provenance: a case study around Taiwan orogen. Chem Geol 248:62–82

Black JR, Kavner A, Schauble EA (2011) Calculation of equilibrium stable isotope partition function ratios for aqueous zinc complexes and metallic zinc. Geochim Cosmochim Acta 75:769–783

Chapman JB, Mason TFD, Weiss DJ, Coles BJ, Wilkinson JJ (2006) Chemical separation and isotopic variations of Cu and Zn from five geological reference materials. Geostand Geoanal Res 30:5–16

Chen H, Savage P, Teng FZ, Helz RT, Moynier F (2013) No zinc isotope fractionation during magmatic differentiation and the isotopic composition of the bulk Earth. Earth Planet Sci Lett 369:34–42

Chen J-B, Gaillardet J, Dessert C, Villemant B, Louvat P, Crispi O, Birck J-L, Wang Y-N (2014) Zn isotope compositions of the thermal spring waters of La Soufrière volcano, Guadeloupe island. Geochim Cosmochim Acta 127:67–82

Cloquet C, Carignan J, Lehmann M, Vanhaecke F (2008) Variation in the isotopic composition of zinc in the natural environment and the use of zinc isotopes in biogeosciences: a review. Anal Bioanal Chem 390:451–463

Conway TM, John SG (2015) The cycling of iron, zinc and cadmium in the North East Pacific Ocean—insights from stable isotopes. Geochim Cosmochim Acta 164:262–283

Cooke DR, Hollings P, Wilkinson JJ, Tosdal RM (2014) Geochemistry of porphyry deposits. In: Treatise on geochemistry, 2nd edn, vol 13, pp 357–381

Cooke DR, Agnew P, Hollings P, Baker M, Chang Z, Wilkinson JJ, Ahmed A, White NC, Zhang L, Thompson J, Gemmell JB, Chen H (2020a) Recent advances in the application of mineral chemistry to exploration for porphyry copper–gold–molybdenum deposits: detecting the geochemical fingerprints and footprints of hypogene mineralization and alteration. Geochem Environ Explor Anal 20:176–188. https://doi.org/10.1144/geochem2019-039

Cooke DR, Wilkinson JJ, Baker M, Agnew P, Phillips J, Chang Z, Chen H, Wilkinson CC, Inglis S, Martin H (2020b) Using mineral chemistry to aid exploration—a case study from the Resolution porphyry Cu-Mo deposit, Arizona. Econ Geol 115:813–840

Crowther HL (2007) A zinc isotope and trace element study of the Irish Zn-Pb orefield. Unpublished PhD thesis, Imperial College, University of London, 361 p

Duan J, Tang J, Lin B (2016) Zinc and lead isotope signatures of the Zhaxikang Pb-Zn deposit, South Tibet: implications for the source of the ore-forming metals. Ore Geol Rev 78:58–68

Ducher M, Blanchard M, Balan E (2018) Equilibrium isotopic fractionation between aqueous Zn and minerals from first-principles calculations. Chem Geol 483:342–350

Fujii T, Moynier F, Pons ML, Albarède F (2011) The origin of Zn isotope fractionation in sulfides. Geochim Cosmochim Acta 75:7632–7643

Gagnevin D, Boyce AJ, Barrie CD, Menuge JF, Blakeman RJ (2012) Zn, Fe and S isotope fractionation in a large hydrothermal system. Geochim Cosmochim Acta 88:183–198

Gao Z, Zhu K, Sun J, Luo Z, Bao C, Tang C, Ma J (2018) Spatial evolution of Zn-Fe-Pb isotopes of sphalerite within a single ore body: a case study from the Dongshengmiao ore deposit, Inner Mongolia, China. Mineral Deposita 53:55–65

John SG, Geis RW, Saito MA, Boyle EA (2007) Zinc isotope fractionation during high-affinity and low-affinity zinc transport by the marine diatom Thalassiosira oceanica. Limnol Oceanog 52(6):2710–2714

John SG, Rouxel OJ, Craddock PR, Engwall AM, Boyle EA (2008) Zinc stable isotopes in seafloor hydrothermal vent fluids and chimneys. Earth Planet Sci Lett 269:17–28

John SG, Helgoe J, Townsend E (2018) Biogeochemical cycling of Zn and Cd and their stable isotopes in the Eastern Tropical South Pacific. Marine Chem 201:256–262

Johnson CM, Beard BL, Albarède F (eds) (2004) Geochemistry of non-traditional stable isotopes. Rev Mineral Geochem 55:454

Kelley KD, Wilkinson JJ, Chapman JB, Crowther HL, Weiss DJ (2009) Zinc isotopes in sphalerite from base metal deposits in the Red Dog district, northern Alaska. Econ Geol 104:767–773

Kunzmann M, Halverson GP, Sossi PA, Raub TD, Payne JL, Kirby J (2013) Zn isotope evidence for immediate resumption of primary productivity after snowball Earth. Geology 41:27–30

Little SH, Sherman DM, Vance D, Hein JR (2014) Molecular controls on Cu and Zn isotopic fractionation in Fe–Mn crusts. Earth Planet Sci Lett 396:213–222

Maréchal CN, Télouk P, Albarède F (1999) Precise analysis of copper and zinc isotopic compositions by plasma-source mass spectrometry. Chem Geol 156:251–273

Maréchal CN, Nicolas E, Douchet C, Albarède F (2000) Abundance of zinc isotopes as a marine biogeochemical tracer. Geochem Geophys Geosyst 1:1015. https://doi.org/10.1029/1999GC000029

Mason TFD, Weiss DJ, Chapman JB, Wilkinson JJ, Tessalina SG, Spiro B, Horstwood MSA, Spratt J, Coles BJ (2005) Zn and Cu isotopic variability in the Alexandrinka volcanic-hosted massive sulfide (VHMS) ore deposit, Urals, Russia. Chem Geol 221:170–187

Mathur R, Wang D (2019) Transition metal isotopes applied to exploration geochemistry: insights from Fe, Cu and Zn. In: Decree S, Robb L (eds) Ore deposits: origin, exploration and exploitation. Wiley, New York, pp 167–183

Mathur R, Zhao Y (2023) Copper isotopes used in mineral exploration. In: Huston DL, Gutzmer J (eds) Isotopes in economic geology, metallogensis and exploration. Springer, Berlin (this volume)

Moeller K, Schoenberg R, Pedersen R-B, Weiss D, Dong S (2012) Calibration of the new certified reference materials ERM-AE633 and ERM-AE647 for copper and IRMM-3702 for zinc isotope amount ratio determinations. Geostand Geoanal Res 36:177–199

Mondillo N, Wilkinson JJ, Boni M, Weiss DJ, Mathur R (2018) A global assessment of Zn isotope fractionation in secondary Zn minerals from sulfide and non-sulfide ore deposits and model for fractionation control. Chem Geol 500:182–193

Moynier F, Vance D, Fujii T, Savage P (2017) The isotope geochemistry of zinc and copper. Rev Mineral Geochem 82:543–600

Pichat S, Douchet C, Albarède F (2003) Zinc isotope variations in deep-sea carbonates from the eastern equatorial Pacific over the last 175 ka. Earth Planet Sci Lett 210:167–178

Ponzevera E, Quételet CR, Berglund M, Taylor PDP, Evans P, Loss RD, Fortunato G (2006) Mass discrimination during MC-ICPMS isotopic ratio measurements: investigation by means of synthetic isotopic mixtures (IRMM-007 series) and application to the calibration of natural-like zinc materials (including IRMM-3702 and IRMM-651). J Am Soc Mass Spectr 17:1413–1428

Rosman KJR, Taylor PDP (1998) Isotopic compositions of the elements. Pure Appl Chem 70:217–235

Sossi PA, Halverson GP, Nebel O, Eggins SM (2015) Combined separation of Cu, Fe and Zn from rock matrices and improved analytical protocols for stable isotope determination. Geostand Geoanal Res 39:129–149

Spinks SC, Uvarova Y (2019) Fractionation of Zn isotopes in terrestrial ferromanganese crusts and implications for tracing isotopically-heterogeneous metal sources. Chem Geol 529:119314

Sverjensky DA (1984) Oilfield brines as ore-forming solutions. Econ Geol 79:23–37

Vance D, Archer C, Bermin J, Kennaway G, Cox EJ, Statham PJ, Lohan MC, Ellwood MJ (2006) Zn isotopes as a new tracer of metal micronutrient usage in the oceans. Geochim Cosmochim Acta 70:A666

Veeramani H, Eagling J, Jamieson-Hanes JH, Kong L, Ptacek CJ, Blowes DW (2015) Zinc isotope fractionation as an indicator of geochemical attenuation processes. Environ Sci Technol Lett 2:314–319

Viers J, Oliva P, Nonelle A, Gelabert A, Sonke J, Freydier R, Gainville R, Dupre B (2007) Evidence of Zn isotopic frationation in a soil-plant system of a pristine tropical watershed (Nsimi, Cameroon). Chem Geol 239:124–137

Wang D, Sun X, Zheng Y, Wu S, Xia S, Chang H, Yu M (2017) Two pulses of mineralization and genesis of the Zhaxikang Sb–Pb–Zn–Ag deposit in southern Tibet: constraints from Fe–Zn isotopes. Ore Geol Rev 84:347–363

Wang D, Zheng Y, Mathur R, Wu S (2018) The Fe–Zn isotopic characteristics and fractionation models: implications for the genesis of the Zhaxikang Sb–Pb–Zn–Ag deposit in southern Tibet. Geofluids 2197891. https://doi.org/10.1155/2018/2197891

Wanty RB, Podda F, De Giudici G, Cidu R, Lattanzi P (2013) Zinc isotope and transition-element dynamics accompanying hydrozincite biomineralization in the Rio Naracauli, Sardinia, Italy. Chem Geol 337–338:1–10

Wilkinson JJ, Weiss DJ, Mason TFD, Coles BJ (2005a) Zinc isotope variation in hydrothermal systems: preliminary evidence from the Irish Midlands ore field. Econ Geol 100:583–590

Wilkinson JJ, Everett CE, Boyce AJ, Gleeson SA, Rye DM (2005b) Intracratonic crustal seawater circulation and the genesis of subseafloor zinc-lead mineralization in the Irish orefield. Geology 33:805–808

Wilkinson JJ, Baker MJ, Cooke DR, Wilkinson CC (2020) Exploration targeting in porphyry Cu systems using propylitic mineral chemistry: a case study of the El Teniente deposit, Chile. Econ Geol 115:771–791

Zhou J-X, Huang Z-L, Zhou M-F, Zhu X-K, Muchez P (2014a) Zinc, sulfur and lead isotopic variations in carbonate-hosted Pb–Zn sulfide deposits, southwest China. Ore Geol Rev 58:41–54

Zhou J-X, Huang Z-L, Lu Z-C, Zhu X-K, Gao J-G, Mirnejad H (2014b) Geology, isotope geochemistry and ore genesis of the Shanshulin carbonate-hosted Pb–Zn deposit, southwest China. Ore Geol Rev 63:209–225

Isotopes in Economic Geology, Metallogeny and Exploration—Future Challenges and Opportunities

David L. Huston and Jens Gutzmer

Abstract

Although the intent of this book is to provide readers with an overview on the current and past usage of isotopes in the broad disciplines of economic geology, metallogenesis and mineral exploration, some of the chapters highlight future challenges and opportunities for the use of both radiogenic and stable isotopes within these disciplines and more broadly. This concluding section identifies and then discusses how some of these challenges might be overcome and the opportunities that might be realized.

1 Challenges to Isotopic Research

As described throughout this book, isotopic research, both on radiogenic and stable isotopes, has been essential to develop models for the genesis of many types of mineral deposits, including information on age and duration of mineralizing events, tectonic and metallogenic setting, fluid, metal and sulfur sources, and alteration and fluid pathways. Despite the usefulness of this information, the application and interpretation of isotopes in economic geology, metallogenesis and exploration faces a number of analytical and interpretational challenges.

1.1 Radiogenic Isotopes and the Geochronology of Mineralizing Systems

Many recent advances in ore genesis resulted from an improving capability to understand mineral systems in the 4th dimension, time, which, in turn, has enabled direct linkages between mineral systems, other geological systems such as tectonic systems, and Earth evolution. With the development of new analytical techniques and the application of existing analytical techniques to new minerals, economic geologists have much better insights into the absolute ages and durations of mineralising events. This insight enables not only a better understanding of mineralizing processes but allows temporal linkages of these processes to other geological events that can be identified and/or tested at scales from the global to the thin section. Despite major advances in capability, many challenges remain to incorporating time into ore genesis and metallogenic models, and exploration practices.

D. L. Huston (✉)
Geocience Australia, GPO Box 378, Canberra, ACT 2601, Australia
e-mail: David.Huston@ga.gov.au

J. Gutzmer
Helmholtz-Institute Freiberg for Resource Technology, Helmholtz-Zentrum Dresden-Rossendorf, Chemnitzer Str. 40, 09599 Freiberg, Germany

D. Huston and J. Gutzmer (eds.), *Isotopes in Economic Geology, Metallogenesis and Exploration*, Mineral Resource Reviews, https://doi.org/10.1007/978-3-031-27897-6_16

One of the most significant challenges is to develop methodologies for dating some classes of mineral deposits and criteria for assessing ages of others. A small proportion of deposits can be robustly dated using ore minerals (i.e., those minerals extracted for metal recovery). Dateable ore minerals, however, are few, for example, molybdenite (using the Re–Os systems: Norman 2023), cassiterite, tantalite and related minerals, uraninite and other uranium minerals (U–Pb: Chelle-Michou and Schaltegger 2023), sphalerite (Rb–Sr: Christensen et al. 1995), scheelite (Sm–Nd; Anglin et al. 1996), and the interpretation of age data from some of these (e.g., uraninite and sphalerite) can be fraught for many reasons, including post-depositional open system behaviour (Chiaradia 2023).

A specific example of these challenges is the fact that while uraninite can produce robust unimodal U–Pb ages (e.g., Cross et al. 2011), it more commonly gives a range of ages that can span hundreds of millions of years (e.g., Polito et al. 2005). Hence, one of the main challenges to ore geochronology is determining the geochronological significance of isotopic data: do the data indicate mineralizing events, subsequent isotopic disturbance events or related processes such as fluid or source mixing? For example, Nash et al. (1981) argued that the formation of some unconformity-related uranium deposits involved repeated introduction of uranium over periods of up to a billion years. More recently, Ehrig et al. (2021) argued for two periods of uranium introduction at the giant Olympic Dam iron-oxide copper–gold deposit in South Australia based on laser ablation-inductively coupled plasma-mass spectrometry (LA-ICP-MS) ages of uraninite—an early event at ca 1590 Ma, and a later event at ca 500 Ma. A challenge to ore geochronology is to determine if multiple ages from uraninite (and other minerals) represent independent metal introduction events, metal redistribution/recrystallization events, or mixing ages.

Some deposit types, for example orogenic gold (Vielreicher et al. 2015), can be dated using non-ore minerals but an assumption must be made (or relationship demonstrated) that the dated non-ore mineral (e.g., monazite or xenotime) is coeval with the ore mineral (gold). Other deposits can be dated using minerals in the associated alteration assemblage; this requires, however, the assumption that the alteration assemblage directly relates to the mineralizing event. A third method to infer mineralization ages is assuming a relationship between a dated rock and mineralization; for example, a skarn deposit can be dated by assuming that its age is the same as the associated granite, or an (assumed) syngenetic deposit can be dated from the age of the host rocks. In many (most) cases, such assumptions are justified and the reported ages reflect the ages of mineralization, but in other cases, the assumption may be wrong. A challenge to ore geochronologists and economic geologists is producing a set of criteria to assess geological relationships and the resulting inferred ages of mineralization.

The greatest challenge to ore geochronology is dating mineralizing events for which dateable ore, ore-related or alteration minerals are not (known to be) present and that cannot be confidently related to other dateable geological events. A good example of this challenge are Mississippi Valley-type deposits, which commonly lack dateable minerals and have ambiguous or controversial relationships to local and/or regional geological events. This challenge may be partly resolved by careful petrographic studies targeted at identifying previously-unidentified but dateable minerals. Other methods of dating, such as paleomagnetic dating (Symons et al. 1998), may offer alternative methods of determining ages.

In addition to challenges specific to mineral deposit studies, geochronology more generally faces continued challenges to produce more precise and more accurate ages, including: (1) precise determination and calibration of decay constants across all isotopic systems used in geochronology, (2) inter-laboratory and inter-method (e.g., secondary ion mass spectrometry (SIMS) versus LA-ICP-MS) calibration, (3) improvements in analytical precision, and (4) improvements in the understanding of isotopic systems. Finally, the integration of "bulk" analyses with low spatial resolution but high analytical precision with in-situ analyses with high spatial resolution but (commonly) low analytical precision is a challenge not only for radiogenic isotopes but also for light and metallic stable isotopes.

1.2 Radiogenic Isotopes in Tracing and Mapping

One of the major challenges facing radiogenic isotopic mapping is developing consistent methodologies for mapping and interpreting the significance of variations in isotopic data. As discussed by Champion and Huston (2023), Huston and Champion (2023) and Waltenberg (2023), there are many options of parameters to map, and multiple models of isotopic evolution complicate these options. To be comparable, isotopic maps must be constructed using similar isotopic growth models, for similar parameters and using similar interpolation methods (e.g., Champion and Huston 2023). Although it is sometimes necessary to use locally constrained models to address local questions, the use of inconsistent methods/models between different regions can produce erroneous maps and interpretations. These issues are in addition to challenges of compiling datasets from different sources and laboratories and the challenges of constructing maps from low-density datasets.

A second challenge to isotopic mapping is to understand and account for processes that affect measured isotopic ratios and derived parameters. These processes, which most strongly affect lead, include pre-mineralizing processes that can modify the source region (e.g., high-grade metamorphism) and post-mineralizing processes that can change initial ratios (e.g., ingrowth and isotopic disturbance). Consistent criteria must be developed to allow isotopic maps to account for such processes, which can produce highly anomalous isotopic signatures. A challenge specific to the Lu–Hf system, in which several tens of zircon spot analyses are acquired per sample (e.g., Waltenberg 2023), is developing consistent methods to determine meaningful parameters (i.e., age and ε_{Hf}) from complex data populations that can be used in isotopic mapping.

The integration of "bulk" and in situ analyses is also a challenge to the use of lead isotopes to trace metal sources and mineralizing processes. This challenge is particularly well illustrated by the work of Gigon et al. (2020), who observed variations in lead isotopic ratios of high spatial resolution but low precision SIMS analyses of galena from the HYC deposit in Australia much greater than the variability observed in low resolution but high precision double-spiked thermal ionization mass spectrometry (DS-TIMS) analyses. Gigon et al. (2020) argued that the in situ SIMS data indicate the mixing of two lead sources, but these relationships are not seen in the DS-TIMS data. Reasons for differences in the datasets are, at this point, unclear.

Finally, like other isotopic systems, access to inexpensive analyses with rapid turnaround is also a challenge for radiogenic isotope analyses used age determinations, source tracing and isotopic mapping. Analyses of most radiogenic isotopic systems still occurs in University or government research laboratories, although commercial geochemical laboratories are starting to offer lead isotope analyses using ICP-MS analyses.

1.3 Light Stable Isotopes

The main challenges to the use and interpretation of stable isotopes are the availability of abundant, low cost and high quality analyses and reconciling differences between bulk and in situ analyses. As indicated by Barker et al. (2013), analyses of light stable isotopes must have a fast turn-around time, be relatively inexpensive and be produced in large number before they are routinely incorporated into mineral exploration programs. The last point is critical, as large populations are required to statistically assess anomalies produced from isotopic data. The availability of inexpensive microanalytical tools such as LA-ICP-MS may achieve rapid production of large numbers of rapid, inexpensive analyses, but as discussed below, interpretation of the results are commonly not simple.

In reviewing stable isotopes in shale-hosted zinc deposits, Williams (2023) found that sulfur isotope values determined using in situ analysis were much more variable that values determined from bulk analyses. Hence, like radiogenic isotopes, the integration of bulk and in situ analyses also presents a challenge to the interpretation of stable isotope data. Even though sample aliquots might only be a few milligrams, bulk analyses homogenize variability seen in in situ analyses, based on much smaller sample volumes. Consequently, the challenge remains to integrate these two broad analytical techniques, with implications for interpretating the scale and process of mineralizing events.

1.4 Metallic Stable Isotopes

As variations in metallic stable isotopes were only discovered in the last two to three decades, research into this field is less mature, and the challenges differ to other fields of isotopic research. The main challenge for metallic stable isotopes is to acquire sufficient data to document natural variability in isotopic ratios and determine processes that cause this variation. The amount of basis data varies according to metal; for zinc, there is a small, but growing, dataset for sediment-hosted deposits, but datasets for other deposit types are very small, in some cases constituting only a handful of analyses (Wilkinson 2023). The datasets for iron and copper are larger (Lobato et al. 2023; Mathur and Zhao 2023), but still require additional data, particularly to understand processes that control isotopic fractionation.

A second challenge is the experimental determination of temperature-dependent mineral-fluid, mineral-melt and mineral–mineral fractionation curves for common Fe-, Cu- and Zn-bearing minerals, including biologically-mediated fractionation. Although fractionation curves for iron and copper isotopes have been determined for some minerals (cf. Lobato et al. 2023; Mathur and Zhao 2023), similar curves for zinc isotopes are limited (Wilkinson 2023). Hence, one of the greatest challenge to interpreting metallic stable isotope data is the acquisition of well-calibrated experimental fractionation curves and understanding application of experimentally determined equilibration relationships to real world ore deposits.

A third challenge for metallic stable isotopes is developing rapid and inexpensive analytical methods, including automation. Like other isotopes, metallic stable isotopes will not be routinely used by the exploration industry until this last challenge is met, although it is important to stress that there is significant interest from industry to use metallic isotopes to resolve specific ore genesis problems.

2 Opportunities for Isotopic Research

Like most other fields of scientific research, there have been important advances in the capability to determine isotopic ratios and understanding processes that cause changes in isotopic ratios. These advances present opportunities to apply the isotopic data to geological and, more specifically, mineral system problems.

In many cases, opportunities in isotopic research have come from unorthodox research, which, in some cases, were in conflict with perceived wisdom at the time—for example, expectations that isotopic fractionation should be mass dependent or that metallic isotopes should not fractionate. Some of these unorthodox opportunities, as described in this book, have provided not only entirely new datasets that can be used to test new and existing models for geological systems, but entirely new ideas about processes involved in mineral systems and, more broadly, Earth systems.

Like in many other fields in geoscience, another important opportunity (and challenge) for isotopic research is data integration. This includes integration of different isotopic systems and/or different analytical techniques, integration of isotopic data with other geoscience data, and integration of isotopic data with data beyond geoscience disciplines. Data integration can occur from thin section to global scales and commonly provides new insights into Earth processes. This opportunity also extends to the rapidly evolving fields of artificial intelligence and machine learning. These fields offer opportunities for comprehensive evaluation of the integrated data using statistical and stochastic approaches that quantify relationships between isotopic and other geoscientific datasets.

2.1 Radiogenic Isotopes and the Geochronology of Mineralizing Events

Despite the challenges described above, ore geochronology has many opportunities, from determining the timing and duration of mineral systems, to the dating of mineralizing events using new minerals and/or isotopic systems, to integrating the geochronology data into process-oriented numerical models of mineralizing events, and to linking these events to other geological events at the district to global scales. As summarized by Chiaradia (2023) and Chelle-Michou and Schaltegger (2023), the duration of individual mineralizing events in porphyry copper systems typically last a few tens of thousands or years, although multiple events may overprint each other to produce to mineralizing systems that last much longer and produce much larger metal endowments. Although well constrained for porphyry mineral systems, the duration of mineralizing events for other systems is poorly known. Moreover, information on the duration and overprinting of mineralizing events can be fed into exploration questions with direct exploration implications: Do highly endowed deposits require long durations? Or multiple events?

Although dating of some deposits is a challenge to ore geochronology, opportunities exist to resolve some of these challenges. These include recognition of dateable minerals in ore assemblages and dating of ore-related minerals not commonly dated at present. Many dateable ore-related minerals are not readily recognized during routine petrography and require systematic microanalytical methods such as scanning electron microscope (SEM)-based image analysis (*aka* automated mineralogy: Sylvester 2012; Schulz 2021) or similar microprobe-based techniques for reliable identification. These techniques not only identify dateable minerals but also place these minerals in textural and paragenetic context. Personal experience indicates that SEM-based image analysis can identify dateable minerals such as phosphates (e.g., apatite, monazite and xenotime) in mineral deposits previously not known to have dateable minerals. In addition, many deposits contain potentially dateable minerals such as fluorite (U–Pb of Sm–Nd), allanite (U–Th–Pb), rare earth element minerals (Sm–Nd), scheelite (U–Pb) and titanite (U–Pb) for which ages are not routinely determined.

Another opportunity for research in ore geochronology is the integration of age data into numerical models of mineral system evolution. For example, Chelle-Michou et al. (2017) used thermal models and Monte Carlo simulations to simulate the evolution of the porphyry copper mineral system during granite emplacement. They combined this modelling with data on endowment and the duration of mineralization from eight porphyry copper deposits around the world to conclude that system duration and the total magma volume are the main controls on copper endowment, and not magma enrichment in sulfur or copper. This illustrates the potential to combine geochronological data with modelling to define the most important controls on endowment for many types of mineral systems.

The growing dataset of high precision and robust ages of mineral deposits also allows linkage of mineralizing events to other geological events at the local to global scales. This allows not only the testing of genetic models, but also incorporating mineralizing systems into Earth evolution. As an example of the former, Phillips et al. (2012) identified two gold mineralizing events in the Victorian goldfields (southeast Australia), an early event at ca 450–440 Ma temporally associated with the Benambran Orogeny but not with magmatism, and a second event at 380–370 Ma temporally associated the

Tabberabberan Orogeny and with granitic magmatism (see also Wilson et al. 2020). Over the last few decades, there has been a debate over the role of magmatism in orogenic gold deposits and the relationship of orogenic to intrusion-related gold deposits. The combination of geochronological (e.g., Philips et al. 2012) and structural (Wilson et al. 2020) syntheses allow for an assessment of the roles of granites in orogenic gold mineral systems (not necessary, at least for the Benambran system in western Victoria) and between orogenic and intrusion-related gold deposits.

Meyer (1981), Lambert and Groves (1981), Lambert et al. (1992), Kerrich et al. (2005) and many others have shown that the distribution of mineral deposits through time is not uniform, with different classes of deposits having distributions that can be related global tectonic events and environmental changes. The early observations by Meyer (1981) and by Lambert and Groves (1981) have largely held up, and the greater availability of high precision geochronological data have refined deposit distributions and demonstrated that the distribution of many deposits are related to global tectonic processes such as the assembly and break-up of supercontinents and global environmental events such as the Great Oxidation Event. Continued acquisition of ore geochronology data will test current ideas on global controls on metallogenesis and generate new ideas.

2.2 Radiogenic Isotopes in Tracing and Mapping

Radiogenic isotope mapping has shown systematic spatial patterns in Sm–Nd, Pb–Pb and Lu–Hf data that appear to be related to continental to province-scale crustal boundaries identified using other datasets (Champion and Huston 2023; Huston and Champion 2023; Waltenberg 2023). In many cases, the tectonic implications of these boundaries are either poorly known or controversial. The changes in isotopic characteristics across these boundaries can provide important constraints on the tectonic processes that produced the boundaries. For example, Champion (2013) interpreted decreasing T_{2DM} from north to south in the North Australian Craton as evidence for a long-lived convergent margin along the southern margin, with implications to metallogenesis of this province.

Armistead et al. (2021), using lead isotope data from volcanic-hosted massive sulfide (VHMS) and orogenic gold deposits, showed that at the global scale the range in μ has changed with time, with the period after 1000 Ma having a more restricted range that the period before. Moreover, S Armistead (pers comm, 2022) suggests that individual terranes have different lead isotope characteristics that may be used as a dataset to provide independent tests of paleotectonic reconstructions and tectonic models, even back into the Paleoarchean. Hence, radiogenic isotope data and derived maps can be used to place constraints on tectonic models, with implications to metallogenic models.

Opportunities also exist to merge data from different isotopic systems into one map. Vervoot et al. (1999) demonstrated that ε_{Hf} and ε_{Nd} strongly correlate for terrestrial rocks, indicating that the two parameters are related by a simple linear relationship. Use of this relationship raises the possibility that isotopic maps from the Sm–Nd and Lu–Hf systems can be combined into one map. Another opportunity is extending isotopic mapping to other isotopic systems, for example the Re–Os system.

Owing to their increasing importance in the energy transition, rare earth elements (REEs) have become critical to the global economy, yet mineral systems that form REE deposits are poorly understood. As the Sm–Nd and Lu–Hf isotopic systems are integral parts of REE mineral systems, these isotopic systems can provide direct constraints on the sources of and processes that enrich REEs.

Finally, Armistead et al. (2022) developed an R tool to automatically calculate μ and other parameters from lead isotope data. Automation of these calculations and isotope mapping methods will allow more widespread use of isotopic data in metallogenic and tectonic studies, and, ultimately, exploration.

2.3 Light Stable Isotopes

Despite being a mature discipline, light stable isotope research has seen several analytical breakthroughs this century, leading to important new insights into process in both Earth and mineral systems (Huston et al. 2023a). These analytical breakthroughs, and breakthroughs in data integration, have and will continue to produce opportunities to apply isotope data to mineral system problems.

A feature of the four chapters on the use of stable isotopes in specific mineral systems is data integration. Huston et al. (2023b) integrate oxygen-hydrogen and sulfur isotope data with temporal data to infer that the fluid temperature and sulfur source in VHMS mineral systems have broadly changed with geological time. Quesnel et al. (2023) integrate a range of isotopic data from orogenic gold deposits with temporal data to show how fluid sources have (and have not) evolved through time, and Hagemann et al. (2023) integrate data from Australia, Brazil and South Africa to show the complexities and similarities of ore fluids that upgraded iron formation to form high-grade iron ore deposits. Finally, Williams (2023) integrates isotopic data with paragenetic observations from major shale-hosted zinc deposits from the North Australian Zinc Belt and the Northern Cordillera in North America to assess differing hypotheses of ore formation and the sources of sulfur and carbon. All four studies illustrate the opportunity of the integration of isotopic data with other data at the global scale to provide insights into mineral system processes not available through the study of individual deposits.

Opportunities for data integration extend to the microscopic scale as analytical capabilities now allow collection of a wide range of isotopic (and other) data from the same thin section and even the same analytical spot. Collection of comprehensive data from the same sample enables a much clearer and more complete view of stable isotopes and the processes that cause their fractionation.

In addition to the opportunities offered by microanalysis described above, the development of effective techniques to analyze multiple isotopes, specifically sulfur isotopes, has offered new insights into the sources of sulfur in mineral deposits and processes that have affected the sulfur cycle through time (Farquhar et al. 2000; Caruso et al. 2022; Huston et al. 2023a,b). Application of multiple sulfur isotope analyses to other mineral systems will continue to provide new constraints on sulfur sources and mineralizing processes.

Unlike multiple sulfur isotopes, clumped isotopes, that is combined variations of isotopes in molecules (isotopologues), have not as yet been used extensively in mineral system studies. Clumped isotope analyses, mostly of CO_2 extracted from carbonate minerals, provides information about the temperature of mineral formation that is independent of the isotopic composition of the fluid (Ghosh et al. 2006). Mering et al. (2018) have shown the potential of clumped isotopes for a number of geothermal systems and mineral deposits to indicate mineralization temperatures and infer $\delta^{18}O$ of the mineralizing fluids. Because the fractionation of clumped isotopes increases with decreasing temperatures, clumped isotopes will be particularly useful in low temperature mineral systems, for example many basin-hosted systems. As temperature calibrations are developed for minerals with higher temperature of closure to isotope reordering, this new tool will have more widespread application (Quesnel et al. 2022). The potential of clumped isotopes is enhanced by the development of new, rapid analytical techniques for very small samples (Sakai et al. 2017). Tunable mid-infrared laser absorption spectrometry (TILDAS), as developed by Sakai et al. (2017), differs from virtually all other methods of isotopic analysis in that it uses infrared spectroscopy rather than mass spectrometry to determine mass ratios.

Finally, recent analytical developments also allow for determination of boron isotopes from minerals in which boron is a minor constituent;

previously boron isotope analyses have largely been restricted to tourmaline. For example, Codeço et al. (2019) determined hydrothermal temperatures and $\delta^{11}B$ of ore fluids at the Panasquiera W-Sn deposit in Portugal using coeval tourmaline and white mica.

2.4 Metallic Stable Isotopes

Being a relatively new discipline, metallic stable isotopes offer a number of opportunities to counterbalance the challenges described above. As both copper and iron occur naturally in multiple valence states, one of the greatest opportunities for the using isotopes of both metals is to understand reactions, in particular redox reactions, involved in hypogene mineralization and supergene enrichment (Lobato et al. 2023; Mathur and Zhao 2023). The major cause of iron isotope fractionation are redox reactions that convert ferric to ferrous iron (or vice versa). These reactions occur in many geological environments and can include both biologically mediated and abiological reactions (Johnson et al. 2008; Lobato et al. 2023). Hence, iron isotopes can be used to better understand processes involved in formation of not only iron ore deposits, but also other deposits in which iron is a major component of the ores.

As discussed by Mathur and Zhao (2023), much of the variability in $\delta^{65}Cu$ in deposits stems from redox and other reactions, either during hypogene ore formation or supergene overprinting. Although $\delta^{65}Cu$ variations occur in high temperature systems such as orthomagmatic mafic-hosted Ni–Cu deposits (Zhao et al. 2017), the greatest fractionations are associated with low temperature systems. Variations in $\delta^{65}Cu$ can track redox reactions and reflect fluid pathways in sediment-hosted copper deposits (Haest et al. 2009), whereas $\delta^{65}Cu$ data can be used to assess the degree of weathering in leached caps that have developed over porphyry copper deposits and distinguish between hypogene versus supergene origins for copper minerals such as chalcocite (Mathur and Zhao 2023).

All three metallic isotope systems discussed in this book have potential as vectors to ore. Lobato et al. (2023) indicate that decreases in $\delta^{56}Fe$ (and $\delta^{18}O$) may vector toward shear zones that have acted as fluid conduits during the upgrading of iron formation to iron ore. Similarly, Mathur and Zhao (2023) show $\delta^{65}Cu$ zonation in a number of deposit types (porphyry copper, epithermal, skarn and layered mafic intrusion), indicating that $\delta^{65}Cu$ may be a useful tool to distinguish ore types and test linkages between deposit types in the same district (e.g., between porphyry copper and epithermal deposits), and assess gossanous exposures. Wilkinson (2023) also notes zonation in $\delta^{66}Zn$ in several sediment-hosted zinc deposits. These variations may have the potential for use as deposit-scale vectors, but more case studies are clearly required.

Finally, based upon current data, metallic stable isotopes have limited opportunity as a tool to identify metal sources. Mathur and Zhao (2023) indicate that variability of $\delta^{65}Cu$ in common rock types is limited, and most variability present in mineral deposits relates to chemical reactions during hypogene mineralization or supergene upgrading. Similarly, although based on a much smaller dataset, Wilkinson (2023) indicates that the variability in $\delta^{66}Zn$ in common rock types is also small. With the exceptions of Precambrian shales and iron formation, $\delta^{56}Fe$ of sedimentary and igneous rocks overlap each other and bulk silicate Earth (Dauphas and Rouxel 2006), limiting the utility of iron isotopes to determine iron sources in most ore deposits.

3 Conclusions

The syntheses of isotopic research related to mineral system science presented in this book highlight the importance of isotopes to develop knowledge of geological processes important to ore formation. Moreover, each chapter has identified important challenges and opportunities for continued contributions of isotopes to ore formation at all scales and to mineral exploration.

Many of the challenges to isotopic science in economic geology, metallogenesis and exploration relate to the availability and quality of data. For a number isotopic systems that have only recently become viable due to analytical developments, the amount of data available is relatively small and gathering sufficient data to determine natural variations in isotopic ratios is a fundamental challenge, but also a major opportunity for further research. For other, more mature isotopic systems, improving analytical quality through inter-laboratory and inter-method calibration is a major challenge. A related challenge is the integration of bulk and in situ analyses. For most isotopic systems, the development of rapid and inexpensive analyses is essential to enable the use of isotopic data in routine mineral exploration, although it must be stressed that the data are used by industry to resolve specific ore genesis questions. New analytical methods, such as TILDAS (Sakai et al. 2017) may offer the opportunity for rapid, high-precision isotopic analyses necessary for routine usage in mineral exploration.

The opportunities for isotopic research are many and varied. They range from the development of new analytical techniques for an increasing number of isotopic systems, through the application of these new developments and more conventional analyses to develop and test concepts of ore systems, and to the integration of data from multiple isotopic systems with other geological data. Perhaps the most exciting opportunity in isotopic research is this integration at scales from the global to the microscopic. Such integration at the global scale has already proved successful in documenting secular changes in the characteristics of mineral systems and the implications of these changes to mineral system processes. Isotopic research continues to be a backbone to the development of models of ore formation, but it faces a major challenge to become widely used in mineral exploration.

Acknowledgements The authors thank Georges Beaudoin, Cyrille Chelle-Michou, Geoff Fraser, Marc Norman and Kathryn Waltenberg for comments on the original draft of the contribution, which is published with permission of the Chief Executive Officer, Geoscience Australia.

References

Anglin CD, Jonasson IR, Franklin JM (1996) Sm-Nd dating of scheelite and tourmaline: implications for the genesis of Archean gold deposits, Val d'Or, Canada. Econ Geol 91:1372–1382

Armistead S, Eglington B, Pehrsson S, Huston D (2021) Pb isotope variability in the Archean: insights from the Superior Province, Canada. Geosci Can 48:153

Armistead S, Eglington B, Pehrsson S (2022) Pbiso: an R package and web app for calculating and plotting Pb isotope data. https://eartharxiv.org/repository/view/2841/. https://doi.org/10.31223/X56G84

Barker SL, Dipple GM, Hickey KA, Lepore WA, Vaughan JR (2013) Applying stable isotopes to mineral exploration: teaching an old dog new tricks. Econ Geol 108:1–9

Caruso S, Fiorentini ML, Champion DC, Lu Y, Ueno Y, Smithies RH (2022) Sulfur isotope systematics of granitoids from the Yilgarn Craton sheds new light on the fluid reservoirs of Neoarchean orogenic gold deposits. Geochim Cosmochim Acta 326:199–213

Champion DC (2013) Neodymium depleted mantle model age map of Australia: explanatory notes and user guide. Geosci Austr Rec 2013/44

Champion DC, Huston DL (2023) Applications of Nd isotopes to ore deposits and metallogenic terranes; using regional isotopic maps and the mineral systems concept. In: Huston DL, Gutzmer J (eds) Isotopes in economic geology, metallogensis and exploration. Springer, Berlin (this volume)

Chelle-Michou C, Schaltegger U (2023) U–Pb dating of mineral deposits: from age constrains to ore-forming processes. In: Huston DL, Gutzmer J (eds) Isotopes in economic geology, metallogensis and exploration. Springer, Berlin (this volume)

Chelle-Michou C, Rottier B, Caricchi L, Simpson G (2017) Tempo of magma degassing and the genesis of porphyry copper deposits. Sci Rep 7:40566

Chiaradia M (2023) Radiometric dating applied to ore deposits: theory and methods. In: Huston DL, Gutzmer J (eds) Isotopes in economic geology, metallogensis and exploration. Springer, Berlin (this volume)

Christensen JN, Halliday AN, Vearncombe JR, Kesler SE (1995) Testing models of large-scale crustal fluid flow using direct dating of sulfides; Rb–Sr evidence for early dewatering and formation of Mississippi Valley-type deposits, Canning Basin, Australia. Econ Geol 90:877–884

Codeço MS, Weis P, Trumbull R, Glodny J, Wiedenbeck M, Romer RL (2019) Boron isotope muscovite-tourmaline geothermometry indicates fluid cooling during magmatic-hydrothermal W-Sn ore formation. Econ Geol 114:153–163

Cross A, Jaireth S, Rapp R, Armstrong R (2011) Reconnaissance-style EPMA chemical U–Th–Pb dating of uraninite. Aust J Earth Sci 58:675–683

Dauphas N, Rouxel O (2006) Mass spectrometry and natural variations of iron isotopes. Mass Spectrom Rev 25:515–590

Ehrig K, Kamenetsky V, McPhie J, Macmillan E, Thompson J, Kamenetsky M, Maas R (2021) Staged formation of the supergiant Olympic Dam uranium deposit, Australia. Geology 49:1312–1316

Farquhar J, Bao H, Thiemens M (2000) Atmospheric influence of Earth's earliest sulphur cycle. Science 289:756–758

Ghosh P, Adkins J, Affek H, Balta B, Guo W, Schauble EA, Schrag D, Eiler JM (2006) ^{13}C–^{18}O bonds in carbonate minerals: a new kind of paleothermometer. Geochim Cosmochim Acta 70:1439–1456

Gigon J, Deloule E, Mercadier J, Huston DL, Richard A, Annesley IR, Wygralak AS, Skirrow RG, Mernagh TP, Masterman K (2020) Tracing metal sources for the giant McArthur River Zn–Pb deposit (Australia) using lead isotopes. Geology 48:478–802

Haest M, Muchez P, Petit JCJ, Vanhaecke F (2009) Cu isotope ratio variations in the Dikulushi Cu–Ag deposit, DRC: of primary origin or induced by supergene reworking. Econ Geol 104:1055–1064

Hagemann S, Hensler A-S, Figueiredo e Silva RC, Tsikos H (2023) Light stable isotope (O, H, C) signatures of BIF-hosted iron ore systems: implications for genetic models and exploration targeting. In: Huston DL, Gutzmer J (eds) Isotopes in economic geology, metallogensis and exploration. Springer, Berlin (this volume)

Huston DL, Champion DC (2023) Applications of lead isotopes to ore geology, metallogenesis and exploration. In: Huston DL, Gutzmer J (eds) Isotopes in economic geology, metallogensis and exploration. Springer, Berlin (this volume)

Huston DL, Trumbull RB, Beaudoin G, Ireland T (2023a) Light stable isotopes (H, B, C, O and S) in ore studies —methods, theory, applications and uncertainties. In: Huston DL, Gutzmer J (eds) Isotopes in economic geology, metallogensis and exploration. Springer, Berlin (this volume)

Huston DL, LaFlamme C, Beaudoin G, Piercey S (2023b) Light stable isotopes in volcanic-hosted massive sulfide ore systems. In: Huston DL, Gutzmer J (eds) Isotopes in economic geology, metallogensis and exploration. Springer, Berlin (this volume)

Johnson CM, Beard BL, Roden EE (2008) The iron isotope fingerprints of redox and biogeochemical cycling in modern and ancient Earth. Annu Rev Earth Planet Sci 36:457–493

Kerrich R, Goldfarb RJ, Richards JP (2005) Metallogenic provinces in an evolving geodynamic framework. Econ Geol 100th Anniv Vol 1097–1136

Lambert IB, Groves DI (1981) Early earth evolution and metallogeny. Handbook Stratabound Stratiform Ore Deposits 8:339–447

Lambert IB, Beukes N, Klein C, Veizer J (1992) Proterozoic mineral deposits through time. In Schopf JW Klein C (eds) The proterozoic biosphere: a multidisciplinary study. Cambridge University Press, Cambridge, pp 59–62

Lobato LM, Figueiredo e Silva RC, Angerer T, Mendes M, Hagemann S (2023) Fe isotopes applied to BIF-hosted iron ore deposits. In: Huston DL, Gutzmer J (eds) Isotopes in economic geology, metallogensis and exploration. Springer, Berlin (this volume)

Mathur R, Zhao Y (2023) Copper isotopes used in mineral exploration. In: Huston DL, Gutzmer J (eds) Isotopes in economic geology, metallogensis and exploration. Springer, Berlin (this volume)

Mering JA, Barker SLL, Huntington KW, Simmons S, Dipple G, Andrew B, Schauer A (2018) Taking the temperature of hydrothermal ore deposits using clumped isotope thermometry. Econ Geol 113:1671–1678

Meyer C (1981) Ore-forming processes in geologic history. Econ Geol 75th Anniv Vol 63–116

Nash JT, Granger HC, Adams SS (1981) Geology and concepts of genesis of important types of uranium deposits. Econ Geol 75th Anniv Vol 63–116

Norman MD (2023) The ^{187}Re-^{187}Os and ^{190}Pt-^{186}Os radiogenic isotope systems: techniques and applications to metallogenic systems. In: Huston DL, Gutzmer J (eds) Isotopes in economic geology, metallogensis and exploration. Springer, Berlin (this volume)

Phillips D, Fu B, Wilson CJL, Kendrick M, Fairmaid A, J. McL Miller JMcL, (2012) Timing of gold mineralization in the western Lachlan Orogen, SE Australia: a critical overview. Austr J Earth Sci 59:495–525

Polito PA, Kyser TK, Rheinberger G, Southgate PN (2005) A paragenetic and isotopic study of the Proterozoic Westmoreland uranium deposits, southern McArthur Basin, Northern Territory, Australia. Econ Geol 100:1243–1260

Quesnel B, Jautzy J, Scheffer C, Raymond G, Beaudoin G, Jørgensen TRC, Pinet N (2022) Clumped isotope geothermometry in Archean mesothermal hydrothermal systems (Augmitto-Bouzan orogenic gold deposit, Abitibi, Québec, Canada): a note of caution and a look forward. Chem Geol 610:121099. https://doi.org/10.1016/j.chemgeo.2022.121099

Quesnel B, Scheffer C, Beaudoin G (2023) The light stable isotope (H, B, C, N, O, Si, S) composition of orogenic gold deposits. In: Huston DL, Gutzmer J (eds) Isotopes in economic geology, metallogensis and exploration. Springer, Berlin (this volume)

Sakai S, Matsuda S, Hikida T, Shimono A, McManus JB, Zahniser M, Nelson D, Dettman DL, Yang D, Ohkouchi N (2017) High-precision simultansous $^{18}O/^{16}O$, $^{13}C/^{12}C$, and $^{17}O/^{16}O$ analyses of microgram quantities of $CaCO_3$ by tunable infrared laser absorption spectrosopy. Anal Chem 89:11846–11852

Schulz B (2021) Monazite microstructures and their interpretation in petrochronology. Front Earth Sci 9:668566

Sylvester PJ (2012) Use of the mineral liberation analyzer (MLA) for mineralogical studies of sediments and

sedimentary rocks. Mineral Assoc Can Short Course Notes 42:1–16
Symons DTA, Lewchuk MT, Leach DL (1998) Age and duration of the Mississippi Valley-type mineraling fluid flow event in the Viburnaum Trend, southeast Missouri, USA, determined from paleomagnetism. Geol Soc Lond Spec Pub 144:27–39
Vervoot JD, Patchett PJ, Blichert-Toft J, Albarede F (1999) Relationships between Lu-Hf and Sm-Nd isotopic systems in the golobal sedimentary system. Earth Planet Sci Lett 168:79–99
Vielreicher N, Groves D, McNaughton N, Fletcher I (2015) The timing of gold mineralization across the eastern Yilgarn craton using U–Pb geochronology of hydrothermal phosphate minerals. Mineral Deposita 50:391–428
Waltenberg K (2023) Application of Lu-Hf isotopes to ore geology, metallogenesis and exploration. In: Huston DL, Gutzmer J (eds) Isotopes in economic geology, metallogensis and exploration. Springer, Berlin (this volume)
Williams N (2023) Light-element stable isotope studies of the clastic-dominated lead-zinc mineral systems of northern Australia & the North American Cordillera: implications for ore genesis and exploration. In: Huston DL, Gutzmer J (eds) Isotopes in economic geology, metallogensis and exploration. Springer, Berlin (this volume)
Wilkinson JJ (2023) The potential of Zn isotopes in the science and exploration of ore deposits. In: Huston DL, Gutzmer J (eds) Isotopes in economic geology, metallogensis and exploration. Springer, Berlin (this volume)
Wilson CJL, Moore DH, Vollgger SA, Madeley HE (2020) Structural evolution of the orogenic gold deposits in central Victoria, Australia: the role of regional stress change and the tectonic regime. Ore Geol Rev 120:103390
Zhao Y, Xue C, Liu S-A, Symons DTA, Zhao X, Yang Y, Ke J (2017) Copper isotope fractionation during sulfide-magma differentiation in the Tulaergen magmatic Ni–Cu deposit, NW China. Lithos 286–287:206–215